AF670647

Space Sciences Series of ISSI

Volume 92

The Space Sciences Series of ISSI books are coherent reports of the findings, discussions, and ideas that result from international scientific workshops regularly held at the International Space Science Institute (ISSI) in Bern, Switzerland. ISSI's main task is to contribute to the achievement of a deeper understanding of the results from space-research missions, adding value to those results through multi-disciplinary research in an atmosphere of international cooperation. The books are reprints of special issues in the Space Science Reviews journal and occasionally of special issues in the Surveys in Geophysics journal.

Joachim Wambsganss · Frédéric Courbin ·
Richard G. McMahon · Priyamvada Natarajan ·
Paul L. Schechter · Sherry H. Suyu ·
Liliya L. R. Williams
Editors

Strong Gravitational Lensing

Previously published in *Space Science Reviews, Collection "Strong Gravitational Lensing"*

Editors
Joachim Wambsganss
Zentrum für Astronomie der Universität Heidelberg
Astronomisches Rechen-Institut
Heidelberg, Germany

Richard G. McMahon
University of Cambridge
Institute of Astronomy
Cambridge, UK

Paul L. Schechter
Kavli Institute for Astrophysics and Space Research
Massachusetts Institute of Technology
Cambridge, MA, USA

Liliya L. R. Williams
Minnesota Institute for Astrophysics
University of Minnesota
Minneapolis, MN, USA

Frédéric Courbin
Universitat de Barcelona
Institut de Ciencies del Cosmos (ICCUB)
Barcelona, Spain

Priyamvada Natarajan
Departments of Astronomy and Physics
Yale University
New Haven, CT, USA

Sherry H. Suyu
Department of Physics
Technical University of Munich School of Natural Sciences
Garching, Germany

ISSN 1385-7525
Space Sciences Series of ISSI
ISBN 978-94-024-2356-3

© The Editor(s) (if applicable) and The Author(s), under exclusive license to Springer Nature B.V. 2026

This work is subject to copyright. All rights are solely and exclusively licensed by the Publisher, whether the whole or part of the material is concerned, specifically the rights of translation, reprinting, reuse of illustrations, recitation, broadcasting, reproduction on microfilms or in any other physical way, and transmission or information storage and retrieval, electronic adaptation, computer software, or by similar or dissimilar methodology now known or hereafter developed.
The use of general descriptive names, registered names, trademarks, service marks, etc. in this publication does not imply, even in the absence of a specific statement, that such names are exempt from the relevant protective laws and regulations and therefore free for general use.
The publisher, the authors and the editors are safe to assume that the advice and information in this book are believed to be true and accurate at the date of publication. Neither the publisher nor the authors or the editors give a warranty, expressed or implied, with respect to the material contained herein or for any errors or omissions that may have been made. The publisher remains neutral with regard to jurisdictional claims in published maps and institutional affiliations.

This Springer imprint is published by the registered company Springer Nature B.V.
The registered company address is: Van Godewijckstraat 30, 3311 GX Dordrecht, The Netherlands

If disposing of this product, please recycle the paper.

Contents

Space Science Reviews (2025) 221:77
https://doi.org/10.1007/s11214-025-01205-2

Strong Gravitational Lensing – Editorial

Joachim Wambsganss[1] · Frédéric Courbin[2,3] · Richard G. McMahon[4,5] · Paul L. Schechter[6,7] · Sherry H. Suyu[8,9,10] · Priyamvada Natarajan[11,12,13] · Liliya L.R. Williams[14]

Received: 5 August 2025 / Accepted: 5 August 2025 / Published online: 19 August 2025
© The Author(s) 2025

The first gravitationally lensed quasar — the double Q0957+561A,B — was discovered more than 45 years ago, by now about 400 multiply imaged quasars are known. Gravitationally lensed quasars have been explored for more than four decades with respect to a variety of cosmological questions: measurement of time delays, determination of the Hubble constant, derivation of the cosmological constant, detection of microlensing, setting limits on compact and smooth dark matter, exploration of quasar size and luminosity profile, modelling of the mass distribution of the lensing galaxy or galaxy cluster, to name just a few.

About 40 years ago, "giant luminous arcs" were discovered, high redshift galaxies gravitationally lensed, magnified and heavily distorted by foreground galaxy clusters. They made Fritz Zwicky's vision from the 1930s come true and opened up a new direction of exploration: Confirming the large masses of clusters of galaxies, and — by their magnification — effectively increasing the diameter of our telescopes.

A few exciting new developments in recent years make the topic of "strong gravitational lensing" very timely and open up new routes for exploration, e.g. the discovery of gravitationally lensed supernovae (as originally suggested by Sjur Refsdal in the early 1960s — before quasars were known!), the new controversy about the Hubble constant, and in particular the data releases of the ESA Gaia mission which allowed to discover many new strongly lensed objects due to the all-sky coverage — opening up a multitude of new roads for exploration.

In the next few years, very wide-ranging cosmological investigations of gravitationally lensed quasars/supernovae/background galaxies will become possible with the NSF-DOE Vera C. Rubin Observatory and in particular the envisioned *Legacy Survey of Space and Time* (LSST): The LSST will open up the time domain for quasar lensing in an unprecedented way which has the potential to revolutionize this field! And in the deep stacked images very many giant luminous arc systems will be discovered in so far unexplored regions of the sky. These very promising developments provide the perfect frame for reviewing the current state of strong gravitational lensing, and allow to discuss new ideas of how to explore the current and near future data treasure with very interesting — even though partly controversial — discussions.

Originally planned for summer 2020, the ISSI Workshop on Strong Gravitational Lensing could not take place due to the corona pandemic. According to the then suggested ISSI "reverse" schedule, about 55 scientists started with an online-kickoff on 18 January 2021 with some review talks and introductory discussions on the important topics to be covered.

Extended author information available on the last page of the article

In the following 14 months, twelve more online meetings were held in which the different groups assigned to the relevant scientific topics discussed and presented their progress on the articles to be written. Finally, the "real" ISSI Workshop took place on 18 – 22 July 2022 in beautiful Bern/Switzerland. Although it was meant to just "wrap up" the papers to be submitted soon thereafter, very lively discussions and new developments lead to further time delays until the last manuscripts became available in a quality which satisfied all parties involved — authors, referees and editors.

Our "product" now is a collection of eleven review articles on various aspects of strong gravitational lensing, covering themes like the search for strong lenses, lensing as a probe for dark matter, lensing and microlensing of supernovae, to name just a few. The topical reviews are framed by two introductory articles on "*Essentials*" and "*Basic Elements*", resp., of strong gravitational lensing — complementary ways to enter the field.

It is a great pleasure to thank the International Space Science Institute (ISSI) for its very generous support of this workshop, and the wonderful ISSI staff for hosting and perfectly organizing this workshop, as well as patiently accompanying the production of this collection. We would also like to thank all authors for their long-lasting support and meticulous work, the referees for very carefully reading, reviewing, commenting and improving all the manuscripts, and the staff of Springer Nature for publishing this Special Issue efficiently. May this collection serve as a good overview for experienced researchers and as an excellent introduction for newcomers to the field of strong gravitational lensing for a few years to come.

Funding Information Open Access funding enabled and organized by Projekt DEAL.

Declarations

Competing Interests The authors declare no competing interests.

Open Access This article is licensed under a Creative Commons Attribution 4.0 International License, which permits use, sharing, adaptation, distribution and reproduction in any medium or format, as long as you give appropriate credit to the original author(s) and the source, provide a link to the Creative Commons licence, and indicate if changes were made. The images or other third party material in this article are included in the article's Creative Commons licence, unless indicated otherwise in a credit line to the material. If material is not included in the article's Creative Commons licence and your intended use is not permitted by statutory regulation or exceeds the permitted use, you will need to obtain permission directly from the copyright holder. To view a copy of this licence, visit http://creativecommons.org/licenses/by/4.0/.

Publisher's Note Springer Nature remains neutral with regard to jurisdictional claims in published maps and institutional affiliations.

Authors and Affiliations

Joachim Wambsganss[1] · Frédéric Courbin[2,3] · Richard G. McMahon[4,5] · Paul L. Schechter[6,7] · Sherry H. Suyu[8,9,10] · Priyamvada Natarajan[11,12,13] · Liliya L.R. Williams[14]

✉ J. Wambsganss
jkw@ari.uni-heidelberg.de

1 Astronomisches Rechen-Institut, Zentrum für Astronomie der Universität Heidelberg, 69120 Heidelberg, Germany

[2] Laboratoire d'Astrophysique Ecole Polytechnique Fédérale de Lausanne (EPFL), Observatoire de Sauverny, 1290 Versoix, Switzerland

[3] Institut de Ciències del Cosmos (ICCUB), 08028 Barcelona, Spain

[4] Institute of Astronomy, University of Cambridge, Cambridge CB3 0HA, UK

[5] Kavli Institute for Cosmology, University of Cambridge, Cambridge, CB3 0HA, UK

[6] Department of Physics, Massachusetts Institute of Technology, Cambridge, 02139, MA, USA

[7] MIT Kavli Institute, Massachusetts Institute of Technology, Cambridge, 02139, MA, USA

[8] TUM School of Natural Sciences, Department of Physics, Technical University of Munich, 85748, Garching, Germany

[9] Max Planck Institute for Astrophysics, 85748, Garching, Germany

[10] Institute of Astronomy and Astrophysics, Academia Sinica, Taipei, 10617, Taiwan

[11] Department of Astronomy, Yale University, New Haven, CT, 06511, USA

[12] Department of Physics, Yale University, New Haven, CT, 06511, USA

[13] Yale Center for Astronomy & Astrophysics, Yale University, New Haven, CT, 06520, USA

[14] Minnesota Institute for Astrophysics, School of Physics and Astronomy, University of Minnesota, Minneapolis, MN, 55455, USA

Space Science Reviews (2025) 221:49
https://doi.org/10.1007/s11214-025-01171-9

Basic Elements of Strong Gravitational Lensing

Paul L. Schechter[1,2] · **Jeremy D. Schnittman**[3,2]

Received: 25 November 2024 / Accepted: 2 May 2025 / Published online: 30 May 2025
This is a U.S. Government work and not under copyright protection in the US; foreign copyright protection may apply 2025

Abstract
Even when used to describe the same phenomenon, equations, graphics and words each give different perspectives and lead to complementary insights. The basic elements of strong gravitational lensing are introduced here favoring words and graphics over equations whenever possible. Fermat's principle is the fundamental driver of strong lensing. Three "D's" encapsulate the essential effects of lensing: **D**elay, **D**eflection and **D**istortion. Gravity and geometry both contribute to the delay of photons from a lensed source. Their interplay determines how the images of a source are deflected and how they are stretched or compressed. Caustics and critical curves are explained. Images of doubly, triply, quadruply and quintuply lensed sources are displayed. A table of symbols, their definitions and distinctions provides a summary of the basic elements of strong lensing.

Keywords Gravitational lensing

1 What Is Strong Gravitational Lensing?

Strong lensing occurs when the gravitational field of an astronomical object is sufficiently strong to produce two or more images of a single background source. The French words "mirage gravitationelle" describe the phenomenon more accurately. The multiple images are distorted; sometimes their handedness is flipped; and as projected onto the sky they are displaced from the actual position of the lensed source. The phenomenon is illustrated schematically in Fig. 1.

Strong lensing has several basic elements that combine in myriad ways to produce complicated strongly lensed systems. These permit the exploration of a number of outstanding questions in astronomy. Our purpose in this article is to equip the reader with the tools needed to carry out such explorations.

✉ P.L. Schechter
schech@mit.edu

J.D. Schnittman
jeremy.d.schnittman@nasa.gov

[1] Department of Physics, Massachusetts Institute of Technology, 77 Massachusetts Avenue, Cambridge, 02139, MA, USA

[2] MIT Kavli Institute, Massachusetts Institute of Technology, 77 Massachusetts Avenue, Cambridge, 02139, MA, USA

[3] Gravitational Astrophysics Lab, NASA Goddard Space Flight Center, Greenbelt, 20771, MD, USA

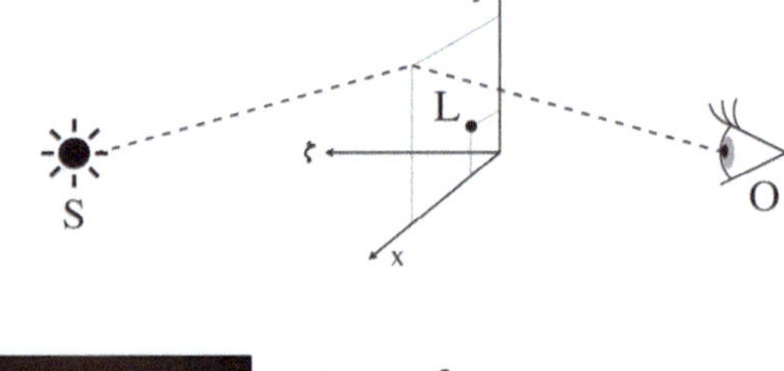

Fig. 1 A gravitational lens L deflects light from a source S. The ζ axis lies along the unperturbed line of sight from the source to the observer O. The lens lies at $\zeta = 0$

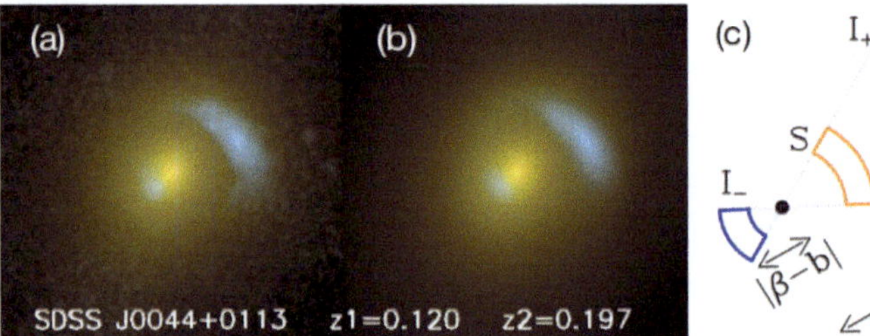

Fig. 2 An archetypical gravitationally lensed system. Panel (a) shows a yellow galaxy lensing a blue galaxy that lies behind it. Panel (b) shows a model for the system. Panel (c) shows a highly idealized schematic of the model. The positive and negative parity images, I_+ and I_-, lie at angular distances $\beta + b$ and $|\beta - b|$ from the center of the lens, where β is the offset of the actual, unlensed, position of the source, S. The strength of the lens model (a singular isothermal spehre) is given by the radius of its Einstein ring, b, which is greater than β

Among the many idealizations that we adopt here for illustative purposes is the assumption that our strong lenses are single objects (as with the case of the elliptical galaxy in Fig. 2). The treatment of "lenses within lenses" is addressed in Natarajan et al. (2024) and Natarajan et al. (2024) in this collection.

2 An Observational Archetype for Strongly Lensed System

In Fig. 2a we see a blue galaxy at redshift 0.197 lensed by a yellowish foreground galaxy with redshift 0.120 (Bolton et al. 2008). This system, SDSS J0044+0113 is among the most ordinary-looking strongly lensed systems known, and we adopt it as our observatinal archetype. The apparent graininess results from statistical noise in the original Hubble Space Telescope (henceforth HST) monochromatic exposure. Figure 2b shows a smooth, theoretical model for this system. The data and model have been "colorized" to conform with our expectations for galaxies at these redshifts.

Figure 2c shows a highly idealized circularly symmetric model drawn from Narayan and Bartelmann (1996) that illustrates some basic elements of strong lensing. The idealized source is shaped like a sector of an annulus, located an angular distance β from the center of the lens.

The idealized lens (known as a **S**ingular **I**sothermal **S**phere; henceforth SIS) has the property that all rays are deflected radially by the same angle, b. One image, I_+, lies at angle $\beta + b$ from the lens and is magnified by a factor $1 + \frac{b}{\beta}$ Another image, I_-, lies at angular distance $|b - \beta|$ from the lens and is (de)magnified by a factor $1 - \frac{b}{\beta}$. Note I_+ has the same handedness as the source (positive image parity), while I_- has the opposite handedness (negative image parity, as indicated by the sign of the magnification).

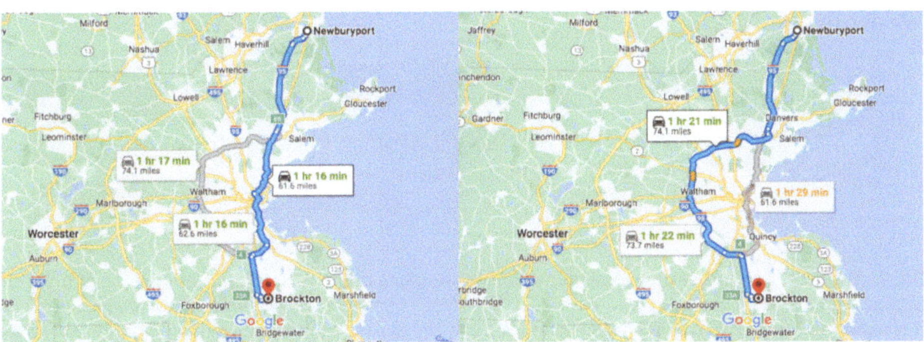

Fig. 3 Automobile traffic plays the same role as an index of refraction, reducing speed and increasing the travel time. In the absence of traffic, the fastest route from Newburyport to Brockton goes through downtown Boston, which lies on a straight line between the two. Panel a) shows that route, chosen by a GPS application. In panel b) traffic is heavy, introducing an effective index of refraction. Though the Fermat route has a longer path length, the total travel time is shorter because traffic is lighter

3 Strong Lensing from Fermat's Principle

Terrestrial mirages are the result of variations with height in the index of refraction n of the Earth's atmosphere. The speed of a photon traveling through a medium with index n is c/n. Fermat's principle tells us that images appear along directions for which a photon's time of flight is a minimum (or a maximum, or a saddle point). Those directions will, in general, be different from the direction along which a source would be seen in the absence of a refractive medium. These deflections from the unlensed light paths increase the path length and consequently delay the arrival of photons. It is therefore the *sum* of the refractive delay and the second, geometric delay, that is subject to Fermat's principle.

The gravitational potentials of astronomical objects (galaxies for the present discussion) create "effective" indices of refraction, which then delay the times of flight of rays passing through them (Binney and Merrifield 1998). Images of a background source appear along directions for which the time of flight is a stationary point.

Something akin to Fermat's principle is at work when one uses GPS enabled software to choose the shortest route from one city to another, as illustrated in Fig. 3. The traffic on the roads acts like an index of refraction, delaying the travel. The software chooses the route that takes the least time.

In the example shown, in the absence of traffic, the quickest route passes through the city of Boston. But during rush hour it takes less time if one bypasses the city. Note that Boston is not small compared to the distance traveled, so one cannot use the "thin lens" approximation described below.

4 The Thin Lens Approximation and the Time Delay Function

The scale on which a galaxy produces a non-trivial effective index of refraction is many orders of magnitude smaller than the distance between the lens and the source or the observer. This permits use of the "thin lens" approximation of conventional optics, greatly simplifying modelling. It is implicit in Fig. 1, in which the direction of travel of photons changes discontinuously in the plane of lens.

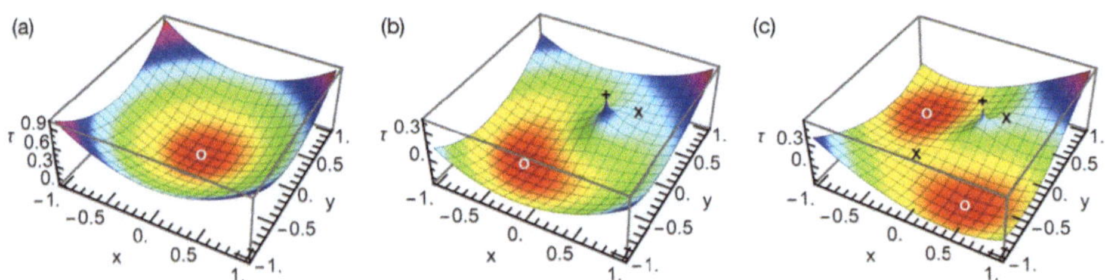

Fig. 4 Time delay plotted as a function of position in the plane of the lens. Minima of the time of flight are marked with a o; saddlepoints by a ×; maxima by a +. Panel (a) shows the time delay in the absence of a lens. It increases quadratically outward from the line of sight, which passes through the origin. Panel (b) shows the effect of a circularly symmetric galaxy offset from the line of sight. Panel (c) shows the effect making that galaxy elliptical. The minimum in panel (b) has been split into two minima and a saddlepoint

The essence of the approximation is to accumulate the time delay of a photon as it passes through the lens and attribute it to a point on a plane at the distance of the lens. Images then form at stationary points of this "time delay function" as shown in Fig. 4.

Figure 4a shows the time delay function in the absence of a lens. As drawn, the line of sight to the source passes through the origin of the lens plane, but for generality in what follows, we take the physical position of the source where it passes through the plane of the lens to be $\vec{s}$. Were an image to appear at sky position $\vec{i}$ (measured in meters), it would be delayed by an amount $\tau_{geom}(\vec{i})$ that reflects the extra path length from the source to the plane of the lens. We have

$$\tau_{geom}(\vec{i}) = \frac{1}{2c}\frac{|\vec{i}-\vec{s}|^2}{\chi_{tran}^{LS}} \quad , \tag{1}$$

where χ_{tran}^{LS} (measured in meters) is the lens-to-source distance, measured in a co-moving coordinate system that expands with the universe The distinctions between this distance measure and several alternatives are discussed at length in Sect. 8.2 below. In the absence of a lens, the time delay function is parabolic. For simplicity we ignore until Sect. 8 the introduction of a *second* parabolic contribution to the delay, generated between the image and the observer.

Figure 4b shows the effect of introducing a circularly symmetric galaxy that is offset from the line of sight. The galaxy induces a gravitational time delay, $\tau_{grav}(\vec{i})$, which added to the geometric delay gives

$$\tau_{sum}(\vec{i}) = \frac{|\vec{i}-\vec{s}|^2}{2c\chi_{tran}^{LS}} + \tau_{grav}(\vec{i}) \quad . \tag{2}$$

We set the gradient of $\tau_{sum}(\vec{i})$ to zero, and then multiply by $c\chi_{tran}^{LS}$ to get

$$\vec{0} = (\vec{i}-\vec{s}) + c\chi_{tran}^{LS}\nabla\tau_{grav}(\vec{i}) \quad . \tag{3}$$

We then solve for the stationary points of the combined delay: $\vec{i}_1$, $\vec{i}_2$ *et cetera*. There are three of these in Fig. 4b – a maximum, a minimum and a saddle point. Dividing the physical displacement in the lens plane of the image from the source, $(\vec{i}-\vec{s})$, by χ_{tran}^{LS} gives the angular deflection of the image, $\vec{\theta}-\vec{\beta}$, where $\vec{\theta}$ is the angular position of the image and $\vec{\beta}$ is the angular position of the source.

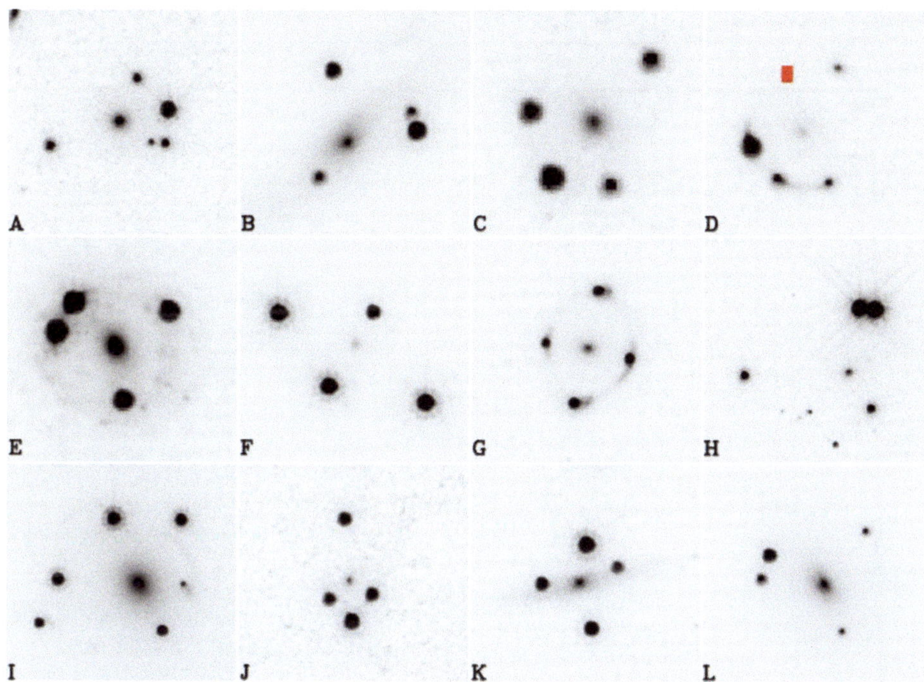

Fig. 5 Negative images of HST exposures of a dozen arbitrarily selected quadruply lensed quasars. The lensing galaxy is always visible at the center of the field, and in most cases is more extended than the four quasar images. There are faint background galaxies in panels A and H. The two faint starlike images in panel I are images of a *second* quasar. A faint charged particle detection in panel D is masked in red. In panels D, E and G, the galaxy that hosts the quasar can be seen forming something of an "Einstein ring"

Figure 4c shows the effect of making the interposed galaxy elliptical (as is expected, at some level, for all galaxies). The minimum has been split into two minima with a saddlepoint in between. For any finite offset of the lens from the source, this splitting occurs *only* if the lens exceeds a minimum ellipticity.

If the lensed source were a supernova or some other transient, one would first see an image at the deeper minimum, then a second at the higher minimum, then a third at the lower saddlepoint, then a fourth at the higher saddlepoint and finally an image at the maximum. In Fig. 5 we show a dozen examples of quadruply lensed quasars formed by elliptically shaped lenses.

5 Coordinates, Distances, and Angles, in Words

The treatment of strong lensing in the preceding sections emphasizes words and drawings. Newcomers to strong lensing may find the plethora of coordinates, distances and angles daunting – more so than any subtlety in the underlying physics.

Before we proceed further, many of the quantities already described need to be given algebraic representations. Yet more quantities, to be introduced in the following sections, are also represented algebraically.

Table 1 is a guide to this exploding collection of symbols.

Table 1 Coordinates, distances, and angles, in words

symbol	ref	definition
S	Fig. 1	source
O	Fig. 1	observer
x	Fig. 1	physical distance from the origin in lens plane
y	Fig. 1	physical distance from the origin in lens plane
ζ	Fig. 1	physical distance along line of sight from lens plane
z	Sect. 8	redshift; not to be confused with ζ
β	Fig. 2	scalar angular displacement of source from lens
$\vec{\beta}$	Sect. 4	angular vector displacement of source from lens
$\vec{\theta}$	Sect. 4	angular vector displacement of image from lens
θ_E	Eq. (25)	the Einstein radius of a lens
$+\&-$	Fig. 2	parities of images for minima and saddle points
b	Fig. 2	angular deflection caused by an isothermal spherical lens
$\vec{i}$	Sect. 4	a physical vector position in the image plane
$\vec{s}$	Sect. 4	the physical vector position in the image plane where the line of sight to the source passes through it
T^{LO}	Sect. 8.3	lens-to-observer time of flight when no lens present
τ	Sect. 4	*time delay* ; extra time of flight w.r.t. no lens present
τ_{geom}	Eq. (1)	the *extra* time of flight due to increased path length
τ_{grav}	Sect. 4	the *extra* time of flight due to gravity
τ_{sum}	Sect. 4	the combined *extra* time of flight
$\tau_{shapiro}$	Eq. (11)	same as τ_{grav} ; calculated from gravitational potential
χ_{long}	Sect. 8.2	a co-moving distance along line of sight
$f_K(\chi_{long})$	Sect. 8.2	a co-moving distance $\perp$ to line of sight
M	Eq. (5)	magnification matrix: M^{-1} defined as $\frac{\partial \vec{s}}{\partial \vec{i}}$
SIEP	Sect. 7	**S**ingular **I**sothermal **E**lliptical (gravitational) **P**otential
SIS+XT	Sect. 7	**S**ingular **I**sothermal **S**pherical potential + e**X**ternal **T**ide
$\Phi(\vec{i},\zeta)$	Sect. 9	Newtonian gravitational potential of lens
$\Phi_{2D}(\vec{i})$	Eq. (10)	projection of $\Phi(\vec{i},\zeta)$ onto lens plane
$\vec{L}$	Sect. 10.1	angular position vector of lens on plane of sky
$\vec{I}$	Sect. 10.1	angular position vector on the plane of the sky of a hypothesized lensed image
$\Psi(\vec{\theta})$	Eq. (13)	Newtonian potential with lens at origin
Γ	Eq. (26)	the strength of the tidal shear flattening the lens potential
η	Sect. 11	the "semi-ellipticity" of an elliptical *potential*
q	Sect. 11	the minor-to-major axis ratio for an ellipse

Strong lensing uses two 2-dimensional coordinate systems with parallel axes, one in the plane of the source, and the other in the plane of the lens. Both planes are perpendicular to the line of sight from the observer to the source. We use ζ as the coordinate along this line of sight in the vicinity of the lens plane, so as not to confuse it with redshift z.

As described below in Sect. 8, the curvature of spacetime leads to a distinction between distances measured along the line of sight, *longitudinally*, and those measured perpendicular to the line of sight, *transversely*.

6 (De-)Magnifications and Distortions

Lensed images are distorted. Were one to change the source position, $\vec{s}$ in Eq. (3), it would produce a change in the image position $\vec{i}$ that differs in both direction and modulus. The matix quantity $\partial\vec{i}/\partial\vec{s}$ measures that distortion and is called the "magnification matrix," M. Its inverse, M^{-1}, can be obtained by taking the gradient of Eq. (3),

$$0 = I - \frac{\partial\vec{s}}{\partial\vec{i}} + c\chi_{tran}^{LS}\nabla\nabla\tau_{grav}(\vec{i}) \quad , \tag{4}$$

where the last term on the right is the Hessian of $\tau_{grav}(\vec{i})$ and I is the identity matrix. The second term on the right of the equation is the inverse magnification matrix, M^{-1}, yielding

$$M^{-1} \equiv \frac{\partial\vec{s}}{\partial\vec{i}} = \begin{pmatrix} 1 + c\chi_{tran}^{LS}\frac{\partial^2\tau_{grav}(\vec{i})}{\partial x^2} & +c\chi_{tran}^{LS}\frac{\partial^2\tau_{grav}(\vec{i})}{\partial x\partial y} \\ +c\chi_{tran}^{LS}\frac{\partial^2\tau_{grav}(\vec{i})}{\partial x\partial y} & 1 + c\chi_{tran}^{LS}\frac{\partial^2\tau_{grav}(\vec{i})}{\partial y^2} \end{pmatrix} \quad . \tag{5}$$

The magnification matrix and its inverse are symmetric for well behaved gravitational time delays, and can be diagonalized. If, after diagonalization, the diagonal terms of M^{-1} are both positive (or both negative), the image is a minimum (or maximum) of the time of flight and have even parity. If the diagonal terms have opposite signs, the image is a saddle point and has the opposite handedness of the source.

For strongly centrally concentrated lenses (as is the case for most galaxies), the second derivatives at the center are large and negative, giving maxima that are highly (and sometimes infinitely) demagnified. High curvature of the time delay surface gives high demagnification. In talking about observations of strong lenses we often ignore these elusive maxima and talk about double and quadruple systems, though in principle the systems in Figs. 4b and 4c are respectively, triple and quintuple.

In Fig. 4c, where three of the images are arranged more-or-less tangentially to the lens, the corresponding component of the inverse magnification matrix is close to zero and the images are highly magnified in the tangential direction.

A useful mnemonic is that there are three "D's" associated with gravitational lensing: *delay, deflection and distortion*, which depend, respectively, on the zeroth, first, and second derivatives of the time delay function.

The magnification matrix might more properly be called the distortion matrix, but we bow to convention on this. The determinant of the magnification matrix gives the scalar factor by which the areas of sources are multiplied, thereby increasing or decreasing the flux received from the source.

7 Close Pairs, Critical Curves, and Caustics

7.1 Close Pairs of Bright Images and Critical Curves

In Fig. 6 we show a slice through the time delay surface along the line connecting a close pair of images. This slice does not tell us which of the two images might be a saddlepoint and whether the other is a minimum or a maximum. A perpendicular slice might show either a maximum or a minimum. But in either case, the maxima and minima within the slice shown are very shallow.

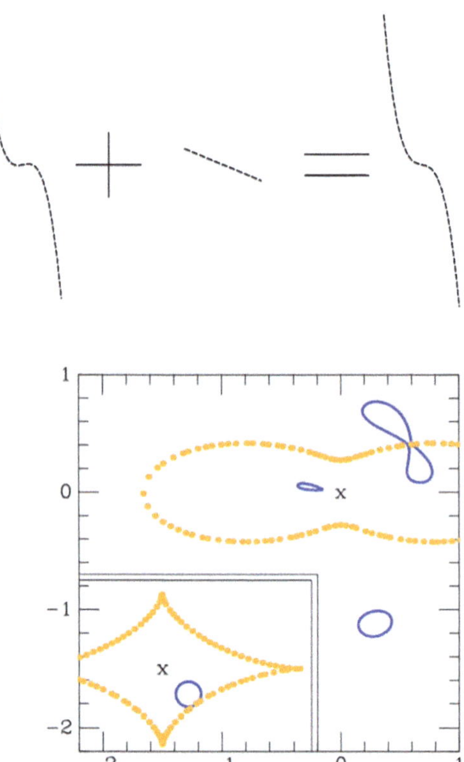

Fig. 6 A section through the time delay surface for a very close pair of images shows very little curvature, and therefore high magnification. Adding a small gradient to the time delay surface eliminates the two stationary points, and hence the two images. But they go out in a blaze of glory, becoming bright before they disappear

Fig. 7 A pair of merging images (upper right) of a circular source (inset) lensed by a **S**ingular **I**sothermal **E**lliptical **P**otential (SIEP). An x marks the center of the potential. The peanut shaped (dotted orange) critical curve shows the locus of points at which images can merge, and along which they have infinite magnification. Sources interior to the (orange dotted) diamond caustic (inset) produce four images. A short arc of the circular source lies outside the caustic and produces only two images

Displacing the source by a small amount or adding mass to the system would produce a correction to the slice through the time delay surface that is linear to first order. If the correction is large enough, we can eliminate these two stationary points. Just prior to their elimination, the two images get arbitrarily close and the curvature along the line connecting them gets arbitrarily small. As magnification is inversely proportional to curvature, that implies infinite magnification along the line connecting the images.

A locus of points along which a pair of images merge is called a "critical curve". One solves for the locus by setting the determinant of the inverse magnification matrix to zero, $|M^{-1}| = 0$. Pairs of images merging at a critical curve are called a "fold" configuration, and are discussed in exhaustive detail by Gaudi and Petters (2002). The pair of images is part of a larger multiplet – most often a quartet or triple – the other images of which can be far from the close pair and less highly magnified. The high magnifications of images close to critical curves draws our attention and result in selection effects that favor systems with close pairs over those with more widely spaced image configurations.

The derivative of the zero eigenvalue of M^{-1} along its eigendirection sets a scale over which the critical curve has significant effect – a kind of "swath of influence". The magnification of an image varies inversely as its distance from the critical curve, measured in units of this scale length.

In Fig. 7 we show two merging images of a circular source. Both images are lopsided – teardrop shaped – and merge along the peanut-shaped critical curve. While images across folds are, to first order, mirror images of each other, that approximation breaks down if the image is not much smaller than the curve's swath of influence.

7.2 Caustics

Critical curves in the image plane map into closed curves in the source plane called "caustics." A source inside a caustic produces two more images than it would outside the caustic." Segments of caustics with finite curvature are called "folds" (Gaudi and Petters 2002). The caustic shown in the inset to Fig. 7, an example of a "diamond caustic" has been calculated for the particular critical curve illustrated, a **S**ingular **I**sothermal **E**lliptical **P**otential (SIEP). The diamond caustic is symmetric about the elongation of the 2 dimensional gravitational potential. Its longer axis is parallel to the shorter axis of the elliptical potential. The diamond caustic is described mathematically by the equation for a stretched astroid. The caustic for a **S**ingular **I**sothermal **S**phere with E**x**ternal tide – SIS+XT – is also a stretched astroid.

Sources that lie near one of the two symmetry axes of a diamond caustic and close to one of its four "cusps" produce a characteristic triplet of images in a short, gentle arc. Cusp configurations are illustrated in Sect. 13.4 below.

8 Distances to and from Lenses and Sources

For simplicity we assume that we live in a ΛCDM universe – one with ordinary matter, "cold dark matter," and a "cosmological constant," Λ. Distances between points are calculated using the Friedman–Robertson–Walker (henceforth FRW) metric of general relativity.

The FRW metric has two salient features that affect strong lensing. First, the spatial separation between points has an overall scale factor, $a(t)$, that, in our universe, increases monotonically as time progresses. The redshift z_s of observed spectral lines in a distant source tell us that this scale factor was smaller when the light was emitted by a factor $1/(1+z_s)$ than it is now. Some of the complications introduced by this expansion are described in Sect. 8.2 below.

The second salient feature of the FRW metric is that distances radiating outward from the pole of the (arbitrarily) adopted coordinate system are calculated differently than those calculated perpendicular to those rays. This is a consequence of the curvature of spacetime, which need not be zero, but is very nearly so (Efstathiou and Gratton 2020).

8.1 Longitudinal and Transverse Distances (on the Earth)

A useful analogy is to imagine rays emanating from two sources at the same latitude $\mathcal{B}$ (expressed in radians) on the Earth's surface that are somehow constrained to propagate along its surface. An observer at the Earth's north pole will see them arriving along lines of constant longitude $\mathcal{L}$ separated by some angle θ (also in radians). The rays arriving at the pole from these sources have traveled a *longitudinal distance* D_{long} proportional to the product of that latitude and the radius of the Earth,

$$D_{long} = (\mathcal{B} - \pi/2) R_{\oplus} \quad . \tag{6}$$

However, the same pair of sources at a common latitude have a *transverse* separation between them, D_{tran}, calculated by multiplying the angular separation by a different factor,

$$D_{tran} = \cos \mathcal{B} R_{\oplus} \theta \quad . \tag{7}$$

This need for both distances is the direct consequence of the curvature of the Earth's surface.

8.2 Co-Moving Distances

The FRW metric is a recipe for computing distances in a coordinate system that is expanding with the universe. By convention, we use a preferred epoch (today) and origin (ourselves). We call these "co-moving coordinates." Distances that are computed in this coordinate system are called "co-moving distances."

We define the "longitudinal co-moving distance" from a source S to ourselves, χ^{OS}_{long} in this preferred coordinate system. As FRW spacetimes are generically curved, there is also a "transverse co-moving distance", denoted $f_K(\chi^{OS}_{long})$, where $K = 0, +1$, and -1 for flat, bowl-like and saddle-like topologies, respectively. It is used to calculate co-moving distances perpendicular to the line of sight.

Both co-moving distances are proportional to $a(t)$. For the special case of a flat universe, we have $f_K(\chi^{AB}_{long}) = \chi^{AB}_{long}$, but for positive and negative curvatures the expression is more complicated, involving, respectively, conventional and hyperbolic sine functions.

Transverse and longitudinal co-moving distances both have the special property that $\chi^{AC} = \chi^{AB} + \chi^{BC}$, which is *not* the case for some alternative kinds of cosmological distance.

8.3 Time of Flight

The time of flight from a source to an observer, T^{OS}, is calculated by taking small increments of the longitudinal co-moving distance, $d\chi$, dividing this by the speed of light to get an increment of time, and dividing each increment of time by the FRW scale factor $a(t)$.

We let $d\zeta$ be an increment of physical distance (measured with a meter stick) along the trajectory. Integrating from the source to the observer gives the time of flight. The scale factor is included to account for special relativistic time dilation, making the time of flight longer than it would otherwise be:

$$T^{LO} = \int_{lens}^{observer} \frac{d\chi^{\zeta}_{long}}{d\zeta}(1 + z_{\zeta})\frac{d\zeta}{c} \quad \text{at } z_{obs} = 0 \quad . \tag{8}$$

Dividing the time-of-flight by the speed of light and integrating we find

$$T^{LO} = \frac{\chi^{OL}_{long}}{c} \tag{9}$$

where χ^{OL}_{long} is the longitudinal co-moving distance.

Hogg (1999) gives a clear, pedagogical roadmap to the different kinds of cosmological distances and the time of flight. We use the same words he does to describe them, but adopt different notation. Hogg illustrates their dependences upon redshift, the cosmological constant Λ and a non-zero curvature, $\Omega_k \neq 0$. His clarity notwithstanding, the expressions he presents are dauntingly non-trivial, even to experienced astronomers.

9 Shapiro Gravitational Time Delay (in the Lens Plane)

The gravitational contribution to an observed time delay is intrinsically general relativistic in nature. But for sources strongly lensed by galaxies and clusters of galaxies, it can be calculated using the Newtonian gravitational potential, $\Phi(\vec{i}, \zeta)$, where $\vec{i}$ is physical position in the lens plane and ζ is physical position perpendicular to it. Since the gravitational delay occurs in or close to the lens, it can be calculated as though it occurs instantaneously as the ray traverses the lens plane, at redshift z_L. This is the driving virtue of the thin lens approximation of Sect. 4.

A great deal of lens modeling is carried out using what is somewhat misleadingly called a two dimensional "potential" $\Phi_{2D}(\vec{i})$, obtained by projecting the actual potential, $\Phi(\vec{i},\zeta)$ onto the lens plane. We define it here as

$$\Phi_{2D}(\vec{i}) \equiv \frac{1}{f_K(\chi_{long}^{OL})} \int_{source}^{observer} \frac{2\Phi(\vec{i},\zeta)}{c^2} d\zeta \quad , \tag{10}$$

where we have the used the transverse co-moving distance, $f_K(\chi_{long}^{OL})$ to render the projected potential dimensionless, a choice that simplifies calculation in what follows.

Following Binney and Merrifield (1998), we derive a general expression for this "Shapiro delay" $\tau_{shapiro}$, by integrating with respect to distance ζ perpendicular to the lens plane,

$$\tau_{shapiro} = \int_{source}^{observer} \frac{2\Phi(\vec{i},\zeta)}{c^2} \frac{d\zeta}{c} \quad , \tag{11}$$

where the use of perpendicular distance ζ is appropriate for small deflection angles. The Shapiro delay is then given by

$$\tau_{shapiro} = \frac{f_K(\chi_{long}^{OL})}{c} \Phi_{2D}(\vec{i}) \quad . \tag{12}$$

In the next section, we solve for the *angular* position of an image on the sky, $\vec{I}$, rather than the corresponding spatial position $\vec{i}$ within the lens plane. To avoid confusion, we represent the 2D potential as a function of angular position,

$$\Psi_{2D}(\vec{I}) = f_K(\chi_{long}^{OL})\Phi_{2D}\left(\frac{f_K(\chi_{long}^{OL})}{1+z_L}\vec{I}\right) \quad , \tag{13}$$

where $f_K(\chi_{long}^{OL})$, the transverse co-moving distance, turns the angle $\vec{I}$ into a co-moving distance and the factor $1+z_L$ accounts for the Hubble expansion.

10 Two Geometric Delays

10.1 Delays as Observed in the Lens Plane

The total time delay τ_{sum} is the sum of three terms: the general relativistic Shapiro delay, and *two* geometric terms, one from the lens plane to the observer and the other from the source to the lens plane. These must be calculated for an observer at rest with respect to the lens plane for consistency with the gravitational delay.

The transverse co-moving distance, $f_K(\chi_{long}^{OL})$ was defined in Sect. 8.2 to transform the angular separation between points perpendicular to the line of sight into co-moving distances. For the lens system illustrated in Fig. 8, it converts the angle between the lens and the image, into χ^{LI}:

$$\chi^{LI} = |\vec{L} - \vec{I}| f_K(\chi_{long}^{OL}) \quad , \tag{14}$$

where $\vec{L}$ and $\vec{I}$ are angular positions on the sky.

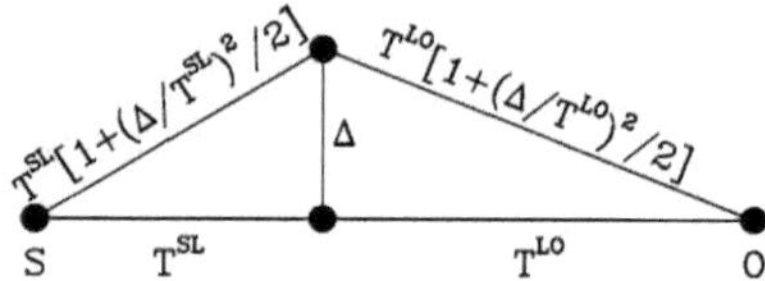

Fig. 8 Times of flight between source and lens, T^{SL}, and between lens and observer, T^{LO}, for a lens directly on the line of sight from the observer to the source. For images that are close to the lens, the Pythagorean theorem and the small angle approximation ($\Delta \ll T^{SL}$ and T^{LO}) give the times of flight from the source to the image, $T^{SL}[1+(\Delta/T^{SL})^2/2]$, and the time of flight from the image to the observer, $T^{LO}[1+(\Delta/T^{LO})^2/2]$

We take our lens to be thin and our deflection angles to be small, so that the lens and the images of the source can be taken to lie in a plane at the same redshift.

As noted in Sect. 8.2, the physical distance between points frozen into the Hubble expansion measured at redshift z_2 is larger (or smaller) than that measured at z_1 by the factor $(1+z_2)/(1+z_1)$. Hence the physical distance Δ^{LI} measured between the lens and an image in the same plane at redshift z_L is smaller than the co-moving distance that would be measured today, χ^{LI}, by a factor $1/(1+z_L)$:

$$\Delta^{LI} = \frac{1}{1+z_L}\chi^{LI} \quad , \tag{15}$$

and the time of flight measured in the lens plane is

$$T^{LI} = \frac{1}{c}\frac{1}{1+z_L}\chi^{LI} \quad . \tag{16}$$

Using the small angle approximation and the Pythagoran Theorem we get the total times of flight from the image to the observer and from the source to the image:

$$T^{IO} \approx T^{LO}\left[1+\frac{1}{2}\left(\frac{T^{LI}}{T^{LO}}\right)^2\right] \quad \text{and} \quad T^{SI} \approx T^{SL}\left[1+\frac{1}{2}\left(\frac{T^{LI}}{T^{SL}}\right)^2\right] \quad . \tag{17}$$

The time delay τ was defined to be the *extra* time of flight compared to the unlensed time of flight. Hence the geometric contribution to the time delay is

$$\tau^{SO}_{geom} = T^{LO}\frac{1}{2}\left(\frac{T^{LI}}{T^{LO}}\right)^2 + T^{SL}\frac{1}{2}\left(\frac{T^{LI}}{T^{SL}}\right)^2 \quad \text{or} \tag{18}$$

$$\tau^{SO}_{geom} = \frac{1}{2}\left[\frac{T^{LO}+T^{SL}}{T^{LO}T^{SL}}\right](T^{LI})^2 \quad . \tag{19}$$

10.2 The "Effective" Time of Flight

The square bracketted term in equation (19) will remind readers of the effective electrical resistance of two resistors connected in parallel in a circuit. By analogy, we define an "effective time of flight"

$$T^{SO}_{eff} \equiv \left[\frac{1}{T^{LO}} + \frac{1}{T^{SL}}\right]^{-1} \quad . \tag{20}$$

The frequency with which this effective quantity appears in strong lensing calculations warrants its being give a special designation, despite its not being the time of flight to anything. We tabulate values of T^{SO}_{eff} in Sect. 10.5 below. The redshifts typical of strongly lensed quasars and the galaxies lensing them are $z_S = 2.0$ and $z_L = 0.5$, giving a $T^{typical}_{eff} \sim \mathcal{O}(1/7)$ the age of the universe.

10.3 Net Time Delay

Summing the gravitational and geometric contributions to the time delay (as measured by an observer in the lens plane) given by equations (12) and (19) we have

$$\tau^{SO}_{sum} = \frac{\chi^{OL}_{long}}{c}\Psi_{2D}(\vec{I}) + \frac{1}{2}\left[\frac{1}{T^{OS}_{eff}}\right]\left(T^{LI}\right)^2 \quad \text{at } z_{lens} \quad . \tag{21}$$

10.4 Images Are Seen at Stationary Points of the Time Delay

As discussed in Sects. 3 and 7, images of the lensed source appear at stationary points of the time delay, where $\nabla_{\vec{I}}\tau^{SO}_{sum} = 0$. The gradients of the gravitational and geometric contributions with respect to angular position of the image on the sky, $\vec{I}$ are equal and opposite and cancel at minima, maxima and saddlepoints.

We have

$$0 = \frac{f_K(\chi^{OL}_{long})}{c}\nabla_{\vec{I}}\Psi_{2D}(\vec{I}) + \frac{1}{2}\left[\frac{1}{T^{OS}_{eff}}\right]\nabla_{\vec{I}}\left(T^{LI}\right)^2 \quad . \tag{22}$$

Substituting the expression for the time of flight T^{LI} in equation (16) and for the angular separation $\vec{I} - \vec{L}$ in equation (14) gives

$$0 = \frac{f_K(\chi^{OL}_{long})}{c}\nabla_{\vec{I}}\Psi_{2D}(\vec{I}) + \frac{1}{2}\left[\frac{1}{T^{OS}_{eff}}\right]\nabla_{\vec{I}}\left(\frac{1}{c}\frac{1}{1+z_L}|\vec{I} - \vec{L}|\, f_K(\chi^{OL}_{long})\right)^2 \quad . \tag{23}$$

Noting that $\nabla_{\vec{I}}|\vec{I} - \vec{L}|$ is a unit vector in the direction $\vec{I} - \vec{L}$ yields

$$0 = \frac{f_K(\chi^{OL}_{long})}{c}\nabla_{\vec{I}}\Psi_{2D}(\vec{I}) + \left[\frac{1}{T^{OS}_{eff}}\right]\left(\frac{1}{c}\frac{1}{1+z_L}f_K(\chi^{OL}_{long})\right)^2(\vec{I} - \vec{L}) \quad . \tag{24}$$

Equation (24) is a variant of what is called "The Lens Equation." It gives the angle between an image and the center of the potential, $\vec{I} - \vec{L}$, in terms of the gradient with respect to image angle of an appropriately defined two dimensional gravitational potential.

Most pedagogical treatments of strong lensing take the lens equation as the starting point for developing subsequent results. It involves quantities called "angular diameter distances" that we have assiduously avoided mentioning. Instead of these, our variant of the lens equation uses times of flight, which is a natural consequence of our taking Fermat's principle as our starting point.

Our hope is that newcomers to the field find it easier to think in terms of time of flight, letting them postpone dealing with the conventional approach until such time as they must communicate their results in presentations.

Table 2 Effective time of flight T_{eff} for respresentative source and lens redshifts

z_L	0.1	0.3	0.5	0.7	1.0
z_S					
0.2	0.041				
0.5	0.064	0.072	0.000		
1.0	0.070	0.115	0.106	0.073	0.000
2.0	0.072	0.127	0.137	0.130	0.106
5.0	0.068	0.117	0.129	0.129	0.120
9.9	0.061	0.097	0.104	0.104	0.098

10.5 Effective Time of Flight Tabulated

In Table 2 of the effective time of flight for a flat, $\Lambda = 0.7$ cosmology for some representative lens and source redshifts. Linear interpolation on the tabulated values gives a good first approximation. One must, however, evaluate the correct expressions if one's ultimate goal is precision cosmology.

11 Pointlike and Extended Sources: QSOs, SNe and Galaxies

Of order ~ 350 quasars have been observed to be strongly lensed (Lemon et al. 2023), of which ~ 60 are quadruple and the rest are double. The optical emission of quasars arises from accretion disks that are much too small to be resolved by today's telescopes – hence the term quasi-stellar objects – so QSOs are effectively point sources. Supernovae are likewise effectively point sources. In both cases the observables are the positions and fluxes of point images.

By contrast the galaxies that host QSOs and SNe can often be spatially resolved. They make it possible to see the distortions (and magnifications) introduced by strong lenses.

For the sake of simplicity, we consider here only two kinds of sources – pointlike and minimally extended. The latter are described by six numbers: two celestial coordinates, a flux, a semi-major axis, an axis ratio, and an orientation. Sources that are more than minimally extended and sufficiently bright can often be decomposed into pointlike and minimally extended subcomponents.

12 An Archetypical Lens Model: The Singular Isothermal Elliptical Potential with a Parallel External Tide

In the same way that our archetypical observed strong lensing system of Sect. 2 is intended as a starting point for thinking about observations, we adopt an archetypical model as a starting point for modelling strong lenses.

In Sect. 2 we modelled our observational archetype with a singular isothermal sphere (SIS), which has the property that images of sources are deflected radially by an angle θ_E.

In Sect. 9 we showed that the gravitational part of the time delay is proportional to a two-dimensional projected gravitational potential given by Equation (10). The 2D potential takes a particularly simple form for the SIS,

$$\Psi_{2D}^{SIS} = |\vec{\theta}|\theta_E \quad , \tag{25}$$

where the Einstein radius, θ_E gives the angular displacement of images from their sources. Note that the potential increases linearrly with distance $|\theta|$ from the center of the lens, giving equipotentials that are equally spaced. Should a source lie directly behind the lens, it is stretched into a circular "Einstein Ring" with radius θ_E.

The SIS model has another property that argues for its use: it produces the well known flat circular velocity curves observed in galaxies (Bosma 1978).

It can be generalized to include non-circular potentials. The singular isothermal elliptical potential (SIEP) has equipotentials that are similar concentric ellipses with "semi-ellipticity" η.[1]

The axis ratios of the equipotentials are $q = (1-\eta)/(1+\eta)$. The SIEP is an SIS potential that has been stretched by a factor $1+\eta$ in one direction and shrunk by a factor $1-\eta$ in the other. As equipotentials are closer along the short axis of the potential, the gradient of the gravitational component of time delay is steeper and minima and stationary points of the total time delay are further from the center of the lens.

Elongated equipotentials are also produced when an SIS lens experiences an e**X**ternal **T**ide from a neighboring galaxy or cluster of galaxies. It is analogous to the production of tides on the Earth's surface and gives a projected potential

$$\Psi_{2D}^{SIS+XT} = |\vec{\theta}|\theta_E - \frac{1}{2}\Gamma(\theta_x^2 - \theta_y^2) \quad . \tag{26}$$

where Γ is a measure of the strength of the tide and where we have chosen to put the perturbing mass on the x-axis of our coordinate system. Whatever shape a lensed image would have had in the absence of the tide, that image is further stretched by $(1+\Gamma)$ along the y-axis and squeezed by $(1-\Gamma)$ along the x-axis.

We use the unconventional word "scronching" (Ellenberg 2021) to describe stretching a circle or an ellipse by a factor of $(1+Q)$ in one direction and squeezing it by $(1-Q)$ in the other direction, where $Q < 1$. We reserve the word "shear" to describe the *net* deformation that results from multiple scronchings by different components of the lensing potential.

Both sources of flattening may contribute simultaneously. Calculation of the combined potential is greatly simplified if one takes them to be parallel to each other, giving the **S**ingular **I**sothermal **E**lliptical **P**otential plus parallel e**X**ternal **T**ide, abbreviated SIEP+$\text{XT}_{\parallel}$ lens (Luhtaru et al. 2021). We postpone discussion of the unusual, geometric perspective it permits until Sect. 13.5.

13 Configurations and Distortions of Single Sources

The defining characteristic of a strongly lensed system is the presence of multiple images of a single (or multiple) source(s). These multiple images are never exact copies either of the source or of each other. They are stretched, sometimes bent, and often have the opposite handedness of the source (though unresolvably so for QSOs and SNe).

We use the word "configuration" to describe the positions on the sky of the multiple images in a strongly lensed system. In this section we describe typical image configurations, working up from not-quite-strongly lensed singles through pairs, triplets, quartets and quintets, and note the characteristic distortions associated with the different configurations.

[1] The mass surface density contours for the SIEP are *not* elliptical, but the equipotentials for elliptical mass models are substantially rounder than the mass itself and are very nearly elliptical (Luhtaru et al. 2021).

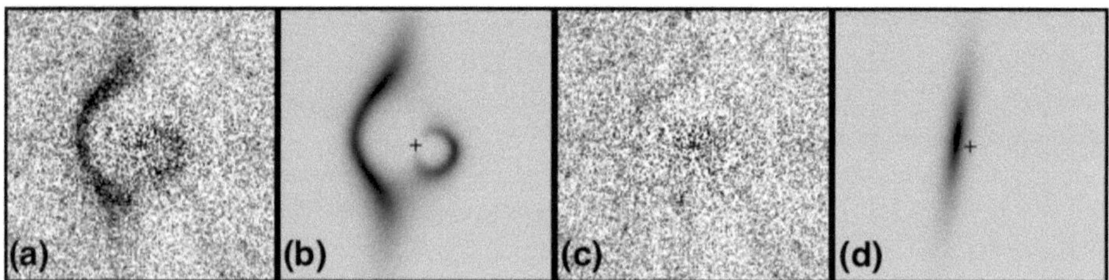

Fig. 9 (a) An HST image of the galaxy-galaxy lens system SDSS J0912+0029. The lensing galaxy (centered on the +) has been subtracted revealing two images of an edge-on background galaxy. (b) A model of the lensed images; (c) residuals from the model fit; (d) the model of the unlensed source

13.1 Not-Quite-Strongly Lensed Single Images

In Fig. 2 we used a source with a highly idealized shape to show that an SIS produces neither magnification nor demagnification in the radial direction but can either stretch or shrink a source in the tangential direction. A source that lies just outside θ_E produces a single image just beyond $2\theta_E$, stretched slightly less than a factor of two in the tangential direction.

This direction-dependent magnification has the effect of transforming an elliptical (minimally) extended source into another elliptical image, with a different semi-major axis, axis ratio and orientation. As the axis ratio will differ by at most a factor of two, a single lensed image is *almost* indistinguishable from an unlensed source. But if the source is more than minimally extended, some curvature will be visible, as described in Sect. 13.2.

13.2 Diametrically Opposed Double Configurations

Images form at minima, maxima and saddle points of the time of flight. One cannot add either a minimum or a maximum without also adding a saddle point. Single images must therefore be minima. When two images are observed, there must be a third image, somewhere. As discussed in Sect. 4, systems that appear to be double and quadruple are in fact triple and quintuple, but with a highly (or infinitely) demagnified image at a maxium of the time of flight.

Our archetypical strongly lensed system of Fig. 2 has images that are diametrically opposed to each other with a lensing galaxy along that diameter. A source at distance $\beta/\theta_E = 0.63$, gives a positive parity image outside the Einstein ring radius θ_E that is stretched by a factor of 2.58 in the tangential direction. It also produces a second, negative parity image (close to the center of the lens) that is squeezed by a factor of 0.73 in the tangential direction.

Had our source been circular, the outer image would have an axis ratio $q \sim 1/2.58$. But our contrived source is more than minimally extended, so it is curved parallel to the Einstein radius, as as seen in Fig. 2. A circular source would give an inner image with $q \sim 0.73$, but with its long axis pointing toward the lens.

Curvature is a generic feature of moderately extended sources, and can be quite dramatic for highly magnified sources. It gives rise to the "arcs" produced in clusters of galaxies that lens background galaxies.

Figure 9 (Bolton et al. 2008) shows the lensed images of a background galaxy that extends well beyond the Einstein ring, θ_E, of the lensing galaxy. Panel a) shows HST data obtained with the F814W filter. Panel b) shows the predictions from a model for the system. Panel c) shows the residual difference between the observations and predictions and panel

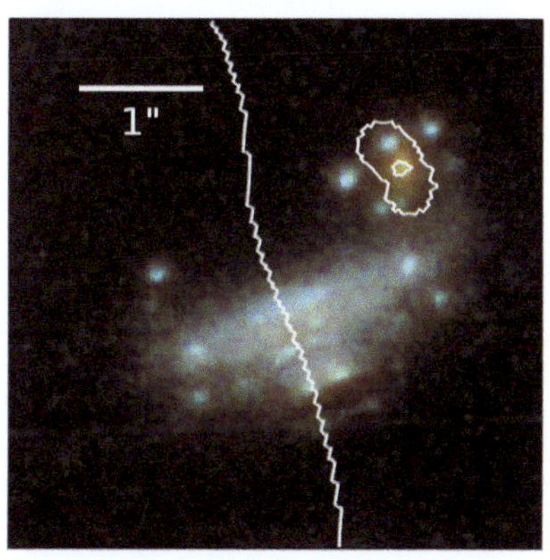

Fig. 10 Images (e)merging from/into a critical curve (white line) in the cluster SDSS J1226+2152. Dai et al. (2020) estimate the stretching perpendicular to the critical curve to be a factor of ~ 20 at 1" from it, but the bright spots are not elongated, indicating they are unresolved

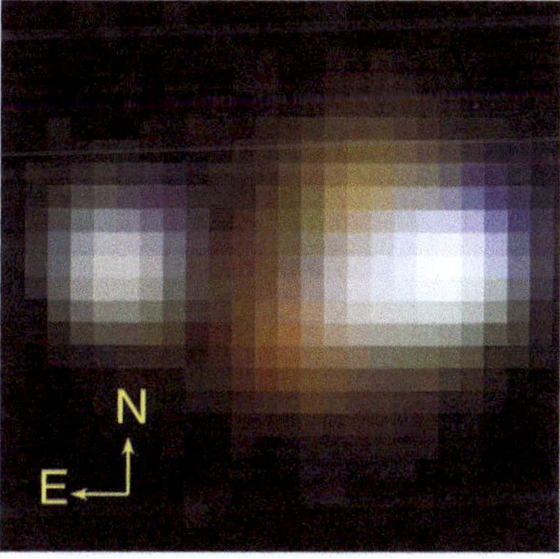

Fig. 11 WGD0316-41, an example of a "straddled saddle," with three bluish images of the quasar running East West, passing between two reddish lensing galaxies. The faintest of the three images is a saddlepoint of the time of flight, and lies close to one of the two minima. The galaxies are separated by $\sim 1.4''$

d) shows the model for the unlensed background galaxy. This image is unlike anything seen in unlensed galaxies, and its interpretation as a strongly lensed system is ironclad.

In Fig. 10 we see a galaxy lensed by the cluster of galaxies SDSS J1226+2152 (Dai et al. 2020) stretched perpendicular to a critical curve derived from a model. Pairs of blue knots, presumably mirror images of each other, straddle the critical curve. They must be unresolved because they are not elongated parallel to their separation.

13.3 Triplets from Straddled Saddles

There is a useful distinction to be made between triplets of lensed images formed between two lensing galaxies, and triplets resulting from "naked cusps," whose origin is described in Sect. 13.4 below. We call the former "straddled saddles" because they have two images that straddle a saddlepoint of the 2D gravitational potential. The inner image is a saddlepoint of the light of the time of flight, while the outer two are minima.

WGD0316-41 (Agnello, private communication), shown in Fig. 11, is an example of a straddled saddle. The middle image does not lie precisely between the two lensing galaxies because the source is offset from the saddlepoint of the 2D potential. Such configurations occur more frequently at the centers of clusters of galaxies (which sometimes have two dark matter concentrations) than between pairs of galaxies.

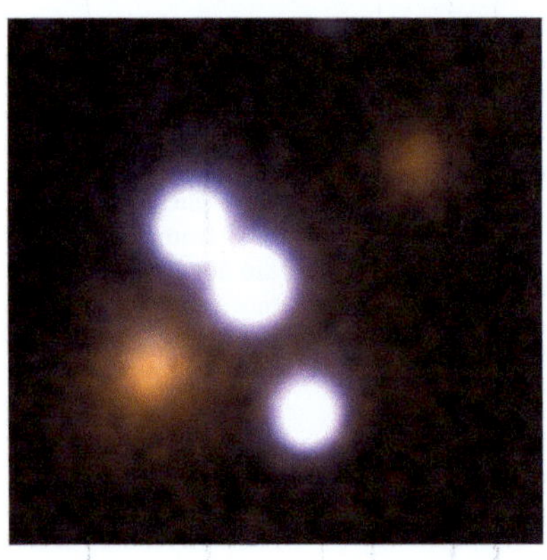

Fig. 12 J0457-78, which seems to be a naked cusp. There is no hint of a fourth image close to the lensing galaxy. A second, fainter galaxy may be seen to the upper right, in which case the potential may be elongated due to a tide rather the flattening of the lensing galaxy. If a tide is indeed the source of elongation, the topology of the lens is the same as for a straddeled saddle, and the distinction becomes one of degree

13.4 Triplets Associated with Cusps

A source that lies along one of the axes of symmetry of a diamond caustic produces a triplet of images that is symmetric around the corresponding axis in the image plane. Mathematically, the diamond caustic is a "scronched" astroid (Falor and Schechter 2022)[2] with four "cusps."

In the same way that the two images near a fold are highly magnified by the same factor (though with opposite parities), the two outer images of a cusp triple are highly magnified and have the same parity. The inner image is twice as highly magnified, and has the opposite parity. Cusps may have either positive or negative net parity.

Starting with a source on one of the symmetry axes and moving perpependicular to it, the cusp relation begins to fail, with two of the images converging and brightening and the other getting fainter. Few, if any investigators explain at what point they chose to call a particular triplet a cusp rather than a fold and a third image.

For most cusp configurations, there is a single diametrically opposed image on the other side of the galaxy whose parity differs from the combined parities of the 3 cusp images.

If one starts with the source at the center of the diamond caustic, the four images form a rhombus. As one moves the source along the longer axis of the diamond caustic, three images begin to converge and get further from the center of the lens. The diametrically opposed image (a saddlepoint) approaches the center of the lens. If the ellipticity of an SIEP is greater than 0.5, the fourth image merges with the central, infinitely demagnified image leaving only the three converging images. This configuration is called a "naked cusp." As with the straddled saddle, two positive parity images flank a negative parity image.

Tides can also cause a naked cusp. The system J0457-7820 (Lemon et al. 2023) shown in Fig. 12 looks at first like a naked cusp configuration. But there is a second, fainter galaxy in roughly the direction of the elongation of the potential. If this second galaxy were too faint to be observed, the system would indeed be a naked cusp. But as it is visible, the topology is that of a straddled saddle. The tidal interpretation is supported by the absence of strong flattening of the lensing galaxy.

13.5 A Taxonomy for Quadruple Image Configurarions

Most four-image configurations arise from flattened potentials, either those of galaxies or those of clusters of galaxies. Discounting translations, rotations and scalings, the space of

[2] A true astroid is the given by the locus of solutions of the equation $x^{\frac{2}{3}} + y^{\frac{2}{3}} = 1$.

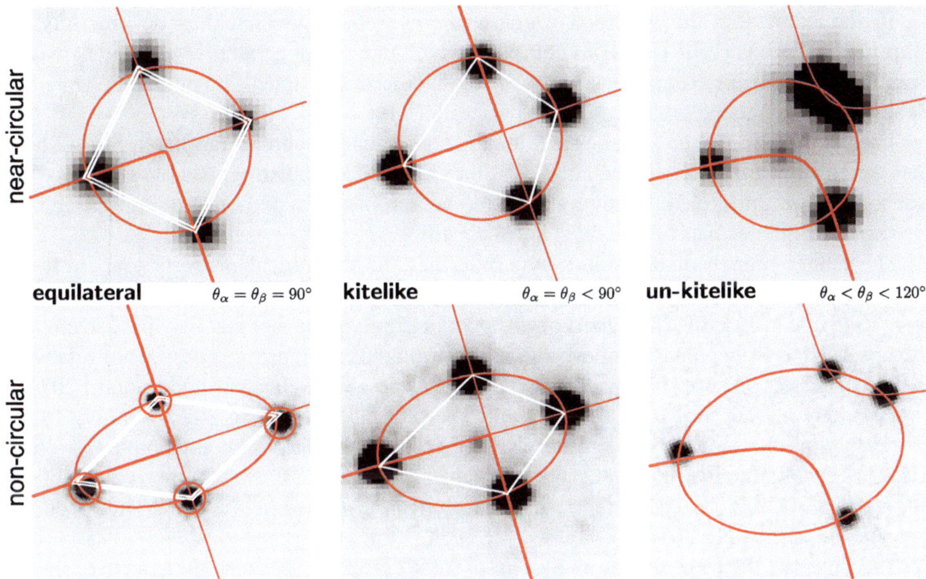

Fig. 13 A classification scheme for quadruple configurations of lensed quasars, illustrated with HST F814W exposures. The "equilateral", "kitelike" and "un-kitelike" configurations are distinguished by their deviations from axisymmetry along one or the other of the two diagonals. "Non-circular" configurations are stretched versions of "nearly circular" configurations

configurations is, at a minimum, 3 dimensional. Schechter (2022) developed a geometric classification scheme for quads that attempts to refine of the qualitative scheme of Saha and Williams (2003).

In this taxonomy, three dimensionless parameters are needed to describe each quad: its non-circularity, its symmetry about one or the other of its diagonals, and the position angle of that diagonal on the sky.

The classification scheme is *intentionally* imprecise. If on visual ("eyeball") examination, without using a ruler, the four images might conceivably lie on a circle, the system is "circular." Similarly, if without measuring, the four images might conceivably be symmetric about one *or* the other diagonals, the system is "kitelike." And if the system is conceivably symmetric about both diagonals, the system is "equilateral."

The scheme is illustrated in Fig. 13. The bottom row shows the clearly non-circular configurations while the top row shows the nearly circular ones. The leftmost column shows the "equilateral" configurations and the central column shows the "kitelike" ones. The rightmost column shows configurations that are "un-kitelike."

Red curves are superposed on the six lenses shown in Fig. 13, with the images lying very close to the four points where the curves intersects. We take the closed curves to be ellipses and the other two curves to be the branches of a hyperbola, which is a geometric representataion of a specific model, the archetypical SIEP+$\mathrm{XT}_{\parallel}$ model, described in Sect. 12.

Other models produce curves that will not, in general, be ellipses or hyperbola. The "lens equation", defined in Sect. 10.4, can be decomposed into a radial component (with respect to the source) and an azimuthal component. Each component of the lens equation gives a locus in the image plane. Images must satisfy both components, and must therefore lie where the loci cross.

To the extent that the loci look *roughly* like hyperbolae, which they do precisely for elliptical potentials (Witt 1996), and ellipses, which they do in general for potentials with a dominant quadrupole component and precisely isothermal elliptical potentials (Wynne and Schechter 2018), the geometric taxonomy is still useful.

The hyperbolae for the symmetric kitelike and equilateral configurations in Fig. 13 have very small semi-major axes – so small that they are almost indistinguishable from crosses. Yet for the two un-kitelike configurations, the branch of the hyperbola passing through the close pair of images barely intersects Wynne's ellipse.

The longer branch of the Witt's hyperbola has the theoretical property that it always passes through the center of the lens. Only the positions of the images (and not their fluxes) were used in constructing the hyperbolae shown in Fig. 13. The fact that they pass so close to the observed lensing galaxy supports the proposition that the lensing potentials of relatively isolated galaxies are very nearly centered on those galaxies (Schechter and Luhtaru 2024).

The non-circular configurations look like "scronched" (Ellenberg 2021) versions of the corresponding the circular configurations, squeezed along one of the asymptotes of Witt's hyperbola and stretched along the other. These asymptotes are the symmetry axes of the underlying gravitational potential, which is elongated *perpendicular* to Wynne's ellipse (Wynne and Schechter 2018).

The fluxes of the four images are not used in the taxonomy because the effects of micro-lensing can be quite large (Weisenbach et al. 2021). Note though that in the absence of micro-lensing, the four images in the equilateral configurations are expected to be nearly equal in flux.

For the un-kitelike configurations, the fluxes for the two closest images are expected to be roughly equal, and brighter than the faintest image by a factor that depends inversely upon their separation, as described in Sect. 7.1.

Readers are invited to judge for themselves the utility of the geometric taxonomy presented at length in this section by first trying to apply it to Fig. 5 and then examining Fig. 14, which shows Witt hyperbolae and Wynne ellipses overlaid. Imprecise as the scheme might be, it puts words to otherwise unarticulated geometric impressions that the images give.

13.6 Monopole + Quadrupole Quintets

The systems described at length in the previous section as quartets might be more accurately described as failed quintets. Had their potentials been slightly less cuspy, they would have produced observable fifth images close to the centers of their lenses. Instead those fifth images are highly or infinitely demagnified (Keeton 2003).

Though this is the rule for quasars lensed by galaxies, clusters of galaxies appear to be less cuspy and *do* show such fifth images.

13.7 Double Lens Quintets

At least three lensed quasars have five images, with the fifth image a saddlepoint attributable to a second lensing galaxy. The five brightest sources in Fig. 15 are images of the quasar PS J0630-1210 (Ostrovski et al. 2018). The faintest and most central of these is a saddle point of the time of flight, and lies between the two faint red galaxies.

13.8 Sextets, Septets *et Cetera*

Starting with a single lens, one can add an arbitrary number of additional images by adding additional lenses. If the added potentials are steeper than isothermal, each additional lens

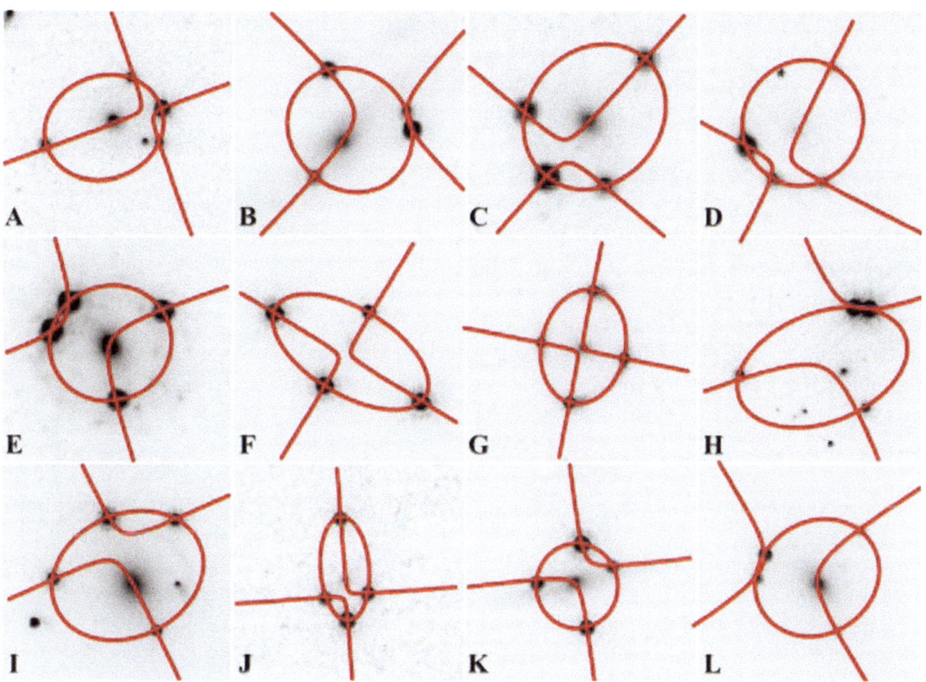

Fig. 14 The red curves show Witt hyperbolae and Wynne ellipses overlaid on the quadruply lensed quasars of Fig. 5. The authors would call systems D, E, K and perhaps B and L circular according to the geometric taxonomy advocated in this section. They would call A, C, D, I, J and K kitelike and F and G equilateral

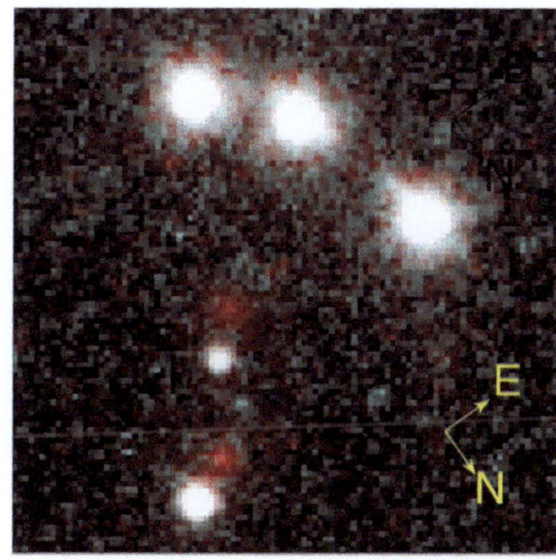

Fig. 15 A double lens quintet. Two galaxies, faint red blobs in this HST composite, produce five blue images of the quasar J0630-1210. The central, odd parity image lies in the saddle of the galaxies' gravitational potential

will add, in the simplest instances, a saddlepoint and an infinitely demagnified maximum. If they are less steep than isothermal, they add a saddlepoint and a maximum of finite magnification, provided that the central surface mass density of the added lens exceeds a value that depends upon the redshifts of the source and the lens.

13.9 The Witt–Wynne Construction: Geometric Not Algebraic

While the archetypical model in introduced in Sect. 2 can be represented by the algebraic expression for it gravitational potential, equation (26) (Luhtaru et al. 2021), we represented it *geometrically* in Sect. 13.5 with the Witt–Wynne construction. One does not need to know *any* physics to use it in modelling strongly lensed systems.

This is analogous to the case of Keplerian ellipses, which are geometric constructs that can be used to model planetary orbits without any understanding of Newton's Universal Law of Gravitation. In both cases one gets a different perspective on the model from the geometric construction than one gets from its algebraic representation. It leads to insights that do not come readily from mathematical expressions.

Readers can gauge for themselves the success of the Witt–Wynne construction in reproducing the positions of quadruply lensed quasars by examining Fig. 14. System "H" (J0818-26), for which it works least well, has a bright galaxy close to Witt's hyperbola and several fainter ones just inside or just outside Wynne's ellipse.

The archetypical SIEP+$XT_{\|}$ also produces algebraic expressions for the fluxes of the four images. These are not represented in the figure, as they are strongly influenced microlensing (Weisenbach et al. 2021).

Though there will be cases where the SIEP+$XT_{\|}$ model fails to capture interesting aspects of some particular system, it has the substantial advantage of permitting the direct, forward calculation of image positions, with no iteration (Falor and Schechter 2022). Comparison of Falor's method with two inverse schemes for solving the lens equation have given speedup factors in excess of twenty thousand.

14 Recapitulation

Three "Ds" capture the principal consequences of strong gravitational lensing: **D**elay, **D**eflection and **D**istortion. Photons traveling from a source to an observer are **D**elayed by the gravitational potential of an intervening lens. They are **D**eflected by gradients in that gravitational potential, increasing the path length. Fermat's principle dictates that images of the source form on trajectories for which the combined time of flight from these two contributions are a minimum, a maximum or a saddlepoint.

A circular source is **D**istorted by *gradients of the gradient* of the potential, which give an inverse magnification matrix. To first order the image of a circular source is stretched and/or squeezed into an ellipse.

Simple circularly symmetric gravitational potentials that are most negative at their centers (e.g. $-GM/r$) produce three images: a minimum, a maximum and a saddlepoint. More complicated potentials add images in pairs. One of these is a saddlepoint, and the other is either a minimum or maximum. Maxima tend to be highly de-magnified, sometimes infinitely, and often unseen.

The **S**ingular **I**sothermal **E**lliptical **P**otential (SIEP) can be treated as the archetypical lens, manifesting most of the basice elements of strong lensing. Adding a parallel **E**xternal **T**ide gives the Witt-Wynne construction, putting the images of a source at the points of intersection of a hyperbola and an ellipse. This model provides a geometric perspective on strong lensing and a good first approximation for many lensed systems.

The generic challenge in strong lensing is to use the shapes and positions of the observed multiple images to infer simultaneously both the photometric structure of the unlensed source and the structure of gravitational potential, which is often dominated by dark matter.

Acknowledgements We thank Caitlin Millard, Aaron Cohen, Luke Weisenbach and Adam Bolton who created and drafted figures for our use. We thank two reviewers for their guidance in clarifying our presentation.

Funding Information Open Access funding provided by the MIT Libraries. Some of the observations presented herein were made with the NASA/ESA Hubble Space Telescope and obtained from the Space Telescope Science Institute, which is operated by the Association of Universities for Research in Astronomy, Inc. under NASA contract NAS 5-26555. They were carried out under program HST-GO-15652.

Declarations

Competing Interests The authors declare that they have no conflicts of interest that are relevant to this article, nor are they affiliated or involvd with any organization or entity with any financial interest or non-financial interest in the subject matter or materials discussed in this manuscript.

Open Access This article is licensed under a Creative Commons Attribution 4.0 International License, which permits use, sharing, adaptation, distribution and reproduction in any medium or format, as long as you give appropriate credit to the original author(s) and the source, provide a link to the Creative Commons licence, and indicate if changes were made. The images or other third party material in this article are included in the article's Creative Commons licence, unless indicated otherwise in a credit line to the material. If material is not included in the article's Creative Commons licence and your intended use is not permitted by statutory regulation or exceeds the permitted use, you will need to obtain permission directly from the copyright holder. To view a copy of this licence, visit http://creativecommons.org/licenses/by/4.0/.

References

Binney J, Merrifield M (1998) Galactic astronomy. Princeton University Press, Princeton

Bolton AS, Burles S, Koopmans LVE, Treu T, Gavazzi R, Moustakas LA, Wayth R, Schlegel DJ (2008) The Sloan Lens ACS Survey. V. The full ACS strong-lens sample. Astrophys J 682(2):964–984. https://doi.org/10.1086/589327. arXiv:0805.1931 [astro-ph]

Bosma A (1978) The distribution and kinematics of neutral hydrogen in spiral galaxies of various morphological types. PhD thesis, University of Groningen, The Netherlands

Dai L, Kaurov AA, Sharon K, Florian M, Miralda-Escudé J, Venumadhav T, Frye B, Rigby JR, Bayliss M (2020) Asymmetric surface brightness structure of caustic crossing arc in SDSS J1226+2152: a case for dark matter substructure. Mon Not R Astron Soc 495(3):3192–3208. https://doi.org/10.1093/mnras/staa1355. arXiv:2001.00261 [astro-ph.GA]

Efstathiou G, Gratton S (2020) The evidence for a spatially flat Universe. Mon Not R Astron Soc 496(1):91–95. https://doi.org/10.1093/mnrasl/slaa093. arXiv:2002.06892 [astro-ph.CO]

Ellenberg J (2021) Shape: the hidden geometry of information, biology, strategy, democracy, and everything else. Penguin Press, New York

Falor C, Schechter PL (2022) The quadruple image configurations of asymptotically circular gravitational lenses. Astron J 164(4):120. https://doi.org/10.3847/1538-3881/ac80bc. arXiv:2205.06269 [gr-qc]

Gaudi BS, Petters AO (2002) Gravitational microlensing near caustics. I. Folds. Astrophys J 574(2):970–984. https://doi.org/10.1086/341063. arXiv:astro-ph/0112531 [astro-ph]

Hogg DW (1999) Distance measures in cosmology. arXiv e-prints arXiv:astro-ph/9905116 [astro-ph]. https://doi.org/10.48550/arXiv.astro-ph/9905116

Keeton CR (2003) Lensing and the centers of distant early-type galaxies. Astrophys J 582(1):17–29. https://doi.org/10.1086/344539. arXiv:astro-ph/0206243 [astro-ph]

Lemon C, Anguita T, Auger-Williams MW, Courbin F, Galan A, McMahon R, Neira F, Oguri M, Schechter P, Shajib A, Treu T, Agnello A, Spiniello C (2023) Gravitationally lensed quasars in Gaia - IV. 150 new lenses, quasar pairs, and projected quasars. Mon Not R Astron Soc 520(3):3305–3328. https://doi.org/10.1093/mnras/stac3721. arXiv:2206.07714 [astro-ph.GA]

Luhtaru R, Schechter PL, de Soto KM (2021) What makes quadruply lensed quasars quadruple?. Astrophys J 915(1):4. https://doi.org/10.3847/1538-4357/abfda1. arXiv:2102.08470 [astro-ph.GA]

Narayan R, Bartelmann M (1996) Lectures on Gravitational Lensing. arXiv e-prints arXiv:astro-ph/9606001 [astro-ph]. https://doi.org/10.48550/arXiv.astro-ph/9606001

Natarajan P, Williams LLR, Bradač M, Grillo C, Ghosh A, Sharon K, Wagner J (2024) Strong Lensing by Galaxy Clusters. Space Sci Rev 220(2):19. https://doi.org/10.1007/s11214-024-01051-8. arXiv:2403.06245[astro-ph.CO]

Ostrovski F, Lemon CA, Auger MW, McMahon RG, Fassnacht CD, Chen GC-F, Connolly AJ, Koposov SE, Pons E, Reed SL, Rusu CE (2018) The discovery of a five-image lensed quasar at z = 3.34 using PanSTARRS1 and Gaia. Mon Not R Astron Soc 473(1):116–120. https://doi.org/10.1093/mnrasl/slx173. arXiv:1709.08975 [astro-ph.GA]

Saha P, Williams LLR (2003) Qualitative theory for lensed QSOs. Astron J 125(6):2769–2782. https://doi.org/10.1086/375204. arXiv:astro-ph/0303261 [astro-ph]

Vernardos G, Sluse D, Pooley D. Schmidt RW, Millon M, Weisenbach L, Motta V, Anguita T, Saha P, O'Dowd M, Peel A, Schechter PL, (2024) Microlensing of Strongly Lensed Quasars. Space Sci Rev 220(2):14. https://doi.org/10.1007/s11214-024-01043-8. arXiv:2312.00931[astro-ph.GA]

Schechter PL (2022) A taxonomy for the configurations of quadruply lensed quasars. Astron J 164(3):113. https://doi.org/10.3847/1538-3881/ac82b5. arXiv:2205.11330 [astro-ph.GA]

Schechter PL, Luhtaru R (2024) Witt's hyperbola is both predicted and observed to pass close to the lensing galaxies in quadruple quasars. Astrophys J 975(1):62. https://doi.org/10.3847/1538-4357/ad7f4d. arXiv:2403.04092 [astro-ph.CO]

Weisenbach L, Schechter PL, Pontula S (2021) "Worst-case" microlensing in the identification and modeling of lensed quasars. Astrophys J 922(1):70. https://doi.org/10.3847/1538-4357/ac2228. arXiv:2105.08690 [astro-ph.GA]

Witt HJ (1996) Using quadruple lenses to probe the structure of the lensing galaxy. Astrophys J 472:1. https://doi.org/10.1086/310358. arXiv:astro-ph/9608197 [astro-ph]

Wynne RA, Schechter PL (2018) Robust modeling of quadruply lensed quasars (and random quartets) using Witt's hyperbola. arXiv e-prints arXiv:1808.06151 [astro-ph.CO]

Publisher's Note Springer Nature remains neutral with regard to jurisdictional claims in published maps and institutional affiliations.

Space Science Reviews (2024) 220:12
https://doi.org/10.1007/s11214-024-01041-w

Essentials of Strong Gravitational Lensing

Prasenjit Saha[1] · Dominique Sluse[2] · Jenny Wagner[3] · Liliya L.R. Williams[4]

Received: 14 April 2023 / Accepted: 30 December 2023 / Published online: 5 February 2024
© The Author(s) 2024

Abstract
Of order one in 10^3 quasars and high-redshift galaxies appears in the sky as multiple images as a result of gravitational lensing by unrelated galaxies and clusters that happen to be in the foreground. While the basic phenomenon is a straightforward consequence of general relativity, there are many non-obvious consequences that make multiple-image lensing systems (aka strong gravitational lenses) remarkable astrophysical probes in several different ways. This article is an introduction to the essential concepts and terminology in this area, emphasizing physical insight. The key construct is the Fermat potential or arrival-time surface: from it the standard lens equation, and the notions of image parities, magnification, critical curves, caustics, and degeneracies all follow. The advantages and limitations of the usual simplifying assumptions (geometrical optics, small angles, weak fields, thin lenses) are noted, and to the extent possible briefly, it is explained how to go beyond these. Some less well-known ideas are discussed at length: arguments using wavefronts show that much of the theory carries over unchanged to the regime of strong gravitational fields; saddle-point contours explain how even the most complicated image configurations are made up of just two ingredients. Orders of magnitude, and the question of why strong lensing is most common for objects at cosmological distance, are also discussed. The challenges of lens modeling, and diverse strategies developed to overcome them, are discussed in general terms, without many technical details.

Keywords Gravitational lensing · Strong gravitational lensing

✉ P. Saha
psaha@physik.uzh.ch

D. Sluse
dsluse@uliege.be

J. Wagner
thegravitygrinch@gmail.com

L.L.R. Williams
llrw@umn.edu

1 Physik-Institut, University of Zurich, Winterthurerstrasse 190, 8057 Zurich, Switzerland

2 STAR Institute, Quartier Agora Allée du six Août, 19c B-4000 Liège, Belgium

3 Bahamas Advanced Study Institute and Conferences, 4A Ocean Heights, Hill View Circle, Stella Maris, Long Island, Bahamas

4 School of Physics and Astronomy, University of Minnesota, 116 Church Street, Minneapolis, MN 55455, USA

Fig. 1 The archetype of strong gravitational lensing: Q0957+561. The two bright bluish objects with diffraction spikes are lensed images of a quasar at redshift $z = 1.41$. The elliptical galaxy near the lower quasar image, and the other galaxies in the field are part of a cluster at $z = 0.36$, which together form the gravitational lens. The field shown is $30'' \times 30''$ with North up and East to the left

1 The General Picture

Some time in centuries past, pieces of glass in shapes resembling lentil seeds (*Lens culinaris*) came to be known as lenses. The name remains with us: the optical elements developed for lighthouses in the 19th century are still called lenses, even though they look more like giant pineapples. A lens, then, is anything that transmits light with some non-trivial optical path or light travel time, causing light deflection. If an interesting optical path is created by a gravitational field, meaning the spacetime metric, we have a gravitational lens.

Light deflection in accordance with Einstein gravity was measured in 1919, and theorists have written about gravitational lensing for even longer,[1] but it was in the Einstein centennial year of 1979, with the discovery of the "double quasar" Q0957+561 by Walsh et al. (1979), that gravitational lensing became part of astronomy. A beautiful image of this system is shown in Fig. 1.

In the decades since 1979 more than a thousand multiply-imaging gravitational-lens systems comparable to Q0957+561 have been discovered, and the discovery rate is accelerating. This article will introduce the essential theory needed to study such systems, which are also called "strong-lensing" systems. The latter term contrasts the phenomenon of multiple image with weak-lensing, where there is light deflection and image distortion but not multiple images — such as in Giblin et al. (2021) on weak lensing by large-scale structure, or Crosta et al. (2017) on weak lensing in the solar system. Note, however, that strong lensing still occurs in weak gravitational fields — we will not consider light propagation near event horizons (as in the well-known images from Akiyama et al. 2022). Some of the theory is also applicable to gravitational microlensing by stars in or near the Milky way (e.g., Mróz et al. 2020), but mainly we will be concerned with lensing by galaxies and clusters at cosmological distances.

[1] For the fascinating scientific history, often at odds with textbook narratives, see Valls-Gabaud (2006) and Kennefick (2012).

1.1 Light Deflection

The practice of strong gravitational lensing involves some simplifying assumptions.

First, geometrical optics is assumed for light sources and the gravitational-lensing processes. Wave optics is essential at the telescope, but we do not consider interference between multiple images. Appendix C briefly discusses how the standard formalism can be extended to wave optics, but for the rest of the present article, light means rays.

Second, gravitational fields are assumed to be weak. As a consequence, light rays are deflected from straight lines only by very small angles. Furthermore, the contributions of many masses can be simply added (or integrated over) without the need to solve Einstein's field equations. Thus, lensing is treated as a first-order perturbation from light travelling along straight lines.

A third assumption is that light deflection occurs over regions much smaller than the distances between light source, lensing mass, and observer. This assumption is expressed in the notions of a *lens plane* and a *source plane*. It is not meant that sources and lenses are flat, but rather that the gradual deflection of light through a gravitational field is integrated along the line of sight through the comparatively small active region, and approximated as a discrete deflection. Distances transverse to the line of sight, however, are treated with respect, and not unceremoniously integrated out.

The well-known expression

$$\frac{4GM_\odot}{c^2 R_\odot} = 1.75'' \simeq 1 \times 10^{-5}\ \text{rad} \tag{1}$$

for gravitational light deflection at the rim of the Sun exemplifies all three assumptions: (i) geometrical optics is understood, (ii) the metric perturbations are of the same order as the deflection in radians, and thus small, and (iii) the Sun is approximated as a flat deflector in the sky. Light deflections by galaxies are coincidentally of the same order as that by the Sun, while clusters of galaxies produce deflections an order of magnitude larger. Thus, the light deflections that will concern us in practice are $\lesssim 1 \times 10^{-4}$ rad and entirely in the regime of weak gravitational fields.

Introductions to gravitational lensing (e.g., Meneghetti 2021) conventionally start by specializing to the regime of geometrical optics, weak fields, and a planar lens. Some of the essential concepts, however, are also valid for strong fields. So let us start a little outside the usual comfort zone, with an example of lensing by a strong gravitational field.

Figure 2 shows light deflection through one of the simplest curved spacetimes. The metric (see Appendix A for details) is similar to a Schwarzschild metric, but has no event horizon (which, though fascinating, would take too long to discuss here). Light rays originate from a point source high above the figure, and are gradually diverging as they approach the strong field in the middle of the figure, and then the lens pulls the rays together. There is no focal point, though there is a sort of focal line along which rays converge symmetrically from paths on opposite sides of the lens. Two such ray intersections on the focal line are shown in the figure. Observers located there will observe a ring of light, provided the lens is circularly symmetric. There are also asymmetric ray intersections (two of which are indicated) in the lower part of the figure, not on the focal line. Observers located there will see the source light coming from two directions in their sky.

The rays near the middle of Fig. 2 undergo very large deflections, even more than 180°. Large deflections occur in strong gravitational fields near neutron stars and black holes, leading to the exotic multiple imaging now observed near the M87 black hole (Broderick et al. 2022). It does not, however, occur in the systems we will consider.

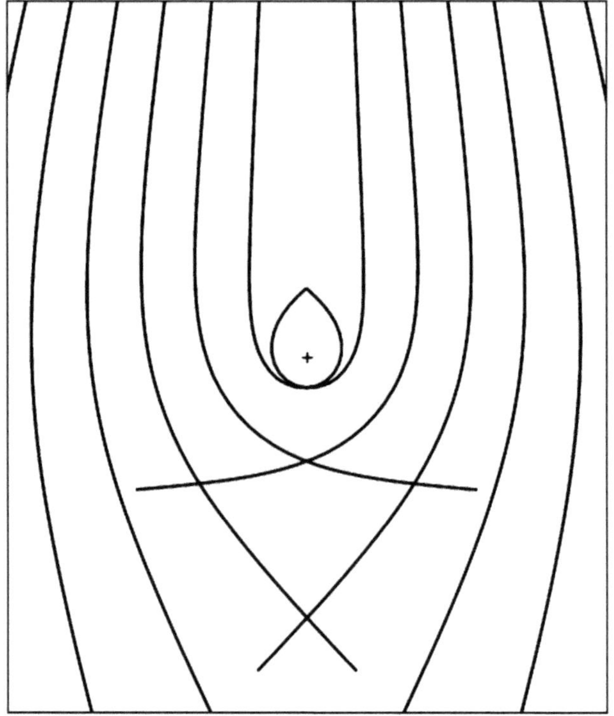

Fig. 2 A representation of the deflection of light rays. The source is far above, outside the figure. There is a strong gravitational field near the center of the figure (marked with a +), which produces large deflections (even > 180°). Farther from the center, the field is weaker and the deflections smaller, though still much larger than the real systems we are considering

Lensing theory conventionally uses observer-centric coordinates. To a given observer, a source appears to be located along a direction (say $\boldsymbol{\theta}$), whereas its true location is along some different direction (say $\boldsymbol{\beta}$). The directions $\boldsymbol{\theta}$ and $\boldsymbol{\beta}$ are conventionally expressed as two-dimensional angles on the sky, measured in radians or arc-seconds. The mapping relating $\boldsymbol{\theta}$ and $\boldsymbol{\beta}$ is called the lens equation, and is conventionally expressed as

$$\boldsymbol{\beta} = \boldsymbol{\theta} - \boldsymbol{\alpha}(\boldsymbol{\theta}) \tag{2}$$

with $\boldsymbol{\alpha}(\boldsymbol{\theta})$ having the interpretation of an apparent deflection angle. Adding angles in this way, one is tacitly assuming that the sky has been mapped to a plane. For large deflections we would need to specify the sky projection, but if all angles are small, at any time, only a very small patch of sky is relevant, and angles can be simply added.

The convention of using observer-centric coordinates seems peculiar to gravitational lensing. In optics, it is usual to put the origin at the lens. Even within astronomy, the study of planetary atmospheres as lenses for occulted stars (see e.g., Elliot and Olkin 1996) uses lens-centric coordinates. Observer-centric coordinates are, however, a good idea, because some useful facts get built into the notation. As we can see from the form of the lens equation (2), the mapping from image plane $\boldsymbol{\theta}$ to source plane $\boldsymbol{\beta}$ is single valued, but the inverse can be multiple valued. Thus each image corresponds to a unique source, whereas each source can have multiple images. Ray-tracing using the lens equation (2) using your favorite deflection field $\boldsymbol{\alpha}(\boldsymbol{\theta})$ is easy and can be entertaining.[2] Each pixel in the image plane is given the bright-

[2]Try visiting

https://phdenzel.github.io/streaming-lens/

on a device with a camera.

ness of the corresponding pixel in the source plane. (Since $\boldsymbol{\beta}$ is a single-valued function of $\boldsymbol{\theta}$, but not the other way round, ray tracing is usually done backwards, from the observer's sky to the source.) Note that ray tracing in this way automatically preserves surface brightness (photons per steradian). A patch of uniform brightness on the source stays the same uniform brightness when lensed, but it can change in size and shape.

The conventional minus sign for the deflection in Eq. (2) means that $\boldsymbol{\alpha}(\boldsymbol{\theta})$ points away from the deflector rather towards it. This is an example of a well-known principle in astrophysics,[3] which is that you first think of how a rational person would do it, and then you *change the sign.*

1.2 Wavefronts and Fermat's Principle

The lens equation (2) would actually be too general for lenses, if we let $\boldsymbol{\alpha}(\boldsymbol{\theta})$ be arbitrary. This is because you cannot send light any way you like using just lenses — for that you would need optical fibers. Imagine a cable consisting of a bundle of twisting optical fibers, and a signal input simultaneously into all the fibers. The overall direction of travel of the light signal along the cable will be different from the direction of travel within individual fibers. Lenses cannot make light travel like this. In lensing, the surface or front described by a bundle of light rays as light travels is always normal to the rays themselves. This front is known as the wavefront. (Despite the name, the wavefront is well-defined in geometrical optics. If the source is small but not effectively a point, the wavefront will be wiggly at the wavelength scale, and will not produce interference.)

Figure 3 is a representation of the wavefront corresponding to the rays in Fig. 2. Here, we imagine the source emitting an omni-directional light flash, and the wavefront shows how far that light flash has travelled in a given time in different directions. Light rays are simply the local normals to the wavefront. Notice that the lens delays the wavefront — drastically in the strong-field region, less in other places. Most interesting for strong gravitational lensing, some observers will be crossed by the wavefront more than once, at different times and from different directions.

Let us now modify the preceding to the scenario shown in Fig. 4. Let us pivot the source (located above and outside the figure) by a small angle $\boldsymbol{\beta}$ about the lens. The incoming wavefront is then inclined by $\boldsymbol{\beta}$. After the wavefront has passed the lens and been delayed by it, we freeze the wavefront. Then we imagine a second backwards wavefront, which emanates from an observer at the bottom of the figure. Any part of this second wavefront can be labelled by the angle $\boldsymbol{\theta}$ on the observer's sky. It is like the observer's sky rising up to meet the frozen wavefront. Let $t(\boldsymbol{\theta})$ be the time at which a point $\boldsymbol{\theta}$ on the rising-sky wavefront meets the original frozen wavefront. Let us now concentrate on any places where the two wavefronts meet tangentially. These correspond to $\nabla t(\boldsymbol{\theta}) = 0$. At these places the normals to the two wavefronts are the same. Since normals to the wavefront are light rays, $\nabla t(\boldsymbol{\theta}) = 0$ corresponds to light rays travelling out from the source with a wavefront and then down the sky to the observer. Moreover, $t(\boldsymbol{\theta})$ at these sky locations is the light travel time, apart from an additive constant. This is Fermat's principle in gravitational lensing.

The two-wavefront picture of lensing (due to Nityananda 1990) is remarkable in that, from the basic postulate that light rays are normal to a wavefront, it leads inexorably to Fermat's principle and the notion of an abstract surface $t(\boldsymbol{\theta})$, even with large deflections and strong fields. We assumed a stationary lens, but in fact Fermat's principle in terms of an arrival time can be extended to arbitrary time-varying gravitational fields in general relativity

[3] Attributed to J. Binney.

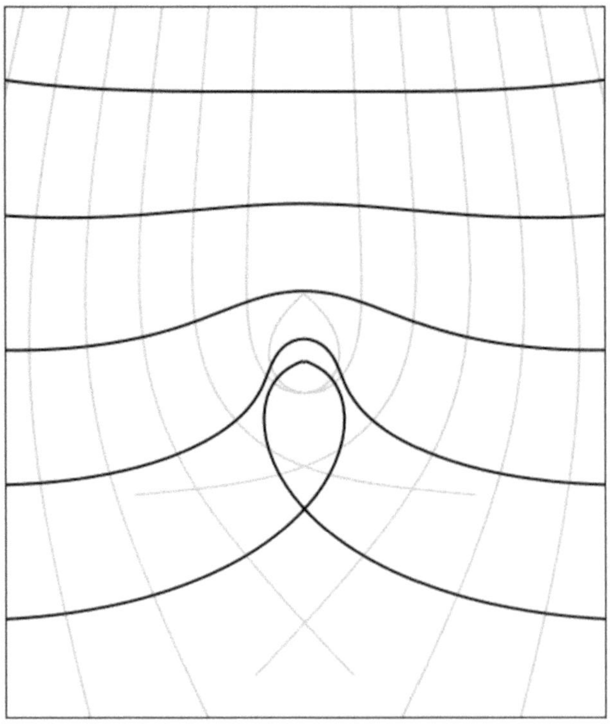

Fig. 3 Wavefronts corresponding to the light rays in Fig. 2. (The source is above outside the figure.) Light gray curves are taken over from that figure. Each of the black curves is the endpoint of all the rays at a particular time. The small loop is the result of very large deflections and would only occur in a strong gravitational field

(Nityananda and Samuel 1992). In the following we will specialize to small deflections, but it is worth emphasizing that the notion of the arrival-time surface $t(\boldsymbol{\theta})$ and its consequences have much wider applicability.

1.3 Geometrical and Gravitational Time Delays

Let us now turn Fig. 4 into equations, under the following simplifying assumptions.

- All angles are taken to be small, in the sense of $\sin\theta \approx \theta$ and $\cos\theta \approx 1 - \frac{1}{2}\theta^2$.
- Fig. 4 has the source far away from the lens. We now assume that the source is at infinity. (We will relax this assumption later.) As before, on the observer's sky the source is located at location $\boldsymbol{\beta}$.
- The gravitational field is assumed localized to the neighborhood of the lens. The backward wavefront is therefore spherical.

Under these assumptions, the frozen wavefront just after passing the lens is a slightly warped plane inclined by $\boldsymbol{\beta}$, while the dashed backward wavefront is $\propto \theta^2$.

Without any lensing mass, the frozen wavefront is simply a plane, and the time taken for backward wavefront to reach it follows from geometry as

$$t_{\rm geom}(\boldsymbol{\theta}) = \frac{D}{c}\left(\tfrac{1}{2}|\boldsymbol{\theta}|^2 - \boldsymbol{\theta}\cdot\boldsymbol{\beta}\right) + \text{const}\,, \tag{3}$$

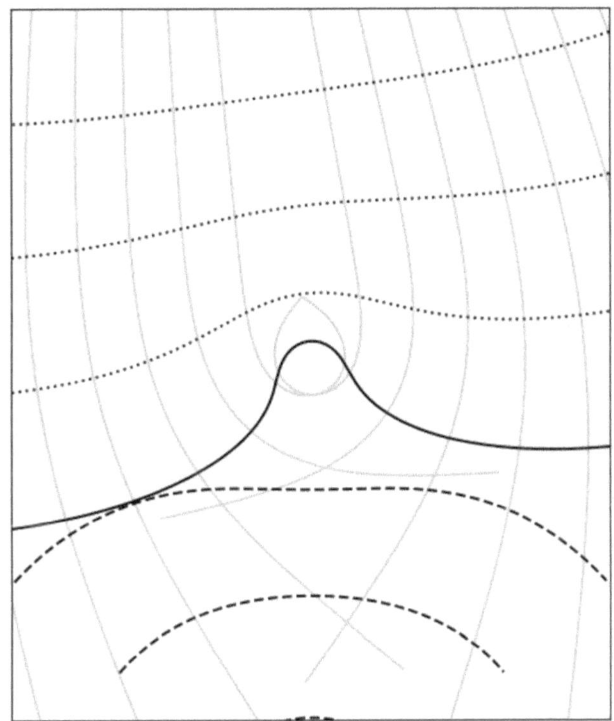

Fig. 4 Modified from Fig. 3. The source (outside the figure) has been moved slightly to the left. The wavefront proceeds as the dotted curves and is then frozen at the solid black curve. Then a second backward wavefront (dashed) emerges from the observer on the ground. Each of the dashed curves is the endpoint at a particular time of rays emanating from the observer, but those rays are not shown

where D is the shortest distance from the observer to the frozen wavefront. Since $\boldsymbol{\beta}$ is constant, we can equivalently write

$$t_{\text{geom}}(\boldsymbol{\theta}) = \frac{D}{2c}|\boldsymbol{\theta} - \boldsymbol{\beta}|^2 . + \text{const} \tag{4}$$

Adding a lensing mass, the frozen wavefront gets warped, and the time taken by the backward wavefront to reach it changes by say $t_{\text{grav}}(\boldsymbol{\theta})$. The light travel time, up to a constant, would then be

$$t(\boldsymbol{\theta}) = t_{\text{grav}}(\boldsymbol{\theta}) + t_{\text{geom}}(\boldsymbol{\theta}) . \tag{5}$$

The form of $t_{\text{grav}}(\boldsymbol{\theta})$ is not obvious, but let us put in the ansatz

$$t_{\text{grav}}(\boldsymbol{\theta}) = -\frac{G}{c^3}\sum_i M_i \ln|\boldsymbol{\theta} - \boldsymbol{\theta}_i| \tag{6}$$

pending verification that it leads to the correct deflection angle. Note here, that, since the angles are all small, the logarithm will always be negative, and because of the minus sign the contribution of the mass term will always be positive. Light rays will reach the observer and images will appear where the gradient of $\nabla t(\boldsymbol{\theta}) = 0$. Working out the gradient we have

$$\boldsymbol{\beta} = \boldsymbol{\theta} - \frac{4G}{c^2 D}\sum_i M_i \frac{\boldsymbol{\theta} - \boldsymbol{\theta}_i}{|\boldsymbol{\theta} - \boldsymbol{\theta}_i|^2} \tag{7}$$

which, as we can see, is the lens equation (2) for a sum of Sun-like deflectors at distance D. As throughout this section, the source is much further away, a restriction we will lift later. This verifies our ansatz (6) for the gravitational time delay.

The combined geometric and gravitational time delay

$$t(\boldsymbol{\theta}) = \frac{D}{2c}|\boldsymbol{\theta} - \boldsymbol{\beta}|^2 - \frac{G}{c^3}\sum_i M_i \ln|\boldsymbol{\theta} - \boldsymbol{\theta}_i|, \tag{8}$$

also variously known as the arrival time or the Fermat potential, is a central concept in gravitational lensing. As we have seen, the lens equation is the zero-gradient condition on the time delay. The second and higher derivatives are also important, as detailed later.

Fermat's principle in gravitational lensing was introduced using different approaches by Schneider (1985) and Blandford and Narayan (1986). Expressions like (8) do, however, appear in earlier work (Cooke and Kantowski 1975; Borgeest 1983) albeit without the interpretation as Fermat's principle. Wavefront arguments were introduced to gravitational lensing by Refsdal (1964). The gravitational time delay (6) is known as the Shapiro delay after its deduction by Shapiro (1964) as a property of null geodesics from general relativity. That time delays also emerge simply as an integral of the deflection angle was shown independently (and actually slightly earlier) by Refsdal (1964).

2 The Cosmological Context

2.1 Why Cosmological Distances?

The lensing galaxy in Q0957+561 is at redshift $z = 0.36$, which is a typical distance for strong lenses. Why don't we see strong lenses nearby?

To see why, let us briefly consider the Sun. The Sun is assuredly a gravitational lens, but we do not observe it as a strong gravitational lens, because the light deflection at its rim (Eq. (1)) is much smaller than its angular size on our sky. If one observed from a much greater distance D, the apparent size of the Sun would become smaller than the deflection angle, and sources directly behind the Sun would be lensed into images on either side. We can express this fact as

$$\frac{4GM}{c^2\xi} = \frac{\xi}{D} \tag{9}$$

together with the condition that ξ (the physical radial distance of the apparent images on the observer's sky) is larger than the solar radius.

What if the Sun were transparent, or we were looking for strongly-lensed neutrinos? Then the requirement that ξ is more than the solar radius would be lifted, and the condition (9) would apply as

$$\frac{4GM(<\xi)}{\xi^2} = \frac{c^2}{D} \tag{10}$$

with the mass enclosed within ξ. Note the latter form — it says that strong lensing can occur if the Newtonian acceleration at the lens is at least c^2/D. (This condition is usually expressed as a critical 2D density M/ξ^2, but let us stay with threshold acceleration for now.) From $D = 1$ au the threshold is too high even for the solar interior, but for $D > 25$ au a transparent Sun will indeed strongly lens neutrinos (Demkov and Puchkov 2000). For $D > 550$ au the optical solar gravity lens would be observable, and is indeed the subject of some space-mission concepts (Turyshev and Toth 2020). But for now, strong lensing needs much larger distances and lower acceleration thresholds.

Let us consider some example cases of the threshold acceleration (10). The values[4]

$$M = M_\odot \qquad \xi \simeq 2\,\mathrm{c\,s} \qquad D \simeq 400\,\mathrm{au} \tag{11}$$

correspond roughly to optical strong lensing by the Sun (the solar gravity lens). If we scale ξ up by 10^3 and D up by 10^6 we get

$$M = M_\odot \qquad \xi \simeq 4\,\mathrm{au} \qquad D \simeq 2\,\mathrm{kpc} \tag{12}$$

which is in the regime of microlensing in the Milky Way. However, the angular scale $\xi/D = 1 \times 10^{-8}$ rad $\simeq 2$ mas which is challenging to image even with optical interferometry (cf. Cassan et al. 2022). If, instead, we scale all three quantities by 10^{12} we have

$$M = 10^{12} M_\odot \qquad R \simeq 20\,\mathrm{kpc} \qquad D \simeq 2\,\mathrm{Gpc}\,, \tag{13}$$

which is typical of massive galaxies at cosmological distances. Moreover the angular scale $\xi/D = 1 \times 10^{-5}$ rad $\simeq 2''$ is comparatively easy to resolve.

Thus, we see that strong lensing is very difficult to observe nearby, but once you can observe galaxies at cosmological distance with arcsecond resolution, strong lensing becomes more frequent.

2.2 Distances in Cosmology

With lenses being at cosmological distances, we need to account for the universe expanding as the light is travelling through it. This is quite straightforward to do, using the usual machinery of modern cosmology, which is covered in many textbooks, and need not be repeated here. Let us nonetheless draw attention to the main concept relevant to lensing, namely the different ways of expressing distance.

The observable associated with distance is the redshift z. While an observed z may contain a small kinematic contribution (of order 10^{-3}), cosmological redshifts are essentially wavelengths expanding with the universe. Thus a redshift of z refers to an epoch when the scale factor of the universe was $1+z$ smaller than now. This applies to any cosmological model that is homogeneous and isotropic on large scales (that is, satisfies the cosmological principle).

In the current standard cosmological model, the universe is spatially flat and its present total energy density is

$$\rho_{\mathrm{cr}} = \frac{3H_0^2}{8\pi G}\,, \tag{14}$$

in which H_0 is the current expansion rate, that is, the Hubble constant.[5] The density at redshift z is ρ_{cr} times

$$E(z) = \Omega_m (1+z)^3 + \Omega_r (1+z)^4 + \Omega_\Lambda\,, \tag{15}$$

[4]Here it is useful to remember that 1 au $\simeq 500$ c s (light-seconds), and 1 pc $\simeq 10^8$ c s (see Appendix B for more precise values).

[5]Distances, times, and angle-size relations needed for studying lenses involve several unit conversions for H_0, which sets the scales. Appendix B gives the relevant conversions.

in which Ω_m is the current matter density fraction, Ω_r is the current relativistic energy-density fraction, and Ω_Λ the "dark energy" density fraction.

The lookback time to some redshift z_d is given by

$$t_\mathrm{d} = \frac{1}{H_0} \int_{z_\mathrm{d}} \frac{dz}{(1+z)\,E(z)} \tag{16}$$

and ct_d is by definition the distance travelled by light from the source to the observer. The comoving distance between two points along the line of sight at redshifts z_d and z_s is given by

$$\chi_\mathrm{ds} = \frac{c}{H_0} \int_{z_\mathrm{d}}^{z_\mathrm{s}} \frac{dz}{E(z)} \tag{17}$$

As the name indicates, the comoving distance scales up distances according to the present scale of the universe.

Gravitationally bound structures like galaxies and clusters do not expand with the universe. But the light coming from them knows[6] nothing of such subtleties, and shows the angular sizes of galaxies and clusters as if they *had* been expanding with the universe. To account for this, angular-size or angular-diameter distances are defined. The angular-diameter distance to the lens (or deflector) and source are

$$D_\mathrm{d} = \frac{\chi_\mathrm{d}}{1+z_\mathrm{d}} \qquad D_\mathrm{s} = \frac{\chi_\mathrm{s}}{1+z_\mathrm{s}}, \tag{18}$$

respectively, with the subscripts having the obvious meanings. Also important is the angular-diameter distance

$$D_\mathrm{ds} = \frac{\chi_\mathrm{s} - \chi_\mathrm{d}}{1+z_\mathrm{s}} \tag{19}$$

from the lens to the source. The latter needs clarification. An observer at the lens would measure the angular distance as $(1+z_\mathrm{d})/(1+z_\mathrm{s})$ times the comoving distance. But that observer's comoving distances are all smaller by $(1+z_\mathrm{d})$ than ours. Hence the expression (19).

In the older literature (notably in the well-known book by Schneider et al. 1992) cosmologies with total density different from ρ_cr are also considered, as are more general considerations of the angular-diameter distances. The expressions given here are restricted to cosmologies that are spatially flat (or equivalently, have mean density equal to ρ_cr), but this is a standard assumption made in nearly all of the recent literature.

Yet another quantity which one occasionally meets in lensing formalism is the conformal time. A conformal time interval is given by $(1+z)^{-1}\,dt$. Using the conformal time, one can trace light rays as if in a static universe. In the previous section, we used some arguments involving running light rays backwards, and the arguments remain valid in an expanding universe if conformal time is understood.

2.3 The Lens and Source Planes

With angular-diameter distances in hand, we can now extend the lens equation and the time delay to the expanding universe, and relax the assumption that sources are at infinity.

[6] This is known as anthropomorphism.

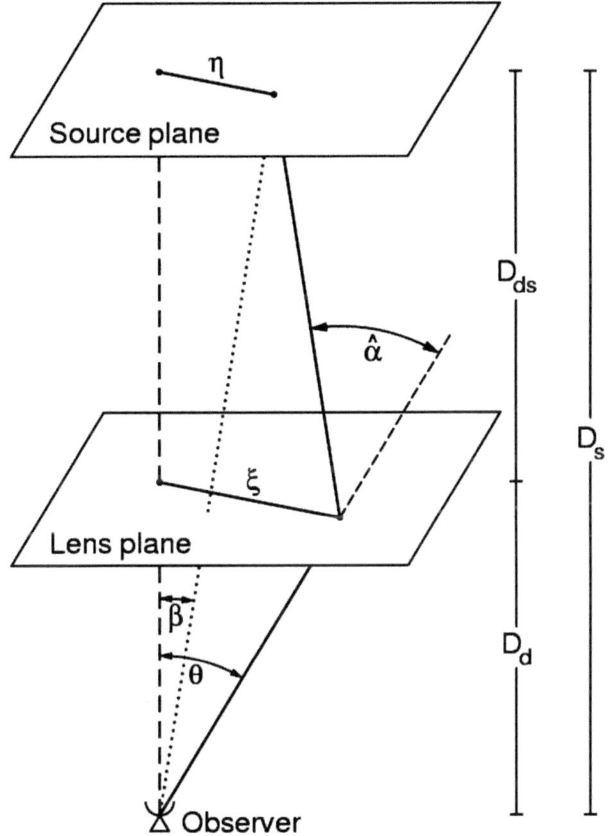

Fig. 5 A diagram (reproduced from Bartelmann and Schneider 2001) of the various lengths and angles used in the lensing formalism. The dashed line is a reference direction, sometimes called the optical axis. The angle $\hat{\alpha}$ is the light deflection by the lens. Note that $D_{\rm d}$, $D_{\rm ds}$, $D_{\rm s}$ are angular-diameter distances, and $D_{\rm d} + D_{\rm ds} > D_{\rm s}$, whereas ξ and η are physical distances

Figure 5 shows the various quantities involved. Through the small-angle approximation, the regions where the source and lens are located, are approximated as planes. The lens plane and source plane have redshifts $z_{\rm d}$ and $z_{\rm s}$ respectively. The lens (deflector) is at angular-diameter distance $D_{\rm d}$ from the observer, so that an angle $\boldsymbol{\theta}$ on the observer's sky corresponds to a physical displacement $\boldsymbol{\xi}$ on the lens plane. Analogously, the source is at angular-diameter distance $D_{\rm s}$ from the observer, and an angle $\boldsymbol{\beta}$ corresponds to a physical displacement $\boldsymbol{\eta}$ on the source plane. Note again that the angular-diameter distance $D_{\rm ds}$ is not equal to $D_{\rm s} - D_{\rm d}$ but larger, because of the way angular-diameter distances are defined.

Projecting distances on the source plane, we have

$$D_{\rm s}\boldsymbol{\beta} = D_{\rm s}\boldsymbol{\theta} - D_{\rm ds}\hat{\boldsymbol{\alpha}}(\boldsymbol{\theta}) \tag{20}$$

and hence

$$\boldsymbol{\beta} = \boldsymbol{\theta} - \frac{D_{\rm ds}}{D_{\rm s}}\hat{\boldsymbol{\alpha}}(\boldsymbol{\theta}) \tag{21}$$

which is our lens equation (2) generalized to an expanding universe. Introducing the apparent (or scaled) deflection angle

$$\boldsymbol{\alpha} \equiv \frac{D_{\rm ds}}{D_{\rm s}} \hat{\boldsymbol{\alpha}} \tag{22}$$

makes the lens equation look exactly as before.

Assembling the lens out of Sun-like deflectors (as in Eq. (7)), we have

$$\boldsymbol{\beta} = \boldsymbol{\theta} - \frac{4G}{c^2} \frac{D_{\rm ds}}{D_{\rm d} D_{\rm s}} \sum_i M_i \frac{\boldsymbol{\theta} - \boldsymbol{\theta}_i}{|\boldsymbol{\theta} - \boldsymbol{\theta}_i|^2}, \tag{23}$$

where we have used $\boldsymbol{\xi} = D_{\rm d}\,\boldsymbol{\theta}$. As is easy to verify, this equation is the zero-gradient condition of the time delay

$$\frac{t\,(\boldsymbol{\theta}, \boldsymbol{\beta})}{1 + z_{\rm d}} = \frac{D_{\rm d} D_{\rm s}}{2c D_{\rm ds}} |\boldsymbol{\theta} - \boldsymbol{\beta}|^2 - \frac{4G}{c^3} \sum_i M_i \ln |\boldsymbol{\theta} - \boldsymbol{\theta}_i| \,. \tag{24}$$

We did not derive the $1 + z_{\rm d}$ factor in the time delay, but one can argue that this factor must be there, because the gravitational time delay occurs at the lens plane and gets redshifted.

One issue we have swept under the carpet here, is that, in Eq. (20), we should have parallel transported the lens plane to the source plane, instead of simply projecting. When one extends to multiple lens planes, as we will do below, disregarding parallel transport produces spurious image rotation at leading order (cf. Grimm and Yoo 2018). Such artifacts are harmless for the applications considered here, but it is good to be aware of them.

2.4 Multiple Lens or Source Planes

Although our equations so far have assumed one source plane and one lens plane, it is not difficult to generalize to multiple planes.

The lensing of a source plane is independent of any other source planes. Hence, for a single lens plane, each source plane $\boldsymbol{\beta}_{\rm s}$ simply has its own lens equation

$$\boldsymbol{\beta}_{\rm s} = \boldsymbol{\theta} - \frac{D_{\rm ds}}{D_{\rm s}} \hat{\boldsymbol{\alpha}}_{\rm d}(\boldsymbol{\theta}) \,. \tag{25}$$

The form of Eq. (25) tells us that a deflection at one $z_{\rm d}$ is equivalent to a scaled deflection at another $z_{\rm d}$. This is useful in that minor lensing masses along the line of sight can be replaced by effective masses on the main lens plane.

It is also possible to generalize the original lens equation (21) explicitly to multiple planes. To do so, let us first rewrite Eq. (21)

$$\frac{\boldsymbol{\xi}}{D_{\rm s}} = \frac{\boldsymbol{\eta}}{D_{\rm d}} - \frac{D_{\rm ds}}{D_{\rm s}} \hat{\boldsymbol{\alpha}} \,. \tag{26}$$

Then let us introduce new subscripts: 0 for the observer plane, 1 for the lens plane, and 2 for the source plane. Let us further introduce $\boldsymbol{x}_0, \boldsymbol{x}_1, \boldsymbol{x}_2$ for comoving positions on these respective planes. This gives

$$\boldsymbol{\xi} = \frac{\boldsymbol{x}_2 - \boldsymbol{x}_1}{1 + z_2} \qquad \boldsymbol{\eta} = \frac{\boldsymbol{x}_1 - \boldsymbol{x}_0}{1 + z_1} \,. \tag{27}$$

Meanwhile we can also write the angular-diameter distances in terms of comoving distances

$$D_{\mathrm{d}} = \frac{\chi_1 - \chi_0}{1+z_1} \quad D_{\mathrm{ds}} = \frac{\chi_2 - \chi_1}{1+z_2} \quad D_{\mathrm{s}} = \frac{\chi_2 - \chi_0}{1+z_2}\,. \tag{28}$$

Here $\boldsymbol{x}_0$ and χ_0 are both zero, as they go from the observer to the same observer, but they will play a useful role presently.

Substituting these various expressions into our rewritten lens equation (26) and simplifying the result, we obtain

$$\frac{\boldsymbol{x}_2 - \boldsymbol{x}_1}{\chi_2 - \chi_1} = \frac{\boldsymbol{x}_1 - \boldsymbol{x}_0}{\chi_1 - \chi_0} - \hat{\boldsymbol{\alpha}}_1\,. \tag{29}$$

This equation relates comoving quantities at three consecutive planes. Since the planes 0, 1, 2 were not special in the derivation, we can generalize to

$$\frac{\boldsymbol{x}_{n+1} - \boldsymbol{x}_n}{\chi_{n+1} - \chi_n} = \frac{\boldsymbol{x}_n - \boldsymbol{x}_{n-1}}{\chi_n - \chi_{n-1}} - \hat{\boldsymbol{\alpha}}_n\,. \tag{30}$$

This is equivalent to Eq. (19) from Petkova et al. (2014). To use it, we need to know the comoving distances χ_n to all the planes, the deflections $\hat{\boldsymbol{\alpha}}_n$ at those planes, and the starting values $\boldsymbol{x}_0 = 0$ (the observer) and $\boldsymbol{x}_1$ (the scaled location on the observer's sky). We can then compute $\boldsymbol{x}_n$ for arbitrarily many lens planes.

So far, the best-known multi-plane system is the "Jackpot lens" J0946+1006 (Gavazzi et al. 2008; Smith and Collett 2021), which has three sources at different redshifts, one of which is also a second lens. Upcoming surveys will surely find more such systems, and the topic is likely to become important.

3 The Standard Formalism of Gravitational Lensing

Several introductions to strong lensing have been published over the years, laying out the basic formalism, though the emphasis and amount of detail differs. Schneider et al. (1992) was the first book on lensing, and was very influential. The historical introduction is quite interesting, and the quantitative portions of the book are very comprehensive and often at an advanced level; however, the portions of the book that deal with observational state of the field have been superseded many times over in the last three decades. Bartelmann and Schneider (2001) is an extension of the book to weak lensing, but can also be used as an introduction to lensing theory in its own right. Blandford and Narayan (1986) give a short but well-explained article, providing a gentle introduction among other things to catastrophe theory as relevant to gravitational lensing (Petters et al. 2001, is a deep dive into that area). They also discuss the similarities with atmospheric mirages, which is further developed in Refsdal and Surdej (1994) who propose an original introduction to the field and explain how to design an optical lens simulator with a piece of glass. (Kochanek 2006) provides an interesting perspective specifically on strong gravitational lensing. Meneghetti (2021) is the most recent review, and includes Jupyter notebooks. The above list is non-exhaustive; in addition to dedicated lensing reviews many text-books today contain chapters on lensing.

Common to all these sources, and indeed to the literature on strong gravitational lensing generally, is a set of concepts and cryptic-sounding terms that do not appear elsewhere in astrophysics. With the physical basis and cosmological context in hand, we now explain what these are.

3.1 Introducing Scaled Quantities

Let us replace the sum over discrete masses in Eq. (24) with an integral over a projected mass distribution

$$\sum_i M_i \ln|\boldsymbol{\theta} - \boldsymbol{\theta}_i| \to D_{\rm d}^2 \int \Sigma(\boldsymbol{\theta}') \ln|\boldsymbol{\theta} - \boldsymbol{\theta}'|\, d^2\boldsymbol{\theta}' . \tag{31}$$

Here $\Sigma(\boldsymbol{\theta})$ is a function of the angular position, but physically, it is a projected density in kg m^{-2} on the lens plane. Hence the $D_{\rm d}^2$ factor.

Next, we introduce two scales

$$\begin{aligned} T_{\rm ds} &\equiv (1+z_{\rm d})\frac{D_{\rm d} D_{\rm s}}{c D_{\rm ds}} , \\ \Sigma_{\rm cr} &\equiv \frac{c^2}{4\pi G} \frac{D_{\rm s}}{D_{\rm d} D_{\rm ds}} \end{aligned} \tag{32}$$

and present their interpretations: As defined, $cT_{\rm ds}$ is a combination of lens and source distances, sometimes called the time-delay distance. Like other cosmological distances, $T_{\rm ds} \propto H_0^{-1}$. In the limit $D_{\rm ds} \to D_{\rm s}$, which is to say sources at infinity, $cT_{\rm ds} \to D_{\rm d}$. Thus $cT_{\rm ds}$ is a replacement for distance D which we used in Sect. 1.3 when introducing the general picture.

The quantity $\Sigma_{\rm cr}$ is known as the critical density for strong lensing, because it turns out to be the threshold for appearance of multiple images. (It is *not* the cosmological critical density, which has different dimensions.) For typical lens and source redshifts $\Sigma_{\rm cr}$ comes to a few kg m^{-2} — about the surface density of window glass (see Appendix B). The proverbial astute reader will have noticed that $G\Sigma_{\rm cr}$ is an acceleration. Indeed the threshold acceleration we mentioned earlier (Eq. (10)) is $4\pi G \Sigma_{\rm cr}$.

Using the scale parameters $T_{\rm ds}$ and $\Sigma_{\rm cr}$ we can replace the time delay and projected mass distribution with the dimensionless quantities $\tau(\boldsymbol{\theta})$ and $\kappa(\boldsymbol{\theta})$

$$\begin{aligned} t(\boldsymbol{\theta},\boldsymbol{\beta}) &= T_{\rm ds}\, \tau(\boldsymbol{\theta},\boldsymbol{\beta}) , \\ \Sigma(\boldsymbol{\theta}) &= \Sigma_{\rm cr}\, \kappa(\boldsymbol{\theta}) \end{aligned} \tag{33}$$

and express the arrival time in the following, commonly used form

$$\begin{aligned} \tau(\boldsymbol{\theta},\boldsymbol{\beta}) &= \tfrac{1}{2}|\boldsymbol{\theta}-\boldsymbol{\beta}|^2 - \psi(\boldsymbol{\theta}) , \\ \psi(\boldsymbol{\theta}) &= \frac{1}{\pi}\int \kappa(\boldsymbol{\theta}') \ln|\boldsymbol{\theta}-\boldsymbol{\theta}'|\, d^2\boldsymbol{\theta}' . \end{aligned} \tag{34}$$

Fermat's principle, $\nabla\tau(\boldsymbol{\theta}) = 0$, gives the lens equation (2) in the form

$$\boldsymbol{\beta} = \boldsymbol{\theta} - \nabla\psi(\boldsymbol{\theta}) . \tag{35}$$

The last term on the right is

$$\boldsymbol{\alpha} = \frac{1}{\pi}\int \frac{\boldsymbol{\theta}-\boldsymbol{\theta}'}{|\boldsymbol{\theta}-\boldsymbol{\theta}'|^2} \kappa(\boldsymbol{\theta}')\, d^2\boldsymbol{\theta}' . \tag{36}$$

Thus $\psi(\boldsymbol{\theta})$ behaves as a scaled version of a gravitational potential. It is known as the lens potential and is the solution of the 2D Poisson equation

$$\nabla^2 \psi(\boldsymbol{\theta}) = 2\kappa(\boldsymbol{\theta}) . \tag{37}$$

Since the logarithmic kernel in Eq. (34) is very broad, the potential will be much smoother than the density.

The scaled density $\kappa(\boldsymbol{\theta})$ is known as the convergence. We will see the reason for this name in Sect. 3.4. There is no single standard name for $\tau(\boldsymbol{\theta}, \boldsymbol{\beta})$, but Fermat potential, (scaled) time delay, or arrival time are all used.

3.2 The Einstein Radius

The quantities κ, ψ, and $\boldsymbol{\alpha}$ behave analogously to the density, classical gravitational potential, and force. In particular, a circular mass distribution behaves like a point mass when observed from outside. This fact has useful consequences, and in particular it helps understand the way κ is defined.

Consider a point mass at the origin, that is

$$\kappa(\boldsymbol{\theta}) = \pi\theta_{\mathrm{E}}^2\, \delta(\boldsymbol{\theta}) , \tag{38}$$

in which θ_{E} is a constant whose meaning will be clear in a moment. The lens equation for this mass is

$$\boldsymbol{\beta} = \left(1 - \theta_{\mathrm{E}}^2/|\boldsymbol{\theta}|^2\right) \boldsymbol{\theta} . \tag{39}$$

A source at $\boldsymbol{\beta} = 0$ will result in a ring image at $|\boldsymbol{\theta}| = \theta_{\mathrm{E}}$ — the well-known Einstein ring. Its radius θ_{E} is known as the Einstein radius.

Because of the aforementioned property of circular mass distributions, the Einstein ring will be the same for any circular distribution of mass within θ_{E} provided the integrated κ equals $\pi\theta_{\mathrm{E}}^2$ — in particular a uniform distribution with $\kappa = 1$ inside the disk.

This property lets us define a notional θ_{E} even when a mass distribution is not circular: as the radius of a region whose mean enclosed κ is unity.

The mass M corresponding to a given Einstein radius will be $\pi\theta_{\mathrm{E}}^2\, \Sigma_{\mathrm{cr}}$ and thus

$$\theta_{\mathrm{E}}^2 = \frac{4GM}{c^2 D_{\mathrm{d}}} \frac{D_{\mathrm{ds}}}{D_{\mathrm{s}}} . \tag{40}$$

For a given mass, θ_{E} decreases with distance. However, the corresponding physical Einstein radius

$$R_{\mathrm{E}} = D_{\mathrm{d}}\theta_{\mathrm{E}} \tag{41}$$

increases with increasing distance.

One further interpretation of θ_{E} follows from rewriting Eq. (40) as

$$\theta_{\mathrm{E}} = \frac{4GM}{c^2 \xi} \frac{D_{\mathrm{ds}}}{D_{\mathrm{s}}} . \tag{42}$$

Recalling that GM/ξ is the squared circular speed v_c^2 at radius ξ we see that θ_{E} is related to kinematics. A circular speed of 300 km s^{-1} corresponds to an Einstein radius of up to $4v_c^2/c^2$ which is roughly an arcsecond.

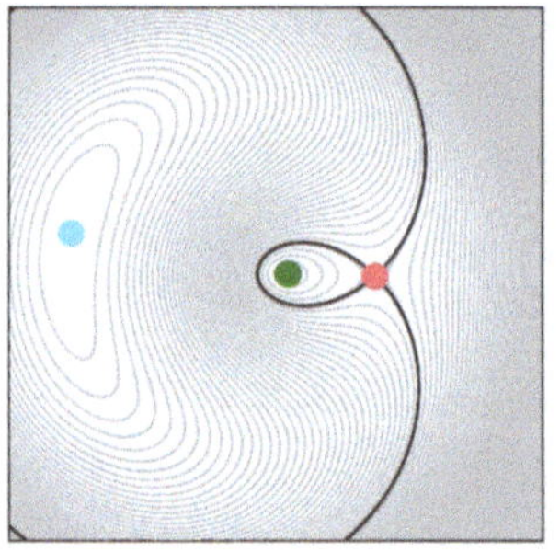
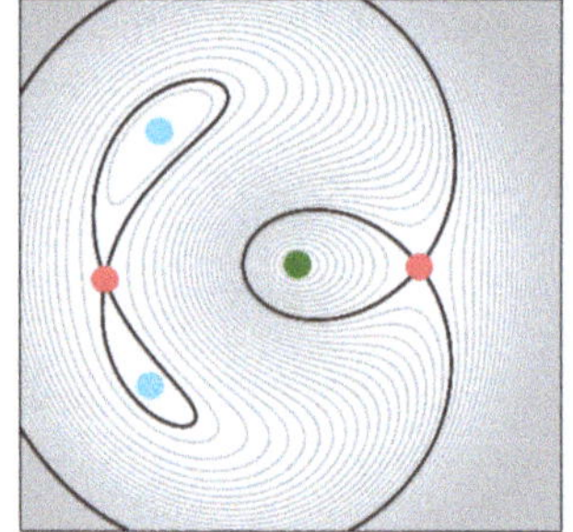
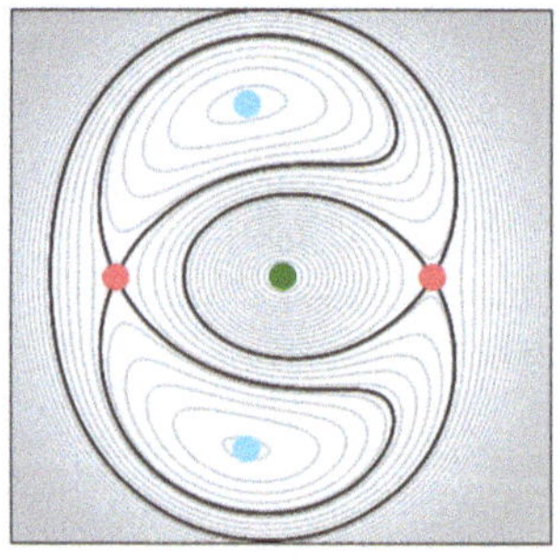

Fig. 6 Contour maps of arrival-time surfaces $\tau(\boldsymbol{\theta})$ image positions marked. Minima are in light-blue, saddle points in orange-red, and maxima in green. Contours through saddle points are highlighted. The potential $\psi(\boldsymbol{\theta})$ is the same in all three panels, but the source positions are different

3.3 The Three Kinds of Images

Let us consider the Fermat potential or arrival time from Eq. (34), that is

$$\tau(\boldsymbol{\theta}, \boldsymbol{\beta}) = \tfrac{1}{2}|\boldsymbol{\theta}|^2 - \psi(\boldsymbol{\theta}) - \boldsymbol{\theta} \cdot \boldsymbol{\beta} \tag{43}$$

plus the constant $\frac{1}{2}|\boldsymbol{\beta}|^2$, which can be discarded. It is interesting and useful to visualize $\tau(\boldsymbol{\theta}, \boldsymbol{\beta})$ as a surface: the time-delay or arrival-time surface.

Without a lens, the arrival-time surface looks like a parabolic well with a minimum at $\boldsymbol{\theta} = \boldsymbol{\beta}$. A lensing mass pushes the surface up at nearby locations. If the lensing mass profile is circular and directly in front of the source, the minimum of the parabola will be replaced by a hill with a circular valley around it. However, unless the lens is perfectly circular and the source is perfectly centered behind it, the valley will not be perfectly symmetric, but will have a bowl on one side. Elongated mass distributions can lead to two bowls, and for more complicated mass distributions, further minima and maxima may appear. Each bowl and hill corresponds to an image location, because, by definition, $\nabla\tau(\boldsymbol{\theta}) = 0$ at minima and maxima. In addition to maxima and minima there are saddle points which also have $\nabla\tau(\boldsymbol{\theta}) = 0$ and hence correspond to images.

The relationship between maxima, minima, and saddle points is conveniently illustrated by contour maps of the arrival-time surface $\tau(\boldsymbol{\theta})$, such as in Fig. 6. The highlighted contours are self-crossing, and the self-intersections are saddle points. The saddle-point contours separate regions of nested simple closed curves, and each such region is either a bowl going down to a minimum, or a hill rising to a maximum. As we can see from the figure, there are two kinds of saddle-point contours. One kind looks like an infinity sign, resembling a Bernoulli lemniscate (which is given in polar coordinates by $r^2 = \cos(2\phi)$). This kind occurs between two bowls, or between two hills. The other kind of saddle-point contours resembles a limaçon trisectrix ($r = 1 - 2\cos\phi$ in polar coordinates), which corresponds to a hill within a bowl.

The lens in Fig. 6 is an elliptical mass distribution (a commonly-used form called a PIEMD — see Appendix A). Each panel shows $\tau(\boldsymbol{\theta}, \boldsymbol{\beta})$ for a different $\boldsymbol{\beta}$. As is evident from the expression (43) we have a function of $\boldsymbol{\theta}$ only plus a tilt due to the $\boldsymbol{\theta} \cdot \boldsymbol{\beta}$ term. In the right-hand panel of Fig. 6 we have a hill in the middle and two bowls above and below it. Turning to the middle panel we see the result of moving the source a little to the left, which tilts the arrival-time surface to the left; the two bowls move to the left. The left panel shows how moving the source further to the left tilts the surface more, and makes the bowls merge into one. A still larger tilt would make the hill disappear too, leaving a single bowl.

More images can form, through nesting of the two cases illustrated in Fig. 6.

1. A minimum is split by a limaçon-like contour into a minimum and a saddle point opposite, or
2. a minimum is split by a lemniscate-like contour into two minima with a saddle point in between.

Less common is

1. the inverted form of the second case, where a maximum is split into two maxima with a saddle point in between.

The inverted form of the first case can occur in ordinary topography, but does not occur in lensing. A circular mass distribution produces case 1; case 2 (whether nested or by itself) is characteristic of elongated mass distributions. For case 3 one needs multiple, very distinct mass peaks. Images always appear and disappear in pairs: a saddle point and either a maximum or minimum. Thus, the number of images is always odd.

3.4 Magnification

Although the number of images is always odd, the number of observable images may not be. This is because different images are variously magnified, and some images can be demagnified so much that they cannot be detected anymore. Magnification and demagnification are implicit in the arrival-time surface, as we will see now.

The contour lines in Fig. 6 hint at magnification. If one looks at this figure from afar (and tries to ignore the heavy saddle-point contours as much as possible) the regions where contour lines are close together look dark, while the regions with widely-spaced contours appear to be bright. Now, consider a solution of the lens equation (2) and take a differential around it as

$$\Delta\boldsymbol{\beta} = (1 - \nabla\nabla\psi(\boldsymbol{\theta}))\,\Delta\boldsymbol{\theta} = \nabla\nabla\tau(\boldsymbol{\theta})\,\Delta\boldsymbol{\theta}\,. \tag{44}$$

If the Hessian $\nabla\nabla\tau(\boldsymbol{\theta})$ is large (small), $\Delta\boldsymbol{\theta}$ will cause us to move off-source quickly (slowly), and meanwhile the contours will be close together (widely spaced). Thus, the apparent brightness between the contour lines gives some qualitative indication of magnification.

The Hessian

$$\mathbf{M}^{-1} \equiv \nabla\nabla\tau(\boldsymbol{\theta},\boldsymbol{\beta}) = 1 - \nabla\nabla\psi(\boldsymbol{\theta}) \tag{45}$$

is an important quantity. Formally inverting Eq. (44) we can write

$$\Delta\boldsymbol{\theta} = \mathbf{M}^{-1}\Delta\boldsymbol{\beta}\,, \tag{46}$$

to express the lensing result of small finite sources. The inverse Hessian (or inverse curvature) of the arrival time is a tensor magnification. It is called the magnification matrix.

The inverse magnification matrix is conventionally separated into a trace and a traceless part. From the Poisson equation (37) it follows that the trace will be $1-\kappa$. The traceless part is called the shear and denoted by γ. Thus we have

$$\mathbf{M}^{-1} = (1-\kappa)\begin{pmatrix}1 & 0\\ 0 & 1\end{pmatrix} - \gamma\begin{pmatrix}\cos 2\varphi & \sin 2\varphi\\ \sin 2\varphi & -\cos 2\varphi\end{pmatrix}, \tag{47}$$

with κ, γ, φ all depending on $\boldsymbol{\theta}$. The meaning of the terms convergence and shear now becomes evident: κ represents an isotropic scaling, while γ represents a shearing oriented about position angle φ. The notation

$$\begin{pmatrix} 1-\kappa-\gamma_1 & \gamma_2 \\ \gamma_2 & 1-\kappa+\gamma_1 \end{pmatrix} \tag{48}$$

for $\mathbf{M}^{-1}$ is also commonly used. Either way, the matrix can be diagonalized (taking $\gamma_1 \to \gamma, \gamma_2 \to 0$) by a suitable rotation of the coordinate system. The resulting diagonal elements, which are the eigenvalues, are $1-\kappa \pm \gamma$. For minima both eigenvalues are positive; for maxima both are negative; at saddle points the eigenvalues have opposite sign. The determinant of $\mathbf{M}$

$$\mu \equiv [(1-\kappa)^2 + \gamma^2]^{-1} \tag{49}$$

is a scalar magnification. It is positive for minima and maxima, and negative for saddle points, the absolute value representing the total size change of a small source. Clearly, if κ or γ become very large, the corresponding image will be demagnified and may be unobservable. At the centers of galaxies and clusters, κ does become extremely large (though γ remains modest). This produces an extreme peak in the arrival time, resulting in a demagnified image (a maximum) at that location. These demagnified maxima are often observed at the centres of lensing clusters. Lensing galaxies, however, nearly always demagnify the central maxima to unobservability.

A curious property of magnification is that there is always at least one image with absolute magnification more than 1. This appears to say that lensing can increase the total amount of light. However, this unphysical property is really an artifact of the small-angle approximation (Wucknitz 2008).

3.5 Critical Curves and Caustics

As we have seen, there are three kinds of images: minima, maxima, and saddle points of the arrival-time surface $\tau(\boldsymbol{\theta})$. As with any topographic surface, the three kinds can be identified according to the two eigenvalues of the curvature matrix $\nabla\nabla\tau(\boldsymbol{\theta})$: minima have both eigenvalues positive; maxima have both eigenvalues negative, whereas saddle points have one of each sign.

Since $\tau(\boldsymbol{\theta})$ has only a quadratic dependence on $\boldsymbol{\beta}$, the curvature matrix $\nabla\nabla\tau(\boldsymbol{\theta})$ is independent of $\boldsymbol{\beta}$. That is to say, although image location depends on $\boldsymbol{\beta}$, the image type —if an image is present— depends only on $\boldsymbol{\theta}$. It follows that image types must be territorial: some regions can have only minima, and so on. If there are regions, there must be boundaries between the regions. The boundaries between regions of different image types are known as critical curves. One can find them by computing where the eigenvalues of the curvature matrix change sign. In general, both eigenvalues will not change sign together, so that minima and maxima do not border each other and there is always a buffer of saddle points between them.

One can map critical curves to the source plane via the lens equation, obtaining curves on the source plane, which are called caustics. Critical curves are abstract curves on the sky, they are no locations in real space. Caustics, on the other hand, are real locations on the source plane. If a point source lies on a caustic, it will light up a point on the critical curve. There may be additional images at other positions, but the image on the critical curve is special because it has formally infinite magnification, since the curvature of the arrival-time surface is zero. This does not create a physical paradox, because (a) physically realistic sources have a finite size, and the singularity can be integrated over, and (b) wave optics would intervene to remove the singularity. If a moving source crosses from one side of a caustic to the other, one of two things happens, depending on the sense of crossing: Either two images are created at the critical curve and leave it in opposite directions, or the converse happens and two images merge and disappear on the critical curve. Positive-parity images always move in the same general direction as the source, while saddle points move the opposite way, so there is no ambiguity about which of these outcomes to expect.

In summary, critical curves are curves of extreme magnification on the sky, whereas caustics are borders on the source plane at which the number of lensed images changes. In ordinary optics, caustics usually refer to bright curves on a screen resulting from lensing of a small bright source; this just means that source and observer are swapped compared to our considerations here. In another difference of terminology, rainbows are sometimes referred to as caustics, but in the gravitational lensing convention they are critical curves.

The consequences of critical curves and caustics are enough to fill a book (Petters et al. 2001) and still leave room for new research. Fortunately, the most common cases are fairly straightforward, and in the following we will concentrate on those.

3.6 Common Image Configurations

The critical curves and caustics of a lens depend only on its potential. The images and saddle-point contours move as the source moves. Figure 7 revisits the lens used for Fig. 6 and illustrates example image configurations. The first panel shows the mass distribution and the corresponding lens potential, the second panel shows the critical curves and caustics. The rest of the figure shows image positions, types, and magnifications for six placements of the source. The lower row has the same source positions as Fig. 6. The outer critical curve separates minima and saddle points, while the inner critical curve separates saddle points and maxima. The outer critical curve corresponds to the inner caustic, and vice versa. The inner caustic has the qualitative form of an astroid ($x^{2/3} + y^{2/3} = 1$ in cartesian coordinates) but is also often called a diamond caustic.

A drastic modification is shown in Fig. 8. The elliptical PIEMD is replaced with a rectangular tile shaped like a candy bar, with a uniform $\kappa = 4$ (for details see Appendix A again). The size is chosen so that the outer critical curve has the same area as before. We see that the potential contours are still ovals, because of the logarithmic kernel in Eq. (34). The caustics are larger, and somewhat differently shaped. The critical curves are shaped very differently. Yet, if we place the source at similar locations with respect to the diamond caustic, the resulting image configurations look similar.

Figure 9 shows six lensed quasars with image configurations similar to those in Figs. 7 and 8. These have been rotated (and some cases also flipped) from the usual North-up/East-left orientation, for better comparison with the panels labeled 1–6 in Figs. 7 and 8. Let us consider each of these examples in turn.

1. Q0957+561 is a typical double (not counting the presumed demagnified image coinciding with the lensing galaxy). Doubles arise when the source is outside the diamond caustic. One image lies outside both of the critical curves, while the other lies between the two critical curves. The former is a minimum, while the latter is a saddle point in the arrival time. The minimum will naturally have a shorter arrival time. The two images form an obtuse angle with respect to the lensing galaxy, which happens when the source location is not aligned with the long or the short axes of the lensing mass. We label it an "inclined double" (cf. Saha and Williams 2003).

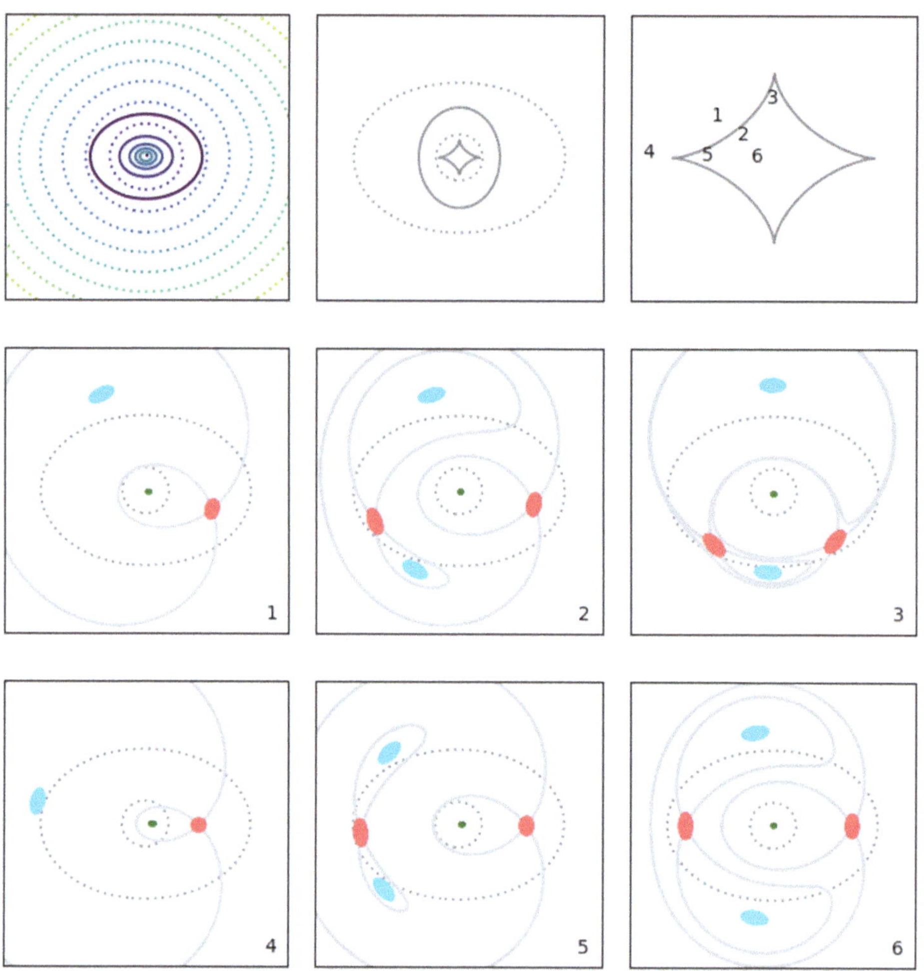

Fig. 7 Lens and image properties for a PIEMD. The upper left panel shows contours of the surface density $\kappa(\boldsymbol{\theta})$ (solid curves) and the $\psi(\boldsymbol{\theta})$ (dotted curves). The upper middle panel shows the critical curves (dotted) and caustic curves (solid). The upper right panel is a detail of the 'diamond' caustic with numbered source positions. The numbers correspond to panels in the middle and lower rows. The numbered panels show critical curves (dotted), saddle-point contours (solid) and images (as before, light blue: minima, orange red: saddle points, green: maxima). The contrast between magnification and demagnification, and the stretching of the lensed images, is actually much more than appears here

2. PG1115+080 is an "inclined quad", which is the type of configuration that arises if the source lies inside the diamond caustic, and is not near any of the cusps. Compared to Q0957+561, two additional images, one a minimum and one a saddle point, form on either side of the outer critical curve. This arrangement is sometimes called a "fold type" because the source lies near a fold caustic.
3. RXJ0911+0551, with three images close together near the lensing galaxy and one image further away on the opposite side, indicates a source near the cusp aligned with the short axis of the lensing mass. It is a "short-axis quad". Three nearly-merging images are characteristic of a source near a cusp.

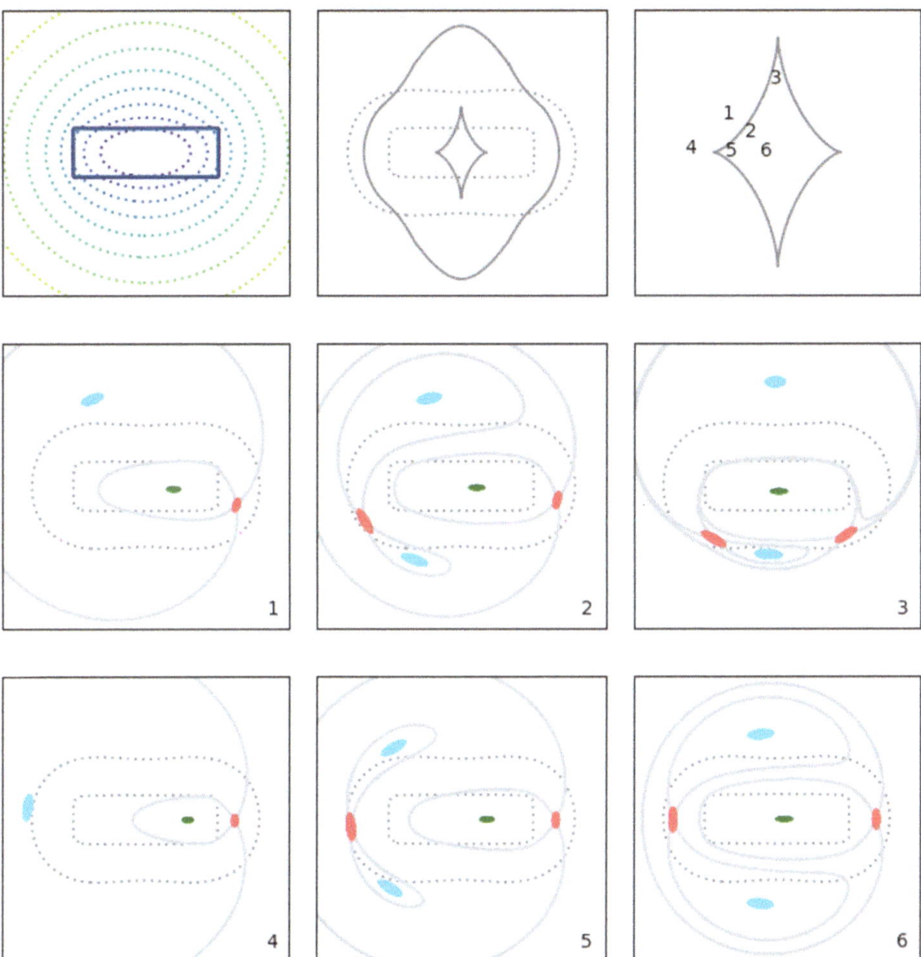

Fig. 8 As Fig. 7 but for a candybar lens

4. HE1104-1805, with two images forming nearly a straight line with the lensing galaxy, is an "axial double". This configuration arises when the source location is aligned with either the long or short axis of the lensing mass.
5. RXJ1131-1231 is a "long-axis quad". It differs from a short-axis quad in that the three nearly-merging images are further from the lensing galaxy than the fourth image. In this case, the source is near a cusp aligned with the long axis of the lensing mass.
6. Q2237+030 with its cross-like arrangement, is a "core quad" and must have its source near the center of the diamond caustic.

With some practice, it is possible to stare at a lens candidate and sketch conjectural saddle-point contours, thereby identifying the type of each image (minimum, maximum, or saddle point) and predicting the time ordering. Figure 10 shows an example.

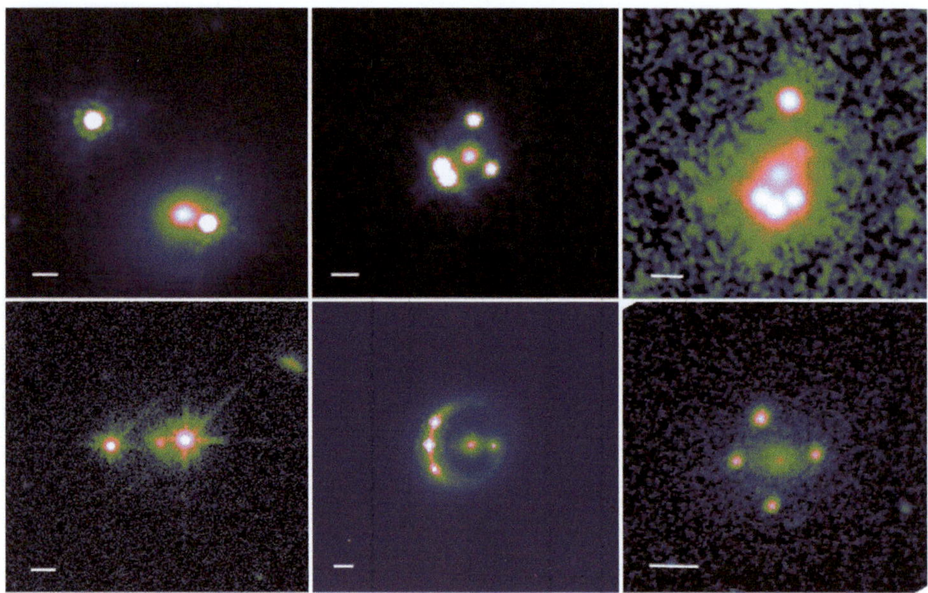

Fig. 9 A collage of HST images of lensed quasars oriented so that the lensed-image positions correspond to the example configurations in Figs. 7 and 8. The systems (from top left in reading order) are Q0957+561, PG1115+080, RXJ0911+0551, HE1104-1805, RXJ1131-1231, and Q2237+030. The horizontal white tick marks $1''$ angular scale. Images were observed in wavelengths corresponding to $I-$, $J-$ or H-band filter

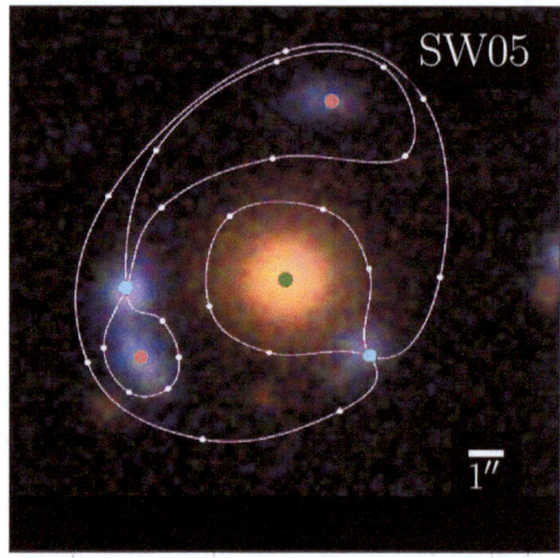

Fig. 10 The lens candidate SW05 J1434+5228 with a conjectural identification of the image structure (taken from Küng et al. 2018) which is similar to those on panel 2 of Figs. 7 and 8, as well as PG1115+080 in Fig. 9. This type of markup can be used as input for lens-modeling, as discussed later in Sect. 4.7

3.7 Degeneracies

The various examples in Figs. 7–10 indicate two contrasting things about image configurations. On the one hand, some basic properties of the lensing mass, the general placement of the source, and the likely time-ordering of the images can be read off directly from the arrangement of images. On the other hand, the lensing mass distribution can be altered quite drastically with little change in the images. In fact, as we will see presently, it is possible to modify a lens while keeping lensing observables exactly fixed. This is the problem of lensing degeneracies.

Let us go back to Fig. 3, and run the light backwards. That is to say, let us consider a single observer above (at the origin of the rays), and imagine sources scattered over the figure. If the sources are densely distributed and their locations are known, the observer can expect to reconstruct the entire wavefront and infer the gravitational field that caused it.

The origin of lensing degeneracies is in that lensing gives information about the gravitational field only along light paths to the observer from lensed sources. If the sources are sparse, the observer's information will be incomplete. Some clusters lenses go a long way towards providing a dense sampling of sources over a range of redshifts, and the situation will improve further with infrared telescopes in space. But for most lenses, one is working with a single source plane, of which only a small part emits light.

A more formal treatment appears in Wagner (2018) and Wagner (2019). The earlier paper characterizes the most general class of degeneracies in the strong lensing formalism for a single source and lens plane without making any assumption about the mass distribution within the lens. The second paper extends the analysis from a single lens plane to additional lenses along the line of sight, connecting the degeneracies in the lensing mass distribution in a single lens plane to degeneracies in the overall matter distribution along the entire line of sight (see especially Fig. 3 in Wagner 2019). These two works explain that one physical cause of these degeneracies is rooted in the fact that lensing only probes the integrated mass density along the entire line of sight from the observer to the source. Consequently, the distribution of the not directly observable dark matter can be redistributed in many ways. At the same time, as shown in Wagner et al. (2018) and Wagner (2019), the observable properties of multiply-imaged galaxies behind strong gravitational lenses uniquely constrain *local* properties of the lens as the shear per mass density at the locations of the multiple images and ratios of mass densities between the different multiple images. This is the maximum information that can be extracted from static multiple images without using any additional information about the overall mass density distribution. All other properties of lenses, especially in the areas far from any multiple image, are determined by the assumptions contained in the lens models.

Among these general degeneracies, there are some simple but still important ones, which we now discuss.

The first is a monopole degeneracy, which we have actually already seen. In Sect. 3.2 we noted that a circular mass distribution behaves from outside like a point mass. From inside it has no deflection. These two properties are analogous to the well-known property of spherical shells in Newtonian gravity, and can be derived using similar methods. It follows that if we replace a mass ring with another mass ring, both rings having the same center and the same mass but different radii, there will be no change in the deflection, except in the annular region between the two rings. The same holds for derivatives of the deflection, including shear and magnification. Thus, images not between the two rings will be unaffected. As a result, we are free to choose any circular or annular region not containing any images, and radially redistribute the mass in that region, with zero effect on images. Moreover, we can repeat this operation in several round regions, as long as they contain no images. At distances outside all the images, we can freely add or remove circular rings of mass.

Our next degeneracy is also very simple, though its consequences are significant. Consider the original lensing equation (2) and multiply both sides by the same factor. This will of course preserve any solution of the equation, but will scale $t(\boldsymbol{\theta})$. That is, the observed image positions will stay the same, but the time delays between different images will get scaled, and so will the implied source sizes. The corresponding transformation of the lensing mass is a little more subtle. To derive it, we first recall Eqs. (36)–(37), and from them we first derive

$$\nabla \cdot \boldsymbol{\alpha} = 2\frac{\Sigma(\boldsymbol{\theta})}{\Sigma_{\mathrm{cr}}} \tag{50}$$

We also have $\nabla \cdot \boldsymbol{\theta} = 2$. Combining these relations, it is clear that multiplying both

$$\boldsymbol{\beta} \text{ and } 1 - \frac{\Sigma(\boldsymbol{\theta})}{\Sigma_{\mathrm{cr}}} \tag{51}$$

by a constant factor amounts to multiplying both sides of the lens equation by the same factor. This degeneracy goes by various names. One is the mass-sheet degeneracy or mass-sheet transformation (or MST), because its limiting case (multiplying the lens equation by zero) makes $\Sigma(\boldsymbol{\theta}) = \Sigma_{\mathrm{cr}}$ everywhere. Another name is the steepness degeneracy, since making the mass distribution less or more like a uniform sheet makes the mass profile steeper or shallower. Yet another name is the magnification transformation, because rescaling $\boldsymbol{\beta}$ with $\boldsymbol{\theta}$ fixed implies changing the magnification.

As expected from the wavefront picture, having more lensed sources reduces the effect of lensing degeneracies. Having more images reduces the scope of the monopole degeneracy, because there is less area available to redistribute mass. The mass-sheet transformation (51) is prevented from being applied globally if sources at different redshifts are present (because Σ_{cr} is not the same for all images), but an extension that stitches together different scalings in different regions of $\boldsymbol{\theta}$ is possible (Liesenborgs et al. 2012; Schneider 2014) There are also more complicated redistributions of mass, coupled with transformations in $\boldsymbol{\beta}$, often constructed numerically in particular cases. These are collectively referred to as shape degeneracies (Saha and Williams 2006), or source-position transformations (Schneider and Sluse 2014).

4 Lens Modeling

4.1 General Considerations

Lens modeling is the process of reconstructing the lensing mass distribution and the unlensed source brightness distribution from the observables. The principal observable is the lensed brightness distribution. Other observables — time delays, or complementary data like stellar kinematics — may be available, depending on the system. Lens and source redshifts enter as parameters, and so does a cosmological model with its parameters to convert these redshifts into distances.

There are several different lens-modeling strategies in use and many codes implementing them are available. Some results, such as measurements of enclosed mass, are robust across different modeling strategies. Other results may depend strongly on the assumptions that go into the lens models. For a quick visual impression, one can compare the mass maps and magnification maps of the lensing cluster Abell 2744[7] obtained by five different groups using the same observations.

[7] https://archive.stsci.edu/prepds/frontier/lensmodels/.

There are many differences of details in how different modeling treat the data. But there is a deeper reason why different modeling methods using exactly the same data can disagree about some of the results, which applies even if the data are noiseless. The problem is lensing degeneracies, which we discussed earlier. Given a gravitational lens and a light source, computing the lensing observables is a well-posed problem. However, the inverse problem of inferring the lensing mass and the unlensed source does not generally have a unique solution, and additional information has to be provided.

The issue is that we only have information about $t(\boldsymbol{\theta})$ where the images are. At these points, we know that $\nabla t(\boldsymbol{\theta}) = 0$ and the values of $\nabla\nabla t(\boldsymbol{\theta})$. We may also know the value of $t(\boldsymbol{\theta})$ (up to a constant), if time delays are measured. Roughly speaking, time delays measure the potential difference between the image locations, the image locations themselves are probes of the gradient of the potential, while fluxes inform about the tidal field, but these pieces of information are available only at image locations. Elsewhere the information from lensing must be supplemented with other data where available, and somehow interpolated where necessary.

In the following we will discuss the essential ideas behind different modeling strategies in the literature.

4.2 Spherical Lens Models

Let us begin with the simplest.

Earlier, we have already briefly considered a point-mass less (see Sect. 3.2 and Eq. (39)). However, while a point-mass lens is useful for modeling microlensing by individual stars within our Galaxy, it is not an adequate approximation for strong lensing by galaxies and clusters. The light paths corresponding to observed multiple images come through the dark halos of lensing galaxies, or even through regions with significant stellar mass. Thus, the first requirement of a realistic model for a galaxy or cluster lens is an extended mass distribution.

An elegant modification of a point mass, that does have an extended mass distribution, is the Plummer model from stellar dynamics,[8] originally proposed a century ago as a model for globular clusters. The Newtonian gravitational potential of a Plummer sphere is

$$\Phi(r) = \frac{GM}{(r^2 + r_c^2)^{1/2}} , \tag{52}$$

with r_c being interpreted as a core radius, which softens the singularity of a point mass. The corresponding density distribution is

$$\rho(r) = \frac{3r_c^2}{4\pi} \frac{M}{(r^2 + r_c^2)^{5/2}} . \tag{53}$$

Projecting on the sky from distance D_d and writing the projected r as $D_\mathrm{d}\theta$ and r_c as $D_\mathrm{d}\theta_c$ we have

$$\Sigma(\theta) = \frac{M}{\pi D_\mathrm{d}^2} \frac{1}{(\theta^2 + \theta_c^2)^2}. \tag{54}$$

[8] This subsection quotes a number of results and models from stellar dynamics, all of which are explained in more detail in Binney and Tremaine (2008).

Dividing by the critical projected density gives

$$\kappa(\theta) = \frac{\theta_{\rm E}^2 \theta_c^2}{(\theta^2 + \theta_c^2)^2} \qquad \theta_{\rm E}^2 = \frac{4GM}{c^2 D_{\rm d}} \frac{D_{\rm ds}}{D_{\rm s}} . \tag{55}$$

The Einstein radius is the same as in Eq. (40) for a point mass. It is easy to verify that the mean κ within a disk of radius $\theta_{\rm E}$ is unity. The lens potential and deflection angle read

$$\begin{aligned} \psi(\theta) &= \tfrac{1}{2}\theta_{\rm E}^2 \ln(\theta^2 + \theta_c^2) , \\ \boldsymbol{\alpha}(\theta) &= \frac{\theta_{\rm E}^2 \boldsymbol{\theta}}{\theta^2 + \theta_c^2} . \end{aligned} \tag{56}$$

As expected, the $\theta_c \to 0$ limit for all the expressions is the point mass.

The Plummer lens is useful as a component in multi-component lens models (which we will discuss below in Sect. 4.6), but by itself, it is not plausible as a galaxy halo. Galaxy halos have circular velocities which are roughly constant. Now, the circular velocity is related to the Newtonian potential by

$$v_c^2 = r \frac{\partial \Phi}{\partial r} \tag{57}$$

and from Eq. (52) we can see that $v_c(r)$ for a Plummer sphere is far from being constant.

A better first approximation to galaxy halos is the isothermal sphere and its variants. An isothermal sphere in stellar dynamics is a collisionless self-gravitating fluid of stars with a Maxwellian velocity distribution. The density is given by

$$G\rho(r) = \frac{\sigma^2}{2\pi r^2} , \tag{58}$$

in which σ is the stellar velocity dispersion which is constant for a given system. The circular velocity is easily derived from the density (58) and is given by

$$v_c^2 = 2\sigma^2 \tag{59}$$

for all r. This is the "flat rotation curve" property. The constituent stars need not be on circular orbits. The velocity distribution could be isotropic or even predominantly radial, but

$$\langle v^2 \rangle = 2\sigma^2 \tag{60}$$

will apply in all cases due to the virial theorem. The observable stellar velocity dispersion will be the average along one direction (the line of sight). Thus

$$\langle v_{\rm los}^2 \rangle \simeq \tfrac{2}{3}\sigma^2 , \tag{61}$$

with equality applying for an isotropic velocity distribution.

To derive the lensing properties of an isothermal sphere, we first project the density (58) in 3D to

$$G\Sigma(\theta) = \frac{\sigma^2}{2D_{\rm d}\theta} \tag{62}$$

on the sky and express it in terms of a new parameter $\theta_{\rm E}$ as

$$\kappa(\theta) = \frac{\theta_{\rm E}}{2\theta}, \qquad \theta_{\rm E} = 4\pi \frac{D_{\rm ds}}{D_{\rm s}} \frac{\sigma^2}{c^2}. \tag{63}$$

As with the Plummer lens, it is easy to verify that the mean κ within a disk of radius $\theta_{\rm E}$ is unity. Hence $\theta_{\rm E}$ really is the Einstein radius. The lens potential and deflection are extraordinarily simple:

$$\begin{aligned} \psi(\theta) &= \theta_{\rm E}\,\theta\,, \\ \boldsymbol{\alpha}(\theta) &= \theta_{\rm E}\,\frac{\boldsymbol{\theta}}{\theta}\,. \end{aligned} \tag{64}$$

The deflection angle is simply $\theta_{\rm E}$ times the radial unit vector. The deflection angle always having the same magnitude $\theta_{\rm E}$ is the lensing analog of the circular speed being the same everywhere.

The simple isothermal lens described so far is often called the SIS, short for "singular isothermal sphere". This refers to the density singularity at the center. The singularity can be removed by putting in a small core, similar to the core in the Plummer sphere, yielding a non-singular isothermal sphere, NSIS for short. In models, however, the singularity is often retained. Redistributing the central density cusp to a circular core has no effect on lensing behavior further out. The only practical effect of a central singular density is to make the central image unobservable (cf. Sect. 3.4 above). Thus, if no central image is observable, modelling the configuration with a central singularity is convenient.

Another singular property of the isothermal sphere is that the total mass is formally infinite. This follows from the density expression (58) and is the reason for M not appearing in the expressions for this model. There are variants of the isothermal sphere (called King models) which have a finite mass. This model class was used in the very first lens models (Young et al. 1980). It is, however, more common to use the SIS as is, tacitly assuming that the density is truncated outside the region of interest.

There are many more circularly symmetric lens models in use. We will not discuss more examples here, as the Plummer and SIS models illustrate the essentials. The relation between kinematics and lensing is especially worth noting. We have seen earlier (Eq. (42)) that for a point lens, $\theta_{\rm E}$ is proportional to the orbital speed squared. We now see that an expression like (63) relating $\theta_{\rm E}$ and the velocity dispersion is a consequence of the virial theorem. A relation of this type can always be set up for any lens in dynamical equilibrium, but the precise coefficient depends on details of the mass distribution.

4.3 Elliptical Lens Models

If a lensing mass is circular, any and all images must be in the same or opposite direction as $\boldsymbol{\beta}$, because there is no other direction in the system. The arrival-time surface will be qualitatively like the left panel of Fig. 6. There will be a minimum on the same side as the source, a maximum near the lens center, and a saddle point on the opposite side. If the lens mass has a sharp peak at the center (as in the SIS), the maximum will be at the center and demagnified to vanishing brightness. Observed multiple-image systems never have all images (including demagnified images at the lens center) in a straight line. Thus, observed lenses are manifestly non-circular.

Generalising a circular lens to an elliptical one is easy, one only has to add a term

$$\gamma_1(\theta_x^2 - \theta_y^2) + 2\gamma_2\theta_x\theta_y \tag{65}$$

to the lens potential ψ of a circular lens. The additional term is called external shear (or XS). By construction, external shear is traceless and hence implies no change in κ, that is, in the mass distribution. Instead, it implies mass *outside* the region of interest. External shear is the lensing analog of a tidal field in orbital dynamics.

Additionally, one can modify the SIS or any other circular lens to make it elliptical. Thus, there is the SIEP (singular isothermal elliptical potential) where the potential of the SIS has elliptical isocontours. Then there is the PIEMD (pseudo-isothermal elliptical mass distribution) where the mass distribution κ of the SIS is both given a core and made elliptical, leading to a quite complicated potential, given below in Appendix A.2. The PIEMD was used for Figs. 6 and 7 earlier. These lens models are discussed in detail in Kassiola and Kovner (1993) and Kormann et al. (1994). A large catalog of commonly-used mass models is given in Keeton (2001).

The disadvantage of ad hoc modifications of the SIS is that they break the connection with dynamics. For circular lenses there are spherical stellar-dynamical models, which make predictions for observable kinematics in a lensing galaxy. For the SIEP or the PIEMD there are no known stellar dynamical models. One then has to either build an ellipsoidal galaxy model numerically, as done by Barnabè and Koopmans (2007) for example (and more recently Shajib 2019, has shown that an arbitrary elliptical lens can be expressed as a sum of stellar-dynamical components) or fall back on spherical models for the kinematics.

How well are real gravitational lenses approximated by elliptical lens models? To answer this question we turn to several studies, which concentrate on quasar quads, because of the precise astrometry provided by quasars. Here there are two contrasting strategies that have been developed: one is to model the observations with elliptical vs more complicated lens models; the other other is to derive and use some diagnostic for departures from ellipticity, without necessarily fitting a model. Let us consider the latter approach first.

One diagnostic for non-elliptic lenses starts from the case of an SIEP, which has

$$\psi(\theta_r, \phi) \propto r(1 + q\cos(\phi - \phi_0)) . \tag{66}$$

This potential exhibits a remarkable relation between the position angles ϕ_i of the images about the lens center. For this lens Kassiola and Kovner (1995) derive what they call a configuration invariant, which can be written as

$$\cos(\phi_1 + \phi_2 - \phi_3) + \cos(\phi_2 + \phi_3 - \phi_1) + \cos(\phi_3 + \phi_1 - \phi_2) = 0 \tag{67}$$

taking the remaining position angle as $\phi_0 = 0$. This expression can be interpreted as a surface in the positional-angle space (ϕ_1, ϕ_2, ϕ_3). Woldesenbet and Williams (2012, 2015) found that about half of the quasars lie close to a polynomial surface, which they call the fundamental surface of quads (FSQ), and image positions from the SIEP satisfy it precisely. Falor and Schechter (2022) argue that the FSQ must be an approximation to Eq. (67).

Another diagnostic for deviations from elliptical symmetry comes from Witt (1996), who considered a general, purely elliptical potential

$$\psi(\theta_x^2 + \theta_y^2/q^2) \tag{68}$$

and showed that all images, as well as the source, all lie on a rectangular hyperbola. It is not difficult to construct Witt's hyperbola. We start with a basic rectangular hyperbola $\theta_x^2 - \theta_y^2 = 1$, which has two branches. There are four simple linear transformations one can apply to the curve: (i) rotate, (ii) shift along θ_x, (iii) shift along θ_y, and (iv) rescale. Together, they transform the curve to

$$\theta_x^2 - \theta_y^2 + c_1\theta_x\theta_y + c_2\theta_x + c_3\theta_y = c_4 , \tag{69}$$

which is a general rectangular hyperbola. Writing down this equation for any four points (θ_x^k, θ_y^k) gives four linear equations for the coefficients, and solving these equations gives a rectangular hyberbola passing through all four points. An important predictive property of Witt's hyperbola is that it applies to any central image as well, and since for a centrally peaked potential the central image is at the lens center, the center of the lensing galaxy will lie on the hyperbola.

Witt's hyperbola itself is independent of a specific choice of an elliptical lens model, but can be used to find lens models via a further construction (Wynne and Schechter 2018). This is to construct an ellipse (Wynne's ellipse) aligned with Witt's hyperbola that passes through the four images of the quad. The four images serve to set the (i) ellipticity, (ii) shift along θ_x, (iii) shift along θ_y, and (iv) scale of Wynne's ellipse. The parameters of the Witt-Wynne construction (see Fig. 11 for an example) translate into parameters for two distinct lens models: SIS+XS (Wynne and Schechter 2018) and SIEP with aligned XS (Luhtaru et al. 2021).

The other approach, that of fitting elliptical and beyond-elliptical models to the data (not only the lensed quasar positions but also extended images) has been pursued by several authors. An early example is the study of PG 1115+080 by Yoo et al. (2005), who concluded that the lens is indeed an ellipse. Sluse et al. (2012) carried out a comparable study of diverse quasar quads, while also identifying perturbers in the form of nearby galaxies. There are also several works (Gomer and Williams 2018; Van de Vyvere et al. 2022a,b) which introduce non-elliptical features (twists, substructures) into simulated lenses, to test how much they affect perturbed models and see how they affect modeling.

While these studies are restricted to quadruply-imaged quasars, they support a more general conclusion that simple elliptical lenses are a good first-order approximation for most lenses, and the counter-examples of large departures from elliptical lenses are usually flagged by the presence of additional lensing galaxies that perturb the main lens. Accurately fitting observations thus requires more features to be added. Schmidt et al. (2023) and Etherington et al. (2022) use elliptical lenses, including components for additional lensing galaxies where necessary, to model 30 and 59 lenses, respectively. In contrast, Collett et al. (2018) and Shajib et al. (2023) each study a single galaxy lens in more detail, using 3D galaxy models and spatially resolved kinematic data. Meanwhile, machine-learning approaches are also being developed (e.g., Gomer et al. 2023) towards more efficient implementations of these models.

4.4 Bayesian Terminology

Lens models, especially if they are rather elaborate with many parameters, are often formulated in Bayesian terms, the key words being *likelihood, prior,* and *posterior,* all three being probability distributions.

To introduce these terms, let us write ϖ for the model parameters and $\mathcal{D}$ for the data. The ϖ are to be understood in a very general sense, they can consist not only of numbers but also of flags specifying the type of model, such as SIEP or PIEMD and so on, and include the location and other properties of the unlensed source. The $\mathcal{D}$ consist of the images of the lensed system and of other available observations (e.g. time-delays). The likelihood is $p(\mathcal{D} \,|\, \varpi)$, meaning the conditional probability of $\mathcal{D}$ being as measured for given ϖ. Basically, the likelihood is the probability of the noise difference between $\mathcal{D}$ and hypothetical

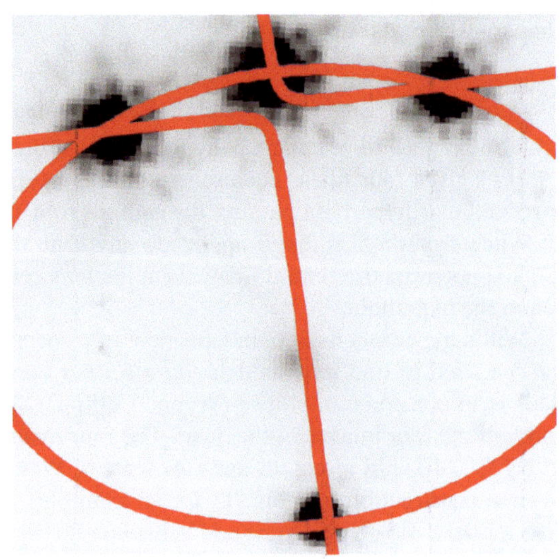

Fig. 11 The Witt-Wynne construction for the quad PS J0147+4630 (from Luhtaru et al. 2021). The image configuration is similar to those on panel 5 of Figs. 7 and 8. The ellipse is similar though not identical to the critical-shear ellipse introduced by Wucknitz (2002) and its short axis is aligned with the long axis of the potential

noiseless data from ϖ. The prior is $p(\varpi)$, and it expresses our knowledge of the parameters before having any data. The posterior

$$p(\varpi \mid \mathcal{D}) \propto p(\mathcal{D} \mid \varpi)\, p(\varpi) \tag{70}$$

is then given by Bayes' theorem in probability, and expresses our knowledge of the parameters given the data.

The likelihood is usually conceptually clear, though it may involve a lot of computation. Typically it has a Gaussian form $p(\mathcal{D} \mid \varpi) \propto \exp\left(-\frac{1}{2}\chi^2\right)$. The prior $p(\varpi)$, on the other hand, may be the subject of debate. If the likelihood factor in Eq. (70) has a sharp peak in ϖ, then the posterior will have the same sharp peak, and the prior will have little influence. If, however, the likelihood does not have a single sharp peak, the choice of prior will strongly influence the posterior.

Actually, the situation with ϖ is more complicated. Because of lensing degeneracies, the likelihood can be insensitive to some parameters. Let us divide the parameters into 'sharp' and 'blunt' $\varpi = (\varpi_s, \varpi_b)$ such that the likelihood $p(\mathcal{D} \mid \varpi_s, \varpi_b)$ is sharply peaked in ϖ_s and blunt in ϖ_b. We can then marginalize out the blunt parameters

$$p(\varpi_s \mid \mathcal{D}) = \int p(\varpi \mid \mathcal{D})\, d\varpi_b \tag{71}$$

to extract the sharp (that is, well-constrained) parameters. As a simple example, consider a system with a clean Einstein ring and no other data. In this example ϖ_s as the enclosed mass and ϖ_b as the radial profile are a possible parameter separation. Suppose, however, that we use a different parametrization that mixes sharp and blunt — say we take $\varpi = (\varpi_1, \varpi_2)$ where ϖ_1 is a power-law slope and ϖ_2 is a normalization — and marginalize out the normalization. We would then get an apparent constraint on the power-law slope, that is really coming from the prior. In this simple example, the problem is easy to diagnose and fix, but the situation gets worse with increasing the number of parameters (distributed between ϖ_s and ϖ_b), and hence the hyper-volume of the parameter space.

4.5 Source Reconstruction

When considering image formation earlier, we assumed that lensed images are small enough to have an unambiguous position. This is clearly valid for lensed quasars and supernovae, and reasonable when speaking of small components in a lensed galaxy, but it is not valid for arcs formed from the lensing of entire galaxies. In such cases one is observing the superposition of a continuous family of lensed sources. In fact, the recent modeling papers we mentioned above have to reconstruct the sources, as well as modeling the lenses.

It is possible to reconstruct a lensed source independently of any lens model, provided there is enough detail in the lensed images. Tessore (2017) and Birrer (2021) develop formalism for doing so, but to explain the general idea, let us consider an example, namely the spectacular five-image system in the lensing cluster CL 0024+1654. Wagner et al. (2018) draw attention to six small components in the source galaxy, all of which appear unambiguously in all five lensed images. They then reconstruct the source in a model-independent way, and compare the results with those obtained by modeling the lens first. Such systems are not common, but nevertheless it is helpful to discuss why model-independent source reconstruction works. Let us think of the six components on the source as being one reference $\boldsymbol{\beta}_0$ and five $\Delta\boldsymbol{\beta}$ vectors. These $\Delta\boldsymbol{\beta}$ are magnified and sheared into five sets of five observable $\Delta\boldsymbol{\theta}$ vectors each by five different magnification matrices. The original $\Delta\boldsymbol{\beta}$ vectors are unknown (5×2 numbers, because we are in 2D). The five magnification matrices are also unknown (3×5 numbers, because magnification matrices are symmetric and have 3 independent components). This yields a total of 25 unknowns. Meanwhile, the $\Delta\boldsymbol{\theta}$ vectors provide $5 \times 5 \times 2 = 50$ constraints. The latter suffice to solve for all $\Delta\boldsymbol{\beta}$ vectors, except for an overall scale, which is degenerate. The reference position $\boldsymbol{\beta}_0$ of the source is not solved for, either, as it also degenerate. Another example, this time for a galaxy lens, appears in the study of B0128+437 by Wagner and Williams (2020), using four images with three components each.

More commonly, however, the source and lens are reconstructed together. Consider an extended source with brightness distribution $s(\boldsymbol{\beta})$. When lensed, this source will produce an extended image system

$$I(\boldsymbol{\theta}) = \int \delta\big[\boldsymbol{\theta} - \boldsymbol{\alpha}(\boldsymbol{\theta}) - \boldsymbol{\beta}\big]\, s(\boldsymbol{\beta})\, d^2\boldsymbol{\beta}\,, \tag{72}$$

where $\delta[\ldots]$ is the delta function. The lens modeling problem is now to recover $\boldsymbol{\alpha}(\boldsymbol{\theta})$ and $s(\boldsymbol{\beta})$ from observations of $I(\boldsymbol{\theta})$ and any other available information.

Let us suppose that a lens model is available. This may be a trial lens model produced in one step of some larger iterative process. But let us say the lens is given, and in particular the deflection field $\boldsymbol{\alpha}(\boldsymbol{\theta})$ is given. Equation (72) then takes the form

$$I(\boldsymbol{\theta}) = \int L(\boldsymbol{\theta}, \boldsymbol{\beta})\, s(\boldsymbol{\beta})\, d^2\boldsymbol{\theta}\,, \tag{73}$$

where $L(\boldsymbol{\theta}, \boldsymbol{\beta})$ is given. In practice, there will be a further convolution by a point spread function (psf) or other instrumental response function, but if the psf is known, it can be absorbed inside $L(\boldsymbol{\theta}, \boldsymbol{\beta})$. Under these assumptions, the source reconstruction becomes a linear image-reconstruction problem.

For optical imaging, noise is generally Gaussian. We can then write the likelihood as $\exp\left(-\frac{1}{2}\chi^2\right)$, where

$$\chi^2 = \int \sigma^{-2}(\boldsymbol{\theta}) \left(I^{\rm data}(\boldsymbol{\theta}) - I(\boldsymbol{\theta})\right)^2 d^2\boldsymbol{\theta}\,, \tag{74}$$

with $\sigma(\boldsymbol{\theta})$ being the noise dispersion. We are writing an integral here for notational convenience, but it is actually a sum over pixels. There are several possible ways to go about solving for $s(\boldsymbol{\alpha})$. One can assume a parametric form for $s(\boldsymbol{\beta})$ and fit for the parameters by minimizing χ^2. Alternatively, one can adopt a basis set for $s(\boldsymbol{\beta})$ and fit for the coefficients in that basis set by minimizing χ^2. Birrer et al. (2015) present a suitable basis set for doing so (shapelets), while Galan et al. (2021) consider wavelets. A third option is to put $s(\boldsymbol{\beta})$ on a grid. In this approach, a simple minimization of χ^2 is actually not recommended, because it tends to amplify the effect of noise, leading to artifacts like negative-brightness pixels or checkerboard patterns. To suppress such artifacts, a penalty function is added to χ^2 and the sum is minimized. The choice of penalty function — or equivalently, the prior on the source light distribution — itself has a large literature in image reconstruction (for example, in interferometry — see Thiébaut and Young 2017) and is known as regularization. Warren and Dye (2003) in an influential paper recommend

$$\left|\nabla^2 s(\boldsymbol{\beta})\right|^2 \tag{75}$$

as the regularization function, which favors smoother $s(\boldsymbol{\beta})$. It can be given a small or large weight, relative to χ^2, for gentle or aggressive smoothing, as desired. Further developments of the idea appear in several later works (Suyu et al. 2006; Vegetti and Koopmans 2009; Vernardos et al. 202). An advantage of a quadratic regularization function like (75) is that the minimization procedure requires solving only linear equations. But it is not a requirement, and source reconstruction using more complicated regularization functions is also possible, notably in the early work by Kochanek et al. (1989) and later in Brewer and Lewis (2006).

4.6 Many-Component Lens Models

Whereas for galaxy lenses a single elliptical component is often used, while multiple components are becoming common, in cluster-lens models multiple components are a must from the start. This is partly because clusters are further from a long-term stable state than galaxies and tend to have more complicated mass distributions, but mainly because cluster lenses present many more lensed images and thus more constraints. An illustrative example are the models of Abell 2744 by Bergamini et al. (2023), consisting of components for individual galaxies, with a larger-scale component for the diffuse matter in the cluster not identified with any individual galaxy.

The various modeling approaches we have discussed so far introduce a lot of parameters to be fitted, but the system remains overdetermined, meaning that there are more data points than parameters. A different category of multi-component methods reconstructs the lens in a basis set such as a grid, where the number of components exceeds the number of data points. By itself, such a free-form reconstruction is underdetermined and non-unique. To solve that problem, regularization can be applied, much as in the source reconstruction discussed above. Using a regularization, one can search, for instance, the smoothest model that fits the data, or the model the most similar to the light distribution. An early example of this technique in lensing is the reconstruction of Abell 370 by Abdelsalam et al. (1998). The mass map of Abell 2744 by Finney et al. (2018) is typical of more recent work on these lines. Hybrids with basic elliptical components and an additional substructure component which is free-form with regularization are also possible (e.g. Vernardos et al. 202; Galan et al. 2022).

4.7 Model Ensembles

What if we specify neither a simple parametric form nor a regularization? In Bayesian terms, this means there are more parameters in ϖ than there are data points in $\mathcal{D}$, while $p(\varpi)$ imposes some requirements on ϖ but has no hierarchy of favored and less-favored models. In practical terms, there is no single optimal model, only an ensemble of models. The ensemble provides confidence regions on any desired quantity, which can be easily extracted by marginalization. Conceptually, lens-model ensembles are like Monte-Carlo Markov chains widely used for likelihood or posterior-probability sampling.

Model ensembles may contain extreme models, but that does not mean the posterior $p(\varpi \mid \mathcal{D})$ is uniform in all properties between the extremes. To give an analogy, consider an ensemble of random walks with a given number and length of steps, and no preference between different random walks. The ensemble will span every random walk from zero net displacement to a straight line (i.e., the maximum displacement possible), but the distribution of displacements will not be uniform between these extremes.

A very general implementation of these ideas for the reconstruction of cluster lenses is the GRALE code (introduced by Liesenborgs et al. 2006, 2007, with many later enhancements). In GRALE each model consists of a large number of Plummer lenses (Eq. (56)) and there is an ensemble or population of models that evolves according to a genetic algorithm to fit the lensing data. The prior is implicitly defined by the algorithm rather than explicit, but is minimal: the mass distribution must be non-negative. The likelihood is not Gaussian, however it does incorporate the data on image locations and shapes, and also the absence of lensed images where they are not observed. The minimal prior means that the model ensemble tends to include a diversity of models degenerate with respect to the observables. Hence the inferred uncertainties are generally larger than in other techniques. The reconstruction of Abell 2744 by Sebesta et al. (2019) is illustrative.

Since galaxy lenses provide much less data than cluster lenses, a more informative prior $p(\kappa)$ is necessary. An explicit but still very weak prior (amounting to requiring the lens to be centrally concentrated) has been used in several modeling codes (Williams and Saha 2000; Saha and Williams 2004; Coles et al. 2014; Küng 2018). Here we mention two contrasting applications. Denzel et al. (2021) used the model-ensemble method to model eight time-delay quasars simultaneous, with H_0 as a shared parameter to be inferred. In (Küng et al. 2018) citizen scientists modeled tens of galaxy lenses by marking up each lens candidate and using a model-ensemble code to quickly explore the range of models consistent with the markup. Figure 10 shows an example of input markup from that work.

Finally, we mention yet another recent strategy, which amounts to finding plausible matches in galaxy-formation simulations instead of trying to reconstruct the lens. That is, the prior $p(\kappa)$ can be a large ensemble of simulated galaxies, which is then filtered to a much smaller ensemble that is consistent with the available data. Any such technique will be computationally expensive, perhaps unrealistically so for clusters, but is starting to become feasible for galaxy lenses (Denzel et al. 2021). What is attractive is the prospect of bypassing lens modeling altogether, and using lensing observations to directly address questions about galaxy structure and formation.

4.8 Model-Comparison Studies

It is acceptable for different lens models (i.e. models that differ in lens/source reconstruction) to disagree on some derived quantities as long as the quantities important for a specific science goal are accurately recovered. Strong lensing very accurately recover the Einstein radius and the mass enclosed within that radius, but other quantities (e.g. the ellipticity of the mass distribution) may be less accurately recovered. Some of these uncertainties on recovered parameters arise because of the degeneracies mentioned earlier, this is what is commonly referred to as "known unknowns", but other differences arise because of "unknown unknowns". The latter are harder to identify and require model comparisons.

We have already mentioned Abell 2744 several times, in connection with different modeling approaches. The system is one of six lensing clusters (the Frontier Field clusters) which have been likewise modeled by several groups. There have further been what one may call meta-analyses (Priewe et al. 2017; Remolina González et al. 2018; Raney et al. 2020) which study and interpret the effect of different choices made by different methods. These comparisons demonstrate a global robustness of models in regions well constrained by the data. They also highlight the complementarity of the parametric and free-form models.

Another kind of model-comparison study are lens modeling challenges, where different groups are invited to reconstruct a lens from mock data. Meneghetti et al. (2017) and (Ding et al. 2021) are two such blind tests of the methods of several groups, covering cluster lenses and galaxy lensed respectively. Such studies are a useful indicator of robustness and uncertainties in the current state of the art.

A third kind of model comparison consists in blindly forecasting some observables with various models. One spectacular case of this exercise happened with the observations of the lensed supernova "SN Refsdal". After the observations of the first four lensed images of the supernova, several groups gathered to predict the reappearance position, brightness of the 5th image (located $\sim 8''$ from the first quadruplet) about hundreds of days after the previous image (Kelly et al. 2023a,b). A fraction of the models successfully reproduced the data, enabling the use of this first multiply imaged lensed supernova for H_0 inference. The success of such a model comparison is very encouraging as numerous supernovae may be discovered in upcoming "all-sky" surveys (see the chapter on transients).

The comparison of models is a useful but difficult exercise because differences between models may arise not only from fundamental degeneracies and prior choices as discussed above, but also from subtle bugs that do not teach much about lensing but reveal a need to fix a code. Importantly, it is also hampered by the difficulty of comparing codes on the same grounds. The comparison of deflection and magnification maps offers a simple mean of comparison but extra work is required to derive differences and biases on specific parameters, such as e.g. the slope of the mean slope of the density profile, or the position angle of the main axis of the mass distribution. An additional hurdle is the difference of conventions in defining some of these quantities. The COOLEST framework (Galan et al. 2023), which provides an infrastructure to gather and compare results of lens models in a painless way, may help and trigger more model comparisons on large ensembles of lenses and ensure a good control of systematic uncertainties.

4.9 Reflections on Lens Modeling

As we remarked earlier (in Sect. 4.1) strong-lensing observations measure the projected gravitational potential its derivatives at the image locations, and the task of lens modeling is, in effect, to interpolate the values elsewhere. Doing so requires additional assumptions, and preferably complementary data as well.

A useful analogy is the density inside the Earth. We can measure gravitational acceleration at the surface or from orbit, and this gives the total mass very accurately, and also provides information on multipoles, but it does not give the radial density profile. That requires theory on the structure of planets, and further data like seismography.

Likewise lens modeling depends on our knowledge of and uncertainties about the structure and dynamics of galaxies and clusters. Are lensing galaxies all isothermal ellipsoids to the few-percent level? Perhaps not. Are dark halos so full of ongoing mergers that all equilibrium models are useless? Perhaps not, either. But between these extremes, there is a diversity of views among researchers in lensing.

The plurality of methods and data used for modeling lens systems may look difficult to unite at first sight. However, the reconstruction of the lensing mass density generally results from a trade-off between model rigidity and flexibility chosen in the light of the data and scientific questions posed, a situation that machine learning scientists may frame in terms of bias and variance arising for under-/over-fitting of the data. But beware, the choice of a model over another is not uniquely governed by those considerations. Inductive biases, namely biases introduced by the choice of a method, also exist. They are inherent to scientific research and evolve hands in hand with scientific results. A diversity of views and methods is often a good sign that our understanding of the world progresses. Strong lensing is no exception to this research walk.

5 Conclusions and Outlook

The introduction to strong gravitational lensing presented here starts by inviting the reader to think how gravitational fields will affect geometrical wavefronts propagating through the Universe. This leads to the main theoretical construct, the Fermat potential or arrival-time surface. The first and second derivatives of the arrival time surface are the lens equation and the distortion (or magnification) matrix, respectively, which are the main players in the quantitative analysis of the problem. The equations are the same as in other approaches that use only geometrical optics to describe light propagation, but many properties of lensing follow from geometrical arguments even without equations. In particular, visualising the arrival-time surface for a specific lensing configuration allows the researcher to gain a qualitative understanding of the problem.

The paper also introduced the reader to the other two essential components of strong lensing: critical curves and caustics. With these theoretical concepts under their belt, the reader is ready to learn about lensing as a probe in astrophysics. As its fundamentals are well understood, the primary application of lensing in astrophysics is to build mass models of the lens, and to recover the structure of the background source. Most of the mass in galaxy and cluster lenses is dark, so lensing is very handy here. Distant, high-redshift sources tend to be faint and small; cosmic lenses make sources larger, allowing for better spatial resolution. We also consider some simple lens models, and briefly describe more sophisticated ones used in the literature.

This short introduction to the essentials of strong lensing should provide the reader with the basic principles and physical intuition as a starting point to dive deeper into this growing field. The applications of lensing are ubiquitous: hardly any aspect of modern astrophysics has been untouched by it. The other papers in this collection cover most, if not all the current, as well as emerging applications of lensing: Search for strong gravitational lensing through serendipitous and deliberate discoveries, Strong lensing by galaxies, Strong lensing by galaxy clusters, Strong lensing and microlensing of quasars, Microcaustics near macrocaustics, Gravitational lensing as a cosmic telescope, Probing the nature of dark matter on (sub-)galactic scales with strong lensing, Time-Delay Cosmography and the estimation of the Hubble Constant and other cosmological parameters, Strong lensing and microlensing of supernovae, Strong lensing of Gamma-Ray Bursts, Fast Radio Bursts, and Gravitational waves, and Diffraction in strong gravitational lensing.

We hope the reader enjoys this and other papers in this series, and finds strong gravitational lensing as enthralling as the authors do!

Appendix A: Example Lens Models Used

A.1 A Softened Schwarzschild Metric

Consider the metric

$$ds^2 = -e^{2\Phi(r)}dt^2 + e^{-2\Phi(r)}dr^2 + r^2\left(d\theta^2 + \sin^2\theta\, d\phi^2\right) \tag{76}$$

If we put

$$2\Phi(r) = \ln\left(1 - \frac{2GM}{c^2 r}\right) \tag{77}$$

we would have a Schwarzschild metric. For the example in Figs. 2–4 we instead we put

$$\Phi(r) = -\frac{GM}{c^2}\frac{1}{\sqrt{r_c^2 + r^2}} \tag{78}$$

where r_c is a small softening length.

This metric has no especially interesting physical interpretation. It is just used as an example of a strong gravitational field without the additional complication of an event horizon.

A.2 PIEMD

The so-called pseudo-isothermal elliptical mass distribution, used as an example in Figs. 6 and 7 is defined via two elliptical radii

$$\begin{aligned} R_e^2 &= q^2(s^2 + \theta_x^2) + \theta_y^2 \\ R_f^2 &= (1 - q^2)\theta_x^2 + (R_e + s)^2 \end{aligned} \tag{79}$$

Then the density is

$$\kappa = \frac{1}{2R_e} \tag{80}$$

the deflection is

$$\begin{aligned} \alpha_x &= \frac{1}{\sqrt{1-q^2}} \arctan\left(\frac{\sqrt{1-q^2}\theta_x}{R_e + s}\right) \\ \alpha_y &= \frac{1}{\sqrt{1-q^2}} \operatorname{arctanh}\left(\frac{\sqrt{1-q^2}\theta_y}{R_e + q^2 s}\right) \end{aligned} \tag{81}$$

while the potential and its second derivatives are

$$\begin{aligned}
\psi &= \theta_x\alpha_x + \theta_y\alpha_y - s\ln R_f \\
\psi_{xx} &= \frac{(qs)^2 + R_e s + \theta_y^2}{R_e R_f} \\
\psi_{yy} &= \frac{s^2 + R_e s + \theta_x^2}{R_e R_f} \\
\psi_{xy} &= -\frac{\theta_x\theta_y}{R_e R_f}
\end{aligned} \tag{82}$$

A.3 A Candy-Bar Lens

The candy-bar lens, used as an example in Fig. 8, is not by itself a serious lens model, but can be a useful component of one. Square mass tiles are used in many free-form models. Another application is in microlensing, where a rectangular mass mass tile is removed from a smooth mass distribution and replaced with point masses. To compute the potential, we need the integral from Eq. (34) for constant κ, that is to say

$$\int \ln\left(\theta_x^2 + \theta_y^2\right) d\theta_x\, d\theta_y \tag{83}$$

over a rectangular region. The result and its derivatives can be expressed in terms of arctans and logs. Expressions for a square are given Saha and Williams (1997) and for a rectangle in Zheng et al. (2022).

Appendix B: Constants and Unit Conversions

Astronomical distances are commonly given in many different units. When contrasting scales are involved, as they often are in gravitational lensing, it is rather convenient to convert everything to light-seconds. Thus, while a parsec is defined as the distance from which an astronomical unit (au) subtends an angle of 1 arcsec, with the au being $1.495978707 \times 10^{11}$ m, the values

$$\begin{aligned}
1\ \text{au} &= c \times 499.0\ \text{s} \\
1\ \text{pc} &= c \times 1.029 \times 10^8\ \text{s}
\end{aligned} \tag{84}$$

are easier to work with, and more digits can be added if necessary. Similarly, the solar mass can be expressed as

$$\frac{GM_\odot}{c^3} = 4.925 \times 10^{-6}\ \text{s} \tag{85}$$

The Hubble constant is commonly given in the form

$$H_0 = h_{70} \times 70\ \text{km s}^{-1} Mpc^{-1} \tag{86}$$

The following conversions may be useful.

$$\begin{aligned} H_0^{-1} &= h_{70}^{-1} \times 14.0 \text{ Gyr} \\ cH_0^{-1} &= h_{70}^{-1} \times 4.28 \text{ Gpc} \\ &= h_{70}^{-1} \times 20.8 \text{ kpc arcsec}^{-1} \end{aligned} \tag{87}$$

The critical density (from Eq. (32)) comes to

$$\frac{cH_0}{4\pi G} = h_{70} \times 0.811 \text{ kg m}^{-2} \tag{88}$$

times a dimensionless factor which is typically $\sim 1/z_{\rm d}$.

The cosmological total density $\rho_{\rm cr}$ (from Eq. (14)) can be expressed in several ways, such as

$$\frac{3H_0^2}{8\pi G}\frac{c^2}{e} = h_{70}^2 \times 5.15 \text{ GeV m}^{-3} \tag{89}$$

Appendix C: Lensed Waves

If a source is large enough, that the distance to the observer from different parts of the source differ by more than a wavelength, radiation from it is incoherent at the observer. This is the case for nearly all astronomical sources. For these, geometrical optics applies. In particular, the arrival-time surface, whether in its scaled form $\tau(\boldsymbol{\theta}, \boldsymbol{\beta})$ or in its dimensional form $t(\boldsymbol{\theta}, \boldsymbol{\beta})$ (see Eqs. (32)–(34)), is defined over the whole sky, but only its zero-gradient points correspond to images.

For sources that are small enough and far enough, however, the whole $t(\boldsymbol{\theta})$ surface interferes. This applies to gravitational waves (see e.g., Dai et al. 2020) and may apply to fast radio bursts (Leung et al. 2022). At frequency ω the resulting electric field of the light (or gravitational wave strain) is given by

$$\int \exp(i\omega t(\boldsymbol{\theta}))\, d^2\boldsymbol{\theta} \tag{90}$$

This is sometimes called the Fresnel-Kirchoff diffraction integral. Of particular interest is the quotient

$$\frac{\int \exp(i\omega t_L(\boldsymbol{\theta}))\, d^2\boldsymbol{\theta}}{\int \exp(i\omega t_{NL}(\boldsymbol{\theta}))\, d^2\boldsymbol{\theta}} \tag{91}$$

of the lens and no-lens cases.

With no lensing

$$t_{NL}(\boldsymbol{\theta}) = \tfrac{1}{2} T_{\rm ds} \left((\theta_x - \beta_x)^2 + (\theta_y - \beta_y)^2\right) \tag{92}$$

with the scale $T_{\rm ds}$ being as defined in from Eq. (32). With a lens present, things are of course more complicated. For one regime, however, namely where the wave period is much shorter than gravitational time delays (that is, $\omega t(\boldsymbol{\theta}) \gg 1$), the integrals can be usefully estimated. The idea is to Taylor-expand $t(\boldsymbol{\theta})$ near a stationary point (say $\boldsymbol{\theta}^*$) where of course an ordinary image would be located. Near a stationary point there is no linear dependence on $\boldsymbol{\theta} - \boldsymbol{\theta}^*$ and the quadratic terms are given by the Hessian (47) matrix. We can assume, without loss of generality, that the Hessian is diagonal. Let us introduce

$$a, b = 1 - \kappa \pm \gamma \qquad \mu = 1/(ab) \tag{93}$$

and then μ will be the scalar magnification. In terms of these we have

$$t_L(\boldsymbol{\theta}) = t(\boldsymbol{\theta}^*) + \tfrac{1}{2} T_{\rm ds} \left(a(\theta_x - \theta_x^*)^2 + b(\theta_y - \theta_y^*)^2 \right) \tag{94}$$

Now both the integrals in (91) have the form

$$\int_{-\infty}^{\infty} \exp\left(\frac{\pm i u^2}{2\sigma^2} \right) du = \frac{e^{\pm i\pi/4}}{\sqrt{2\pi}\,\sigma} \tag{95}$$

Note changing the sign in integrand changes the phase of the integral by $\pi/2$. We need to pay attention to the signs of ω, a, and b. Bearing this mind and simplifying, we get

$$|\mu|^{1/2} \exp\left(i\omega t(\boldsymbol{\theta}^*) - i\Delta \right) \tag{96}$$

where

$$\Delta = \tfrac{1}{2}\pi \,\mathrm{sgn}\,\omega \times \begin{cases} 0 \text{ for minima} \\ 1 \text{ for saddle points} \\ 2 \text{ for maxima} \end{cases} \tag{97}$$

The Δ term here is known as the Morse phase.

Acknowledgements We would like to thank the International Space Science Institute in Bern, Switzerland (ISSI) for their hospitality, and the conveners for organizing the stimulating workshop on "Strong Gravitational Lensing". We thank our co-participants at the workshop for many discussions and suggestions, and the reviewers for their comments. We especially thank P. L. Schechter and P. Schneider for sharing many insights.

Funding Open access funding provided by University of Zurich.

Declarations

Competing Interests The authors declare no competing interests.

Open Access This article is licensed under a Creative Commons Attribution 4.0 International License, which permits use, sharing, adaptation, distribution and reproduction in any medium or format, as long as you give appropriate credit to the original author(s) and the source, provide a link to the Creative Commons licence, and indicate if changes were made. The images or other third party material in this article are included in the article's Creative Commons licence, unless indicated otherwise in a credit line to the material. If material is not included in the article's Creative Commons licence and your intended use is not permitted by statutory regulation or exceeds the permitted use, you will need to obtain permission directly from the copyright holder. To view a copy of this licence, visit http://creativecommons.org/licenses/by/4.0/.

References

Abdelsalam HM, Saha P, Williams LLR (1998) Mon Not R Astron Soc 294:734. https://doi.org/10.1046/j.1365-8711.1998.01356.x

Akiyama K, Alberdi A, Alef W et al (2022) Astrophys J Lett 930:L17. https://doi.org/10.3847/2041-8213/ac6756

Barnabè M, Koopmans LVE (2007) Astrophys J 666:726. https://doi.org/10.1086/520495

Bartelmann M, Schneider P (2001) Phys Rep 340:291. https://doi.org/10.1016/S0370-1573(00)00082-X

Bergamini P, Acebron A, Grillo C et al (2023) Astrophys J 952:84. https://doi.org/10.3847/1538-4357/acd643

Binney J, Tremaine S (2008) Galactic dynamics, 2nd edn. Princeton University Press, Princeton

Birrer S (2021) Astrophys J 919:38. https://doi.org/10.3847/1538-4357/ac1108

Birrer S, Amara A, Refregier A (2015) Astrophys J 813:102. https://doi.org/10.1088/0004-637X/813/2/102

Blandford R, Narayan R (1986) Astrophys J 310:568. https://doi.org/10.1086/164709

Borgeest U (1983) Astron Astrophys 128:162

Brewer BJ, Lewis GF (2006) Astrophys J 637:608. https://doi.org/10.1086/498409

Broderick AE, Pesce DW, Gold R et al (2022) Astrophys J 935:61. https://doi.org/10.3847/1538-4357/ac7c1d

Cassan A, Ranc C, Absil O et al (2022) Nat Astron 6:121. https://doi.org/10.1038/s41550-021-01514-w

Coles JP, Read JI, Saha P (2014) Mon Not R Astron Soc 445:2181. https://doi.org/10.1093/mnras/stu1781

Collett TE, Oldham LJ, Smith RJ et al (2018) Science 360:1342. https://doi.org/10.1126/science.aao2469

Cooke JH, Kantowski R (1975) Astrophys J Lett 195:L11. https://doi.org/10.1086/181697

Crosta M, Geralico A, Lattanzi MG, Vecchiato A (2017) Phys Rev D 96:104030. https://doi.org/10.1103/PhysRevD.96.104030

Dai L, Zackay B, Venumadhav T, Roulet J, Zaldarriaga M (2020). ArXiv:e-prints. arXiv:2007.12709

Demkov YN, Puchkov AM (2000) Phys Rev D 61:083001. https://doi.org/10.1103/PhysRevD.61.083001

Denzel P, Coles JP, Saha P, Williams LLR (2021) Mon Not R Astron Soc 501:784. https://doi.org/10.1093/mnras/staa3603

Denzel P, Mukherjee S, Saha P (2021) Mon Not R Astron Soc 506:1815. https://doi.org/10.1093/mnras/stab1716

Ding X, Treu T, Birrer S et al (2021) Mon Not R Astron Soc 503:1096. https://doi.org/10.1093/mnras/stab484

Elliot JL, Olkin CB (1996) Annu Rev Earth Planet Sci 24:89. https://doi.org/10.1146/annurev.earth.24.1.89

Etherington A, Nightingale JW, Massey R et al (2022) Mon Not R Astron Soc 517:3275. https://doi.org/10.1093/mnras/stac2639

Falor C, Schechter PL (2022) Astron J 164:120. https://doi.org/10.3847/1538-3881/ac80bc

Finney EQ, Bradač M, Huang K-H et al (2018) Astrophys J 859:58. https://doi.org/10.3847/1538-4357/aabf97

Galan A, Peel A, Joseph R, Courbin F, Starck JL (2021) Astron Astrophys 647:A176. https://doi.org/10.1051/0004-6361/202039363

Galan A, Van de Vyvere L, Gomer M, Vernardos G, Sluse D (2023) J Open Sour Softw 8:5567. https://doi.org/10.21105/joss.05567

Galan A, Vernardos G, Peel A, Courbin F, Starck JL (2022) Astron Astrophys 668:A155. https://doi.org/10.1051/0004-6361/202244464

Gavazzi R, Treu T, Koopmans LVE et al (2008) Astrophys J 677:1046. https://doi.org/10.1086/529541

Giblin B, Heymans C, Asgari M et al (2021) Astron Astrophys 645:A105. https://doi.org/10.1051/0004-6361/202038850

Gomer MR, Williams LLR (2018) Mon Not R Astron Soc 475:1987. https://doi.org/10.1093/mnras/stx3294

Gomer MR, Ertl S, Biggio L et al (2023) Astron Astrophys 679:A59. https://doi.org/10.1051/0004-6361/202347507

Grimm N, Yoo J (2018) J Cosmol Astropart Phys 2018:067. https://doi.org/10.1088/1475-7516/2018/07/067

Kassiola A, Kovner I (1993) Astrophys J 417:450. https://doi.org/10.1086/173325

Kassiola A, Kovner I (1995) Mon Not R Astron Soc 272:363. https://doi.org/10.1093/mnras/272.2.363

Keeton CR (2001). ArXiv e-prints. arXiv:astro-ph/0102341

Kelly PL, Rodney S, Treu T et al (2023a) Astrophys J 948:93. https://doi.org/10.3847/1538-4357/ac4ccb

Kelly PL, Rodney S, Treu T et al (2023b) Science 380:abh1322. https://doi.org/10.1126/science.abh1322

Kennefick D (2012) In: Lehner C, Renn J, Schemmel M (eds) Einstein and the changing worldviews of physics. Birkhäuser, Boston, pp 201–232. https://doi.org/10.1007/978-0-8176-4940-1_9

Kochanek CS (2006) In: Meylan G, Jetzer P, North P, Schneider P, Kochanek CS, Wambsganss J (eds) Gravitational lensing: strong, weak and micro. Saas-fee advanced courses, vol. 33. Springer, Berlin, Heidelberg, pp 91–268. https://doi.org/10.1007/978-3-540-30310-7_2

Kochanek CS, Blandford RD, Lawrence CR, Narayan R (1989) Mon Not R Astron Soc 238:43. https://doi.org/10.1093/mnras/238.1.43

Kormann R, Schneider P, Bartelmann M (1994) Astron Astrophys 284:285

Küng R (2018) Astron Comput 23:115. https://doi.org/10.1016/j.ascom.2018.02.007

Küng R, Saha P, Ferreras I et al (2018) Mon Not R Astron Soc 474:3700. https://doi.org/10.1093/mnras/stx3012

Leung C, Kader Z, Masui KW et al (2022) Phys Rev D 106:043017. https://doi.org/10.1103/PhysRevD.106.043017
Liesenborgs J, De Rijcke S (2012) Mon Not R Astron Soc 425:1772. https://doi.org/10.1111/j.1365-2966.2012.21751.x
Liesenborgs J, De Rijcke S, Dejonghe H (2006) Mon Not R Astron Soc 367:1209. https://doi.org/10.1111/j.1365-2966.2006.10040.x
Liesenborgs J, de Rijcke S, Dejonghe H, Bekaert P (2007) Mon Not R Astron Soc 380:1729. https://doi.org/10.1111/j.1365-2966.2007.12236.x
Luhtaru R, Schechter PL, de Soto KM (2021) Astrophys J 915:4. https://doi.org/10.3847/1538-4357/abfda1
Meneghetti M (2021) Introduction to gravitational lensing; with Python examples. Lecture Notes in Physics, vol 956. Springer, Cham. https://doi.org/10.1007/978-3-030-73582-1
Meneghetti M, Natarajan P, Coe D et al (2017) Mon Not R Astron Soc 472:3177. https://doi.org/10.1093/mnras/stx2064
Mróz P, Udalski A, Szymański MK et al (2020) Astrophys J Supp 249:16. https://doi.org/10.3847/1538-4365/ab9366
Nityananda R (1990) Curr Sci 59:1044. http://www.jstor.org/stable/24094893
Nityananda R, Samuel J (1992) Phys Rev D 45:3862. https://doi.org/10.1103/PhysRevD.45.3862
Petkova M, Metcalf RB, Giocoli C (2014) Mon Not R Astron Soc 445:1954. https://doi.org/10.1093/mnras/stu1860
Petters AO, Levine H, Wambsganss J (2001) Singularity theory and gravitational lensing. Progress in mathematical physics, vol 21. Birkhäuser, Boston. https://doi.org/10.1007/978-1-4612-0145-8
Priewe J, Williams LLR, Liesenborgs J, Coe D, Rodney SA (2017) Mon Not R Astron Soc 465:1030. https://doi.org/10.1093/mnras/stw2785
Raney CA, Keeton CR, Brennan S, Fan H (2020) Mon Not R Astron Soc 494:4771. https://doi.org/10.1093/mnras/staa921
Refsdal S (1964) Mon Not R Astron Soc 128:295. https://doi.org/10.1093/mnras/128.4.295
Refsdal S (1964) Mon Not R Astron Soc 128:307. https://doi.org/10.1093/mnras/128.4.307
Refsdal S, Surdej J (1994) Rep Prog Phys 57:117. https://doi.org/10.1088/0034-4885/57/2/001
Remolina González JD, Sharon K, Mahler G (2018) Astrophys J 863:60. https://doi.org/10.3847/1538-4357/aacf8e
Saha P, Williams LLR (1997) Mon Not R Astron Soc 292:148. https://doi.org/10.1093/mnras/292.1.148
Saha P, Williams LLR (2003) Astron J 125:2769. https://doi.org/10.1086/375204
Saha P, Williams LLR (2004) Astron J 127:2604. https://doi.org/10.1086/383544
Saha P, Williams LLR (2006) Astrophys J 653:936. https://doi.org/10.1086/508798
Schmidt T, Treu T, Birrer S et al (2023) Mon Not R Astron Soc 518:1260. https://doi.org/10.1093/mnras/stac2235
Schneider P (1985) Astron Astrophys 143:413
Schneider P (2014) Astron Astrophys 568:L2. https://doi.org/10.1051/0004-6361/201424450
Schneider P, Ehlers J, Falco EE (1992) Gravitational lenses. Astronomy and astrophysics library. Springer, New York. https://doi.org/10.1007/978-3-662-03758-4
Schneider P, Sluse D (2014) Astron Astrophys 564:A103. https://doi.org/10.1051/0004-6361/201322106
Sebesta K, Williams LLR, Liesenborgs J, Medezinski E, Okabe N (2019) Mon Not R Astron Soc 488:3251. https://doi.org/10.1093/mnras/stz1950
Shajib AJ (2019) Mon Not R Astron Soc 488:1387. https://doi.org/10.1093/mnras/stz1796
Shajib AJ, Mozumdar P, Chen GCF et al (2023) Astron Astrophys 673:A9. https://doi.org/10.1051/0004-6361/202345878
Shapiro II (1964) Phys Rev Lett 13:789. https://doi.org/10.1103/PhysRevLett.13.789
Sluse D, Chantry V, Magain P, Courbin F, Meylan G (2012) Astron Astrophys 538:A99. https://doi.org/10.1051/0004-6361/201015844
Smith RJ, Collett TE (2021) Mon Not R Astron Soc 505:2136. https://doi.org/10.1093/mnras/stab1399
Suyu SH, Marshall PJ, Hobson MP, Blandford RD (2006) Mon Not R Astron Soc 371:983. https://doi.org/10.1111/j.1365-2966.2006.10733.x
Tessore N (2017) Astron Astrophys 597:L1. https://doi.org/10.1051/0004-6361/201629947
Thiébaut É, Young J (2017) J Opt Soc Am A 34:904. https://doi.org/10.1364/JOSAA.34.000904
Turyshev SG, Toth VT (2020) Phys Rev D 102:024038. https://doi.org/10.1103/PhysRevD.102.024038
Valls-Gabaud D (2006) In: Alimi J-M, Füzfa A (eds) Albert Einstein century international conference. AIP Conf Proc, vol. 861, pp 1163–1163. https://doi.org/10.1063/1.2399715
Van de Vyvere L, Gomer MR, Sluse D et al (2022a) Astron Astrophys 659:A127. https://doi.org/10.1051/0004-6361/202141551
Van de Vyvere L, Sluse D, Gomer MR, Mukherjee S (2022b) Astron Astrophys 663:A179. https://doi.org/10.1051/0004-6361/202243382

Vegetti S, Koopmans LVE (2009) Mon Not R Astron Soc 392:945. https://doi.org/10.1111/j.1365-2966.2008.14005.x

Vernardos G, Koopmans LVE (2022) Mon Not R Astron Soc 516:1347. https://doi.org/10.1093/mnras/stac1924

Wagner J (2018) Astron Astrophys 620:A86. https://doi.org/10.1051/0004-6361/201834218

Wagner J (2019) Mon Not R Astron Soc 487:4492. https://doi.org/10.1093/mnras/stz1587

Wagner J (2019) Universe 5:177. https://doi.org/10.3390/universe5070177

Wagner J, Liesenborgs J, Tessore N (2018) Astron Astrophys 612:A17. https://doi.org/10.1051/0004-6361/201731932

Wagner J, Williams LLR (2020) Astron Astrophys 635:A86. https://doi.org/10.1051/0004-6361/201936628

Walsh D, Carswell RF, Weymann RJ (1979) Nature 279:381. https://doi.org/10.1038/279381a0

Warren SJ, Dye S (2003) Astrophys J 590:673. https://doi.org/10.1086/375132

Williams LLR, Saha P (2000) Astron J 119:439. https://doi.org/10.1086/301234

Witt HJ (1996) Astrophys J Lett 472:L1. https://doi.org/10.1086/310358

Woldesenbet AG, Williams LLR (2012) Mon Not R Astron Soc 420:2944. https://doi.org/10.1111/j.1365-2966.2011.20110.x

Woldesenbet AG, Williams LLR (2015) Mon Not R Astron Soc 454:862. https://doi.org/10.1093/mnras/stv1960

Wucknitz O (2002) Mon Not R Astron Soc 332:951. https://doi.org/10.1046/j.1365-8711.2002.05426.x

Wucknitz O (2008) Mon Not R Astron Soc 386:230. https://doi.org/10.1111/j.1365-2966.2008.13017.x

Wynne RA, Schechter PL (2018). ArXiv e-prints. arXiv:1808.06151

Yoo J, Kochanek CS, Falco EE, McLeod BA (2005) Astrophys J 626:51. https://doi.org/10.1086/429959

Young P, Gunn JE, Kristian J, Oke JB, Westphal JA (1980) Astrophys J 241:507. https://doi.org/10.1086/158365

Zheng W, Chen X, Li G, Chen H-Z (2022) Astrophys J 931:114. https://doi.org/10.3847/1538-4357/ac68ea

Publisher's Note Springer Nature remains neutral with regard to jurisdictional claims in published maps and institutional affiliations.

Space Science Reviews (2024) 220:87
https://doi.org/10.1007/s11214-024-01105-x

Strong Lensing by Galaxies

A.J. Shajib[1,2] · G. Vernardos[3,4,5] · T.E. Collett[6] · V. Motta[7] · D. Sluse[8] · L.L.R. Williams[9] · P. Saha[10] · S. Birrer[11,12,13] · C. Spiniello[14,15] · T. Treu[16]

Received: 17 April 2023 / Accepted: 26 August 2024 / Published online: 8 November 2024
© The Author(s) 2024

Abstract
Strong gravitational lensing at the galaxy scale is a valuable tool for various applications in astrophysics and cosmology. Some of the primary uses of galaxy-scale lensing are to study elliptical galaxies' mass structure and evolution, constrain the stellar initial mass function, and measure cosmological parameters. Since the discovery of the first galaxy-scale lens in the 1980s, this field has made significant advancements in data quality and modeling techniques. In this review, we describe the most common methods for modeling lensing observables, especially imaging data, as they are the most accessible and informative source of lensing observables. We then summarize the primary findings from the literature on the astrophysical and cosmological applications of galaxy-scale lenses. We also discuss the current limitations of the data and methodologies and provide an outlook on the expected improvements in both areas in the near future.

Keywords Gravitational lensing: strong · Galaxies: elliptical and lenticular, cD · Galaxies: structure · Galaxies: evolution · Cosmological parameters

1 Introduction

This review article discusses applications of galaxy-scale strong lenses to study the properties of the deflector galaxies, that is, the central lensing galaxies. These have so far typically been massive elliptical galaxies at $0.1 \lesssim z \lesssim 1$ and strong lensing has been chiefly applied to study their internal structure and composition. However, a large sample of strong lenses with deflectors other than massive ellipticals is expected to be discovered in this decade from the upcoming deep sky surveys. We can also gain important insights into the formation and evolution of the deflectors by comparing their structural properties (e.g., the logarithmic slope of the density profile and the dark matter fraction) across cosmic times. Furthermore, in this review article, we present cosmological applications of the galaxy-scale lenses that do not require time delay information – that is, measuring cosmological parameters such as the matter density parameter Ω_{m} and the dark energy equation-of-state parameter w_{de}. Cosmological application involving the time delay measurements, that is, measuring primarily the Hubble constant, is reviewed by Birrer et al. (2024). Additionally, Vegetti et al. (2023) review the application of galaxy-scale lensing to study sub-galactic structures of dark matter.

A.J. Shajib is NHFP Einstein Fellow.

Extended author information available on the last page of the article

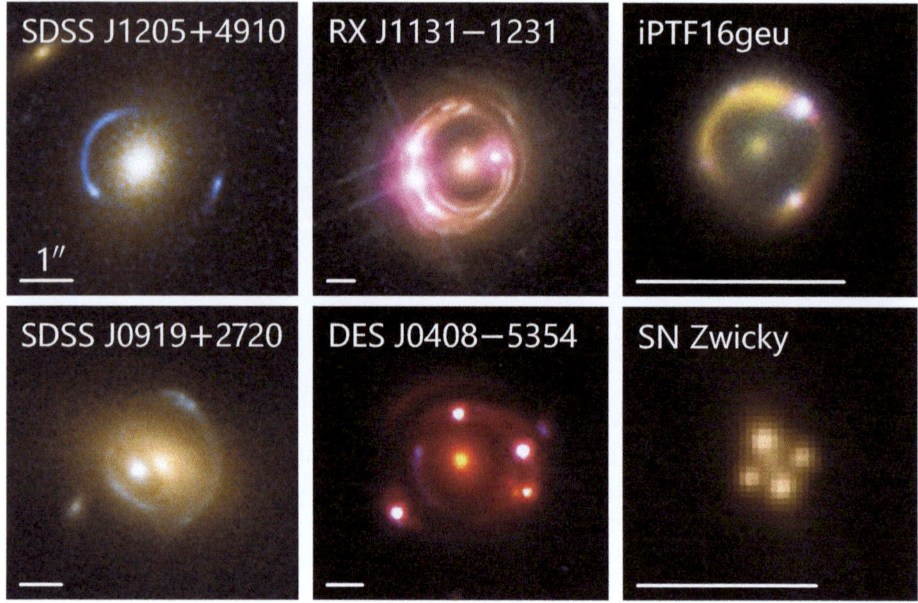

Fig. 1 Examples of galaxy-scale lenses with different types of background sources. These are false color images created from multi-band *HST* imaging, and in some cases, also combined with the *Chandra* X-ray data (RX J1131−1231) and the Keck Observatory IR imaging (iPTF16geu). The first column shows two lenses with background galaxies without any resolved point source (Bolton et al. 2006; Courbin et al. 2012). The second column shows two lensed quasar systems (Suyu et al. 2013; Shajib et al. 2020). The third column shows two lensed supernovae (Goobar et al. 2017; Pierel et al. 2023). The white bar in each panel represents 1″. *Image credits: NASA, ESA, A. Bolton, the SLACS team,* Chandra, *A. J. Shajib, W. M. Keck Observatory, T. Li, and J. Pierel*

In this introductory section, we provide a brief description of the lensing phenomenology (Sect. 1.1) and discuss the advantages of strong-lensing observables in comparison with other probes of galaxy mass, such as stellar dynamics (Sect. 1.2). The remainder of this review article is organized as follows. In Sect. 2, we highlight the significant historical results involving galaxy-scale lenses and introduce several prominent lens samples. In Sect. 3, we describe the strong-lensing observables and their modeling and analysis methods. We discuss the application of galaxy-scale strong lensing to study galaxy properties and evolution in Sect. 4 and to constrain cosmological parameters in Sect. 5. Next, in Sect. 6, we discuss open issues – both in technical aspects and scientific questions – and provide future outlooks. We conclude the review article in Sect. 7.

1.1 Brief Description of Lensing Phenomenology at the Galaxy-Scale

The background extended source in a galaxy-scale strong lens can be lensed into multiple arcs or a complete Einstein ring. Multiple point images will also appear if there is a point source within the background galaxy, for example, an active galactic nucleus (AGN) or quasar, or a supernova (see Fig. 1). The former type of system is called galaxy–galaxy lenses. In contrast, the latter is usually referred to as 'quads' (for the case of four detected point images) or 'doubles' (for the case of two detected point images). The different manifestations of strong lenses – the appearance of arcs or a full Einstein ring, or the number of point images – depend on the position of the source with respect to the lens caustics (as introduced in Saha et al. 2024). Strong lensing of point sources can provide three types of observables: image positions, image magnification ratios, and time delays between the images. It follows from the lensing theory that all three are properties of the Fermat potential or the arrival-time surface. Images form at the local extrema – minima, saddle points, and maxima – of this potential. Magnification is inversely proportional to the determinant of the Fermat potential's Jacobian matrix. Lastly, time delays are the differences in the Fermat potential at the image locations (for a detailed explanation, see Saha et al. 2024).

However, not all of these observables are available for every lens. Image positions are almost always observable for point-source lens systems. Image flux ratios, while also always observable, can only be turned into actual magnifications when the intrinsic source flux is known. However, interpreting these magnifications during modeling requires extra care, as they can be affected by more complex features in the galaxy-scale mass distribution, for example, baryonic disks (Hsueh et al. 2016, 2017), microlensing by individual stars and planets in the lensing galaxy (see Vernardos et al. 2024), intermediate-mass-scale structures like dark matter subhalos (see Vegetti et al. 2023), or dust extinction by the lens galaxy (e.g., Motta et al. 2002; Mediavilla et al. 2005). Time delays can be obtained only for variable point sources, like quasars, supernovae, and, in the future, gravitational waves and fast radio bursts (see Birrer et al. 2024). Measuring time delays can be a difficult task, especially for quasars, which require long-term monitoring spanning from a few seasons to years (e.g., Eigenbrod et al. 2005; Bonvin et al. 2016; Millon et al. 2020).

The above description of point-source lensing must be slightly modified for extended sources comparable in size to the lens caustics. Instead of point-like image positions, the light from such a source is spread into an extended area in the image plane and then further smeared by the point spread function. As a result, it is hard to know *a priori* how the observed flux of the multiple images of the source on the lens plane traces back to the same location on the source plane. Therefore, even if the actual source brightness is known, a lens model is required to compute the magnification field (i.e., the Jacobian of the lens potential) across the lens plane. Due to their size, extended sources are not variable on time scales relevant for lensing. Thus, time delays are not observable without a variable point source. Note that extended arcs from the point source's host galaxy are also usually present for a lens system that includes a point source. Therefore, we can constrain the lens model by simultaneously utilizing the lensed arcs' flux distribution and the point-image positions.

1.2 Unique Advantages of Lensing as a Probe of Galaxy Structure

Other than lensing, the only commonly used probe of the mass distribution in galaxies is kinematics, that is, the velocity dispersion or streaming motions of stars, gas, or globular clusters. For this to be fully informative of the mass distribution, however, spatially resolved measurements are necessary (Cappellari 2016), which is often limited to nearby galaxies at $z \leq 0.5$ due to the sizeable observational cost, or due to the smaller size on the sky of objects at higher redshifts. Using individual stars' motions, for example, from *Gaia*, to map a galaxy's mass is only limited to our Milky Way (e.g., Nitschai et al. 2020). Thus, an aperture-integrated velocity dispersion measurement (from either long-slit or integral field spectroscopy) is the only possible dynamical observable for galaxies other than ours. However, strong-lensing observables are obtained from high-resolution imaging data, which are much more informative than the dynamical observables for galaxies beyond the local Universe ($z \gtrsim 0.03$, e.g., Smith et al. 2015). This is especially advantageous for studying the lens galaxies since the lensing signal does not depend on the surface brightness of the galaxy being studied, unlike stellar kinematics, but on the combination of the lens galaxy mass and the source brightness. Strong lensing can provide ∼1–2% constraints on the mass enclosed within the Einstein radius from imaging data alone. To obtain a similarly precise mass constraint from dynamical observations at high redshift, either integration times longer by a factor $\mathcal{O}(10)$ are needed on current facilities, or we must wait for better quality adaptive optics systems planned for future extremely large telescopes.

Furthermore, the dynamical observables have their own intrinsic degeneracies, that is, the mass–anisotropy degeneracy (e.g., Treu and Koopmans 2002). There is an important complementarity between lensing and dynamical observables, which can be used to break their corresponding degeneracies (e.g., Courteau et al. 2014). Finally, gravitational lensing responds to both baryonic and dark matter without any assumptions on their dynamical state, regardless of whether it is in equilibrium or not.

2 Historical Background

This section provides a historical note on the initial discoveries of galaxy-scale strong lenses (Sect. 2.1). Then, in Sect. 2.2, we briefly introduce several prominent samples of galaxy-scale lenses that have contributed to the major science applications described in Sect. 4 and Sect. 5. Additional references can be found in the review by Treu (2010).

2.1 Initial Discoveries of Strong Lensing Systems

The first strong lens, the double quasar 0957+561A, B, was discovered in 1979 by Walsh et al. (1979). The system consists of two images of the quasar separated by $5\farcs7$. The authors first offered a 'conventional' interpretation that the two images are different, individual quasars that happen to be close to each other and share the same physical characteristics. Since no gravitational lens was known before that, their less conventional view was that the two are multiple images of the same source. Subsequent work showed that the lensing hypothesis was correct. For example, the structure of the radio jets emanating from the two images of the quasar is consistent with them being mirror imaged, as one would expect for a minimum and a saddle-point image (Gorenstein et al. 1988; Garrett et al. 1994). The discovery of the first quadruply imaged quasar, PG 1115+080, was announced the following year (Weymann et al. 1980). It was initially called a 'triple' because the second arriving minimum and its neighboring saddle point were too close to be resolved. These two lenses' discoveries opened up a new field in astrophysics: observations of multiply-imaged, 'strong' gravitational lenses. Discoveries of other types of lensed sources followed: for example, the first dust-obscured Seyfert 2 AGN, IRAS F10214+4724, was detected in the infrared (IR), and later identified as being lensed (Eisenhardt et al. 1996; Lehar and Broadhurst 1996).

The increasing number of detections of such lensed systems spurred a lens modeling effort. The initial studies that used more detailed models beyond a point-like lens mass appeared in the early 1980s (Young et al. 1980, 1981b,a), and already recognized "that there are several plausible ways to reproduce the observations", foreshadowing the importance of lens model degeneracies. The role of the lens mass granularity due to individual stars in the lensing galaxy was also recognized very early on (Chang and Refsdal 1979; Young 1981) and later grew into the rich sub-field of extragalactic stellar microlensing (see Vernardos et al. 2024 for a review).

2.2 Prominent Samples of Galaxy-Scale Lenses

In this subsection, we briefly introduce some of the most prominent samples of galaxy-scale lenses that had an impact on the science applications presented in Sect. 4 and Sect. 5. Note that this is not a complete list of all the discovered galaxy-scale lenses. The specifics of lens searching and discovery to build samples like these are reviewed by Lemon et al. (2024), which we refer the reader to for recent developments in the search methods and discoveries.

2.2.1 MG-VLA Survey-Based Samples

The first systematic search for strongly lensed systems at radio wavelengths took place in the eighties within the MIT–Greenbank–Very Large Array (MG-VLA) survey (Lawrence et al. 1986). This survey discovered a few famous radio-loud lensed quasars among thousands of radio sources scrutinized with high resolution by the Very Large Array (VLA). The Jodrell Bank–VLA Astromtric Survey (JVAS; Patnaik et al. 1992; King et al. 1996), and its successor, the Cosmic Lens All Sky Survey (CLASS; Myers et al. 1995) was the largest survey carried out for a long time. This survey targeted the whole northern sky ($0° < \mathrm{Dec} < 75°$) for multiple images (separated by $0''.3 < \Delta\theta < 6''.0$) among flat spectrum radio sources brighter than 30 mJy. CLASS discovered 22 new systems, among which twelve are doubles, nine are quads, and one displays six images (Browne et al. 2003).

2.2.2 The CASTLES Sample

The CfA-Arizona Space Telescope LEns Survey (CASTLES[1]) is a follow-up *Hubble Space Telescope* (*HST*) imaging survey of ∼100 galaxy-scale lenses[2] known at the time, some from previous surveys such as CLASS (Muñoz et al. 1998) and others from serendipity or targeted surveys (for details, see Lemon et al. 2024). This survey collected the first uniform ensemble of high-resolution images of known galaxy-scale lens systems, including both galaxy–quasar and galaxy–galaxy lenses. An account of early systematic lens searches, which generally unveiled small samples of less than six systems, can be found in Claeskens and Surdej (2002).

2.2.3 Samples of Lensed Sub-Mm Galaxies

Lens searches in the sub-mm have proved to be efficient in finding hundreds of lensed dusty star-forming galaxies at high redshift ($z \sim 1$–4) with high purity in the candidate sample, using the sharp cutoff in the luminosity function for these galaxies (see Lemon et al. 2024). The initial samples detected using this technique came from the *Herschel* Astrophysical Terahertz Large Area Survey (HATLAS) and the *Herschel* Multi-tiered Extragalactic Survey (HerMES; Negrello et al. 2010, 2017). Since these samples of lenses are source-selected, the selection function of the lens galaxies is less affected than for lens-selected samples, thus providing an advantageous avenue to study galaxy properties (e.g., Dye et al. 2014, 2018; Amvrosiadis et al. 2018; Maresca et al. 2022).

[1] https://lweb.cfa.harvard.edu/castles/.

[2] There are a handful cluster-scale lenses included in this sample as well.

2.2.4 SDSS-Based Samples

The Sloan Lens ACS (SLACS) survey discovered 85 galaxy–galaxy lenses from the Sloan Digital Sky Survey (SDSS) spectroscopic data by identifying multiple redshifts in the fiber (with $3''$ diameter) spectra. The SLACS survey also followed these systems up with multi-band *HST* imaging (Bolton et al. 2006; Auger et al. 2009). The sample was expanded with 40 new systems with smaller deflector masses by the SLACS for the Masses (S4TM) sample (Shu et al. 2015). In addition to galaxy–galaxy lenses, the SDSS Quasar Lens Search (SQLS) discovered a sample of 28 galaxy–quasar lenses using SDSS multicolor imaging data (Oguri et al. 2006). More galaxy–quasar lens systems were discovered from joint SDSS and UKIRT Infrared Deep Sky Survey (UKIDSS) data by the Major UKIDSS–SDSS Cosmic Lens Survey (MUSCLES; Jackson et al. 2012).

2.2.5 CFHTLS-Based Samples

The Strong Lensing Legacy Survey (SL2S; Gavazzi et al. 2012) discovered a sample of $\sim$35 galaxy–galaxy lenses from the Canada–France–Hawaii Telescope Legacy Survey (CFHTLS) data. Some newer lens samples have also been discovered from this survey (More et al. 2016a; Paraficz et al. 2016).

2.2.6 BOSS-Based Samples

The Baryon Oscillation Spectroscopic Survey Emission-Line Lens Survey (BELLS) discovered $\sim$30 galaxy–galaxy systems from the Baryon Oscillation Spectroscopic Survey (BOSS) and obtained *HST* imaging for them (Brownstein et al. 2012). This survey was later expanded into the BELLS for the GALaxy-Lyα EmitteR sYstems (BELLS GALLERY) survey, where the source galaxies are specifically Lyα emitters (Shu et al. 2016). A sample of 13 strongly lensed quasars has also been discovered from the BOSS data (More et al. 2016b).

2.2.7 Other Samples and Ongoing Efforts from Recent Surveys

Numerous large-area sky surveys have recently discovered several other lens samples. The STRong-Lensing Insights into the Dark Energy Survey (STRIDES) collaboration has discovered $\sim$30 quadruply lensed quasar systems from the Dark Energy Survey (DES) data – often in combination with data from other sky surveys – and obtained multi-band *HST* imaging of them in IR, optical, and ultra-violet (UV) bands (Shajib et al. 2019; Schmidt et al. 2023). Many galaxy–galaxy lens candidates have also been identified in the DES data (Jacobs et al. 2019b,a; Rojas et al. 2021; Tran et al. 2022). In addition to the DES, surveys such as the Canada–France Imaging Survey (CFIS), *Gaia*, the Hyper Suprime-Cam (HSC) survey, the Kilo Degree Survey (KiDS), and the Panoramic Survey Telescope and Rapid Response System (Pan-STARRS) have provided a plethora of newly discovered galaxy-scale strong lenses (e.g., Petrillo et al. 2017, 2019; Agnello et al. 2018; Krone-Martins et al. 2018; Delchambre et al. 2018; Lemon et al. 2018, 2020; Cañameras et al. 2021; Savary et al. 2022; Li et al. 2021; Lemon et al. 2022; Wong et al. 2022). Most of these samples still contain candidate lenses and require spectroscopic confirmation (by measuring the redshifts) and high-resolution imaging (to perform lens modeling) for the science applications described in Sect. 4 and Sect. 5. See Lemon et al. (2024) for a detailed discussion of the recent and ongoing efforts.

3 Observables and Analysis Methods

This section describes the strong-lensing observables (Sect. 3.1) and the analysis techniques to constrain galaxy properties from them. We describe the lens modeling methods in Sect. 3.2 and the commonly used models in Sect. 3.3. We outline the Bayesian hierarchical framework in Sect. 3.4, which allows inferring population characteristics of galaxies from a sample of lenses. Lastly, non-lensing observables most commonly combined with strong lensing ones are presented in Sect. 3.5.

3.1 Lensing Observables

The two types of lensing observables for galaxy-scale lenses are imaging of the lensing system (Sect. 3.1.1) and the time delay between a pair of images from a point source (Sect. 3.1.2).

3.1.1 Imaging of the Lens System

The most common and informative lensing observables result from imaging data with angular resolution much better than the Einstein radius. An extended source can be lensed into clearly identifiable arcs that can form partial or complete Einstein rings (see Fig. 1). The conjugate points on the arcs, that is, locations that are traced back to the same location in the source plane, can be used to simultaneously constrain a lens model (through its deflection angles and magnification) and reconstruct the surface brightness of the source that is *a priori* unknown. Although simple models can be constrained from any image that displays lensing features with sufficient signal-to-noise ratio, high-resolution imaging from space- or ground-based telescopes offers many more observational constraints (i.e., conjugate pixels). This is crucial for exploring more sophisticated models, which are required for precise science applications (e.g., see Sect. 4 and Sect. 5).

If the background source contains a point-like emitting region – for example, a quasar or a supernova – the positions of its multiple images (i.e., conjugate points) can be extracted from the data and used as constraints for a lens model. Although the simple addition of two or four conjugate points may initially seem insignificant compared to the hundreds of pixels that correspond to the extended source, the flux contained in the corresponding pixels may be brighter than the entire host galaxy, for example, in the case of an AGN, and their positions can be constrained with sub-pixel accuracy (e.g., Shajib et al. 2019; Schmidt et al. 2023). Hence, these conjugate points have a strong effect on the resulting model and must be treated separately. In addition to the astrometry of the point-like images, their magnification ratio (or flux ratio) can be used as an observational constraint. However, image magnifications are susceptible to micro- and milli-lensing, dust extinction,[3] and the effect of higher order moments in the mass distribution of the lens usually attributed to the complex, non-linear physics of baryons, for example, galactic disks and bars (Hsueh et al. 2016, 2017). Therefore, many lensing analyses involving systems with such point-like sources choose not to use magnification ratios as constraints (e.g., Shajib et al. 2019). However, suppose the microlensing and dust extinction effects can be incorporated and quantified within a lens model. In that case, any residual flux ratio anomaly would signal a departure from the smooth macro-model for the deflector galaxy and thus could indicate the presence of sub-galactic dark matter structures within the galaxy-scale halo (e.g., Mao and Schneider

[3] Lensed arcs from extended sources are also vulnerable to dust extinction.

1998; Nierenberg et al. 2017; Gilman et al. 2020). Such detection of dark substructures can provide important insights into the nature of the dark matter, as described in Vegetti et al. (2023).

Lens modeling by simultaneously fitting the imaging data in multiple bands (from radio wavelengths to UV) has become commonly employed in the literature when such data are available (e.g., Dye et al. 2014; Oldham and Auger 2018; Shajib et al. 2019; Young et al. 2022; Tan et al. 2024). Such multi-band modeling has the advantage of adding constraints in regions of the images that may be poorly detected in some wavelength ranges and deblending the lensing galaxy from the images and arcs. The main drawback is the increase in model complexity due to the wavelength dependence of the lens galaxy and source morphology. However, such multi-band lens modeling is expected to be ubiquitous in the upcoming decade, with multi-band data being more available thanks to the current and upcoming facilities such as the *JWST* and the Vera Rubin Observatory (Shajib et al. 2024). Several automated modeling pipelines (for both single-band and multi-band data) are being developed to tackle the computational aspect of modeling very large lens samples (Oguri and Marshall 2010; Collett 2015) to be discovered by the Rubin Observatory, *Euclid*, and the *Roman Space Telescope* (e.g., Chan et al. 2015; Nightingale et al. 2018; Shajib et al. 2021; Etherington et al. 2022; Schmidt et al. 2023; Tan et al. 2024).

3.1.2 Time Delay

If the background source is a variable point source, for example, a quasar or a supernova, the delay between the arrival times of photons at its multiple different images can be measured through long-term monitoring that spans from a few months to decades (e.g., Eigenbrod et al. 2005; Bonvin et al. 2016; Millon et al. 2020). The time delays are variant under the well-known mass-sheet degeneracy (MSD) in lensing, unlike the imaging observables (Falco et al. 1985; Schneider and Sluse 2014). Thus, they can be combined with the imaging observables to break the MSD when constraining the potential of the lensing galaxy. However, such a combination of these observables requires a fiducial cosmological model since the time delays depend on the cosmology, particularly the Hubble constant. A more detailed discussion on the measurement of time delays is provided in Birrer et al. (2024).

3.2 Lens Modeling Methods

Lens modeling is the process of constraining properties of the lens galaxy and the source from the lensing observables. Traditional methods are based on reconstructing the source light and lens potential to fit the data under some assumptions, such as regularization, and are described in Sect. 3.2.1. More recently, machine learning methods are being developed for these purposes, which we present in Sect. 3.2.2. It is beyond the scope of this review to provide a beginner's guide to modeling galaxy-scale lenses, but we refer the reader to Saha et al. (2024) for an introduction to lens modeling. Readers interested in the science applications without needing a technical discussion on lens modeling and analysis techniques may go directly to Sect. 4 and refer back to the rest of this Sect. 3 as needed.

3.2.1 Likelihood-Based Inference

Such methods require a likelihood function that leads to an optimized model that can reproduce an observed multiply-imaged system down to the noise level. We note that whereas an optimized forward model aims to reproduce the data to the noise level, it is often difficult to achieve that in practice, often owing to the simplifying assumptions made in the model. As a result, fine-tuning the model complexity is commonly required through trial-and-error to meet the accuracy requirement for a given science case (e.g., Shajib et al. 2019; Schmidt et al. 2023). In general, the lens model consists of three main components: the background source's flux distribution, the mass distribution in the lensing galaxy (or galaxies), and the flux distribution in the lensing galaxy (or galaxies). The likelihood function that measures the goodness-of-fit of the model to the data can be defined as

$$\mathcal{L}(\boldsymbol{d} \mid \boldsymbol{m}) \propto \exp\left[-\frac{1}{2}(\boldsymbol{d}-\boldsymbol{m})^T\,\Sigma_d^{-1}\,(\boldsymbol{d}-\boldsymbol{m})\right], \tag{1}$$

where $\boldsymbol{d}$ is the vector of pixel values in the data and

$$\boldsymbol{m} \equiv \boldsymbol{m}(\boldsymbol{\xi}_{\rm mass},\ \boldsymbol{\xi}_{\rm source},\ \boldsymbol{\xi}_{\rm light}) \tag{2}$$

is the model-computed flux in the data pixels. Here, $\boldsymbol{\xi}_{\rm mass}$ is the set of model parameters defining the lens mass distribution, $\boldsymbol{\xi}_{\rm source}$ is the set of model parameters defining the source's flux distribution, $\boldsymbol{\xi}_{\rm light}$ is the set of model parameters defining the lens galaxy's flux distribution, and Σ_d is the covariance matrix of the data. Some studies choose to subtract the lens galaxy's flux distribution from the imaging data before lens modeling (e.g., Bolton et al. 2008b). Given that the lens flux is loosely related to the lensing phenomenon only through mass-follows-light arguments that are not strictly required, we omit it in the following discussion for brevity. This likelihood function can be extended in a Bayesian framework to include prior (or regularization) terms on the source (Warren and Dye 2003; Treu and Koopmans 2004; Koopmans 2005), or used to compute the Bayesian evidence (Suyu et al. 2006; Shajib et al. 2020; Vernardos and Koopmans 2022) and perform model comparisons.

When the lensed arcs from an extended source are resolved, each pixel is a constraint for the lens model. Given a deflection field $\boldsymbol{\alpha}(\boldsymbol{\theta})$, which depends on the lens mass distribution through $\boldsymbol{\xi}_{\rm mass}$, the lens equation

$$\boldsymbol{\beta}(\boldsymbol{\theta}) = \boldsymbol{\theta} - \boldsymbol{\alpha}(\boldsymbol{\theta}) \tag{3}$$

can be used to map any position $\boldsymbol{\theta}$ on the image plane to the corresponding position $\boldsymbol{\beta}$ on the source plane (for a detailed explanation of the strong lensing formalism, see Schneider et al. 1992, Meneghetti 2021, or Saha et al. 2024). We can then easily compute the lensed flux at any location on the image plane as

$$I(\boldsymbol{\theta}) = S[\boldsymbol{\beta}(\boldsymbol{\theta})], \tag{4}$$

where S is the light distribution of the source that depends on $\boldsymbol{\xi}_{\rm source}$, and we use the fact that lensing conserves surface brightness. The dependence of $I(\boldsymbol{\theta})$ on the mass through $\boldsymbol{\beta}(\boldsymbol{\theta})$ is almost always non-linear, which means that we cannot directly (i.e., through a linear inversion) solve Eq. (3) to obtain the true parameters $\boldsymbol{\xi}_{\rm mass}$, even when S (and equivalently the values of $\xi_{\rm source}$) is perfectly known.

In practice, S is an unknown that must be solved simultaneously with the lens potential and requires special attention. The most straightforward choice for it is an analytic function, for example, a Sérsic profile (Sérsic 1968), whose parameters are treated in the same forward-modeling way as for the lens potential. However, several more advanced techniques have been developed over the years that allow a free-form reconstruction of the source, each

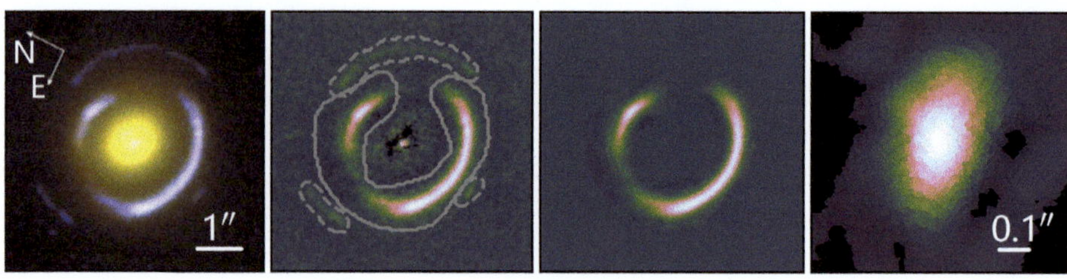

Fig. 2 *First panel:* False-color image of the system SDSS J0946+1006 combining three *HST* filters (Sonnenfeld et al. 2012). *Second panel: HST* image in the F814W filter of the system with the galaxy light subtracted. This system has lensed arcs from multiple source galaxies at different redshifts, which are grouped with solid and dashed contours. *Third panel:* Model of the lensed arc from only the brightest source. *Fourth panel:* Corresponding source reconstructed on an adaptive grid using every data pixel and curvature regularization (for the analysis of the data, see Chap. 4 of Bayer 2021 and for the modeling method, see Vernardos and Koopmans 2022)

with its advantages and disadvantages. The semi-linear inversion technique of Warren and Dye (2003) uses a regular grid of pixels to approximate the source brightness. Although the degrees of freedom are much higher in this case, the use of regularization or prior terms of a specific form can greatly facilitate obtaining the best-fit solution. If a prior is used with quadrature terms of the source pixel brightness values, then the derivative of the posterior probability function[4] can be obtained analytically given $\boldsymbol{\xi}_{\rm mass}$. As a result, the source reconstruction turns into a linear inversion problem for a given set of non-linear parameters within $\{\boldsymbol{\xi}_{\rm mass}, \boldsymbol{\xi}_{\rm source}\}$, where $\boldsymbol{\xi}_{\rm source}$ now refers only to non-linear parameters of the source that are not solved through the linear inversion.

Typical choices of source regularization are gradient and curvature that impose smoothness on the source solution through its derivatives (a standard approach in this kind of problem, e.g., Press et al. 1992). Alternatively, Vegetti et al. (2014) and Nightingale et al. (2018) use adaptive regularization, which changes the degree of smoothing based on the source brightness, and both studies discuss that this scheme is vital for reconstructing compact sources. Other choices include a covariance kernel prior, which is based on observations of the galaxy brightness power spectrum and thus a more physically justifiable choice (Vernardos and Koopmans 2022), or a multi-scale regularization through the use of sparsity constraints on a wavelet representation of the source (Galan et al. 2021). Choosing a different basis to represent the source can in itself significantly reduce the number of degrees of freedom while still having enough flexibility to represent complex light profiles across different scales (Birrer et al. 2015; Tagore and Jackson 2016). This can be similarly achieved by reconstructing the source on an irregular, adaptive grid that can have increased resolution in the most magnified areas of the source, that is, near the caustics (Vegetti and Koopmans 2009; Nightingale and Dye 2015; Vernardos and Koopmans 2022). Figure 2 illustrates an example of lens modeling based on high-resolution *HST* imaging of a lensing system with a bright, extended source component. The source is reconstructed on an adaptive grid using curvature regularization (Chap. 4 of Bayer 2021).

If there is a point source within the extended source galaxy, then the positions of its point-like images provide additional constraints on the mass model as conjugate points. One approach to model the point-like source is to include the location of its multiple images as free parameters and then require the mass model to trace them back to the same location

[4]This is also referred to as the penalty or loss function.

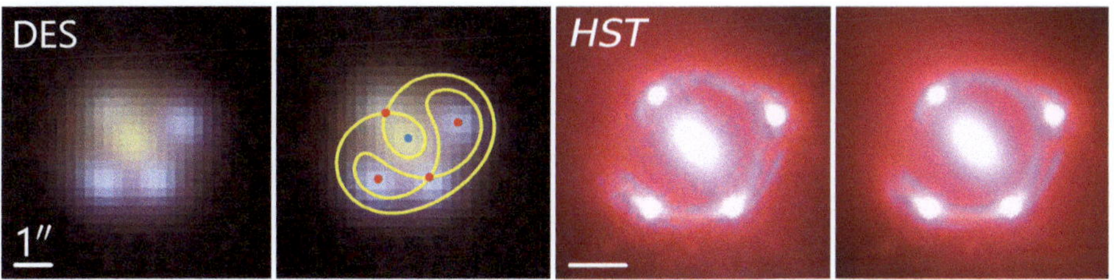

Fig. 3 Lens modeling of a galaxy-scale lens system – here, the lensed quasar WGD J2038−4008. *First panel:* The false-color image of the system from the DES *giy*-bands, where only the point images are resolved (Agnello et al. 2018). *Second panel:* Illustration of a lens model based only on the quasar image positions (red points; Agnello et al. 2018). The blue point shows the center of the lensing galaxy, and the yellow lines trace the saddle-point contours on the arrival-time surface. *Third panel:* The false-color image of the system from 3-band *HST* imaging (in F160W, F814W, and F475X filters), where the lensed arcs from the quasar host galaxy can be seen in greater detail. *Fourth panel:* The pixel-level model of the *HST* imaging from Shajib et al. (2022). The white bars represent 1″

on the source plane through the lens equation. Alternatively, the point-source location can be free, and the predicted image positions must match the observed ones. In practice, the choice between the two approaches is based on computational and sub-grid effect arguments (see Keeton 2001b, for a more detailed presentation on the topic). Figure 3 illustrates an example of lens modeling based on high-resolution *HST* imaging of a lensing system with both a point-like and an extended source component, that is, a quasar and its host galaxy. In this case, the model for the extended host galaxy assumes an analytic Sérsic profile and an additional free-form component described by a basis set of shapelets (Birrer et al. 2015; Shajib et al. 2022).

3.2.2 Machine-Learning Based Parameter Extraction

Recently, machine learning (ML) based methods have been developed to extract lensing parameters – for example, the Einstein radius, power-law index, ellipticity, and shear parameters – from the imaging data (e.g., Hezaveh et al. 2017; Morningstar et al. 2019; Adam et al. 2022; Schuldt et al. 2023). Some studies also explored reconstructing the lens mass and source flux distributions using machine learning algorithms (e.g., Chianese et al. 2020; Karchev et al. 2022; Mishra-Sharma and Yang 2022; Biggio et al. 2023). In these approaches, a machine learning algorithm – usually a neural network – is trained using synthetic data since real examples of lens systems are not adequate in number for such data-intensive training. This leads to the critical caveat that the mass density profile in the simulated galaxies is based on empirical priors (e.g., Hezaveh et al. 2017) or cosmological simulations (e.g., Adam et al. 2022). Thus, the inferred parameters are prior-dependent – either empirical or physical – in a similar manner that the forward modeling approach depends on the adopted mass model and the associated priors.

Although ML-based inferences have yet to demonstrate their potential in science applications with real data, they are essential for quickly extracting lensing parameters with minimum human supervision. This is crucial for modeling huge samples of lenses like those expected from several large surveys upcoming in the 2020s (e.g., Park et al. 2021; Wagner-Carena et al. 2021). For such large datasets, the traditional forward modeling techniques would be unfeasible due to either computational or human time restrictions. Finally, ML-based inference provides a direct and fast way to constrain quantities of interest that can be otherwise too cumbersome to infer from the data through a traditional approach, for example, detecting individual or populations of dark subhalos or constraining the subhalo-mass function (Brehmer et al. 2019; Diaz Rivero and Dvorkin 2020; Coogan et al. 2020, 2022; Ostdiek et al. 2022a,b; Vernardos et al. 2020; Wagner-Carena et al. 2021; Zhang et al. 2022; Anau Montel et al. 2023). Note that this particular science application is discussed in detail in Vegetti et al. (2023).

3.3 Lens Mass Models

Lens mass models can vary in complexity and eventually in the number of free parameters, reflecting that the mass distribution in galaxies is a non-trivial problem for which methodologies and algorithms are still evolving. Although it is possible to extract lensing information with free-form models and directly connect them to galaxy properties of interest (see Saha et al. 2024), lens modeling with simple parametrization has been the most common practice in the literature (e.g., Ritondale et al. 2019; Schmidt et al. 2023). We briefly introduce some commonly used models with simple parametrization in Sect. 3.3.1, discuss the degeneracies that impact them in Sect. 3.3.2, and describe free-form models in Sect. 3.3.3.

3.3.1 Simply Parametrized Mass Models

Keeton (2001a) provides a large catalog of simply parametrized models, which includes the commonly used ones such as the singular isothermal sphere or ellipsoid (SIS or SIE; Kormann et al. 1994), the pseudo-isothermal elliptical mass distribution (PIEMD; Kassiola and Kovner 1993), the softened power-law elliptical mass distribution (SPEMD; Barkana 1998), and the Navarro–Frenk–White (NFW; Navarro et al. 1996, 1997) profile. Defining the ellipticity in the potential makes lensing computation very efficient, even for complex radial profiles, since all the lensing quantities can be obtained either from the potential itself (e.g., for the time delay) or through numerical differentiation of the potential (e.g., for the deflection, convergence, or magnification; Kovner 1987; Golse and Kneib 2002). However, moderately elliptical potentials (e.g., with axis ratio $q \lesssim 0.6$) can lead to unphysical shapes in the convergence (Kassiola and Kovner 1993) and also introduce implicit azimuthal variation or ellipticity gradient in the convergence (Gomer et al. 2023). For simply parametrized profiles with ellipticity defined in the convergence, computing deflection angle becomes computationally expensive without an analytical solution due to needing a 2D numerical integration. Among such elliptical convergence profiles, Tessore and Metcalf (2015) provide an analytical solution for the power-law radial form. Alternative parameterizations of the elliptical NFW convergence profile have also been devised (Oguri 2021; Heyrovský and Karamazov 2024). Shajib (2019) provides a computationally efficient general solution for any radial form using a superposition of elliptical Gaussian components (i.e., the multi-Gaussian expansion; Emsellem et al. 1994; van de Ven et al. 2010).

On top of such a simply parametrized profile describing the primary lens mass distribution, it is also often necessary to include a constant shear field (often shortened as XS) lensing potential to accurately model the distortions in the lensed arcs or the Einstein ring. These additional distortions can arise from nearby perturbers and large-scale structures (Keeton et al. 1997) or the additional angular structure in the central lensing galaxy or galaxies (Witt 1996; Hilbert et al. 2007; Gomer and Williams 2018, 2021; Barrera et al. 2021; Van de Vyvere et al. 2022a,b; Etherington et al. 2024). This constant shear field is commonly referred to as 'external shear' in the literature, pointing to the former of the two origins mentioned above. However, we recommend using the term 'residual shear' instead as a more general terminology. The magnitude of residual shear commonly exceeds 0.1, which is difficult to explain if this shear originates solely from the line-of-sight structures. Until the mid-2010s, the SIE+XS model has been the most popular choice to model large samples of galaxy-scale lenses (e.g., Bolton et al. 2008a; Sonnenfeld et al. 2013), adequate for the science application requirements at the time. Although simple SIE models can sufficiently constrain the Einstein radius, obtaining other important properties, such as the radial slope of the mass profile, requires models with additional degrees of freedom. Improved data quality and analysis techniques have allowed the use of such models (e.g., Ritondale et al. 2019; Shajib et al. 2021).

Simply parametrized lens models with larger degrees of freedom, for example, a superposition of a stellar component with a constant or varying mass-to-light ratio and a dark matter component usually described by the NFW profile, have been adopted by some studies (e.g., Treu et al. 2010; Sonnenfeld et al. 2015; Oldham and Auger 2018; Shajib et al. 2021). Although necessary for the addressed science questions, such mass profiles with more free parameters amplify the impact of degeneracies inherent to lensing. It is thus often necessary to constrain these additional parameters by incorporating non-strong-lensing observables such as stellar kinematics and weak lensing, or by incorporating informative priors (e.g., Sonnenfeld 2018; Shajib et al. 2021). We note that Sonnenfeld (2018) and Shajib et al. (2021) combined strong-lensing information with stellar kinematics or weak lensing using 'summary observables', such as the Einstein radius or the reduced slope $\xi_{\rm rad}$ (see Sect. 3.3.2 for definition), instead of simultaneously fitting the abovementioned complex model to the full lensing information in the imaging data.

While the additional degrees of freedom beyond the simple SIE model focus on the radial profile of the lens in most cases, azimuthal structures such as disky-ness, boxiness, ellipticity gradients, or isodensity twists may leave noticeable imprints in the lensed images. For point-like sources, the image flux ratios are the most susceptible to perturbations (Möller et al. 2003; Keeton et al. 2003, 2005). In contrast, for extended sources, the imprint is generally more subtle and detectable only from high-resolution imaging data (Van de Vyvere et al. 2022a,b). Fortunately, most of those structures arise from baryonic physics and may also be detectable in the luminosity profile of the lens (Van de Vyvere et al. 2022b).

3.3.2 Common Degeneracies in Simple Parametric Modeling

The degeneracies in lens modeling – both intrinsic in the data and stemming from the parametrization scheme – are discussed in Saha et al. (2024). Here, we summarize the common degeneracies that largely impact simply parametrized lens models for the convenience of the readers.

The MSD, which is intrinsic to imaging observables in lensing, originates from the mass-sheet transform (MST, Falco et al. 1985; Saha 2000)

$$\kappa \rightarrow \kappa' = \lambda\kappa + 1 - \lambda, \tag{5}$$

where λ is a constant. This equation implies a source position transformation (e.g., Schneider and Sluse 2014), where the unknown source position is altered as

$$\beta \rightarrow \beta' = \lambda\beta. \tag{6}$$

A more general but approximate degeneracy happens when λ is not a constant anymore but depends on the position, that is, $\lambda \equiv \lambda(\boldsymbol{\theta})$ (Unruh et al. 2017; Wertz et al. 2018).

Most simply parametrized mass profiles artificially limit the MSD. For example, this is easily demonstrable for the power-law mass model, as the MST of a power law is mathematically not a power law anymore. Although MST-invariant quantities exist that the imaging observables can constrain (e.g., Wagner 2017; Wagner and Tessore 2018), commonly employed mass models are not parametrized based on those quantities. The standard practice in the literature to assume simply parametrized models, such as the power law, is usually validated on numerous non-lensing constraints demonstrating that the power-law model is a 'good' approximation for elliptical galaxy mass profiles (e.g., Thomas et al. 2007; Tortora et al. 2014; Bellstedt et al. 2018). The model-independent radial quantities constrained by the imaging data are the Einstein radius $\theta_{\rm E}$ and the MST-invariant 'reduced slope' defined as

$$\xi_{\rm rad} \propto \frac{\theta_{\rm E}\alpha''(\theta_{\rm E})}{1-\kappa(\theta_{\rm E})}, \tag{7}$$

where α'' is the second derivative of the deflection angle (see Eq. 42 of Birrer 2021 for the full definition of $\xi_{\rm rad}$, also Kochanek 2021). For a power-law convergence profile $\kappa(\theta) \propto \theta^{-\gamma_{\rm PL}+1}$, this quantity becomes $\xi_{\rm rad} = \gamma_{\rm PL} - 2$. As a result, the choice of a SIE+XS lens model fixes $\xi_{\rm rad} = 0$ and only extracts $\theta_{\rm E}$ from the imaging data.

Degeneracies in lens modeling can also arise from particular parametrizations of the lens model. One such example is the shear–ellipticity degeneracy, as the total shear can be redistributed between the 'internal' shear, arising from the ellipticity of the central deflector, and the external shear (Kassiola and Kovner 1993). Specifically, a SIE+XS model can be modeled with only an SIE model that has a quadrupole moment equaling $1/3^{\rm rd}$ of the shear (An 2005). This degeneracy is particularly apparent when modeling with point-like image positions as the only constraints (Witt 1996).

3.3.3 Free-Form Models

In the most commonly used lens models, the main focus is on capturing the radial shape of the mass profile, while any azimuthal structure beyond an elliptical shape with residual shear is of secondary importance. Real galaxies can have more complicated mass profiles, with higher order moments present, such as disky-ness, boxiness, or bar- or disk-like components (e.g., Trotter et al. 2000; Claeskens et al. 2006; Hsueh et al. 2017; Frigo et al. 2019), radial dependence of the ellipticity or the orientation of the isodensity contours (i.e., ellipticity gradients, twists, or lopsidedness, e.g., Hao et al. 2006; Nightingale et al. 2019; Barrera et al. 2021), or even features that do not fit into a simple parametric description, such as merger products that are not yet completely relaxed and populations of substructures (e.g., satellite subhalos or perturbers along the line of sight). Detecting such deviations from the simple parametric profiles depends on numerous factors, such as their alignment with the smooth potential (Van de Vyvere et al. 2022b), the complexity of the brightness profile of the source (Vernardos and Koopmans 2022), the signal-to-noise ratio, etc. However, multipole components in the mass potential beyond the combined effect of ellipticity and residual shear have been recently detected in very-long-baseline interferometric observations of a strong lens (Powell et al. 2022). Although not accounting for such structures can bias the mass model by up to several percent – which is mostly acceptable except for time-delay cosmography (see Birrer et al. 2024) and for detecting dark matter substructure (see Vegetti et al. 2023) – their detection holds valuable information on the formation history and evolution of galaxies.

To this extent, free-form techniques have been developed that either entirely dismiss any parametric mass component and employ a grid of mass pixels to describe the lens potential (Saha and Williams 1997), or retain a parametric model as a first-order smooth component and combine it with a similar pixel grid that now focuses specifically on capturing higher order deviations (Koopmans 2005; Suyu et al. 2010). In both cases, regularization assumptions or priors on the free-form pixel grid are necessary to obtain a solution and to prevent the appearance of unphysical mass distributions. Existing techniques are based on forward models or extensions of the semi-linear inversion technique (Vernardos and Koopmans 2022).

The specific form of the regularization priors plays an important role in the quality of the obtained solutions. It has been shown that purely mathematically motivated priors (e.g., curvature) can lead to biased potentials as opposed to more physically driven ones, based on the observed light properties of real galaxies (e.g., Vernardos and Koopmans 2022). Galan et al. (2022) proposed a wavelet-based regularization technique that finds solutions that satisfy sparsity constraints. Biggio et al. (2023) completely replaced the pixel grid with a neural network. These approaches allow the use of purely data-driven regularization that is the most compatible with the data (without computing and comparing the Bayesian evidence). It remains to be seen how well these new and promising techniques can perform on real data and robustly recover deviations from smooth, parametric models that encode galaxy evolution.

One way to allow the prior to be less informative for free-form models is to marginalize over an ensemble of solutions. Such ensembles were first introduced for free-form models made up of mass tiles or pixels (Williams and Saha 2000; Saha and Williams 2004; Coles et al. 2014). In these and related works, the mass distribution is required to be non-negative and centrally concentrated in a broad sense. Within these prior conditions, models that correctly reproduce the observed positions of point-like image features are randomly sampled to form the ensemble of solutions. This ensemble of solutions is then effectively the posterior of the model parameters, which include all the mass pixel values, and the posteriors of model-predicted quantities can also be obtained from this ensemble (e.g., Williams and Saha 2000). The ensemble can be further filtered according to how well the whole image or the pixels on the lensed arcs from the extended source can be fitted (e.g., Denzel et al. 2021b).

Free-form models naturally allow a broad range of mass profile shapes, both radially and azimuthally, thus exploring the degenerate space of the mass profile shapes. The same effect can be obtained with simply parametrized models by combining posteriors from models with different parametric forms, albeit to the limited extent allowed by the variety of the adopted parametrizations (e.g., Suyu et al. 2014; Birrer et al. 2019; Shajib et al. 2022).

3.4 Bayesian Hierarchical Framework

The Bayesian hierarchical framework can be used to constrain the population properties from a sample of strong lenses (e.g., Sonnenfeld et al. 2015). This framework also allows one to incorporate a selection function of the lensing galaxies and generalize the sample properties to the population of all galaxies that are of the same type as the lensing ones (e.g., Sonnenfeld et al. 2019). Within the hierarchical analysis, there are two levels of parameters: hyper-parameters that dictate the distribution of the parent population of the lens galaxies and parameters pertaining to individual lens galaxies sampled from the parent population. The hierarchical framework connects the population-level hyper-parameters to the observed data through the individual-galaxy-level parameters. According to the Bayes' theorem, the posterior probability distribution of the hyper-parameters u is given by

$$p(u \mid \mathcal{D}) \propto p(\mathcal{D} \mid u)\ p(u), \tag{8}$$

where $\mathcal{D}$ is the dataset, $p(\mathcal{D} \mid u)$ is the likelihood, and $p(u)$ is the prior. For model parameters w_i pertaining to individual galaxies, the above equation can be expressed as

$$p(u \mid \mathcal{D}) \propto p(u) \prod_i \int \mathrm{d}w_i \; p(\mathcal{D}_i \mid w_i) \; p(w_i \mid u), \tag{9}$$

where $\mathcal{D}_i$ is the data from to the i-th individual lens galaxy. This approach can infer any property at the population level with the associated mean and scatter values. See, for example, Sonnenfeld and Cautun (2021) for a detailed presentation on the hierarchical framework with specific examples of hyper-parameters u.

3.5 Incorporating Non-strong-Lensing Observables

Incorporating non-strong-lensing observables can be used to break the degeneracies in strong lensing analysis. For example, stellar kinematics data can be used to break the MSD (e.g., Romanowsky and Kochanek 1999; Treu and Koopmans 2002, 2004; Shajib et al. 2018; Birrer et al. 2020; Shajib et al. 2021; Tan et al. 2024), whereas spectroscopic stellar population analysis (e.g., Spiniello et al. 2011), weak lensing (e.g., Gavazzi et al. 2007; Sonnenfeld et al. 2018; Shajib et al. 2021), and microlensing (e.g., Schechter et al. 2014; Oguri et al. 2014) information can help mitigate the degeneracy between the stellar and dark matter distributions. Here, we briefly describe combining strong lensing with stellar kinematics (Sect. 3.5.1) and weak lensing (Sect. 3.5.2).

3.5.1 Combining Stellar Kinematics with Strong Lensing

Imaging observables probe the 2D mass distribution of the lens projected on the plane of the sky, whereas stellar kinematics probe its full 3D mass distribution. Thus, a combination of the two helps break the MSD to robustly constrain the mass distribution in galaxies. Although elliptical galaxies are triaxial, assuming spherical symmetry has been a standard practice for the case of a single aperture-integrated stellar velocity dispersion measurements (see Sonnenfeld et al. 2012, for a discussion on the impact of this assumption). Then, the stellar velocity dispersion is obtained by solving the spherical Jeans equation

$$\frac{\mathrm{d}\left(l(r)\,\sigma_\mathrm{r}^2\right)}{\mathrm{d}r} + \frac{2\beta_\mathrm{ani}(r)\,l(r)\,\sigma_\mathrm{r}^2}{r} = -l(r)\,\frac{\mathrm{d}\Phi}{\mathrm{d}r}. \tag{10}$$

Here, $l(r)$ is the 3D luminosity density of the stars, σ_r is the intrinsic radial velocity dispersion, and $\beta_\mathrm{ani}(r)$ is the anisotropy parameter relating σ_r with the tangential velocity dispersion σ_t given by

$$\beta_\mathrm{ani}(r) \equiv 1 - \frac{\sigma_\mathrm{t}^2}{\sigma_\mathrm{r}^2}. \tag{11}$$

By solving the Jeans equation, the line-of-sight velocity dispersion, which is the kinematic observable, is obtained as

$$\sigma_\mathrm{los}^2(R) = \frac{2G}{I(R)} \int_R^\infty \mathcal{K}_\beta\left(\frac{r}{R}\right) \frac{l(r)\,M(r)}{r}\,\mathrm{d}r \tag{12}$$

(Mamon and Łokas 2005). Here, $M(r)$ is the 3D enclosed mass within radius r. The function $\mathcal{K}_\beta(u)$ depends on the parameterization of $\beta(r)$ (see Mamon and Łokas 2005 for specific forms of $\mathcal{K}_\beta(u)$ corresponding to different $\beta(r)$). Thus, the observed velocity dispersion in Eq. (12) can be written as a function of the lens model parameters as

$$\sigma_{\text{los}}^2 = \frac{D_{\text{s}}}{D_{\text{ds}}} c^2 \, J(\boldsymbol{\xi}_{\text{mass}}, \boldsymbol{\xi}_{\text{light}}, \beta_{\text{ani}}, \lambda), \tag{13}$$

where $\boldsymbol{\xi}_{\text{mass}}$ are the deflector's mass model parameters, $\boldsymbol{\xi}_{\text{light}}$ are the deflector's light model parameters (Birrer et al. 2016). In this form, the function J is independent of cosmology. It only depends on the lens model parameters $\{\boldsymbol{\xi}_{\text{mass}}, \boldsymbol{\xi}_{\text{light}}\}$, the anisotropy profile $\beta_{\text{ani}}(r)$, and the MST parameter λ. All the cosmological dependence of σ_{los} is contained in the distance ratio $D_{\text{s}}/D_{\text{ds}}$. To reproduce the observed velocity dispersion integrated within an aperture, the computed luminosity-weighted velocity dispersion needs to be blurred with the PSF $\mathcal{P}$ as

$$\sigma_{\text{ap}}^2 = \frac{\int_{\text{ap}} \left[I(R)\, \sigma_{\text{los}}^2(R)\right] * \mathcal{P}\; R \,\mathrm{d}R\mathrm{d}\theta}{\int_{\text{ap}} I(R) * \mathcal{P}\; R\,\mathrm{d}R\mathrm{d}\theta}, \tag{14}$$

where the $*$ symbol denotes the convolution operation. To obtain analytic solutions for specific choices of mass, light, and anisotropy profiles, see Koopmans (2006) for the case of power-law mass and light profiles with constant anisotropy, and Suyu et al. (2010) for the case with power-law mass profile, the Hernquist light profile (Hernquist 1990), and isotropic stellar orbits.

The constraints from the velocity dispersion measurement can be folded in the lens model posterior with a multiplicative likelihood term

$$\mathcal{L}_{\text{kin}} \propto \exp\left[-\frac{(\sigma_{\text{ap}}^{\text{obs}} - \sigma_{\text{ap}}^{\text{model}})^2}{2\sigma_{\sigma_{\text{ap}}^{\text{obs}}}^2}\right], \tag{15}$$

where $\sigma_{\sigma_{\text{ap}}^{\text{obs}}}$ is the uncertainty in the observed velocity dispersion $\sigma_{\text{ap}}^{\text{obs}}$. Whereas the imaging observables from strong lensing cannot constrain the MST parameter λ, the stellar velocity dispersion constrains λ through the likelihood $\mathcal{L}_{\text{kin}}$ (see Birrer et al. 2024, for a detailed discussion within a hierarchical Bayesian framework).

Although integrated velocity dispersion from long-slit spectra is the most commonly used kinematic observable in strong lensing studies, a few studies have also incorporated spatially resolved velocity dispersion, mainly from integral field unit (IFU) spectra (e.g., Barnabè and Koopmans 2007; Barnabè et al. 2011; Czoske et al. 2012; Spiniello et al. 2015).

3.5.2 Combining Weak Lensing with Strong Lensing

Weak lensing measures the excess shear quantity ($\Delta\Sigma$) from tidal distortions of galaxies far away ($\gtrsim$10 arcsec) from the central lensing galaxy. Thus, weak lensing provides information on the mass distribution of the lensing galaxy's outer region, that is, where the dark matter halo dominates. In contrast, strong-lensing observables provide information on the enclosed mass within the Einstein radius. Still, there remains a degeneracy between the luminous and dark matter fractions within the total enclosed mass. If a specific profile is assumed for the dark matter distribution, weak lensing data can break this degeneracy between luminous and dark component normalizations (Sonnenfeld 2018; Shajib et al. 2021).

The weak lensing information is often convenient or appropriate to be incorporated within the hierarchical framework, although there are examples of such combination without using a hierarchical framework (e.g., Gavazzi et al. 2007). Since the weak lensing signal from one single lensing galaxy does not usually have enough constraining power on the dark matter normalization, it is often required to stack weak lensing signals from a large sample of elliptical galaxies, which are not-necessarily lensing galaxies. If it is justified to assume that this large sample of elliptical galaxies and the lens galaxy sample under consideration are subsamples of the same parent population, then the weak lensing observables $\mathcal{D}_{\text{weak}}$ and the strong-lensing observables $\mathcal{D}_{\text{strong}}$ can be jointly considered in the posterior probability function of the population hyper-parameters u as

$$p(u \mid \mathcal{D}_{\text{strong}}, \mathcal{D}_{\text{weak}}) \propto p(\mathcal{D}_{\text{strong}} \mid u)\ p(\mathcal{D}_{\text{weak}} \mid u)\ p(u). \tag{16}$$

Here, the strong lensing likelihood $p(\mathcal{D}_{\text{strong}} \mid u)$ can then be expanded using the single system likelihoods similar to the product term in Eq. (9). Since from Bayes' theorem, we have

$$p(\mathcal{D}_{\text{weak}} \mid u)\ p(u) \propto p(u \mid \mathcal{D}_{\text{weak}}), \tag{17}$$

it is also valid for numerical convenience to first obtain the posterior of u from weak lensing observables only and then fold this posterior as the prior of u in the hierarchical analysis with only strong-lensing data.

4 Applications in Galaxy Properties and Evolution

This section describes what we can learn about galaxy structure and evolution using the lensing galaxy properties. Since strong-lensing galaxies are typically massive ellipticals, most of the strong-lensing studies in the field relate to this type of galaxy (Sect. 4.1–Sect. 4.3). However, we briefly discuss strong lensing by spiral galaxies at the end of this section in Sect. 4.4.

4.1 Galaxy Mass Density Profile

All galaxies are believed to form and grow inside their dark matter halos. Thus, a massive galaxy's total mass density profile comprises two components: the baryonic matter distribution, which includes stars and gas, and the dark matter distribution. As seen from the comparison of the observed galaxy luminosity function and the distribution of simulated dark matter halos (i.e., abundance matching, see Moster et al. 2010), the stellar-mass fraction decreases with mass, for the mass range of typical lensing galaxies. This is attributed to AGN feedback and is evident from simple comparisons of the stellar and total mass in lensing galaxies (Auger et al. 2009; Küng et al. 2018).

A power-law model $\rho(r) \propto r^{-\gamma_{\text{pl}}}$ close to the isothermal case (i.e., $\gamma_{\text{pl}} \sim 2$) has been found to be sufficient to describe several lensing and non-lensing observables to the noise level; for example, from strong lensing only or in combination with stellar dynamics (Kochanek 1995; Treu and Koopmans 2004; Dye and Warren 2005; Koopmans et al. 2006, 2009; Barnabè et al. 2009; Auger et al. 2010b; Dutton and Treu 2014; Ritondale et al. 2019; Powell et al. 2022; Tan et al. 2024), from the combination of strong and weak lensing (Gavazzi et al. 2007), from stellar dynamics only (Bertin and Stiavelli 1993; Gerhard et al. 2001; Thomas et al. 2007; Tortora et al. 2014; Cappellari et al. 2015; Bellstedt et al.

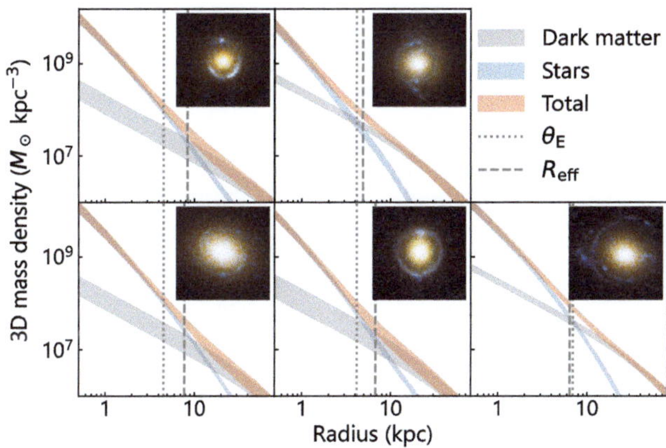

Fig. 4 Examples of 3D mass distribution in dark matter and stars (i.e., baryons) in lensing elliptical galaxies (Shajib et al. 2021). The false-color image created from the *HST* imaging of each system is illustrated in the inset (*image credit: NASA/ESA, A. Bolton, and the SLACS team*). Shajib et al. (2021) combined this *HST* imaging data with the stellar kinematics and weak lensing information to decouple the stellar (teal) and dark matter (grey) components of the total mass density profile (red). Only five examples are illustrated here out of the 23 analyzed lenses from the SLACS sample. The vertical dotted and dashed lines mark the Einstein radius θ_E and the half-light or effective radius R_{eff}, respectively. In most cases, the total density profile is close to the power-law profile, with occasional deviations appearing near the center or far outside the Einstein radius

2018; Derkenne et al. 2021), and from X-ray luminosity (Humphrey and Buote 2010). This phenomenon – that the total mass profile in ellipticals approximately follows the power law, whereas neither the baryonic nor the dark matter component individually follows the power law – is referred to as the 'bulge–halo conspiracy' (Treu and Koopmans 2004), similar to the 'disk–halo conspiracy' in spiral galaxies (van Albada and Sancisi 1986). Figure 4 illustrates the dark and luminous components constrained by combining kinematic and weak lensing information with strong lensing for five SLACS lens galaxies (Shajib et al. 2021). In most cases, the total density profile is very close to a power-law form, with deviations only being prominent far from the half-light radius. Galaxy formation simulations suggest that the close-to-isothermal nature of the total density profile originates from rearranging the mass distribution through collisionless accretion in gas-poor mergers (Johansson et al. 2009; Remus et al. 2013).

Analysis of the SLACS lenses finds the mean logarithmic slope $\langle \gamma_{pl} \rangle = 2.08 \pm 0.03$ with an intrinsic scatter of 0.16 ± 0.02 from a sample of 85 galaxy–galaxy lenses (Koopmans et al. 2009; Auger et al. 2010b). The median redshift of the SLACS lenses is $\langle z_{SLACS} \rangle \simeq 0.19$. The SIE lens model was adopted in this analysis to constrain θ_E from the imaging data. Then, the power-law index γ_{pl} was obtained from the stellar velocity dispersion measured by the SDSS. Shajib et al. (2021) re-analyzed 23 systems from the SLACS sample with a power-law model instead of the SIE model to obtain the mean logarithmic slope $\langle \gamma_{pl} \rangle = 2.08 \pm 0.03$ with an intrinsic scatter of 0.13 ± 0.02. Ritondale et al. (2019) analyzed 17 galaxy–galaxy lens systems from the BELLS GALLERY sample with a power-law mass model to find the average logarithmic slope $\langle \gamma_{pl} \rangle = 2.00 \pm 0.01$. The mean redshift of the BELLS GALLERY sample is approximately $\langle z_{BG} \rangle \simeq 0.5$.

Project Dinos (Tan et al. 2024) reanalyzed multi-band *HST* imaging for a sample of ~50 lenses from the SLACS and SL2S samples and then combined the lensing constraints with the stellar kinematics to directly constrain any potential deviation from the power-law profile, where the deviation is parametrized with the internal MST parameter λ_{int} after

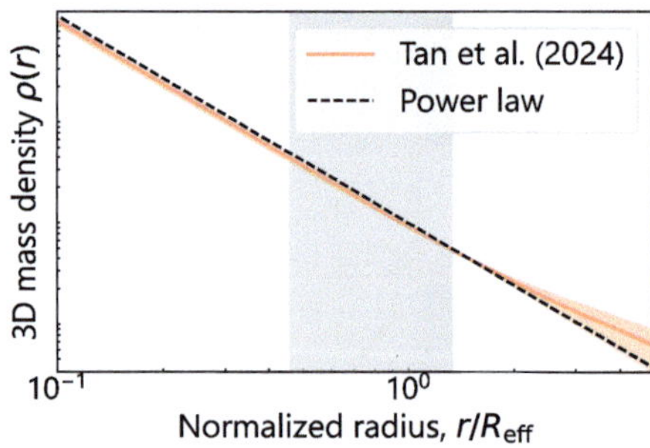

Fig. 5 The shape of the mean 3D total mass profile for the lens galaxies from the SLACS and SL2S samples constrained by Tan et al. (2024, *cf.* Fig. 11 therein) from a joint lensing–dynamics analysis. The orange line shows the population mean with the shaded region representing the 68% (1σ) credible region. The dashed black line traces the fiducial power-law mass distribution. The vertical grey shaded region shows the 1σ range of the Einstein radius distribution of the sample. For the dynamical modeling, these authors adopted a spatially constant stellar anisotropy profile, consistent with the spatially resolved velocity dispersion measurements of local elliptical galaxies (see references therein). The 3D mass density profile $\rho(r)$ along the vertical axis is normalized at a reference radius and thus has no units in this illustration. The population mean of the mass density profile is consistent with the power-law model within 1σ

correcting for the line-of-sight effects (i.e., the external convergence). The 'internal' MST is referred to as such to differentiate it from the effect of the external convergence that acts as an 'external' mass sheet. These authors find the power-law model is consistent with both the lensing and kinematic observables within 1σ for the baseline choice of spatially constant anisotropy profile that is informed by exquisite IFU kinematics of local elliptical galaxies (Fig. 5).

The surface mass in elliptical galaxies constrained from strong lensing can explain the origin of the so-called tilt of the fundamental plane, that is, the tight correlation between the effective radius, the effective surface brightness, and the central velocity dispersion (Ciotti et al. 1996). If the surface mass is used instead of the surface brightness, then the resulting mass plane (in place of the fundamental plane) is not tilted (Bolton et al. 2008a). This result implies that the tilt of the fundamental plane stems from the increase in the dark matter fraction with increasing velocity dispersion or dynamical mass (Auger et al. 2009).

In summary, the total mass profile in galaxies seems to be well described by a power law. However, there is also a recent indication of a potential departure from the power law suggested by the non-correlation between the logarithmic slopes from lensing-only and lensing–dynamics analyses (Etherington et al. 2023). The implications of this finding for the individual baryonic and dark matter components are discussed separately below.

4.1.1 Luminous (Baryonic) Mass Profile

In recent years, spatially resolved spectroscopic surveys of nearby ellipticals using IFUs and high-resolution *HST* imaging have considerably advanced our understanding of the structure and evolution of these galaxies (Cappellari 2016). Such detailed studies of intermediate redshift galaxies ($z \sim 0.2 - 0.7$) using direct observations are impossible. However, the mass of galaxies acting as strong lenses at these redshifts can be mapped out in detail, thereby providing critical information about the progenitors of present-day galaxies and their evolution.

In the last decade, imaging and kinematic data have led to a revision of galaxy classification. The modern analyses classify ellipticals into fast and slow rotators, and those with and without central cores, resulting in four classes (Cappellari 2016; Krajnović et al. 2020). Massive ellipticals ($M \gtrsim 10^{11} M_\odot$) in the local Universe tend to have small cores with size 0.02–0.50 kpc (Krajnović et al. 2020), that is, much smaller than the Einstein radius of typical lenses. In contrast, less massive ones appear to have cuspy central light profiles. Cores are believed to result from in-spiraling super-massive black hole (SMBH) binaries, which transfer their angular momentum outward to stars and leave a flatter density core at the center. The population of massive elliptical galaxies is also more likely to be composed of slow rotators, being also morphologically rounder than the lower mass fast rotator counter-part (e.g., Weijmans et al. 2014; van de Sande et al. 2017).

Although it is well known that the baryons and the dark matter do not follow the same radial profile, it is unknown *a priori* whether they have the same angular structure or not. Strong lensing allows us to compare the azimuthal distributions of the light and the total mass distributions, thus detecting any possible difference between the angular structures of the dark matter and the baryons. Several studies report a strong correlation in the ellipticity between the matter and light distributions (e.g., Koopmans 2006; Gavazzi et al. 2012; Sluse et al. 2012; Kostrzewa-Rutkowska et al. 2014). In contrast, some other studies only report weak or no correlation (e.g., Keeton et al. 1998; Ferreras et al. 2008; Rusu et al. 2016; Shajib et al. 2019, 2021). Some differences can be attributed to data quality, modeling procedure, or selection effects (Shajib et al. 2021). The major axes are usually well aligned (with position angle difference $\lesssim 10°$) between the mass and light distributions (e.g., Keeton et al. 1998; Kochanek 2022; Koopmans 2006; Treu et al. 2009; Sluse et al. 2012; Bruderer et al. 2016; Shajib et al. 2019, 2021). The cases where the major axes of the mass and light do not align within $\sim 10°$ also generally have large residual shear. These findings suggest that stellar orbits highly misaligned with the potential can only be sustained in non-isolated galaxies (indicated by the large residual shear if interpreted as originating from the presence of nearby galaxies), which is consistent with what is found in simulations (Heiligman and Schwarzschild 1979; Martinet and de Zeeuw 1988; Adams et al. 2007; Debattista et al. 2015). Although the stellar initial mass function (IMF) pertains to the luminous structure, we present the strong lensing results on the stellar IMF separately in Sect. 4.2.

4.1.2 Dark Matter Profile

Numerical simulations show that the dark matter halos and the baryonic mass within them initially follow the NFW profile before star formation begins. However, the baryonic gas has to cool down and fall inward for star formation to begin. The contraction in the baryonic matter deepens the gravitational potential. Thus, the dark matter distribution also contracts in response, for which the adiabatic contraction scenario can work reasonably well (Blumenthal et al. 1986; Cautun et al. 2020). In the process of adiabatic contraction of a spherical mass distribution, the initial radius r_i of a dark matter particle and its final radius r_f is related as

$$r_\mathrm{i}\, M_\mathrm{i}(r_\mathrm{i}) = r_\mathrm{f}\, M_\mathrm{f}(r_\mathrm{f}), \tag{18}$$

where $M(r)$ is the enclosed 3D mass within radius r (Blumenthal et al. 1986). However, numerical simulations find that dark matter does not fully respond to the baryonic infall according to the theoretical model of Blumenthal et al. (1986) (e.g., Gnedin et al. 2004; Abadi et al. 2010). Dutton et al. (2007) prescribe a formalism defining a halo response parameter ν to adjust the degree of contraction (i.e., the response to the baryonic infall) as

$$r_\mathrm{i} \equiv \Gamma^{-\nu}(r_\mathrm{f})\, r_\mathrm{f}, \tag{19}$$

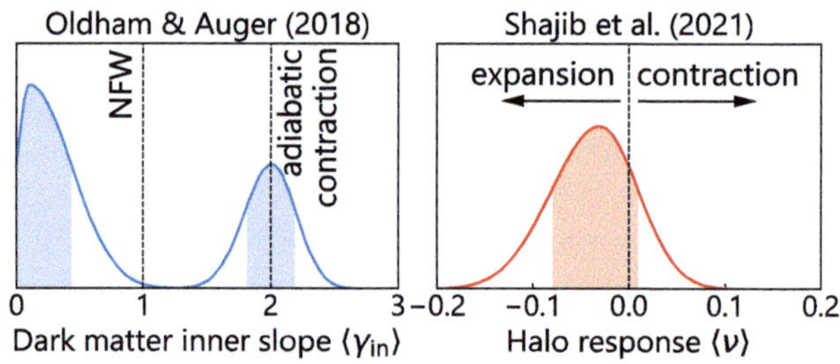

Fig. 6 Results on the dark matter distribution from joint lensing–dynamics analyses. *Left-hand panel:* the probability density function for the sample mean of the dark matter inner logarithmic slope $\langle\gamma_{\rm in}\rangle$. This result was obtained from a hierarchical analysis performed on the lensing–dynamics data for 12 strong lenses. The probability distribution shows a bimodality, where one mode is consistent with the adiabatic contraction scenario (i.e., $\gamma_{\rm in} \approx 2$) and the other mode is consistent with an expanded halo (i.e., $\gamma_{\rm in} \approx 0$). *Right-hand panel:* the probability density function of the sample mean of the halo response parameter $\langle\nu\rangle$ obtained from a SLACS subsample of 23 lenses (Shajib et al. 2021). This result is consistent with no contraction or expansion from the regular NFW profile (i.e., $\nu = 0$ marked by the vertical dashed line). Although these two results disagree, potential systematics stemming from different parametrizations of the adiabatic contraction or the stellar anisotropy profile are yet to be ruled out as the source of this discrepancy

where $\Gamma(r_{\rm f}) \equiv r_{\rm f}/r_{\rm i}$ is the contraction factor. In this formalism, $\nu = 0$ corresponds to no contraction, and $\nu = 1$ corresponds to fully responsive contraction according to the model of Blumenthal et al. (1986). The simulations of Gnedin et al. (2004) point to $\nu \sim 0.8$, and those of Abadi et al. (2010) to $\nu \sim 0.4$ (Dutton and Treu 2014).

Dye and Warren (2005) constrained the dark matter profile based on lensing analysis of one lens system and found the inner slope of the dark matter halo to be consistent with the NFW profile. Several studies have combined additional information, for example, stellar kinematics and weak lensing, with the strong-lensing data to decompose the dark matter and baryonic components from the total density profile constrained by strong lensing. Dutton and Treu (2014) adopted multiple dark matter contraction models with fixed ν to values between -0.5 and 1. These authors find that $\nu = 0$, that is, no contraction, best matches the data from the SLACS sample. Shajib et al. (2021) allowed for a fully variable ν parameter within -0.5 and 1 in their dark matter model. These authors find the average contraction in their sample to be $\langle\nu\rangle = -0.03^{+0.04}_{-0.05}$, which is consistent with no contraction and rules out the contraction results from simulations with high statistical significance (Fig. 6 right-hand panel). In contrast, Oldham and Auger (2018) find a cuspier inner logarithmic slope than the NFW profile for the majority of their sample of 12 lens systems, with a smaller subset having shallower ones, pointing to the impact of the environment in the evolution of these galaxies (Fig. 6 left-hand panel). However, the potential systematic dependence for all of the above results on modeling choices – for example, dark matter profile and anisotropy profile parameterizations – is yet to be investigated thoroughly.

4.2 Stellar Initial Mass Function

When combined with ancillary data, strong lensing provides a method to infer the stellar mass-to-light ratio ($M_\star/L$; Treu et al. 2010), which can then be related to the low-mass end of the stellar IMF slope in the lensing galaxies (for a review, see Smith 2020). This is mainly because low-mass stars ($M_\star < 0.5 M_\odot$) contribute only by a few percent to the integrated light in the optical, but they give a much larger contribution to the mass (Conroy and van Dokkum 2012).

As described in Sect. 4.1, the total mass distribution in the lensing galaxy can be decomposed into the stellar and dark components by combining lensing and dynamical observables. Thus, the total stellar mass $M_\star^{\rm LD}$ can be obtained. An independent method to get the stellar mass $M_\star^{\rm SPS}$ of the galaxy is the stellar population synthesis (SPS) method applied on the lens galaxy's photometric or spectroscopic data (Spiniello et al. 2011, 2012). The $M_\star^{\rm SPS}/L$ computed via SPS analysis depends on the choice of the IMF slope. In particular, a bottom-heavier IMF implies a larger $M_\star^{\rm SPS}/L$ because dwarf stars contribute more to the mass than light. Hence, the mismatch parameter, $\alpha_{\rm IMF} \equiv M_\star^{\rm LD}/M_\star^{\rm SPS}$, can be used to infer the lightness or heaviness of the IMF (Treu et al. 2010). For example, if a light IMF – such as the Chabrier IMF (Chabrier 2003) – is adopted in the SPS method, then $\alpha_{\rm IMF} \sim 1$ would point to the Chabrier IMF being consistent with the lensing and dynamical observables. However, $\alpha_{\rm IMF} \sim 2$ would indicate that the IMF in the lens galaxies is bottom-heavier (i.e., characterized by a larger number of dwarf stars), with a slope more similar to that of a Salpeter IMF (Salpeter 1955), or even steeper. Note that $M_\star^{\rm LD}$ may depend on lens modeling assumptions, such as the choice of dark matter density profile, the stellar mass-to-light ratio being spatially constant or varying, and the assumed anisotropy profile of the stellar orbits in the dynamical modeling (e.g., see the discussions in Sonnenfeld et al. 2018, 2019).

Although the IMF slope and the low-mass cutoff are degenerate with respect to the lensing data, both can be constrained when combined with dynamics and stellar population analysis. Barnabè et al. (2013) show this by studying two strong lenses from the X-Shooter Lens Survey (Spiniello et al. 2011) for which both *HST* imaging (for precise lens modeling) and X-Shooter spectra (for stellar population analysis) are available. Chromatic, microlensing-induced flux anomalies in a galaxy–quasar strong lens can also be used to constrain the stellar IMF (Schechter et al. 2014). This technique is described in Vernardos et al. (2024), which discusses the theory and applications of microlensing.

Whereas the IMF within the Milky Way is light – that is, consistent with the Chabrier IMF regardless of the stellar population age and environment (Chabrier 2003; Bastian et al. 2010) – the majority of strong lensing studies on elliptical galaxies report consistency with a heavier IMF (e.g., Spiniello et al. 2011; Sonnenfeld et al. 2012; Oldham and Auger 2018). We note, however, that the stellar IMF is degenerate with the choice of the dark matter density profile in most of these studies (Auger et al. 2010a). For instance, the SLACS analysis – by combining lensing and dynamics – finds the IMF in elliptical galaxies at mean redshift $\langle z \rangle \sim 0.2$ to be consistent with the Salpeter IMF, that is, $\alpha_{\rm IMF} \sim 2$ (see Fig. 7; Treu et al. 2010, also, Grillo et al. 2009). This result is reproduced with more flexible models for the same SLACS systems or a subset of them (Auger et al. 2010b; Shajib et al. 2021). This is also in agreement with the more general findings, based on dynamics or SPS analysis only, that the IMF is bottom-heavier for more massive galaxies in general (Fig. 7; Cappellari et al. 2012; La Barbera et al. 2013; Spiniello et al. 2014).The general picture is that the low-mass end of the IMF might not be universal across all galaxies, as generally assumed in the last thirty years. However, a consensus on the physical mechanisms responsible for its variation has not yet been reached. According to theoretical work (e.g., Hopkins 2013; Chabrier et al. 2014), high density, temperature, and turbulence of the gas are key parameters that drive the fragmentation of molecular clouds. Higher density and temperature make the fragmentation easier, forming more dwarf stars, that is, a bottom-heavier IMF.

In contrast with the studies mentioned above on massive, local elliptical galaxies, Sonnenfeld et al. (2019) find $\log_{10} \alpha_{\rm IMF} = -0.04 \pm 0.11$ with respect to the Chabrier IMF in the SPS method for a sample of strong lensing galaxies at $z \sim 0.6$, which is in tension with Shetty and Cappellari (2014) that reports consistency with a Salpeter IMF for a galaxy at $z \sim 1$ from a dynamical analysis. However, allowing spatial gradients in the $M_\star/L$ can alleviate this tension (Sonnenfeld et al. 2018, 2019). Indeed, a radial gradient in the $M_\star/L$, or

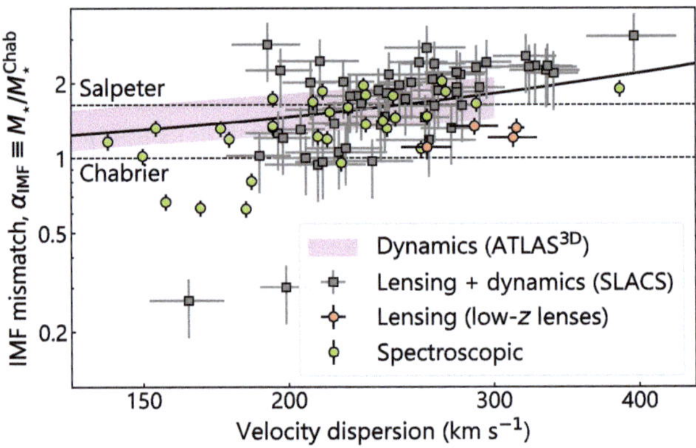

Fig. 7 Measurements of the stellar IMF from various probes. The pink shaded region traces the 1σ intrinsic scatter of the fitted relation from dynamical constraints of the ATLAS^{3D} sample (Cappellari et al. 2013; McDermid et al. 2014). The grey points are based on joint lensing–dynamics analysis for a subsample of the SLACS lenses (Posacki et al. 2015). The orange points are lensing-only measurements from a sample of low-redshift ($z = 0.031$–0.066) lenses; 3 from the SINFONI Nearby Elliptical Lens Locator Survey (SNELLS) and one discovered from a search on the publicly available MUSE data (Smith et al. 2015; Collier et al. 2018). The green points show the fully spectroscopy-based measurements for a sample of 34 elliptical galaxies from Conroy and van Dokkum (2012). The two horizontal dashed lines mark the values expected for Salpeter and Chabrier IMFs. The solid black line illustrates the dependency of the IMF on the velocity dispersion, fitted by Posacki et al. (2015) using the ATLAS^{3D} and the SLACS samples

equivalently a radially varying IMF, has been reported by studies based on the SPS method applied on local massive ellipticals (Martín-Navarro et al. 2015b; van Dokkum et al. 2017; Sarzi et al. 2018; Barbosa et al. 2021a,b). These authors find that the central region within $\sim$2 kpc has a heavy IMF (even super-Salpeter), and the IMF in the outer regions gradually becomes light (i.e., Milky-Way-like). The current belief is that the IMF is bottom-heavy for stars formed very early in cosmic time via a quick and violent SF burst. These stars usually form a 'red nugget' ($z \sim 2$, Damjanov et al. 2011; Oldham et al. 2017): an ultra-compact red-and-dead massive core. Then, with a second and more time-extended phase, red nuggets merge, interact with other structures in the Universe, and accrete gas. This process causes a growth in size up to a factor of $\sim$5, and only slightly in mass, transforming them into giant local, massive ellipticals. Depending on the merger history of each single galaxy, the red nugget can remain almost untouched in the innermost region, dominating the light there. In this case, this region would have a bottom-heavy IMF. This seems to be the case for NGC 3311, the central galaxy in the Hydra cluster (Barbosa et al. 2021a), M87 (Sarzi et al. 2018), and many other very massive low-z elliptical galaxies. However, the red nugget can also be destroyed or contaminated by accreted or lately formed stars. In that case, one would measure a Milky-Way-like IMF, which is the characteristic for stars formed later on and through more time-extended star formation channels. This 'two-phase formation scenario' (Naab et al. 2009) is also supported by the discovery of 'relic galaxies' (Trujillo et al. 2014; Spiniello et al. 2021), the local counterparts of red nuggets that somehow wholly missed the size-growth and evolved passively and undisturbed across cosmic time. The IMF for these peculiar and rare objects has been measured to be steep everywhere up to at least one effective radius (Ferré-Mateu et al. 2017). Finally, simulations have lately shown that not all elliptical galaxies formed via this two-phase formation scenario (e.g., Pulsoni et al. 2021), although this becomes more and more common with increasing stellar mass.

The finding of the Salpeter IMF from a combination of lensing and dynamics in the near-by Universe is consistent with this scenario, as strong-lensing information is sensitive to the galaxy's inner region (typically $\lesssim$6 kpc). However, at higher redshift, lensing probes larger and larger regions, which explains the results presented in Sonnenfeld et al. (2019). A radially decreasing $M_{\star}/L$ also explains or alleviates the reported tension between lensing-based studies themselves (Sonnenfeld et al. 2018, 2019; Shajib et al. 2021).

Furthermore, studies of few local massive lenses for which the Einstein radius is much smaller than the effective radius, and hence where the stellar mass dominates the lensing inference, indicate that bottom-heavy IMFs are excluded by lensing (Ferreras et al. 2010; Smith et al. 2015; Leier et al. 2016). Allowing for a variable cut-off on the low-mass end of the IMF can reconcile the $M_{\star}/L$ measurement from strong lensing of Smith et al. (2015) with the IMF-sensitive absorption line measurements of Conroy and van Dokkum (2012). Still, the discrepancy with other lensing–dynamics measurements remains.

In conclusion, the currently preferred scenario sees the majority of massive galaxies having a bottom-heavy IMF in their innermost region, where the pristine stellar population dominates that formed at $z > 2$ through a star formation burst, while stars in outskirt are distributed by a Milky-Way-like IMF. However, depending on the single galaxy's detailed merger tree and cosmic evolution, the IMF can differ from system to system.

In the future, IFU-based stellar kinematics and population analysis of strong lensing galaxies in combination with lensing constraints can shed light on the presence or absence of IMF variations and spatial gradients in ellipticals at intermediate redshifts. However, to properly track any evolution in the IMF properties of elliptical galaxies across redshift, it would be essential to mitigate systematic impacts through a uniform choice of models and to account for selection differences between samples.

4.3 Constraints on the Very Central Densities and SMBH Mass from Central Images

Gravitational lensing theory predicts that the number of multiple images must always be odd. In systems with three (or five) images, two (or four) are formed roughly at the Einstein radius from the lens center, which is $\lesssim 2''$ for galaxies. The odd 3rd (or 5th) image is formed very near the center of the lens. It is always demagnified, usually significantly, and superimposed on the light of the lensing galaxy, making it hard to detect. Since the demagnification of the central image depends on the central density profile, the detections (or the lack thereof) of the central image can constrain or put an upper limit (or lower limit) on the steepness of the inner density profile or the SMBH mass (e.g., Winn et al. 2004), with the degeneracy between the two broken with stellar light distribution informing the stellar mass profile (e.g., Wong et al. 2015; Tamura et al. 2015).

Most existing searches for central images of strongly lensed quasars (given that they are much brighter sources) rely on optical, or radio wavelengths. The radio wavelengths are most favorable as the lensing galaxy is generally transparent in that range, but these investigations are limited by the fact that quasars are usually radio-quiet. Out of ~200 doubles discovered to date, only two have observed central images where the lens is a single galaxy:[5] PMN J1632$-$0033, with the central image demagnified by a factor of 0.004 or 6 magnitudes compared to the brightest image (Winn et al. 2004), and PKS 1830$-$211, with the central image demagnified by a factor of 0.007 or 5.4 magnitudes (Muller et al. 2020). No reliable detection exists of the ~50 known quads with a single lensing galaxy. Upper limits have also been placed. For example, Quinn et al. (2016) place an upper limit of $\sim 10^{-4}$ on the

[5]There are at least five cases of central images where the lens has two or more main lensing galaxies.

magnification of the central image with respect to the brightest visible image in the double B1030+074.

Upper limits on the flux of the central image have also been placed at X-ray wavelengths for several quads and doubles. The stacking of X-ray monitoring data allows effective exposures of several hundreds of kilo-seconds without any contamination from the lens. However, the data are still too shallow to strongly constrain the lens galaxy's inner density profile. One of the deepest upper limits has been achieved for HE 0435−1123 (e.g., Chen et al. 2012; Guerras et al. 2017).

Detection of central images, or the lack thereof, has been used to place constraints on the central mass density of the lensing galaxies and the mass of the central SMBH (Mao et al. 2001; Rusin and Ma 2001; Wong et al. 2015; Tamura et al. 2015; Quinn et al. 2016; Perera et al. 2023). Even without a central image detected, a lensed image that is sufficiently close to the center can be used to measure the SMBH mass. For example, such an SMBH mass ($M_{\rm SMBH} = 3.27 \pm 2.12 \times 10^{10}$ $M_\odot$; 3σ confidence limit) was measured using the image at $\sim$1 kpc distance from the center in Abel 1201 ($z_{\rm lens} = 0.169$), providing the first lensing-based measurement of an SMBH mass with limits placed on both sides (Nightingale et al. 2023). Additionally, Millon et al. (2023) demonstrated the usefulness of strong lensing *by* a quasar (SDSS J0919+2720, shown in Fig. 1) to measure its host galaxy mass to robustly probe the SMBH–host mass relation, where the SMBH mass was measured through conventional methods based on spectroscopic data.

For other kinds of SMBHs, prospects for detecting binary SMBHs due to their lensing effects are discussed in Li et al. (2012), Hezaveh et al. (2015). Free-floating SMBHs, which could be of primordial origin or formed through co-evolution with their previous host galaxies, can be detected through the kinks they produce on razor-thin arcs with sub-mas width and resolved in radio observations (Banik et al. 2019). The detection (or the lack thereof) of such free-floating SMBHs can place constraints (or upper limits) on their mass density and fractional contribution to the dark matter.

4.4 Spiral Galaxies

Due to their lower mass and presence of a disk, spiral galaxies have a substantially lower lensing cross-section than elliptical ones (Keeton and Kochanek 1998). Only a handful of spiral galaxies lensing quasars have been discovered to date. The most well-known and the first example of such a system is the Einstein Cross (2237+0305), which is lensed by the bulge of a nearby spiral galaxy (Huchra et al. 1985). Targeted searches for galaxies lensed by a spiral have been carried out, allowing the discovery of several dozens of systems (Treu et al. 2011, and reference therein). Complications in studying those systems arise from the dust and the disk mass component. Correcting for the reddening by dust is needed to model extended lensed images, which rely on conserving the surface brightness between the lens and the source plane. On the other hand, the disk component yields strong discontinuities in the gravitational potential and needs to be explicitly modeled using, for example, an exponential mass density profile. Whereas disk-like features can yield flux anomalies in lensed quasars (Hsueh et al. 2016, 2017; Gilman et al. 2017), they can be disentangled from the dust using multi-band data (Möller et al. 2003). Once these ingredients are accounted for, the combination of kinematics and lensing information can be used to break the degeneracy – that exists in dynamical studies alone – between the disk and halo components (Maller et al. 2000; Dutton et al. 2011; Suyu et al. 2012).

5 Applications in Cosmology

In this section, we present applications of galaxy-scale lens systems to measure cosmological parameters without using time-delay information. Measurement of the Hubble constant (H_0) and other cosmological parameters based on the time delays are discussed in detail in Birrer et al. (2024).

Here, we briefly establish some necessary definitions for use in this section. More detailed explanations of the cosmological connection with strong lensing formalism are given in Saha et al. (2024). The angular diameter distance between two objects at redshifts z_1 and z_2 (with $z_1 < z_2$) for a flat universe is given by

$$D_{\rm ang}(z_1, z_2) = \frac{c}{H_0(1+z_2)} \int_{z_1}^{z_2} \frac{{\rm d}z'}{E(z')}, \tag{20}$$

where $E(z) \equiv H(z)/H_0$ is the dimensionless Friedman equation given by

$$E(z) = \sqrt{\Omega_{\rm m}(1+z)^3 + \Omega_{\rm r}(1+z)^4 + \Omega_{\rm de}(1+z)^{3(1+w_{\rm de})}}. \tag{21}$$

Here, $\Omega_{\rm m}$, $\Omega_{\rm r}$, and $\Omega_{\rm de}$ represent the matter, radiation, and dark energy density parameters, respectively, at $z = 0$. The parameter $w_{\rm de}$ is the equation-of-state parameter of the dark energy given by $w_{\rm de} \equiv p_{\rm de}/\rho_{\rm de}c^2$, where $p_{\rm de}$ and $\rho_{\rm de}$ denote the pressure and density of the dark energy, respectively. In the ΛCDM model, $w_{\rm de} = -1$ is assumed, that is, the dark energy density stays constant through cosmic time. The wCDM model is one natural extension of the ΛCDM model, where $w_{\rm de} \neq -1$ is allowed; however, $w_{\rm de} < -1/3$ should still be satisfied to reproduce an accelerated Universe.

In the following subsections, we discuss methods to estimate cosmological parameters (primarily, $\Omega_{\rm m}$, $\Omega_{\rm de}$, and $w_{\rm de}$) using multiple-source-plane lenses (Sect. 5.1), using the stellar kinematics of strong lenses (Sect. 5.2), and using galaxy–galaxy lensing statistics (Sect. 5.3).

5.1 Utilizing Multiple Sources at Different Redshifts

Strong lens systems with multiple sources (i.e., compound lenses) at different redshifts can be used as cosmographic probes. Currently, only a tiny sample of galaxy-scale compound lenses are known (e.g., Lewis et al. 2002; Gavazzi et al. 2008; Collett and Smith 2020), but at the *HST* snapshot depth they are expected to occur in about 1% of galaxy-scale strong lenses (Gavazzi et al. 2008). Indeed, if one were to stare deeply at any single plane lens, other sources would almost inevitably be discovered, as is spectacularly demonstrated by the discovery of a third source (a $z \approx 6$ Lyman-α emitter) behind the Jackpot lens in deep Multi Unit Spectroscopic Explorer (MUSE) data from the Very Large Telescope (Collett and Smith 2020). Forthcoming surveys are expected to discover $\mathcal{O}(1000)$ compound lenses.

Since the Einstein radius is a function of the lens mass and the cosmological distances, the ratio of Einstein radii in a compound lens with sources at two or more redshifts is independent of the mass (Gavazzi et al. 2008; Collett et al. 2012). In practice, the method also requires a complete understanding of the lens density profile and additional lensing by the source galaxies and other perturbing masses along the line of sight.

For a two-source-plane system with one primary lens, the lens equation can be written as

$$\begin{aligned} \mathbf{y} &= \mathbf{x} - \beta_{12}\boldsymbol{\alpha}_{\rm d}(\mathbf{x}), \\ \mathbf{z} &= \mathbf{x} - \boldsymbol{\alpha}_{\rm d}(\mathbf{x}) - \boldsymbol{\alpha}_{\rm s1}\left(\mathbf{x} - \beta_{12}\boldsymbol{\alpha}_{\rm d}(\mathbf{x})\right), \end{aligned} \tag{22}$$

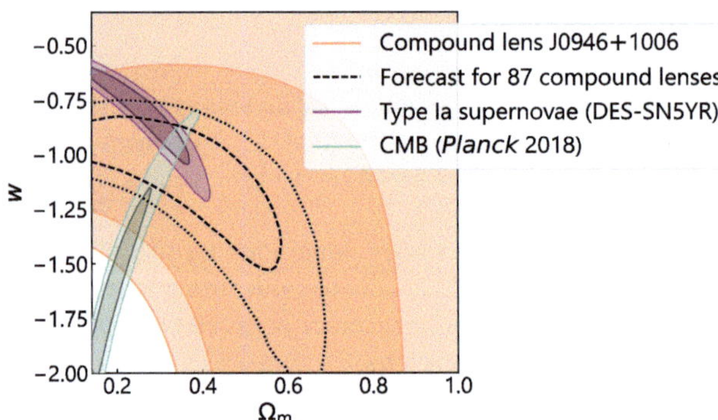

Fig. 8 Cosmological parameters constrained from the compound lens system SDSS J0946+1006 (shown in Fig. 2). The 68% and 95% credible regions on the $\Omega_{\rm m}$–$w_{\rm de}$ plane from this compound plane system are shown in orange (Collett and Auger 2014). The black dashed and dotted contours illustrate the forecasted 68% and 95% credible regions, respectively, for a sample of 87 compound lenses to be discovered by the Rubin Observatory LSST (Sharma et al. 2023). The purple and emerald contours are the constraints from the DES Type Ia supernovae sample (DES Collaboration 2024) and the CMB (Planck Collaboration 2020), respectively, illustrating the complementarity of compound lenses to these probes

where $\mathbf{x}$ are positions on the image plane, $\mathbf{y}$ and $\mathbf{z}$ are the unlensed positions of the first and second source, respectively, $\boldsymbol{\alpha}_{\rm d}$ is the deflection caused by the primary lens, $\boldsymbol{\alpha}_{\rm s1}$ is the deflection caused by the closer of the two sources, and β_{ij} is the cosmological scaling factor

$$\beta_{ij} = \frac{D_{ij} D_{\rm s}}{D_j D_{is}}. \tag{23}$$

For realistic redshifts and cosmologies, β_{ij} is sensitive to the matter density parameter $\Omega_{\rm m}$ and the equation-of-state parameter $w_{\rm de}$ but has no dependence on the Hubble constant H_0.

Amongst galaxy-scale compound lenses, only the Jackpot lens (Gavazzi et al. 2008), shown in Fig. 2, has been used to precisely constrain cosmology since this system has a favorable redshift configuration – other compound lenses have multiple sources at similar redshifts, thus having $\beta \approx 1$ regardless of the cosmology. Collett and Auger (2014) modeled the *HST* imaging performing a pixellated reconstruction of both sources to make a 1.1% measurement on β^{-1}. Converting this into constraints on the dark energy, this single compound lens with a cosmic microwave background (CMB) prior from Planck Collaboration (2014) constrains $w_{\rm de}$ to 0.2 precision. With hundreds of compound lenses expected in the Rubin Observatory Legacy Survey of Space and Time (LSST) and the *Euclid* surveys, constraints on both $w_{\rm de}$ and its redshift derivative are expected to be comparable with established cosmological probes (Fig. 8; Sharma et al. 2023).

5.2 Utilizing Stellar Kinematics of Single Source Lenses

The enclosed projected mass inside the Einstein radius is independent of the mass profile (Schneider et al. 1992). Similarly, the speeds of the stars within a galaxy are sensitive to the total mass enclosed within their orbits. Whilst we cannot measure the speeds of individual stars in a lens, we can measure the velocity dispersion of the ensemble. For an SIS lens with stars on isotropic orbits, these two quantities are related by

$$\theta_{\rm E} = 4\pi \left(\frac{\sigma_{\rm los}}{c}\right)^2 \frac{D_{\rm ds}}{D_{\rm s}}, \tag{24}$$

where $\sigma_{\rm los}$ is the velocity dispersion of the deflector. Ofek et al. (2003) estimated that deviations from isothermality and orbital isotropy can cause the observed velocity dispersion to differ from Eq. (24) by up to 20%.

Equation (24) becomes far more complicated if the density profile is not isothermal or the orbits of the stars are not isotropic, but the fundamental relationship remains that one can constrain the distance ratio $d^{\rm obs} \equiv D_{\rm ds}/D_{\rm s}$ using the observed Einstein radius $\theta_{\rm E}$ and observed velocity dispersion. Thus, a sample of single-source lenses with measured $\sigma_{\rm los}$ can be used to estimate cosmological parameters (Grillo et al. 2008) by maximizing the likelihood function

$$\mathcal{L}\left(\boldsymbol{\Theta} \mid \mathcal{D}\right) \propto \exp\left[-\frac{1}{2}\sum_{i=1}^{N_{\rm SL}} \frac{\left[d_i^{\rm th}\left(z_{\rm d}, z_{\rm s}; \boldsymbol{\Theta}\right) - d_i^{\rm obs}\left(\theta_{{\rm E},i}, \sigma_{{\rm los},i}\right)\right]^2}{(\delta d_i^{\rm obs})^2}\right], \tag{25}$$

where $\boldsymbol{\Theta}$ is a set consisting of the free parameters in the assumed cosmological model and the free parameters that describe the density profile and anisotropy profile of the lenses, $\mathcal{D}$ is the data, $N_{\rm SL}$ is the number of lens systems, and $\delta d_i^{\rm obs}$ is the uncertainty of each $d_i^{\rm obs}$, which depends on the $\sigma_{{\rm los},i}$ and $\theta_{{\rm E},i}$ uncertainties.

Similar to the multiple-source lens systems described in Sect. 5.1, this method is also independent of H_0. If the cosmological parameters and the lens population parameters are inferred simultaneously for a large sample of lenses (e.g., $N_{\rm lens} \sim$10 000) discovered by the Rubin Observatory LSST and *Euclid*, it will be possible to achieve very competitive precision with other probes such as Type Ia supernovae and the CMB (Li et al. 2024).

The primary systematics in this method can potentially arise from these assumptions: (i) the measured $\theta_{\rm E}$ is independent of the choice of the lens mass profile (e.g., Cao et al. 2015), and (ii) the measured line-of-sight velocity dispersion is equal to that for an SIS profile. Recent modeling methods provide robust $\theta_{\rm E}$ measurement within a few percent regardless of the mass profile choice (Birrer 2021). Treu et al. (2006) argue that $\sigma_{\rm los} \simeq \sigma_{\rm SIS}$ for the lens elliptical galaxies with velocity dispersion in the range 200–300 km s^{-1}. These results are further confirmed by analyzing other samples (Bolton et al. 2006; Auger et al. 2010b).

Ultimately, the only way to move forward with lensing and dynamics as a precision cosmological probe is to simultaneously infer the astrophysical parameters of the lens population and the cosmological parameters. By building up a sample of many lenses it should be possible to investigate how the Einstein radius grows with source redshift regardless of the underlying density profile of strong lenses. Exploiting the fact that lenses at the same redshift can be expected to be somewhat self-similar, Li et al. (2024) showed that with 10 000 lenses, it is possible to disentangle lens population properties and cosmological parameters. These authors assumed that lenses have the same intrinsic scatter as Auger et al. (2010b) found for the SLACS lenses. The method is fundamentally limited by how self-similar lenses are, and if their properties evolve with redshift.

5.3 Utilizing Galaxy–Galaxy Lensing Statistics

The statistics of strong lensing were initially expected to be powerful for probing the cosmological parameters (Fukugita et al. 1990). At the most basic level, matter clusters can collapse to the densities required to form strong lenses, whereas the cosmological constant does not cluster and cannot form lenses. Therefore, it was expected that lensing rates should be suppressed for larger values of Ω_Λ. Further information is contained in the Einstein radius distribution and the lens and source redshift distributions. In practice, this topic has

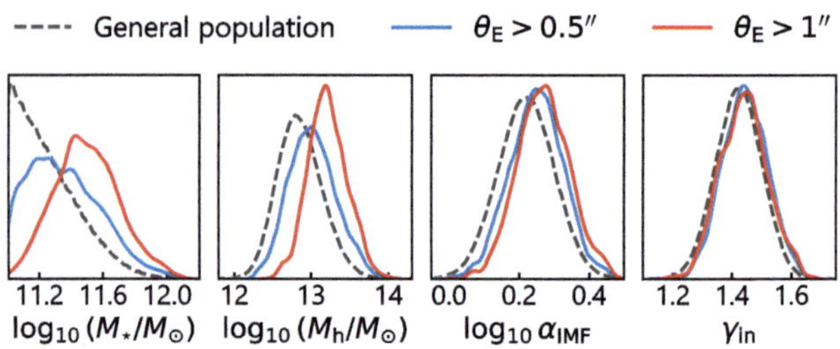

Fig. 9 Impact of strong lensing selection function, estimated by Sonnenfeld et al. (2023), on various galaxy properties: from left to right, stellar mass $M_\star$, halo mass M_h, IMF mismatch parameter α_{IMF} with $\alpha_{IMF} = 1$ corresponding to the Chabrier IMF, and the dark matter's inner logarithmic slope γ_{in}. The grey dashed line shows the distribution of the general population of galaxies. The blue and red lines show the distributions of galaxy–galaxy strong lenses with $\theta_E > 0''.5$ and $\theta_E > 1''$, respectively. Strong lenses are expectedly biased toward more massive galaxies. The mean IMF mismatch parameter in strong lensing galaxies is biased by 10% from the general population, and the mean inner logarithmic slope is biased by 5%

fallen out of fashion due to the cosmological sensitivity being overwhelmed by astrophysical uncertainties of the unlensed source population, the lens discovery selection function, and the lensing properties of typical galaxies (e.g., Mitchell et al. 2005; Chae 2010).

6 Open Problems and Future Outlook

In this section, we discuss the current open problems that are expected to be tackled in this decade: the selection function (Sect. 6.1), the self-similarity assumption (Sect. 6.2), degeneracies in strong-lensing (Sect. 6.3) and non-strong-lensing observables (Sect. 6.4), and comparison with galaxy simulations (Sect. 6.5). We also provide future outlooks on these issues whenever appropriate.

6.1 Selection Function

Although strong lensing is a powerful probe to study galaxy properties, the lensing phenomenon is a rare occurrence requiring a serendipitous alignment of two line-of-sight objects separated by a large cosmological distance. Thus, samples of strong lensing galaxies inherently occupy a tiny fraction of the population of all galaxies. When strong lensing studies aim to infer properties of the general population of galaxies based on such a small fraction of galaxies, the lens sample's selection function must be considered. The strong lensing samples are inherently biased towards lensing galaxies that are more massive and concentrated (Mandelbaum et al. 2009; Sonnenfeld et al. 2023). Although triaxial galaxies with the major axis more aligned along the line of sight also have larger lensing cross-sections, interestingly for a given mass and shape, the effect of viewing angle does not affect the selection function when averaged over (Mandelbaum et al. 2009, we note that this study only considered point sources). Sonnenfeld et al. (2023) estimate that the mean of the IMF mismatch parameter α_{IMF} measured from a sample of lens galaxies is only biased by 10% and the mean of the inner slope of the dark matter by 5% (Fig. 9). These bias levels are dependent on the completeness in the Einstein radius distribution of the lens sample but independent of the source properties, with the galaxy–galaxy lenses and galaxy–quasar lenses having the same levels of bias.

The lens samples to date often had highly complex selection functions, largely due to the selection or discovery procedure being highly tuned to maximize the number of discovered systems. As a result, treatment of selection function on actual lens samples has been rare except for a handful of studies (Arneson et al. 2012; Sonnenfeld et al. 2015, 2019). For the ideal case of a known selection function, Sonnenfeld (2022) provides the formally correct solution to account for selection effects in the lens samples. However, several technical challenges still remain to implement this formal solution on real samples, for example, the requirement to well characterize the lens-detection efficiency of the survey and also the efficiency in obtaining follow-up spectroscopy if that was taken into account for sample selection. The treatment of the selection effect with parametric functions by Sonnenfeld et al. (2019) can be considered as an approximation to this formal solution. To keep the selection function easily treatable, it can be advisable to pre-emptively mitigate the complexity of the selection procedure. Alternatively, it can be possible to form a subsample of lens systems that has a well-characterized selection function from a much larger sample of discovered systems, for example, those discovered by current and future surveys such as the *Euclid*, the Rubin Observatory LSST, and the *Roman Space Telescope* (Collett 2015; Sonnenfeld 2022).

6.2 Assumption of Self-Similarity

When hierarchically inferring population properties of lenses by combining multiple strong lensing systems (Sect. 3.4), it is crucial to describe and quantify potential differences between subsets of the considered sample for accurate population-level inference. Differences in the population may arise from the sample selection due to search criteria and techniques (Sect. 6.1) or intrinsic differences in the sources. Such differences in secondary selection might impact and bias population-level constraints when assuming that two different samples can be described with an identical underlying population. For example, Birrer et al. (2020) present a hierarchical analysis under the assumptions that the same population level parameters describe the SLACS and the Time-Delay COSMOgraphy (TDCOSMO) lenses to constrain the mass density profiles of the time-delay lenses better. There are different ways to mitigate such assumptions. The first is to select a purified sample of lenses as self-similar as possible when performing hierarchical analyses, also suggested in Sect. 6.1. However, although the currently small number of known lenses will increase through future surveys and enable this approach, it is still unclear in which parameter space these lenses can be considered self-similar and whether there are important latent variables to consider. The second way is to model and describe differences between populations on a first-principle level, folding in differential selection effects and other aspects into the analysis (as stated in Sect. 6.1).

6.3 Degeneracies in Strong-Lensing Observables

The lensing data's ability to constrain the lens's radial structure, particularly in disentangling dark and luminous mass components, strongly depends on the data quality. As discussed in Saha et al. (2024), the modeling of point-source astrometry primarily encodes information on the lens's quadrupole moment. Still, it provides limited constraints on the monopole (the total mass). This, however, does not mean that any monopole model combined with a quadrupole component will accurately reproduce a set of lensed image positions. For some choice of the monopole, extreme or unphysical values of the quadrupole may be required, naturally excluding some mathematically correct solutions. This explains why, for example, a single-component, constant mass-to-light ratio model generally yields large shear amplitudes to reproduce the observed astrometry of lensed systems to high accuracy (e.g., Sluse et al. 2012).

While extended lensed images may provide detailed azimuthal information and allow one to constrain the ratio of radial magnifications at different galactocentric distances, their effective constraint on the density profile remains sensitive to the MSD (e.g., Sonnenfeld 2018). While the MSD provides mathematically large leverage to modify the results, its impact may remain in practice generally small, with a typical change on the total density profile that may not exceed a few tens percent. The prior on the mass profile (i.e., choice of mass distribution families or a free-form model with some regularization) may further limit the impact of degeneracies. One may, however, need to be careful when choosing a mass distribution, as a model that is too rigid compared to the true mass density may yield biased posteriors or underestimated parameter uncertainties (e.g., Sonnenfeld 2018; Kochanek 2021). Finally, it is important to note that point images with measured time delays limit the impact of degeneracies in lensing-only observables for an assumed cosmology (Saha and Williams 2001; Kochanek 2002).

6.4 Degeneracies in Non-strong-Lensing Observables

The modeling of stellar kinematics data requires an assumption on the anisotropy profile of stellar orbits. Integrated velocity dispersions obtained from single-slit spectra cannot constrain the anisotropy profile, which leads to the so-called mass–anisotropy degeneracy (Treu and Koopmans 2002) — typically adopted anisotropy profiles are either isotropic or the Osipkov–Merritt profile (Osipkov 1979; Merritt 1985a,b). Whereas the isotropy assumption does not entail any free parameter, the Osipkov–Merritt profile depends on a scale radius. Due to the mass–anisotropy degeneracy, the posterior of the anisotropy scale radius is dominated by the adopted prior (e.g., Shajib et al. 2018). Particular choices of the anisotropy profile and the associated prior may lead to systematic differences between studies involving strong lensing and kinematics data. For example, Sonnenfeld et al. (2018) find the existence of a mass-to-light ratio gradient in the SLACS lenses assuming isotropic orbits, whereas Shajib et al. (2021) find consistency with a constant mass-to-light ratio assuming the Osipkov–Merritt anisotropy profile for a subsample of SLACS. Additionally, the unknown 3D structures of the mass distribution and the tracer distribution are also potential sources of systematics (Cappellari 2008).

Spatially resolved velocity dispersion measurements can better constrain the anisotropy profile by breaking the mass–anisotropy degeneracy. However, spatially resolved kinematics data from IFU spectroscopy are more expensive than an integrated measurement from long-slit spectroscopy. Thus, usage of such data in lensing studies has been limited (see van de Ven et al. 2010; Barnabè et al. 2011; Shajib et al. 2023, for example of IFU data being combined with strong lensing).

6.5 Comparison with Galaxy Simulations

The past decade has seen significant progress in understanding galaxies' structure and formation. Within the ΛCDM paradigm, there is general agreement regarding the gravitational aspect of galaxy formation. The 'gastrophysics' is less well understood and requires subgrid models to simulate, but still, the galaxies formed in simulations like Illustris (Vogelsberger et al. 2014), FIRE (Hopkins et al. 2014), EAGLE (Crain et al. 2015), IllustrisTNG (Nelson et al. 2019) are much more credible than previous generations of simulated galaxies.

There are also equilibrium galaxy models, including stars, gas, and dark matter, of which the AGAMA (Vasiliev 2019) simulations are arguably the most sophisticated.

Despite the great advances in the fidelity of the simulations, there have been discrepancies between the simulated predictions and the observed properties of galaxies. In particular, simulations have not been successful yet in reproducing the observed distributions of the logarithmic slope γ_{pl} and the dark matter fraction f_{dm} simultaneously. A no-feedback or weak feedback prescription was required in some of the simulations to reproduce the γ_{pl} distribution, which, however, led to underestimating f_{dm} compared to the observations due to overestimating the star formation efficiency (e.g., Naab et al. 2007; Duffy et al. 2010; Johansson et al. 2009). Similarly, matching the f_{dm} distribution required strong feedback prescriptions, but these produce too shallow γ_{pl} compared to the observed distribution. More recently, the IllustrisTNG simulation reproduced a f_{dm}–γ_{pl} distribution that is consistent with the strong lensing observations if the stellar IMF corresponds to the Salpeter IMF (Wang et al. 2020; Shajib et al. 2021). In contrast, Mukherjee et al. (2022), find that the EAGLE and SLACS f_{dm}–γ_{pl} distributions agree while using a Chabrier IMF, supporting the important role played by feedback and sub-grid physics in reproducing this relation.

Furthermore, there has been an apparent tension in the redshift evolution of the logarithmic slope γ_{pl} between observations and simulations. Strong lensing observations report a steepening of γ_{pl} with decreasing redshift at $z < 1$ (Ruff et al. 2011; Sonnenfeld et al. 2013). Such a steepening would require dissipative processes through wet mergers along the evolutionary track of elliptical galaxies. Simulations instead find no evidence for redshift evolution of γ_{pl} below $z = 1$, or a slightly shallowing trend in γ_{pl} with decreasing redshift (see Fig. 10; Xu et al. 2017; Remus et al. 2017; Wang et al. 2020). Strong lensing selection effects could be a potential source of this discrepancy (Sonnenfeld et al. 2015). Moreover, this tension vanishes if the same strong lensing analysis is applied to the simulated galaxies from Illustris and Magneticum to extract γ_{pl} by combining lensing and kinematic information (Xu et al. 2017; Remus et al. 2017). Therefore, the modeling systematics in the joint analysis of lensing and dynamical observables cannot be ruled out as another potential source of the above tension. For an example of modeling systematic, if the true mass distribution in lensing elliptical galaxies is not an accurate power law, then lensed images are formed at different galactocentric radii as the lens and source redshifts vary (for a relevant test of systematic, see Gomer et al. 2022). In that case, minor deviations from the adopted power-law model may mimic an (absence of) evolution of the galaxy's total density profile with redshift. To account for this effect, Dutton and Treu (2014) suggest using a mass-weighted slope.

An alternative approach to compare observed galaxy properties to theoretical models is the semi-empirical one adopted by Shankar et al. (2017, 2018). These authors find their semi-empirical model for massive elliptical galaxies to be consistent with un-contracted NFW halos and the Salpeter IMF, corroborating with the majority of the previous literature (see Sect. 4.1.2 and Sect. 4.2). These authors also investigate the redshift dependence of the logarithmic slope γ_{pl} mentioned above and find that a redshift dependence of the Sérsic index is necessary to explain it. In contrast, the selection function does not contribute to producing the redshift-dependent trend in γ_{pl}.

The specific importance of various baryonic processes in the evolution of ellipticals is still an open question. The simulations would require further fine-tuning to consistently reproduce all of their observable properties. However, modeling systematics and the selection function must be appropriately considered for accurate comparison between simulations and observations. The first results from a statistical framework jointly considering simulations and observations to infer the galaxy evolution scenario indicate that AGN feedback is essential (Denzel et al. 2021a), but application to a large sample of lenses is required.

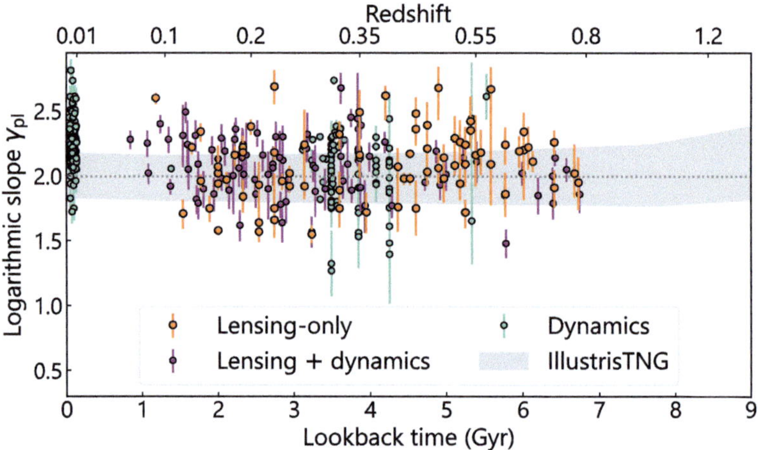

Fig. 10 Comparison of measured logarithmic slopes $\gamma_{\rm pl}$ of the total density profile at different redshifts between lensing-only measurements (orange points), lensing–dynamics measurements (purple points), dynamical measurements (emerald points), and the IllustrisTNG simulation (grey shaded region; Wang et al. 2020). The lensing-only measurements are from the SLACS, SL2S, and BELLS samples (Tan et al. 2024). The lensing–dynamics measurements are from the SLACS and SL2S samples (Auger et al. 2010b; Sonnenfeld et al. 2013). The dynamical measurements are from the ATLAS$^{\rm 3D}$, the Frontier Fields, and the Middle Ages Galaxy Properties with Integral field spectroscopy (MAGPI) surveys (Poci et al. 2017; Derkenne et al. 2021, 2023). The horizontal dotted line traces $\gamma_{\rm pl} = 2$, the isothermal case

An additional usage of simulated galaxies is to take them as the deflector galaxies in synthesizing strong-lensing observables for testing and validating the assumptions made in the simply parametrized mass models (e.g., Enzi et al. 2020; Ding et al. 2021a). These investigations often find the simple power-law parametrization to be inadequate to accurately describe the data (e.g., He et al. 2023). However, numerical inadequacies in synthesizing the strong-lensing observables can also potentially hamper such investigations (Van de Vyvere et al. 2020). Alternatively, simply parametrized lens models can also be tested based on independent empirical observations, for example, using high-resolution imaging of nearby ellipticals (Gilman et al. 2017), or using dynamical models from highly resolved IFU spectroscopy of ellipticals (e.g., Cao et al. 2022; Poci and Smith 2022). In the future, exquisite high-resolution imaging from the *JWST* or extremely large telescopes, or advanced dynamical models, such as Schwarzschild models (extending the original method of Schwarzschild 1979), will provide powerful means to carry out these important validation tests.

7 Concluding Remarks

In this review article, we have provided a review of the applications of galaxy-scale strong lensing in astrophysics and cosmology. Inevitably, some special sub-topics within the field of galaxy-scale strong lensing have evolved into proper research fields, having acquired a methodology and literature extensive enough to warrant a dedicated review. These are: detecting dark matter substructures and linking their properties to the dark matter particle, and measuring H_0 and other cosmological parameters through time delays, examined in Vegetti et al. (2023) and Birrer et al. (2024), respectively.

We started with a brief historical overview in Sect. 2. Then, in Sect. 3, we have discussed both strong-lensing and complementary non-strong-lensing observables and methodologies to model and extract meaningful results from such data. The most available and informative data for galaxy-scale strong lenses come from imaging. We reviewed the most common modeling methods found in the literature to model such data and constrain galaxy properties from lensed arcs, and in some cases with the inclusion of multiple images of a point-like source. Next, we reviewed the main scientific results from the literature on the astrophysics of galaxies in Sect. 4 and on cosmology in Sect. 5. We then discussed the currently open questions and provided future outlooks in Sect. 6.

The open questions presented in Sect. 6 provide exciting opportunities for the near future. Several large-area sky surveys – namely the *Rubin*, *Euclid*, and *Roman* observatories – will discover thousands of new galaxy-scale lensing systems. These treasure troves of data will provide the necessary statistical power to shed light on the open questions on galaxy evolution and cosmology.

Acknowledgements The authors thank the anonymous referees for helpful comments that improved the content and quality of the manuscript. The authors also thank Caro Derkenne and Nandini Sahu for providing a compilation of measurements from the literature, which was used in Fig. 10. The authors thank the International Space Science Institute in Bern (ISSI) for their hospitality and the conveners for organizing the stimulating workshop on "Strong Gravitational Lensing". Support for this work was provided by NASA through the NASA Hubble Fellowship grant HST-HF2-51492 awarded to AJS by the Space Telescope Science Institute, which is operated by the Association of Universities for Research in Astronomy, Inc., for NASA, under contract NAS5-26555. This work was also supported by the U.S. Department of Energy (DOE) Office of Science Distinguished Scientist Fellow Program. GV has received funding from the European Union's Horizon 2020 research and innovation program under the Marie Skłodowska-Curie grant agreement No 897124. This research was made possible by the generosity of Eric and Wendy Schmidt by recommendation of the Schmidt Futures program. VM acknowledges partial support from Centro de Astrofísica de Valparaíso and project "Fortalecimiento del Sistema de Investigación e Innovación de la Universidad de Valparaíso" (UVA20993). This article made use of LENSTRONOMY (Birrer and Amara 2018; Birrer et al. 2021), ASTROPY (Astropy Collaboration 2013, 2018, 2022), NUMPY (Oliphant 2015), SCIPY (Jones et al. 2001), MATPLOTLIB (Hunter 2007), SEABORN (Waskom et al. 2014), and CHAINCONSUMER (Hinton 2016).

Funding Open access funding provided by EPFL Lausanne.

Declarations

Competing Interests The authors have no conflicts of interest to report.

Open Access This article is licensed under a Creative Commons Attribution 4.0 International License, which permits use, sharing, adaptation, distribution and reproduction in any medium or format, as long as you give appropriate credit to the original author(s) and the source, provide a link to the Creative Commons licence, and indicate if changes were made. The images or other third party material in this article are included in the article's Creative Commons licence, unless indicated otherwise in a credit line to the material. If material is not included in the article's Creative Commons licence and your intended use is not permitted by statutory regulation or exceeds the permitted use, you will need to obtain permission directly from the copyright holder. To view a copy of this licence, visit http://creativecommons.org/licenses/by/4.0/.

References

Abadi MG, Navarro JF, Fardal M, Babul A, Steinmetz M (2010) Galaxy-induced transformation of dark matter haloes. Mon Not R Astron Soc 407(1):435–446. https://doi.org/10.1111/j.1365-2966.2010.16912.x

Adam A, Perreault-Levasseur L, Hezaveh Y (2022) Pixelated Reconstruction of Gravitational Lenses using Recurrent Inference Machines. ArXiv e-prints, arXiv:2207.01073

Adams FC, Bloch AM, Butler SC, Druce JM, Ketchum JA (2007) Orbital instabilities in a triaxial cusp potential. Astrophys J 670(2):1027–1047. https://doi.org/10.1086/522581

Agnello A, Lin H, Kuropatkin N, Buckley-Geer E, Anguita T, Schechter PL, Morishita T, Motta V, Rojas K, Treu T, Amara A, Auger MW, Courbin F, Fassnacht CD, Frieman J, More A, Marshall PJ, McMahon RG, Meylan G, Suyu SH, Glazebrook K, Morgan N, Nord B, Abbott TMC, Abdalla FB, Annis J, Bechtol K, Benoit-Lévy A, Bertin E, Bernstein RA, Brooks D, Burke DL, Rosell AC, Carretero J, Cunha CE, D'Andrea CB, da Costa LN, Desai S, Drlica-Wagner A, Eifler TF, Flaugher B, García-Bellido J, Gaztanaga E, Gerdes DW, Gruen D, Gruendl RA, Gschwend J, Gutierrez G, Honscheid K, James DJ, Kuehn K, Lahav O, Lima M, Maia MAG, March M, Menanteau F, Miquel R, Ogando RLC, Plazas AA, Sanchez E, Scarpine V, Schindler R, Schubnell M, Sevilla-Noarbe I, Smith M, Soares-Santos M, Sobreira F, Suchyta E, Swanson MEC, Tarle G, Tucker D, Wechsler R (2018) DES meets Gaia: discovery of strongly lensed quasars from a multiplet search. Mon Not R Astron Soc 479(4):4345–4354. https://doi.org/10.1093/mnras/sty1419

Amvrosiadis A, Eales SA, Negrello M, Marchetti L, Smith MWL, Bourne N, Clements DL, De Zotti G, Dunne L, Dye S, Furlanetto C, Ivison RJ, Maddox SJ, Valiante E, Baes M, Baker AJ, Cooray A, Crawford SM, Frayer D, Harris A, Michałowski MJ, Nayyeri H, Oliver S, Riechers DA, Serjeant S, Vaccari M (2018) ALMA observations of lensed Herschel sources: testing the dark matter halo paradigm. Mon Not R Astron Soc 475(4):4939–4952. https://doi.org/10.1093/mnras/sty138

An JH (2005) Gravitational lens under perturbations: symmetry of perturbing potentials with invariant caustics. Mon Not R Astron Soc 356(4):1409–1428. https://doi.org/10.1111/j.1365-2966.2004.08581.x

Anau Montel N, Coogan A, Correa C, Karchev K, Weniger C (2023) Estimating the warm dark matter mass from strong lensing images with truncated marginal neural ratio estimation. Mon Not R Astron Soc 518(2):2746–2760. https://doi.org/10.1093/mnras/stac3215

Arneson RA, Brownstein JR, Bolton AS (2012) Quantifying the biases of spectroscopically selected gravitational lenses. Astrophys J 753(1):4. https://doi.org/10.1088/0004-637X/753/1/4

Astropy Collaboration (2013) Astropy: a community Python package for astronomy. Astron Astrophys 558:33. https://doi.org/10.1051/0004-6361/201322068

Astropy Collaboration (2018) The Astropy Project: building an open-science project and status of the v2.0 core package. Astron J 156:123. https://doi.org/10.3847/1538-3881/aabc4f

Astropy Collaboration (2022) The astropy project: sustaining and growing a community-oriented open-source project and the latest major release (v5.0) of the core package. Astrophys J 935(2):167. https://doi.org/10.3847/1538-4357/ac7c74

Auger MW, Treu T, Bolton AS, Gavazzi R, Koopmans LVE, Marshall PJ, Bundy K, Moustakas LA (2009) The Sloan Lens ACS Survey. IX. Colors, lensing, and stellar masses of early-type galaxies. Astrophys J 705(2):1099–1115. https://doi.org/10.1088/0004-637X/705/2/1099

Auger MW, Treu T, Bolton AS, Gavazzi R, Koopmans LVE, Marshall PJ, Moustakas LA, Burles S (2010b) The Sloan Lens ACS Survey. X. Stellar, dynamical, and total mass correlations of massive early-type galaxies. Astrophys J 724:511–525. https://doi.org/10.1088/0004-637X/724/1/511

Auger MW, Treu T, Gavazzi R, Bolton AS, Koopmans LVE, Marshall PJ (2010a) Dark matter contraction and the stellar content of massive early-type galaxies: disfavoring "light" initial mass functions. Astrophys J Lett 721(2):163–167. https://doi.org/10.1088/2041-8205/721/2/L163

Banik U, van den Bosch FC, Tremmel M, More A, Despali G, More S, Vegetti S, McKean JP (2019) Constraining the mass density of free-floating black holes using razor-thin lensing arcs. Mon Not R Astron Soc 483(2):1558–1573. https://doi.org/10.1093/mnras/sty3267

Barbosa CE, Spiniello C, Arnaboldi M, Coccato L, Hilker M, Richtler T (2021a) A preserved high-z compact progenitor in the heart of NGC 3311 revealed with MUSE 2D stellar population analysis. Astron Astrophys 649:93. https://doi.org/10.1051/0004-6361/202039809

Barbosa CE, Spiniello C, Arnaboldi M, Coccato L, Hilker M, Richtler T (2021b) What does (not) drive the variation of the low-mass end of the stellar initial mass function of early-type galaxies. Astron Astrophys 645:1. https://doi.org/10.1051/0004-6361/202039810

Barkana R (1998) Fast calculation of a family of elliptical mass gravitational lens models. Astrophys J 502(2):531–537. https://doi.org/10.1086/305950

Barnabè M, Koopmans LVE (2007) A unifying framework for self-consistent gravitational lensing and stellar dynamics analyses of early-type galaxies. Astrophys J 666(2):726–746. https://doi.org/10.1086/520495

Barnabè M, Czoske O, Koopmans LVE, Treu T, Bolton AS, Gavazzi R (2009) Two-dimensional kinematics of SLACS lenses - II. Combined lensing and dynamics analysis of early-type galaxies at z = 0.08-0.33. Mon Not R Astron Soc 399(1):21–36. https://doi.org/10.1111/j.1365-2966.2009.14941.x

Barnabè M, Czoske O, Koopmans LVE, Treu T, Bolton AS (2011) Two-dimensional kinematics of SLACS lenses - III. Mass structure and dynamics of early-type lens galaxies beyond $z \simeq 0.1$. Mon Not R Astron Soc 415(3):2215–2232. https://doi.org/10.1111/j.1365-2966.2011.18842.x

Barnabè M, Spiniello C, Koopmans LVE, Trager SC, Czoske O, Treu T (2013) A low-mass cut-off near the hydrogen burning limit for Salpeter-like initial mass functions in early-type galaxies. Mon Not R Astron Soc 436(1):253–258. https://doi.org/10.1093/mnras/stt1727

Barrera B, Williams LLR, Coles JP, Denzel P (2021) Bridging the Gap Between Simply Parametrized and Free-Form Pixelated Models of Galaxy Lenses: the Case of WFI 2033-4723 Quad. Open J Astrophys 4. https://doi.org/10.21105/astro.2108.04348

Bastian N, Covey KR, Meyer MR (2010) A universal stellar initial mass function? A critical look at variations. Annu Rev Astron Astrophys 48:339–389. https://doi.org/10.1146/annurev-astro-082708-101642

Bayer D (2021) Probing the nature of dark matter with the power spectrum of small-scale mass structure in massive elliptical lens galaxies. PhD thesis, University of Groningen. https://doi.org/10.33612/diss.155450412

Bellstedt S, Forbes DA, Romanowsky AJ, Remus R-S, Stevens ARH, Brodie JP, Poci A, McDermid R, Alabi A, Chevalier L, Adams C, Ferré-Mateu A, Wasserman A, Pandya V (2018) The SLUGGS survey: a comparison of total-mass profiles of early-type galaxies from observations and cosmological simulations, to $\sim$4 effective radii. Mon Not R Astron Soc 476(4):4543–4564. https://doi.org/10.1093/mnras/sty456

Bertin G, Stiavelli M (1993) Structure and dynamics of elliptical galaxies. Rep Prog Phys 56(4):493–556. https://doi.org/10.1088/0034-4885/56/4/001

Biggio L, Vernardos G, Galan A, Peel A, Courbin F (2023) Modeling lens potentials with continuous neural fields in galaxy-scale strong lenses. Astron Astrophys 675:125. https://doi.org/10.1051/0004-6361/202245126

Birrer S (2021) Gravitational lensing formalism in a curved arc basis: a continuous description of observables and degeneracies from the weak to the strong lensing regime. Astrophys J 919(1):38. https://doi.org/10.3847/1538-4357/ac1108

Birrer S, Amara A (2018) Lenstronomy: multi-purpose gravitational lens modelling software package. Phys Dark Universe 22:189–201. https://doi.org/10.1016/j.dark.2018.11.002

Birrer S, Amara A, Refregier A (2015) Gravitational lens modeling with basis sets. Astrophys J 813:102. https://doi.org/10.1088/0004-637X/813/2/102

Birrer S, Amara A, Refregier A (2016) The mass-sheet degeneracy and time-delay cosmography: analysis of the strong lens RXJ1131-1231. J Cosmol Astropart Phys 8:020. https://doi.org/10.1088/1475-7516/2016/08/020

Birrer S, Treu T, Rusu CE, Bonvin V, Fassnacht CD, Chan JHH, Agnello A, Shajib AJ, Chen GC-F, Auger M, Courbin F, Hilbert S, Sluse D, Suyu SH, Wong KC, Marshall P, Lemaux BC, Meylan G (2019) H0LiCOW - IX. Cosmographic analysis of the doubly imaged quasar SDSS 1206+4332 and a new measurement of the Hubble constant. Mon Not R Astron Soc 484(4):4726–4753. https://doi.org/10.1093/mnras/stz200

Birrer S, Shajib AJ, Galan A, Millon M, Treu T, Agnello A, Auger M, Chen GC-F, Christensen L, Collett T, Courbin F, Fassnacht CD, Koopmans LVE, Marshall PJ, Park J-W, Rusu CE, Sluse D, Spiniello C, Suyu SH, Wagner-Carena S, Wong KC, Barnabè M, Bolton AS, Czoske O, Ding X, Frieman JA, Van de Vyvere L (2020) TDCOSMO. IV. Hierarchical time-delay cosmography – joint inference of the Hubble constant and galaxy density profiles. Astron Astrophys 643:165. https://doi.org/10.1051/0004-6361/202038861

Birrer S, Shajib A, Gilman D, Galan A, Aalbers J, Millon M, Morgan R, Pagano G, Park J, Teodori L, Tessore N, Ueland M, Van de Vyvere L, Wagner-Carena S, Wempe E, Yang L, Ding X, Schmidt T, Sluse D, Zhang M, Amara A (2021) Lenstronomy II: a gravitational lensing software ecosystem. J Open Sour Softw 6(62):3283. https://doi.org/10.21105/joss.03283

Birrer S, Millon M, Sluse D, Shajib AJ, Courbin F, Erickson S, Koopmans LVE, Suyu SH, Treu T (2024) Time-delay cosmography: measuring the Hubble constant and other cosmological parameters with strong gravitational lensing. Space Sci Rev 220(5):48. https://doi.org/10.1007/s11214-024-01079-w

Blumenthal GR, Faber SM, Flores R, Primack JR (1986) Contraction of dark matter galactic halos due to baryonic infall. Astrophys J 301:27. https://doi.org/10.1086/163867

Bolton AS, Burles S, Koopmans LVE, Treu T, Moustakas LA (2006) The Sloan Lens ACS Survey. I. A large spectroscopically selected sample of massive early-type lens galaxies. Astrophys J 638:703–724. https://doi.org/10.1086/498884

Bolton AS, Burles S, Koopmans LVE, Treu T, Gavazzi R, Moustakas LA, Wayth R, Schlegel DJ (2008b) The Sloan Lens ACS Survey. V. The full ACS strong-lens sample. Astrophys J 682(2):964–984. https://doi.org/10.1086/589327

Bolton AS, Treu T, Koopmans LVE, Gavazzi R, Moustakas LA, Burles S, Schlegel DJ, Wayth R (2008a) The Sloan Lens ACS Survey. VII. Elliptical galaxy scaling laws from direct observational mass measurements. Astrophys J 684(1):248–259. https://doi.org/10.1086/589989

Bonvin V, Tewes M, Courbin F, Kuntzer T, Sluse D, Meylan G (2016) COSMOGRAIL: the COSmological MOnitoring of GRAvItational lenses. XV. Assessing the achievability and precision of time-delay measurements. Astron Astrophys 585:88. https://doi.org/10.1051/0004-6361/201526704

Brehmer J, Mishra-Sharma S, Hermans J, Louppe G, Cranmer K (2019) Mining for dark matter substructure: inferring subhalo population properties from strong lenses with machine learning. Astrophys J 886(1):49. https://doi.org/10.3847/1538-4357/ab4c41

Browne IWA, Wilkinson PN, Jackson NJF, Myers ST, Fassnacht CD, Koopmans LVE, Marlow DR, Norbury M, Rusin D, Sykes CM, Biggs AD, Blandford RD, de Bruyn AG, Chae K-H, Helbig P, King LJ, McKean JP, Pearson TJ, Phillips PM, Readhead ACS, Xanthopoulos E, York T (2003) The cosmic lens all-sky survey - II. Gravitational lens candidate selection and follow-up. Mon Not R Astron Soc 341(1):13–32. https://doi.org/10.1046/j.1365-8711.2003.06257.x

Brownstein JR, Bolton AS, Schlegel DJ, Eisenstein DJ, Kochanek CS, Connolly N, Maraston C, Pandey P, Seitz S, Wake DA, Wood-Vasey WM, Brinkmann J, Schneider DP, Weaver BA (2012) The BOSS emission-line lens survey (BELLS). I. A large spectroscopically selected sample of lens galaxies at redshift ~0.5. Astrophys J 744(1):41. https://doi.org/10.1088/0004-637X/744/1/41

Bruderer C, Read JI, Coles JP, Leier D, Falco EE, Ferreras I, Saha P (2016) Light versus dark in strong-lens galaxies: dark matter haloes that are rounder than their stars. Mon Not R Astron Soc 456(1):870–884. https://doi.org/10.1093/mnras/stv2582

Cañameras R, Schuldt S, Shu Y, Suyu SH, Taubenberger S, Meinhardt T, Leal-Taixé L, Chao DC-Y, Inoue KT, Jaelani AT, More A (2021) HOLISMOKES. VI. New galaxy-scale strong lens candidates from the HSC-SSP imaging survey. Astron Astrophys 653:6. https://doi.org/10.1051/0004-6361/202141758

Cao S, Biesiada M, Gavazzi R, Piórkowska A, Zhu Z-H (2015) Cosmology with strong-lensing systems. Astrophys J 806(2):185. https://doi.org/10.1088/0004-637X/806/2/185

Cao X, Li R, Nightingale JW, Massey R, Robertson A, Frenk CS, Amvrosiadis A, Amorisco NC, He Q, Etherington A, Cole S, Zhu K (2022) Systematic errors induced by the elliptical power-law model in galaxy-galaxy strong lens modeling. Res Astron Astrophys 22(2):025014. https://doi.org/10.1088/1674-4527/ac3f2b

Cappellari M (2008) Measuring the inclination and mass-to-light ratio of axisymmetric galaxies via anisotropic jeans models of stellar kinematics. Mon Not R Astron Soc 390(1):71–86. https://doi.org/10.1111/j.1365-2966.2008.13754.x

Cappellari M (2016) Structure and kinematics of early-type galaxies from integral field spectroscopy. Annu Rev Astron Astrophys 54:597–665. https://doi.org/10.1146/annurev-astro-082214-122432

Cappellari M, McDermid RM, Alatalo K, Blitz L, Bois M, Bournaud F, Bureau M, Crocker AF, Davies RL, Davis TA, de Zeeuw PT, Duc P-A, Emsellem E, Khochfar S, Krajnović D, Kuntschner H, Lablanche P-Y, Morganti R, Naab T, Oosterloo T, Sarzi M, Scott N, Serra P, Weijmans A-M, Young LM (2012) Systematic variation of the stellar initial mass function in early-type galaxies. Nature 484(7395):485–488. https://doi.org/10.1038/nature10972

Cappellari M, McDermid RM, Alatalo K, Blitz L, Bois M, Bournaud F, Bureau M, Crocker AF, Davies RL, Davis TA, de Zeeuw PT, Duc P-A, Emsellem E, Khochfar S, Krajnović D, Kuntschner H, Morganti R, Naab T, Oosterloo T, Sarzi M, Scott N, Serra P, Weijmans A-M, Young LM (2013) The ATLAS^{3D} project - XX. Mass-size and mass-σ distributions of early-type galaxies: bulge fraction drives kinematics, mass-to-light ratio, molecular gas fraction and stellar initial mass function. Mon Not R Astron Soc 432(3):1862–1893. https://doi.org/10.1093/mnras/stt644

Cappellari M, Romanowsky AJ, Brodie JP, Forbes DA, Strader J, Foster C, Kartha SS, Pastorello N, Pota V, Spitler LR, Usher C, Arnold JA (2015) Small scatter and nearly isothermal mass profiles to four half-light radii from two-dimensional stellar dynamics of early-type galaxies. Astrophys J Lett 804(1):21. https://doi.org/10.1088/2041-8205/804/1/L21

Cautun M, Benítez-Llambay A, Deason AJ, Frenk CS, Fattahi A, Gómez FA, Grand RJJ, Oman KA, Navarro JF, Simpson CM (2020) The Milky Way total mass profile as inferred from Gaia DR2. Mon Not R Astron Soc 494(3):4291–4313. https://doi.org/10.1093/mnras/staa1017

Chabrier G (2003) Galactic stellar and substellar initial mass function. Publ Astron Soc Pac 115(809):763–795. https://doi.org/10.1086/376392

Chabrier G, Hennebelle P, Charlot S (2014) Variations of the stellar initial mass function in the progenitors of massive early-type galaxies and in extreme starburst environments. Astrophys J 796(2):75. https://doi.org/10.1088/0004-637X/796/2/75

Chae K-H (2010) Galaxy evolution from strong-lensing statistics: the differential evolution of the velocity dispersion function in Concord with the Λ cold dark matter paradigm. Mon Not R Astron Soc 402(3):2031–2048. https://doi.org/10.1111/j.1365-2966.2009.16073.x

Chan JHH, Suyu SH, Chiueh T, More A, Marshall PJ, Coupon J, Oguri M, Price P (2015) Chitah: strong-gravitational-lens hunter in imaging surveys. Astrophys J 807(2):138. https://doi.org/10.1088/0004-637X/807/2/138

Chang K, Refsdal S (1979) Flux variations of QSO 0957 + 561 A, B and image splitting by stars near the light path. Nature 282(5739):561–564. https://doi.org/10.1038/282561a0

Chen B, Dai X, Kochanek CS, Chartas G, Blackburne JA, Morgan CW (2012) X-ray monitoring of gravitational lenses with Chandra. Astrophys J 755(1):24. https://doi.org/10.1088/0004-637X/755/1/24

Chianese M, Coogan A, Hofma P, Otten S, Weniger C (2020) Differentiable strong lensing: uniting gravity and neural nets through differentiable probabilistic programming. Mon Not R Astron Soc 496(1):381–393. https://doi.org/10.1093/mnras/staa1477

Ciotti L, Lanzoni B, Renzini A (1996) The tilt of the fundamental plane of elliptical galaxies - I. Exploring dynamical and structural effects. Mon Not R Astron Soc 282(1):1–12. https://doi.org/10.1093/mnras/282.1.1

Claeskens J-F, Surdej J (2002) Gravitational lensing in quasar samples. Astron Astrophys Rev 10(4):263–311. https://doi.org/10.1007/s001590000010

Claeskens J-F, Sluse D, Riaud P, Surdej J (2006) Multi wavelength study of the gravitational lens system RXS J1131-1231. II. Lens model and source reconstruction. Astron Astrophys 451(3):865–879. https://doi.org/10.1051/0004-6361:20054352

Coles JP, Read JI, Saha P (2014) Gravitational lens recovery with GLASS: measuring the mass profile and shape of a lens. Mon Not R Astron Soc 445(3):2181–2197. https://doi.org/10.1093/mnras/stu1781

Collett TE (2015) The population of galaxy-galaxy strong lenses in forthcoming optical imaging surveys. Astrophys J 811:20. https://doi.org/10.1088/0004-637X/811/1/20

Collett TE, Auger MW (2014) Cosmological constraints from the double source plane lens SDSSJ0946+1006. Mon Not R Astron Soc 443(2):969–976. https://doi.org/10.1093/mnras/stu1190

Collett TE, Smith RJ (2020) A triple rollover: a third multiply imaged source at z = 6 behind the Jackpot gravitational lens. Mon Not R Astron Soc 497(2):1654–1660. https://doi.org/10.1093/mnras/staa1804

Collett TE, Auger MW, Belokurov V, Marshall PJ, Hall AC (2012) Constraining the dark energy equation of state with double-source plane strong lenses. Mon Not R Astron Soc 424(4):2864–2875. https://doi.org/10.1111/j.1365-2966.2012.21424.x

Collier WP, Smith RJ, Lucey JR (2018) A new strong-lensing galaxy at z= 0.066: another elliptical galaxy with a lightweight IMF. Mon Not R Astron Soc 478(2):1595–1600. https://doi.org/10.1093/mnras/sty1188

Conroy C, van Dokkum PG (2012) The stellar initial mass function in early-type galaxies from absorption line spectroscopy. II. Results. Astrophys J 760(1):71. https://doi.org/10.1088/0004-637X/760/1/71

Coogan A, Karchev K, Weniger C (2020) Targeted Likelihood-Free Inference of Dark Matter Substructure in Strongly-Lensed Galaxies. ArXiv e-prints, arXiv:2010.07032

Coogan A, Anau Montel N, Karchev K, Grootes MW, Nattino F, Weniger C (2022) One never walks alone: the effect of the perturber population on subhalo measurements in strong gravitational lenses. ArXiv e-prints, arXiv:2209.09918

Courbin F, Faure C, Djorgovski SG, Rérat F, Tewes M, Meylan G, Stern D, Mahabal A, Boroson T, Dheeraj R, Sluse D (2012) Three quasi-stellar objects acting as strong gravitational lenses. Astron Astrophys 540:36. https://doi.org/10.1051/0004-6361/201118015

Courteau S, Cappellari M, de Jong RS, Dutton AA, Emsellem E, Hoekstra H, Koopmans LVE, Mamon GA, Maraston C, Treu T, Widrow LM (2014) Galaxy masses. Rev Mod Phys 86(1):47–119. https://doi.org/10.1103/RevModPhys.86.47

Crain RA, Schaye J, Bower RG, Furlong M, Schaller M, Theuns T, Dalla Vecchia C, Frenk CS, McCarthy IG, Helly JC, Jenkins A, Rosas-Guevara YM, White SDM, Trayford JW (2015) The EAGLE simulations of galaxy formation: calibration of subgrid physics and model variations. Mon Not R Astron Soc 450(2):1937–1961. https://doi.org/10.1093/mnras/stv725

Czoske O, Barnabè M, Koopmans LVE, Treu T, Bolton AS (2012) Two-dimensional kinematics of SLACS lenses - IV. The complete VLT-VIMOS data set. Mon Not R Astron Soc 419(1):656–668. https://doi.org/10.1111/j.1365-2966.2011.19726.x

Damjanov I, Abraham RG, Glazebrook K, McCarthy PJ, Caris E, Carlberg RG, Chen H-W, Crampton D, Green AW, Jørgensen I, Juneau S, Le Borgne D, Marzke RO, Mentuch E, Murowinski R, Roth K, Savaglio S, Yan H (2011) Red nuggets at high redshift: structural evolution of quiescent galaxies over 10 gyr of cosmic history. Astrophys J Lett 739:44. https://doi.org/10.1088/2041-8205/739/2/L44

Debattista VP, van den Bosch FC, Roškar R, Quinn T, Moore B, Cole DR (2015) Internal alignments of red versus blue discs in dark matter haloes. Mon Not R Astron Soc 452(4):4094–4110. https://doi.org/10.1093/mnras/stv1563

Delchambre L, Krone-Martins A, Wertz O, Ducourant C, Galluccio L, Klüter J, Mignard F, Teixeira R, Djorgovski SG, Stern DP, Graham MJ, Bastian U, Surdej J, Wambsganss J, Le Campion J-F, Slezak E (2018) Gaia gral iii - gaia dr2 gravitational lens systems: a systematic blind search for new lensed systems. ArXiv e-prints, http://adsabs.harvard.edu/abs/2018arXiv180702845D

Denzel P, Coles JP, Saha P, Williams LLR (2021b) The Hubble constant from eight time-delay galaxy lenses. Mon Not R Astron Soc 501(1):784–801. https://doi.org/10.1093/mnras/staa3603

Denzel P, Mukherjee S, Saha P (2021a) A new strategy for matching observed and simulated lensing galaxies. Mon Not R Astron Soc 506(2):1815–1831. https://doi.org/10.1093/mnras/stab1716

Derkenne C, McDermid RM, Poci A, Remus R-S, Jørgensen I, Emsellem E (2021) Total mass density slopes of early-type galaxies using jeans dynamical modelling at redshifts 0.29 < z < 0.55. Mon Not R Astron Soc 506(3):3691–3716. https://doi.org/10.1093/mnras/stab1996

Derkenne C, McDermid RM, Poci A, Mendel JT, D'Eugenio F, Jeon S, Remus R-S, Bellstedt S, Battisti AJ, Bland-Hawthorn J, Ferré-Mateu A, Foster C, Harborne KE, Lagos CDP, Peng Y, Sharda P, Sharma G, Sweet S, Tran K-VH, Valenzuela LM, Vaughan S, Wisnioski E, Yi SK (2023) The MAGPI survey: impact of environment on the total internal mass distribution of galaxies in the last 5 Gyr. Mon Not R Astron Soc 522(3):3602–3626. https://doi.org/10.1093/mnras/stad1079

DES Collaboration (2024) tHe Dark Energy Survey: Cosmology Results with ~1500 New High-redshift Type Ia Supernovae Using the Full 5-year Dataset. ArXiv e-prints, arXiv:2401.02929. https://doi.org/10.48550/arXiv.2401.02929

Diaz Rivero A, Dvorkin C (2020) Direct detection of dark matter substructure in strong lens images with convolutional neural networks. Phys Rev D 101(2):023515. https://doi.org/10.1103/PhysRevD.101.023515

Ding X, Treu T, Birrer S, Chen GC-F, Coles J, Denzel P, Frigo M, Galan A, Marshall PJ, Millon M, More A, Shajib AJ, Sluse D, Tak H, Xu D, Auger MW, Bonvin V, Chand H, Courbin F, Despali G, Fassnacht CD, Gilman D, Hilbert S, Kumar SR, Lin JY-Y, Park JW, Saha P, Vegetti S, Van de Vyvere L, Williams LLR (2021a) Time delay lens modelling challenge. Mon Not R Astron Soc 503(1):1096–1123. https://doi.org/10.1093/mnras/stab484

Duffy AR, Schaye J, Kay ST, Dalla Vecchia C, Battye RA, Booth CM (2010) Impact of baryon physics on dark matter structures: a detailed simulation study of halo density profiles. Mon Not R Astron Soc 405(4):2161–2178. https://doi.org/10.1111/j.1365-2966.2010.16613.x

Dutton AA, Treu T (2014) The bulge-halo conspiracy in massive elliptical galaxies: implications for the stellar initial mass function and halo response to baryonic processes. Mon Not R Astron Soc 438:3594–3602. https://doi.org/10.1093/mnras/stt2489

Dutton AA, van den Bosch FC, Dekel A, Courteau S (2007) A revised model for the formation of disk galaxies: low spin and dark halo expansion. Astrophys J 654(1):27–52. https://doi.org/10.1086/509314

Dutton AA, Brewer BJ, Marshall PJ, Auger MW, Treu T, Koo DC, Bolton AS, Holden BP, Koopmans LVE (2011) The SWELLS survey - II. Breaking the disc-halo degeneracy in the spiral galaxy gravitational lens SDSS J2141-0001. Mon Not R Astron Soc 417(3):1621–1642. https://doi.org/10.1111/j.1365-2966.2011.18706.x

Dye S, Warren SJ (2005) Decomposition of the visible and dark matter in the Einstein ring 0047-2808 by semilinear inversion. Astrophys J 623(1):31–41. https://doi.org/10.1086/428340

Dye S, Negrello M, Hopwood R, Nightingale JW, Bussmann RS, Amber S, Bourne N, Cooray A, Dariush A, Dunne L, Eales SA, Gonzalez-Nuevo J, Ibar E, Ivison RJ, Maddox S, Valiante E, Smith M (2014) Herschel-ATLAS: modelling the first strong gravitational lenses. Mon Not R Astron Soc 440(3):2013–2025. https://doi.org/10.1093/mnras/stu305

Dye S, Furlanetto C, Dunne L, Eales SA, Negrello M, Nayyeri H, van der Werf PP, Serjeant S, Farrah D, Michałowski MJ, Baes M, Marchetti L, Cooray A, Riechers DA, Amvrosiadis A (2018) Modelling high-resolution ALMA observations of strongly lensed highly star-forming galaxies detected by Herschel. Mon Not R Astron Soc 476(4):4383–4394. https://doi.org/10.1093/mnras/sty513

Eigenbrod A, Courbin F, Vuissoz C, Meylan G, Saha P, Dye S (2005) COSMOGRAIL: the COSmological MOnitoring of GRAvItational lenses. I. How to sample the light curves of gravitationally lensed quasars to measure accurate time delays. Astron Astrophys 436(1):25–35. https://doi.org/10.1051/0004-6361:20042422

Eisenhardt PR, Armus L, Hogg DW, Soifer BT, Neugebauer G, Werner MW (1996) Hubble Space Telescope observations of the luminous IRAS source FSC 10214+4724: a gravitationally lensed infrared quasar. Astrophys J 461:72. https://doi.org/10.1086/177038

Emsellem E, Monnet G, Bacon R (1994) The multi-Gaussian expansion method: a tool for building realistic photometric and kinematical models of stellar systems I. The formalism. Astron Astrophys 285:723–738

Enzi W, Vegetti S, Despali G, Hsueh J-W, Metcalf RB (2020) Systematic errors in strong gravitational lensing reconstructions, a numerical simulation perspective. Mon Not R Astron Soc 496(2):1718–1729. https://doi.org/10.1093/mnras/staa1224

Etherington A, Nightingale JW, Massey R, Cao X, Robertson A, Amorisco NC, Amvrosiadis A, Cole S, Frenk CS, He Q, Li R, Tam S-I (2022) Automated galaxy-galaxy strong lens modelling: no lens left behind. Mon Not R Astron Soc 517(3):3275–3302. https://doi.org/10.1093/mnras/stac2639

Etherington A, Nightingale JW, Massey R, Robertson A, Cao X, Amvrosiadis A, Cole S, Frenk CS, He Q, Lagattuta DJ, Lange S, Li R (2023) Beyond the bulge-halo conspiracy? Density profiles of early-type galaxies from extended-source strong lensing. Mon Not R Astron Soc 521(4):6005–6018. https://doi.org/10.1093/mnras/stad582

Etherington A, Nightingale JW, Massey R, Tam S-I, Cao X, Niemiec A, He Q, Robertson A, Li R, Amvrosiadis A, Cole S, Diego JM, Frenk CS, Frye BL, Harvey D, Jauzac M, Koekemoer AM, Lagattuta DJ, Lange S, Limousin M, Mahler G, Sirks E, Steinhardt CL (2024) Strong gravitational lensing's 'external shear' is not shear. Mon Not R Astron Soc 531(3):3684–3697. https://doi.org/10.1093/mnras/stae1375

Falco EE, Gorenstein MV, Shapiro II (1985) On model-dependent bounds on h(0) from gravitational images application of q0957 + 561a, b. Astrophys J Lett 289:1–4. https://doi.org/10.1086/184422

Ferré-Mateu A, Trujillo I, Martín-Navarro I, Vazdekis A, Mezcua M, Balcells M, Domínguez L (2017) Two new confirmed massive relic galaxies: red nuggets in the present-day universe. Mon Not R Astron Soc 467:1929–1939. https://doi.org/10.1093/mnras/stx171

Ferreras I, Saha P, Burles S (2008) Unveiling dark haloes in lensing galaxies. Mon Not R Astron Soc 383(3):857–863. https://doi.org/10.1111/j.1365-2966.2007.12606.x

Ferreras I, Saha P, Leier D, Courbin F, Falco EE (2010) Constraining the low-mass end of the initial mass function with gravitational lensing. Mon Not R Astron Soc 409(1):30–34. https://doi.org/10.1111/j.1745-3933.2010.00941.x

Frigo M, Naab T, Hirschmann M, Choi E, Somerville RS, Krajnovic D, Davé R, Cappellari M (2019) The impact of AGN on stellar kinematics and orbits in simulated massive galaxies. Mon Not R Astron Soc 489(2):2702–2722. https://doi.org/10.1093/mnras/stz2318

Fukugita M, Futamase T, Kasai M (1990) A possible test for the cosmological constant with gravitational lenses. Mon Not R Astron Soc 246:24

Galan A, Peel A, Joseph R, Courbin F, Starck J-L (2021) SLITRONOMY: towards a fully wavelet-based strong lensing inversion technique. Astron Astrophys 647:176. https://doi.org/10.1051/0004-6361/202039363

Galan A, Vernardos G, Peel A, Courbin F, Starck J-L (2022) Using wavelets to capture deviations from smoothness in galaxy-scale strong lenses. Astron Astrophys 668:A155. https://doi.org/10.1051/0004-6361/202244464

Garrett MA, Calder RJ, Porcas RW, King LJ, Walsh D, Wilkinson PN (1994) Global VLBI observations of the gravitational lens system 0957+561 A, B. Mon Not R Astron Soc 270:457–464. https://doi.org/10.1093/mnras/270.2.457

Gavazzi R, Treu T, Rhodes JD, Koopmans LVE, Bolton AS, Burles S, Massey RJ, Moustakas LA (2007) The Sloan Lens ACS Survey. IV. The mass density profile of early-type galaxies out to 100 effective radii. Astrophys J 667:176–190. https://doi.org/10.1086/519237

Gavazzi R, Treu T, Koopmans LVE, Bolton AS, Moustakas LA, Burles S, Marshall PJ (2008) The Sloan Lens ACS Survey. VI. Discovery and analysis of a double Einstein ring. Astrophys J 677(2):1046–1059. https://doi.org/10.1086/529541

Gavazzi R, Treu T, Marshall PJ, Brault F, Ruff A (2012) The SL2S galaxy-scale gravitational lens sample. I. The alignment of mass and light in massive early-type galaxies at z = 0.2-0.9. Astrophys J 761(2):170. https://doi.org/10.1088/0004-637X/761/2/170

Gerhard O, Kronawitter A, Saglia RP, Bender R (2001) Dynamical family properties and dark halo scaling relations of giant elliptical galaxies. Astron J 121(4):1936–1951. https://doi.org/10.1086/319940

Gilman D, Agnello A, Treu T, Keeton CR, Nierenberg AM (2017) Strong lensing signatures of luminous structure and substructure in early-type galaxies. Mon Not R Astron Soc 467(4):3970–3992. https://doi.org/10.1093/mnras/stx158

Gilman D, Birrer S, Nierenberg A, Treu T, Du X, Benson A (2020) Warm dark matter chills out: constraints on the halo mass function and the free-streaming length of dark matter with eight quadruple-image strong gravitational lenses. Mon Not R Astron Soc 491(4):6077–6101. https://doi.org/10.1093/mnras/stz3480

Gnedin OY, Kravtsov AV, Klypin AA, Nagai D (2004) Response of dark matter halos to condensation of baryons: cosmological simulations and improved adiabatic contraction model. Astrophys J 616(1):16–26. https://doi.org/10.1086/424914

Golse G, Kneib J-P (2002) Pseudo elliptical lensing mass model: application to the NFW mass distribution. Astron Astrophys 390:821–827. https://doi.org/10.1051/0004-6361:20020639

Gomer MR, Williams LLR (2018) The impact of ΛCDM substructure and baryon-dark matter transition on the image positions of quad galaxy lenses. Mon Not R Astron Soc 475(2):1987–2002. https://doi.org/10.1093/mnras/stx3294

Gomer MR, Williams LLR (2021) Galaxy-lens determination of H_0: the effect of the ellipse + shear modelling assumption. Mon Not R Astron Soc 504(1):1340–1354. https://doi.org/10.1093/mnras/stab930

Gomer MR, Sluse D, Van de Vyvere L, Birrer S, Courbin F (2022) TDCOSMO. VIII. A key test of systematics in the hierarchical method of time-delay cosmography. Astron Astrophys 667:86. https://doi.org/10.1051/0004-6361/202244324

Gomer MR, Sluse D, Van de Vyvere L, Birrer S, Shajib AJ, Courbin F (2023) The ellipticity parameterization for an NFW profile: an overlooked angular structure in strong lens modeling. Astron Astrophys 679:128. https://doi.org/10.1051/0004-6361/202346308

Goobar A, Amanullah R, Kulkarni SR, Nugent PE, Johansson J, Steidel C, Law D, Mörtsell E, Quimby R, Blagorodnova N, Brandeker A, Cao Y, Cooray A, Ferretti R, Fremling C, Hangard L, Kasliwal M, Kupfer T, Lunnan R, Masci F, Miller AA, Nayyeri H, Neill JD, Ofek EO, Papadogiannakis S, Petrushevska T, Ravi V, Sollerman J, Sullivan M, Taddia F, Walters R, Wilson D, Yan L, Yaron O (2017) iPTF16geu: a multiply imaged, gravitationally lensed type Ia supernova. Science 356(6335):291–295. https://doi.org/10.1126/science.aal2729

Gorenstein MV, Cohen NL, Shapiro II, Rogers AEE, Bonometti RJ, Falco EE, Bartel N, Marcaide JM (1988) VLBI observations of the gravitational lens system 0957+561: structure and relative magnification of the A and B images. Astrophys J 334:42. https://doi.org/10.1086/166816

Grillo C, Lombardi M, Bertin G (2008) Cosmological parameters from strong gravitational lensing and stellar dynamics in elliptical galaxies. Astron Astrophys 477:397–406. https://doi.org/10.1051/0004-6361:20077534

Grillo C, Gobat R, Lombardi M, Rosati P (2009) Photometric mass and mass decomposition in early-type lens galaxies. Astron Astrophys 501(2):461–474. https://doi.org/10.1051/0004-6361/200811604

Guerras E, Dai X, Steele S, Liu A, Kochanek CS, Chartas G, Morgan CW, Chen B (2017) Extended X-ray monitoring of gravitational lenses with Chandra and joint constraints on X-ray emission regions. Astrophys J 836(2):206. https://doi.org/10.3847/1538-4357/aa5728

Hao CN, Mao S, Deng ZG, Xia XY, Wu H (2006) Isophotal shapes of elliptical/lenticular galaxies from the Sloan Digital Sky Survey. Mon Not R Astron Soc 370(3):1339–1350. https://doi.org/10.1111/j.1365-2966.2006.10545.x

He Q, Nightingale J, Robertson A, Amvrosiadis A, Cole S, Frenk CS, Massey R, Li R, Amorisco NC, Metcalf RB, Cao X, Etherington A (2023) Testing strong lensing subhalo detection with a cosmological simulation. Mon Not R Astron Soc 518(1):220–239. https://doi.org/10.1093/mnras/stac2779

Heiligman G, Schwarzschild M (1979) On the nonexistence of three-dimensional tube orbits around the intermediate axis in a triaxial galaxy model. Astrophys J 233:872–876. https://doi.org/10.1086/157449

Hernquist L (1990) An analytical model for spherical galaxies and bulges. Astrophys J 356:359. https://doi.org/10.1086/168845

Heyrovský D, Karamazov M (2024) Gravitational lensing by an ellipsoidal Navarro–Frenk–White dark-matter halo: an analytic solution and its properties. ArXiv e-prints, arXiv:2404.00169. https://doi.org/10.48550/arXiv.2404.00169

Hezaveh YD, Marshall PJ, Blandford RD (2015) Probing the inner kiloparsec of massive galaxies with strong gravitational lensing. Astrophys J Lett 799(2):22. https://doi.org/10.1088/2041-8205/799/2/L22

Hezaveh YD, Levasseur LP, Marshall PJ (2017) Fast automated analysis of strong gravitational lenses with convolutional neural networks. Nature 548:555–557. https://doi.org/10.1038/nature23463

Hilbert S, White SDM, Hartlap J, Schneider P (2007) Strong lensing optical depths in a ΛCDM universe. Mon Not R Astron Soc 382(1):121–132. https://doi.org/10.1111/j.1365-2966.2007.12391.x

Hinton SR (2016) ChainConsumer. J Open Sour Softw 1:00045. https://doi.org/10.21105/joss.00045

Hopkins PF (2013) Variations in the stellar cmf and imf: from bottom to top. Mon Not R Astron Soc 433(1):170–177. https://doi.org/10.1093/mnras/stt713

Hopkins PF, Kereš D, Oñorbe J, Faucher-Giguère C-A, Quataert E, Murray N, Bullock JS (2014) Galaxies on FIRE (feedback in realistic environments): stellar feedback explains cosmologically inefficient star formation. Mon Not R Astron Soc 445(1):581–603. https://doi.org/10.1093/mnras/stu1738

Hsueh J-W, Fassnacht CD, Vegetti S, McKean JP, Spingola C, Auger MW, Koopmans LVE, Lagattuta DJ (2016) SHARP - II. Mass structure in strong lenses is not necessarily dark matter substructure: a flux ratio anomaly from an edge-on disc in B1555+375. Mon Not R Astron Soc 463(1):51–55. https://doi.org/10.1093/mnrasl/slw146

Hsueh J-W, Oldham L, Spingola C, Vegetti S, Fassnacht CD, Auger MW, Koopmans LVE, McKean JP, Lagattuta DJ (2017) SHARP - IV. An apparent flux-ratio anomaly resolved by the edge-on disc in B0712+472. Mon Not R Astron Soc 469(3):3713–3721. https://doi.org/10.1093/mnras/stx1082

Huchra J, Gorenstein M, Kent S, Shapiro I, Smith G, Horine E, Perley R (1985) 2237+0305: a new and unusual gravitational lens. Astron J 90:691–696. https://doi.org/10.1086/113777

Humphrey PJ, Buote DA (2010) The slope of the mass profile and the tilt of the fundamental plane in early-type galaxies. Mon Not R Astron Soc 403(4):2143–2151. https://doi.org/10.1111/j.1365-2966.2010.16257.x

Hunter JD (2007) Matplotlib: a 2d graphics environment. Comput Sci Eng 9(3):90–95. https://doi.org/10.1109/MCSE.2007.55

Jackson N, Rampadarath H, Ofek EO, Oguri M, Shin M-S (2012) New lensed quasars from the MUSCLES survey. Mon Not R Astron Soc 419(3):2014–2024. https://doi.org/10.1111/j.1365-2966.2011.19857.x

Jacobs C, Collett T, Glazebrook K, Buckley-Geer E, Diehl HT, Lin H, McCarthy C, Qin AK, Odden C, Caso Escudero M, Dial P, Yung VJ, Gaitsch S, Pellico A, Lindgren KA, Abbott TMC, Annis J, Avila S, Brooks D, Burke DL, Carnero Rosell A, Carrasco Kind M, Carretero J, da Costa LN, De Vicente J, Fosalba P, Frieman J, García-Bellido J, Gaztanaga E, Goldstein DA, Gruen D, Gruendl RA, Gschwend J, Hollowood DL, Honscheid K, Hoyle B, James DJ, Krause E, Kuropatkin N, Lahav O, Lima M, Maia MAG, Marshall JL, Miquel R, Plazas AA, Roodman A, Sanchez E, Scarpine V, Serrano S, Sevilla-Noarbe I, Smith M, Sobreira F, Suchyta E, Swanson MEC, Tarle G, Vikram V, Walker AR, Zhang Y, DES Collaboration (2019a) An extended catalog of galaxy-galaxy strong gravitational lenses discovered in DES using convolutional neural networks. Astrophys J Suppl Ser 243(1):17. https://doi.org/10.3847/1538-4365/ab26b6

Jacobs C, Collett T, Glazebrook K, McCarthy C, Qin AK, Abbott TMC, Abdalla FB, Annis J, Avila S, Bechtol K, Bertin E, Brooks D, Buckley-Geer E, Burke DL, Carnero Rosell A, Carrasco Kind M, Carretero J, da Costa LN, Davis C, De Vicente J, Desai S, Diehl HT, Doel P, Eifler TF, Flaugher B, Frieman J, García-Bellido J, Gaztanaga E, Gerdes DW, Goldstein DA, Gruen D, Gruendl RA, Gschwend J, Gutierrez G, Hartley WG, Hollowood DL, Honscheid K, Hoyle B, James DJ, Kuehn K, Kuropatkin N, Lahav O, Li TS, Lima M, Lin H, Maia MAG, Martini P, Miller CJ, Miquel R, Nord B, Plazas AA, Sanchez E, Scarpine V, Schubnell M, Serrano S, Sevilla-Noarbe I, Smith M, Soares-Santos M, Sobreira F, Suchyta E, Swanson MEC, Tarle G, Vikram V, Walker AR, Zhang Y, Zuntz J, DES Collaboration (2019b) Finding high-redshift strong lenses in DES using convolutional neural networks. Mon Not R Astron Soc 484(4):5330–5349. https://doi.org/10.1093/mnras/stz272

Johansson PH, Naab T, Burkert A (2009) Equal- and unequal-mass mergers of disk and elliptical galaxies with black holes. Astrophys J 690(1):802–821. https://doi.org/10.1088/0004-637X/690/1/802

Jones E, Oliphant T, Peterson P (2001) Others, SciPy: open source scientific tools for Python. http://www.scipy.org/

Karchev K, Coogan A, Weniger C (2022) Strong-lensing source reconstruction with variationally optimized Gaussian processes. Mon Not R Astron Soc 512(1):661–685. https://doi.org/10.1093/mnras/stac311

Kassiola A, Kovner I (1993) Elliptic mass distributions versus elliptic potentials in gravitational lenses. Astrophys J 417:450. https://doi.org/10.1086/173325

Keeton CR (2001a) A Catalog of Mass Models for Gravitational Lensing. ArXiv e-prints, arXiv:astro-ph/0102341. https://doi.org/10.48550/arXiv.astro-ph/0102341

Keeton CR (2001b) Computational Methods for Gravitational Lensing. ArXiv e-prints, arXiv:astro-ph/0102340

Keeton CR, Kochanek CS (1998) Gravitational lensing by spiral galaxies. Astrophys J 495(1):157–169. https://doi.org/10.1086/305272

Keeton CR, Kochanek CS, Seljak U (1997) Shear and ellipticity in gravitational lenses. Astrophys J 482(2):604–620. https://doi.org/10.1086/304172

Keeton CR, Kochanek CS, Falco EE (1998) The optical properties of gravitational lens galaxies as a probe of galaxy structure and evolution. Astrophys J 509(2):561–578. https://doi.org/10.1086/306502

Keeton CR, Gaudi BS, Petters AO (2003) Identifying lenses with small-scale structure. I. Cusp lenses. Astrophys J 598(1):138–161. https://doi.org/10.1086/378934

Keeton CR, Gaudi BS, Petters AO (2005) Identifying lenses with small-scale structure. II. Fold lenses. Astrophys J 635(1):35–59. https://doi.org/10.1086/497324

King LJ, Browne IWA, Wilkinson PN (1996) Observations and predictions of the ratio of 3-image to 5-image systems in JVAS. In: Kochanek CS, Hewitt JN (eds) Astrophysical applications of gravitational lensing, IAU Symposium, vol 173. Springer, Dordrecht, pp 191–194. https://doi.org/10.1007/978-94-009-0221-3_50

Kochanek CS (1995) Evidence for dark matter in MG 1654+134. Astrophys J 445:559. https://doi.org/10.1086/175721

Kochanek CS (2002) What do gravitational lens time delays measure? Astrophys J 578(1):25–32. https://doi.org/10.1086/342476

Kochanek CS (2021) Overconstrained models of time delay lenses redux: how the angular tail wags the radial dog. Mon Not R Astron Soc 501(4):5021–5028. https://doi.org/10.1093/mnras/staa4033

Kochanek CS (2022) Mass Follows Light. In: The Shapes of Galaxies and Their Dark Halos, pp 62–71. https://doi.org/10.1142/9789812778017_0010.

Koopmans LVE (2005) Gravitational imaging of cold dark matter substructures. Mon Not R Astron Soc 363(4):1136–1144. https://doi.org/10.1111/j.1365-2966.2005.09523.x

Koopmans LVE (2006) Gravitational lensing & stellar dynamics. In: Mamon GA, Combes F, Deffayet C, Fort B (eds) Mass profiles and shapes of cosmological structures. EAS Publications Series, vol 20. pp 161–166. https://doi.org/10.1051/eas:2006064

Koopmans LVE, Treu T (2003) The structure and dynamics of luminous and dark matter in the early-type lens galaxy of 0047-281 at z = 0.485. Astrophys J 583:606–615. https://doi.org/10.1086/345423

Koopmans LVE, Treu T, Bolton AS, Burles S, Moustakas LA (2006) The Sloan Lens ACS Survey. III. The structure and formation of early-type galaxies and their evolution since z ~1. Astrophys J 649(2):599–615. https://doi.org/10.1086/505696

Koopmans LVE, Bolton A, Treu T, Czoske O, Auger MW, Barnabè M, Vegetti S, Gavazzi R, Moustakas LA, Burles S (2009) The structure and dynamics of massive early-type galaxies: on homology, isothermality, and isotropy inside one effective radius. Astrophys J Lett 703(1):51–54. https://doi.org/10.1088/0004-637X/703/1/L51

Kormann R, Schneider P, Bartelmann M (1994) Isothermal elliptical gravitational lens models. Astron Astrophys 284:285–299

Kostrzewa-Rutkowska Z, Wyrzykowski Ł, Auger MW, Collett TE, Belokurov V (2014) The evolution of late-type galaxies from CASSOWARY lensing systems. Mon Not R Astron Soc 441(4):3238–3248. https://doi.org/10.1093/mnras/stu783

Kovner I (1987) Marginal gravitational lenses of large separation: probing superclusters. Nature 325(6104):507–509. https://doi.org/10.1038/325507a0

Krajnović D, Ural U, Kuntschner H, Goudfrooij P, Wolfe M, Cappellari M, Davies R, de Zeeuw TP, Duc P-A, Emsellem E, Karick A, McDermid RM, Mei S, Naab T (2020) Formation channels of slowly rotating early-type galaxies. Astron Astrophys 635:129. https://doi.org/10.1051/0004-6361/201937040

Krone-Martins A, Delchambre L, Wertz O, Ducourant C, Mignard F, Teixeira R, Klüter J, Le Campion J-F, Galluccio L, Surdej J, Bastian U, Wambsganss J, Graham MJ, Djorgovski SG, Slezak E (2018) Gaia GraL: Gaia DR2 gravitational lens systems. I. New quadruply imaged quasar candidates around known quasars. Astron Astrophys 616:11. https://doi.org/10.1051/0004-6361/201833337

Küng R, Saha P, Ferreras I, Baeten E, Coles J, Cornen C, Macmillan C, Marshall P, More A, Oswald L, Verma A, Wilcox JK (2018) Models of gravitational lens candidates from space warps CFHTLS. Mon Not R Astron Soc 474(3):3700–3713. https://doi.org/10.1093/mnras/stx3012

La Barbera F, Ferreras I, Vazdekis A, de la Rosa IG, de Carvalho RR, Trevisan M, Falcón-Barroso J, Ricciardelli E (2013) SPIDER VIII - constraints on the stellar initial mass function of early-type galaxies from a variety of spectral features. Mon Not R Astron Soc 433(4):3017–3047. https://doi.org/10.1093/mnras/stt943

Lawrence CR, Bennett CL, Hewitt JN, Langston GI, Klotz SE, Burke BF, Turner KC (1986) 5 GHz radio structure and optical identifications of sources from the MG survey. II. Maps and finding charts. Astrophys J Suppl Ser 61:105. https://doi.org/10.1086/191109

Lehar J, Broadhurst T (1996) Is IRAS F102144724 gravitationally lensed? In: Kochanek CS, Hewitt JN (eds) Astrophysical applications of gravitational lensingl. IAU Symposium, vol 173. Springer, Dordrecht, pp 247–252 https://doi.org/10.1007/978-94-009-0221-3_65

Leier D, Ferreras I, Saha P, Charlot S, Bruzual G, La Barbera F (2016) Strong gravitational lensing and the stellar IMF of early-type galaxies. Mon Not R Astron Soc 459(4):3677–3692. https://doi.org/10.1093/mnras/stw885

Lemon CA, Auger MW, McMahon RG, Ostrovski F (2018) Gravitationally lensed quasars in Gaia - II. Discovery of 24 lensed quasars. Mon Not R Astron Soc 479(4):5060–5074. https://doi.org/10.1093/mnras/sty911

Lemon C, Auger MW, McMahon R, Anguita T, Apostolovski Y, Chen GC-F, Fassnacht CD, Melo AD, Motta V, Shajib A, Treu T, Agnello A, Buckley-Geer E, Schechter PL, Birrer S, Collett T, Courbin F, Rusu CE, Abbott TMC, Allam S, Annis J, Avila S, Bertin E, Brooks D, Burke DL, Carnero Rosell A, Carrasco Kind M, Carretero J, Costanzi M, da Costa LN, De Vicente J, Desai S, Eifler TF, Flaugher B, Frieman J, García-Bellido J, Gaztanaga E, Gerdes DW, Gruen D, Gruendl RA, Gschwend J, Gutierrez G, Honscheid K, James DJ, Kim A, Krause E, Kuehn K, Kuropatkin N, Lahav O, Lima M, Lin H, Maia MAG, March M, Marshall JL, Menanteau F, Miquel R, Palmese A, Paz-Chinchón F, Plazas AA, Roodman A, Sanchez E, Schubnell M, Serrano S, Smith M, Soares-Santos M, Suchyta E, Tarle G, Walker AR (2020) The STRong lensing insights into the dark energy survey (STRIDES) 2017/2018 follow-up campaign: discovery of 10 lensed quasars and 10 quasar pairs. Mon Not R Astron Soc 494(3):3491–3511. https://doi.org/10.1093/mnras/staa652

Lemon C, Anguita T, Auger M, Courbin F, Galan A, McMahon R, Neira F, Oguri M, Schechter P, Shajib A, Treu T (2022) Gravitationally lensed quasars in Gaia – IV. 150 new lenses, quasar pairs, and projected quasars. ArXiv e-prints, arXiv:2206.07714

Lemon C, Courbin F, More A, Schechter P, Cañameras R, Delchambre L, Leung C, Shu Y, Spiniello C, Hezaveh Y, Klüter J, McMahon R (2024) Searching for strong gravitational lenses. Space Sci Rev 220(2):23. https://doi.org/10.1007/s11214-024-01042-9

Lewis GF, Ibata RA, Ellison SL, Aracil B, Petitjean P, Pettini M, Srianand R (2002) Spatially resolved STIS spectra of the gravitationally lensed broad absorption line quasar APM08279+5255: the nature of component C and evidence for microlensing. Mon Not R Astron Soc 334(1):7–10. https://doi.org/10.1046/j.1365-8711.2002.05700.x

Li N, Mao S, Gao L, Loeb A, di Stefano R (2012) Effects of supermassive binary black holes on gravitational lenses. Mon Not R Astron Soc 419(3):2424–2432. https://doi.org/10.1111/j.1365-2966.2011.19893.x

Li R, Napolitano NR, Spiniello C, Tortora C, Kuijken K, Koopmans LVE, Schneider P, Getman F, Xie L, Long L, Shu W, Vernardos G, Huang Z, Covone G, Dvornik A, Heymans C, Hildebrandt H, Radovich M, Wright AH (2021) High-quality strong lens candidates in the final Kilo-Degree Survey footprint. Astrophys J 923(1):16. https://doi.org/10.3847/1538-4357/ac2df0

Li T, Collett TE, Krawczyk CM, Enzi W (2024) Cosmology from large populations of galaxy-galaxy strong gravitational lenses. Mon Not R Astron Soc 527(3):5311–5323. https://doi.org/10.1093/mnras/stad3514

Maller AH, Simard L, Guhathakurta P, Hjorth J, Jaunsen AO, Flores RA, Primack JR (2000) Breaking the disk/halo degeneracy with gravitational lensing. Astrophys J 533(1):194–202. https://doi.org/10.1086/308641

Mamon GA, Łokas EL (2005) Dark matter in elliptical galaxies - II. Estimating the mass within the virial radius. Mon Not R Astron Soc 363:705–722. https://doi.org/10.1111/j.1365-2966.2005.09400.x

Mandelbaum R, van de Ven G, Keeton CR (2009) Galaxy density profiles and shapes - II. Selection biases in strong lensing surveys. Mon Not R Astron Soc 398(2):635–657. https://doi.org/10.1111/j.1365-2966.2009.15166.x

Mao S, Schneider P (1998) Evidence for substructure in lens galaxies? Mon Not R Astron Soc 295(3):587–594. https://doi.org/10.1046/j.1365-8711.1998.01319.x

Mao S, Witt HJ, Koopmans LVE (2001) The influence of central black holes on gravitational lenses. Mon Not R Astron Soc 323(2):301–307. https://doi.org/10.1046/j.1365-8711.2001.04143.x

Maresca J, Dye S, Amvrosiadis A, Bendo G, Cooray A, De Zotti G, Dunne L, Eales S, Furlanetto C, González-Nuevo J, Greener M, Ivison R, Lapi A, Negrello M, Riechers D, Serjeant S, Tergolina M, Wardlow J (2022) Modelling high-resolution ALMA observations of strongly lensed dusty star-forming galaxies detected by Herschel. Mon Not R Astron Soc 512(2):2426–2438. https://doi.org/10.1093/mnras/stac585

Martín-Navarro I, La Barbera F, Vazdekis A, Falcón-Barroso J, Ferreras I (2015b) Radial variations in the stellar initial mass function of early-type galaxies. Mon Not R Astron Soc 447(2):1033–1048. https://doi.org/10.1093/mnras/stu2480

Martinet L, de Zeeuw T (1988) Orbital stability in rotating triaxial stellar systems. Astron Astrophys 206:269–278

McDermid RM, Cappellari M, Alatalo K, Bayet E, Blitz L, Bois M, Bournaud F, Bureau M, Crocker AF, Davies RL, Davis TA, de Zeeuw PT, Duc P-A, Emsellem E, Khochfar S, Krajnović D, Kuntschner H, Morganti R, Naab T, Oosterloo T, Sarzi M, Scott N, Serra P, Weijmans A-M, Young LM (2014) Connection between dynamically derived initial mass function normalization and stellar population parameters. Astrophys J Lett 792(2):37. https://doi.org/10.1088/2041-8205/792/2/L37

Mediavilla E, Muñoz JA, Kochanek CS, Falco EE, Arribas S, Motta V (2005) The first precise determination of an optical-far-ultraviolet extinction curve beyond the local group (z=0.83). Astrophys J 619(2):749–754. https://doi.org/10.1086/426579

Meneghetti M (2021) Introduction to gravitational lensing: with Python example. Lecture Notes in Physics, vol 956. Springer, Cham. https://doi.org/10.1007/978-3-030-73582

Merritt D (1985a) Distribution functions for spherical galaxies. Mon Not R Astron Soc 214:25–28. https://doi.org/10.1093/mnras/214.1.25P

Merritt D (1985b) Spherical stellar systems with spheroidal velocity distributions. Astron J 90:1027–1037. https://doi.org/10.1086/113810

Millon M, Courbin F, Bonvin V, Paic E, Meylan G, Tewes M, Sluse D, Magain P, Chan JHH, Galan A, Joseph R, Lemon C, Tihhonova O, Anderson RI, Marmier M, Chazelas B, Lendl M, Triaud AHMJ, Wyttenbach A (2020) COSMOGRAIL. XIX. Time delays in 18 strongly lensed quasars from 15 years of optical monitoring. Astron Astrophys 640:105. https://doi.org/10.1051/0004-6361/202037740

Millon M, Courbin F, Galan A, Sluse D, Ding X, Tewes M, Djorgovski SG (2023) Strong gravitational lensing by AGNs as a probe of the quasar-host relations in the distant Universe. Nat Astron 7:959–966. https://doi.org/10.1038/s41550-023-01982-2

Mishra-Sharma S, Yang G (2022) Strong Lensing Source Reconstruction Using Continuous Neural Fields. ArXiv e-prints, arXiv:2206.14820

Mitchell JL, Keeton CR, Frieman JA, Sheth RK (2005) Improved cosmological constraints from gravitational lens statistics. Astrophys J 622(1):81–98. https://doi.org/10.1086/427910

Möller O, Hewett P, Blain AW (2003) Discs in early-type lensing galaxies: effects on magnification ratios and measurements of H_0. Mon Not R Astron Soc 345(1):1–15. https://doi.org/10.1046/j.1365-8711.2003.06758.x

More A, Oguri M, Kayo I, Zinn J, Strauss MA, Santiago BX, Mosquera AM, Inada N, Kochanek CS, Rusu CE, Brownstein JR, da Costa LN, Kneib J-P, Maia MAG, Quimby RM, Schneider DP, Streblyanska A, York DG (2016b) The SDSS-III BOSS quasar lens survey: discovery of 13 gravitationally lensed quasars. Mon Not R Astron Soc 456(2):1595–1606. https://doi.org/10.1093/mnras/stv2813

More A, Verma A, Marshall PJ, More S, Baeten E, Wilcox J, Macmillan C, Cornen C, Kapadia A, Parrish M, Snyder C, Davis CP, Gavazzi R, Lintott CJ, Simpson R, Miller D, Smith AM, Paget E, Saha P, Küng R, Collett TE (2016a) SPACE WARPS- II. New gravitational lens candidates from the CFHTLS discovered through citizen science. Mon Not R Astron Soc 455(2):1191–1210. https://doi.org/10.1093/mnras/stv1965

Morningstar WR, Perreault Levasseur L, Hezaveh YD, Blandford R, Marshall P, Putzky P, Rueter TD, Wechsler R, Welling M (2019) Data-driven reconstruction of gravitationally lensed galaxies using recurrent inference machines. Astrophys J 883(1):14. https://doi.org/10.3847/1538-4357/ab35d7

Moster BP, Somerville RS, Maulbetsch C, van den Bosch FC, Macciò AV, Naab T, Oser L (2010) Constraints on the relationship between stellar mass and halo mass at low and high redshift. Astrophys J 710(2):903–923. https://doi.org/10.1088/0004-637X/710/2/903

Motta V, Mediavilla E, Muñoz JA, Falco E, Kochanek CS, Arribas S, García-Lorenzo B, Oscoz A, Serra-Ricart M (2002) Detection of the 2175 Å extinction feature at z = 0.83. Astrophys J 574(2):719–725. https://doi.org/10.1086/341118

Mukherjee S, Koopmans LVE, Tortora C, Schaller M, Metcalf RB, Schaye J, Vernardos G (2022) SEAGLE - III: towards resolving the mismatch in the dark-matter fraction in early-type galaxies between simulations and observations. Mon Not R Astron Soc 509(1):1245–1251. https://doi.org/10.1093/mnras/stab3014

Muller S, Jaswanth S, Horellou C, Martí-Vidal I (2020) All good things come in threes: the third image of the lensed quasar PKS 1830-211. Astron Astrophys 641:2. https://doi.org/10.1051/0004-6361/202038978

Muñoz JA, Falco EE, Kochanek CS, Lehár J, McLeod BA, Impey CD, Rix H-W, Peng CY (1998) The castles project. Astrophys Space Sci 263:51–54. https://doi.org/10.1023/A:1002120921330

Myers ST, Fassnacht CD, Djorgovski SG, Blandford RD, Matthews K, Neugebauer G, Pearson TJ, Readhead ACS, Smith JD, Thompson DJ, Womble DS, Browne IWA, Wilkinson PN, Nair S, Jackson N, Snellen IAG, Miley GK, de Bruyn AG, Schilizzi RT (1995) 1608+656: a quadruple-lens system found in the CLASS gravitational lens survey. Astrophys J Lett 447:5. https://doi.org/10.1086/309556

Naab T, Johansson PH, Ostriker JP, Efstathiou G (2007) Formation of early-type galaxies from cosmological initial conditions. Astrophys J 658(2):710–720. https://doi.org/10.1086/510841

Naab T, Johansson PH, Ostriker JP (2009) Minor mergers and the size evolution of elliptical galaxies. Astrophys J Lett 699:178–182. https://doi.org/10.1088/0004-637X/699/2/L178

Navarro JF, Frenk CS, White SDM (1996) The structure of cold dark matter halos. Astrophys J 462:563. https://doi.org/10.1086/177173

Navarro JF, Frenk CS, White SDM (1997) A universal density profile from hierarchical clustering. Astrophys J 490:493–508. https://doi.org/10.1086/304888

Negrello M, Hopwood R, De Zotti G, Cooray A, Verma A, Bock J, Frayer DT, Gurwell MA, Omont A, Neri R, Dannerbauer H, Leeuw LL, Barton E, Cooke J, Kim S, da Cunha E, Rodighiero G, Cox P, Bonfield DG, Jarvis MJ, Serjeant S, Ivison RJ, Dye S, Aretxaga I, Hughes DH, Ibar E, Bertoldi F, Valtchanov I, Eales S, Dunne L, Driver SP, Auld R, Buttiglione S, Cava A, Grady CA, Clements DL, Dariush A, Fritz J, Hill D, Hornbeck JB, Kelvin L, Lagache G, Lopez-Caniego M, Gonzalez-Nuevo J, Maddox S, Pascale E, Pohlen M, Rigby EE, Robotham A, Simpson C, Smith DJB, Temi P, Thompson MA, Woodgate BE, York DG, Aguirre JE, Beelen A, Blain A, Baker AJ, Birkinshaw M, Blundell R, Bradford CM, Burgarella D, Danese L, Dunlop JS, Fleuren S, Glenn J, Harris AI, Kamenetzky J, Lupu RE, Maddalena RJ, Madore BF, Maloney PR, Matsuhara H, Michałowski MJ, Murphy EJ, Naylor BJ, Nguyen H, Popescu C, Rawlings S, Rigopoulou D, Scott D, Scott KS, Seibert M, Smail I, Tuffs RJ, Vieira JD, van der Werf PP, Zmuidzinas J (2010) The detection of a population of submillimeter-bright, strongly lensed galaxies. Science 330(6005):800. https://doi.org/10.1126/science.1193420

Negrello M, Amber S, Amvrosiadis A, Cai Z-Y, Lapi A, Gonzalez-Nuevo J, De Zotti G, Furlanetto C, Maddox SJ, Allen M, Bakx T, Bussmann RS, Cooray A, Covone G, Danese L, Dannerbauer H, Fu H, Greenslade J, Gurwell M, Hopwood R, Koopmans LVE, Napolitano N, Nayyeri H, Omont A, Petrillo CE, Riechers DA, Serjeant S, Tortora C, Valiante E, Verdoes Kleijn G, Vernardos G, Wardlow JL, Baes M, Baker AJ, Bourne N, Clements D, Crawford SM, Dye S, Dunne L, Eales S, Ivison RJ, Marchetti L, Michałowski MJ, Smith MWL, Vaccari M, van der Werf P (2017) The Herschel-ATLAS: a sample of 500 μm-selected lensed galaxies over 600 deg^2. Mon Not R Astron Soc 465(3):3558–3580. https://doi.org/10.1093/mnras/stw2911

Nelson D, Springel V, Pillepich A, Rodriguez-Gomez V, Torrey P, Genel S, Vogelsberger M, Pakmor R, Marinacci F, Weinberger R, Kelley L, Lovell M, Diemer B, Hernquist L (2019) The IllustrisTNG simulations: public data release. Comput Astrophys Cosmol 6(1):2. https://doi.org/10.1186/s40668-019-0028-x

Nierenberg AM, Treu T, Brammer G, Peter AHG, Fassnacht CD, Keeton CR, Kochanek CS, Schmidt KB, Sluse D, Wright SA (2017) Probing dark matter substructure in the gravitational lens HE 0435-1223 with the WFC3 grism. Mon Not R Astron Soc 471(2):2224–2236. https://doi.org/10.1093/mnras/stx1400

Nightingale JW, Dye S (2015) Adaptive semi-linear inversion of strong gravitational lens imaging. Mon Not R Astron Soc 452(3):2940–2959. https://doi.org/10.1093/mnras/stv1455

Nightingale JW, Dye S, Massey RJ (2018) AutoLens: automated modeling of a strong lens's light, mass, and source. Mon Not R Astron Soc 478(4):4738–4784. https://doi.org/10.1093/mnras/sty1264

Nightingale JW, Massey RJ, Harvey DR, Cooper AP, Etherington A, Tam S-I, Hayes RG (2019) Galaxy structure with strong gravitational lensing: decomposing the internal mass distribution of massive elliptical galaxies. Mon Not R Astron Soc 489(2):2049–2068. https://doi.org/10.1093/mnras/stz2220

Nightingale JW, Smith RJ, He Q, O'Riordan CM, Kegerreis JA, Amvrosiadis A, Edge AC, Etherington A, Hayes RG, Kelly A, Lucey JR, Massey RJ (2023) Abell 1201: detection of an ultramassive black hole in a strong gravitational lens. Mon Not R Astron Soc 521(3):3298–3322. https://doi.org/10.1093/mnras/stad587

Nitschai MS, Cappellari M, Neumayer N (2020) First Gaia dynamical model of the Milky Way disc with six phase space coordinates: a test for galaxy dynamics. Mon Not R Astron Soc 494(4):6001–6011. https://doi.org/10.1093/mnras/staa1128

Ofek EO, Rix H-W, Maoz D (2003) The redshift distribution of gravitational lenses revisited: constraints on galaxy mass evolution. Mon Not R Astron Soc 343:639. https://doi.org/10.1046/j.1365-8711.2003.06707.x

Oguri M (2021) Fast calculation of gravitational lensing properties of elliptical Navarro-frenk-white and hernquist density profiles. Publ Astron Soc Pac 133(1025):074504. https://doi.org/10.1088/1538-3873/ac12db

Oguri M, Marshall PJ (2010) Gravitationally lensed quasars and supernovae in future wide-field optical imaging surveys. Mon Not R Astron Soc 405:2579–2593. https://doi.org/10.1111/j.1365-2966.2010.16639.x

Oguri M, Inada N, Pindor B, Strauss MA, Richards GT, Hennawi JF, Turner EL, Lupton RH, Schneider DP, Fukugita M, Brinkmann J (2006) The Sloan Digital Sky Survey quasar lens search. I. Candidate selection algorithm. Astron J 132(3):999–1013. https://doi.org/10.1086/506019

Oguri M, Rusu CE, Falco EE (2014) The stellar and dark matter distributions in elliptical galaxies from the ensemble of strong gravitational lenses. Mon Not R Astron Soc 439(3):2494–2504. https://doi.org/10.1093/mnras/stu106

Oldham LJ, Auger MW (2018) Dark matter contraction and stellar-mass-to-light ratio gradients in massive early-type galaxies. Mon Not R Astron Soc 476(1):133–150. https://doi.org/10.1093/mnras/sty065

Oldham L, Auger MW, Fassnacht CD, Treu T, Brewer BJ, Koopmans LVE, Lagattuta D, Marshall P, McKean J, Vegetti S (2017) Red nuggets grow inside-out: evidence from gravitational lensing. Mon Not R Astron Soc 465(3):3185–3202. https://doi.org/10.1093/mnras/stw2832

Oliphant TE (2015) Guide to NumPy, 2nd edn. CreateSpace Independent Publishing Platform, USA.

Osipkov LP (1979) Spherical systems of gravitating bodies with an ellipsoidal velocity distribution. Pisma v Astronom Zh 5:77–80

Ostdiek B, Diaz Rivero A, Dvorkin C (2022a) Extracting the subhalo mass function from strong lens images with image segmentation. Astrophys J 927:83. https://doi.org/10.3847/1538-4357/ac2d8d

Ostdiek B, Diaz Rivero A, Dvorkin C (2022b) Image segmentation for analyzing galaxy-galaxy strong lensing systems. Astron Astrophys 657:14. https://doi.org/10.1051/0004-6361/202142030

Paraficz D, Courbin F, Tramacere A, Joseph R, Metcalf RB, Kneib J-P, Dubath P, Droz D, Filleul F, Ringeisen D, Schäfer C (2016) The PCA lens-finder: application to CFHTLS. Astron Astrophys 592:75. https://doi.org/10.1051/0004-6361/201527971

Park JW, Wagner-Carena S, Birrer S, Marshall PJ, Lin JY-Y, Roodman A, LSST Dark Energy Science Collaboration (2021) Large-scale gravitational lens modeling with Bayesian neural networks for accurate and precise inference of the Hubble constant. Astrophys J 910(1):39. https://doi.org/10.3847/1538-4357/abdfc4

Patnaik AR, Browne IWA, Walsh D, Chaffee FH, Foltz CB (1992) B 1422+231: a new gravitationally lensed system at Z = 3.62. Mon Not R Astron Soc 259:1–4. https://doi.org/10.1093/mnras/259.1.1P

Perera D, Williams LLR, Scarlata C (2023) Constraints on the inner regions of lensing galaxies from central images using a recent AGN offset distribution. Mon Not R Astron Soc 522(2):1863–1880. https://doi.org/10.1093/mnras/stad1110

Petrillo CE, Tortora C, Chatterjee S, Vernardos G, Koopmans LVE, Verdoes Kleijn G, Napolitano NR, Covone G, Schneider P, Grado A, McFarland J (2017) Finding strong gravitational lenses in the kilo degree survey with convolutional neural networks. Mon Not R Astron Soc 472(1):1129–1150. https://doi.org/10.1093/mnras/stx2052

Petrillo CE, Tortora C, Vernardos G, Koopmans LVE, Verdoes Kleijn G, Bilicki M, Napolitano NR, Chatterjee S, Covone G, Dvornik A, Erben T, Getman F, Giblin B, Heymans C, de Jong JTA, Kuijken K, Schneider P, Shan H, Spiniello C, Wright AH (2019) LinKS: discovering galaxy-scale strong lenses in the kilo-degree survey using convolutional neural networks. Mon Not R Astron Soc 484(3):3879–3896. https://doi.org/10.1093/mnras/stz189

Pierel JDR, Arendse N, Ertl S, Huang X, Moustakas LA, Schuldt S, Shajib AJ, Shu Y, Birrer S, Bronikowski M, Hjorth J, Suyu SH, Agarwal S, Agnello A, Bolton AS, Chakrabarti S, Cold C, Courbin F, Della Costa JM, Dhawan S, Engesser M, Fox OD, Gall C, Gomez S, Goobar A, Jha SW, Jimenez C, Johansson J, Larison C, Li G, Marques-Chaves R, Mao S, Mazzali PA, Perez-Fournon I, Petrushevska T, Poidevin F, Rest A, Sheu W, Shirley R, Silver E, Storfer C, Strolger LG, Treu T, Wojtak R, Zenati Y (2023) LensWatch. I. Resolved HST observations and constraints on the strongly lensed type Ia supernova 2022qmx ("SN Zwicky"). Astrophys J 948(2):115. https://doi.org/10.3847/1538-4357/acc7a6

Planck Collaboration (2014) Planck 2013 results. XVI. Cosmological parameters. Astron Astrophys 571:16. https://doi.org/10.1051/0004-6361/201321591

Planck Collaboration (2020) Planck 2018 results. VI. Cosmological parameters. Astron Astrophys 641:6. https://doi.org/10.1051/0004-6361/201833910

Poci A, Smith RJ (2022) Comparing lensing and stellar orbital models of a nearby massive strong-lens galaxy. Mon Not R Astron Soc 512(4):5298–5310. https://doi.org/10.1093/mnras/stac776

Poci A, Cappellari M, McDermid RM (2017) Systematic trends in total-mass profiles from dynamical models of early-type galaxies. Mon Not R Astron Soc 467(2):1397–1413. https://doi.org/10.1093/mnras/stx101

Posacki S, Cappellari M, Treu T, Pellegrini S, Ciotti L (2015) The stellar initial mass function of early-type galaxies from low to high stellar velocity dispersion: homogeneous analysis of Atlas^{3d} and Sloan Lens ACS galaxies. Mon Not R Astron Soc 446(1):493–509. https://doi.org/10.1093/mnras/stu2098

Powell DM, Vegetti S, McKean JP, Spingola C, Stacey HR, Fassnacht CD (2022) A lensed radio jet at milli-arcsecond resolution I: Bayesian comparison of parametric lens models. ArXiv e-prints, arXiv:2207.03375

Press WH, Rybicki GB, Hewitt JN (1992) The time delay of gravitational lens 0957+561. I. Methodology and analysis of optical photometric data. Astrophys J 385:404. https://doi.org/10.1086/170951

Pulsoni C, Gerhard O, Arnaboldi M, Pillepich A, Rodriguez-Gomez V, Nelson D, Hernquist L, Springel V (2021) The stellar halos of ETGs in the IllustrisTNG simulations. II. Accretion, merger history, and dark halo connection. Astron Astrophys 647:95. https://doi.org/10.1051/0004-6361/202039166

Quinn J, Jackson N, Tagore A, Biggs A, Birkinshaw M, Chapman S, De Zotti G, McKean J, Pérez-Fournon I, Scott D, Serjeant S (2016) A new VLA/e-MERLIN limit on central images in the gravitational lens system CLASS B1030+074. Mon Not R Astron Soc 459(3):2394–2407. https://doi.org/10.1093/mnras/stw773

Remus R-S, Burkert A, Dolag K, Johansson PH, Naab T, Oser L, Thomas J (2013) The dark halo-spheroid conspiracy and the origin of elliptical galaxies. Astrophys J 766(2):71. https://doi.org/10.1088/0004-637X/766/2/71

Remus R-S, Dolag K, Naab T, Burkert A, Hirschmann M, Hoffmann TL, Johansson PH (2017) The co-evolution of total density profiles and central dark matter fractions in simulated early-type galaxies. Mon Not R Astron Soc 464(3):3742–3756. https://doi.org/10.1093/mnras/stw2594

Ritondale E, Auger MW, Vegetti S, McKean JP (2019) Resolving on 100 pc scales the uv-continuum in Lyman-αemitters between redshift 2 and 3 with gravitational lensing. Mon Not R Astron Soc 482(4):4744–4762. https://doi.org/10.1093/mnras/sty2833

Rojas K, Savary E, Clément B, Maus M, Courbin F, Lemon C, Chan JHH, Vernardos G, Joseph R, Cañameras R, Galan A (2021) Strong lens systems search in the Dark Energy Survey using Convolutional Neural Networks. ArXiv e-prints, arXiv:2109.00014

Romanowsky AJ, Kochanek CS (1999) Constraints on H_0 from the central velocity dispersions of lens galaxies. Astrophys J 516(1):18–26. https://doi.org/10.1086/307077

Ruff AJ, Gavazzi R, Marshall PJ, Treu T, Auger MW, Brault F (2011) The SL2S galaxy-scale lens sample. II. Cosmic evolution of dark and luminous mass in early-type galaxies. Astrophys J 727(2):96. https://doi.org/10.1088/0004-637X/727/2/96

Rusin D, Ma C-P (2001) Constraints on the inner mass profiles of lensing galaxies from missing odd images. Astrophys J Lett 549(1):33–37. https://doi.org/10.1086/319129

Rusu CE, Oguri M, Minowa Y, Iye M, Inada N, Oya S, Kayo I, Hayano Y, Hattori M, Saito Y, Ito M, Pyo T-S, Terada H, Takami H, Watanabe M (2016) Subaru telescope adaptive optics observations of gravitationally lensed quasars in the Sloan Digital Sky Survey. Mon Not R Astron Soc 458(1):2–55. https://doi.org/10.1093/mnras/stw092

Saha P (2000) Lensing degeneracies revisited. Astron J 120:1654–1659. https://doi.org/10.1086/301581

Saha P, Williams LLR (1997) Non-parametric reconstruction of the galaxy lens in PG 1115+080. Mon Not R Astron Soc 292(1):148–156. https://doi.org/10.1093/mnras/292.1.148

Saha P, Williams LLR (2001) Beware the nonuniqueness of Einstein rings. Astron J 122:585–590

Saha P, Williams LLR (2004) A portable modeler of lensed quasars. Astron J 127(5):2604–2616. https://doi.org/10.1086/383544

Saha P, Sluse D, Wagner J, Williams LLR (2024) Essentials of strong gravitational lensing. Space Sci Rev 220(1):12. https://doi.org/10.1007/s11214-024-01041-w

Salpeter EE (1955) The luminosity function and stellar evolution. Astrophys J 121:161. https://doi.org/10.1086/145971

Sarzi M, Spiniello C, La Barbera F, Krajnović D, van den Bosch R (2018) MUSE observations of M87: radial gradients for the stellar initial-mass function and the abundance of sodium. Mon Not R Astron Soc 478(3):4084–4100. https://doi.org/10.1093/mnras/sty1092

Savary E, Rojas K, Maus M, Clément B, Courbin F, Gavazzi R, Chan JHH, Lemon C, Vernardos G, Cañameras R, Schuldt S, Suyu SH, Cuillandre J-C, Fabbro S, Gwyn S, Hudson MJ, Kilbinger M, Scott D, Stone C (2022) A search for galaxy-scale strong gravitational lenses in the Ultraviolet Near Infrared Optical Northern Survey (UNIONS).. Astron Astrophys 666:A1. https://doi.org/10.1051/0004-6361/202142505

Schechter PL, Pooley D, Blackburne JA, Wambsganss J (2014) A calibration of the stellar mass fundamental plane at z ~0.5 using the micro-lensing-induced flux ratio anomalies of macro-lensed quasars. Astrophys J 793(2):96. https://doi.org/10.1088/0004-637X/793/2/96

Schmidt T, Treu T, Birrer S, Shajib AJ, Lemon C, Millon M, Sluse D, Agnello A, Anguita T, Auger-Williams MW, McMahon RG, Motta V, Schechter P, Spiniello C, Kayo I, Courbin F, Ertl S, Fassnacht CD, Frieman JA, More A, Schuldt S, Suyu SH, Aguena M, Andrade-Oliveira F, Annis J, Bacon D, Bertin E, Brooks D, Burke DL, Carnero Rosell A, Carrasco Kind M, Carretero J, Conselice C, Costanzi M, da Costa LN, Pereira MES, De Vicente J, Desai S, Doel P, Everett S, Ferrero I, Friedel D, García-Bellido J, Gaztanaga E, Gruen D, Gruendl RA, Gschwend J, Gutierrez G, Hinton SR, Hollowood DL, Honscheid K, James DJ, Kuehn K, Lahav O, Menanteau F, Miquel R, Palmese A, Paz-Chinchón F, Pieres A, Plazas Malagón AA, Prat J, Rodriguez-Monroy M, Romer AK, Sanchez E, Scarpine V, Sevilla-Noarbe I, Smith M, Suchyta E, Tarle G, To C, Varga TN, DES Collaboration (2023) STRIDES: automated uniform models for 30 quadruply imaged quasars. Mon Not R Astron Soc 518(1):1260–1300. https://doi.org/10.1093/mnras/stac2235

Schneider P, Sluse D (2014) Source-position transformation: an approximate invariance in strong gravitational lensing. Astron Astrophys 564:103. https://doi.org/10.1051/0004-6361/201322106

Schneider P, Ehlers J, Falco EE (1992) Gravitational Lenses. https://doi.org/10.1007/978-3-662-03758-4

Schuldt S, Cañameras R, Shu Y, Suyu SH, Taubenberger S, Meinhardt T, Leal-Taixé L (2023) HOLISMOKES – IX. Neural network inference of strong-lens parameters and uncertainties from ground-based images. Astron Astrophys 671:A147 https://doi.org/10.1051/0004-6361/202244325

Schwarzschild M (1979) A numerical model for a triaxial stellar system in dynamical equilibrium. Astrophys J 232:236–247. https://doi.org/10.1086/157282

Sérsic JL (1968) Atlas de Galaxias Australes

Shajib AJ (2019) Unified lensing and kinematic analysis for any elliptical mass profile. Mon Not R Astron Soc 488(1):1387–1400. https://doi.org/10.1093/mnras/stz1796

Shajib AJ, Treu T, Agnello A (2018) Improving time-delay cosmography with spatially resolved kinematics. Mon Not R Astron Soc 473:210–226. https://doi.org/10.1093/mnras/stx2302

Shajib AJ, Birrer S, Treu T, Auger MW, Agnello A, Anguita T, Buckley-Geer EJ, Chan JHH, Collett TE, Courbin F, Fassnacht CD, Frieman J, Kayo I, Lemon C, Lin H, Marshall PJ, McMahon R, More A, Morgan ND, Motta V, Oguri M, Ostrovski F, Rusu CE, Schechter PL, Shanks T, Suyu SH, Meylan G, Abbott TMC, Allam S, Annis J, Avila S, Bertin E, Brooks D, Carnero Rosell A, Carrasco Kind M, Carretero J, Cunha CE, da Costa LN, De Vicente J, Desai S, Doel P, Flaugher B, Fosalba P, García-Bellido J, Gerdes DW, Gruen D, Gruendl RA, Gutierrez G, Hartley WG, Hollowood DL, Hoyle B, James DJ, Kuehn K, Kuropatkin N, Lahav O, Lima M, Maia MAG, March M, Marshall JL, Melchior P, Menanteau F, Miquel R, Plazas AA, Sanchez E, Scarpine V, Sevilla-Noarbe I, Smith M, Soares-Santos M, Sobreira F, Suchyta E, Swanson MEC, Tarle G, Walker AR (2019) Is every strong lens model unhappy in its own way? Uniform modelling of a sample of 13 quadruply+ imaged quasars. Mon Not R Astron Soc 483:5649–5671. https://doi.org/10.1093/mnras/sty3397

Shajib AJ, Birrer S, Treu T, Agnello A, Buckley-Geer EJ, Chan JHH, Christensen L, Lemon C, Lin H, Millon M, Poh J, Rusu CE, Sluse D, Spiniello C, Chen GC-F, Collett T, Courbin F, Fassnacht CD, Frieman J, Galan A, Gilman D, More A, Anguita T, Auger MW, Bonvin V, McMahon R, Meylan G, Wong KC, Abbott TMC, Annis J, Avila S, Bechtol K, Brooks D, Brout D, Burke DL, Carnero Rosell A, Carrasco Kind M, Carretero J, Castander FJ, Costanzi M, da Costa LN, De Vicente J, Desai S, Dietrich JP, Doel P, Drlica-Wagner A, Evrard AE, Finley DA, Flaugher B, Fosalba P, García-Bellido J, Gerdes DW, Gruen D, Gruendl RA, Gschwend J, Gutierrez G, Hollowood DL, Honscheid K, Huterer D, James DJ, Jeltema T, Krause E, Kuropatkin N, Li TS, Lima M, MacCrann N, Maia MAG, Marshall JL, Melchior P, Miquel R, Ogando RLC, Palmese A, Paz-Chinchón F, Plazas AA, Romer AK, Roodman A, Sako M, Sanchez E, Santiago B, Scarpine V, Schubnell M, Scolnic D, Serrano S, Sevilla-Noarbe I, Smith M, Soares-Santos M, Suchyta E, Tarle G, Thomas D, Walker AR, Zhang Y (2020) STRIDES: a 3.9 per cent measurement of the Hubble constant from the strong lens system DES J0408-5354. Mon Not R Astron Soc 494(4):6072–6102. https://doi.org/10.1093/mnras/staa828

Shajib AJ, Treu T, Birrer S, Sonnenfeld A (2021) Dark matter haloes of massive elliptical galaxies at z ∼ 0.2 are well described by the Navarro-Frenk-White profile. Mon Not R Astron Soc 503(2):2380–2405. https://doi.org/10.1093/mnras/stab536

Shajib AJ, Wong KC, Birrer S, Suyu SH, Treu T, Buckley-Geer EJ, Lin H, Rusu CE, Poh J, Palmese A, Agnello A, Auger-Williams MW, Galan A, Schuldt S, Sluse D, Courbin F, Frieman J, Millon M (2022) TDCOSMO. IX. Systematic comparison between lens modelling software programs: time-delay prediction for WGD 2038−4008. Astron Astrophys 667:123. https://doi.org/10.1051/0004-6361/202243401

Shajib AJ, Mozumdar P, Chen GC-F, Treu T, Cappellari M, Knabel S, Suyu SH, Bennert VN, Frieman JA, Sluse D, Birrer S, Courbin F, Fassnacht CD, Villafaña L, Williams PR (2023) TDCOSMO. XII. Improved Hubble constant measurement from lensing time delays using spatially resolved stellar kinematics of the lens galaxy. Astron Astrophys 673:9. https://doi.org/10.1051/0004-6361/202345878

Shajib AJ, Smith GP, Birrer S, Verma A, Arendse N, Collett TE (2024) Strong gravitational lenses from the Vera C. Rubin Observatory. ArXiv e-prints, arXiv:2406.08919. https://doi.org/10.48550/arXiv.2406.08919

Shankar F, Sonnenfeld A, Mamon GA, Chae K-H, Gavazzi R, Treu T, Diemer B, Nipoti C, Buchan S, Bernardi M, Sheth R, Huertas-Company M (2017) Revisiting the Bulge-Halo conspiracy. I. Dependence on galaxy properties and halo mass. Astrophys J 840(1):34. https://doi.org/10.3847/1538-4357/aa66ce

Shankar F, Sonnenfeld A, Grylls P, Zanisi L, Nipoti C, Chae K-H, Bernardi M, Petrillo CE, Huertas-Company M, Mamon GA, Buchan S (2018) Revisiting the bulge-halo conspiracy - II. Towards explaining its puzzling dependence on redshift. Mon Not R Astron Soc 475(3):2878–2890. https://doi.org/10.1093/mnras/stx3086

Sharma D, Collett TE, Linder EV (2023) Testing cosmology with double source lensing. J Cosmol Astropart Phys 2023(4):001. https://doi.org/10.1088/1475-7516/2023/04/001

Shetty S, Cappellari M (2014) Salpeter normalization of the stellar initial mass function for massive galaxies at z ~1. Astrophys J Lett 786(2):10. https://doi.org/10.1088/2041-8205/786/2/L10

Shu Y, Bolton AS, Brownstein JR, Montero-Dorta AD, Koopmans LVE, Treu T, Gavazzi R, Auger MW, Czoske O, Marshall PJ, Moustakas LA (2015) The Sloan Lens ACS Survey. XII. Extending strong lensing to lower masses. Astrophys J 803(2):71. https://doi.org/10.1088/0004-637X/803/2/71

Shu Y, Bolton AS, Mao S, Kochanek CS, Pérez-Fournon I, Oguri M, Montero-Dorta AD, Cornachione MA, Marques-Chaves R, Zheng Z, Brownstein JR, Ménard B (2016) The BOSS emission-line lens survey. IV. Smooth lens models for the BELLS GALLERY sample. Astrophys J 833(2):264. https://doi.org/10.3847/1538-4357/833/2/264

Sluse D, Chantry V, Magain P, Courbin F, Meylan G (2012) COSMOGRAIL: the COSmological MOnitoring of GRAvItational lenses. X. Modeling based on high-precision astrometry of a sample of 25 lensed quasars: consequences for ellipticity, shear, and astrometric anomalies. Astron Astrophys 538:99. https://doi.org/10.1051/0004-6361/201015844

Smith RJ (2020) Evidence for initial mass function variation in massive early-type galaxies. Annu Rev Astron Astrophys 58:577–615. https://doi.org/10.1146/annurev-astro-032620-020217

Smith RJ, Lucey JR, Conroy C (2015) The SINFONI nearby elliptical lens locator survey: discovery of two new low-redshift strong lenses and implications for the initial mass function in giant early-type galaxies. Mon Not R Astron Soc 449(4):3441–3457. https://doi.org/10.1093/mnras/stv518

Sonnenfeld A (2018) On the choice of lens density profile in time delay cosmography. Mon Not R Astron Soc 474(4):4648–4659. https://doi.org/10.1093/mnras/stx3105

Sonnenfeld A (2022) Statistical strong lensing. III. Inferences with complete samples of lenses. Astron Astrophys 659:132. https://doi.org/10.1051/0004-6361/202142301

Sonnenfeld A, Cautun M (2021) Statistical strong lensing. I. Constraints on the inner structure of galaxies from samples of a thousand lenses. Astron Astrophys 651:18. https://doi.org/10.1051/0004-6361/202140549

Sonnenfeld A, Treu T, Gavazzi R, Marshall PJ, Auger MW, Suyu SH, Koopmans LVE, Bolton AS (2012) Evidence for dark matter contraction and a Salpeter initial mass function in a massive early-type galaxy. Astrophys J 752(2):163. https://doi.org/10.1088/0004-637X/752/2/163

Sonnenfeld A, Gavazzi R, Suyu SH, Treu T, Marshall PJ (2013) The SL2S galaxy-scale lens sample. III. Lens models, surface photometry, and stellar masses for the final sample. Astrophys J 777:97. https://doi.org/10.1088/0004-637X/777/2/97

Sonnenfeld A, Treu T, Gavazzi R, Suyu SH, Marshall PJ, Auger MW, Nipoti C (2013) The SL2S galaxy-scale lens sample. IV. The dependence of the total mass density profile of early-type galaxies on redshift, stellar mass, and size. Astrophys J 777(2):98. https://doi.org/10.1088/0004-637X/777/2/98

Sonnenfeld A, Treu T, Marshall PJ, Suyu SH, Gavazzi R, Auger MW, Nipoti C (2015) The sl2s galaxy-scale lens sample. V. Dark matter halos and stellar imf of massive early-type galaxies out to redshift 0.8. Astrophys J 800:94. https://doi.org/10.1088/0004-637X/800/2/94

Sonnenfeld A, Leauthaud A, Auger MW, Gavazzi R, Treu T, More S, Komiyama Y (2018) Evidence for radial variations in the stellar mass-to-light ratio of massive galaxies from weak and strong lensing. Mon Not R Astron Soc 481:164–184. https://doi.org/10.1093/mnras/sty2262

Sonnenfeld A, Jaelani AT, Chan J, More A, Suyu SH, Wong KC, Oguri M, Lee C-H (2019) Survey of gravitationally-lensed objects in hsc imaging (sugohi). III. Statistical strong lensing constraints on the stellar imf of cmass galaxies. Astron Astrophys 630:71. https://doi.org/10.1051/0004-6361/201935743

Sonnenfeld A, Li S-S, Despali G, Gavazzi R, Shajib AJ, Taylor EN (2023) Strong lensing selection effects. Astron Astrophys 678:4. https://doi.org/10.1051/0004-6361/202346026

Spiniello C, Koopmans LVE, Trager SC, Czoske O, Treu T (2011) The x-shooter lens survey - I. Dark matter domination and a Salpeter-type initial mass function in a massive early-type galaxy. Mon Not R Astron Soc 417(4):3000–3009. https://doi.org/10.1111/j.1365-2966.2011.19458.x

Spiniello C, Trager SC, Koopmans LVE, Chen YP (2012) Evidence for a mild steepening and bottom-heavy initial mass function in massive galaxies from sodium and titanium-oxide indicators. Astrophys J Lett 753(2):32. https://doi.org/10.1088/2041-8205/753/2/L32

Spiniello C, Trager S, Koopmans LVE, Conroy C (2014) The stellar IMF in early-type galaxies from a non-degenerate set of optical line indices. Mon Not R Astron Soc 438(2):1483–1499. https://doi.org/10.1093/mnras/stt2282

Spiniello C, Koopmans LVE, Trager SC, Barnabè M, Treu T, Czoske O, Vegetti S, Bolton A (2015) The X-Shooter Lens Survey - II. Sample presentation and spatially-resolved kinematics. Mon Not R Astron Soc 452(3):2434–2444. https://doi.org/10.1093/mnras/stv1490

Spiniello C, Tortora C, D'Ago G, Coccato L, La Barbera F, Ferré-Mateu A, Pulsoni C, Arnaboldi M, Gallazzi A, Hunt L, Napolitano NR, Radovich M, Scognamiglio D, Spavone M, Zibetti S (2021) INSPIRE: INvestigating stellar population in RElics. II. First data release (DR1). Astron Astrophys 654:136. https://doi.org/10.1051/0004-6361/202140856

Suyu SH, Marshall PJ, Hobson MP, Blandford RD (2006) A Bayesian analysis of regularized source inversions in gravitational lensing. Mon Not R Astron Soc 371(2):983–998. https://doi.org/10.1111/j.1365-2966.2006.10733.x

Suyu SH, Marshall PJ, Auger MW, Hilbert S, Blandford RD, Koopmans LVE, Fassnacht CD, Treu T (2010) Dissecting the gravitational lens B1608+656. II. Precision measurements of the Hubble constant, spatial curvature, and the dark energy equation of state. Astrophys J 711(1):201–221. https://doi.org/10.1088/0004-637X/711/1/201

Suyu SH, Hensel SW, McKean JP, Fassnacht CD, Treu T, Halkola A, Norbury M, Jackson N, Schneider P, Thompson D, Auger MW, Koopmans LVE, Matthews K (2012) Disentangling baryons and dark matter in the spiral gravitational lens B1933+503. Astrophys J 750(1):10. https://doi.org/10.1088/0004-637X/750/1/10

Suyu SH, Auger MW, Hilbert S, Marshall PJ, Tewes M, Treu T, Fassnacht CD, Koopmans LVE, Sluse D, Blandford RD, Courbin F, Meylan G (2013) Two accurate time-delay distances from strong lensing: implications for cosmology. Astrophys J 766(2):70. https://doi.org/10.1088/0004-637X/766/2/70

Suyu SH, Treu T, Hilbert S, Sonnenfeld A, Auger MW, Blandford RD, Collett T, Courbin F, Fassnacht CD, Koopmans LVE, Marshall PJ, Meylan G, Spiniello C, Tewes M (2014) Cosmology from gravitational lens time delays and Planck data. Astrophys J Lett 788(2):35. https://doi.org/10.1088/2041-8205/788/2/L35

Tagore AS, Jackson N (2016) On the use of shapelets in modelling resolved, gravitationally lensed images. Mon Not R Astron Soc 457(3):3066–3075. https://doi.org/10.1093/mnras/stw057

Tamura Y, Oguri M, Iono D, Hatsukade B, Matsuda Y, Hayashi M (2015) High-resolution ALMA observations of SDP.81. I. The innermost mass profile of the lensing elliptical galaxy probed by 30 milli-arcsecond images. Publ Astron Soc Jpn 67(4):72. https://doi.org/10.1093/pasj/psv040

Tan CY, Shajib AJ, Birrer S, Sonnenfeld A, Treu T, Wells P, Williams D, Buckley-Geer EJ, Drlica-Wagner A, Frieman J (2024) Project dinos I: a joint lensing-dynamics constraint on the deviation from the power law in the mass profile of massive ellipticals. Mon Not R Astron Soc. https://doi.org/10.1093/mnras/stae884

Tessore N, Metcalf RB (2015) The elliptical power law profile lens. Astron Astrophys 580:79. https://doi.org/10.1051/0004-6361/201526773

Thomas J, Saglia RP, Bender R, Thomas D, Gebhardt K, Magorrian J, Corsini EM, Wegner G (2007) Dynamical modelling of luminous and dark matter in 17 coma early-type galaxies. Mon Not R Astron Soc 382(2):657–684. https://doi.org/10.1111/j.1365-2966.2007.12434.x

Tortora C, Napolitano NR, Saglia RP, Romanowsky AJ, Covone G, Capaccioli M (2014) Evolution of central dark matter of early-type galaxies up to z $\sim$0.8. Mon Not R Astron Soc 445(1):162–174. https://doi.org/10.1093/mnras/stu1712

Tran K-VH, Harshan A, Glazebrook K, Keerthi Vasan GC, Jones T, Jacobs C, Kacprzak GG, Barone TM, Collett TE, Gupta A, Henderson A, Kewley LJ, Lopez S, Nanayakkara T, Sanders RL, Sweet SM (2022)

The AGEL survey: spectroscopic confirmation of strong gravitational lenses in the DES and DECaLS fields selected using convolutional neural networks. Astron J 164:148. https://doi.org/10.3847/1538-3881/ac7da2

Treu T (2010) Strong lensing by galaxies. Annu Rev Astron Astrophys 48:87–125. https://doi.org/10.1146/annurev-astro-081309-130924

Treu T, Koopmans LVE (2002) The internal structure of the lens PG1115+080: breaking degeneracies in the value of the Hubble constant. Mon Not R Astron Soc 337:6–10. https://doi.org/10.1046/j.1365-8711.2002.06107.x

Treu T, Koopmans LVE (2004) Massive dark matter halos and evolution of early-type galaxies to z $\sim$ 1. Astrophys J 611:739–760. https://doi.org/10.1086/422245

Treu T, Koopmans LV, Bolton AS, Burles S, Moustakas LA (2006) The Sloan Lens ACS Survey. II. Stellar populations and internal structure of early-type lens galaxies. Astrophys J 640:662–672. https://doi.org/10.1086/500124

Treu T, Gavazzi R, Gorecki A, Marshall PJ, Koopmans LVE, Bolton AS, Moustakas LA, Burles S (2009) The slacs survey. VIII. The relation between environment and internal structure of early-type galaxies. Astrophys J 690(1):670–682. https://doi.org/10.1088/0004-637X/690/1/670

Treu T, Auger MW, Koopmans LVE, Gavazzi R, Marshall PJ, Bolton AS (2010) The initial mass function of early-type galaxies. Astrophys J 709(2):1195–1202. https://doi.org/10.1088/0004-637X/709/2/1195

Treu T, Dutton AA, Auger MW, Marshall PJ, Bolton AS, Brewer BJ, Koo DC, Koopmans LVE (2011) The SWELLS survey - I. A large spectroscopically selected sample of edge-on late-type lens galaxies. Mon Not R Astron Soc 417(3):1601–1620. https://doi.org/10.1111/j.1365-2966.2011.19378.x

Trotter CS, Winn JN, Hewitt JN (2000) A multipole-Taylor expansion for the potential of the gravitational lens MG J0414+0534. Astrophys J 535(2):671–691. https://doi.org/10.1086/308861

Trujillo I, Ferré-Mateu A, Balcells M, Vazdekis A, Sánchez-Blázquez P (2014) NGC 1277: a massive compact relic galaxy in the nearby universe. Astrophys J Lett 780:20. https://doi.org/10.1088/2041-8205/780/2/L20

Unruh S, Schneider P, Sluse D (2017) Ambiguities in gravitational lens models: the density field from the source position transformation. Astron Astrophys 601:77. https://doi.org/10.1051/0004-6361/201629048

van Albada TS, Sancisi R (1986) Dark matter in spiral galaxies. Philos Trans R Soc Lond Ser A 320(1556):447–464. https://doi.org/10.1098/rsta.1986.0128

van de Sande J, Bland-Hawthorn J, Brough S, Croom SM, Cortese L, Foster C, Scott N, Bryant JJ, d'Eugenio F, Tonini C, Goodwin M, Konstantopoulos IS, Lawrence JS, Medling AM, Owers MS, Richards SN, Schaefer AL, Yi SK (2017) The SAMI galaxy survey: revising the fraction of slow rotators in IFS galaxy surveys. Mon Not R Astron Soc 472(2):1272–1285. https://doi.org/10.1093/mnras/stx1751

van de Ven G, Falcón-Barroso J, McDermid RM, Cappellari M, Miller BW, de Zeeuw PT (2010) The Einstein cross: constraint on dark matter from stellar dynamics and gravitational lensing. Astrophys J 719(2):1481–1496. https://doi.org/10.1088/0004-637X/719/2/1481

Van de Vyvere L, Sluse D, Mukherjee S, Xu D, Birrer S (2020) The impact of mass map truncation on strong lensing simulations. Astron Astrophys 644:108. https://doi.org/10.1051/0004-6361/202038942

Van de Vyvere L, Gomer MR, Sluse D, Xu D, Birrer S, Galan A, Vernardos G (2022b) TDCOSMO. VII. Boxyness/discyness in lensing galaxies: detectability and impact on H_0. Astron Astrophys 659:127. https://doi.org/10.1051/0004-6361/202141551

Van de Vyvere L, Sluse D, Gomer MR, Mukherjee S (2022a) Consequences of the lack of azimuthal freedom in the modeling of lensing galaxies. Astron Astrophys 663:179. https://doi.org/10.1051/0004-6361/202243382

van Dokkum P, Conroy C, Villaume A, Brodie J, Romanowsky AJ (2017) The stellar initial mass function in early-type galaxies from absorption line spectroscopy. III. Radial gradients. Astrophys J 841(2):68. https://doi.org/10.3847/1538-4357/aa7135

Vasiliev E (2019) AGAMA: action-based galaxy modelling architecture. Mon Not R Astron Soc 482(2):1525–1544. https://doi.org/10.1093/mnras/sty2672

Vegetti S, Koopmans LVE (2009) Bayesian strong gravitational-lens modelling on adaptive grids: objective detection of mass substructure in galaxies. Mon Not R Astron Soc 392(3):945–963. https://doi.org/10.1111/j.1365-2966.2008.14005.x

Vegetti S, Koopmans LVE, Auger MW, Treu T, Bolton AS (2014) Inference of the cold dark matter substructure mass function at z = 0.2 using strong gravitational lenses. Mon Not R Astron Soc 442(3):2017–2035. https://doi.org/10.1093/mnras/stu943

Vegetti S, Birrer S, Despali G, Fassnacht CD, Gilman D, Hezaveh Y, Perreault Levasseur L, McKean JP, Powell DM, O'Riordan CM, Vernardos G (2023) Strong gravitational lensing as a probe of dark matter. ArXiv e-prints, arXiv:2306.11781. https://doi.org/10.48550/arXiv.2306.11781

Vernardos G, Koopmans LVE (2022) The Very Knotty Lenser: exploring the role of regularization in source and potential reconstructions using Gaussian Process Regression. Mon Not R Astron Soc 516(1):1347–1372. https://doi.org/10.1093/mnras/stac1924

Vernardos G, Tsagkatakis G, Pantazis Y (2020) Quantifying the structure of strong gravitational lens potentials with uncertainty-aware deep neural networks. Mon Not R Astron Soc 499(4):5641–5652. https://doi.org/10.1093/mnras/staa3201

Vernardos G, Sluse D, Pooley D, Schmidt RW, Millon M, Weisenbach L, Motta V, Anguita T, Saha P, O'Dowd M, Peel A, Schechter PL (2024) Microlensing of strongly lensed quasars. Space Sci Rev 220(1):14. https://doi.org/10.1007/s11214-024-01043-8

Vogelsberger M, Genel S, Springel V, Torrey P, Sijacki D, Xu D, Snyder G, Nelson D, Hernquist L (2014) Introducing the illustris project: simulating the coevolution of dark and visible matter in the Universe. Mon Not R Astron Soc 444(2):1518–1547. https://doi.org/10.1093/mnras/stu1536

Wagner J (2017) Generalised model-independent characterisation of strong gravitational lenses. I. Theoretical foundations. Astron Astrophys 601:131. https://doi.org/10.1051/0004-6361/201630200

Wagner J, Tessore N (2018) Generalised model-independent characterisation of strong gravitational lenses. II. Transformation matrix between multiple images. Astron Astrophys 613:6. https://doi.org/10.1051/0004-6361/201730947

Wagner-Carena S, Park JW, Birrer S, Marshall PJ, Roodman A, Wechsler RH, LSST Dark Energy Science Collaboration (2021) Hierarchical inference with Bayesian neural networks: an application to strong gravitational lensing. Astrophys J 909(2):187. https://doi.org/10.3847/1538-4357/abdf59

Walsh D, Carswell RF, Weymann RJ (1979) 0957+561 A, B: twin quasistellar objects or gravitational lens? Nature 279:381–384. https://doi.org/10.1038/279381a0

Wang Y, Vogelsberger M, Xu D, Mao S, Springel V, Li H, Barnes D, Hernquist L, Pillepich A, Marinacci F, Pakmor R, Weinberger R, Torrey P (2020) Early-type galaxy density profiles from IllustrisTNG - I. Galaxy correlations and the impact of baryons. Mon Not R Astron Soc 491(4):5188–5215. https://doi.org/10.1093/mnras/stz3348

Warren SJ, Dye S (2003) Semilinear gravitational lens inversion. Astrophys J 590(2):673–682. https://doi.org/10.1086/375132

Waskom M, Botvinnik O, Hobson P, Cole JB, Halchenko Y, Hoyer S, Miles A, Augspurger T, Yarkoni T, Megies T, Coelho LP, Wehner D, cynddl, Ziegler E, diego0020, Zaytsev YV, Hoppe T, Seabold S, Cloud P, Koskinen M, Meyer K, Qalieh A, Allan D (2014) seaborn: v0.5.0 (November 2014). https://doi.org/10.5281/zenodo.12710

Weijmans A-M, de Zeeuw PT, Emsellem E, Krajnović D, Lablanche P-Y, Alatalo K, Blitz L, Bois M, Bournaud F, Bureau M, Cappellari M, Crocker AF, Davies RL, Davis TA, Duc P-A, Khochfar S, Kuntschner H, McDermid RM, Morganti R, Naab T, Oosterloo T, Sarzi M, Scott N, Serra P, Verdoes Kleijn G, Young LM (2014) The ATLAS ^{3D} project - XXIV. The intrinsic shape distribution of early-type galaxies. Mon Not R Astron Soc 444(4):3340–3356. https://doi.org/10.1093/mnras/stu1603

Wertz O, Orthen B, Schneider P (2018) Ambiguities in gravitational lens models: impact on time delays of the source position transformation. Astron Astrophys 617:140. https://doi.org/10.1051/0004-6361/201732240

Weymann RJ, Latham D, Angel JRP, Green RF, Liebert JW, Turnshek DA, Turnshek DE, Tyson JA (1980) The triple QSO PG1115 + 08: another probable gravitational lens. Nature 285(5767):641–643. https://doi.org/10.1038/285641a0

Williams LLR, Saha P (2000) Pixelated lenses and H_0 from time-delay quasars. Astron J 119(2):439–450. https://doi.org/10.1086/301234

Winn JN, Rusin D, Kochanek CS (2004) The central image of a gravitationally lensed quasar. Nature 427(6975):613–615. https://doi.org/10.1038/nature02279

Witt HJ (1996) Using quadruple lenses to probe the structure of the lensing galaxy. Astrophys J Lett 472:1. https://doi.org/10.1086/310358

Wong KC, Suyu SH, Matsushita S (2015) The innermost mass distribution of the gravitational lens SDP.81 from ALMA observations. Astrophys J 811(2):115. https://doi.org/10.1088/0004-637X/811/2/115

Wong KC, Chan JHH, Chao DC-Y, Jaelani AT, Kayo I, Lee C-H, More A, Oguri M (2022) Survey of Gravitationally-lensed Objects in HSC Imaging (SuGOHI). VIII. New galaxy-scale lenses from the HSC SSP. ArXiv e-prints, arXiv:2203.12805

Xu D, Springel V, Sluse D, Schneider P, Sonnenfeld A, Nelson D, Vogelsberger M, Hernquist L (2017) The inner structure of early-type galaxies in the illustris simulation. Mon Not R Astron Soc 469(2):1824–1848. https://doi.org/10.1093/mnras/stx899

Young P (1981) Q0957+561: effects of random stars on the gravitational lens. Astrophys J 244:756–767. https://doi.org/10.1086/158752

Young P, Gunn JE, Kristian J, Oke JB, Westphal JA (1980) The double quasar Q0957+561 A, B: a gravitational lens image formed by a galaxy at z=0.39. Astrophys J 241:507–520. https://doi.org/10.1086/158365

Young P, Deverill RS, Gunn JE, Westphal JA, Kristian J (1981b) The triple quasar Q1115+080A, B, C: a quintuple gravitational lens image. Astrophys J 244:723–735. https://doi.org/10.1086/158750

Young P, Gunn JE, Kristian J, Oke JB, Westphall JA (1981a) Q0957+561: detrailed models of the gravitational lens effect. Astrophys J 244:736–755. https://doi.org/10.1086/158751

Young AJ, Keeton CR, Baker AJ (2022) A new method for the reconstruction of strongly lensed galaxies with resolved kinematics. Astrophys J 929(1):6. https://doi.org/10.3847/1538-4357/ac59af

Zhang G, Mishra-Sharma S, Dvorkin C (2022) Inferring subhalo effective density slopes from strong lensing observations with neural likelihood-ratio estimation. Mon Not R Astron Soc. https://doi.org/10.1093/mnras/stac3014

Publisher's Note Springer Nature remains neutral with regard to jurisdictional claims in published maps and institutional affiliations.

Authors and Affiliations

A.J. Shajib[1,2] · G. Vernardos[3,4,5] · T.E. Collett[6] · V. Motta[7] · D. Sluse[8] · L.L.R. Williams[9] · P. Saha[10] · S. Birrer[11,12,13] · C. Spiniello[14,15] · T. Treu[16]

✉ G. Vernardos
georgios.vernardos@epfl.ch

A.J. Shajib
ajshajib@uchicago.edu

1 Department of Astronomy and Astrophysics, University of Chicago, Chicago, IL, 60637, USA

2 Kavli Institute for Cosmological Physics, University of Chicago, Chicago, IL, 60637, USA

3 Institute of Physics, Laboratory of Astrophysics, Ecole Polytechnique Fédérale de Lausanne (EPFL), Observatoire de Sauverny, 1290, Versoix, Switzerland

4 Department of Astrophysics, American Museum of Natural History, Central Park West and 79th Street, New York, NY, 10024-5192, USA

5 Department of Physics and Astronomy, Lehman College of the CUNY, Bronx, NY, 10468, USA

6 Institute of Cosmology and Gravitation, University of Portsmouth, Burnaby Rd, Portsmouth, PO1 3FX, UK

7 Instituto de Fisica y Astronomia, Facultad de Ciencias, Universidad de Valparaiso, Avda. Gran Bretana 1111, Valparaiso, Chile

8 STAR Institute, Quartier Agora - Allée du six Aout, 19c B-4000, Liége, Belgium

9 School of Physics and Astronomy, University of Minnesota, 116 Church Street SE, Minneapolis, MN, 55455, USA

10 Physik-Institut, University of Zurich, Winterthurerstrasse 190, 8057, Zurich, Switzerland

11 Kavli Institute for Particle Astrophysics and Cosmology and Department of Physics, Stanford University, Stanford, CA, 94305, USA

12 SLAC National Accelerator Laboratory, Menlo Park, CA, 94025, USA

13 Department of Physics and Astronomy, Stony Brook University, Stony Brook, NY, 11794, USA

14 Department of Physics, University of Oxford, Denys Wilkinson Building, Keble Road, Oxford, OX1 3RH, UK

15 INAF – Osservatorio Astronomico di Capodimonte, Via Moiariello 16, 80131, Naples, Italy

16 Department of Physics and Astronomy, University of California, Los Angeles, 430 Portola Plaza, Los Angeles, CA 90095, USA

Space Science Reviews (2024) 220:19
https://doi.org/10.1007/s11214-024-01051-8

Strong Lensing by Galaxy Clusters

P. Natarajan[1,2,3] · L.L.R. Williams[4] · M. Bradač[5,6] · C. Grillo[7] · A. Ghosh[4] · K. Sharon[8] · J. Wagner[9]

Received: 18 April 2023 / Accepted: 23 January 2024 / Published online: 15 February 2024
© The Author(s) 2024

Abstract
Galaxy clusters as gravitational lenses play a unique role in astrophysics and cosmology: they permit mapping the dark matter distribution on a range of scales; they reveal the properties of high and intermediate redshift background galaxies that would otherwise be unreachable with telescopes; they constrain the particle nature of dark matter and are a powerful probe of global cosmological parameters, like the Hubble constant. In this review we summarize the current status of cluster lensing observations and the insights they provide, and offer a glimpse into the capabilities that ongoing, and the upcoming next generation of telescopes and surveys will deliver. While many open questions remain, cluster lensing promises to remain at the forefront of discoveries in astrophysics and cosmology.

Keywords Gravitational lensing · Strong gravitational lensing · Galaxy clusters

1 Introduction

In hierarchical structure formation in the Universe, per the standard cold dark paradigm, galaxy clusters are now understood to be the most recently assembled and most massive

✉ P. Natarajan
priyamvada.natarajan@yale.edu

✉ M. Bradač
Marusa.Bradac@fmf.uni-lj.si

1 Department of Astronomy, Yale University, New Haven, CT, 06511, USA

2 Department of Physics, Yale University, New Haven, CT, 06511, USA

3 Yale Center for Astronomy & Astrophysics, Yale University, New Haven, CT, 06520, USA

4 Minnesota Institute for Astrophysics, School of Physics and Astronomy, University of Minnesota, Minneapolis, MN, 55455, USA

5 Department of Physics, University of California, Davis, CA, 95616, USA

6 University of Ljubljana, 1000, Ljubljana, Slovenia

7 Department of Physics, University of Milan, 20122, Milano, MI, Italy

8 Department of Astronomy, University of Michigan, Ann Arbor, MI, 48109, USA

9 Bahamas Advanced Study Institute and Conferences, 4A Ocean Heights, Hill View Circle, Stella Maris, Long Island, Bahamas

structures to form. Clusters are therefore also expected to be the largest repositories of dark matter. As a result, clusters are some of the most efficient gravitational lenses known to date. Zwicky (1937) was the first to point out that gravitational lensing by galaxy clusters might serve as a valuable tool to measure the amount of this unseen mass component and permit study of distant, magnified objects that lie behind clusters. Despite his bold prediction, the lack of appropriate resolution and imaging facilities at the time and the lack of theoretical understanding of how structure assembles deferred progress in gravitational lensing studies. Several decades later, the presence of a foreground cluster was proposed by Young et al. (1980) as the explanation for a larger-than-expected separation between the images of a double quasar in the first detected double quasar Q0957+561 (Walsh et al. 1979). With this observational impetus and hence renewed interest, the lensing properties of clusters were revisited and explored in detail by Narayan et al. (1984).

The first observational detection of strong lensing by a cluster—a giant arc in Abell 370—was reported in Lynds and Petrosian (1989) and independently by Soucail et al. (1987). These extremely elongated distorted shapes seen in the core of the massive cluster Abell 370 were recognized by Paczynski (1987) as merging images of a strongly lensed background galaxy, and were soon confirmed as such with the measurement of the redshift of the arc by Soucail et al. (1988). With this discovery of giant arcs, the strong lensing regime was established as a property of the deflection produced by dense cluster cores.

With the discovery of the weak, systematic alignment in the shapes of distant background galaxies produced in the outer regions of clusters as reported in Tyson et al. (1990), observational evidence for both strong and weak lensing were confirmed. Several significant early papers (Schneider 1984; Blandford and Narayan 1986; Blandford et al. 1989; Kochanek 1990; Miralda-Escude 1991; Kaiser 1992; Kaiser and Squires 1993) helped establish the formalism for "inverting" observed lensing effects into mapping the mass distributions of galaxy clusters. Subsequent technological advances in the CCD imaging that allowed deeper and sharper images as well as the development and availability of spectrographs drove the rapid discovery of cluster lenses.

The truly transformative revolution in lensing, and strong lensing studies in particular, came with imaging from space with the deployment of the *Hubble Space Telescope*. After correction of its initially blurry images with the repair mission that installed the WFPC-2 camera in 1993, Hubble recovered its image sharpness, and one of the first post-repair image releases was the now iconic image of the massive cluster lens Abell 2218 (Kneib et al. 1996), replete with strong lensing features that are discernible by eye. High resolution images combined with a large field of view have been a game changer for studies of cluster lenses (e.g., Smail and Dickinson 1995), for example, for tracing out the circularly averaged density profile of equilibrium clusters and comparing these to the predictions of cosmological models (Sect. 5.1), and for studying the most distant galaxies using the magnifying power of cluster lenses (Sect. 5.3). Strong lensing features in clusters are now detected in the inner one arcminute central region and on arc-second scales (see left panel of Fig. 2 for a gallery of HST detected dramatic cluster arcs). On smaller scales, multiple images produced by individual cluster galaxies have also been detected in the deepest HST images (Natarajan et al. 1998, 2017; Meneghetti et al. 2020).

The *James Webb Space Telescope* (JWST), launched late 2021, looked at a galaxy cluster, SMACS J0723.3-7327, as one of its first targets, offering an unprecedented and detailed view of a massive lens in infra-red wavelengths. Figure 1 compares the HST images to the newly obtained JWST images (top two panels), and the bottom panel shows the X-ray image superimposed on the JWST image. The improvement in the quality of the JWST image, compared to an already spectacular HST image is visually apparent and striking.

Fig. 1 The galaxy cluster lens SMACS J0723.3-7327 at $z = 0.39$. *Top left:* HST image of the cluster showing multiple strongly lensed features visible by eye. *Top right:* JWST image of the same cluster that in turn reveals a significantly larger number of strongly lensed features. The improvement in the quality of the image is evident upon visual inspection. *Bottom:* Chandra X-ray image showing the spatial distribution of the hot X-ray gas (blue) overlaid over the JWST image. The images span nearly the same field of view and scale. The top right and bottom panels are about 2.35 arcmin wide, or 750 kpc at the redshift of the cluster, $z = 0.388$

2 Strong Lensing as a Probe of Cluster Properties

Clusters are powerful astrophysical laboratories as they are objects where the interplay of dark matter and baryons can be effectively probed on multiple physical scales (see reviews by Kneib and Natarajan 2011; Umetsu 2020). On spatial scales greater than about 1-few kpc, clusters are dominated by dark matter, even at their centers. The baryonic component in clusters is mostly in the form of very hot plasma with $T \sim 10^7 K$, the Intra-Cluster Medium (ICM) detected through its bremsstrahlung, or free-free, emission observed in X-rays. The stellar component is mostly confined to cluster member galaxies, including, in many cases, a massive central dominant (cD) galaxy or the brightest cluster galaxy (BCG) that has been known to have an extended stellar distribution (Schombert 1986; Uson et al. 1991) and deeper recent observations have helped delineate this diffuse emission from a population of intra-cluster stars more clearly, referred to as the Intra-Cluster Light (ICL) component, in the form a diffuse cloud of light, which appears to trace the overall dark matter distribution with great fidelity (Gonzalez et al. 2005; Montes and Trujillo 2019). Notably, as discussed first by Uson et al. (1991), disentangling the BCG and the extended cD envelope is challenging even today. Even including all the baryonic components in the mass inventory in clusters, they still stand as dark matter dominated cosmic objects.

As noted in Sect. 1, the central $\lesssim 0.3$ Mpc of galaxy clusters are the regions that contain a high concentration of matter (dark + baryonic); these are the spatial scales on which the gravitational coupling between baryons and dark matter manifests. Currently the dark matter component of the Universe is believed to be cold and collision-less, rendering it inert in terms of interactions aside from gravity with the baryonic components. Cosmological simulations of structure formation over cosmic time find that dark matter settles into a near universal density profile across several decades in mass—the Navarro, Frenk and White profile (NFW)—with an inner slope of $\rho \sim r^{-1}$, that is expected in clusters of galaxies as well. However, if for example, dark matter interacts weakly with itself, this unusual, additional self-interaction, would preferentially produce discernible effects in this central dense region (see Sect. 5.5). While the mass distribution in the innermost regions is typically dominated by baryons, fortunately, these regions are also the ones that are best probed by strong lensing observations.

One of the key measured properties for clusters is their mass and the detailed spatial distribution of the mass. The radial mass distribution in clusters can be probed and measured using 3 independent methods that all map the underlying gravitational potential: (i) the dynamics of cluster member galaxies assuming they are in equilibrium and robustly trace the gravitational potential, (ii) the X-ray emission from the hot plasma, also assuming it is in hydrostatic equilibrium with the cluster potential, and (iii) gravitational lensing, that makes no prior assumptions about the gravitational potential. The advantage of lensing as a technique is that light deflection is achromatic, and is also entirely independent of the dynamical state of the cluster. Lensing observations therefore permit study of merging and out-of-equilibrium clusters as well. Besides, lensing delivers not just the projected, *radially averaged* mass profiles for cluster, but also permits mapping of the *detailed* two-dimensional mass distribution.

The power of cluster lensing derives from the fact that deeper imaging data not only permit probing the detailed spatial distribution of dark matter in the cluster but that it also simultaneously acts as a cosmic telescope allowing us to probe faint sources in the background Universe. Strong lensing by galaxy clusters brings into view distant, faint background galaxies that would otherwise remain inaccessible to even the most powerful telescopes on the ground or in space; the high magnification of intermediate and high redshift star-forming galaxies allows us to study the building blocks of these galaxies at an unprecedented spatial scale and at a fraction of the observing time otherwise necessary. Modeling and characterizing the mass distribution of cluster lenses therefore stands to open a new window into the early Universe by magnifying the most distant first sources that assemble, a prospect keenly awaited for with JWST data. However, as massive clusters are rare objects, defining systematic criteria to characterize cluster lens samples has proved to be challenging. Furthermore, given the dependence of the strength of lensing on the underlying geometry of the Universe, lensing has also been established as a powerful method to constrain cosmological parameters that characterize the underlying world model (Gilmore and Natarajan 2009; Jullo et al. 2010, 2015; Magana et al. 2018; Caminha et al. 2022a). There are synergies in the general modeling approaches for characterizing the lensing properties of individual galaxies and those adopted for clusters. However, the existence of a richer phenomenology and more extensive data available in both the strong and weak lensing regimes, warrants more sophisticated models tailored for lensing clusters. We outline the range of mass modeling techniques that have been developed and applied to cluster lenses in Sect. 3.

The Hubble Frontier Fields program (HFF thereafter), a key Director's Discretionary large program (Lotz et al. 2017) dedicated 840 HST orbits to ultra-deep strong lensing imaging of six selected massive cluster lenses. These fields were also observed across several wavelengths, adding to wealth of information for modeling and deeper understanding of clusters and high redshift Universe. These coordinated multi-wavelength surveys include

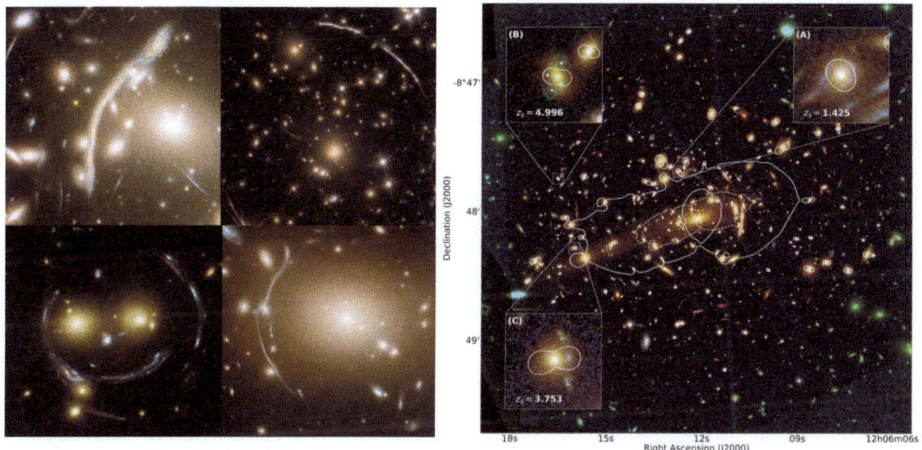

Fig. 2 Left Panel: Gallery of HST detected cluster arcs seen in the cluster Abell 370 (top-left); SDSS J1038+4849 (bottom-left); PSZ1 G311.65−18.48 (top-right) and Abell 611 (bottom-right). **Right Panel:** GGSL in the central region of the HFF galaxy cluster MACSJ1206: (Meneghetti et al. 2020). The image is a composite and combines HST observations in multiple bands: F105W, F110W, F125W, F140W, F160W (red channel), F606W, F625W, F775W, F814W, F850LP (green channel), F435W and F475W (blue channel). The dashed and solid lines in the right panel show the lens critical curves for potential sources at redshifts of 1 and 7, respectively. Panels A, B, C are zoom-ins of three individual GGSL events produced by sources measured to be at redshifts $z = 1.425$, $z = 4.996$, and $z = 3.753$. The GGSL sources in panel B were not visible in the HST image and were discovered in an observation with the Multi-Unit-Spectroscopic-Explorer (MUSE) spectrograph of the VLT

ALMA (González-López et al. 2017), the Spitzer Frontier Fields [1] and Chandra, and JVLA Frontier Fields Campaigns van Weeren et al. (2017). HFF has yet again transformed our understanding of the mass distribution and evolutionary states of clusters. Chosen as the most efficient lenses, these six massive cluster lenses permit probing high mass end of the cluster mass function. Intriguingly, with deeper HFF imaging, in addition to the giant arcs and multiple image families observed in cluster lenses, additional strong lensing features produced by individual cluster members, referred to as Galaxy-Galaxy Strong Lensing (GGSL), have also been detected (see the right panel of Fig. 2 for some notable GGSL examples).

3 Modeling the Mass Distribution in Cluster Lenses

The strong distortion of space-time by a massive cluster is so significant, that its effect on the shapes of background galaxies is unambiguously detectable, as evident in HST and even in ground-based images. Background sources, strongly lensed by foreground galaxy clusters result in the production of magnified, extremely distorted images referred to as giant arcs; and multiple images of the same source, that exhibit similar color and morphology, that albeit may have different levels of distortion and mirror symmetry. This wide range of phenomena—ranging from their geometry, brightness and observed spatial locations in the cluster—can and is used as input observational constraints to reconstruct the total mass of the cluster lens. Utilizing the observed distorted images to recover the original undistorted

[1] https://irsa.ipac.caltech.edu/data/SPITZER/Frontier/overview.html.

shapes enables constraining the mass distribution of the cluster lens, yielding "lens models". Lensing observations and the derived foreground cluster mass distributions have revealed that the bulk of the gravitational potential in clusters comprises the undetected "dark matter".

Several independent lens modeling algorithms to recover cluster mass distributions have been developed, widely tested and used by the strong lensing community. Regardless of the specific technique, all strong lensing modeling algorithms use as input constraints the positions of multiple images of the same background sources, sometimes referred to as "families", and ideally, when available the measured spectroscopic redshifts of the sources. Although the magnification in the strong lensing regime is conducive to follow-up spectroscopy to obtain redshifts, it is challenging, and often photometric redshifts are used instead.

The purpose of strong lens modeling algorithms is to find a representation of the projected mass distribution on the lens plane that offers the smallest scatter between the observed and model-predicted lensing observables. In the source plane (a.k.a. "source plane minimization"), this is accomplished by minimizing the scatter of the model-predicted (unobserved) source position: for a set of N images of the source, the source position β_i is calculated for each image i using the lens equation:

$$\beta_i = \theta_i - \alpha_i = \theta_i - \frac{D_{\rm ls}}{D_{\rm os}} \hat{\alpha}_i, \tag{1}$$

where α_i is the deflection angle at the observer, $\hat{\alpha}$ is the deflection angle at the lens plane, and $D's$ are angular diameter distances connecting lens, source and observer; and θ is observed source position (a schematic of the lensing geometry can be found in Fig. 3 of the review by Kneib and Natarajan 2011). In the image plane, the predicted positions of counter-images are calculated for each observed image, and compared to the observed ones. The latter, commonly referred to as "image plane minimization", is significantly more computationally intensive since the lensing equation cannot be inverted. The best-fit model is typically identified by maximizing a likelihood function, which usually means minimizing a χ^2 function. Finally, missing information can be dealt with in different ways, such as setting priors on the source redshifts based on photometric information using what are now standard methods, or allowing the model to predict the unknown redshifts of sources.

The different approaches to lens modeling are generally divided into two broad categories, referred to as "parametric" and "free-form" (the latter is also often referred to as "non parametric") methods. Parametric approaches involve representing the mass distribution of the cluster lens with a set of physically-motivated profiles for the surface mass density, which are well described by analytic functional forms, and profile specific properties such as the ellipticity, and positional angles. Parametric models are conceptually motivated by cosmological simulations of structure formation in which cluster scale masses are observed to comprise of large and small scale dark matter halos that are gravitationally bound. Cluster mass distributions are modeled as a superposition of one or more large scale components, corresponding to the smooth cluster and/or group scale dark matter halo seen in simulations and several small scale sub-halos that are associated with the positions of individual cluster member galaxies. The overall gravitational potential of a cluster is therefore modeled as a sum of contributions from large scale dark matter halos and smaller scale sub-halos:

$$\phi_{\rm cluster} = \phi_{\rm smooth} + \Sigma_i\, \phi_{\rm sub}. \tag{2}$$

Fig. 3 Left: HST Frontier Fields image of the cluster MACS 0416 at $z = 0.397$. Right: Overlaid on HST image is reconstructed dark matter distribution of the cluster recovered using `Lenstool`shown as a blue haze. Image Credit: NASA/Hubble, ESA, HST Frontier Fields and J. Lotz, M. Mountain, A. Koekemoer, and the HFF Team (STScI). Acknowledgement for overlaid reconstructed dark matter map for Abell 2744: Mathilde Jauzac (Durham University, UK), Jean-Paul Kneib (Ecole Polytechnique Federale de Lausanne, Switzerland) & Priyamvada Natarajan (Yale University, USA). Image taken by the Advanced Camera for Surveys (ACS) and WFC3 including B, V, I, Y, J, H & W filters, with a FOV that is 2.08 x 2.32 arcminutes

As a result, at each position in the cluster, we can explicitly write the convergence, $\kappa(\mathbf{x})$, and the shear, $\gamma(\mathbf{x})$, as a sum over these components (Natarajan and Kneib 1997),

$$\kappa(\mathbf{x}) = \kappa_{\text{smooth}}(\mathbf{x}) + \sum_i \kappa_{\text{sub},i}(\mathbf{x}) \tag{3}$$

$$\gamma(\mathbf{x}) = \gamma_{\text{smooth}}(\mathbf{x}) + \sum_i \gamma_{\text{sub},i}(\mathbf{x}) \tag{4}$$

where κ_{smooth}; γ_{smooth} and $\kappa_{\text{sub},i}$; $\gamma_{\text{sub},i}$ are the contributions to the convergence and shear from the smooth component and from the sub-halos respectively.

Examples of commonly used analytic density profiles used to model the gravitational potentials are: a generalized Navarro et al. (1997), and truncated pseudo isothermal models (PIEMDs) that permit the potential to have finite mass and non-zero ellipticity or cored isothermal models. Parametric models have therefore been particularly powerful for comparing properties of observed cluster lenses with theoretical predictions from cold dark matter simulations. An example of parametric reconstruction of the mass distribution of the HFF cluster MACS 0416 with the publicly available software `Lenstool` is shown in Fig. 3.

On the other hand, free-form approaches describe the gravitational lens without assuming a predetermined mass density profile. For example, the lens plane is divided into a fixed or adaptive grid. The lens plane is then populated with some basis functions set, each grid cell is assigned one member of that set. Depending on the method, the number of grid cells can range between a few dozen and many thousands. The amplitude, i.e., mass of each basis function is determined directly by image properties. The motivation for free-form methods is that the mass distribution need not follow simple parametric forms, especially on cluster-wide scales and in merging clusters.

Modeling approaches vary by the degree to which they assume that mass and light distribution in clusters are correlated. Most, if not all, of the parametric approaches assign mass to observed individual cluster-member galaxies, motivated by empirical correlations like the Faber–Jackson relation derived from observations of early-type galaxies in clusters (Natarajan and Kneib 1997). Some algorithms make no such assumption, and rely only on the lensing evidence to reveal the presence of mass. Finally, several algorithms adopt a hybrid approach: some mass components are inserted as physically-motivated parametric halos, while other components within the same model are set as free-form functions (e.g., Niemiec et al. 2020).

As we note in the following section, when sufficient lensing constraints from rich observational data are available the different approaches outlined above perform equally well. However, they do vary in their strengths and weaknesses, and the decision of which approach to adopt depends primarily on the science goal and motivating question. The primary advantage of the parametric approach is that it can produce satisfactory results even with a small number of constraints that sparsely sample the image plane, as the assumed functional form that describes the lens plane fills in the gaps. And it is particularly well suited for comparison of the properties of observed cluster lenses with simulated clusters. Therefore, for tests of the CDM paradigm, where large cosmological simulations are the only available theoretical test-bed, parametric lensing models have been particularly useful and insightful (Natarajan et al. 2007, 2017; Meneghetti et al. 2020). The high flexibility[2] of free-form models, on the other hand, require rich data sets as a high density of lensing constraints in all regions of the lens plane. However, given sufficient constraints, free-form models can be used to compare properties like the radial slope of the projected mass density against theoretical predictions, and have the flexibility to uncover substructure that may be potentially not be associated with light (Ghosh et al. 2021). Models that do not include a prior assumption about the correlation between mass and light are particularly useful for investigating this exact question—to what extent does mass follow light in the Universe as a function of physical scale (Sebesta et al. 2016; Ghosh et al. 2023).

Development of the formalism and first cluster-scale lens modeling attempts appeared in the literature nearly 25 years ago (Narasimha and Chitre 1989; Kovner 1989; Kneib et al. 1993, 1996; Natarajan and Kneib 1997; Abdelsalam et al. 1998). As noted above, at present, with the availability of richer data from deeper HST imaging surveys over the past two decades, a number of reliable cluster lens mapping methods are available as tabulated below. While the most commonly used methods tend to be parametric, recent years have seen the introduction of several new free-form methods.

- `Lenstool` (parametric; Kneib et al. 1996; Jullo et al. 2007a; Jullo and Kneib 2009; Niemiec et al. 2020). An extension of `Lenstool` includes free-form B-splines in addition to parametric functions (Beauchesne et al. 2021; Limousin et al. 2022).
- Light-Traces-Mass (LTM, parametric; Broadhurst et al. 2005a; Zitrin et al. 2009a, 2015a), which closely ties the total mass to the observed light distribution.
- `GLAFIC` (parametric; Oguri 2010; Ishigaki et al. 2015; Kawamata et al. 2016).
- `GRAVLENS` (parametric; Keeton 2010; McCully et al. 2014).
- GLEE (parametric; Suyu and Halkola 2010; Suyu et al. 2012).
- Weak & Strong Lensing Analysis Package (WSLAP+, hybrid; Diego et al. 2005, 2007, 2016b), which combines parametrized galaxy-scale mass distributions with a free-form distribution on larger scales.
- Strong and Weak Lensing United (SWUnited, free-form; Bradač et al. 2006, 2009), which, unlike most other methods reconstructs lensing potential, not projected mass.

[2]It may be somewhat confusing that free-form models are also called "non-parametric", since they often have *more* free parameters than parametric models.

- PixeLens (free-form; Williams and Saha 2004; Saha et al. 2006; Köhlinger and Schmidt 2014), which is primarily designed for modeling systems with small number of images, but has been used for cluster reconstruction.
- `Grale` (free-form; Liesenborgs et al. 2006; Williams and Liesenborgs 2019; Liesenborgs et al. 2020; Ghosh et al. 2021), which uses an adaptive grid and a genetic algorithm to find solutions in a high dimensional parameter space.
- JPEG parametrization (free-form; Lam 2019), which takes inspiration from JPEG compression to do strong lensing inversion.
- `relensing` (free-form; Torres-Ballesteros and Castañeda 2023), which reconstructs gravitational deflection potential on an adaptive irregular grid.
- `MARS` (free-form; Cha and Jee 2022), which uses a MaxEnt-regularized strong lensing inversion.

3.1 Comparison of Mass Modeling Methods Using Synthetic and Real Clusters

The independent cluster lens modeling methodologies outlined above have been carefully compared and contrasted in order to better understand their relative advantages and limitations. Below, we describe significant community-wide comparison exercises conducted to date and their conclusions. The first (Sect. 3.1.1) involved invitation to multiple teams to reconstruct the mass distribution of two simulated lensing clusters from their produced images wherein all teams were provided with the same input constraints on the identified families of multiple images and their redshifts. By construction, the details of the true underlying mass distributions in this case were known. This exercise was in preparation for the production of publicly available magnification maps from multiple teams for the HFF sample.

For a second exercise (Sect. 3.1.2), since it involved deep HFF imaging data sets, the mapping teams first arrived at a consensus set of input constraints, i.e., mutually agreed upon the identification of multiple lensed image families and either their inferred or measured redshifts to use uniformly for lens mass modeling. And from this cooperative and collaborative effort, magnification maps generated from various independent codes and teams were then made publicly available for the larger astronomical community to use. These maps have been successfully deployed for studies beyond lensing since.

The third exercise (Sect. 3.1.3) was an unanticipated continuation of the previous one, with the serendipitous discovery of a lensed supernovae in the field of two HFF clusters MACS J1149.6+2223 and Abell 2744. Multiple independent teams with their respective methodologies predicted magnification (for the cluster Abell 2744), and the arrival time and magnification of the final image (in the case of MACS J1149) as its observation was awaited.

3.1.1 Comparison of Lens Modeling Methods: Reconstructing the Mass and Magnification of Simulated Cluster Lenses

In this detailed comparison exercise Meneghetti et al. (2017) compared mass distributions derived for the simulated clusters *"Ares"* and *"Hera"* from ten independent modeling groups. Simulated observations of these two simulated clusters that mocked the depth and resolution of HFF observations, were used as the reference sample. *Ares* was generated using the semi-analytical code MOKA, as a superposition of analytical elliptical profiles representing the main cluster and its member galaxies consistent with cosmological initial conditions. *Hera*, on the other hand was more realistic, as it was directly taken from the output of a large volume, high resolution cosmological N-body simulation of the cold dark matter Universe. A concise summary of the comparison results are presented in a figure in Meneghetti et al.

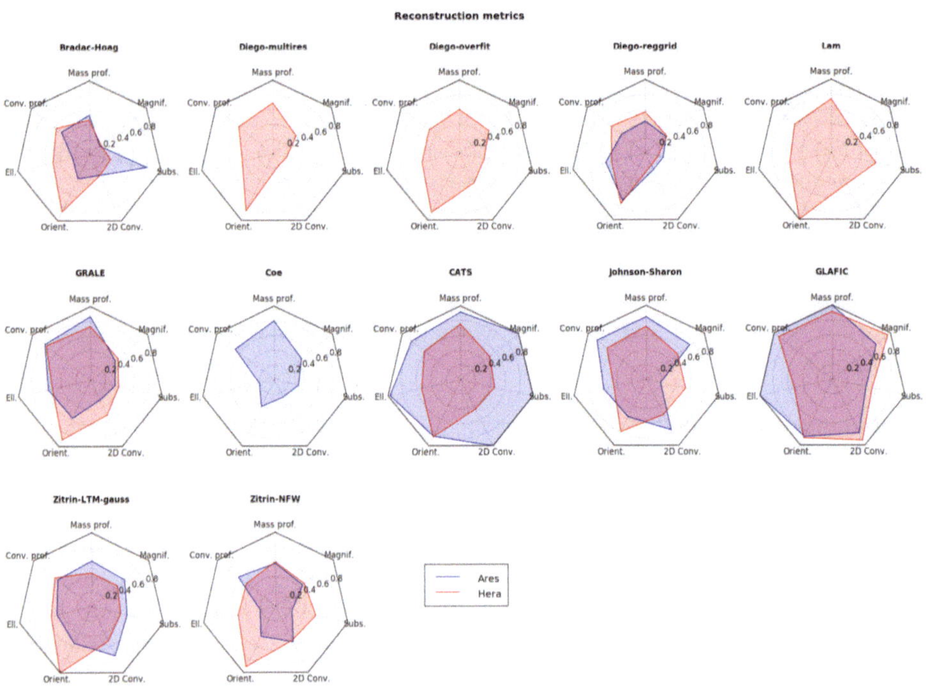

Fig. 4 Reproduced from Meneghetti et al. (2017), an HFF cluster lens mass model comparison project. Their caption reads: Radar plot showing the scores of each model for all metrics discussed in the paper. Larger polygons correspond to better overall performance. Each chart corresponds to a different lens model (see labels on the top) and shows results for both *Ares* (blue) and *Hera* (red), or whichever is available. The seven metrics are shown on the vertices of each chart. For each metric, the scores range from 0 (worst; plotted at the centre of the chart) to 1 (best; plotted at the vertex), normalized to the maximum value recorded by all models. A filled polygon is obtained by connecting the plotted scores of all metrics for each reconstruction

(2017), which we reproduce here as Fig. 4. It is a so-called radar plot; each of the 7 spokes represents a particular cluster property, such as mass profile slope, ellipticity, magnification, etc. If a property is reconstructed well (poorly) by a model the corresponding point is close to the vertex (center) of the black and white polygon. Reconstructed models of *Ares* and *Hera* are represented by blue and red colored polygons respectively. The larger the colored area the better that model reproduces the properties of the synthetic cluster. Upon detailed comparison it was found that several lensing properties that relied on integrated quantities were recovered equally well by the different lens modeling approaches using completely independent methodologies. For instance, the reconstructed mass profiles of the simulated clusters across methods matched at the level of $\sim 10\%$ for both parametric and free-form methods; and the agreement on the mass enclosed within the Einstein radius, was found to be extremely robust for all methods. The results also showed that the parametric modeling methods tend to be generally better at capturing two-dimensional properties of the lens core. This was true in particular for *Ares*, because its true mass distribution resembled assumptions made by parametric methods. In addition, parametric models that permit comparison of additional quantities derivable directly from simulations like the subhalo mass function, and the radial distribution of subhalos, fared well.

Predictably, the uncertainties in reconstructed magnification maps grew as a function of the magnification itself, resulting with largest uncertainties around the critical lines where the magnification diverges. The accuracy in the magnification estimates dropped by a factor of three when the magnification increased from $\mu_{\rm true} = 3$ to $\mu_{\rm true} = 10$. Another interesting result from this work was that even groups that used the same mass reconstruction algorithms with slightly different priors (e.g., CATS and Johnson-Sharon teams that both used the code `Lenstool`) and very similar inputs, obtained somewhat different maps, revealing the impact of the choice of priors and enabling quantifying their downstream effect.

3.1.2 Comparison of Lens Modeling Methods: Magnification Maps for the HFF Sample

The six clusters of the HFF sample were selected to be amongst the most efficient lensing clusters on the sky and as they were legacy fields observed with Directors Discretionary time, the data was released immediately to the larger astronomical community. However, for exploitation of this unique data set, to interpret the properties of faint background lensed galaxies that these clusters bring into view for instance, reliable lensing maps for each cluster were required to be made available publicly as well. To do so, five independent lens modeling groups were selected to provide preliminary magnification maps for each of the HFF clusters prior to the observing campaign in order to facilitate rapid analysis of the HFF data by all members of the community. These models were initially based on a common set of input data, derived from shallower pre-HFF archival HST imaging, early spectroscopic campaigns, and a common shared catalog of lensed background galaxies. Once the data were taken, these models were continually collectively updated by each of the groups. As noted before the techniques adopted by groups spanned both parametric and free-form methods.

The public HFF lens models include maps of mass (κ) and shear (γ) from which magnifications can be derived for sources at any redshift using scripts provided on the HST site. The models cover regions constrained by strongly lensed, multiply-imaged galaxies, within the HST ACS fields of view of the cluster cores. Some models extend to larger areas, including the HFF parallel fields, and also incorporate ground-based weak lensing data. All data, models and detailed description of individual methods and maps are available at the following website: https://archive.stsci.edu/prepds/frontier/lensmodels/.

Since cluster magnification is of paramount importance for the HFF project, several papers compared magnification distributions as reconstructed by various independent groups. Figure 5 shows magnification maps of Abell 2744 from 6 HFF teams, analyzed in Priewe et al. (2017). It was found that in the broadest terms the reconstructed magnification maps more or less agreed with each other. However, differences are also seen, and often exceed their quoted statistical errors. The resulting dispersion ranged from 30% at low magnifications, $\mu \sim 2$, to 70% at high magnifications, $\mu \sim 40$. This implies that the true uncertainties are probably currently underestimated in most of the modeling methods. These differences are mainly attributed to lensing degeneracies, suggesting that a small lens plane rms value is not a sufficient condition for a model to robustly and reliably predict the true magnifications.

The more recent magnification comparison study by Raney et al. (2020b) also arrives at similar conclusions, finding $\sim 40\%$ dispersion at low magnifications ($\mu \sim 3$) that increases to $\sim 65\%$ dispersion at high magnifications ($\mu \sim 10$). In contrast, excellent agreement exists for the derived integrated quantities like the cumulative mass distribution and it is significantly well constrained within $1'$ radius from the cluster core, with $< 5\%$ dispersion. Additionally, Remolina González, Sharon and Mahler (2018) examined source plane scatter in 10 mass reconstruction models for the HFF cluster MACS J0416. Using source plane rms as a metric of diagnose lens model performance, they quantified the ability of different models to predict unknown multiple images. They found that while free-form reconstructions rely heavily on the quality and quantity of the lensing data, parametric models are not as

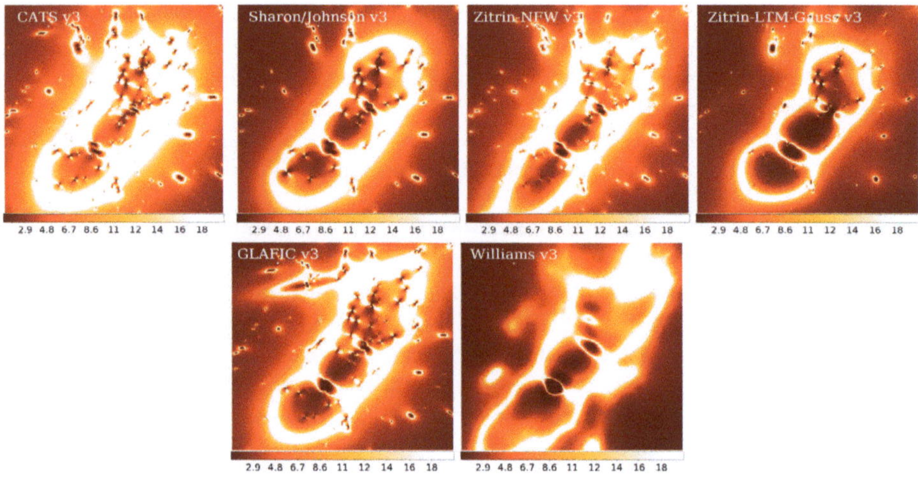

Fig. 5 Magnification maps of 6 different reconstructions of galaxy cluster Abell 2744. The reconstruction team names appear in the top left of each panel. Magnification scale is the same for all panels, and goes from $\mu = 1$ (darkest color) to $\mu = 20$ (white). Each panel is a square 100 arcsec on the side. Taken from Priewe et al. (2017)

susceptible to uncertainties in the data. However, parametric models do not necessarily benefit from adding more constraints with the inclusion of photometric redshift measurements when the existing amount of spectroscopic data available is large (of the order of measured redshifts for a hundred or so lensed images), as in the case of the HFF clusters.

3.1.3 Comparison of Lens Modeling Methods: Using Time Delays and Magnifications of Quasars and Supernovae

After the serendipitous discovery of the Supernova Refsdal in 2014, multiply imaged by a cluster member galaxy in the HFF cluster MACS J1149.6+2223 at $z = 0.544$, it was predicted that there would be the rare opportunity to see the supernova again in about one year, after the four images had faded away Kelly et al. (2015). This is because the initially observed four-image pattern was only one component of the lensing configuration. This supernova might have appeared as a single image some 10-20 years prior in the cluster field, and it reappeared at the predicted position between mid-November 2015 and December 11, 2015 (the approximately one month uncertainty is the interval between two consecutive Hubble observations). This was in good agreement with the blinded model predictions made by several independent groups deploying the range of lens reconstruction techniques prior to its observation Sharon and Johnson (2015), Oguri (2015), Diego et al. (2016a), Treu et al. (2016). The time delay between the original quadruplet observed in 2014 and the latest appearance of the supernova in 2015 was used to infer the value of the Hubble constant. This is the first time this technique, originally proposed by Refsdal, was successfully applied to supernovae Vega-Ferrero et al. (2018).

In the case of another supernova detected behind the HFF lensing cluster Abell 2744 at $z = 0.308$, once again, specific predictions for the magnification derived from different lensing models have been compared in detail. Rodney et al. (2015) computed the magnification predictions from various lens models for the location and redshift of the Type-Ia supernova HFF14Tom, leveraging the idea that the true luminosity of a SN-Ia can be inferred from

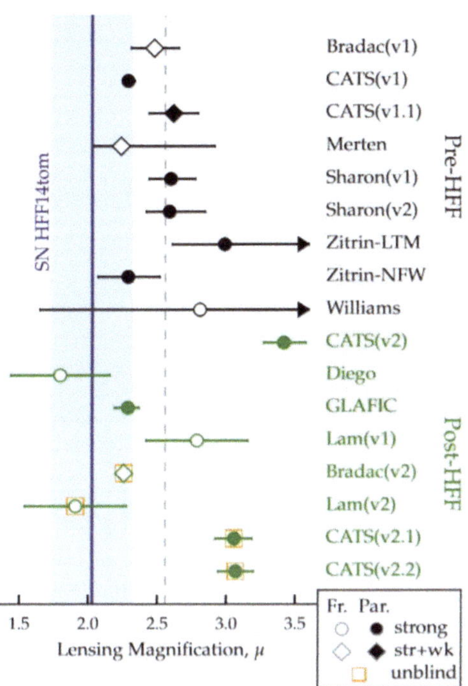

Fig. 6 The observed lensing magnification is compared to predictions from multiple independent lens models. The vertical blue line shows the constraints from SN HFF14Tom, with the shaded region marking the total uncertainty. Markers with horizontal error bars show the median magnification and 68% confidence region from each model. Circles indicate models that use only strong-lensing constraints, while diamonds denote those that also incorporate weak-lensing measurements. Free-form (parametric) models are shown as open (filled) markers. The top half, with points in black, shows the 9 models that were constructed using only data available before the start of the HFF observations. The lower 8 models in green use additional input constraints, including new multiply imaged systems and redshifts. The final four points, with square orange outlines, are the "unblind" models that were generated after the magnification of the SN was known. The black dashed line marks the unweighted mean for all 17 models, at $\mu = 2.6$. Image reproduced with permission from Rodney et al. (2015), copyright by AAS

its lightcurve (see Fig. 6). They report that the median of the lensing model magnification predictions is 25% higher than the magnifications deduced from the observed brightness of the supernova, and that many model predictions disagreed by more than their stated uncertainties. A possible reason for the discrepancy is that there are not enough spectroscopically confirmed multiple image systems to constrain the lens models: accurate and precise image redshifts are as important for lens reconstructions as image positions.

3.1.4 Comparison of Lens Modeling Methods: Detailed Comparison of Spatial Regions Within Clusters

In addition to the exercises described above, it is instructive to compare regions of special interest in the reconstructed mass maps generated by various lens modeling methods. These could include regions around main cluster galaxies, or those that have excess reconstructed mass but are deficient in light. Ghosh et al. (2021) show that reconstructions with distinct modelling philosophies – parametric vs. free-form – often lead to similar results regarding specific mass features, while at the same time, models using the same algorithm can draw

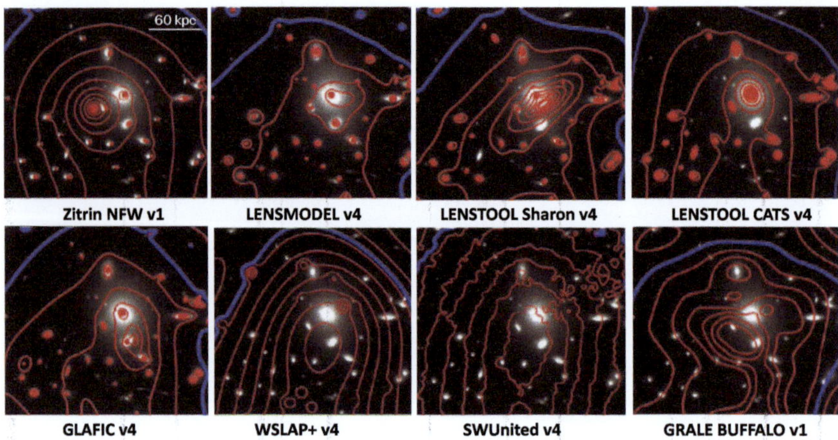

Fig. 7 Comparison between different lens inversion methods mass maps of the same region in the HFF cluster Abell 370, centered on the northern of the 2 brightest cluster galaxies. Each panel is a square of 45 arcseconds on the side, or ~ 250 kpc. The gray scale is the HST image in ACS/F814W filter. The panels show eight different reconstructions by various teams. The contours are those of equal projected dimensionless surface mass density, κ, with the blue contour corresponding to $\kappa = 1$. All but 3 (LENSMODEL HFFv4, LENSTOOL CATS HFFv4, Glafic HFFv4) show that the local center of mass and light do not coincide, with the center of mass offset the center of the BCG. For maps with offset centers, the offset is generally to the South of the BCG, though the morphology and the degree of displacement differs between models

contrasting conclusions, just by using somewhat different model priors and data constraints. Figure 7 presents a region in an HFF cluster Abell 370, centered on the northern BCG. The red contours show the projected surface mass density, κ, with the blue contour indicating $\kappa = 1$. Most models show that the center of mass appears to be displaced, generally towards the South, compared to the location of the center of light. This inferred feature of misalignment between mass and light is consistent with the merging, disturbed nature of this cluster. The fact that not all mass reconstructions agree on the detailed mass distribution in that region, including the misalignment between mass and light, is mostly due to lensing degeneracies, which are present not just on cluster scale, like the mass sheet degeneracy, but also on smaller scales (Lasko et al. 2023). A larger number of multiply imaged sources than currently available will help to map out that region more accurately (Lin et al. 2023) in the future. Other interesting examples are presented in Ghosh et al. (2021) and Ghosh et al. (2023). The high spatial resolution with which strong lensing can measure the mass distribution can test predictions of dark matter properties (e.g., Harvey et al. 2019). In summary, what these results imply is the importance of continued stress-testing of all lens mass modeling methods and continued comparison between them to delineate systematics that need to be understood in greater detail in cluster lensing studies.

3.2 Connecting the Strong and Weak Lensing Regimes

While strong-lensing driven mass modeling techniques have been used to obtain high signal-to-noise reconstructions in the central regions of clusters, the power of weak lensing is critical to extend the derived mass distributions out to the virial radius (e.g., Clowe et al. 2006b; Umetsu et al. 2014; Umetsu 2020). Connecting strong lensing and weak lensing regimes has therefore long been seen as a powerful way to recover the detailed and extended mass distribution of clusters from the central region to the outskirts. The method also has the advantage of breaking the so-called mass-sheet degeneracy (MSD), which says that the surface mass density κ is invariant under the following global transformation $\kappa(\theta) \to \lambda\kappa(\theta) + (1 - \lambda)$, where λ is between 0 and 1. Non-zero λ rescales the original surface mass density, and adds a constant density sheet to the lens. As a result, the overall density slope becomes less steep (Schneider and Seitz 1995). Weak lensing helps break MSD because it is reasonable to assume that beyond the weak lensing regime, i.e., far from the cluster center, cluster density is negligible.

Many of the lens modeling methods described above leverage the potency of this combination and therefore include both weak and strong lensing data, and some methods also include higher order information from flexion as well. Flexion parameters are similar to shear, but instead of the second derivatives of the lensing potential, they encode the third derivatives (e.g., Goldberg and Natarajan 2002; Bacon et al. 2006; Lanusse et al. 2016). When both shear and flexion are important, image shapes deviate from purely elliptical distortion and can look 'banana'-shaped. Flexion has been used to reconstruct mass distribution in some clusters (e.g., Okura et al. 2008).

The simplest adopted approach is to combine strong and weak lensing in a parametric fashion (e.g., `Lenstool` has included strong and weak lensing out to large cluster radii; Natarajan et al. 2009; Niemiec et al. 2020; Schäfer et al. 2020). These and other methods that were used to model cluster cores previously have an advantage that they are numerically stable, they are, however, inherently limited to the fitting a set of parametrized models Natarajan et al. (1998). Recently, free-form methods have also been used for combined analysis (Diego et al. 2005; Bradač et al. 2006; Merten 2016; Liesenborgs et al. 2020). Modeling that include combining the two lensing regimes has been successful in predicting mass profiles on all scales and in determining other observables like the mass-concentration relation in clusters (Merten et al. 2015). They have also proven useful in reconstructing magnification fields for high-redshift galaxies beyond the regime where strongly lensed images at lower redshifts are typically found as this pertains to the regime and region where magnification is poorly constrained (Priewe et al. 2017). Inclusion of higher order constraints like flexion can also add useful information in this intermediate regime as demonstrated by Cain et al. (2016), Leonard et al. (2007).

Another successful example of combining strong and weak lensing mass reconstruction has been the case of the Bullet Cluster (Fig. 8, Clowe et al. 2006a; Bradač et al. 2006). The Bullet cluster comprises a pair of plane-of-the-sky merging sub-clusters, where the dominant baryonic mass in the form of X-ray emitting gas, and the total mass, dominated by dark matter are not aligned post collision on the sky, providing strong evidence that dark matter is collisionless. In particular strong lensing provides useful constraints – for instance, the offset between the peak of the baryonic mass, as traced by X-rays, and the peak of the total mass, as traced by strong and weak lensing, is over an arcminute, and was measured at $> 10\sigma$ (Bradač et al. 2006), while the limits with only weak lensing were 8σ (Clowe et al. 2006a). The Bullet Cluster configuration – the location of the mass peaks, the offset between baryonic and dark matter distributions and the inferred relative velocity of the colliding sub-clusters was a challenge to reproduce for cold dark matter simulations Springel and Farrar (2007). The strong constraints on the inferred positions of the peaks of mass distribution provides a challenge for simulations that assume non-negligible self-interaction cross sections for dark matter (Randall et al. 2008; Lage and Farrar 2015). In addition, results from combined analysis also provide useful and meaningful constraints on several alternative gravity models as they fail to predict large mass concentration needed to produce strong lensing constraints (e.g., Angus et al. 2006).

Many clusters now have successful measurements of mass and detailed mass distributions in combined regimes. HFF clusters have been reconstructed using combined strong

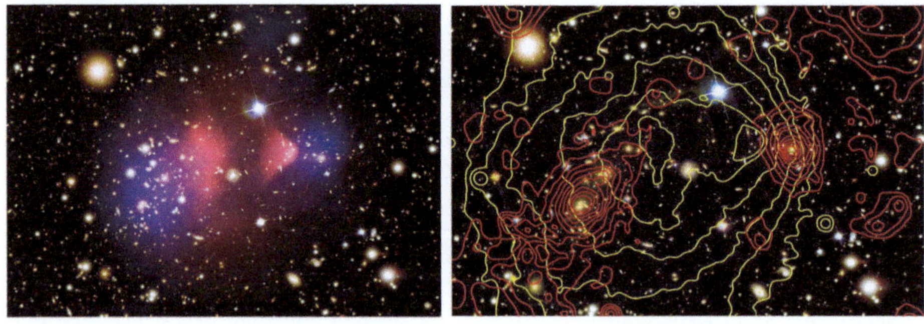

Fig. 8 Left: The color composite of the Bullet Cluster using only weak lensing measurement. Overlaid in *blue* shade is the surface mass density map from the weak lensing mass reconstruction. The X-ray emitting plasma is shown in *red*. Credit NASA/CXC/CfA/STScI/Magellan, Clowe et al. 2006a. Right: Strong and weak lensing reconstruction for the Bullet Cluster. Overlaid in red contours is the surface mass density κ from the combined weak and strong lensing mass reconstruction. The X-ray brightness contours from the 500 ks Chandra ACIS-I observations are overlaid in yellow. Bradač et al. 2006, 2009

and weak lensing analysis; see for example (Jauzac et al. 2016; Merten et al. 2011; Wang et al. 2015; Hoag et al. 2016; Finney et al. 2018; Strait et al. 2018). The scientific results from these studies are discussed in Sect. 5.

4 Surveys for Cluster Lenses

4.1 HST Cluster Lens Samples: CLASH, RELICS, HFF, BUFFALO Surveys

HST has devoted significant resources to cluster lensing observations via multiple programs including CLASH, RELICS, HFF and BUFFALO. Significant progress in the field of galaxy cluster physics has been made thanks to the data collected within the HST Multi-Cycle Treasury program Cluster Lensing And Supernova survey with Hubble (CLASH; Postman et al. 2012), complemented with a comprehensive and coordinated spectroscopic campaign carried out with the ground-based telescopes (e.g., the CLASH-VLT Large Programme; Rosati et al. 2014).

The CLASH survey was awarded 524 orbits of HST time (GO 12066; PI: M. Postman) to observe 25 massive (virial mass $M_{\rm vir} \approx$ 5-30 $\times 10^{14}\, M_{\odot}$, and X-ray temperature $T_{\rm X} \geq$ 5 keV) galaxy clusters in 16 broadband filters, ranging from approximately 2000 to 17000 Å, with the Wide Field Camera 3 (WFC3) and the Advanced Camera for Surveys (ACS). The sample, spanning a wide redshift range ($z = 0.18 - 0.90$), was carefully chosen to be largely free of lensing bias and representative of relaxed clusters, on the basis of their symmetric and smooth X-ray emission profiles. CLASH had four main scientific goals: 1) to measure the total mass profiles of cluster over a wide radial range by using strong and weak lensing observations; 2) to detect new Type Ia supernovae out to redshift $z \sim 2.5$ to improve the constraints on the dark energy equation of state; 3) to discover and study the highly magnified first galaxies that formed after the Big Bang ($z > 7$) and are brought into view by the cluster lens; and 4) to study galaxy evolution by comparing cluster member properties with those of and background galaxies (for a thorough overview, see Postman et al. 2012).

A Large Programme (186.A-0798, PI: P. Rosati; Rosati et al. 2014) of 225 hours with the VIMOS instrument at the VLT was also approved to perform a panoramic spectroscopic survey of the 14 CLASH clusters that are visible from ESO-Paranal. This observational campaign was aimed at measuring in each cluster the redshifts of 1) approximately 500 cluster members within a radius of 3 Mpc; 2) determine spectroscopic redshifts for 10-30 lensed multiple images inside the HST field of view, including possible highly-magnified candidates out to $z \approx 7$ (e.g., Balestra et al. 2013); 3) possible supernova hosts. In the first CLASH cluster that was followed-up (i.e., MACS J1206.2−0847), the spectroscopic redshifts were exploited to build robust strong lensing models (Zitrin et al. 2012a, Umetsu et al. 2012, Eichner et al. 2013, Grillo et al. 2014); to obtain an independent total mass estimate from the spatial distribution and kinematics of the cluster members (Biviano et al. 2013, Sartoris et al. 2014); to study the properties of the intracluster light (Presotto et al. 2014), and to characterize the galaxy stellar masses as a function of environment and the stellar mass density profiles (Annunziatella et al. 2014).

Using the unprecedented data set—panchromatic HST imaging and VLT spectroscopic data, astronomers have measured the gravitational potential in the inner regions of the cluster (i.e., the total mass profile in the core) for some of the CLASH clusters, by modeling the multiple images of several newly discovered strong lensing systems (e.g., Zitrin et al. 2015b, Coe et al. 2012, Monna et al. 2014, Grillo et al. 2015). Moreover, the combination of the HST observations with deep, multi-band, wide-field imaging from Subaru has allowed the weak lensing detection and therefore extension of the mapping of the total mass profile of several clusters out to their virial radii (Medezinski et al. 2013, Umetsu et al. 2014, Merten et al. 2015).

The advent of the MUSE instrument at the VLT has revolutionized strong lensing studies in galaxy clusters. MUSE capabilities permit simultaneous identification and study of cluster members and multiply imaged background sources out to $z = 6.6$. The exceptional suitability of MUSE to reach these goals has been demonstrated by several successful programs (e.g., Karman et al. 2015, Richard et al. 2015, Grillo et al. 2016, Caminha et al. 2019, Bergamini et al. 2021), which have enabled the construction of detailed mass models of the cores of cluster lenses.

The results from systematic studies of lensing clusters with HST, including those from CLASH, led to the deeper Hubble Frontier Fields Initiative (HFF; Lotz et al. 2017) that targeted six massive galaxy clusters, for a total of 140 HST orbits per cluster in 7 broad-band filters, achieving in all of them unprecedented depth of $\sim$29 mag (AB) using Director Discretionary Time. This program has detected some of the highest redshift galaxies and enabled the first characterization of this sample of star-forming galaxies in a statistically meaningful way (e.g., Bouwens et al. 2017a). The HFF data have also provided a great opportunity to study the structure of the dark matter halos hosting these clusters (e.g., Caminha et al. 2017, Natarajan et al. 2017, Bergamini et al. 2019, Meneghetti et al. 2020).

The Reionization Lensing Cluster Survey (RELICS; PI: D. Coe; Coe et al. 2019) consisted of 190 orbits of HST time to observe the cores of 41 galaxy clusters in 7 broadband filters, ranging from approximately 0.4 to 1.7 μm with the WFC3 and the ACS. This sample, spanning a wide redshift range (z = 0.18-0.97), was carefully chosen to be representative of the most massive clusters ($M_{500} \gtrsim 8 \times 10^{14}$ M$_\odot$; thus also many of the most efficient cluster lenses) discovered so far, on the basis of their Sunyaev-Zel'dovich (SZ) mass estimates released by the Planck collaboration, and other known strong lenses. These include ACT-CLJ0102−49151 (El Gordo) (Menanteau et al. 2012, Zitrin et al. 2013), PLCK G287.0+32.9 with arcs 80$''$ from the core (Gruen et al. 2014) and CL 0152−1357, a massive merging cluster at $z = 0.83$ (Acebron et al. 2019). The HST imaging of the HFF combined with Spitzer imaging (390 hours; PIs: M. Bradac & T. Soifer) yielded strongly lensed high-z candidates among the best and brightest known.

The latest contribution to the lensing data comes from The Beyond Ultra-deep Frontier Fields and Legacy Observations (BUFFALO) Program, with 101 orbits and 101 parallels as a part of Hubble Space Telescope's Treasury program, taking data from 2018-2020 (Steinhardt et al. 2020). BUFFALO expanded the existing coverage of the HFF fields in additonal filters: WFC3/IR F105W, F125W, and F160W and ACS/WFC F606W and F814W around each of the six HFF clusters and flanking fields. These fields are already covered by additional deep multi-wavelength datasets, including Spitzer and Chandra. Further important additions to the data are provided by the spectroscopic surveys, such as the Pilot-WINGS (Lagattuta et al. 2022).

4.2 X-ray and SZ Surveys for Cluster Lenses

X-ray selected clusters have been invaluable in identifying potential strong lensing clusters. Horesh et al. (2010) concluded that clusters from the MAssive Cluster Survey (MACS; Ebeling et al. 2001) are significantly more—by a factor of ~ 6—efficient as lenses than optically selected clusters. Many of the MACS clusters are now household names among those doing cluster lens modeling. X-ray clusters are also used in another capacity. Several efforts have been underway at X-ray wavelengths to obtain complementary information on the cluster hot gas component (e.g., Ogrean et al. 2015). Owing to the adopted selection criteria, in all the clusters of the CLASH sample, measurements of the X-ray emission of the hot intracluster medium, taken with the Chandra telescope (and in some cases also with the XMM-Newton telescope) were available. These observations are in most cases sufficiently deep to estimate the hot gas and total mass profiles, under the assumption of hydrostatic equilibrium (Donahue et al. 2014). In some clusters, a combined strong lensing, photometric and X-ray analysis has allowed a precise decomposition of the cluster total mass profiles into their stellar, hot-gas and dark-matter components (Annunziatella et al. 2017, Bonamigo et al. 2017, 2018, Granata et al. 2022). Systematic searches for clusters at high redshifts, $z \sim 1$, has become possible using their Sunyev-Zeldovich (SZ) emission. The important surveys were done by the Planck CMB satellite and the South Pole Telescope (SPT). The masses of these clusters can be calibrated through weak lensing (Sereno et al. 2017; Raghunathan et al. 2019; Schrabback et al. 2021).

4.3 Ground-Based Imaging Surveys for Cluster-Scale Strong Lenses

Wide optical and spectroscopic surveys such as the Sloan Digital Sky Survey (SDSS; Blanton et al. 2017) have been a fruitful hunting ground for several strong lensing surveys; from galaxy-lensed quasars, through galaxy-galaxy lenses, to the most massive cluster lenses. Teams have used different methods to identify candidate lenses in these public data. Some of the most well studied cluster lenses (e.g., the "poster child" cluster lens Abell 1689) are hardly unidentifiable as such in shallow SDSS imaging. The numerous multiple images in Abell 1689 are too faint to be robustly identified. Most shallow, ground-based surveys for cluster lenses aim to discover the lensing effect itself, i.e., an occurrence of highly distorted and magnified galaxy in the form of one of more giant arcs. The Sloan Giant Arcs Survey (SGAS; PI: M. Gladders; Bayliss et al. 2011; Sharon et al. 2020) yielded hundreds of cluster lenses by mining the SDSS DR7 imaging data. Starting with the SDSS public catalog, SGAS first identified clusters and groups using the red sequence technique (Gladders and Yee 2000). g, r, i, and z cutouts around these fields were then examined by eye to identify lines of sight with bright arcs, and given scores. Lens and arc candidates were confirmed through imaging and spectroscopy with larger telescopes, reaching close to 100% followup completeness of high-scoring candidates. The selection process was designed such that purity and completeness could be quantified, making the results useful for statistical studies of lensing occurrence (a.k.a arcs statistics, see Sect. 5.4 and cluster studies. Due to its selection function, the SGAS sample provided a bountiful supply of highly magnified background sources at "cosmic noon" – those galaxies that reside in the epoch when the Universe formed most of its stars. The high magnification makes feasible detailed spectroscopic studies of the physical properties of these galaxies (see Sharon et al. 2020, and references therein), also leading to offshoot surveys such as MEGaSaURA (Rigby et al. 2018). Similar methods for discovery and confirmation of strong lensing clusters were employed in searching through the RCS-2 survey (Gilbank et al. 2011) for lenses, and most recently in the COOL-LAMPS (Khullar et al. 2021) survey.

4.4 Cosmological and Large Area Surveys

There are multiple on-going and planned surveys that will garner large samples of cluster lenses. While these are expected to be shallower than the HST cluster lens samples discussed above, the majority of them stand to detect and exploit weak lensing. One of recently concluded ground-based surveys is the Dark Energy Survey (DES).

DES is a large survey of distant galaxies including clusters that aims to uncover the nature of dark energy that drives cosmic acceleration using multiple techniques, that include its impact on the abundance of galaxy clusters; weak gravitational lensing signals; type Ia supernovae and the detection of large-scale correlations between galaxies. The survey uses a powerful wide-field imaging camera called the Dark Energy Camera, or DECam, installed on the 4-meter Blanco telescope in Chile. During its five-year campaign (2013-2018), the DECam imaged approximately 5000 square degrees of the sky using five broadband filters, taking advantage of the excellent viewing conditions available on Cerro Tololo. The final DES dataset will consist of precise photometric and morphological information for over 200 million galaxies out to redshifts of 2.0. The survey is also periodically revisiting smaller patches of sky to find and study over 2500 Type Ia supernovae. Forecasts for expected cluster weak lensing results from the DES and other ground based surveys with the Rubin Observatory as well as from the Nancy Grace Roman Space Telescope are promising for constraining cosmological parameters Wu et al. (2021).

The recently successfully launched and deployed Euclid mission consists of a 1.2 m space telescope that is dedicated to study the imprint of dark energy and gravity via two powerful, complementary cosmological probes: weak gravitational lensing and galaxy clustering (via baryonic acoustic oscillations and redshift space distortion). These two complementary probes will capture signatures of the expansion rate of the Universe and the assembly of cosmic structures (Amendola et al. 2013). The key instruments aboard the telescope are a high quality panoramic visible imager (VIS), a near infrared 3-filter (Y, J and H) photometer (NISP-P) and a slitless spectrograph (NISP-S). These instruments will permit accurate shape measurements and therefore quantify weak gravitational lensing effects of dark matter as well as three-dimensional mapping of structures with spectroscopic redshifts of galaxies and clusters of galaxies (Sartoris et al. 2016). Euclid will observe 15,000 $\deg^2$ and is expected to find of the order of 60,000 galaxy clusters in the redshift range $z = 0.2 - 2$. It is forecast that $\sim$ 50,000 of them will have background galaxy densities $>$ 15 galaxies per sq. arcmin allowing accurate determination of cluster masses from weak lensing studies out to the virial radius and beyond (Euclid Collaboration et al. 2019). The imaging and spectroscopic capabilities of Euclid will enable internal mass calibration from weak lensing and the study of the dynamics of cluster galaxies, in combination with ground based cluster surveys like DES (Tutusaus et al. 2020). Euclid is also expected to detect several thousand lensed arcs, as predicted from the analysis with the Skylens simulation package, which has catalyzed the development of new automated arc-finder algorithms. Three "Euclid Deep Fields" covering around 40 deg^2 in total will be also observed extending the scientific scope of the mission to access the very high-redshift universe.

5 Science Results from Modeling Observed Cluster Lenses

5.1 Insights on Larger-Scales

One of the first clusters to have the mass profile derived from observed lensing was Abell 1689 (Broadhurst et al. 2005a). Its profile is consistent with that expected in the LCDM universe, but with a somewhat larger concentration (Broadhurst et al. 2005b; Coe et al. 2010), which can be explained by the lensing selection bias because higher concentration makes a cluster a stronger lens. A number of studies over the next decade looked at the distribution of properties of large sets of clusters (Richard et al. 2010; Zitrin et al. 2011). More recently, Fox et al. (2022) analysed a large sample (74) of RELICS clusters using their strong lensing data. Not all of these were found to be in equilibrium and relaxed, and consequently they display a large diversity in their density profiles, especially in the radial range 50-200 kpc.

In order to test the standard LCDM cosmological model, one needs to select clusters that are in equilibrium, because simulations (Navarro et al. 1997, 2004), and theory (Hjorth and Williams 2010; Williams et al. 2010; Wagner 2020) make specific predictions for the spherically averaged density profiles of relaxed dark matter halos. To determine whether the cluster radial mass distribution is in good agreement with the predictions of, one needs a large radial range of the mass reconstructions. The strong lensing region of clusters, wherein the surface mass density exceeds the critical value, is relatively small, spanning at most one decade in radius. Therefore, it needs to be supplemented with weak lensing data. Umetsu et al. (2016) performed a joint analysis of 20 CLASH clusters using their strong, weak and magnification data, over two decades in radius, from $\sim 30 - 3000$ kpc, corresponding to approximately the virial radius of these clusters. They found that the stacked density profile in these clusters is well described by the fitting formulae from the cold dark matter simulations, i.e., NFW (Navarro et al. 1997) and Einasto (Navarro et al. 2004) profiles, as well theoretical DARKexp (Hjorth and Williams 2010) profiles for relaxed, collisionless, self-gravitating systems. We re-iterate that massive galaxy clusters are uniquely suited for this important test of dark matter properties because their mass is dominated by dark matter over a wide range of spatial scales. In addition to radially averaged density profile, one can also use the median ellipticity of a set of relaxed clusters as a test of standard cosmology and the nature of dark matter. Using the same set of 20 CLASH clusters, Umetsu et al. (2018) found that the median projected axis ratio of clusters' dark matter halos is in agreement with the predictions from recent numerical simulations of the standard collisionless cold dark matter model.

On much smaller scales than discussed above, the density profile in the very central regions of clusters, $\lesssim 50$ kpc, or a few percent of the virial radius, are often hard to estimate because the de-magnified central images (maxima in the Fermat potential) are usually not detected. A detection of two central images has been claimed in the cluster Abell 3827 by (Massey et al. 2018a). Yet, the question of the central profile—whether clusters have flat density cores or cusps—is very important for constraining of the nature of dark matter (also see Sect. 5.5).

Limousin et al. (2022) modeled two unimodal clusters (i.e., with a single dominant center) AS 1063, and MACS J1206, and one bimodal cluster MACS J0416. While the last cluster is an ongoing merger, the dynamical status of the unimodal clusters is not clear; they show some evidence of disturbance, and so are probably not relaxed. A new metric utilizing the computation of the power spectrum of fluctuations in the mass and X-ray gas maps reveals that unimodal clusters that appear smooth may still be out of equilibrium Cerini et al. (2023). The authors conclude that cored profiles interior to $\sim$ 50 kpc are favored over cusped ones in these 3 clusters. However, for the purposes of comparison with cosmological model predictions, and stress-testing the underlying structure formation model itself more clusters, and preferably relaxed ones, need to be analysed to make a conclusive statement about the mass distribution in the very central regions of clusters.

5.2 Mapping Dark Matter Substructure & Galaxy-Galaxy Lensing in Clusters

The standard paradigm that describes structure formation in the Universe is the concordance cold dark matter model with a non-zero cosmological constant (LCDM). In simulations of the LCDM model, where dark matter is collisionless, the internal density distribution in dark-matter halos, as noted previously, converges to a roughly "universal" and cuspy density profile, the NFW profile (Navarro et al. 1996, 1997) over several decades in mass. Moreover, the degree of central concentration of a halo in LCDM depends on its formation epoch, and hence on its total mass (Wechsler et al. 2002, Zhao et al. 2003). In this scenario, objects that virialize early are dense and compact if they get accreted and or bound into a larger halo during hierarchical assembly. These smaller scale collapsed structures are usually referred to as sub-halos that are held within the gravitational potential well of the more massive host halo. Spiraling in toward the center owing to dynamical friction, while they are truncated or disrupted by tidal forces, leads to changes in sub-halo masses, angular momentum and energy (Ghigna et al. 1998; De Lucia et al. 2004).

On cluster scales, observational tests of several concrete LCDM predictions using lensing mass models have been performed. The abundance of sub-halos in clusters, also referred to the mass function of substructure (or the sub-halo mass function) is predicted to be a power law of the form $dn/dm \propto m^{-1.8}$. This prediction has been robustly tested with increasingly higher resolution HST data of cluster lenses and generations of state-of-the-art cosmological simulations ranging from the Millenium simulation (Natarajan and Springel 2004; Natarajan et al. 2007) to the Illustris simulation suite (Natarajan et al. 2017). Predictions of tidal stripping of subhalos in LCDM have also been tested (Natarajan et al. 2002a; Sand et al. 2002, 2004; Natarajan et al. 2009) with tidal radii estimated from lensing mass models. Overall, properties of cluster lenses and their sub-halo mass functions were found to be in good agreement with LCDM predictions. Recently more sophisticated and detailed comparisons with high fidelity mass reconstructions for clusters have become possible, so much so that we are now in a position to stress-test the LCDM model (Grillo et al. 2015, Natarajan et al. 2017, Meneghetti et al. 2020). In fact, Natarajan et al. (2017) report tension between the observed radial distribution of subhalos in the massive HFF cluster lenses compared to simulated clusters in LCDM, an issue that is replicated on galaxy scales as well, wherein the radial distribution of observed satellites is discrepant with theoretical predictions Carlsten et al. (2020).

Meneghetti et al. (2020) reported that the observed number of Galaxy-Galaxy Strong Lensing (GGSL; see the right panels of Fig. 2) events, i.e., strong lensing events observed on the scale of individual cluster galaxies on mass scales of $\sim 10^{11}\,M_\odot$ in cluster lenses, exceeds the expectations from LCDM simulations by more than one order of magnitude. These significant differences in the spatial and mass distributions of the cluster sub-halos do not seem to be fully fixed by the inclusion in cosmological simulations of baryonic physics, in the form of cooling, star formation, and feedback by active galactic nuclei. Robertson (2021) and Bahé (2021) argue that the reported under-abundance of GGSL events in simulations is due to their low resolution. These authors use considerably higher mass resolution runs, and also include several baryonic physical processes. They report that this results in a higher number density of substructures in lensing clusters, leading to comparable simulated and observed occurrence of GGSL events. GGSL is computed using the area enclosed within smaller scale secondary caustics. However, the majority of the contribution to the GGSL cross section computed from the simulations reported in Robertson (2021), Bahé (2021) arises from in-falling group-scale substructures that are more massive than those reported in Meneghetti et al. (2020). In addition, in recent work, Ragagnin et al. (2022) note that the stellar masses of cluster galaxies in the suite of simulations studied by Bahé (2021), Robertson (2021) vastly exceed that of observed cluster galaxies which in turn artificially enhances their computed GGSL cross section.

In continuing work pursuing the origin of this discrepancy Meneghetti et al. (2022) quantify the impact of the numerical resolution and AGN feedback scheme adopted on the predicted GGSL probability to alleviate the gap with observations. Improving the mass resolution by a factor of ten and twenty-five, while using the same galaxy formation model that includes AGN feedback does not affect the GGSL probability. Adopting an AGN feedback scheme that is less efficient at suppressing gas cooling and star formation leads to an increase in the GGSL probability by a factor between three and six, however, while such simulations form overly massive subhalos whose contribution to the GGSL would be higher, their Einstein radii are too large to be consistent with the observations. The primary contributors to the observed GGSL cross-sections are compact subhalos with lower masses (in the mass range of $\sim 10^{10-11.5}\,M_\odot$), which are not present in the appropriate abundance as inferred from observations even in the current highest resolution LCDM simulations. Meanwhile, Ragagnin et al. (2022) find that regardless of the resolution and galaxy formation model adopted, simulations are unable to simultaneously reproduce the observed stellar masses and compactness (or maximum circular velocities) of cluster galaxies and therefore account for GGSL as measured from observed lenses.

Substructure also exists further away from the centers of clusters and their distribution can also be compared to LCDM simulations. An excess of substructures in Abell 2744 as reconstructed from lensing effects compared to the galaxy clusters in the Millenium XXL N-body simulation, as discussed in Schwinn et al. (2017), suggested a mis-match. However, this disagreement can be brought into concordance by more carefully matching the definitions of substructure in simulations and observations. Applying the same definition, reconciles their abundances in Abell 2744 (Schwinn et al. 2018).

An alternative way to look at substructure within clusters is to analyse their projected mass power spectra. This recently proposed approach eliminates the need to identify individual substructures, and relies instead on characterizing the power spectrum of fluctuations as a function of scale to compare the lensing reconstructed mass maps with that of simulated clusters (Mohammed et al. 2016; Cerini et al. 2023).

Strong lensing when combined with kinematics has also made a significant impact on our understanding of the central galaxies in clusters, BCGs, and the dark matter profile slopes in these very central regions of clusters Natarajan and Kneib (1996). Sand et al. (2004, 2008), Newman et al. (2013) found that once the BCG stellar mass of the BCG is subtracted, the dark matter slope is significantly shallower than NFW within $\sim$ 30kpc, comparable to the effective radius of BCGs. Newman et al. (2015a) placed these findings in a larger context of central profile slopes as a function of halo mass. For groups and individual galaxies the central slope get progressively steeper. They conclude that dissipation-less processes are more important in the assembly of BCGs in larger halos, like clusters, leading to shallower slopes. Baryons, through gas cooling, are more important in lower mass halos, like groups and individual galaxies, and lead to steeper central density profiles.

5.3 Cluster Lenses as Nature's Telescopes

One of the key benefits of gravitational lensing by clusters is the magnification they induce on the background galaxy population. Bringing into view faint background sources that would otherwise not be accessible even with the most powerful space-based telescopes, cluster lenses act as cosmic telescopes, that permit detailed studies of the magnified galaxies behind them. Lensing magnification stretches the area behind clusters by the magnification factor, thereby increasing the flux we receive from distant galaxies as lensing conserves surface brightness. Acting as nature's telescopes, clusters are expected to serendipitously reveal the earliest forming galaxies as magnification enhances their reach to probe the very early Universe.

At intermediate redshifts, the high lensing magnification by galaxy clusters allows us to peer *into* galaxies. A particularly valuable utility is in studying star forming galaxies at "Cosmic Noon", $z \sim 1-3$, a critical epoch for the assembly of the massive galaxy population in the Universe. Galaxies that would otherwise be too faint for detailed spectroscopy to be practical, even with the largest telescopes, are magnified by factors of tens to hundreds, making their detailed study feasible (Rigby et al. 2018). This is thanks to the overall magnification, which significantly reduces the required observing time for mid- and high-resolution spectroscopy. The high linear magnification, which is manifested as (usually tangential) distortion, resolves the internal structure of galaxies, allowing us to measure the sizes and physical properties of even individual star forming clumps within them (e.g., Johnson et al. 2017). Detailed lens models are then deployed to place the measurements in context of their intrinsic (source plane) values, reverse the magnification effect, and reconstruct the galaxy morphology (e.g., Sharon et al. 2022a).

A key output that lens mass modeling provides is the calculation of the area in the source plane that has been magnified by a given factor μ. The area in the source plane that has (unsigned) magnification greater than μ scales as μ^{-2}. As an example, we present in Fig. 9, taken from Johnson et al. (2014), the magnification power of the six HFF clusters modeled by `Lenstool`, plotted as the cumulative area (left axis) and volume (right axis) magnified above a certain value as a function of magnification. The source plane area in this illustrative case was computed by mapping background sources at $z = 9$, and summing the area that is magnified by more than a given magnification factor. To estimate the volume, the area is multiplied by the co-moving distance between $z = 8.5$ and $z = 9.5$. On average, the total $z = 9$ area that is observed through each cluster lens is about 20% of the WFC3/IR FOV. Approximately 10% of this area is magnified by more than a factor of six, which is equivalent to lowering the limiting magnitude by at least 2 magnitudes.

Cluster lenses have brought into view intrinsically faint and extremely high redshift sources, including a candidate Lyman-break galaxy MACS0647-JD at $z = 11$ in the CLASH survey Coe et al. (2013) and most recently galaxy candidates detected by JWST at $z \sim 9-13$ magnified by modest factors of 3–8 by the foreground cluster WHL0137 that was observed in 8 JWST filters spanning 0.8–5.0 microns and previously in 9 HST filters spanning 0.4–1.7

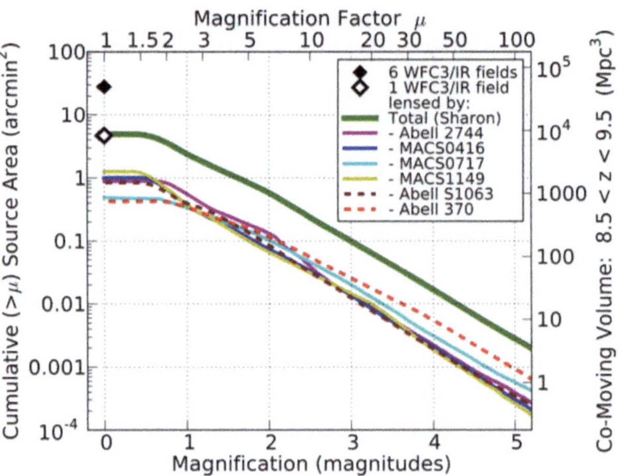

Fig. 9 The cumulative area (left axis) and co-moving volume (right axis) of background sources at $8.5 < z < 9.5$ lensed by each of the HFF clusters to magnification μ and higher. We show, for reference, the corresponding area and co-moving volume of one and six WFC3/IR field of view regions (diamonds). Image reproduced with permission from Johnson et al. (2014), copyright by AAS

microns. Recent data from the JWST UNCOVER & GLASS surveys have also revealed the presence of a high density of bright galaxies behind the cluster lens Abell 2744 (Castellano et al. 2023). Other cluster surveys like RELICS have uncovered several highly magnified sources at $z \sim 6-7$ (Acebron et al. 2018). In addition, the number counts of highly magnified sources obtained from cluster lenses permit constraining the faint end of the galaxy luminosity function from moderate to high redshifts (see for example results from the use of HFF cluster lenses: Alavi et al. 2016; Bouwens et al. 2017b; Livermore et al. 2017; Atek et al. 2018; Yue et al. 2018; Bouwens et al. 2022). The faint end slope encodes important information about the efficiency of star formation, a key unknown in our current understanding of galaxy formation. JWST with its access to the highest redshift, early forming galaxies is anticipated to transform these studies in the near future.

5.4 Statistics of Giant Arcs in Clusters

Bright, merging multiple images of extended background sources lensed by clusters often appear as curved, highly elongated arcs, called giant arcs. Originally, arc statistics were introduced in Grossman and Narayan (1988) to investigate the abundance of giant arcs as a test of the hypothesis that these features were actually produced by gravitationally lensing. The abundance and the observable properties of giant arcs depend on the total mass and density distribution of clusters, their redshift evolution, the properties and evolution of sources, the nature of dark matter, as well as global cosmology (e.g., Bartelmann et al. 1998; Williams et al. 1999; Wambsganss et al. 2004; D'Aloisio and Natarajan 2011; Mahdi et al. 2014). In recent years the global cosmological parameters are estimated from other observations, mostly the CMB and baryonic acoustic oscillations (BOA), however giant arcs can be potentially used as a probe of redshift evolution of non-linear structure in the Universe, and the nature of dark matter.

The impact of the evolution of structure formation, in particular for different types of dark matter, on strong gravitational lensing observables is detailed in Mahdi et al. (2014) and Elahi et al. (2014) in which warm and cold dark matter structure growth is compared with respect to their giant arc statistics. Both works conclude that warm dark matter (WDM) clusters show more extended and massive substructures than those of their cold dark matter (CDM) counterparts, thus enhancing the shear and convergence. In addition, these WDM substructures reach farther out into the cluster halo than the CDM ones. These effects lead to WDM clusters having enhanced giant arc production and not a reduced one as one would naively expect, if one considers only that small-scale structures are smoothed out in WDM cosmologies. Beyond the single lens plane, the impact of deflecting structures along the line of sight which are uncorrelated with the main lens are analysed in, e.g., Puchwein and Hilbert (2009), Bayliss et al. (2014), Li et al. (2019). They conclude that such structures need to be taken into account in order not to underestimate the abundance of giant arcs, because line-of-sight structures can boost the lensing cross-section for individual clusters by up to 50%.

Complementary to using giant arcs, highly magnified but undistorted images (e.g., Zitrin and Broadhurst 2009) can also be used to probe the structure of clusters (Williams and Lewis 1998; Morioka and Futamase 2015). Another approach to arc statistics is to search for observations of uncommon and irregular multiple image configurations deviating from the theoretically predicted ones (usually fold and cusp configurations) (Orban de Xivry and Marshall 2009; Meena and Bagla 2021). Finding extremely rare image properties can refute the underlying assumptions, e.g., about the statistical distribution of dark matter in a given background cosmology. Though a potentially powerful diagnostic, statistics of giant arcs still needs further improvement in order to be turned into a reliable probe. The main difficulties—accounting for selection effects in sufficiently realistic lensing mass models (e.g., Li et al. 2016; Plazas Malagón 2020)—will need to be tackled to make use the upcoming data: a 30-fold increase in the number of giant arcs from the full sky surveys like *Euclid* (Meneghetti et al. 2013). Extensive reviews of the development of arc statistics are given in the introduction of Caminha et al. (2013), and in the overview paper by Meneghetti et al. (2013).

5.5 Constraints on the Nature of Dark Matter

Though in the standard cosmological model dark matter particles are cold and collisionless, and none of the existing observations rule that out, the possibility remains that dark matter, or some fraction thereof, is either warm (WDM) or self-interacting (SIDM). Cluster strong lensing can be used to test the properties of dark matter particles and constrain the level of self-interaction.

WDM: predicts the existence of numerous sub-galactic scale dark matter halos, which would be largely invisible through in electromagnetic emission (Diemand et al. 2007; Springel et al. 2008). In WDM cosmologies, these low mass subhalos would be erased by the large streaming length of relativistic particles (Bose et al. 2017; Rossi 2020). Specifically, and WDM predict very different abundances of subhalos with masses $\lesssim 10^8 - 10^9 \mathrm{M}_\odot$. Therefore, measuring subhalo abundances and properties down to this mass range will help indicate what cosmological model we live in. Isolated subhalos are very hard to detect gravitationally, however, in galaxies and clusters of galaxies their small individual lensing effects can be detected because they are superimposed and are amplified by the magnification effect of the underlying larger scale lens Şengül et al. (2023). In clusters, their presence will be especially easy to detect near the critical curves created by the cluster as whole, where magnifications are very high over ~arcsec scales. These probes hold great promise (Kaurov et al. 2019; Dai et al. 2020; Griffiths et al. 2021; Diego et al. 2022), and as suggested in recent work, new lens modeling methodologies like the use of the curved arc bases might permit future detection Şengül et al. (2023).

SIDM: The recent interest in self-interacting dark matter was spurred by apparent discrepancies with the predictions of on galaxy scales (Spergel and Steinhardt 2000) – the so-called cusp-core problem. And even though most of these observations can now probably be accounted for with collisionless dark matter, the interest in dark matter self-interactions persists. There are a few cluster-scale measurements that can be performed using strong lensing to place constraints on dark matter particle self-interaction cross-section per unit mass, σ/m. Some of the earliest proposed tests were cluster ellipticity (Miralda-Escudé 2002): SIDM should result in rounder cluster cores and the sizes of tidal truncation radii for subhalos, that provide an estimate on the interaction cross-section (Natarajan et al. 2002b). However, it was later shown (Peter et al. 2013) that ellipticity is likely subject to other dynamical effects, and cluster core sizes may likely provide a better diagnostic. The size of constant density cores in clusters can also place constraints on SIDM; a recent core estimate of $\lesssim$ 4kpc in Abell 611 implies $\sigma/m \lesssim 0.1\,\mathrm{g/cm^2}$ (Andrade et al. 2019). Using results of BAHAMAS-SIDM hydrodynamic simulations Robertson et al. (2019) discuss lensing-based test of SIDM for future cluster surveys. For the current CLASH survey they conclude that cross-section $\sigma/m \gtrsim 1\,\mathrm{g/cm^2}$ (for systems with velocity dispersions of $1000\,\mathrm{km\,s^{-1}}$) can be ruled out. Baryons play an important role in determining the properties of clusters' central regions. Taking their distribution into account, Despali et al. (2019) conclude that future wide-field surveys might be able to distinguish between CDM and SIDM models based on the distribution of the Einstein radii of cluster lenses. Because SIDM leads to flat density cores in clusters, the central bright galaxies are expected to wobble (i.e., wander away from the cluster center) and result in larger cluster-mass dependent offsets in SIDM cosmologies, compared to LCDM. Based on BAHAMAS-SIDM simulations, Harvey et al. (2019) conclude that their sample of 10 observed clusters is consistent with CDM at 1.5σ level. One of the additional predictions of SIDM is that dark matter halos can undergo gravothermal collapse late in their evolution, similar to that of Globular Clusters, thereby increasing the central halo densities by a factor of several (Yang and Yu 2021). In that case one could expect individual galaxies in galaxy clusters to act as exceptionally strong lenses, producing $\sim$ arcsecond scale multiple images. As noted previously many such GGSLs have recently been observed in galaxy clusters (Meneghetti et al. 2020), and though their high abundance still remains an open question within collisionless, SIDM accompanied by gravothermal collapse may offer an alternative if warranted.

Clusters offer yet another powerful test of the nature of dark matter. Similar to gas ram pressure stripping, self-interactions between dark matter particles would tend to transfer momentum between two cluster-scale halos in merging clusters, or between cluster-scale halo and embedded galaxy-scale halo, resulting in offsets between baryonic and dark matter peaks. These effects have been investigated in simulations (Kahlhoefer et al. 2015), and signatures of these interactions were searched for in galaxy clusters. After initial reports about finding offsets (Massey et al. 2015), better data acquired later showed that the offsets were consistent with zero interaction cross-section (Massey et al. 2018b). Robertson et al. (2017) use the Bullet Cluster and estimated that even an interaction cross-section as large as $\sigma/m = 2\ \mathrm{cm^2 g^{-1}}$ is not ruled out by this cluster. However, they also caution that measuring offsets between baryonic and dark matter peaks must be done consistently in simulations and observations and this remains a challenge.

5.6 Cosmography with Cluster Lenses

If a source is strongly lensed into two images, the difference in time that light takes to reach the observer from the two different directions, i.e., the time delay (Δt) between the two images due to the difference in path length, is related to the gravitational potential of the lens ($\phi[\theta]$) and the so-called time-delay distance ($D_{\Delta t}$). The latter scales with the three angular-diameter distances, observer-deflector (D_d), observer-source (D_s), and deflector-source (D_ds): $D_{\Delta t} \propto D_\mathrm{d} \times D_\mathrm{s}/D_\mathrm{ds}$. Since each of the distances is proportional to c/H_0, it implies that $D_{\Delta t} \propto 1/H_0$. The uncertainty in $D_{\Delta t}$ (and thus on H_0) is approximately the sum in quadrature of the uncertainty in time-delay ($\delta\Delta t$) and the uncertainty in differences in gravitational potential ($\delta\phi$). The time delays observed in the multiple images of time-varying sources strongly lensed by galaxy clusters have typical values of several months/years, so they can be measured with an approximate 1-3% precision (e.g., Fohlmeister et al. 2013, Dahle et al. 2015, Kelly et al. 2016). Most recently, Muñoz et al. (2022) report $< 0.05\%$ precision on the 3 time delays of the quasar in cluster SDSS J1004+4114, which has been monitored for 14.5 years. Supernova, with their known lightcurves can provide precise estimates with considerably shorter monitoring. The main remaining uncertainty lies in the potential of the cluster lens. With sufficient photometric and spectroscopic observational data, it may be possible that $\delta\phi \lesssim 5\%$ can be reached in lens galaxy clusters (e.g., Grillo et al. 2015, 2016). This corresponds to a remarkable precision of $\sim$6% on the value of H_0 from a single lens cluster with measured time delays (e.g., Grillo et al. 2018, 2020).

Time-delay distances are primarily sensitive to the value of H_0, and more mildly on those of other cosmological parameters. If a lens produces multiple images of two sources (1 and 2), located at different redshifts, the observed positions of the multiple images provide information about ϕ and the so-called family ratio $(D_{\rm ds1} \times D_{\rm s2})/(D_{\rm s1} \times D_{\rm ds2})$. In galaxy clusters, usually showing several multiple images, different values of this ratio can be used at the same time to estimate the values of the cosmological matter ($\Omega_{\rm m}$) and dark-energy (Ω_Λ) density parameters, and the dark-energy equation-of state (w) parameter (e.g. Soucail et al. 2004; Gilmore and Natarajan 2009), defining the global geometry of the Universe. Using observed cluster lenses Jullo et al. (2010), Caminha et al. (2016), Magana et al. (2018) have demonstrated that cluster strong lensing cosmography could be competitive with other cosmological probes once systematics are better quantified and controlled.

Multiple images of supernova Refsdal in HFF cluster MACS J1149 was recently used to constraint H_0 (Kelly et al. 2023). In addition to supernova, variable quasars can also be used as sources behind clusters. One of these is SDSS J1004+4112. The time delay of the fourth arriving image was predicted by different groups some years ago. After a nearly 15 year observing campaign that delay was finally measured to be 6.73 ± 0.003 years (Muñoz et al. 2022), and agrees well with some of the predictions (Liesenborgs et al. 2009; Mohammed et al. 2015). Using the measured time delays in three clusters with background quasars and archival HST imaging, Napier et al. (2023) measured the value of H_0 which agrees with the one from Refsdal within 1σ, but has somewhat larger modeling uncertainties, because the 3 clusters have fewer lensed images than HFF clusters. They estimate that of order of 50 cluster-lensed quasar systems would help reduce the uncertainties on H_0 to below the 1% level, which is expected to be detected in the next decade (Robertson et al. 2020). These results show that galaxy clusters are competitive in the measurement of the Hubble constant, once sufficient systems are discovered and analyzed.

Most of the baryons in galaxy clusters are in the form of hot, X-ray emitting gas. Massive galaxy clusters can be assumed to have formed from representative regions of the Universe, and hence the ratio of the mass in baryons they contain to that of dark matter (i.e., the gas mass fraction) should be a constant, not evolving with redshift. This assumption can be used to constrain global cosmological parameters, like $h^2\Omega_b$, and Ω_m (Applegate et al. 2016; Allen et al. 2011; Mantz et al. 2014). Lensing plays a key role in this determination as it measures the cluster masses. One can also turn the argument around: assume cosmological parameters and determine if the baryon fraction in clusters evolves with redshift. Holanda et al. (2017) estimate that there is no statistically significant evolution of the baryon mass fraction with z.

5.7 Impact of Line-of-Sight Structures and Multiple-Lens Planes

Traditional single-plane lensing analyses assume that the mass of the deflector is concentrated in a single plane, at the redshift of a cluster. More advanced multi-plane lensing studies, taking into account the observed mass distributions at redshifts different from that of the cluster, are becoming possible, thanks to some recent software developments. In the former case, the light rays emitted from a distant source and received by an observer are deflected only once (i.e., when crossing the galaxy cluster); in the latter, multiple deflections have to be added properly (see Blandford and Narayan 1986; Schneider 2014). The observed positions and fluxes of each galaxy contributing to the total deflection has to be re-scaled iteratively in the modeling process, according to the deflection and magnification values associated with the mass of the objects in the foreground.

Studies by Bayliss et al. (2014), D'Aloisio et al. (2014), Chirivì et al. (2018), Williams et al. (2018), McCully et al. (2017) found that line-of-sight (LOS) galaxies can have a significant impact on the reconstruction of some multiple image positions, but the inclusion of LOS mass distributions in the strong lensing models of galaxy clusters can reduce only modestly the rms offset between the observed and model-predicted positions of all multiple images. These studies also noted that the perturbing contribution of foreground galaxies is more important than that of the background ones and that accounting for LOS mass structures, as if they were at the cluster redshift, can partially lower this offset. Recently, Raney et al. (2020a) confirmed that LOS galaxies can account for a significant fraction of the typical image position residuals of current state-of-the-art strong lensing models of galaxy clusters and that the values of the image magnifications can be affected by a few percent.

5.8 Local Properties of Clusters Around Extended Images

Most of the work on cluster lensing presented in the literature involves reconstruction of the mass distribution over the whole lens plane region of the cluster. As described in earlier sections, this reconstruction can use simply parametrized models, free-form models, or hybrid ones. However, a very different type of analysis is also possible, one that does not involve modeling the whole contiguous central region of the cluster, and recovers cluster properties only *at* the locations of extended images.

The strong gravitational lensing formalism provides only *local* information about the deflecting mass distribution in the lens plane at the positions of the multiple images. Assuming that the surface mass density and shear are constant over the area considered for evaluation around the multiple image locations, we can use the image morphology to retrieve ratios of convergences, $f^{(ij)}$ between multiple images i and j arising from the same individual source, $i, j = 1, \ldots, n_{\mathrm{i}}$ (with n_{i} being the total number of images from the same source), and the reduced shear components at the positions of all n_{i} images

$$f^{(ij)} \equiv \frac{1-\kappa(\boldsymbol{\theta}^{(i)})}{1-\kappa(\boldsymbol{\theta}^{(j)})}\,, \quad g_1^{(i)} = \frac{\gamma_1(\boldsymbol{\theta}^{(i)})}{1-\kappa(\boldsymbol{\theta}^{(i)})}\,, \quad g_2^{(i)} = \frac{\gamma_2(\boldsymbol{\theta}^{(i)})}{1-\kappa(\boldsymbol{\theta}^{(i)})}\,. \tag{5}$$

These convergence ratios and reduced shear components can be determined by features in the brightness distributions of the multiple images, most often star forming regions. At least three multiple images with three features that are not aligned in each image are required. Wagner (2019) details how to uniquely determine the $f^{(ij)}$ and $g^{(i)}$ from these feature vectors, Lin et al. (2022) and Lin et al. (2023) show the impact of non-unique feature matching when one image contains multiple candidate features that can be matched with one single feature showing in other images. Lin et al. (2022) also set up an algorithm to identify the positions of these features in the presence of noise. Alternatively, the quadrupole moment of a featureless brightness profile around its centre of light can be employed.[3]

Inserting Eq. (5) into the distortion matrix for each multiple image

$$\mathrm{A} = \begin{pmatrix} 1-\kappa-\gamma_1 & \gamma_2 \\ \gamma_2 & 1-\kappa+\gamma_1 \end{pmatrix}, \tag{6}$$

ratios of magnifications can also be determined based on these local lens properties. If flux measurements covering the same area of the multiple images are available, a comparison between the flux ratios and the magnification ratios yields a consistency check for the local lens properties in Eq. (5). Deviations may hint at dust extinction, additional micro-lensing, or lensing effects beyond convergence and shear.

To investigate the impact of higher-order lensing effects in galaxy clusters, gradients of reduced shear and convergence maps in simulated clusters, like *Ares* and *Hera* (Meneghetti et al. 2017), are analysed, as further detailed in Wagner (2022). On the resolution scale of the pixels used in these simulations, the results reveal that, in more than 90% of all pixels covering the lensing area, the gradients in θ_1- and θ_2-directions are less than 10% of the convergence and reduced shear values at these pixels. More than 67% of all pixels in the *Ares* and *Hera* lensing area have gradients of less than 2% of their respective distortion matrix entries. Hence, at current observational precision, neglecting higher-order lensing effects on galaxy-cluster scales is valid to a good approximation.

In addition, assuming a smooth deflecting mass density in the vicinity of the multiple images that straddle a critical curve, we can use the observables in the multiple images to constrain the position of the critical curve between these images. To leading order, points on the critical curve between two images in a fold-configuration can be determined as the mid-point dividing the connection line of corresponding features in the two multiple images. Retrieving local lens properties is very useful in newly detected galaxy clusters in which only few spectroscopically confirmed multiple images have been identified. Using a synthetic cluster *Ares*, Johnson and Sharon (2016) estimate that one needs > 10 spectroscopically confirmed image sets to precisely determine the mass enclosed within the Einstein radius of a cluster-lens. Thus, the lens modelling algorithms introduced in Sect. 3 cannot be applied to galaxy clusters like RM J223013.1-080853.1 (Rykoff et al. 2016), which was identified after a serendipitous discovery of a multiple-image system. So far, it only shows three spectroscopically confirmed multiple images for a common background galaxy at $z_\mathrm{s} = 0.8200$. Nevertheless, Griffiths et al. (2021) succeeded in constraining the smoothness scale of dark matter by applying the approach outlined above to the three highly resolved images. Figure 10 shows the galaxy cluster at $z_\mathrm{d} = 0.526$, determined by its $\sim$60 cluster member galaxies, and details of the confirmed multiple images.

Using the `ptmatch`-program[4] Wagner and Tessore (2018), Wagner et al. (2018) determine the properties in Eq. (5). The fact that the algorithm obtains a solution supports the assumption of constant entries in the distortion matrix in the areas spanned by the brightness features shown in Fig. 10 (left). The observed flux ratios in all five available *HST* filters (F110W, F140W, F160W, F606W, F814W) also show a high degree of agreement with the magnification ratios of `ptmatch`, which is a second consistency check corroborating the assumption. Each lens model has its underlying assumptions, for instance that light traces mass

[3] An example for the usage of the quadrupole ellipticity and direction is given in Wagner and Williams (2020) albeit for a galaxy-scale lens.

[4] available at https://github.com/ntessore/imagemap.

Fig. 10 Galaxy cluster RM J223013.1-080853.1 at $z_d = 0.526 \pm 0.018$ with a cusp configuration of highly resolved images from a clumpy disk galaxy at $z_s = 0.8200 \pm 0.0005$ (right). The features in the brightness profiles (numbered yellow circles in the details on the left) are used to reconstruct the local lens properties in Eq. (5)

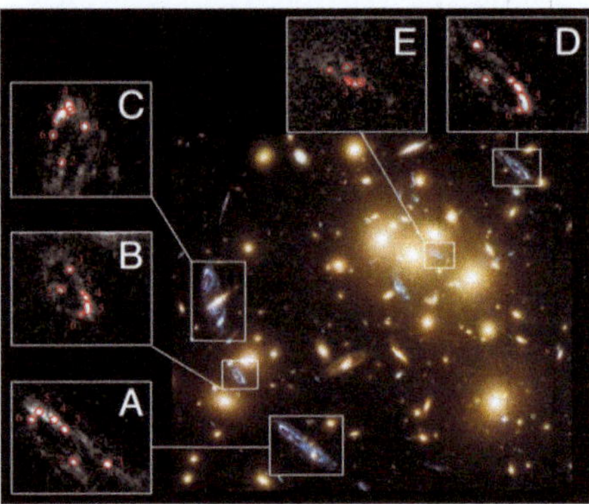

Fig. 11 The galaxy cluster strong gravitational lens CL0024+1654 at $z_d = 0.39$ (background) and the five multiple images of a blue spiral background galaxy at $z_s = 0.1675$ (details around CL0024) with their six brightness features, star-forming regions, that can be used to extract the local lens properties in Eq. (5). (See Lin et al. 2022, for details.)

(see Jullo et al. 2007b; Zitrin et al. 2009b), or regularisation constraints on the smoothness of the mass density. Such assumptions are not employed in the observation-based approach outlined in Sect. 5.8. A comparison between the local lens properties (Eq. (5)) extracted from the lens-model reconstructions at the multiple image positions and the same properties directly obtained from the multiple-image observables (according to the procedure described in Sect. 5.8) allows us to corroborate or refute lens model assumptions.

As an example, Wagner et al. (2018) use the strongly-lensing galaxy cluster CL0024+1654 at $z_d = 0.39$ to investigate the differences between the lens-model-independent local lens properties, those obtained in a `Lenstool` and a `Grale` reconstruction. The local lens properties are extracted at the positions of five highly resolved images of a blue spiral background galaxy at $z_s = 1.675$. In addition to these images, the `Lenstool` and `Grale` reconstructions employ the same set of 14 multiple image positions of 5 other background sources. Images' extended nature is not taken into account, only their centres of light are used as image positions.

The six brightness features in the five resolved multiple images shown in Fig. 11 constrain the local lens properties of Eq. (5) at the five positions in the cluster area for several filter bands, which shows overall consistency of these local lens properties (Lin et al. 2022). Extracting the same local lens properties determined by the `Lenstool` and `Grale` reconstructions, the authors find a high degree of agreement between the $f^{(ij)}$- and $g^{(j)}$-values of all three approaches. This implies that in the vicinity of the highly resolved multiple images, (i) the light-traces-mass assumption and the hypothesis of a constant mass-to-light-ratio for the brightest cluster member galaxies as used in `Lenstool` can both be corroborated for CL0024+1654; (ii) `Grale` arrives at similar conclusions via a completely different route.

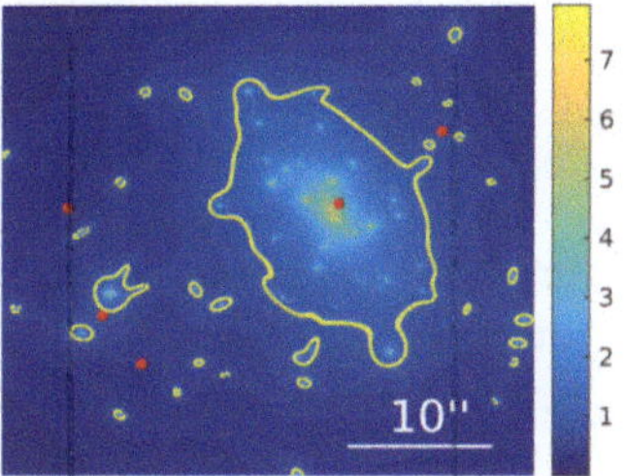

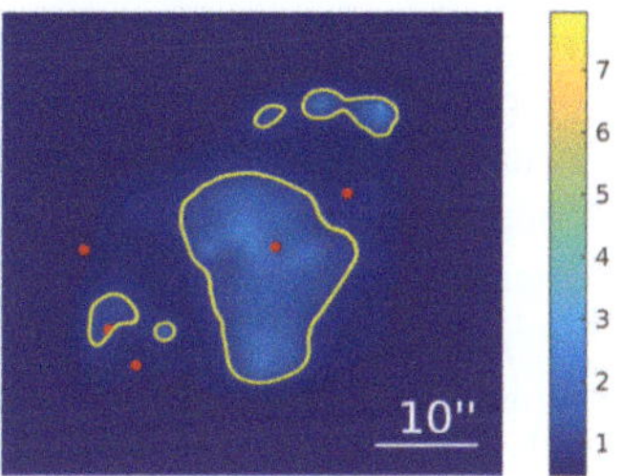

Fig. 12 Convergence maps for CL0024+1654 obtained from `Lenstool` (left) and `Grale` (right) using the five images of the highly resolved spiral galaxy (red dots) and 14 additional images from five other source. The yellow curves mark the isocontour $\kappa = 1$. Image reproduced with permission from Wagner et al. (2018), copyright by ESO

Despite the fact that `Lenstool` and `Grale` generate the same *local* lens properties at the five image positions, their *global* convergence maps greatly differ, as shown in Fig. 12. This is to be expected because the lens models are dominated by the multiple image observables only at their positions. Farther away from the data points, lens-model specific additional assumptions determine the global morphology of the reconstruction, causing the differences in the resulting convergence maps.

In the regime of several tens of multiple-image systems, the differences in the convergence maps between model-based lens reconstructions, as described above, occur because the sparse multiple-image constraints are not sufficient, and need to be complemented by additional assumptions to obtain a global lens morphology. The observation-based, model-free method discussed in this Section can be used to compare and cross-check additional modeling assumptions. In the future, observations with JWST will increase the number of multiple images per cluster by an order of magnitude from about 100 to 1000 (Ghosh et al. 2020). Accordingly, the impact of model assumptions will decrease and the area covered by multiple images which is reconstructed with an accuracy above 90% will increase from about 40-50% to about 65%. The use of model-free methods can also be extended to more images, further confirming the recovered mass distribution in clusters.

6 Lensing by Galaxy Groups

Galaxy groups are similar to clusters, but have lower masses, $\sim 10^{13} - 10^{15} M_{\odot}$, and so 'fill' the mass gap between individual galaxies and clusters. Given that they are more numerous, galaxy groups can be important contributors to the study of background sources, the mass distribution in lenses, or cosmological parameters. For that reason, Fox and Pen (2001) suggested that groups can be used to test for the NFW profile. They analytically compared the lensing efficiency of NFW vs. singular isothermal sphere (SIS) profiles, and concluded that NFW dark matter halos are ~ 200 less efficient than SIS, therefore strong lensing statistics by groups could provide a sensitive test of the inner density region of virialized structures. Their finding of low group cross-section to strong lensing is consistent with there being no instances of strong lensing by groups in the Hubble Deep Field. Newman et al. (2015b) addressed the same question of group's density profile. They used a sample of 10 strong lensing groups and found that the central regions are consistent with NFW, but are possibly somewhat steeper, which they ascribed to the contribution from the central elliptical galaxy. Verdugo et al. (2014) analyzed how the Einstein radius of galaxy groups in the Strong Lensing Legacy Survey (Limousin et al. 2009; More et al. 2012) relates to the group's properties, like velocity dispersion. They showed that all correlations between group properties have a considerable amount of scatter, but there is some evidence that the Einstein radius is decreasing with increasing redshift. Foëx et al. (2014) used weak lensing data for 80 strong lensing galaxy groups to estimate the groups' mass vs. concentration relation. The relation they found is steeper than that seen in simulations. They attribute this to the fact that strong lensing selected groups are preferentially aligned along our line of sight because that way they present a larger strong lensing cross-section, and are more easily detected. Detailed study of one group, SL2S J02140-0535 by Verdugo et al. (2011), whose measured concentration parameter is somewhat larger than predicted by simulations, is in agreement with this. Therefore galaxy groups selected by their strong lensing features may constitute a biased sample of groups in general.

In the recent years there have been a few dedicated searches for strong lensing in galaxy groups that aim to find and analyze or order 100 or more groups, instead of 10 or so used in earlier works. Jaelani et al. (2020) carried out a Survey of Gravitationally-lensed Objects in HSC Imaging using Hyper Suprime-Cam Subaru Strategic Program (HSC-SSP) Survey (Aihara et al. 2018), covering $\sim 1114\text{deg}^2$. They discovered 641 candidate lens systems, of which 47 are almost certainly bonafide lenses; see Fig. 13 for some examples. The future of group-scale lensing will definitely make use of such large surveys where the observational selection effects are well quantified, and resulting catalogues of groups can be used for statistical studies. But the other aspect of groups, detailed study of individual groups that have abundant data on both lensing and kinematics also promises to yield interesting results. For example, Wang et al. (2022) studies the mass distribution in CASSOWARY 31, using strong lensing data from HST imaging and kinematics from MUSE integral-field spectroscopy. They identified 5 sets of multiple images, on par with some clusters, and obtained detailed mass maps of the system, concluding that it is peculiar fossil group, strongly dominated by dark matter in its central regions. Systems like this can finally realize the prediction from 20 years ago (Fox and Pen 2001) that groups are excellent systems to probe the structure of dark matter halos.

7 Prospects for Cluster Lensing from Upcoming Instruments and Surveys

JWST observations are expected to revolutionize the study of cluster lensing. The early results are very promising and exciting. Caminha et al. (2022b), Golubchik et al. (2022), Mahler et al. (2023), Sharon et al. (2022b) all present mass models of the first strong gravitational lens observed by the JWST, SMACS J0723.3-7327. In Fig. 1 we show the lens model presented in Mahler et al. (2023). The first deep-field observations of the JWST have yielded a surprisingly large number of very high redshift galaxy candidates, $z = 10 - 16$ (Furtak et al. 2023). The JWST Prime Extragalactic Areas for Reionization and Lensing Science (PEARLS) project will look for high redshift galaxies behind 7 lensing clusters (Windhorst et al. 2023). Meanwhile Laporte et al. (2022) report finding a lensed protocluster candidate at $z = 7.66$ behind a lensing galaxy cluster SMACS0723-7327. JWST data will allow testing our current cosmological paradigm via the detailed studies of substructure that it will provide for direct comparison with LCDM simulations and predictions.

The upcoming Euclid space mission with the expectation of detecting upto 60,000 galaxy clusters from $z \sim 0.2 - 2$, and of the order of several thousand strongly lensed arcs stands to transform our understanding of the assembly history of clusters. With accurate weak lensing

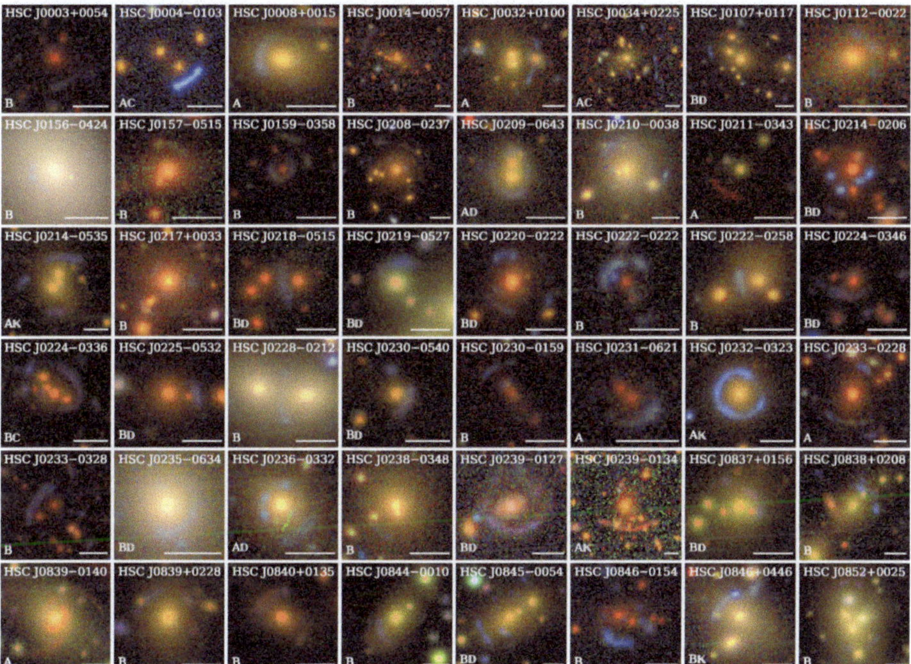

Fig. 13 Examples of group-scale lensing candidates from the Hyper Supreme Cam on the *Subaru Telescope* (Jaelani et al. 2020). The letters A or B at the bottom of each panel indicate the confidence grade of this being a lens. The white horizontal bar at the bottom of each panel is 5 arcseconds long. North is up, East is to the left

mass estimates out to the virial radius, Euclid will enable tracing the mass build-up and growth of structure in the Universe. Euclid is expected to provide a stringent test of the cold dark matter paradigm as data from the mission will permit probing the radial density profile and the ellipticity of clusters, both of which are predicted by LCDM. The anticipated data deluge has catalyzed the development of many new automatic lensing detection algorithms. In combination with DES, Euclid data will provide tighter cosmological constraints from clusters. On the modeling side, improved existing lens inversion methods, as well as new reconstruction methodologies under development currently will rise to meet the challenge of JWST data and data from other planned surveys are expected to improve our understanding of the clusters and the high redshift Universe.

Acknowledgements We thank the International Space Science Institute in Bern (ISSI) for their hospitality for organizing the stimulating workshop on "Strong Gravitational Lensing". We thank Masamune Oguri and Anupreeta More for helpful comments and discussions.

Declarations

Competing Interests The authors declare no competing interests.

Open Access This article is licensed under a Creative Commons Attribution 4.0 International License, which permits use, sharing, adaptation, distribution and reproduction in any medium or format, as long as you give appropriate credit to the original author(s) and the source, provide a link to the Creative Commons licence, and indicate if changes were made. The images or other third party material in this article are included in the article's Creative Commons licence, unless indicated otherwise in a credit line to the material. If material is not included in the article's Creative Commons licence and your intended use is not permitted by statutory regulation or exceeds the permitted use, you will need to obtain permission directly from the copyright holder. To view a copy of this licence, visit http://creativecommons.org/licenses/by/4.0/.

References

Abdelsalam HM, Saha P, Williams LLR (1998) Nonparametric reconstruction of Abell 2218 from combined weak and strong lensing. Astron J 116(4):1541–1552. https://doi.org/10.1086/300546. arXiv:astro-ph/9806244 [astro-ph]

Acebron A, Cibirka N, Zitrin A, Coe D, Agulli I, Sharon K, Bradač M, Frye B, Livermore RC, Mahler G, Salmon B, Umetsu K, Bradley L, Andrade-Santos F, Avila R, Carrasco D, Cerny C, Czakon NG, Dawson WA, Hoag AT, Huang KH, Johnson TL, Jones C, Kikuchihara S, Lam D, Lovisari L, Mainali R, Oesch PA, Ogaz S, Ouchi M, Past M, Paterno-Mahler R, Peterson A, Ryan RE, Sendra-Server I, Stark DP, Strait V, Toft S, Trenti M, Vulcani B (2018) RELICS: strong-lensing analysis of the massive clusters MACS J0308.9+2645 and PLCK G171.9-40.7. Astrophys J 858(1):42. https://doi.org/10.3847/1538-4357/aabe29. arXiv:1803.00560 [astro-ph.CO]

Acebron A, Alon M, Zitrin A, Mahler G, Coe D, Sharon K, Cibirka N, Bradač M, Trenti M, Umetsu K, Andrade-Santos F, Avila RJ, Bradley L, Carrasco D, Cerny C, Czakon NG, Dawson WA, Frye B, Hoag AT, Huang KH, Johnson TL, Jones C, Kikuchihara S, Lam D, Livermore RC, Lovisari L, Mainali R, Oesch PA, Ogaz S, Ouchi M, Past M, Paterno-Mahler R, Peterson A, Ryan RE, Salmon B, Sendra-Server I, Stark DP, Strait V, Toft S, Vulcani B (2019) RELICS: high-resolution constraints on the inner mass distribution of the z = 0.83 merging cluster RXJ0152.7-1357 from strong lensing. Astrophys J 874(2):132. https://doi.org/10.3847/1538-4357/ab0adf. arXiv:1810.08122 [astro-ph.CO]

Aihara H, Arimoto N, Armstrong R et al (2018) The hyper suprime-cam SSP survey: overview and survey design. Publ Astron Soc Jpn 70:S4. https://doi.org/10.1093/pasj/psx066. arXiv:1704.05858 [astro-ph.IM]

Alavi A, Siana B, Richard J, Rafelski M, Jauzac M, Limousin M, Freeman WR, Scarlata C, Robertson B, Stark DP, Teplitz HI, Desai V (2016) The evolution of the faint end of the UV luminosity function during the peak epoch of star formation ($1 < z < 3$). Astrophys J 832(1):56. https://doi.org/10.3847/0004-637X/832/1/56. arXiv:1606.00469 [astro-ph.GA]

Allen SW, Evrard AE, Mantz AB (2011) Cosmological parameters from observations of galaxy clusters. Annu Rev Astron Astrophys 49(1):409–470. https://doi.org/10.1146/annurev-astro-081710-102514. arXiv:1103.4829 [astro-ph.CO]

Amendola L, Appleby S, Bacon D, Baker T, Baldi M, Bartolo N, Blanchard A, Bonvin C, Borgani S, Branchini E, Burrage C, Camera S, Carbone C, Casarini L, Cropper M, de Rham C, Di Porto C, Ealet A, Ferreira PG, Finelli F, García-Bellido J, Giannantonio T, Guzzo L, Heavens A, Heisenberg L, Heymans C, Hoekstra H, Hollenstein L, Holmes R, Horst O, Jahnke K, Kitching TD, Koivisto T, Kunz M, La Vacca G, March M, Majerotto E, Markovic K, Marsh D, Marulli F, Massey R, Mellier Y, Mota DF, Nunes NJ, Percival W, Pettorino V, Porciani C, Quercellini C, Read J, Rinaldi M, Sapone D, Scaramella R, Skordis C, Simpson F, Taylor A, Thomas S, Trotta R, Verde L, Vernizzi F, Vollmer A, Wang Y, Weller J, Zlosnik T (2013) Cosmology and fundamental physics with the Euclid satellite. Living Rev Relativ 16(1):6. https://doi.org/10.12942/lrr-2013-6. arXiv:1206.1225 [astro-ph.CO]

Andrade KE, Minor Q, Nierenberg A, Kaplinghat M (2019) Detecting dark matter cores in galaxy clusters with strong lensing. Mon Not R Astron Soc 487(2):1905–1926. https://doi.org/10.1093/mnras/stz1360. arXiv:1901.00507 [astro-ph.GA]

Angus GW, Famaey B, Zhao HS (2006) Can MOND take a bullet? Analytical comparisons of three versions of MOND beyond spherical symmetry. Mon Not R Astron Soc 371(1):138–146. https://doi.org/10.1111/j.1365-2966.2006.10668.x. arXiv:astro-ph/0606216 [astro-ph]

Annunziatella M, Biviano A, Mercurio A, Nonino M, Rosati P, Balestra I, Presotto V, Girardi M, Gobat R, Grillo C, Kelson D, Medezinski E, Postman M, Scodeggio M, Brescia M, Demarco R, Fritz A, Koekemoer A, Lemze D, Lombardi M, Sartoris B, Umetsu K, Vanzella E, Bradley L, Coe D, Donahue M, Infante L, Kuchner U, Maier C, Regős E, Verdugo M, Ziegler B (2014) CLASH-VLT: the stellar mass function and stellar mass density profile of the z = 0.44 cluster of galaxies MACS J1206.2-0847. Astron Astrophys 571:A80. https://doi.org/10.1051/0004-6361/201424102. arXiv:1408.6356 [astro-ph.GA]

Annunziatella M, Bonamigo M, Grillo C, Mercurio A, Rosati P, Caminha G, Biviano A, Girardi M, Gobat R, Lombardi M, Munari E (2017) Mass profile decomposition of the frontier fields cluster MACS J0416-2403: insights on the dark-matter inner profile. Astrophys J 851(2):81. https://doi.org/10.3847/1538-4357/aa9845. arXiv:1711.02109 [astro-ph.CO]

Applegate DE, Mantz A, von Allen SW, der Linden A, Morris RG, Hilbert S, Kelly PL, Burke DL, Ebeling H, Rapetti DA, Schmidt RW (2016) Cosmology and astrophysics from relaxed galaxy clusters - IV. Robustly calibrating hydrostatic masses with weak lensing. Mon Not R Astron Soc 457(2):1522–1534. https://doi.org/10.1093/mnras/stw005. arXiv:1509.02162 [astro-ph.CO]

Atek H, Richard J, Kneib JP, Schaerer D (2018) The extreme faint end of the UV luminosity function at z $\sim$ 6 through gravitational telescopes: a comprehensive assessment of strong lensing uncertainties. Mon Not R Astron Soc 479(4):5184–5195. https://doi.org/10.1093/mnras/sty1820. arXiv:1803.09747 [astro-ph.GA]

Bacon DJ, Goldberg DM, Rowe BTP, Taylor AN (2006) Weak gravitational flexion. Mon Not R Astron Soc 365(2):414–428. https://doi.org/10.1111/j.1365-2966.2005.09624.x. arXiv:astro-ph/0504478 [astro-ph]

Bahé YM (2021) Strongly lensed cluster substructures are not in tension with ΛCDM. Mon Not R Astron Soc. https://doi.org/10.1093/mnras/stab1392. arXiv:2101.12112 [astro-ph.GA]

Balestra I, Vanzella E, Rosati P, Monna A, Grillo C, Nonino M, Mercurio A, Biviano A, Bradley L, Coe D, Fritz A, Postman M, Seitz S, Scodeggio M, Tozzi P, Zheng W, Ziegler B, Zitrin A, Annunziatella M, Bartelmann M, Benitez N, Broadhurst T, Bouwens R, Czoske O, Donahue M, Ford H, Girardi M, Infante L, Jouvel S, Kelson D, Koekemoer A, Kuchner U, Lemze D, Lombardi M, Maier C, Medezinski E, Melchior P, Meneghetti M, Merten J, Molino A, Moustakas L, Presotto V, Smit R, Umetsu K (2013) CLASH-VLT: spectroscopic confirmation of a z = 6.11 quintuply lensed galaxy in the frontier fields cluster RXC J2248.7-4431. Astron Astrophys 559:L9. https://doi.org/10.1051/0004-6361/201322620. arXiv:1309.1593 [astro-ph.CO]

Bartelmann M, Huss A, Colberg JM, Jenkins A, Pearce FR (1998) Arc statistics with realistic cluster potentials. IV. Clusters in different cosmologies. Astron Astrophys 330:1–9. arXiv:astro-ph/9707167 [astro-ph]

Bayliss MB, Hennawi JF, Gladders MD, Koester BP, Sharon K, Dahle H, Oguri M (2011) Gemini/GMOS spectroscopy of 26 strong-lensing-selected galaxy cluster cores. Astrophys J Suppl Ser 193(1):8. https://doi.org/10.1088/0067-0049/193/1/8. arXiv:1010.2714 [astro-ph.CO]

Bayliss MB, Johnson T, Gladders MD, Sharon K, Oguri M (2014) Line-of-sight structure toward strong lensing galaxy clusters. Astrophys J 783(1):41. https://doi.org/10.1088/0004-637X/783/1/41. arXiv:1312.3637 [astro-ph.CO]

Beauchesne B, Clément B, Richard J, Kneib JP (2021) Improving parametric mass modelling of lensing clusters through a perturbative approach. Mon Not R Astron Soc. https://doi.org/10.1093/mnras/stab1684. arXiv:2106.05029 [astro-ph.CO]

Bergamini P, Rosati P, Mercurio A, Grillo C, Caminha GB, Meneghetti M, Agnello A, Biviano A, Calura F, Giocoli C, Lombardi M, Rodighiero G, Vanzella E (2019) Enhanced cluster lensing models with measured galaxy kinematics. Astron Astrophys 631:A130. https://doi.org/10.1051/0004-6361/201935974. arXiv:1905.13236 [astro-ph.GA]

Bergamini P, Rosati P, Vanzella E, Caminha GB, Grillo C, Mercurio A, Meneghetti M, Angora G, Calura F, Nonino M, Tozzi P (2021) A new high-precision strong lensing model of the galaxy cluster MACS J0416.1-2403. Robust characterization of the cluster mass distribution from VLT/MUSE deep observations. Astron Astrophys 645:A140. https://doi.org/10.1051/0004-6361/202039564. arXiv:2010.00027 [astro-ph.GA]

Biviano A, Rosati P, Balestra I, Mercurio A, Girardi M, Nonino M, Grillo C, Scodeggio M, Lemze D, Kelson D, Umetsu K, Postman M, Zitrin A, Czoske O, Ettori S, Fritz A, Lombardi M, Maier C, Medezinski E, Mei S, Presotto V, Strazzullo V, Tozzi P, Ziegler B, Annunziatella M, Bartelmann M, Benitez N, Bradley L, Brescia M, Broadhurst T, Coe D, Demarco R, Donahue M, Ford H, Gobat R, Graves G, Koekemoer A, Kuchner U, Melchior P, Meneghetti M, Merten J, Moustakas L, Munari E, Regős E, Sartoris B, Seitz S, Zheng W (2013) CLASH-VLT: the mass, velocity-anisotropy, and pseudo-phase-space density profiles of the z = 0.44 galaxy cluster MACS J1206.2-0847. Astron Astrophys 558:A1. https://doi.org/10.1051/0004-6361/201321955. arXiv:1307.5867 [astro-ph.CO]

Blandford R, Narayan R (1986) Fermat's principle, caustics, and the classification of gravitational lens images. Astrophys J 310:568. https://doi.org/10.1086/164709

Blandford RD, Kochanek CS, Kovner I, Narayan R (1989) Gravitational lens optics. Science 245(4920):824–830. https://doi.org/10.1126/science.245.4920.824

Blanton MR, Bershady MA, Abolfathi B et al (2017) Sloan digital sky survey IV: mapping the Milky Way, nearby galaxies, and the distant universe. Astron J 154(1):28. https://doi.org/10.3847/1538-3881/aa7567. arXiv:1703.00052 [astro-ph.GA]

Bonamigo M, Grillo C, Ettori S, Caminha GB, Rosati P, Mercurio A, Annunziatella M, Balestra I, Lombardi M (2017) Joining X-ray to lensing: an accurate combined analysis of MACS J0416.1-2403. Astrophys J 842(2):132. https://doi.org/10.3847/1538-4357/aa75cc. arXiv:1705.10322 [astro-ph.GA]

Bonamigo M, Grillo C, Ettori S, Caminha GB, Rosati P, Mercurio A, Munari E, Annunziatella M, Balestra I, Lombardi M (2018) Dissection of the collisional and collisionless mass components in a mini sample of CLASH and HFF massive galaxy clusters at z $\approx$ 0.4. Astrophys J 864(1):98. https://doi.org/10.3847/1538-4357/aad4a7. arXiv:1807.10286 [astro-ph.GA]

Bose S, Hellwing WA, Frenk CS, Jenkins A, Lovell MR, Helly JC, Li B, Gonzalez-Perez V, Gao L (2017) Substructure and galaxy formation in the Copernicus Complexio warm dark matter simulations. Mon Not R Astron Soc 464(4):4520–4533. https://doi.org/10.1093/mnras/stw2686. arXiv:1604.07409 [astro-ph.CO]

Bouwens RJ, Illingworth GD, Oesch PA, Atek H, Lam D, Stefanon M (2017a) Extremely small sizes for faint z $\sim$ 2-8 galaxies in the Hubble frontier fields: a key input for establishing their volume density and UV emissivity. Astrophys J 843(1):41. https://doi.org/10.3847/1538-4357/aa74e4. arXiv:1608.00966 [astro-ph.GA]

Bouwens RJ, Oesch PA, Illingworth GD, Ellis RS, Stefanon M (2017b) The z $\sim$ 6 luminosity function fainter than -15 mag from the Hubble frontier fields: the impact of magnification uncertainties. Astrophys J 843(2):129. https://doi.org/10.3847/1538-4357/aa70a4. arXiv:1610.00283 [astro-ph.GA]

Bouwens RJ, Illingworth G, Ellis RS, Oesch P, Stefanon M (2022) z 2-9 galaxies magnified by the Hubble frontier field clusters. II. Luminosity functions and constraints on a faint-end turnover. Astrophys J 940(1):55. https://doi.org/10.3847/1538-4357/ac86d1. arXiv:2205.11526 [astro-ph.GA]

Bradač M, Clowe D, Gonzalez AH, Marshall P, Forman W, Jones C, Markevitch M, Randall S, Schrabback T, Zaritsky D (2006) Strong and weak lensing united. III. Measuring the mass distribution of the merging galaxy cluster 1ES 0657-558. Astrophys J 652:937–947. https://doi.org/10.1086/508601. arXiv:astro-ph/0608408 [astro-ph]

Bradač M, Treu T, Applegate D, Gonzalez AH, Clowe D, Forman W, Jones C, Marshall P, Schneider P, Zaritsky D (2009) Focusing cosmic telescopes: exploring redshift $z \sim 5$–6 galaxies with the bullet cluster 1E0657 - 56. Astrophys J 706(2):1201–1212. https://doi.org/10.1088/0004-637X/706/2/1201. arXiv:0910.2708 [astro-ph.CO]

Broadhurst T, Benítez N, Coe D, Sharon K, Zekser K, White R, Ford H, Bouwens R, Blakeslee J, Clampin M, Cross N, Franx M, Frye B, Hartig G, Illingworth G, Infante L, Menanteau F, Meurer G, Postman M, Ardila DR, Bartko F, Brown RA, Burrows CJ, Cheng ES, Feldman PD, Golimowski DA, Goto T, Gronwall C, Herranz D, Holden B, Homeier N, Krist JE, Lesser MP, Martel AR, Miley GK, Rosati P, Sirianni M, Sparks WB, Steindling S, Tran HD, Tsvetanov ZI, Zheng W (2005a) Strong-lensing analysis of A1689 from deep advanced camera images. Astrophys J 621(1):53–88. https://doi.org/10.1086/426494. arXiv:astro-ph/0409132 [astro-ph]

Broadhurst T, Takada M, Umetsu K, Kong X, Arimoto N, Chiba M, Futamase T (2005b) The surprisingly steep mass profile of A1689, from a lensing analysis of subaru images. Astrophys J Lett 619(2):L143–L146. https://doi.org/10.1086/428122. arXiv:astro-ph/0412192 [astro-ph]

Cain B, Bradač M, Levinson R (2016) Reconstruction of small-scale galaxy cluster substructure with lensing flexion. Mon Not R Astron Soc 463(4):4287–4300. https://doi.org/10.1093/mnras/stw2270. arXiv:1503.08218 [astro-ph.CO]

Caminha GB, Estrada J, Makler M (2013) Magnification Bias in Gravitational Arc Statistics. ArXiv e-prints arXiv:1308.6569 [astro-ph.CO]

Caminha GB, Grillo C, Rosati P, Balestra I, Karman W, Lombardi M, Mercurio A, Nonino M, Tozzi P, Zitrin A, Biviano A, Girardi M, Koekemoer AM, Melchior P, Meneghetti M, Munari E, Suyu SH, Umetsu K, Annunziatella M, Borgani S, Broadhurst T, Caputi KI, Coe D, Delgado-Correal C, Ettori S, Fritz A, Frye B, Gobat R, Maier C, Monna A, Postman M, Sartoris B, Seitz S, Vanzella E, Ziegler B (2016) CLASH-VLT: a highly precise strong lensing model of the galaxy cluster RXC J2248.7-4431 (Abell S1063) and prospects for cosmography. Astron Astrophys 587:A80. https://doi.org/10.1051/0004-6361/201527670. arXiv:1512.04555 [astro-ph.CO]

Caminha GB, Grillo C, Rosati P, Balestra I, Mercurio A, Vanzella E, Biviano A, Caputi KI, Delgado-Correal C, Karman W, Lombardi M, Meneghetti M, Sartoris B, Tozzi P (2017) A refined mass distribution of the cluster MACS J0416.1-2403 from a new large set of spectroscopic multiply lensed sources. Astron Astrophys 600:A90. https://doi.org/10.1051/0004-6361/201629297. arXiv:1607.03462 [astro-ph.GA]

Caminha GB, Rosati P, Grillo C, Rosani G, Caputi KI, Meneghetti M, Mercurio A, Balestra I, Bergamini P, Biviano A, Nonino M, Umetsu K, Vanzella E, Annunziatella M, Broadhurst T, Delgado-Correal C, Demarco R, Koekemoer AM, Lombardi M, Maier C, Verdugo M, Zitrin A (2019) Strong lensing models of eight CLASH clusters from extensive spectroscopy: accurate total mass reconstructions in the cores. Astron Astrophys 632:A36. https://doi.org/10.1051/0004-6361/201935454. arXiv:1903.05103 [astro-ph.GA]

Caminha GB, Suyu SH, Grillo C, Rosati P (2022a) Galaxy cluster strong lensing cosmography. Cosmological constraints from a sample of regular galaxy clusters. Astron Astrophys 657:A83. https://doi.org/10.1051/0004-6361/202141994. arXiv:2110.06232 [astro-ph.CO]

Caminha GB, Suyu SH, Mercurio A, Brammer G, Bergamini P, Acebron A, Vanzella E (2022b) First JWST observations of a gravitational lens. Mass model from new multiple images with near-infrared observations of SMACS J0723.3−7327. Astron Astrophys 666:L9. https://doi.org/10.1051/0004-6361/202244517. arXiv:2207.07567 [astro-ph.GA]

Carlsten SG, Greene JE, Peter AHG, Greco JP, Beaton RL (2020) Radial distributions of dwarf satellite systems in the local volume. Astrophys J 902(2):124. https://doi.org/10.3847/1538-4357/abb60b. arXiv:2006.02444 [astro-ph.GA]

Castellano M, Fontana A, Treu T, Merlin E, Santini P, Bergamini P, Grillo C, Rosati P, Acebron A, Leethochawalit N, Paris D, Bonchi A, Belfiori D, Calabrò A, Correnti M, Nonino M, Polenta G, Trenti M, Boyett K, Brammer G, Broadhurst T, Caminha GB, Chen W, Filippenko AV, Fortuni F, Glazebrook K, Mascia S, Mason CA, Menci N, Meneghetti M, Mercurio A, Metha B, Morishita T, Nanayakkara T, Pentericci L, Roberts-Borsani G, Roy N, Vanzella E, Vulcani B, Yang L, Wang X (2023) Early results from GLASS-JWST. XIX. A high density of bright galaxies at $z \approx 10$ in the A2744 region. Astrophys J Lett 948(2):L14. https://doi.org/10.3847/2041-8213/accea5. arXiv:2212.06666 [astro-ph.GA]

Cerini G, Cappelluti N, Natarajan P (2023) New metrics to probe the dynamical state of galaxy clusters. Astrophys J 945(2):152. https://doi.org/10.3847/1538-4357/acbccb. arXiv:2209.06831 [astro-ph.CO]

Cha S, Jee MJ (2022) MARS: a new maximum-entropy-regularized strong lensing mass reconstruction method. Astrophys J 931(2):127. https://doi.org/10.3847/1538-4357/ac69df. arXiv:2202.10489 [astro-ph.CO]

Chirivì G, Suyu SH, Grillo C, Halkola A, Balestra I, Caminha GB, Mercurio A, Rosati P (2018) MACS J0416.1-2403: impact of line-of-sight structures on strong gravitational lensing modelling of galaxy clusters. Astron Astrophys 614:A8. https://doi.org/10.1051/0004-6361/201731433. arXiv:1706.07815 [astro-ph.CO]

Clowe D, Bradač M, Gonzalez AH, Markevitch M, Randall SW, Jones C, Zaritsky D (2006a) A direct empirical proof of the existence of dark matter. Astrophys J Lett 648(2):L109–L113. https://doi.org/10.1086/508162. arXiv:astro-ph/0608407 [astro-ph]

Clowe D, Schneider P, Aragón-Salamanca A, Bremer M, De Lucia G, Halliday C, Jablonka P, Milvang-Jensen B, Pelló R, Poggianti B, Rudnick G, Saglia R, Simard L, White S, Zaritsky D (2006b) Weak lensing mass reconstructions of the ESO distant cluster survey. Astron Astrophys 451(2):395–408. https://doi.org/10.1051/0004-6361:20041787. arXiv:astro-ph/0511746 [astro-ph]

Coe D, Benítez N, Broadhurst T, Moustakas LA (2010) A high-resolution mass map of galaxy cluster substructure: LensPerfect analysis of A1689. Astrophys J 723(2):1678–1702. https://doi.org/10.1088/0004-637X/723/2/1678. arXiv:1005.0398 [astro-ph.CO]

Coe D, Umetsu K, Zitrin A, Donahue M, Medezinski E, Postman M, Carrasco M, Anguita T, Geller MJ, Rines KJ, Diaferio A, Kurtz MJ, Bradley L, Koekemoer A, Zheng W, Nonino M, Molino A, Mahdavi A, Lemze D, Infante L, Ogaz S, Melchior P, Host O, Ford H, Grillo C, Rosati P, Jiménez-Teja Y, Moustakas J, Broadhurst T, Ascaso B, Lahav O, Bartelmann M, Benítez N, Bouwens R, Graur O, Graves G, Jha S, Jouvel S, Kelson D, Moustakas L, Maoz D, Meneghetti M, Merten J, Riess A, Rodney S, Seitz S (2012) CLASH: precise new constraints on the mass profile of the galaxy cluster A2261. Astrophys J 757(1):22. https://doi.org/10.1088/0004-637X/757/1/22. arXiv:1201.1616 [astro-ph.CO]

Coe D, Zitrin A, Carrasco M, Shu X, Zheng W, Postman M, Bradley L, Koekemoer A, Bouwens R, Broadhurst T, Monna A, Host O, Moustakas LA, Ford H, Moustakas J, van der Wel A, Donahue M, Rodney SA, Benítez N, Jouvel S, Seitz S, Kelson DD, Rosati P (2013) CLASH: three strongly lensed images of a candidate $z \approx 11$ galaxy. Astrophys J 762(1):32. https://doi.org/10.1088/0004-637X/762/1/32. arXiv:1211.3663 [astro-ph.CO]

Coe D, Salmon B, Bradač M et al (2019) RELICS: reionization lensing cluster survey. Astrophys J 884(1):85. https://doi.org/10.3847/1538-4357/ab412b. arXiv:1903.02002 [astro-ph.GA]

Dahle H, Gladders MD, Sharon K, Bayliss MB, Rigby JR (2015) Time delay measurements for the cluster-lensed sextuple quasar SDSS J2222+2745. Astrophys J 813(1):67. https://doi.org/10.1088/0004-637X/813/1/67. arXiv:1505.06187 [astro-ph.CO]

Dai L, Kaurov AA, Sharon K, Florian M, Miralda-Escudé J, Venumadhav T, Frye B, Rigby JR, Bayliss M (2020) Asymmetric surface brightness structure of caustic crossing arc in SDSS J1226+2152: a case for dark matter substructure. Mon Not R Astron Soc 495(3):3192–3208. https://doi.org/10.1093/mnras/staa1355. arXiv:2001.00261 [astro-ph.GA]

D'Aloisio A, Natarajan P (2011) The effects of primordial non-Gaussianity on giant-arc statistics. Mon Not R Astron Soc 415(2):1913–1927. https://doi.org/10.1111/j.1365-2966.2011.18837.x. arXiv:1102.5097 [astro-ph.CO]

D'Aloisio A, Natarajan P, Shapiro PR (2014) The effect of large-scale structure on the magnification of high-redshift sources by cluster lenses. Mon Not R Astron Soc 445(4):3581–3591. https://doi.org/10.1093/mnras/stu1931. arXiv:1311.1614 [astro-ph.CO]

De Lucia G, Kauffmann G, Springel V, White SDM, Lanzoni B, Stoehr F, Tormen G, Yoshida N (2004) Substructures in cold dark matter haloes. Mon Not R Astron Soc 348(1):333–344. https://doi.org/10.1111/j.1365-2966.2004.07372.x. arXiv:astro-ph/0306205 [astro-ph]

Despali G, Sparre M, Vegetti S, Vogelsberger M, Zavala J, Marinacci F (2019) The interplay of self-interacting dark matter and baryons in shaping the halo evolution. Mon Not R Astron Soc 484(4):4563–4573. https://doi.org/10.1093/mnras/stz273. arXiv:1811.02569 [astro-ph.GA]

Diego JM, Protopapas P, Sandvik HB, Tegmark M (2005) Non-parametric inversion of strong lensing systems. Mon Not R Astron Soc 360(2):477–491. https://doi.org/10.1111/j.1365-2966.2005.09021.x. arXiv:astro-ph/0408418 [astro-ph]

Diego JM, Tegmark M, Protopapas P, Sandvik HB (2007) Combined reconstruction of weak and strong lensing data with WSLAP. Mon Not R Astron Soc 375(3):958–970. https://doi.org/10.1111/j.1365-2966.2007.11380.x. arXiv:astro-ph/0509103 [astro-ph]

Diego JM, Broadhurst T, Chen C, Lim J, Zitrin A, Chan B, Coe D, Ford HC, Lam D, Zheng W (2016a) A free-form prediction for the reappearance of supernova Refsdal in the Hubble Frontier Fields cluster MACSJ1149.5+2223. Mon Not R Astron Soc 456(1):356–365. https://doi.org/10.1093/mnras/stv2638. arXiv:1504.05953 [astro-ph.CO]

Diego JM, Broadhurst T, Wong J, Silk J, Lim J, Zheng W, Lam D, Ford H (2016b) A free-form mass model of the Hubble Frontier Fields cluster AS1063 (RXC J2248.7-4431) with over one hundred constraints. Mon Not R Astron Soc 459(4):3447–3459. https://doi.org/10.1093/mnras/stw865. arXiv:1512.07916 [astro-ph.CO]

Diego JM, Pascale M, Kavanagh BJ, Kelly P, Dai L, Frye B, Broadhurst T (2022) Godzilla, a monster lurks in the Sunburst galaxy. Astron Astrophys 665:A134. https://doi.org/10.1051/0004-6361/202243605. arXiv:2203.08158 [astro-ph.GA]

Diemand J, Kuhlen M, Madau P (2007) Formation and evolution of galaxy dark matter halos and their substructure. Astrophys J 667(2):859–877. https://doi.org/10.1086/520573. arXiv:astro-ph/0703337 [astro-ph]

Donahue M, Voit GM, Mahdavi A, Umetsu K, Ettori S, Merten J, Postman M, Hoffer A, Baldi A, Coe D, Czakon N, Bartelmann M, Benitez N, Bouwens R, Bradley L, Broadhurst T, Ford H, Gastaldello F, Grillo C, Infante L, Jouvel S, Koekemoer A, Kelson D, Lahav O, Lemze D, Medezinski E, Melchior P, Meneghetti M, Molino A, Moustakas J, Moustakas LA, Nonino M, Rosati P, Sayers J, Seitz S, Van der Wel A, Zheng W, Zitrin A (2014) CLASH-X: a comparison of lensing and X-ray techniques for measuring the mass profiles of galaxy clusters. Astrophys J 794(2):136. https://doi.org/10.1088/0004-637X/794/2/136. arXiv:1405.7876 [astro-ph.CO]

Ebeling H, Edge AC, Henry JP (2001) MACS: a quest for the most massive galaxy clusters in the universe. Astrophys J 553(2):668–676. https://doi.org/10.1086/320958. arXiv:astro-ph/0009101 [astro-ph]

Eichner T, Seitz S, Suyu SH, Halkola A, Umetsu K, Zitrin A, Coe D, Monna A, Rosati P, Grillo C, Balestra I, Postman M, Koekemoer A, Zheng W, Høst O, Lemze D, Broadhurst T, Moustakas L, Bradley L, Molino A, Nonino M, Mercurio A, Scodeggio M, Bartelmann M, Benitez N, Bouwens R, Donahue M, Infante L, Jouvel S, Kelson D, Lahav O, Medezinski E, Melchior P, Merten J, Riess A (2013) Galaxy halo truncation and giant arc surface brightness reconstruction in the cluster MACSJ1206.2-0847. Astrophys J 774(2):124. https://doi.org/10.1088/0004-637X/774/2/124. arXiv:1306.5240 [astro-ph.CO]

Elahi PJ, Mahdi HS, Power C, Lewis GF (2014) Warm dark haloes accretion histories and their gravitational signatures. Mon Not R Astron Soc 444(3):2333–2345. https://doi.org/10.1093/mnras/stu1614. arXiv:1406.3413 [astro-ph.CO]

Euclid Collaboration, Adam R, Vannier M, Maurogordato S, Biviano A, Adami C, Ascaso B, Bellagamba F, Benoist C, Cappi A, Díaz-Sánchez A, Durret F, Farrens S, Gonzalez AH, Iovino A, Licitra R, Maturi M, Mei S, Merson A, Munari E, Pelló R, Ricci M, Rocci PF, Roncarelli M, Sarron F, Amoura Y, Andreon S, Apostolakos N, Arnaud M, Bardelli S, Bartlett J, Baugh CM, Borgani S, Brodwin M, Castander F, Castignani G, Cucciati O, De Lucia G, Dubath P, Fosalba P, Giocoli C, Hoekstra H, Mamon GA, Melin JB, Moscardini L, Paltani S, Radovich M, Sartoris B, Schultheis M, Sereno M, Weller J, Burigana C, Carvalho CS, Corcione L, Kurki-Suonio H, Lilje PB, Sirri G, Toledo-Moreo R, Zamorani G (2019) Euclid preparation. III. Galaxy cluster detection in the wide photometric survey, performance and algorithm selection. Astron Astrophys 627:A23. https://doi.org/10.1051/0004-6361/201935088. arXiv:1906.04707 [astro-ph.CO]

Finney EQ, Bradač M, Huang KH, Hoag A, Morishita T, Schrabback T, Treu T, Borello Schmidt K, Lemaux BC, Wang X, Mason C (2018) Mass modeling of frontier fields cluster MACS J1149.5+2223 using strong and weak lensing. Astrophys J 859:58. https://doi.org/10.3847/1538-4357/aabf97

Foëx G, Motta V, Jullo E, Limousin M, Verdugo T (2014) SARCS strong-lensing galaxy groups. II. Mass-concentration relation and strong-lensing bias. Astron Astrophys 572:A19. https://doi.org/10.1051/0004-6361/201424706. arXiv:1409.5905 [astro-ph.CO]

Fohlmeister J, Kochanek CS, Falco EE, Wambsganss J, Oguri M, Dai X (2013) A two-year time delay for the lensed quasar SDSS J1029+2623. Astrophys J 764(2):186. https://doi.org/10.1088/0004-637X/764/2/186. arXiv:1207.5776 [astro-ph.CO]

Fox DC, Pen UL (2001) Gravitational lensing by galaxy groups in the Hubble deep field. Astrophys J 546(1):35–46. https://doi.org/10.1086/318242. arXiv:astro-ph/0009068 [astro-ph]

Fox C, Mahler G, Sharon K, Remolina González JD (2022) The strongest cluster lenses: an analysis of the relation between strong gravitational lensing strength and the physical properties of galaxy clusters. Astrophys J 928(1):87. https://doi.org/10.3847/1538-4357/ac5024. arXiv:2104.05585 [astro-ph.CO]

Furtak LJ, Shuntov M, Atek H, Zitrin A, Richard J, Lehnert MD, Chevallard J (2023) Constraining the physical properties of the first lensed $z \sim 9$–16 galaxy candidates with JWST. Mon Not R Astron Soc 519(2):3064–3075. https://doi.org/10.1093/mnras/stac3717. arXiv:2208.05473 [astro-ph.GA]

Ghigna S, Moore B, Governato F, Lake G, Quinn T, Stadel J (1998) Dark matter haloes within clusters. Mon Not R Astron Soc 300(1):146–162. https://doi.org/10.1046/j.1365-8711.1998.01918.x. arXiv:astro-ph/9801192 [astro-ph]

Ghosh A, Williams LLR, Liesenborgs J (2020) Free-form grale lens inversion of galaxy clusters with up to 1000 multiple images. Mon Not R Astron Soc 494(3):3998–4014. https://doi.org/10.1093/mnras/staa962. arXiv:2004.01724 [astro-ph.CO]

Ghosh A, Williams LLR, Liesenborgs J, Acebron A, Jauzac M, Koekemoer AM, Mahler G, Niemiec A, Steinhardt C, Faisst AL, Lagattuta D, Natarajan P (2021) Further support for a trio of mass-to-light deviations in Abell 370: free-form grale lens inversion using BUFFALO strong lensing data. Mon Not R Astron Soc 506(4):6144–6158. https://doi.org/10.1093/mnras/stab1196. arXiv:2104.11781 [astro-ph.CO]

Ghosh A, Adams D, Williams LLR, Liesenborgs J, Alavi A, Scarlata C (2023) An excursion into the core of the cluster lens Abell 1689. Mon Not R Astron Soc 525(2):2519–2534. https://doi.org/10.1093/mnras/stad2418. arXiv:2206.08584 [astro-ph.CO]

Gilbank DG, Gladders MD, Yee HKC, Hsieh BC (2011) The red-sequence cluster survey-2 (RCS-2): survey details and photometric catalog construction. Astron J 141(3):94. https://doi.org/10.1088/0004-6256/141/3/94. arXiv:1012.3470 [astro-ph.CO]

Gilmore J, Natarajan P (2009) Cosmography with cluster strong lensing. Mon Not R Astron Soc 396(1):354–364. https://doi.org/10.1111/j.1365-2966.2009.14612.x. arXiv:astro-ph/0605245 [astro-ph]

Gladders MD, Yee HKC (2000) A new method for galaxy cluster detection. I. The algorithm. Astron J 120(4):2148–2162. https://doi.org/10.1086/301557. arXiv:astro-ph/0004092 [astro-ph]

Goldberg DM, Natarajan P (2002) The galaxy octopole moment as a probe of weak-lensing shear fields. Astrophys J 564(1):65–72. https://doi.org/10.1086/324202. arXiv:astro-ph/0107187 [astro-ph]

Golubchik M, Furtak LJ, Meena AK, Zitrin A (2022) HST strong-lensing model for the first JWST galaxy cluster SMACS J0723.3-7327. Astrophys J 938(1):14. https://doi.org/10.3847/1538-4357/ac8ff1. arXiv:2207.05007 [astro-ph.CO]

Gonzalez AH, Zabludoff AI, Zaritsky D (2005) Intracluster light in nearby galaxy clusters: relationship to the halos of brightest cluster galaxies. Astrophys J 618(1):195–213. https://doi.org/10.1086/425896. arXiv:astro-ph/0406244 [astro-ph]

González-López J, Bauer FE, Romero-Cañizales C, Kneissl R, Villard E, Carvajal R, Kim S, Laporte N, Anguita T, Aravena M, Bouwens RJ, Bradley L, Carrasco M, Demarco R, Ford H, Ibar E, Infante L, Messias H, Muñoz Arancibia AM, Nagar N, Padilla N, Treister E, Troncoso P, Zitrin A (2017) The ALMA frontier fields survey. I. 1.1 mm continuum detections in Abell 2744, MACS J0416.1-2403 and MACS J1149.5+2223. Astron Astrophys 597:A41. https://doi.org/10.1051/0004-6361/201628806. arXiv:1607.03808 [astro-ph.GA]

Granata G, Mercurio A, Grillo C, Tortorelli L, Bergamini P, Meneghetti M, Rosati P, Caminha GB, Nonino M (2022) Improved strong lensing modelling of galaxy clusters using the fundamental plane: detailed mapping of the baryonic and dark matter mass distribution of Abell S1063. Astron Astrophys 659:A24. https://doi.org/10.1051/0004-6361/202141817. arXiv:2107.09079 [astro-ph.GA]

Griffiths RE, Rudisel M, Wagner J, Hamilton T, Huang PC, Villforth C (2021) Hamilton's object – a clumpy galaxy straddling the gravitational caustic of a galaxy cluster: constraints on dark matter clumping. Mon Not R Astron Soc 506(2):1595–1608. https://doi.org/10.1093/mnras/stab1375. arXiv:2105.04562 [astro-ph.CO]

Grillo C, Gobat R, Presotto V, Balestra I, Mercurio A, Rosati P, Nonino M, Vanzella E, Christensen L, Graves G, Biviano A, Lemze D, Bartelmann M, Benitez N, Bouwens R, Bradley L, Broadhurst T, Coe D, Donahue M, Ford H, Infante L, Jouvel S, Kelson D, Koekemoer A, Lahav O, Medezinski E, Melchior P, Meneghetti M, Merten J, Molino A, Monna A, Moustakas J, Moustakas LA, Postman M, Seitz S, Umetsu K, Zheng W, Zitrin A (2014) CLASH: extending galaxy strong lensing to small physical scales with distant sources highly magnified by galaxy cluster members. Astrophys J 786(1):11. https://doi.org/10.1088/0004-637X/786/1/11. arXiv:1403.0573 [astro-ph.CO]

Grillo C, Suyu SH, Rosati P, Mercurio A, Balestra I, Munari E, Nonino M, Caminha GB, Lombardi M, De Lucia G, Borgani S, Gobat R, Biviano A, Girardi M, Umetsu K, Coe D, Koekemoer AM, Postman M, Zitrin A, Halkola A, Broadhurst T, Sartoris B, Presotto V, Annunziatella M, Maier C, Fritz A, Vanzella E, Frye B (2015) CLASH-VLT: insights on the mass substructures in the frontier fields cluster MACS J0416.1-2403 through accurate strong lens modeling. Astrophys J 800(1):38. https://doi.org/10.1088/0004-637X/800/1/38. arXiv:1407.7866 [astro-ph.CO]

Grillo C, Karman W, Suyu SH, Rosati P, Balestra I, Mercurio A, Lombardi M, Treu T, Caminha GB, Halkola A, Rodney SA, Gavazzi R, Caputi KI (2016) The story of supernova "Refsdal" told by MUSE. Astrophys J 822(2):78. https://doi.org/10.3847/0004-637X/822/2/78. arXiv:1511.04093 [astro-ph.GA]

Grillo C, Rosati P, Suyu SH, Balestra I, Caminha GB, Halkola A, Kelly PL, Lombardi M, Mercurio A, Rodney SA, Treu T (2018) Measuring the value of the Hubble constant "à la refsdal". Astrophys J 860(2):94. https://doi.org/10.3847/1538-4357/aac2c9. arXiv:1802.01584 [astro-ph.CO]

Grillo C, Rosati P, Suyu SH, Caminha GB, Mercurio A, Halkola A (2020) On the accuracy of time-delay cosmography in the frontier fields cluster MACS J1149.5+2223 with supernova refsdal. Astrophys J 898(1):87. https://doi.org/10.3847/1538-4357/ab9a4c. arXiv:2001.02232 [astro-ph.CO]

Grossman SA, Narayan R (1988) Arcs from gravitational lensing. Astrophys J Lett 324:L37. https://doi.org/10.1086/185086

Gruen D, Seitz S, Brimioulle F, Kosyra R, Koppenhoefer J, Lee CH, Bender R, Riffeser A, Eichner T, Weidinger T, Bierschenk M (2014) Weak lensing analysis of SZ-selected clusters of galaxies from the SPT and Planck surveys. Mon Not R Astron Soc 442(2):1507–1544. https://doi.org/10.1093/mnras/stu949. arXiv:1310.6744 [astro-ph.CO]

Harvey D, Robertson A, Massey R, McCarthy IG (2019) Observable tests of self-interacting dark matter in galaxy clusters: BCG wobbles in a constant density core. Mon Not R Astron Soc 488(2):1572–1579. https://doi.org/10.1093/mnras/stz1816. arXiv:1812.06981 [astro-ph.CO]

Hjorth J, Williams LLR (2010) Statistical mechanics of collisionless orbits. I. Origin of central cusps in dark-matter halos. Astrophys J 722(1):851–855. https://doi.org/10.1088/0004-637X/722/1/851. arXiv:1010.0265 [astro-ph.CO]

Hoag A, Huang KH, Treu T, Bradač M, Schmidt KB, Wang X, Brammer GB, Broussard A, Amorin R, Castellano M, Fontana A, Merlin E, Schrabback T, Trenti M, Vulcani B (2016) The grism lens-amplified survey from space (GLASS). VI. Comparing the mass and light in MACS J0416.1-2403 using frontier field imaging and GLASS spectroscopy. Astrophys J 831:182. https://doi.org/10.3847/0004-637X/831/2/182. arXiv:1603.00505

Holanda RFL, Busti VC, Gonzalez JE, Andrade-Santos F, Alcaniz JS (2017) Cosmological constraints on the gas depletion factor in galaxy clusters. J Cosmol Astropart Phys 2017(12):016. https://doi.org/10.1088/1475-7516/2017/12/016. arXiv:1706.07321 [astro-ph.CO]

Horesh A, Maoz D, Ebeling H, Seidel G, Bartelmann M (2010) The lensing efficiencies of MACS X-ray-selected versus RCS optically selected galaxy clusters. Mon Not R Astron Soc 406(2):1318–1336. https://doi.org/10.1111/j.1365-2966.2010.16763.x. arXiv:1004.2048 [astro-ph.CO]

Ishigaki M, Kawamata R, Ouchi M, Oguri M, Shimasaku K, Ono Y (2015) Hubble frontier fields first complete cluster data: faint galaxies at $z \sim 5$–10 for UV luminosity functions and cosmic reionization. Astrophys J 799(1):12. https://doi.org/10.1088/0004-637X/799/1/12. arXiv:1408.6903 [astro-ph.GA]

Jaelani AT, More A, Oguri M, Sonnenfeld A, Suyu SH, Rusu CE, Wong KC, Chan JHH, Kayo I, Lee CH, Chao DCY, Coupon J, Inoue KT, Futamase T (2020) Survey of Gravitationally lensed Objects in HSC Imaging (SuGOHI) - V. Group-to-cluster scale lens search from the HSC-SSP survey. Mon Not R Astron Soc 495(1):1291–1310. https://doi.org/10.1093/mnras/staa1062. arXiv:2002.01611 [astro-ph.GA]

Jauzac M, Eckert D, Schwinn J, Harvey D, Baugh CM, Robertson A, Bose S, Massey R, Owers M, Ebeling H, Shan HY, Jullo E, Kneib JP, Richard J, Atek H, Clément B, Egami E, Israel H, Knowles K, Limousin M, Natarajan P, Rexroth M, Taylor P, Tchernin C (2016) The extraordinary amount of substructure in the Hubble Frontier Fields cluster Abell 2744. Mon Not R Astron Soc 463(4):3876–3893. https://doi.org/10.1093/mnras/stw2251. arXiv:1606.04527 [astro-ph.CO]

Johnson TL, Sharon K (2016) The systematics of strong lens modeling quantified: the effects of constraint selection and redshift information on magnification, mass, and multiple image predictability. Astrophys J 832(1):82. https://doi.org/10.3847/0004-637X/832/1/82. arXiv:1608.08713 [astro-ph.CO]

Johnson TL, Sharon K, Bayliss MB, Gladders MD, Coe D, Ebeling H (2014) Lens models and magnification maps of the six Hubble frontier fields clusters. Astrophys J 797(1):48. https://doi.org/10.1088/0004-637X/797/1/48. arXiv:1405.0222 [astro-ph.CO]

Johnson TL, Rigby JR, Sharon K, Gladders MD, Florian M, Bayliss MB, Wuyts E, Whitaker KE, Livermore R, Murray KT (2017) Star formation at $z = 2.481$ in the lensed galaxy SDSS J1110+6459: star formation down to 30 pc scales. Astrophys J Lett 843(2):L21. https://doi.org/10.3847/2041-8213/aa7516. arXiv:1707.00706 [astro-ph.GA]

Jullo E, Kneib JP (2009) Multiscale cluster lens mass mapping - I. Strong lensing modelling. Mon Not R Astron Soc 395(3):1319–1332. https://doi.org/10.1111/j.1365-2966.2009.14654.x. arXiv:0901.3792 [astro-ph.CO]

Jullo E, Kneib JP, Limousin M, Elíasdóttir Á., Marshall PJ, Verdugo T (2007a) A Bayesian approach to strong lensing modelling of galaxy clusters. New J Phys 9(12):447. https://doi.org/10.1088/1367-2630/9/12/447. arXiv:0706.0048 [astro-ph]

Jullo E, Kneib JP, Limousin M, Elíasdóttir Á., Marshall PJ, Verdugo T (2007b) A Bayesian approach to strong lensing modelling of galaxy clusters. New J Phys 9(12):447. https://doi.org/10.1088/1367-2630/9/12/447. arXiv:0706.0048 [astro-ph]

Jullo E, Natarajan P, Kneib JP, D'Aloisio A, Limousin M, Richard J, Schimd C (2010) Cosmological constraints from strong gravitational lensing in clusters of galaxies. Science 329:924–927. https://doi.org/10.1126/science.1185759. arXiv:1008.4802 [astro-ph.CO]

Jullo E, Acebron A, Limousin M, Giocoli C, Despali G, Bonamigo M (2015) Strong lensing cosmography in the frontier fields. Proc Int Astron Union 11(A29B):801–803. https://doi.org/10.1017/S1743921316006888

Kahlhoefer F, Schmidt-Hoberg K, Kummer J, Sarkar S (2015) On the interpretation of dark matter self-interactions in Abell 3827. Mon Not R Astron Soc 452(1):L54–L58. https://doi.org/10.1093/mnrasl/slv088. arXiv:1504.06576 [astro-ph.CO]

Kaiser N (1992) Weak gravitational lensing of distant galaxies. Astrophys J 388:272. https://doi.org/10.1086/171151

Kaiser N, Squires G (1993) Mapping the dark matter with weak gravitational lensing. Astrophys J 404:441. https://doi.org/10.1086/172297

Karman W, Caputi KI, Grillo C, Balestra I, Rosati P, Vanzella E, Coe D, Christensen L, Koekemoer AM, Krühler T, Lombardi M, Mercurio A, Nonino M, van der Wel A (2015) MUSE integral-field spectroscopy towards the frontier fields cluster Abell S1063. I. Data products and redshift identifications. Astron Astrophys 574:A11. https://doi.org/10.1051/0004-6361/201424962. arXiv:1409.3507 [astro-ph.GA]

Kaurov AA, Dai L, Venumadhav T, Miralda-Escudé J, Frye B (2019) Highly magnified stars in lensing clusters: new evidence in a galaxy lensed by MACS J0416.1-2403. Astrophys J 880(1):58. https://doi.org/10.3847/1538-4357/ab2888. arXiv:1902.10090 [astro-ph.GA]

Kawamata R, Oguri M, Ishigaki M, Shimasaku K, Ouchi M (2016) Precise strong lensing mass modeling of four Hubble frontier field clusters and a sample of magnified high-redshift galaxies. Astrophys J 819(2):114. https://doi.org/10.3847/0004-637X/819/2/114. arXiv:1510.06400 [astro-ph.GA]

Keeton CR (2010) On modeling galaxy-scale strong lens systems. Gen Relativ Gravit 42(9):2151–2176. https://doi.org/10.1007/s10714-010-1041-1

Kelly PL, Rodney SA, Treu T, Foley RJ, Brammer G, Schmidt KB, Zitrin A, Sonnenfeld A, Strolger LG, Graur O, Filippenko AV, Jha SW, Riess AG, Bradac M, Weiner BJ, Scolnic D, von Malkan MA, der Linden A, Trenti M, Hjorth J, Gavazzi R, Fontana A, Merten JC, McCully C, Jones T, Postman M, Dressler A, Patel B, Cenko SB, Graham ML, Tucker BE (2015) Multiple images of a highly magnified supernova formed by an early-type cluster galaxy lens. Science 347(6226):1123–1126. https://doi.org/10.1126/science.aaa3350. arXiv:1411.6009 [astro-ph.CO]

Kelly PL, Rodney SA, Treu T, Strolger LG, Foley RJ, Jha SW, Selsing J, Brammer G, Bradač M, Cenko SB, Graur O, Filippenko AV, Hjorth J, McCully C, Molino A, Nonino M, Riess AG, Schmidt KB, Tucker B, von der Linden A, Weiner BJ, Zitrin A (2016) Deja vu all over again: the reappearance of supernova refsdal. Astrophys J Lett 819(1):L8. https://doi.org/10.3847/2041-8205/819/1/L8. arXiv:1512.04654 [astro-ph.CO]

Kelly PL, Rodney S, Treu T, Oguri M, Chen W, Zitrin A, Birrer S, Bonvin V, Dessart L, Diego JM, Filippenko AV, Foley RJ, Gilman D, Hjorth J, Jauzac M, Mandel K, Millon M, Pierel J, Sharon K, Thorp S, Williams L, Broadhurst T, Dressler A, Graur O, Jha S, McCully C, Postman M, Schmidt KB, Tucker BE, von der Linden A (2023) Constraints on the Hubble constant from supernova Refsdal's reappearance Science 380:6649:abh1322. https://doi.org/10.1126/science.abh1322

Khullar G, Gozman K, Lin JJ, Martinez MN, Matthews Acuña OS, Medina E, Merz K, Sanchez JA, Sisco EE, Kavin Stein DJ, Sukay EO, Tavangar K, Bayliss MB, Bleem LE, Brownsberger S, Dahle H, Florian MK, Gladders MD, Mahler G, Rigby JR, Sharon K, Stark AA (2021) COOL-LAMPS. I. An extraordinarily bright lensed galaxy at redshift 5.04. Astrophys J 906(2):107. https://doi.org/10.3847/1538-4357/abcb86. arXiv:2011.06601 [astro-ph.GA]

Kneib JP, Natarajan P (2011) Cluster lenses. Astron Astrophys Rev 19:47. https://doi.org/10.1007/s00159-011-0047-3. arXiv:1202.0185 [astro-ph.CO]

Kneib JP, Mellier Y, Fort B, Mathez G (1993) The distribution of dark matter in distant cluster-lenses: modelling A 370. Astron Astrophys 273:367.

Kneib JP, Ellis RS, Smail I, Couch WJ, Sharples RM (1996) Hubble space telescope observations of the lensing cluster Abell 2218. Astrophys J 471:643. https://doi.org/10.1086/177995. arXiv:astro-ph/9511015 [astro-ph]

Kochanek CS (1990) Inverting cluster gravitational lenses. Mon Not R Astron Soc 247:135.

Köhlinger F, Schmidt RW (2014) Strong lensing in RX J1347.5-1145 revisited. Mon Not R Astron Soc 437(2):1858–1871. https://doi.org/10.1093/mnras/stt2017. arXiv:1310.0021 [astro-ph.CO]

Kovner I (1989) Diagnostics of compact clusters of galaxies by giant luminous arcs. Astrophys J 337:621. https://doi.org/10.1086/167133

Lagattuta DJ, Richard J, Bauer FE, Cerny C, Claeyssens A, Guaita L, Jauzac M, Jeanneau A, Koekemoer AM, Mahler G, Prieto Lyon G, Acebron A, Meneghetti M, Niemiec A, Zitrin A, Bianconi M, Connor T, Cen R, Edge A, Faisst AL, Limousin M, Massey R, Sereno M, Sharon K, Weaver JR (2022) Pilot-WINGS: an extended MUSE view of the structure of Abell 370. Mon Not R Astron Soc 514(1):497–517. https://doi.org/10.1093/mnras/stac418. arXiv:2202.04663 [astro-ph.GA]

Lage C, Farrar GR (2015) The bullet cluster is not a cosmological anomaly. J Cosmol Astropart Phys 2015(2):038. https://doi.org/10.1088/1475-7516/2015/02/038. arXiv:1406.6703 [astro-ph.GA]

Lam D (2019) A new approach to free-form cluster lens modeling inspired by the JPEG image compression method. Publ Astron Soc Pac 131(1005):114505. https://doi.org/10.1088/1538-3873/ab35c0. arXiv:1906.00006 [astro-ph.CO]

Lanusse F, Starck JL, Leonard A, Pires S (2016) High resolution weak lensing mass mapping combining shear and flexion. Astron Astrophys 591:A2. https://doi.org/10.1051/0004-6361/201628278. arXiv:1603.01599 [astro-ph.CO]

Laporte N, Zitrin A, Dole H, Roberts-Borsani G, Furtak LJ, Witten C (2022) A lensed protocluster candidate at z = 7.66 identified in JWST observations of the galaxy cluster SMACS0723−7327. Astron Astrophys 667:L3. https://doi.org/10.1051/0004-6361/202244719. arXiv:2208.04930 [astro-ph.GA]

Lasko K, Williams LLR, Ghosh A (2023) What multiple images say about the large-scale mass maps of galaxy clusters. Mon Not R Astron Soc 525(4):5423–5436. https://doi.org/10.1093/mnras/stad2622. arXiv:2309.05730

Leonard A, Goldberg DM, Haaga JL, Massey R (2007) Gravitational shear, flexion, and strong lensing in Abell 1689. Astrophys J 666(1):51–63. https://doi.org/10.1086/520109. arXiv:astro-ph/0702242 [astro-ph]

Li N, Gladders MD, Rangel EM, Florian MK, Bleem LE, Heitmann K, Habib S, Fasel P (2016) PICS: simulations of strong gravitational lensing in galaxy clusters. Astrophys J 828(1):54. https://doi.org/10.3847/0004-637X/828/1/54. arXiv:1511.03673 [astro-ph.CO]

Li N, Gladders MD, Heitmann K, Rangel EM, Child HL, Florian MK, Bleem LE, Habib S, Finkel HJ (2019) The importance of secondary halos for strong lensing in massive galaxy clusters across redshift. Astrophys J 878(2):122. https://doi.org/10.3847/1538-4357/ab1f74. arXiv:1810.13330 [astro-ph.CO]

Liesenborgs J, De Rijcke S, Dejonghe H (2006) A genetic algorithm for the non-parametric inversion of strong lensing systems. Mon Not R Astron Soc 367(3):1209–1216. https://doi.org/10.1111/j.1365-2966.2006.10040.x. arXiv:astro-ph/0601124 [astro-ph]

Liesenborgs J, de Rijcke S, Dejonghe H, Bekaert P (2009) Non-parametric strong lens inversion of SDSS J1004+4112. Mon Not R Astron Soc 397(1):341–349. https://doi.org/10.1111/j.1365-2966.2009.14912.x. arXiv:0904.2382 [astro-ph.CO]

Liesenborgs J, Williams LLR, Wagner J, De Rijcke S (2020) Extended lens reconstructions with grale: exploiting time-domain, substructural, and weak lensing information. Mon Not R Astron Soc 494(3):3253–3274. https://doi.org/10.1093/mnras/staa842. arXiv:2003.10377 [astro-ph.CO]

Limousin M, Cabanac R, Gavazzi R, Kneib JP, Motta V, Richard J, Thanjavur K, Foex G, Pello R, Crampton D, Faure C, Fort B, Jullo E, Marshall P, Mellier Y, More A, Soucail G, Suyu S, Swinbank M, Sygnet JF, Tu H, Valls-Gabaud D, Verdugo T, Willis J (2009) A new window of exploration in the mass spectrum: strong lensing by galaxy groups in the SL2S. Astron Astrophys 502(2):445–456. https://doi.org/10.1051/0004-6361/200811473. arXiv:0812.1033 [astro-ph]

Limousin M, Beauchesne B, Jullo E (2022) Dark matter in galaxy clusters: parametric strong-lensing approach. Astron Astrophys 664:A90. https://doi.org/10.1051/0004-6361/202243278. arXiv:2202.02992 [astro-ph.CO]

Lin J, Wagner J, Griffiths RE (2022) Generalised model-independent characterisation of strong gravitational lenses VIII. Automated multi-band feature detection to constrain local lens properties. Mon Not R Astron Soc. https://doi.org/10.1093/mnras/stac2576

Lin J, Wagner J, Griffiths RE (2023) Much ado about no offset – characterizing the anomalous multiple-image configuration and the model-driven displacement between light and mass in the multiplane strong lens Abell 3827. Mon Not R Astron Soc 526(2):2776–2794. https://doi.org/10.1093/mnras/stad2800

Livermore RC, Finkelstein SL, Lotz JM (2017) Directly observing the galaxies likely responsible for reionization. Astrophys J 835(2):113. https://doi.org/10.3847/1538-4357/835/2/113. arXiv:1604.06799 [astro-ph.GA]

Lotz JM, Koekemoer A, Coe D, Grogin N, Capak P, Mack J, Anderson J, Avila R, Barker EA, Borncamp D, Brammer G, Durbin M, Gunning H, Hilbert B, Jenkner H, Khandrika H, Levay Z, Lucas RA, MacKenty J, Ogaz S, Porterfield B, Reid N, Robberto M, Royle P, Smith LJ, Storrie-Lombardi LJ, Sunnquist B, Surace J, Taylor DC, Williams R, Bullock J, Dickinson M, Finkelstein S, Natarajan P, Richard J, Robertson B, Tumlinson J, Zitrin A, Flanagan K, Sembach K, Soifer BT, Mountain M (2017) The frontier fields: survey design and initial results. Astrophys J 837(1):97. https://doi.org/10.3847/1538-4357/837/1/97. arXiv:1605.06567 [astro-ph.GA]

Lynds R, Petrosian V (1989) Luminous arcs in clusters of galaxies. Astrophys J 336:1. https://doi.org/10.1086/166989

Magana J, Acebron A, Motta V, Verdugo T, Jullo E, Limousin M (2018) Strong lensing modeling in galaxy clusters as a promising method to test cosmography. I. Parametric dark energy models. Astrophys J 865(2):122. https://doi.org/10.3847/1538-4357/aada7d

Mahdi HS, van Beek M, Elahi PJ, Lewis GF, Power C, Killedar M (2014) Gravitational lensing in WDM cosmologies: the cross-section for giant arcs. Mon Not R Astron Soc 441(3):1954–1963. https://doi.org/10.1093/mnras/stu705. arXiv:1404.1644 [astro-ph.CO]

Mahler G, Jauzac M, Richard J, Beauchesne B, Ebeling H, Lagattuta D, Natarajan P, Sharon K, Atek H, Claeyssens A, Clément B, Eckert D, Edge A, Kneib JP, Niemiec A (2023) Precision modeling of JWST's first cluster lens SMACS J0723.3-7327. Astrophys J 945(1):49. https://doi.org/10.3847/1538-4357/acaea9. arXiv:2207.07101 [astro-ph.GA]

Mantz AB, Allen SW, Morris RG, Rapetti DA, Applegate DE, von Kelly PL, der Linden A, Schmidt RW (2014) Cosmology and astrophysics from relaxed galaxy clusters - II. Cosmological constraints. Mon Not R Astron Soc 440(3):2077–2098. https://doi.org/10.1093/mnras/stu368. arXiv:1402.6212 [astro-ph.CO]

Massey R, Williams L, Smit R, Swinbank M, Kitching TD, Harvey D, Jauzac M, Israel H, Clowe D, Edge A, Hilton M, Jullo E, Leonard A, Liesenborgs J, Merten J, Mohammed I, Nagai D, Richard J, Robertson A, Saha P, Santana R, Stott J, Tittley E (2015) The behaviour of dark matter associated with four bright cluster galaxies in the 10 kpc core of Abell 3827. Mon Not R Astron Soc 449(4):3393–3406https://doi.org/10.1093/mnras/stv467. arXiv:1504.03388 [astro-ph.CO]

Massey R, Harvey D, Liesenborgs J, Richard J, Stach S, Swinbank M, Taylor P, Williams L, Clowe D, Courbin F, Edge A, Israel H, Jauzac M, Joseph R, Jullo E, Kitching TD, Leonard A, Merten J, Nagai D, Nightingale J, Robertson A, Romualdez LJ, Saha P, Nagai T, Smit R, Tam SI Tittley E (2018b) Dark matter dynamics in Abell 3827: new data consistent with standard cold dark matter. Mon Not R Astron Soc 477(1):669–677, https://doi.org/10.1093/mnras/sty630. arXiv:1708.04245 [astro-ph.CO]

Massey R, Harvey D, Liesenborgs J, Richard J, Stach S, Swinbank M, Taylor P, Williams L, Clowe D, Courbin F, Edge A, Israel H, Jauzac M, Joseph R, Jullo E, Kitching TD, Leonard A, Merten J, Nagai D, Nightingale J, Robertson A, Romualdez LJ, Saha P, Smit R, Tam SI, Tittley E (2018a) Dark matter dynamics in Abell 3827: new data consistent with standard cold dark matter. Mon Not R Astron Soc 477(1):669–677. https://doi.org/10.1093/mnras/sty630. arXiv:1708.04245 [astro-ph.CO]

McCully C, Keeton CR, Wong KC, Zabludoff AI (2014) A new hybrid framework to efficiently model lines of sight to gravitational lenses. Mon Not R Astron Soc 443(4):3631–3642. https://doi.org/10.1093/mnras/stu1316. arXiv:1401.0197 [astro-ph.CO]

McCully C, Keeton CR, Wong KC, Zabludoff AI (2017) Quantifying environmental and line-of-sight effects in models of strong gravitational lens systems. Astrophys J 836(1):141. https://doi.org/10.3847/1538-4357/836/1/141. arXiv:1601.05417 [astro-ph.CO]

Medezinski E, Umetsu K, Nonino M, Merten J, Zitrin A, Broadhurst T, Donahue M, Sayers J, Waizmann JC, Koekemoer A, Coe D, Molino A, Melchior P, Mroczkowski T, Czakon N, Postman M, Meneghetti M, Lemze D, Ford H, Grillo C, Kelson D, Bradley L, Moustakas J, Bartelmann M, Benítez N, Biviano A, Bouwens R, Golwala S, Graves G, Infante L, Jiménez-Teja Y, Jouvel S, Lahav O, Moustakas L, Ogaz S, Rosati P, Seitz S, Zheng W (2013) CLASH: complete lensing analysis of the largest cosmic lens MACS J0717.5+3745 and surrounding structures. Astrophys J 777(1):43. https://doi.org/10.1088/0004-637X/777/1/43. arXiv:1304.1223 [astro-ph.CO]

Meena AK, Bagla JS (2021) Exotic image formation in strong gravitational lensing by clusters of galaxies - I. Cross-section. Mon Not R Astron Soc 503(2):2097–2107. https://doi.org/10.1093/mnras/stab577. arXiv:2009.13418 [astro-ph.CO]

Menanteau F, Hughes JP, Sifón C, Hilton M, González J, Infante L, Barrientos LF, Baker AJ, Bond JR, Das S, Devlin MJ, Dunkley J, Hajian A, Hincks AD, Kosowsky A, Marsden D, Marriage TA, Moodley K, Niemack MD, Nolta MR, Page LA, Reese ED, Sehgal N, Sievers J, Spergel DN, Staggs ST, Wollack E (2012) The Atacama cosmology telescope: ACT-CL J0102-4915 "el gordo," a massive merging cluster at redshift 0.87. Astrophys J 748(1):7. https://doi.org/10.1088/0004-637X/748/1/7. arXiv:1109.0953 [astro-ph.CO]

Meneghetti M, Bartelmann M, Dahle H, Limousin M (2013) Arc statistics. Space Sci Rev 177(1–4):31–74. https://doi.org/10.1007/s11214-013-9981-x. arXiv:1303.3363 [astro-ph.CO]

Meneghetti M, Natarajan P, Coe D, Contini E, De Lucia G, Giocoli C, Acebron A, Borgani S, Bradac M, Diego JM, Hoag A, Ishigaki M, Johnson TL, Jullo E, Kawamata R, Lam D, Limousin M, Liesenborgs J, Oguri M, Sebesta K, Sharon K, Williams LLR, Zitrin A (2017) The Frontier Fields lens modelling comparison project. Mon Not R Astron Soc 472(3):3177–3216. https://doi.org/10.1093/mnras/stx2064. arXiv:1606.04548 [astro-ph.CO]

Meneghetti M, Davoli G, Bergamini P, Rosati P, Natarajan P, Giocoli C, Caminha GB, Metcalf RB, Rasia E, Borgani S, Calura F, Grillo C, Mercurio A, Vanzella E (2020) An excess of small-scale gravitational lenses observed in galaxy clusters. Science 369(6509):1347–1351. https://doi.org/10.1126/science.aax5164. arXiv:2009.04471 [astro-ph.GA]

Meneghetti M, Ragagnin A, Borgani S, Calura F, Despali G, Giocoli C, Granato GL, Grillo C, Moscardini L, Rasia E, Rosati P, Angora G, Bassini L, Bergamini P, Caminha GB, Granata G, Mercurio A, Metcalf RB, Natarajan P, Nonino M, Pignataro GV, Ragone-Figueroa C, Vanzella E, Acebron A, Dolag K, Murante G, Taffoni G, Tornatore L, Tortorelli L, Valentini M (2022) The probability of galaxy-galaxy strong lensing events in hydrodynamical simulations of galaxy clusters. Astron Astrophys 668:A188. https://doi.org/10.1051/0004-6361/202243779. arXiv:2204.09065 [astro-ph.CO]

Merten J (2016) Mesh-free free-form lensing - I. Methodology and application to mass reconstruction. Mon Not R Astron Soc 461(3):2328–2345. https://doi.org/10.1093/mnras/stw1413. arXiv:1412.5186 [astro-ph.CO]

Merten J, Coe D, Dupke R, Massey R, Zitrin A, Cypriano ES, Okabe N, Frye B, Braglia FG, Jiménez-Teja Y, Benítez N, Broadhurst T, Rhodes J, Meneghetti M, Moustakas LA, Sodré LJ, Krick J, Bregman JN (2011) Creation of cosmic structure in the complex galaxy cluster merger Abell 2744. Mon Not R Astron Soc 417(1):333–347. https://doi.org/10.1111/j.1365-2966.2011.19266.x. arXiv:1103.2772 [astro-ph.CO]

Merten J, Meneghetti M, Postman M, Umetsu K, Zitrin A, Medezinski E, Nonino M, Koekemoer A, Melchior P, Gruen D, Moustakas LA, Bartelmann M, Host O, Donahue M, Coe D, Molino A, Jouvel S, Monna A, Seitz S, Czakon N, Lemze D, Sayers J, Balestra I, Rosati P, Benítez N, Biviano A, Bouwens R, Bradley L, Broadhurst T, Carrasco M, Ford H, Grillo C, Infante L, Kelson D, Lahav O, Massey R, Moustakas J, Rasia E, Rhodes J, Vega J, Zheng W (2015) CLASH: the concentration-mass relation of galaxy clusters. Astrophys J 806(1):4. https://doi.org/10.1088/0004-637X/806/1/4. arXiv:1404.1376 [astro-ph.CO]

Miralda-Escude J (1991) Gravitational lensing by clusters of galaxies: constraining the mass distribution. Astrophys J 370:1. https://doi.org/10.1086/169789

Miralda-Escudé J (2002) A test of the collisional dark matter hypothesis from cluster lensing. Astrophys J 564(1):60–64. https://doi.org/10.1086/324138. arXiv:astro-ph/0002050 [astro-ph]

Mohammed I, Saha P, Liesenborgs J (2015) Lensing time delays as a substructure constraint: a case study with the cluster SDSS J1004+4112. Publ Astron Soc Jpn 67(2):21. https://doi.org/10.1093/pasj/psu155. arXiv:1412.3464 [astro-ph.CO]

Mohammed I, Saha P, Williams LLR, Liesenborgs J, Sebesta K (2016) Quantifying substructures in Hubble Frontier Field clusters: comparison with ΛCDM simulations. Mon Not R Astron Soc 459(2):1698–1709. https://doi.org/10.1093/mnras/stw727. arXiv:1507.01532 [astro-ph.CO]

Monna A, Seitz S, Greisel N, Eichner T, Drory N, Postman M, Zitrin A, Coe D, Halkola A, Suyu SH, Grillo C, Rosati P, Lemze D, Balestra I, Snigula J, Bradley L, Umetsu K, Koekemoer A, Kuchner U, Moustakas L, Bartelmann M, Benítez N, Bouwens R, Broadhurst T, Donahue M, Ford H, Host O, Infante L, Jimenez-Teja Y, Jouvel S, Kelson D, Lahav O, Medezinski E, Melchior P, Meneghetti M, Merten J, Molino A, Moustakas J, Nonino M, Zheng W (2014) CLASH: $z \sim 6$ young galaxy candidate quintuply lensed by the frontier field cluster RXC J2248.7-4431. Mon Not R Astron Soc 438(2):1417–1434. https://doi.org/10.1093/mnras/stt2284. arXiv:1308.6280 [astro-ph.CO]

Montes M, Trujillo I (2019) Intracluster light: a luminous tracer for dark matter in clusters of galaxies. Mon Not R Astron Soc 482(2):2838–2851. https://doi.org/10.1093/mnras/sty2858. arXiv:1807.11488 [astro-ph.GA]

More A, Cabanac R, More S, Alard C, Limousin M, Kneib JP, Gavazzi R, Motta V (2012) The CFHTLS-Strong Lensing Legacy Survey (SL2S): Investigating the group-scale lenses with the SARCS sample. Astrophys J 749(1):38. https://doi.org/10.1088/0004-637x/749/1/38

Morioka M, Futamase T (2015) Lens statistics with gravitationally lensed yet morphologically regular images. Astrophys J 805(2):184. https://doi.org/10.1088/0004-637X/805/2/184

Muñoz JA, Kochanek CS, Fohlmeister J, Wambsganss J, Falco E, Forés-Toribio R (2022) The longest delay: a 14.5 yr campaign to determine the third time delay in the lensing cluster SDSS J1004+4112. Astrophys J 937(1):34. https://doi.org/10.3847/1538-4357/ac8877. arXiv:2206.08597 [astro-ph.GA]

Napier K, Sharon K, Dahle H, Bayliss M, Gladders MD, Mahler G, Rigby JR, Florian M (2023) Hubble constant measurement from three large-separation quasars strongly lensed by galaxy clusters. Astrophys J 959(2):134. https://doi.org/10.3847/1538-4357/ad045a. arXiv:2301.11240 [astro-ph.CO]

Narasimha D, Chitre SM (1989) Lensing of extended sources by dark galactic halos. Astron J 97:327. https://doi.org/10.1086/114983

Narayan R, Blandford R, Nityananda R (1984) Multiple imaging of quasars by galaxies and clusters. Nature 310(5973):112–115. https://doi.org/10.1038/310112a0

Natarajan P, Kneib JP (1996) Probing the dynamics of cluster-lenses. Mon Not R Astron Soc 283(3):1031–1046. https://doi.org/10.1093/mnras/283.3.1031. arXiv:astro-ph/9602035 [astro-ph]

Natarajan P, Kneib JP (1997) Lensing by galaxy haloes in clusters of galaxies. Mon Not R Astron Soc 287(4):833–847. https://doi.org/10.1093/mnras/287.4.833. arXiv:astro-ph/9609008 [astro-ph]

Natarajan P, Springel V (2004) Abundance of substructure in clusters of galaxies. Astrophys J Lett 617(1):L13–L16. https://doi.org/10.1086/427079. arXiv:astro-ph/0411515 [astro-ph]

Natarajan P, Kneib JP, Smail I, Ellis RS (1998) The mass-to-light ratio of early-type galaxies: constraints from gravitational lensing in the rich cluster AC 114. Astrophys J 499(2):600–607. https://doi.org/10.1086/305660. arXiv:astro-ph/9706129 [astro-ph]

Natarajan P, Kneib JP, Smail I (2002a) Evidence for tidal stripping of dark matter halos in massive cluster lenses. Astrophys J Lett 580(1):L11–L15. https://doi.org/10.1086/345399. arXiv:astro-ph/0207049 [astro-ph]

Natarajan P, Loeb A, Kneib JP, Smail I (2002b) Constraints on the collisional nature of the dark matter from gravitational lensing in the cluster A2218. Astrophys J 580(1):L17. https://doi.org/10.1086/345547

Natarajan P, De Lucia G, Springel V (2007) Substructure in lensing clusters and simulations. Mon Not R Astron Soc 376(17):180–192. https://doi.org/10.1111/j.1365-2966.2007.11399.x. arXiv:astro-ph/0604414 [astro-ph]

Natarajan P, Kneib JP, Smail I, Treu T, Ellis R, Moran S, Limousin M, Czoske O (2009) The survival of dark matter halos in the cluster cl 0024+16. Astrophys J 693(1):970–983. https://doi.org/10.1088/0004-637X/693/1/970. arXiv:0711.4587 [astro-ph]

Natarajan P, Chadayammuri U, Jauzac M, Richard J, Kneib JP, Ebeling H, Jiang F, van den Bosch F, Limousin M, Jullo E, Atek H, Pillepich A, Popa C, Marinacci F, Hernquist L, Meneghetti M, Vogelsberger M (2017) Mapping substructure in the HST Frontier Fields cluster lenses and in cosmological simulations. Mon Not R Astron Soc 468(2):1962–1980. https://doi.org/10.1093/mnras/stw3385. arXiv:1702.04348 [astro-ph.GA]

Navarro JF, Frenk CS, White SDM (1996) The structure of cold dark matter halos. Astrophys J 462:563. https://doi.org/10.1086/177173. arXiv:astro-ph/9508025 [astro-ph]

Navarro JF, Frenk CS, White SDM (1997) A universal density profile from hierarchical clustering. Astrophys J 490(2):493–508. https://doi.org/10.1086/304888. arXiv:astro-ph/9611107 [astro-ph]

Navarro JF, Hayashi E, Power C, Jenkins AR, Frenk CS, White SDM, Springel V, Stadel J, Quinn TR (2004) The inner structure of ΛCDM haloes – III. Universality and asymptotic slopes. Mon Not R Astron Soc 349(3):1039–1051. https://doi.org/10.1111/j.1365-2966.2004.07586.x. arXiv:astro-ph/0311231 [astro-ph]

Newman AB, Treu T, Ellis RS, Sand DJ (2013) The density profiles of massive, relaxed galaxy clusters. II. Separating luminous and dark matter in cluster cores. Astrophys J 765(1):25. https://doi.org/10.1088/0004-637X/765/1/25. arXiv:1209.1392 [astro-ph.CO]

Newman AB, Ellis RS, Treu T (2015a) Luminous and dark matter profiles from galaxies to clusters: bridging the gap with group-scale lenses. Astrophys J 814(1):26. https://doi.org/10.1088/0004-637X/814/1/26. arXiv:1503.05282 [astro-ph.GA]

Newman AB, Ellis RS, Treu T (2015b) Luminous and dark matter profiles from galaxies to clusters: bridging the gap with group-scale lenses. Astrophys J 814(1):26. https://doi.org/10.1088/0004-637X/814/1/26. arXiv:1503.05282 [astro-ph.GA]

Niemiec A, Jauzac M, Jullo E, Limousin M, Sharon K, Kneib JP, Natarajan P, Richard J (2020) Hybrid-LENSTOOL: a self-consistent algorithm to model galaxy clusters with strong- and weak-lensing simultaneously. Mon Not R Astron Soc 493(3):3331–3340. https://doi.org/10.1093/mnras/staa473. arXiv:2002.04635 [astro-ph.CO]

Ogrean GA, van Weeren RJ, Jones C, Clarke TE, Sayers J, Mroczkowski T, Nulsen PEJ, Forman W, Murray SS, Pandey-Pommier M, Randall S, Churazov E, Bonafede A, Kraft R, David L, Andrade-Santos F, Merten J, Zitrin A, Umetsu K, Goulding A, Roediger E, Bagchi J, Bulbul E, Donahue M, Ebeling H, Johnston-Hollitt M, Mason B, Rosati P, Vikhlinin A (2015) Frontier Fields clusters: Chandra and JVLA view of the pre-merging cluster MACS J0416.1-2403. Astrophys J 812(2):153. https://doi.org/10.1088/0004-637X/812/2/153. arXiv:1505.05560 [astro-ph.CO]

Oguri M (2010) The mass distribution of SDSS J1004+4112 revisited. Publ Astron Soc Jpn 62:1017. https://doi.org/10.1093/pasj/62.4.1017. arXiv:1005.3103 [astro-ph.CO]

Oguri M (2015) Predicted properties of multiple images of the strongly lensed supernova SN Refsdal. Mon Not R Astron Soc 449:L86–L89. https://doi.org/10.1093/mnrasl/slv025. arXiv:1411.6443 [astro-ph.CO]

Okura Y, Umetsu K, Futamase T (2008) A method for weak-lensing flexion analysis by the HOLICs moment approach. Astrophys J 680(1):1–16. https://doi.org/10.1086/587676. arXiv:0710.2262 [astro-ph]

Orban de Xivry G, Marshall P (2009) An atlas of predicted exotic gravitational lenses. Mon Not R Astron Soc 399(1):2–20. https://doi.org/10.1111/j.1365-2966.2009.14925.x. arXiv:0904.1454 [astro-ph.CO]

Paczynski B (1987) Giant luminous arcs discovered in two clusters of galaxies. Nature 325(6105):572–573. https://doi.org/10.1038/325572a0

Peter AHG, Rocha M, Bullock JS, Kaplinghat M (2013) Cosmological simulations with self-interacting dark matter - II. Halo shapes versus observations. Mon Not R Astron Soc 430(1):105–120. https://doi.org/10.1093/mnras/sts535. arXiv:1208.3026 [astro-ph.CO]

Plazas Malagón AA (2020) Image simulations for strong and weak gravitational lensing. Symmetry 12(4):494. https://doi.org/10.3390/sym12040494. arXiv:2003.06090 [astro-ph.CO]

Postman M, Coe D, Benítez N et al (2012) The cluster lensing and supernova survey with Hubble: an overview. Astrophys J Suppl Ser 199(2):25. https://doi.org/10.1088/0067-0049/199/2/25. arXiv:1106.3328 [astro-ph.CO]

Presotto V, Girardi M, Nonino M, Mercurio A, Grillo C, Rosati P, Biviano A, Annunziatella M, Balestra I, Cui W, Sartoris B, Lemze D, Ascaso B, Moustakas J, Ford H, Fritz A, Czoske O, Ettori S, Kuchner U, Lombardi M, Maier C, Medezinski E, Molino A, Scodeggio M, Strazzullo V, Tozzi P, Ziegler B, Bartelmann M, Benitez N, Bradley L, Brescia M, Broadhurst T, Coe D, Donahue M, Gobat R, Graves G, Kelson D, Koekemoer A, Melchior P, Meneghetti M, Merten J, Moustakas LA, Munari E, Postman M, Regős E, Seitz S, Umetsu K, Zheng W, Zitrin A (2014) Intracluster light properties in the CLASH-VLT cluster MACS J1206.2-0847. Astron Astrophys 565:A126. https://doi.org/10.1051/0004-6361/201323251. arXiv:1403.4979 [astro-ph.CO]

Priewe J, Williams LLR, Liesenborgs J, Coe D, Rodney SA (2017) Lens models under the microscope: comparison of Hubble Frontier Field cluster magnification maps. Mon Not R Astron Soc 465(1):1030–1045. https://doi.org/10.1093/mnras/stw2785. arXiv:1605.07621 [astro-ph.CO]

Puchwein E, Hilbert S (2009) Cluster strong lensing in the Millennium simulation: the effect of galaxies and structures along the line-of-sight. Mon Not R Astron Soc 398(3):1298–1308. https://doi.org/10.1111/j.1365-2966.2009.15227.x. arXiv:0904.0253 [astro-ph.CO]

Ragagnin A, Meneghetti M, Bassini L, Ragone-Figueroa C, Granato GL, Despali G, Giocoli C, Granata G, Moscardini L, Bergamini P, Rasia E, Valentini M, Borgani S, Calura F, Dolag K, Grillo C, Mercurio A, Murante G, Natarajan P, Rosati P, Taffoni G, Tornatore L, Tortorelli L (2022) Galaxies in the central regions of simulated galaxy clusters. Astron Astrophys 665:A16. https://doi.org/10.1051/0004-6361/202243651. arXiv:2204.09067 [astro-ph.CO]

Raghunathan S, Patil S, Baxter E et al (2019) Mass calibration of optically selected DES clusters using a measurement of CMB-cluster lensing with SPTpol data. Astrophys J 872(2):170. https://doi.org/10.3847/1538-4357/ab01ca. arXiv:1810.10998 [astro-ph.CO]

Randall SW, Markevitch M, Clowe D, Gonzalez AH, Bradač M (2008) Constraints on the self-interaction cross section of dark matter from numerical simulations of the merging galaxy cluster 1E 0657-56. Astrophys J 679(2):1173–1180. https://doi.org/10.1086/587859. arXiv:0704.0261 [astro-ph]

Raney CA, Keeton CR, Brennan S (2020a) Exploring effects on magnifications due to line-of-sight galaxies in the Hubble Frontier Fields. Mon Not R Astron Soc 492(1):503–527. https://doi.org/10.1093/mnras/stz3116. arXiv:1911.02101 [astro-ph.CO]

Raney CA, Keeton CR, Brennan S, Fan H (2020b) Systematic versus statistical uncertainties in masses and magnifications of the Hubble Frontier Fields. Mon Not R Astron Soc 494(4):4771–4793. https://doi.org/10.1093/mnras/staa921. arXiv:2004.05952 [astro-ph.CO]

Remolina González JD Sharon K, Mahler G (2018) An evaluation of 10 lensing models of the frontier fields cluster MACS J0416.1-2403. Astrophys J 863(1):60. https://doi.org/10.3847/1538-4357/aacf8e. arXiv:1807.03291 [astro-ph.CO]

Richard J, Smith GP, Kneib JP, Ellis RS, Sanderson AJR, Pei L, Targett TA, Sand DJ, Swinbank AM, Dannerbauer H, Mazzotta P, Limousin M, Egami E, Jullo E, Hamilton-Morris V, Moran SM (2010) LoCuSS: first results from strong-lensing analysis of 20 massive galaxy clusters at z = 0.2. Mon Not R Astron Soc 404(1):325–349. https://doi.org/10.1111/j.1365-2966.2009.16274.x. arXiv:0911.3302 [astro-ph.CO]

Richard J, Patricio V, Martinez J, Bacon R, Clement B, Weilbacher P, Soto K, Wisotzki L, Vernet J, Pello R, Schaye J, Turner M, Martinsson T (2015) MUSE observations of the lensing cluster SMACSJ2031.8-4036: new constraints on the mass distribution in the cluster core. Mon Not R Astron Soc 446:L16–L20. https://doi.org/10.1093/mnrasl/slu150. arXiv:1409.2488 [astro-ph.CO]

Rigby JR, Bayliss MB, Sharon K, Gladders MD, Chisholm J, Dahle H, Johnson T, Paterno-Mahler R, Wuyts E, Kelson DD (2018) The Magellan evolution of galaxies spectroscopic and ultraviolet reference atlas (MegaSaura). I. The sample and the spectra. Astron J 155(3):104. https://doi.org/10.3847/1538-3881/aaa2ff. arXiv:1710.07294 [astro-ph.GA]

Robertson A (2021) The galaxy-galaxy strong lensing cross-sections of simulated ΛCDM galaxy clusters. Mon Not R Astron Soc 504(1):L7–L11. https://doi.org/10.1093/mnrasl/slab028. arXiv:2101.12067 [astro-ph.GA]

Robertson A, Massey R, Eke V (2017) What does the Bullet Cluster tell us about self-interacting dark matter? Mon Not R Astron Soc 465(1):569–587. https://doi.org/10.1093/mnras/stw2670. arXiv:1605.04307 [astro-ph.CO]

Robertson A, Harvey D, Massey R, Eke V, McCarthy IG, Jauzac M, Li B, Schaye J (2019) Observable tests of self-interacting dark matter in galaxy clusters: cosmological simulations with SIDM and baryons. Mon Not R Astron Soc 488(3):3646–3662. https://doi.org/10.1093/mnras/stz1815. arXiv:1810.05649 [astro-ph.CO]

Robertson A, Smith GP, Massey R, Eke V, Jauzac M, Bianconi M, Ryczanowski D (2020) What does strong gravitational lensing? The mass and redshift distribution of high-magnification lenses. Mon Not R Astron Soc 495(4):3727–3739. https://doi.org/10.1093/mnras/staa1429. arXiv:2002.01479 [astro-ph.CO]

Rodney SA, Patel B, Scolnic D, Foley RJ, Molino A, Brammer G, Jauzac M, Bradač M, Broadhurst T, Coe D, Diego JM, Graur O, Hjorth J, Hoag A, Jha SW, Johnson TL, Kelly P, Lam D, McCully C, Medezinski E, Meneghetti M, Merten J, Richard J, Riess A, Sharon K, Strolger LG, Treu T, Wang X, Williams LLR, Zitrin A (2015) Illuminating a dark lens: a type Ia supernova magnified by the frontier fields galaxy cluster Abell 2744. Astrophys J 811(1):70. https://doi.org/10.1088/0004-637X/811/1/70. arXiv:1505.06211 [astro-ph.CO]

Rosati P, Balestra I, Grillo C, Mercurio A, Nonino M, Biviano A, Girardi M, Vanzella E, Clash-VLT Team (2014) CLASH-VLT: a VIMOS large programme to map the dark matter mass distribution in galaxy clusters and probe distant lensed galaxies. Messenger 158:48–53

Rossi G (2020) The sejong suite: cosmological hydrodynamical simulations with massive neutrinos, dark radiation, and warm dark matter. Astrophys J Suppl Ser 249(2):19. https://doi.org/10.3847/1538-4365/ab9d1e. arXiv:2007.15279 [astro-ph.CO]

Rykoff ES, Rozo E, Hollowood D, DES Collaboration et al (2016) The RedMaPPer galaxy cluster catalog from DES science verification data. Astrophys J Suppl Ser 224(1):1. https://doi.org/10.3847/0067-0049/224/1/1. arXiv:1601.00621 [astro-ph.CO]

Saha P, Read JI, Williams LLR (2006) Two strong-lensing clusters confront universal dark matter profiles. Astrophys J Lett 652(1):L5–L8. https://doi.org/10.1086/509782. arXiv:astro-ph/0610011 [astro-ph]

Sand DJ, Treu T, Ellis RS (2002) The dark matter density profile of the lensing cluster MS 2137-23: a test of the cold dark matter paradigm. Astrophys J Lett 574(2):L129–L133. https://doi.org/10.1086/342530. arXiv:astro-ph/0207048 [astro-ph]

Sand DJ, Treu T, Smith GP, Ellis RS (2004) The dark matter distribution in the central regions of galaxy clusters: implications for cold dark matter. Astrophys J 604(1):88–107. https://doi.org/10.1086/382146. arXiv:astro-ph/0309465 [astro-ph]

Sand DJ, Treu T, Ellis RS, Smith GP, Kneib JP (2008) Separating baryons and dark matter in cluster cores: a full two-dimensional lensing and dynamic analysis of Abell 383 and MS 2137-23. Astrophys J 674(2):711–727. https://doi.org/10.1086/524652. arXiv:0710.1069 [astro-ph]

Sartoris B, Biviano A, Rosati P, Borgani S, Umetsu K, Bartelmann M, Girardi M, Grillo C, Lemze D, Zitrin A, Balestra I, Mercurio A, Nonino M, Postman M, Czakon N, Bradley L, Broadhurst T, Coe D, Medezinski E, Melchior P, Meneghetti M, Merten J, Annunziatella M, Benitez N, Czoske O, Donahue M, Ettori S, Ford H, Fritz A, Kelson D, Koekemoer A, Kuchner U, Lombardi M, Maier C, Moustakas LA, Munari E, Presotto V, Scodeggio M, Seitz S, Tozzi P, Zheng W, Ziegler B (2014) CLASH-VLT: constraints on the dark matter equation of state from accurate measurements of galaxy cluster mass profiles. Astrophys J Lett 783(1):L11. https://doi.org/10.1088/2041-8205/783/1/L11. arXiv:1401.5800 [astro-ph.CO]

Sartoris B, Biviano A, Fedeli C, Bartlett JG, Borgani S, Costanzi M, Giocoli C, Moscardini L, Weller J, Ascaso B, Bardelli S, Maurogordato S, Viana PTP (2016) Next generation cosmology: constraints from the Euclid galaxy cluster survey. Mon Not R Astron Soc 459(2):1764–1780. https://doi.org/10.1093/mnras/stw630. arXiv:1505.02165 [astro-ph.CO]

Schäfer C, Fourestey G, Kneib JP (2020) Lenstool-HPC: a high performance computing based mass modelling tool for cluster-scale gravitational lenses. Astron Comput 30:100360. https://doi.org/10.1016/j.ascom.2019.100360. arXiv:2004.06352 [astro-ph.IM]

Schneider P (1984) The amplification caused by gravitational bending of light. Astron Astrophys 140(1):119–124

Schneider P (2014) Can one determine cosmological parameters from multi-plane strong lens systems? Astron Astrophys 568:L2. https://doi.org/10.1051/0004-6361/201424450. arXiv:1406.6152 [astro-ph.CO]

Schneider P, Seitz C (1995) Steps towards nonlinear cluster inversion through gravitational distortions. I. Basic considerations and circular clusters. Astron Astrophys 294:411–431. arXiv:astro-ph/9407032 [astro-ph]

Schombert JM (1986) The structure of brightest cluster members. I. Surface photometry. Astrophys J Suppl Ser 60:603. https://doi.org/10.1086/191100

Schrabback T, Bocquet S, Sommer M, Zohren H, van den Busch JL, Hernández-Martín B, Hoekstra H, Raihan SF, Schirmer M, Applegate D, Bayliss M, Benson BA, Bleem LE, Dietrich JP, Floyd B, Hilbert S, Hlavacek-Larrondo J, McDonald M, Saro A, Stark AA, Weissgerber N (2021) Mass calibration of distant SPT galaxy clusters through expanded weak-lensing follow-up observations with HST, VLT & Gemini-south. Mon Not R Astron Soc 505(3):3923–3943. https://doi.org/10.1093/mnras/stab1386. arXiv:2009.07591 [astro-ph.CO]

Schwinn J, Jauzac M, Baugh CM, Bartelmann M, Eckert D, Harvey D, Natarajan P, Massey R (2017) Abell 2744: too much substructure for ΛCDM?. Mon Not R Astron Soc 467(3):2913–2923. https://doi.org/10.1093/mnras/stx277. arXiv:1611.02790 [astro-ph.CO]

Schwinn J, Baugh CM, Jauzac M, Bartelmann M, Eckert D (2018) Uncovering substructure with wavelets: proof of concept using Abell 2744. Mon Not R Astron Soc 481(4):4300–4310. https://doi.org/10.1093/mnras/sty2566. arXiv:1804.07401 [astro-ph.CO]

Sebesta K, Williams LLR, Mohammed I, Saha P, Liesenborgs J (2016) Testing light-traces-mass in Hubble Frontier Fields cluster MACS-J0416 1-2403. Mon Not R Astron Soc 461(2):2126–2134. https://doi.org/10.1093/mnras/stw1433. arXiv:1507.08960 [astro-ph.CO]

Şengül A, Birrer S, Natarajan P, Dvorkin C (2023) Detecting low-mass perturbers in cluster lenses using curved arc bases. Mon Not R Astron Soc 526(2):2525–2541. https://doi.org/10.1093/mnras/stad2784. arXiv:2303.14786 [astro-ph.GA]

Sereno M, Covone G, Izzo L, Ettori S, Coupon J, Lieu M (2017) PSZ2LenS. Weak lensing analysis of the Planck clusters in the CFHTLenS and in the RCSLenS. Mon Not R Astron Soc 472(2):1946–1971. https://doi.org/10.1093/mnras/stx2085. arXiv:1703.06886 [astro-ph.CO]

Sharon K, Johnson TL (2015) Revised lens model for the multiply imaged lensed supernova, "SN refsdal" in MACS J1149+2223. Astrophys J Lett 800(2):L26. https://doi.org/10.1088/2041-8205/800/2/L26. arXiv:1411.6933 [astro-ph.CO]

Sharon K, Bayliss MB, Dahle H, Dunham SJ, Florian MK, Gladders MD, Johnson TL, Mahler G, Paterno-Mahler R, Rigby JR, Whitaker KE, Akhshik M, Koester BP, Murray K, Remolina González JD, Wuyts E (2020) Strong lens models for 37 clusters of galaxies from the SDSS giant arcs survey. Astrophys J Suppl Ser 247(1):12. https://doi.org/10.3847/1538-4365/ab5f13. arXiv:1904.05940 [astro-ph.GA]

Sharon K, Mahler G, Rivera-Thorsen TE, Dahle H, Gladders MD, Bayliss MB, Florian MK, Kim KJ, Khullar G, Mainali R, Napier KA, Navarre A, Rigby JR, Remolina González JD, Sharma S (2022a) The cosmic telescope that lenses the sunburst arc, PSZ1 G311.65-18.48: strong gravitational lensing model and source plane analysis. Astrophys J 941(2):203. https://doi.org/10.3847/1538-4357/ac927a. arXiv:2209.03417 [astro-ph.GA]

Sharon K, Mahler G, Rivera-Thorsen TE, Dahle H, Gladders MD, Bayliss MB, Florian MK, Kim KJ, Khullar G, Mainali R, Napier KA, Navarre A, Rigby JR, Remolina González JD Sharma S (2022b) The cosmic telescope that lenses the sunburst arc, PSZ1 G311.65-18.48: strong gravitational lensing model and source plane analysis. Astrophys J 941(2):203. https://doi.org/10.3847/1538-4357/ac927a. arXiv:2209.03417 [astro-ph.GA]

Smail I, Dickinson M (1995) Lensing by distant clusters: HST observations of weak shear in the field of 3C 324. Astrophys J Lett 455:L99. https://doi.org/10.1086/309842. arXiv:astro-ph/9510050 [astro-ph]

Soucail G, Fort B, Mellier Y, Picat JP (1987) A blue ring-like structure in the center of the A 370 cluster of galaxies. Astron Astrophys 172:L14–L16.

Soucail G, Mellier Y, Fort B, Cailloux M (1988) Spectroscopic observations of the distant cluster of galaxies Abell 370: a catalogue of 84 spectra. Astron Astrophys 73:471–514

Soucail G, Kneib JP, Golse G (2004) Multiple-images in the cluster lens Abell 2218: constraining the geometry of the Universe? Astron Astrophys 417:L33–L37. https://doi.org/10.1051/0004-6361:20040077. arXiv:astro-ph/0402658 [astro-ph]

Spergel DN, Steinhardt PJ (2000) Observational evidence for self-interacting cold dark matter. Phys Rev Lett 84(17):3760–3763. https://doi.org/10.1103/PhysRevLett.84.3760. arXiv:astro-ph/9909386 [astro-ph]

Springel V, Farrar GR (2007) The speed of the 'bullet' in the merging galaxy cluster 1E0657-56. Mon Not R Astron Soc 380(3):911–925. https://doi.org/10.1111/j.1365-2966.2007.12159.x. arXiv:astro-ph/0703232 [astro-ph]

Springel V, Wang J, Vogelsberger M, Ludlow A, Jenkins A, Helmi A, Navarro JF, Frenk CS, White SDM (2008) The aquarius project: the subhaloes of galactic haloes. Mon Not R Astron Soc 391(4):1685–1711. https://doi.org/10.1111/j.1365-2966.2008.14066.x. arXiv:0809.0898 [astro-ph]

Steinhardt CL, Jauzac M, Acebron A et al (2020) The BUFFALO HST survey. Astrophys J Suppl Ser 247(2):64. https://doi.org/10.3847/1538-4365/ab75ed. arXiv:2001.09999 [astro-ph.GA]

Strait V, Bradač M, Hoag A, Huang KH, Treu T, Wang X, Amorin R, Castellano M, Fontana A, Lemaux BC, Merlin E, Schmidt KB, Schrabback T, Tomczack A, Trenti M, Vulcani B (2018) Mass and light of Abell 370: a strong and weak lensing analysis. Astrophys J 868:129. https://doi.org/10.3847/1538-4357/aae834. arXiv:1805.08789

Suyu SH, Halkola A (2010) The halos of satellite galaxies: the companion of the massive elliptical lens SL2S J08544-0121. Astron Astrophys 524:A94. https://doi.org/10.1051/0004-6361/201015481. arXiv:1007.4815 [astro-ph.CO]

Suyu SH, Hensel SW, McKean JP, Fassnacht CD, Treu T, Halkola A, Norbury M, Jackson N, Schneider P, Thompson D, Auger MW, Koopmans LVE, Matthews K (2012) Disentangling baryons and dark matter in the spiral gravitational lens B1933+503. Astrophys J 750(1):10. https://doi.org/10.1088/0004-637X/750/1/10. arXiv:1110.2536 [astro-ph.CO]

Torres-Ballesteros DA, Castañeda L (2023) RELENSING: reconstructing the mass profile of galaxy clusters from gravitational lensing. Mon Not R Astron Soc 518(3):4494–4516. https://doi.org/10.1093/mnras/stac3253. arXiv:2201.10076 [astro-ph.CO]

Treu T, Brammer G, Diego JM, Grillo C, Kelly PL, Oguri M, Rodney SA, Rosati P, Sharon K, Zitrin A, Balestra I, Bradač M, Broadhurst T, Caminha GB, Halkola A, Hoag A, Ishigaki M, Johnson TL, Karman W, Kawamata R, Mercurio A, Schmidt KB, Strolger LG, Suyu SH, Filippenko AV, Foley RJ, Jha SW, Patel B (2016) "Refsdal" meets popper: comparing predictions of the re-appearance of the multiply imaged supernova behind MACSJ1149.5+2223. Astrophys J 817(1):60. https://doi.org/10.3847/0004-637X/817/1/60. arXiv:1510.05750 [astro-ph.CO]

Tutusaus I, Martinelli M, Cardone VF, Camera S et al (2020) Euclid: the importance of galaxy clustering and weak lensing cross-correlations within the photometric Euclid survey. Astron Astrophys 643:A70. https://doi.org/10.1051/0004-6361/202038313. arXiv:2005.00055 [astro-ph.CO]

Tyson JA, Valdes F, Wenk RA (1990) Detection of systematic gravitational lens galaxy image alignments: mapping dark matter in galaxy clusters. Astrophys J Lett 349:L1. https://doi.org/10.1086/185636

Umetsu K (2020) Cluster-galaxy weak lensing. Astron Astrophys Rev 28(1):7. https://doi.org/10.1007/s00159-020-00129-w. arXiv:2007.00506 [astro-ph.CO]

Umetsu K, Medezinski E, Nonino M, Merten J, Zitrin A, Molino A, Grillo C, Carrasco M, Donahue M, Mahdavi A, Coe D, Postman M, Koekemoer A, Czakon N, Sayers J, Mroczkowski T, Golwala S, Koch PM, Lin KY, Molnar SM, Rosati P, Balestra I, Mercurio A, Scodeggio M, Biviano A, Anguita T, Infante L, Seidel G, Sendra I, Jouvel S, Host O, Lemze D, Broadhurst T, Meneghetti M, Moustakas L, Bartelmann M, Benítez N, Bouwens R, Bradley L, Ford H, Jiménez-Tej Y, Kelson D, Lahav O, Melchior P, Moustakas J, Ogaz S, Seitz S, Zheng W (2012) CLASH: mass distribution in and around MACS J1206.2-0847 from a full cluster lensing analysis. Astrophys J 755(1):56. https://doi.org/10.1088/0004-637X/755/1/56. arXiv:1204.3630 [astro-ph.CO]

Umetsu K, Medezinski E, Nonino M, Merten J, Postman M, Meneghetti M, Donahue M, Czakon N, Molino A, Seitz S, Gruen D, Lemze D, Balestra I, Benítez N, Biviano A, Broadhurst T, Ford H, Grillo C, Koekemoer A, Melchior P, Mercurio A, Moustakas J, Rosati P, Zitrin A (2014) CLASH: weak-lensing shear-and-magnification analysis of 20 galaxy clusters. Astrophys J 795(2):163. https://doi.org/10.1088/0004-637X/795/2/163. arXiv:1404.1375 [astro-ph.CO]

Umetsu K, Zitrin A, Gruen D, Merten J, Donahue M, Postman M (2016) CLASH: joint analysis of strong-lensing, weak-lensing shear, and magnification data for 20 galaxy clusters. Astrophys J 821(2):116. https://doi.org/10.3847/0004-637X/821/2/116. arXiv:1507.04385 [astro-ph.CO]

Umetsu K, Sereno M, Tam SI, Chiu IN, Fan Z, Ettori S, Gruen D, Okumura T, Medezinski E, Donahue M, Meneghetti M, Frye B, Koekemoer A, Broadhurst T, Zitrin A, Balestra I, Benítez N, Higuchi Y, Melchior P, Mercurio A, Merten J, Molino A, Nonino M, Postman M, Rosati P, Sayers J, Seitz S (2018) The projected dark and baryonic ellipsoidal structure of 20 CLASH galaxy clusters. Astrophys J 860(2):104. https://doi.org/10.3847/1538-4357/aac3d9. arXiv:1804.00664 [astro-ph.CO]

Uson JM, Boughn SP, Kuhn JR (1991) Diffuse light in dense clusters of galaxies. I. R-band observations of Abell 2029. Astrophys J 369:46. https://doi.org/10.1086/169737

van Weeren RJ, Ogrean GA, Jones C, Forman WR, Andrade-Santos F, Pearce CJJ, Bonafede A, Brüggen M, Bulbul E, Clarke TE, Churazov E, David L, Dawson WA, Donahue M, Goulding A, Kraft RP, Mason B, Merten J, Mroczkowski T, Nulsen PEJ, Rosati P, Roediger E, Randall SW, Sayers J, Umetsu K, Vikhlinin A, Zitrin A (2017) Chandra and JVLA observations of HST frontier fields cluster MACS J0717.5+3745. Astrophys J 835(2):197. https://doi.org/10.3847/1538-4357/835/2/197. arXiv:1701.04096 [astro-ph.CO]

Vega-Ferrero J, Diego JM, Miranda V, Bernstein GM (2018) The Hubble constant from SN refsdal. Astrophys J Lett 853(2):L31. https://doi.org/10.3847/2041-8213/aaa95f. arXiv:1712.05800 [astro-ph.CO]

Verdugo T, Motta V, Muñoz RP, Limousin M, Cabanac R, Richard J (2011) Gravitational lensing and dynamics in SL2S J02140-0535: probing the mass out to large radius. Astron Astrophys 527:A124. https://doi.org/10.1051/0004-6361/201014965. arXiv:1005.1566 [astro-ph.CO]

Verdugo T, Motta V, Foëx G, Forero-Romero JE, Muñoz RP, Pello R, Limousin M, More A, Cabanac R, Soucail G, Blakeslee JP, Mejía-Narváez AJ, Magris G, Fernández-Trincado JG (2014) Characterizing

SL2S galaxy groups using the Einstein radius. Astron Astrophys 571:A65. https://doi.org/10.1051/0004-6361/201423696. arXiv:1409.2900 [astro-ph.CO]

Wagner J (2019) A model-independent characterisation of strong gravitational lensing by observables. Universe 5(7):177. https://doi.org/10.3390/universe5070177. arXiv:1906.05285 [astro-ph.CO]

Wagner J (2020) Cosmic structures from a mathematical perspective 1: dark matter halo mass density profiles. Gen Relativ Gravit 52(6):61. https://doi.org/10.1007/s10714-020-02715-w. arXiv:2002.00960 [astro-ph.CO]

Wagner J (2022) Generalised model-independent characterisation of strong gravitational lenses - VII. Impact of source properties and higher-order lens properties on the local lens reconstruction. Astron Astrophys 663:A157. https://doi.org/10.1051/0004-6361/202243562

Wagner J, Tessore N (2018) Generalised model-independent characterisation of strong gravitational lenses. II. Transformation matrix between multiple images. Astron Astrophys 613:A6. https://doi.org/10.1051/0004-6361/201730947. arXiv:1704.01822 [astro-ph.CO]

Wagner J, Williams LLR (2020) Model-independent and model-based local lensing properties of B0128+437 from resolved quasar images. Astron Astrophys 635:A86. https://doi.org/10.1051/0004-6361/201936628. arXiv:1909.01349 [astro-ph.GA]

Wagner J, Liesenborgs J, Tessore N (2018) Model-independent and model-based local lensing properties of CL0024+1654 from multiply imaged galaxies. Astron Astrophys 612:A17. https://doi.org/10.1051/0004-6361/201731932. arXiv:1709.03531

Walsh D, Carswell RF, Weymann RJ (1979) 0957+561 A, B: twin quasistellar objects or gravitational lens?. Nature 279:381–384. https://doi.org/10.1038/279381a0

Wambsganss J, Bode P, Ostriker JP (2004) Giant arc statistics in Concord with a concordance lambda cold dark matter universe. Astrophys J Lett 606(2):L93–L96. https://doi.org/10.1086/421459. arXiv:astro-ph/0306088 [astro-ph]

Wang X, Hoag A, Huang KH, Treu T, Bradač M, Schmidt KB, Brammer GB, Vulcani B, Jones TA, Ryan RE, Amorín R, Castellano M, Fontana A, Merlin E, Trenti M (2015) The grism lens-amplified survey from space (GLASS). IV. Mass reconstruction of the lensing cluster abell 2744 from frontier field imaging and GLASS spectroscopy. Astrophys J 811:29. https://doi.org/10.1088/0004-637X/811/1/29. arXiv:1504.02405

Wang H, Cañameras R, Caminha GB, Suyu SH, Yıldırım A, Chirivì G, Christensen L, Grillo C, Schuldt S (2022) Constraining the multi-scale dark-matter distribution in CASSOWARY 31 with strong gravitational lensing and stellar dynamics. Astron Astrophys 668:A162. https://doi.org/10.1051/0004-6361/202243600. arXiv:2203.13759 [astro-ph.GA]

Wechsler RH, Bullock JS, Primack JR, Kravtsov AV, Dekel A (2002) Concentrations of dark halos from their assembly histories. Astrophys J 568(1):52–70. https://doi.org/10.1086/338765. arXiv:astro-ph/0108151 [astro-ph]

Williams LLR, Lewis GF (1998) Undistorted lensed images in galaxy clusters. Mon Not R Astron Soc 294:299–306. https://doi.org/10.1046/j.1365-8711.1998.01179.x. arXiv:astro-ph/9710042 [astro-ph]

Williams LLR, Liesenborgs J (2019) The role of multiple images and model priors in measuring H_0 from supernova Refsdal in galaxy cluster MACS J1149 5+2223. Mon Not R Astron Soc 482(4):5666–5677. https://doi.org/10.1093/mnras/sty3113. arXiv:1806.11113 [astro-ph.CO]

Williams LLR, Saha P (2004) Models of the Giant Quadruple Quasar SDSS J1004+4112. Astron J 128(6):2631–2641. https://doi.org/10.1086/426007. arXiv:astro-ph/0409418 [astro-ph]

Williams LLR, Navarro JF, Bartelmann M (1999) The core structure of galaxy clusters from gravitational lensing. Astrophys J 527(2):535–544. https://doi.org/10.1086/308127. arXiv:astro-ph/9905134 [astro-ph]

Williams LLR, Hjorth J, Wojtak R (2010) Statistical mechanics of collisionless orbits. III. Comparison with N-body simulations. Astrophys J 725(1):282–287. https://doi.org/10.1088/0004-637X/725/1/282. arXiv:1010.0267 [astro-ph.CO]

Williams LLR, Sebesta K, Liesenborgs J (2018) Evidence for the line-of-sight structure in the Hubble Frontier Field cluster MACSJ0717 5+3745. Mon Not R Astron Soc 480(3):3140–3151. https://doi.org/10.1093/mnras/sty2088. arXiv:1711.05265 [astro-ph.CO]

Windhorst RA, Cohen SH, Jansen RA, Summers J, Tompkins S, Conselice CJ, Driver SP, Yan H, Coe D, Frye B, Grogin N, Koekemoer A, Marshall MA, O'Brien R, Pirzkal N, Robotham A, Ryan RE, Willmer CNA, Carleton T, Diego JM, Keel WC, Porto P, Redshaw C, Scheller S, Wilkins SM, Willner SP, Zitrin A, Adams NJ, Austin D, Arendt RG, Beacom JF, Bhatawdekar RA, Bradley LD, Broadhurst T, Cheng C, Civano F, Dai L, Dole H, D'Silva JCJ, Duncan KJ, Fazio GG, Ferrami G, Ferreira L, Finkelstein SL, Furtak LJ, Gim HB, Griffiths A, Hammel HB, Harrington KC, Hathi NP, Holwerda BW, Honor R, Huang JS, Hyun M, Im M, Joshi BA, Kamieneski PS, Kelly P, Larson RL, Li J, Lim J, Ma Z, Maksym P, Manzoni G, Meena AK, Milam SN, Nonino M, Pascale M, Petric A, Pierel JDR, del Carmen Polletta M, Röttgering HJA, Rutkowski MJ, Smail I, Straughn AN, Strolger LG, Swirbul A, Trussler JAA, Wang L,

Welch B, Wyithe JS, Yun M, Zackrisson E, Zhang J, Zhao X (2023) JWST PEARLS. Prime extragalactic areas for reionization and lensing science: project overview and first results. Astron J 165(1):13. https://doi.org/10.3847/1538-3881/aca163. arXiv:2209.04119 [astro-ph.CO]
Wu HY, Weinberg DH, Salcedo AN, Wibking BD (2021) Cosmology with galaxy cluster weak lensing: statistical limits and experimental design. Astrophys J 910(1):28. https://doi.org/10.3847/1538-4357/abdc23
Yang D, Yu HB (2021) Self-interacting dark matter and small-scale gravitational lenses in galaxy clusters. Phys Rev D 104(10):103031. https://doi.org/10.1103/PhysRevD.104.103031. arXiv:2102.02375 [astro-ph.GA]
Young P, Gunn JE, Kristian J, Oke JB, Westphal JA (1980) The double quasar Q0957+561 A, B: a gravitational lens image formed by a galaxy at $z = 0.39$. Astrophys J 241:507–520. https://doi.org/10.1086/158365
Yue B, Castellano M, Ferrara A, Fontana A, Merlin E, Amorín R, Grazian A, Mármol-Queralto E, Michalowski MJ, Mortlock A, Paris D, Parsa S, Pilo S, Santini P, Criscienzo MD (2018) On the faint end of the galaxy luminosity function in the epoch of reionization: updated constraints from the hst frontier fields. Astrophys J 868(2):115. https://doi.org/10.3847/1538-4357/aae77f
Zhao DH, Mo HJ, Jing YP, Börner G (2003) The growth and structure of dark matter haloes. Mon Not R Astron Soc 339(1):12–24. https://doi.org/10.1046/j.1365-8711.2003.06135.x. arXiv:astro-ph/0204108 [astro-ph]
Zitrin A, Broadhurst T (2009) Discovery of the largest known lensed images formed by a critically convergent lensing cluster. Astrophys J Lett 703(2):L132–L136. https://doi.org/10.1088/0004-637X/703/2/L132. arXiv:0906.5079 [astro-ph.CO]
Zitrin A, Broadhurst T, Umetsu K, Coe D, Benítez N, Ascaso B, Bradley L, Ford H, Jee J, Medezinski E, Rephaeli Y, Zheng W (2009a) New multiply-lensed galaxies identified in ACS/NIC3 observations of Cl0024+1654 using an improved mass model. Mon Not R Astron Soc 396(4):1985–2002. https://doi.org/10.1111/j.1365-2966.2009.14899.x. arXiv:0902.3971 [astro-ph.CO]
Zitrin A, Broadhurst T, Umetsu K, Coe D, Benítez N, Ascaso B, Bradley L, Ford H, Jee J, Medezinski E, Rephaeli Y, Zheng W (2009b) New multiply-lensed galaxies identified in ACS/NIC3 observations of Cl0024+1654 using an improved mass model. Mon Not R Astron Soc 396(4):1985–2002. https://doi.org/10.1111/j.1365-2966.2009.14899.x. arXiv:0902.3971 [astro-ph.CO]
Zitrin A, Broadhurst T, Barkana R, Rephaeli Y, Benítez N (2011) Strong-lensing analysis of a complete sample of 12 MACS clusters at $z > 0.5$: mass models and Einstein radii. Mon Not R Astron Soc 410(3):1939–1956. https://doi.org/10.1111/j.1365-2966.2010.17574.x. arXiv:1002.0521 [astro-ph.CO]
Zitrin A, Rosati P, Nonino M, Grillo C, Postman M, Coe D, Seitz S, Eichner T, Broadhurst T, Jouvel S, Balestra I, Mercurio A, Scodeggio M, Benítez N, Bradley L, Ford H, Host O, Jimenez-Teja Y, Koekemoer A, Zheng W, Bartelmann M, Bouwens R, Czoske O, Donahue M, Graur O, Graves G, Infante L, Jha S, Kelson D, Lahav O, Lazkoz R, Lemze D, Lombardi M, Maoz D, McCully C, Medezinski E, Melchior P, Meneghetti M, Merten J, Molino A, Moustakas LA, Ogaz S, Patel B, Regoes E, Riess A, Rodney S, Umetsu K, der Van Wel A (2012b) CLASH: new multiple images constraining the inner mass profile of MACS J1206.2-0847. Astrophys J 749(2):97. https://doi.org/10.1088/0004-637X/749/2/97. arXiv:1107.2649 [astro-ph.CO]
Zitrin A, Menanteau F, Hughes JP, Coe D, Barrientos LF, Infante L, Mandelbaum R (2013) A highly elongated prominent lens at $z = 0.87$: first strong-lensing analysis of El Gordo. Astrophys J Lett 770(1):L15. https://doi.org/10.1088/2041-8205/770/1/L15. arXiv:1304.0455 [astro-ph.CO]
Zitrin A, Fabris A, Merten J, Melchior P, Meneghetti M, Koekemoer A, Coe D, Maturi M, Bartelmann M, Postman M, Umetsu K, Seidel G, Sendra I, Broadhurst T, Balestra I, Biviano A, Grillo C, Mercurio A, Nonino M, Rosati P, Bradley L, Carrasco M, Donahue M, Ford H, Frye BL, Moustakas J (2015a) Hubble space telescope combined strong and weak lensing analysis of the CLASH sample: mass and magnification models and systematic uncertainties. Astrophys J 801(1):44. https://doi.org/10.1088/0004-637X/801/1/44. arXiv:1411.1414 [astro-ph.CO]
Zitrin A, Fabris A, Merten J, Melchior P, Meneghetti M, Koekemoer A, Coe D, Maturi M, Bartelmann M, Postman M, Umetsu K, Seidel G, Sendra I, Broadhurst T, Balestra I, Biviano A, Grillo C, Mercurio A, Nonino M, Rosati P, Bradley L, Carrasco M, Donahue M, Ford H, Frye BL Moustakas J (2015b) Hubble space telescope combined strong and weak lensing analysis of the CLASH sample: mass and magnification models and systematic uncertainties. Astrophys J 801(1):44. https://doi.org/10.1088/0004-637X/801/1/44. arXiv:1411.1414 [astro-ph.CO]
Zwicky F (1937) On the masses of nebulae and of clusters of nebulae. Astrophys J 86:217. https://doi.org/10.1086/143864

Publisher's Note Springer Nature remains neutral with regard to jurisdictional claims in published maps and institutional affiliations.

Space Science Reviews (2024) 220:23
https://doi.org/10.1007/s11214-024-01042-9

Searching for Strong Gravitational Lenses

Cameron Lemon[1] · Frédéric Courbin[1] · Anupreeta More[2,3] · Paul Schechter[4] · Raoul Cañameras[5,6] · Ludovic Delchambre[7] · Calvin Leung[4] · Yiping Shu[5,8] · Chiara Spiniello[9,10] · Yashar Hezaveh[11,12] · Jonas Klüter[13] · Richard McMahon[14,15]

Received: 14 June 2023 / Accepted: 30 December 2023 / Published online: 21 February 2024
© The Author(s) 2024

Abstract
Strong gravitational lenses provide unique laboratories for cosmological and astrophysical investigations, but they must first be discovered – a task that can be met with significant contamination by other astrophysical objects and asterisms. Here we review strong lens searches, covering various sources (quasars, galaxies, supernovae, FRBs, GRBs, and GWs), lenses (early- and late-type galaxies, groups, and clusters), datasets (imaging, spectra, and lightcurves), and wavelengths. We first present the physical characteristics of the lens and source populations, highlighting relevant details for constructing targeted searches. Search techniques are described based on the main lensing feature that is required for the technique to work, namely one of: (i) an associated magnification, (ii) multiple spatially-resolved images, (iii) multiple redshifts, or (iv) a non-zero time delay between images. To use the current lens samples for science, and for the design of future searches, we list several selection biases that exist due to these discovery techniques. We conclude by discussing the future of lens searches in upcoming surveys and the new population of lenses that will be discovered.

Keywords Gravitational lensing: strong

1 Introduction

The first measurement of the deflection of light by a gravitational field was the angular displacement of stars close to the Solar limb. The experiment, carried out during the 7-minute-long Solar eclipse visible on 29 May 1919, measured a small displacement of stars in the Taurus constellation, of the order of 1.75 arcseconds (Dyson et al. 1920), supporting the factor of two increase in the deflection angle prediction of general relativity over Newtonian gravity. Should the Sun be counted as the first lens "discovery"? Throughout this chapter we will be concerned with strong gravitational lensing, in which multiple images of the source are observed. Though the 1919 experiment would not count as the first strong lens, it is the first genuine observation of the gravitational lensing phenomenon. The first strong lens discovery would wait 60 years.

The relevant calculations of multiple imaging were first published by Chwolson (1924), but a flurry of work was sparked a decade later by Einstein's 1936 publication on the subject (Einstein 1936). In fact, Einstein had made similar unpublished calculations as early as 1912, possibly to explain the intensity of a newly discovered Nova, Nova Geminorum 1912, but

Extended author information available on the last page of the article

eventually concluded that the low probability, symmetrical lightcurve, and achromaticity of lensing were an incompatible explanation (Sauer 2008). As Einstein had only considered stars as both the lens and source, he correctly concluded that such image separations were unobservable. In 1937, Zwicky noted that galaxies or clusters are much more likely lenses to provide observable lens systems (Zwicky 1937). After the discovery of quasars in 1963, they were quickly recognised as a possible variable lensed source by Refsdal for measuring the Hubble constant, alongside supernovae (Refsdal 1964). Indeed, the first lens discovered in 1979 was a doubly imaged lensed quasar, QJ0957+561, a.k.a. "The Twin Quasar" (Walsh et al. 1979).

The 1980s saw a few more serendipitous discoveries, including the first quadruply imaged lensed quasars: PG1115+080 (Weymann et al. 1980) and Q2237+030 (a.k.a. the Einstein Cross, Huchra et al. 1985). The same decade saw the discovery of giant luminous arcs in the galaxy clusters Abell 370 and 2244–02 (Lynds and Petrosian 1986; Soucail et al. 1987). The arcs were soon spectroscopically confirmed to be behind the cluster (Soucail et al. 1988; Lynds and Petrosian 1989), confirming Zwicky's prescient suggestion that clusters were likely gravitational lenses.

Systematic lens searches soon began, as high-resolution imaging with the *Hubble Space Telescope* (*HST*) and at radio wavelengths was used to look for multiple imaging of quasars. The first successful search in the optical was the doubly imaged quasar UM 673 (Surdej et al. 1987), quickly followed by the quadruply imaged quasar H1413+117 a.k.a. the "Cloverleaf" (Magain et al. 1988). In the radio, systematic searches found the first full Einstein ring (Hewitt et al. 1988) and multiply-imaged radio sources such as MG 2016+112 (Lawrence et al. 1984; Narasimha et al. 1984). By the early 2000s, the Cosmic Lens All-Sky Survey (CLASS; Myers et al. 2003) found 22 lenses within a sample of 13,500 radio sources. These searches are often called "source-selected", because the sample pre-selection is made on the properties of the background, lensed source.

One difficulty with source-selected lens candidates resides in their confirmation as real lenses: the glare of the bright quasar images masks the often faint lensing galaxy. The task is particularly challenging with ground-based data, although image deconvolution techniques have been effective for deblending in optical imaging (e.g. Courbin et al. 1998; Burud et al. 1998) and spectroscopy (e.g. Eigenbrod et al. 2007). High-resolution data from *HST* or from ground-based Adaptive Optics (AO) remains the best way to confirm lenses. Source-selected lens samples are mainly limited to quasars, but there are $\sim$1000 times more galaxies than quasars and therefore about as many more galaxy-galaxy strong lenses than galaxy-quasar strong lenses. In the former, the lens galaxy light typically dominates, even taking into account the magnification of the source. Searches for galaxy-galaxy lenses have thus focused on a pre-selection of galaxies potentially acting as lenses and are therefore called "lens-selected".

Cluster lenses are also typically lens-selected, as deeper imaging is obtained for known clusters to search for multiple images. The clusters are first found through their X-ray emission (e.g., Ebeling et al. 2001), or optical imaging surveys (e.g. Gladders and Yee 2005), and subsequent *HST* follow-up reveals, on average, at least 1 giant arc per cluster (Horesh et al. 2010) for the most massive (X-ray-bright) clusters, let alone many more compact sets of multiple images. While the discovery of multiple images in any cluster is very likely, it is a question of the depth and resolution of available imaging, as indeed deep observations of the most massive isolated galaxies can also often uncover faint background lensed sources (e.g., Collett and Smith 2020).

For both lens- and source- selected searches, wide-field imaging and spectroscopic surveys at various wavelengths are now the main catalyst for developing new searches, as the depth and quality of the data allows not only for many more lenses to be discovered, but also new types of lenses. The Sloan Digital Sky Survey (SDSS) was the first such survey, with targeted campaigns yielding new lensed galaxies from imaging (e.g., Belokurov et al. 2007) and spectra (e.g., Bolton et al. 2004), and lensed quasars from a combination of spectra and imaging (e.g., Pindor et al. 2004; Oguri et al. 2006). As modern surveys have ever increasing depth, area, and quality, both machine learning and citizen science have played a pivotal role in dealing with the enormous number of candidates from the increased depth and area (e.g., Marshall et al. 2016; Jacobs et al. 2019a; Rojas et al. 2021). Entirely unique datasets, such as the high-resolution catalogue of *Gaia*, have allowed for the efficient identification of bright lensed quasars across the whole sky (e.g., Stern et al. 2021; Lemon et al. 2023).

The aim of this chapter is to present a detailed review of the various search methods presented in the literature for finding strong gravitational lenses of all source and lens types, and all image separations. The above dichotomy between source- and lens-selected samples often represents only the first filtering of a survey's catalogue; the subsequent selection algorithms are where most of the filtering happens. Data types, wavelengths, and source types are all possible ways to subdivide the many lens searches in the literature. We find it more helpful, however, to avoid repetition by broadly sorting the search methods into those targeting a common lensing feature, that is to say, either a magnification, multiple imaging, multiple redshifts, or an associated time delay.

This chapter is outlined as follows: in Sects. 2 and 3 we describe the properties of the lens and source populations respectively. In Sect. 4, we provide an overview of the literature in terms of selection methods following the categories described above. Possible biases and selection effects from these searches are described in Sect. 5, and in Sect. 6 we discuss how lens finding may evolve in the coming decade, in a context where key facilities such as *Euclid*, Rubin-LSST, and Roman, will be delivering actual data.

2 Deflector Classes

Gravitational lenses span a wide range of characteristics, with various brightnesses, image configurations, separations, lens types, source types, and environments. Figure 1 shows a gallery of gravitational lenses to demonstrate this variety of possibilities. Before describing the lens search techniques (Sect. 4), it is first prudent to describe the physical characteristics of the possible lens systems themselves. For simplicity, we split the descriptions between the deflector (this Section) and source populations (Sect. 3). We limit our discussion to *macro-lenses*, in which the multiple images of the source are separated by the order of arcseconds, and can be readily resolved by current observatories. Such separations are only possible for deflectors and sources at cosmological distances, and with masses above 10^8 $M_{\odot}$ so the relevant lensing objects are isolated galaxies (early- and late-type), galaxy groups, and galaxy clusters.

For lens-selected searches, an initial catalogue of systems is based on finding examples of the expected lens type. For example, high-redshift galaxies lensed by galaxy clusters might be found through an all-sky selection of galaxy clusters. Below we describe the physical properties of these deflector classes to better understand the lens search techniques and their possible biases.

2.1 Galaxies

Galaxy families are often split by morphology; we can broadly categorise them into ellipticals (early-type), and spirals (late-type) via the Hubble sequence. Ellipticals are featureless

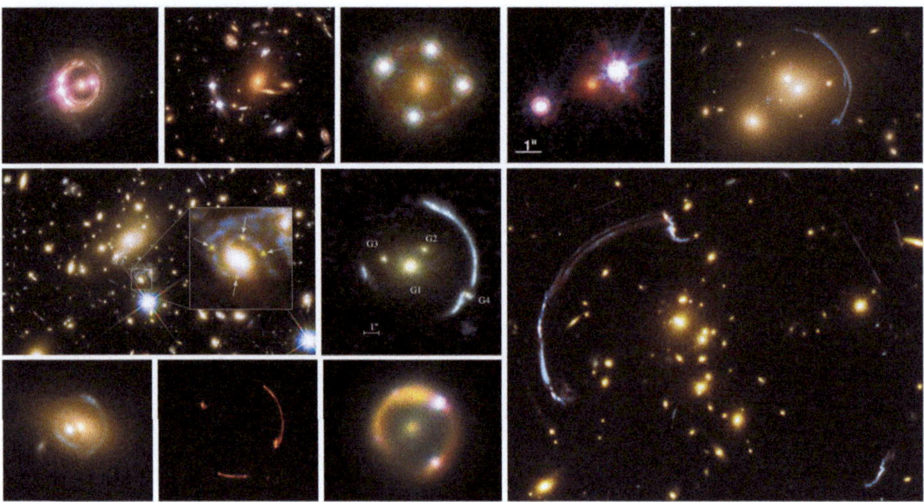

Fig. 1 Various strong lens systems with quasar, galaxy, and supernova sources. Images are composed from multi-band *HST* imaging, a mix of *HST* and ground-based adaptive optics, or Very Long Baseline Interferometry (VLBI). Sources without optical, infrared, sub-mm, or radio imaging counterparts should not be forgotten, however no definite localised examples have yet been discovered. Credit: NASA/ESA/NRAO/AUI/NSF

quiescent galaxies with old stellar populations and no star formation, representing the final product of hierarchical galaxy formation. Spirals, on the other hand, show prominent star formation and coherent rotation, in contrast to pressure-supported ellipticals. Due to their larger numbers at high masses, ellipticals dominate over spirals as lensing galaxies (approximately 7 to 1), with the latter only contributing at image separations smaller than 2 arcseconds (Turner et al. 1984; Kochanek 1996). For more details on lensing by galaxies, we refer the reader to the Galaxy Lensing Chapter.

2.1.1 Early-Type Lenses

Elliptical galaxies, often found within groups and clusters, have ellipsoidal isophotes, are composed of old stellar populations, and lack significant star formation, dust, or gas. They are accordingly red, with the most massive examples termed Luminous Red Galaxies (LRGs). Due to both selection effects of brighter galaxies, and their increased lensing efficiency, massive LRGs constitute the majority of known high-confidence galaxy-scale lenses, and dominate the system's flux (and therefore colours) in ground-based imaging surveys, as shown in Fig. 2. Ellipticals' spectra contain clear absorption lines from old stars (CaII H and K, Mg-b, and Na-D), and are void of emission lines. A typical feature used to measure redshifts of ellipticals is the characteristic 4000Å break immediately redward of the CaII H and K lines. The light profiles of LRGs are often well-described by a single Sersic profile (with Sersic index $n \approx 4$), however there is evidence for a dichotomy of formation pathways for massive ellipticals, with one coreless population with extra central light and lower Sersic index fits (Kormendy et al. 2009).

As ellipticals dominate the galaxy-scale lensing population, several lensing studies have constrained their mass distributions, revealing standard NFW halos, heavy initial mass functions (e.g. Sonnenfeld et al. 2015; Shajib et al. 2021a), and an alignment between total mass and light (e.g. Keeton et al. 1997; Shajib et al. 2019; Schmidt et al. 2022). Their total mass

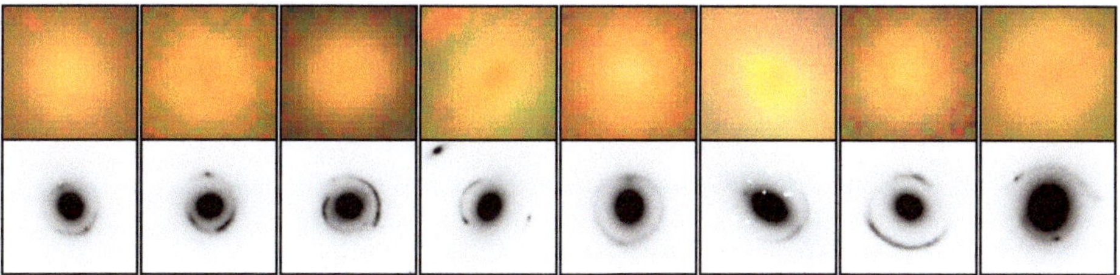

Fig. 2 Ground-based *grz* images from the Legacy Survey (top) and *HST*-ACS WFC F814W (bottom) images of eight galaxy-galaxy strong lenses from the SLACS sample (Bolton et al. 2008). The images are ordered by Einstein radius from left to right. Each cutout is $5''$ across

profiles are approximately isothermal, despite the relative non-isothermal contributions from baryons and dark matter (the so-called 'bulge-halo conspiracy', e.g., Treu and Koopmans 2004; Auger et al. 2010; Dutton and Treu 2014), however detailed individual lens studies prefer multiple baryonic components or departure from purely elliptical masses (e.g., Nightingale et al. 2019; Powell et al. 2022). The velocity-dispersion function (VDF) has been found to be consistent with no redshift evolution below $z \sim 0.5$ (Shu et al. 2012), however, the evolution beyond that is unclear, as large samples of spectroscopic measurements are currently limited to below $z \approx 1$. Complete gravitational lens samples, through their image separation statistics, are a sensitive probe of the elliptical VDF and its evolution (Oguri et al. 2012).

2.1.2 Late-Type Lenses

The mass distribution of spiral galaxies can be decomposed into their disk, bulge, and halo. The relative mass contributions of each component is unclear from kinematic data alone due to the so-called 'disc-halo degeneracy' (van Albada and Sancisi 1986). Gravitational lensing offers a unique method to break this degeneracy given the additional mass constraints from lensing, and constrain both the initial mass function and halo properties (Barnabè et al. 2012).

The efficiency (or cross-section) for lensing by these systems is strongly dependent on the inclination, as this significantly changes the projected mass density – a key quantity for lensing probabilities and configurations. Keeton and Kochanek (1998) show that the bulge component is necessary for strong lensing to occur in face-on systems, which would otherwise not surpass the surface mass density required for multiple imaging, and quad systems are expected almost exclusively from edge-on systems. A handful of serendipitous discoveries and a concerted search within SDSS spectra – SWELLS (Sloan WFC Edge-on Late-type Lens Survey, Treu et al. 2011, see also Sect. 4.4.1) – now provide ~ 20 such examples. Figure 3 shows a selection of these lenses. The presence of dust is common in these galaxies, and the extinction of both point sources and extended sources must be accounted for during searches and modelling.

2.2 Groups and Clusters

Galaxy groups are loosely defined as comprising several to ~ 50 galaxies within ~ 1 Mpc, and are thought to be embedded within their dark matter halos. Galaxy clusters contain anywhere upwards from $\sim$50 members, and can span up to 10 Mpc. Equivalently, groups have masses $\sim 10^{13-14}$ $M_{\odot}$, and clusters have masses over $\sim 10^{14}$ $M_{\odot}$. Such clusters can lens

Fig. 3 Examples of lenses with late-type galaxy deflectors; the first four are from the SWELLS sample (Treu et al. 2011), and the last one is a serendipitously discovered lensed quasar (Castander et al. 2006)

many dozens of high-redshift sources, making them efficient tools for studying high-redshift galaxies individually or in a statistical sense. While these extra sources provide more constraints (e.g., 30 sources with a total of 90 multiple images in Abell 2744, Bergamini et al. 2023), the mass distributions are naturally also more complex than isolated galaxies, and even state-of-the-art models are limited to reproducing image positions of multiply sourced systems to within a few tenths of an arcsecond in the image plane (e.g., Acebron et al. 2022). As a result, the lensing configurations are more often non-standard (e.g., Orban de Xivry and Marshall 2009), often leading to larger source magnifications than isolated galaxies (e.g., Robertson et al. 2020), and the lensed images have wider image separations than produced by galaxy-scale lenses (e.g., More et al. 2012). Complex configurations are particularly common in disturbed clusters with ongoing mergers, reflective of the turbulent mass distributions. At cluster scales, the lack of baryonic cooling beyond a certain mass scale implies fundamentally different mass profiles relative to isolated galaxies. X-ray surface brightness and strong and weak lensing analyses show that the total density profiles of relaxed clusters are well-described by Navarro-Frenk-White (NFW) profiles, namely an inner density slope of $\alpha = 1$ ($\rho(r) \propto r^{-\alpha}$; Schmidt and Allen 2007; Newman et al. 2013). This leads to less efficient lensing in clusters, and explains well the paucity of observed cluster-scale lensed quasars (Keeton 1998; Kochanek and White 2001). The associated shallower mass profiles lead to a predicted surplus of cusp geometries, in which only three bright images are apparent, and coupled with triaxiality, this can reduce the lensing efficiency by up to a factor of 4 (Oguri and Keeton 2004).

We briefly discuss the discovery methods for clusters, since this is often a significant step towards finding new giant arcs. The hot intracluster gas emits X-ray emission, which has been used to discover clusters in all-sky X-ray surveys, most notably the MAssive Cluster Survey (MACS, Ebeling et al. 2001). In optical imaging surveys, clusters are found as overdensities in the 4-dimensional space of position, colour, and magnitude, for example in the Red-Sequence Cluster Sequence (RCS, Gladders and Yee 2005) and the SDSS Giant Arcs Survey (SGAS, Sharon et al. 2020). The available datasets for these catalogues are key to the discovery of multiply imaged sources. The deepest datasets – such as those of the Frontier Fields (Lotz et al. 2017) and the Reionization Lensing Cluster Survey (RELICS, Coe et al. 2019) – reveal sources as intrinsically faint as 33 mag.

For a full description of lensing by clusters, we point the reader to the Cluster Lensing Chapter.

3 Lensed Source Classes

We briefly review the various sources of gravitational lenses, highlighting the physical characteristics that are important for their discovery and application to lens searching methods.

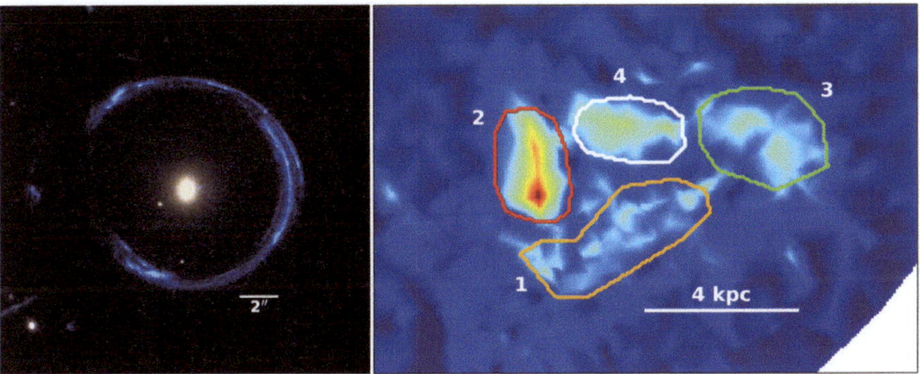

Fig. 4 *Left*: *HST* F475W/F606W/F814W colour image of the Cosmic Horseshoe; *right*: reconstructed source ($z = 2.38$) from the F160W observation, showing four distinct star-forming regions. Reproduced from James et al. (2018)

3.1 Galaxies

The deflector classes from Sect. 2 can naturally be the sources of lower redshift deflectors. However, ETG sources are intrinsically rare, with only 15 known examples of 'Early-type Early-type Lenses' (EELs, systems in which both the lens and source are ETGs, Oldham et al. 2017). While the mass function was more important when discussing the deflector population, the luminosity functions and source shapes are the key metrics for the source populations. We cover two general classes of common source galaxies: unobscured, blue star-forming galaxies and dust-obscured star-forming galaxies (or sub-mm galaxies).

3.1.1 Blue Star-Forming Galaxies

In the hierarchical picture of galaxy formation, galaxies are built up through mergers and accretion of smaller galaxies, accompanied by the onset of star formation. This star formation is seen to peak around 'Cosmic Noon' ($z \approx 2$), in which the merging galaxies are predominantly low-mass, late-type galaxies – irregulars and spirals. Such high-surface-brightness, compact galaxies dominate the galaxy population, and particularly the observed-frame optical luminosity function, and thus represent one of the most common lensed sources of gravitational lenses.

The spectrum of a star-forming galaxy composes a blue continuum with absorption lines from gas proximate to the star-forming regions, and any associated emission/nebular lines, such as [OII] 3727 and [OIII] 5007. These features are particularly useful for identification in spectra (see Sect. 4.4.1). At optical wavelengths, requiring the detection of multiple common emission lines typically limits the redshifts of such sources to $z \lesssim 1$. Beyond this, single emission lines can only suggest a redshift, however at $z \gtrsim 2$, Lyα enters the optical, and may be distinguished at sufficient signal-to-noise as its profile often has a sharp edge towards the blue, and a tail towards the red (Shu et al. 2016a).

The shapes of these galaxies compose multiple star-forming regions perhaps due to recent mergers, with several compact and diffuse components (Ritondale et al. 2019). For example, Fig. 4 shows the well-known lensed star-forming galaxy, the Cosmic Horseshoe (Belokurov et al. 2007), with the reconstructed source showing several typical clumps.

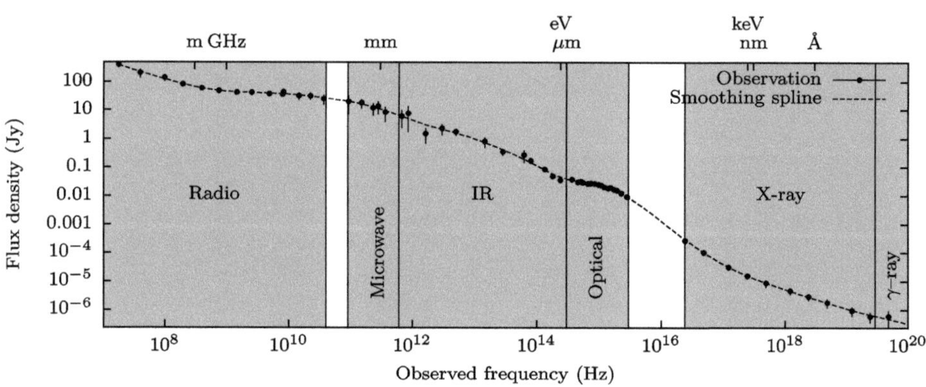

Fig. 5 Multi-wavelength observations of the 3C273 quasar at $z = 0.158$ (data taken from Türler et al. 1999)

3.1.2 Dusty Star-Forming Galaxies

Dusty star-forming galaxies (DSFGs), also known as submillimeter galaxies, are high redshift (median $z \sim 4$) galaxies with remarkably bright blackbody emission in the rest-frame far infrared, which is redshifted into the millimeter and submillimeter bands. These galaxies are some of the most active star-forming galaxies in the Universe, with star-formation rates exceeding thousands of solar masses per year (e.g., Casey et al. 2014). Their star-forming regions are obscured by dust, which, after being heated by UV radiation, are responsible for the bright far-infrared emission. In addition to copious amounts of dust, these galaxies contain vast reservoirs of molecular gas, which are commonly observed through their emission lines (e.g., Weiss et al. 2013). In contrast, these galaxies are notably difficult to detect in rest-frame optical or UV-bands, as such emission is absorbed and scattered by dust.

One of the unique properties of DSFGs is the fact that their (unlensed) luminosity function has a sharp cut-off at the bright end (e.g., Negrello et al. 2010). This simply means that (without lensing) there are large numbers of lower-flux galaxies and almost no extremely bright sources. Therefore if an unresolved extremely bright DSFG is observed, it is highly likely to be strongly lensed due to the associated magnification (see Sect. 4.2.1). Since most lensing galaxies are faint in the submillimeter, such lensed DSFGs do not suffer from lens-source deblending problems that are common at optical and infrared wavelengths.

3.2 Quasars

Supermassive black holes are thought to exist in the central potential of all massive galaxies, and can have luminosities of up to $10^{15} L_{\odot}$, due to the thermal emission associated with friction in the accretion disk. While in this actively accreting state, they are known as Active Galactic Nuclei (AGN), and the brightest high-redshift (typically $z > 1$) examples are quasars, which offer a common source population for gravitational lenses, given their brightness and luminosity function peak at high-redshift.

Quasars emit over the whole electromagnetic spectrum, as shown in Fig. 5 in the case of 3C273, the apparently brightest quasar on the sky. We point the reader towards Beckmann and Shrader (2012) and Padovani et al. (2017) for explanations of the physical processes that cause the multi-wavelength emission of quasars. In brief, the accretion disk contributes a continuum power-law in the rest-frame UV to infrared, and surrounding broad and narrow line regions of ionised gas add broad and narrow lines attributed to specific ions. Strong radio emission is found in about 10% of quasars, and is attributed to the presence of powerful jets that can extend out to many times the size of the host galaxy. A variety of observed quasar 'types' are possible depending on viewing angle: when viewed face-on, sources have prominent blue continua and broad emission lines (type-I); edge-on viewing leads to an obscured accretion disk and narrow lines (type-II). This continuum of viewing angles leads to a variety of observed quasar properties, and is further divided by levels of dust in the host galaxy, the presence of radio jets, and absorption due to outflows.

A typical scale for an accretion disk is several light days, thus quasars are unresolved at cosmological distances, and appear as point sources often outshining their host galaxies. To select these systems without contamination from stars, several wavelength-specific selection techniques have been used: UV-excess selection (Richards et al. 2009), unique infrared colours (Stern et al. 2012a), and radio and X-ray detections. Quasars are also seen to be significantly variable at all wavelengths on timescales above a week. This can be used as a selection method, with the variability amplitude being found to positively correlate with bluer rest-frame wavelengths and smaller black hole masses (MacLeod et al. 2010). We note that UV-excess selections are not reliable for higher quasar redshifts, as the Lyman alpha dropout appears in the UV filters, leading to colours similar to stars. Equivalently for infrared colour selections, the local minimum at 1 μm is redshifted into redder filters, increasing stellar contamination.

3.3 Transients

Transients cover all variable astrophysical phenomena which eventually fade away on human timescales. While supernovae have been known for millennia, several new classes of transient have only been discovered in recent years, and our current understanding suggests that they should exist at high-redshift offering sources for gravitational lenses. The discovery of the first definite lensed examples is only a matter of time as observatories open up new volumes and resolving capabilities. For complete reviews of lensed transients, we direct the reader to Oguri (2019a) and Liao et al. (2022).

3.3.1 Supernovae

Supernovae are intrinsically rare events, constituting the final moments in the evolution of only the most massive stars or specific binary star systems. The former are known as Type Ia supernovae and the latter are core-collapse supernovae. A wide array of supernova subtypes exist based on the presence, or lack thereof, of various narrow absorption lines and exact lightcurve shape. All supernova lightcurves are smoothly varying with a maximum brightness at 15-25 days after explosion, followed by a slow fade, and typical supernovae rates are roughly once per galaxy per century. Refsdal (1964) first suggested supernovae as sources in gravitational lens systems, highlighting that their time variations allowed precise determination of a time delay, and in turn, a measurement of the Hubble constant.

For detection of such transients, the Zwicky Transient Facility has played a pivotal role by surveying the entire sky every two nights down to a magnitude of $r \sim 20$ thanks to its mosaic camera, capable of imaging 47 square degrees of sky at once. From 2016 to 2023, over 100,000[1] transients consistent with supernovae have been reported by ZTF and various cadenced imaging surveys. However only 10% of these have spectra, often limited to those brighter than ~18.5 mag.

[1] Based on searches of the Transient Name Server: https://www.wis-tns.org/stats-maps

The standard-candle nature of Type Ia supernovae makes them valuable not only for cosmological probes (see, e.g., Goobar and Leibundgut 2011, for a review), but also as a method for detecting an associated magnification from strong lensing. Only a spectroscopic redshift, classification, and peak magnitude are needed (see Sect. 4.2.1). We refer to the reader to the Lensed Supernova Chapter for a full description.

3.3.2 Fast Radio Bursts

Fast Radio Bursts (FRBs) are a new class of radio transients at cosmological distances whose durations range from nano- to milli-seconds. Around 100 FRBs have been discovered, and their population-inferred redshift peaks at $z = 0.4$ (Shin et al. 2022). The individual redshifts of most bursts are unknown since they are unlocalised, though a handful have been localised to $z > 0.5$ galaxies (e.g., Ryder et al. 2022). The physical origin of FRBs is unknown, however they are divided into repeating and non-repeating sources. We point readers to Petroff et al. (2019) and Cordes and Chatterjee (2019) for detailed reviews.

3.3.3 Gamma Ray Bursts

Gamma Ray Bursts (GRBs) are brief highly energetic explosions that appear at gamma ray frequencies and display afterglows at all wavelengths. They are divided into long (70%) and short (30%) GRBs, separated by durations around $\sim$2 seconds (Kouveliotou et al. 1993). Long GRBs are associated with supernovae, and short GRBs with neutron star mergers.

Since GRBs are particularly bright and occur at cosmological redshifts, they are expected to be gravitationally lensed (Oguri 2019b). Of order 10^4 GRBs have been observed, one third of those with the Fermi Gamma-ray Space Telescope. Were the redshift distribution for GRBs the same as that for quasars, one would expect several of these to have been gravitationally lensed. The search techniques and potential discoveries are presented in Sect. 4.5.2.

3.3.4 Gravitational Waves

Since the first detection of gravitational waves (GW) from the merger of binary black holes in 2015 by the Laser Interferometer Gravitational-wave Observatory (LIGO), nearly 100 GW events from binary systems have been detected. These binaries comprise either a pair of black holes (up to $100M_{\odot}$), neutron stars ($\sim 2\mathrm{M}_{\odot}$), or a neutron star and a black hole. The current ground-based GW observatories, LIGO–Virgo–Kagra, are sensitive to GW signals in the frequency range $10 - 10^3$ Hz and the angular accuracy of the sky localisation contours spans an area of $\mathcal{O}(100)$ deg^2. Supermassive black hole mergers (up to $10^{10}M_{\odot}$) will emit in the nano-Hz frequency range, in which Pulsar Timing Arrays (PTAs) are sensitive (Nanograv Collaboration et al. 2023; Xu et al. 2023). Gravitational lensing of GWs is speculated to be discovered within the next few years with much theory already laid down a few decades ago (e.g. Deguchi and Watson 1986; Nakamura 1998).

Binary black holes (BBHs) are expected to be the first lensed GW detections owing to their cosmological distances. The exquisite precision on the time delay between multiply lensed BBH will make them an attractive probe of cosmological parameters such as the Hubble constant (e.g. Liao et al. 2017). This is feasible via their distance measurements when combined with the complementary electromagnetic (EM) data constraints, for instance, optical redshifts, accurate lensed image positions and lensing configuration. If the binary has a direct EM counterpart as is expected in the case of neutron stars (e.g. GW170817, the first BNS), lensing can shed light on the physical processes right before the merger through an early warning study (Magare et al. 2023).

4 Selection Methods

One single search method for all lenses is impossible, given the wide variety of lens types, source types, image configurations, colours, available surveys, etc. Therefore lens searches are tailored to target specific lenses in specific datasets. Search techniques can be fully described by: (i) the type of lens being searched for, (ii) the data being used, and (iii) the algorithm that prioritises or selects lens candidates. These algorithms often target one of four unique features of gravitational lenses: source magnification, multiple well-separated images, multiple redshifts, or time delays. We divide this section by these lensing attributes. For example, a discovery method that searches for multiple blue arcs around a red galaxy requires only the second feature, multiple imaging, and is independent of the other three. We remind the reader, however, that modern selection techniques can rely on a combination of these methods. We describe serendipitous discoveries in their own subsection, and citizen science searches within the multiple imaging section. Machine learning techniques are given their own section though principally rely on the resolved-images feature of lensing. A summary of the various discovery methods, divided by source type, are listed in Table 1 with approximate numbers of known systems.

4.1 Serendipitous Discoveries

Nearly all of the first dozen lenses were serendipitous discoveries, and even now lenses continue to be found fortuitously, and merit single publications. We do not aim to describe all these discoveries but highlight some of the most interesting and important examples.

The first known lens, Q0956+561, was part of a follow-up programme for quasars (at that time defined as radio sources) from the Jodrell Bank radio survey, to identify optical counterparts, and determine redshifts (Walsh et al. 1979). They obtained spectra of two blue point sources that might be associated to the radio emission, separated from one another by 6 arcseconds. The spectra were close to identical (at $z = 1.405$) and soon after the lensing galaxy was discovered (Young et al. 1980). Within a year of this first lens discovery, a known quasar, PG1115+080, was being followed up for high resolution spectroscopy, when two further stellar objects were observed on the TV guiding monitor within ~3 arcseconds. Weymann et al. (1980) proposed this as a lensed triple (at $z = 1.722$), and later observations showed that the brightest image A was in fact a close pair of images (Hege et al. 1981). A few more years would pass before a third gravitational lens was discovered, also by serendipity. Q2237+0305, also known as the Einstein Cross, was discovered by Huchra et al. (1985), during a spectroscopic survey of bright galaxies. The spectrum showed a quasar at $z = 1.7$ at the nucleus of a $z = 0.039$ spiral galaxy, and later imaging revealed the four individual quasar images (Yee 1988). Lenses that are outliers in terms of their extreme properties, such as the very low redshift lens of Q2237+0305, are, perhaps unsurprisingly, often found serendipitously. For example, one of the brightest known lensed quasars, APM08279+5255, at $z = 3.87$, was a serendipitous discovery within a survey of distant cool carbon stars (Irwin et al. 1998), which share similar infrared colours to high-redshift quasars. Also, the lensed quasar with the lowest source redshift, RXJ1131−1231, was discovered serendipitously while measuring polarisations of unlensed quasars (Sluse et al. 2005). Even in the last few years, bright quadruply lensed quasars are discovered as a byproduct of visual inspection – the two diamonds of Lucey et al. (2018) were identified when removing contaminants for the target catalogue of the Taipan Galaxy Survey.

Many lensed galaxies have also been discovered unexpectedly, including many of the earliest examples. The giant arc in the galaxy cluster Abell 370 was noticed on the CCD images

Table 1 Summary of discovery methods and numbers of known lens systems. Text in italics implies that the method is a prospective idea for a search

Source	Magnification Selection	Multiple Imaging Selection	Multiple Redshift Selection	Time-delay Selection	Number
Galaxy	very pure for DSFG sources given steep bright-end of luminosity function	visual inspection, citizen science, ring-finders, neural networks	narrow emission lines in galaxy fibre searches / IFU searches	N/A	$\sim$1500
Quasar	UV-X-ray luminosity relation / proximity zone size from Lyα transmission / Eddington ratio	catalogue-only from *Gaia* / deblended SED similarity / extended variability	rare, serendipitous broad emission lines in galaxy fibre spectra	rare, one discovery from OGLE lightcurves	$\sim$300
Supernova	spectroscopic supernovae that are too bright / *inconsistent with photometric redshift of nearby elliptical*	multiple detections in difference images of *HST* monitoring of clusters	N/A	*monitor known supernovae for re-brightening*	5
GW FRB GRB	N/A	*resolved images of EM counterparts*	N/A	*require similar waveform and sky localisation of time-separated detections*	0

during a spectroscopic campaign of the galaxy members (Soucail et al. 1987). Originally it was thought to be due to tidal interactions or star formation, but soon after, spectroscopy placed the source at $z = 0.59$, i.e. lensed by the $z = 0.374$ cluster. Galaxy-scale complete Einstein rings are striking and thus relatively easy to spot even within a wide field of data, and are thus regular serendipitous discoveries in the literature (e.g., Tanaka et al. 2016; Bettinelli et al. 2016), and similarly for bright galaxy-scale arcs (e.g., Allam et al. 2007).

4.2 Magnification Selection

A general property of strong gravitational lensing is the *magnification theorem*: at least one image is as bright as the source (Schneider 1984). Therefore a magnification > 1 is expected in any strong lens system, implying that the brightest sources will contain an enhanced fraction of lenses. Some systems will show anomalously bright sources that would disobey limits imposed by physics if not lensed. We discuss these two consequences as lens search techniques below.

4.2.1 The Brightest Objects

Simply obtaining high-resolution imaging of the brightest quasars was an early strategic search for lensed quasars above $z \approx 1$ (for example with *HST*; Bahcall et al. 1992; Maoz et al. 1992), and is still a potentially fruitful method for the brightest high-redshift quasars (McGreer 2013; Fan et al. 2019). The bright quasar catalogues of the Hamburg Quasar and Hamburg/ESO surveys have continued to provide small-separation lensed quasars even long after their initial identification (e.g. Wisotzki et al. 1996; Blackburne et al. 2008). This increased fraction of lensed sources at the brightest end of the observed luminosity function is even stronger for sources with particularly steep luminosity functions – namely very few bright sources and many faint sources. This is apparent at submillimeter wavelengths with dusty star-forming galaxies (DSFGs) (Coppin et al. 2006). Negrello et al. (2010) took advantage of this by selecting bright sources within the 550 deg^2 of the Herschel Astrophysical Terahertz Large Area Survey (H-ATLAS), providing both a large enough area and high enough resolution to select five new lensed dusty star-forming galaxies without contamination from unlensed DSFGs, but still requiring low-redshift interlopers to be removed through colour cuts. Extensions of this technique have been used to build sizeable samples of strongly lensed DSFGs from the Herschel and South Pole Telescope surveys (e.g., Vieira et al. 2013; Wardlow et al. 2013), and to carry out the selection at even brighter magnitude limits across wider areas, using the Planck all-sky survey (e.g., Cañameras et al. 2015).

4.2.2 Outliers from Physical Relations

The associated magnification of strong lensing can be used to identify outliers from non-linear physical relations. The brightest sub-mm galaxies (Sect. 4.2.1) can be thought of as outliers from an underlying true luminosity function, however in this section we highlight objects with other physical properties or relations that are disobeyed when magnification changes the observed brightness.

The X-ray – UV luminosity relation for quasars has a clear flattening at higher luminosities. As outlined in Stern and Walton (2020), if a source is strongly lensed, the UV and X-ray luminosities are increased by the same factor, taking them above the known relation, suggestive of lensing, as they demonstrated for the system MG1131+0456 (see Fig. 6). It has also been used to argue against strong lensing for high-redshift lensed quasar candidates

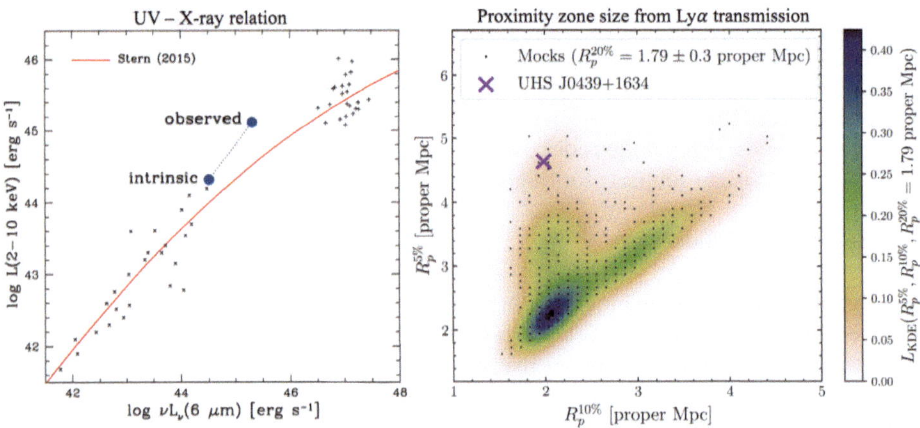

Fig. 6 Examples of lens discovery as outliers from physical relations, which can otherwise be explained by an associated magnification from lensing. *Left*: MG1131+0456 UV - X-ray luminosity before and after the magnification against the observed relation, reproduced from Stern and Walton (2020). *Right*: proximity zone size for the lensed quasar UHSJ0439+1634 appearing as an outlier among mock systems with similar *apparent* luminosities, reproduced from Davies et al. (2020)

(Connor et al. 2021): the physical size of the proximity zone of a quasar at high-redshift should be related to the intrinsic luminosity of the ionising quasar source. However, strong gravitational lensing will change the inferred luminosity, while the inferred size remains the same, and a discrepancy between the size-luminosity relation could support the lensing hypothesis. Davies et al. (2020) showed that this technique can successfully recover the known $z = 6.5$ lensed quasar, UHSJ0439+1634, while it conclusively rules out lensing for the candidate SDSSJ0100+2802 (see Fig. 6).

Another physical limit is that of accretion rates onto SMBHs, i.e. an Eddington ratio of unity. This has been used to propose an associated magnification to a quasar, in the case of J2329–0522 at $z = 4.85$ (Yue et al. 2023). In this case, however, no multiple imaging has been observed with *HST* data, and thus has been dubbed an intermediately-lensed system, where the weak lensing magnification is still significant.

The magnification associated with lensing is most apparent when the source has a specific absolute luminosity. This is the case with type Ia supernovae – they are *standard candles*, all having similar peak absolute brightnesses. This feature is notably used to probe cosmological parameters, as the redshift and luminosity distances can both be measured. Assuming a cosmology, simply measuring the redshift of a type Ia supernova allows the prediction of the observed peak brightness. In the case of gravitational lensing, the additional magnification will lead to an observed peak brightness brighter than predicted. This is how the first galaxy-scale lensed supernova, iPTF16geu, was discovered (Goobar et al. 2017). By obtaining spectroscopic redshifts and classifications of supernovae discovered in the intermediate Palomar Transient Factory (iPTF), iPTF16geu – at $z = 0.409$ – was found to be 4 magnitudes (a factor of 50) too bright. High-resolution follow-up imaging clearly revealed the 4 multiple images, with a fraction of this extra magnification associated to microlensing (e.g., More et al. 2017). The discovery spectra also showed clear absorption signatures associated to the lensing galaxy, presenting a further sign of strong lensing. A second example, SN Zwicky, has also been found in similar fashion (Goobar et al. 2023), with an Einstein radius below $0.2''$, demonstrating that this selection technique can uniquely probe the full mass range of lensing galaxies. Even without a spectroscopic redshift, Quimby et al. (2014a) show how irregular SN Ia colours relative to observed magnitude can be used to infer a higher redshift and hence associated magnification.

Requiring a spectroscopic redshift creates a strong dependence on magnitude, as the brightest supernovae candidates are followed up preferentially. For current imaging surveys – such as the Zwicky Transient Facility – very few lensed supernova are expected to be brighter than this magnitude cutoff, due to the cross-section to lensing being small at low redshift. Future surveys, such as the LSST, will discover several hundreds of lensed supernovae. However, many of these cases will not require spectra for confirmation, since the LSST imaging quality will be able to resolve individual images. (Goldstein and Nugent 2017) proposed a method that relies neither on resolving the multiple images, nor a spectroscopic redshift, nor a classification of the supernova: by considering supernovae near ellipticals to be hosted by them, one can calculate the apparent magnitude if the supernova were type Ia using the photometric redshift of the elliptical. Since the vast majority of other supernovae are fainter than type Ia, a surplus in observed flux should then be associated to lensing magnification of a source supernova (not necessarily even type Ia when the magnification is large).

4.3 Multiple-Imaging Selection

In this section, we discuss the techniques that identify lenses as systems containing spatially separated components as potential images of a background source. Typically this requires the analysis of optical and infrared imaging data (i.e., pixels), but in some cases catalogued detections can be used directly. In most cases, the starting data for a search are derived from wide-field ground-based imaging surveys, and higher resolution follow-up data are needed to not only confirm the lens nature, but also for most scientific analyses. Figure 7 shows the difference in imaging quality between discovery data and subsequent higher-resolution imaging for four quadruply imaged lensed quasars.

4.3.1 Citizen Science

Citizen science is one of the novel techniques that has emerged in the last two decades as a systematic means of doing science in Astronomy. Typically, it involves citizens from all over the world collecting and/or analysing data to collectively achieve a scientific goal. Repetitive tasks requiring special skills, such as pattern recognition, and that are difficult to carry out with computer algorithms are deemed suitable for citizen science. One of the first citizen science projects, Galaxy Zoo, was successfully used to study the morphology of galaxies (Lintott et al. 2008). Another major citizen science initiative – The Zooniverse (Simpson et al. 2014) – was setup to support many more citizen science projects, both within and outside of astronomy, including Planet Hunters, Bat Detectives, and Weather Rescue. Searching for gravitational lenses in modern wide-field, deep, multi-band imaging datasets is well suited to citizen science.

Space Warps (Marshall et al. 2016) was the first citizen science project designed specifically to discover strong lenses from astronomical imaging surveys. The classification interface has tools to aid visual analysis, a spotter's guide which displays typical examples of lenses and non-lenses and the data stream is interspersed with training images unknown to the users. On the backend, a Bayesian pipeline combines classifications from multiple users to assign probabilities for an image to contain a lens. The end product is not only a high-confidence catalogue of lens candidates with calibrated probabilities, but also a vetted sample of contaminants which can be used by the community to improve their lens

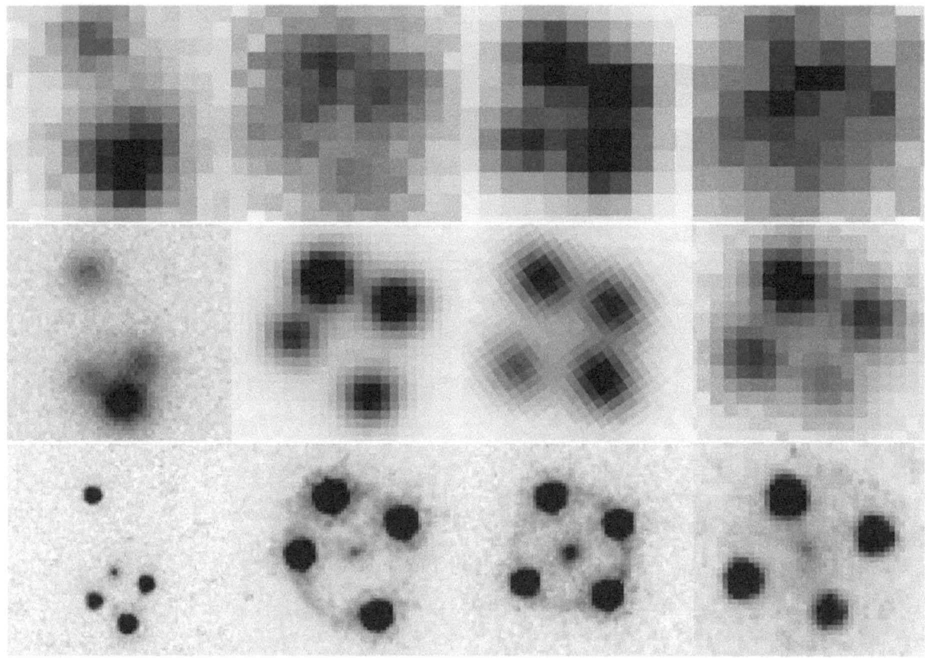

Fig. 7 *Top*: Discovery images of four lensed quasars – from left to right: DESJ0029–3814, ATLASJ0259–1608, DESJ0405–3308, and ATLASJ2344–3056. The pixel sizes of 0.26″ (DES) and 0.21″ (ATLAS) are typical of current ground-based surveys. *Middle*: Magellan images obtained with pixel sizes of 0.07″ (left three) and 0.11″ (right) in excellent seeing. *Bottom*: *HST* F814W imaging

search algorithms, and for specific studies of spiral galaxies, red star-forming galaxies or ring galaxies.

Space Warps has to-date conducted lens searches in VISTA-CFHT Stripe 82 Survey (Geach et al. 2015), the CFHT Legacy Survey (More et al. 2016a) and the Hyper Suprime-Cam Survey (Sonnenfeld et al. 2020). A unique feature of Space Warps is the supervised learning approach implemented in the analysis of the image classification wherein realistic training images are used. As a result, the final sample of candidate lenses can be well-characterised in terms of completeness and purity which is useful for the statistical analyses of lenses. Recently, a byproduct of the Zooniverse Hubble Asteroid Hunter project has been to identify strong lenses in archival *HST* observations, identifying ∼200 new high-quality candidates (Garvin et al. 2022). Citizen science will remain an important way to test automated lens discovery algorithms in future surveys, such as LSST and *Euclid*.

4.3.2 Visual Inspection

Before citizen science projects, the brute force approach of inspecting all available imaging to look for lenses was only possible with small area surveys. In particular *HST* has been extensively searched thanks to its depth and data quality, in the Hubble Deep Field (Hogg et al. 1996; Barkana et al. 1999), GOODS/ACS fields (Fassnacht et al. 2004), the COSMOS fields (Faure et al. 2008; Jackson 2008), the fields of known lenses (Fassnacht et al. 2006; Newton et al. 2009), and various other fields (e.g., Moustakas et al. 2007; More et al. 2011).

High-resolution imaging can also be obtained at radio wavelengths: the Cosmic Lens All-Sky Survey (CLASS) imaged over 13,500 radio sources at a resolution of $0.2''$ with the Very Large Array (VLA) at 8.4 GHz (Myers et al. 2003; Browne et al. 2003). Together with the $\sim$2400 images from the Jodrell Bank VLA Astrometic Survey (JVAS, King et al. 1999), the sample was inspected for systems with multiple components, and promising candidates were followed-up for higher-resolution imaging with the Multi-Element Radio Linked Interferometer Network (MERLIN) and Very Long Baseline Array (VLBA), each at 5GHz. These efforts led to the discovery of 22 new lensed quasars. Similar investigations have been carried out with the VLA in the Southern hemisphere, which uncovered 4 further radio-loud lensed quasars (e.g., Winn et al. 2002), and again more recently with the VLBA using a different parent sample (Spingola et al. 2019).

In lower-resolution imaging, the lensing nature is often not obvious, except for the widest separation systems. The 14,000 square degrees of SDSS imaging fuelled a variety of such searches for wider-separation lensed galaxies (typically with images separated by $>3''$ from the lensing galaxy), through visual inspection of red galaxies with nearby blue catalogued detections (e.g., Shin et al. 2008; Kubo et al. 2009; Diehl et al. 2009; Stark et al. 2013). Before visual inspection, images can be pre-processed to remove the bright lensing galaxies that often hide faint background arcs, either by simply subtracting a rotated version of the system (Anguita et al. 2012), or through Principal Component Analysis (Joseph et al. 2014; Paraficz et al. 2016). Similarly, giant arc searches have been performed by visual inspection of pre-identified clusters in both *HST* (Sand et al. 2005), SDSS (e.g., Estrada et al. 2007; Hennawi et al. 2008), the Dark Energy Survey (DES, Diehl et al. 2017; O'Donnell et al. 2022), and at group-to-cluster scales in the Hyper Suprime Cam Subaru Strategic Program Survey (HSC-SSP, Jaelani et al. 2020).

However, for those lenses with smaller separations, comparable to the Point Spread Function (PSF) width of the survey, contaminant systems resembling lenses quickly become overwhelming and outnumber the real lenses. For lensed quasars, these contaminants are most often quasars projected close to foreground stars, or compact star-forming galaxies. Thus, such searches are designed at specifically removing these contaminants.

In the remainder of this section, we will discuss techniques that apply some algorithms for reducing the number of candidates to a small fraction of the input catalogue. However, these searches still require a final step of visual inspection of cutouts, sometimes many tens of thousands. Biases from individual inspector's subjectivity can naturally lead to biases towards certain types of lenses (see Sect. 5), and it is common practice to have multiple authors inspect the same stamps, or inject true positives into the sample to assess this bias statistically. Rojas et al. (2023) investigated the effects of visual classification by using a labelled training set of 1500 mock and real images of lenses and non-lenses. 55 classifiers with varying levels of experience in lens discovery graded the systems through visual inspection of their colour images. As expected, arcs with low signal-to-noise or Einstein radii less than or equal to the PSF width are rarely recovered. They found that substantial variations are found between individual classifiers which can be mitigated by combining scores from 6 or more individual classifiers.

4.3.3 Catalogue Searches

Wide-field imaging and spectroscopic surveys are now available across various wavelengths, meaning that all-sky classification and cataloguing of galaxies and quasars is possible down to a few arcseconds resolution. Simple queries and self-cross-matches within these catalogues can be used to look for nearby sources with similar colours as lens candidates. One of the first dedicated catalogue-based lens searches of this type was the SDSS Quasar Lens Search (SQLS), which obtained deeper imaging and/or long-slit spectroscopy of spectroscopically confirmed quasars from SDSS which were either extended or had multiple nearby components in the SDSS imaging catalogue. Colour cuts on companions reduced the many contaminants, and the multi-year campaign discovered 49 new lensed quasars (Oguri et al. 2006; Inada et al. 2012). These searches have been extended by including additional infrared imaging for candidate selection (Jackson et al. 2012), or simply by application to newer spectroscopic samples (More et al. 2016b).

Lensed quasars can also be discovered as outliers from the unlensed quasar population in their catalogued magnitudes, due to the additional contribution from the lens galaxy, which leads to an apparent near-infrared flux excess. Ofek et al. (2007) applied this colour cut on SDSS spectroscopic quasars with red $g - H$ colours using the Two Micron All Sky Survey (2MASS), successfully identifying a new double quasar.

Even without spectra, quasars can be selected photometrically thanks to their unique infrared colours with the Wide-field Infrared Survey Explorer all-sky survey (WISE, Wright et al. 2010; Stern et al. 2012a), and various radio and X-ray surveys (see, e.g., Flesch 2021). Searches using WISE photometric quasar candidates coupled with resolved or extended detections in ground-based imaging data such as DES and VST-ATLAS have yielded a handful of new lenses (e.g., Agnello et al. 2015; Ostrovski et al. 2017), but typically with significant contamination from quasar-star projections and star-forming galaxies which mimic the infrared colours of quasars (e.g., Williams et al. 2018). WISE alone is limited due to its resolution of $\sim 5''$, resulting in the photometry of any quasar-star projection becoming blended and appearing very similar to that of a lensed quasar. Significant visual inspection is often required as a final step in these processes (Rusu et al. 2019; Dawes et al. 2023; He et al. 2023).

The high level of contamination in photometric searches was partially addressed by *Gaia*, which provided an all-sky, space-based-resolution map of bright point sources (*Gaia* Collaboration et al. 2016). For known lens systems, the multiple lensed images were often all catalogued in *Gaia*, and contaminant systems could be removed with various catalogue parameters (e.g., Lemon et al. 2018). Cross-matches between the aforementioned photometric quasar catalogues and *Gaia* yielded several dozen new lenses (Agnello et al. 2018; Spiniello et al. 2018; Desira et al. 2022), aided by the proper motion and parallax parameters and improved completeness in the second *Gaia* data release. For those systems with only one image bright enough to be detected or simply with undetected images, the single *Gaia* detection was useful when combined with ground-based imaging. By looking for astrometric offsets between the resolved *Gaia* detection of one image and the blended weighted centroid of the ground-based image, coupled with a difference in the overall photometry, Lemon et al. (2017) discovered several candidate lensed quasars which were otherwise only seen as single point sources in ground-based imaging. This is similar to the method employed by Jackson and Browne (2007), who looked for offsets between radio positions from FIRST and optical SDSS galaxy positions. Methods making use of *Gaia* data alone and its improved completeness and various catalogued parameters has unveiled a potentially large population of new lenses and dual quasar candidates, thanks to high-resolution *HST* follow-up imaging (Shen et al. 2021; Chen et al. 2022b; Mannucci et al. 2022). However, contamination from quasar-star projections still dominates, as detections with close neighbours either do not have recorded proper motions or parallax, or these are consistent with zero for most faint stars. Simultaneous use of *Gaia* and pixel-fitting of other imaging surveys has overcome this remaining contamination (Sect. 4.3.4). Quadruply imaged point-like sources occupy specific regions in the position/flux-ratio space, such that any experienced observer can easily recognize lens systems from spurious star/galaxy asterisms by only looking at the image positions and their relative magnifications. Indeed, depending on the position of the source compared to the tangential caustics, one usually ends up with three type of configurations: (i) cusps, when the sources stands near the intersection of two caustics, (ii) folds, when close to a single caustic and (iii) crosses, when the source lies close to the centre of the caustics.

The position of four postulated images can be forward modelled by a simple lensing potential (e.g., Falor and Schechter 2022), such as the Singular Isothermal Ellipsoid Potential (SIEP), which often recovers image positions for most known lensed quasars, and requires seven free parameters for the fit (2 for the source position, 2 for the galaxy position, and 3 for galaxy parameters, i.e., Einstein radius and shape). Witt (1996) demonstrated that the galaxy, source, and images of a SIEP must all lie on a hyperbola with asymptotes along the major and minor axes of the potential. This reduces the dimensionality of the model to three degrees of freedom, while the constraints are simply the positions of the four images. Wynne and Schechter (2018) also shows that the images of an SIEP must lie on an ellipse centred on the source and axes parallel to the hyperbola's asymptotes. They subsequently suggest a two-dimensional minimisation (source position along the hyperbola and potential axis ratio) to find the best fit model, from which one can report a figure of merit which determines the disagreement from a SIEP model. They report that for a certain threshold of this figure of merit, 98% of random quartets can be removed while only removing 20% of known quad quasars – namely those with peculiar potentials such as double lensing galaxies. Schechter and Wynne (2019) present an even faster determination of the best-fit model by insisting that the two brightest images fall on a shifted copy of the Witt hyperbola, ensuring always a four-image solution and no need for solving a minimization problem.

The application to lens finding simply requires four positions, which promises an extremely quick lens discovery technique in the catalogues of high-resolution imaging surveys. Indeed, this method recovers the majority of known lensed quasars from *Gaia*, suggesting that there are no bright missing quads. Machine learning has also been applied to this problem, and is discussed in Sect. 4.6.2.

We conclude our review of catalogue-based methods with cluster lensed quasars. Typically, giant arcs in clusters can be difficult to detect due to their low signal-to-noise and require pixel-level search algorithms (see Sect. 4.3.4). Quasar sources, however, are often bright enough to be available in ground-based imaging catalogues. Shu et al. (2019) cross-matched their *Gaia* AGN catalogue to known bright cluster galaxies, and discovered a new $21''$separation doubly lensed quasar. Similarly Yantovski-Barth et al. (2023) created a catalogue of clusters from the Legacy Surveys imaging, and found 5 new high-grade lens candidates (of which 2 were known) by searching for multiple DESI quasar targets within the Einstein radius of their clusters.

4.3.4 Pixel-Level Searches

For lensed quasar discovery, the astrometric lensing information is entirely in two (doubles) or four (quads) relative image positions (often without a lensing galaxy visible). Doubles are often indistinguishable from common contaminant quasar-star or star-star pairs, and thus careful pixel-level analysis that looks for colour similarity through the use of multiple survey datasets remains the leading discovery methodology. For lensed galaxy discovery however, the state-of-the-art is machine learning. We defer the discussion of these searches to their own section (see Sect. 4.6), but here describe the earlier techniques that were applied to look for arcs.

A major limitation of catalogue searches in ground-based imaging datasets is related to deblending – for any lens system with images separated by less than the size of the PSF, the separate images and lensing galaxy risk being catalogued as a single object. Other methods, such as identifying multiple redshifts in spectra (e.g., Bolton et al. 2004, see Sect. 4.4.1), offer ways to overcome this limitation, however it is still possible to detect the multiple images by analysing the pixels themselves. Schechter et al. (2017) fit multiple components to the multi-band pixels of VST-ATLAS for photometric quasar candidates, prioritising systems containing components with similar quasar-like colours, discovering four quasar pairs and three lensed quasars. They remove star-forming galaxies, a common source of contamination for WISE-selected photometric quasars, as systems with components incompatible with a given image's PSF. This technique has also been extended to DES pixels (Anguita et al. 2018). Chan et al. (2022) have applied the same principle to the $0.6''$ median-seeing survey of the Ultraviolet Near Infrared Optical Northern Survey (UNIONS), identifying nearby point sources with similar $u - r$ colours via a convolution of the Laplacian of the point spread function.

Given the precise positions of the *Gaia* detections, Lemon et al. (2019) analysed the pixels of WISE, to extract the photometry at the known positions of the individual point sources from *Gaia*. Simple colour cuts on the two components were able to remove 80% of previously identified quasar/star projections and retain 99% of lensed quasars. Applying this selection, alongside inspection of Pan-STARRS and DECaLS imaging, long-slit spectroscopic follow-up for 208 candidates was obtained with 2m- to 4m-class telescopes (ISIS on the WHT, ALFOSC on the NOT, and EFOSC2 on the NTT), with a success rate (quasar pairs or lenses) of 88% (Lemon et al. 2019, 2023). *Gaia*'s limiting magnitude is $\sim$21, while ground-based imaging surveys are now reaching beyond this, and should resolve many fainter lensed quasars.

Pixel-level analysis for compatibility with lens configurations has also been used for final candidate selection. Typical features in lensed galaxies at both galaxy- and cluster-scales are arcs, which can be targeted through algorithms built to identify such elongated and curved features. Several 'arc-enchancing' algorithms have been published, which simply require detection of significant structures in the resulting images: Horesh et al. (2005) targeted giant arcs in clusters by applying SEXTRACTOR multiple times, repeatedly removing non-elliptical sources from the segmentation map; Lenzen et al. (2004) applied anisotropic diffusion to enhance arcs along their direction while keeping them the same thickness and reducing noise; and Alard (2006) defined a local elongation map based on second order moments within a few times the effective PSF size. The latter was successfully applied to the Canada-France-Hawaii Telescope Legacy Survey (CFHTLS) by More et al. (2012). Seidel and Bartelmann (2007) similarly use moments to determine local elongation but in cells of a grid, which are subsequently grouped with nearby cells sharing similar local elongation directions, resulting in groups of arc-like structures which can have curvature, and do not require a subsequent detection algorithm. This arcfinder has since been extended to require colour information for the arcs (Maturi et al. 2014). Gavazzi et al. (2014) presented and applied their RINGFINDER algorithm to CFHTLS; by analysing PSF-matched $g - \alpha i$ images of galaxies (i.e., scaled i-band images subtracted from g-band images to remove the galaxy light), RINGFINDER finds connected tangential pixels between $0.5''$ and $2.7''$ from the galaxy, requiring at least two detections or one elongated detection. Lee (2017) use a circular Hough transform to identify overdensities of light lying on rings in SDSS LRG cutouts, recovering the four known Einstein rings in the sample, with 40% purity.

Requiring a more strict compatibility with a source lensed by a simple mass model has been applied at the full pixel-level light distribution (Marshall et al. 2009) and on extracted image positions (CHITAH, Chan et al. 2015; Sonnenfeld et al. 2018; Chan et al. 2020), which results in high purities, but at the risk of missing systems with more complex lensing potentials. YATTALENS combines arc detection and lens modelling for lens discovery

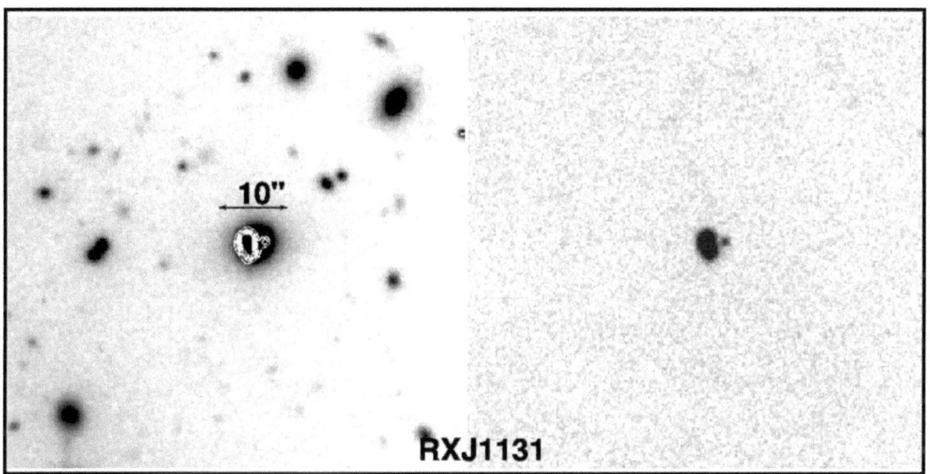

Fig. 8 Fig. 3 from Kochanek et al. (2006) showing the field of RXJ1131-1231 (left) and the variability map built from several epochs (right). The lens system is clearly the only extended variable source in the field

(Sonnenfeld et al. 2018). Arc candidates are first found as tangential detections in galaxy-subtracted g-band images of the HSC-SSP survey; subsequently the pixels are fit with a lens model (Sersic for lensing galaxy + lensed Sersic for arcs) and with a non-lens model consisting of only Sersic profiles. The systems with a better fit from a lens model were kept as candidates. The algorithm has been further applied to find new lenses in more recent data releases of HSP-SSP (Sonnenfeld et al. 2020; Wong et al. 2022).

4.3.5 Variability

In recent years, cadenced wide-field optical imaging surveys have opened up the possibility of finding variables and transients by comparing different epochs of imaging, via a method called *difference imaging*. If the source quasars are sufficiently variable, lensed quasars appear as unique sources in such difference images, since they consist of multiple crowded variable point sources, unlike most variable objects, as first proposed by Kochanek et al. (2006). They predicted that the dominant source of extended variable objects as seen by examining combined difference images would be lensed quasars, with some small contamination from binary quasars, quasar/variable star projections and variable star pairs. Figure 8 shows such an extended stacked variability map of a known lensed quasar. Lacki et al. (2009) first demonstrated the efficacy of this technique on 20,536 candidates in the SDSS Supernova Survey region, finding only eight extended variables recovering the only known lensed quasar in the sample. Kostrzewa-Rutkowska et al. (2018) searched 670 square degrees of OGLE (Optical Gravitational Lensing Experiment; an optical imaging survey of the Magellanic Clouds region with cadence of $\sim$3 days) for red WISE sources, i.e. photometric quasar candidates, with at least two variable components within 6 arcseconds yielding 63 candidates, of which 3 showed similar lightcurves. Two were spectroscopically confirmed as quasar-star projections, while OGLEJ0218-7335 was confirmed as a new lensed quasar. Within this area and to the single epoch depth of OGLE, approximately 10 lensed quasars are expected. Lemon et al. (2020) used component modelling of DES epoch data to measure variability properties of lensed quasars and quasar-star projections, further demonstrating variability as a powerful selection and prioritisation tool. They also uncovered a *variability bias* associated with finding lensed quasars through variability alone, further discussed in Sect. 5. More recently attempts have been made to implement such a variability search in HSC by Chao et al. (2021), however the implementation of the difference imaging and selection criteria can limit the purity and efficiency of such a method, and many artefacts can lead to large numbers of false positives. Dux et al. (2023b) use the multiple epochs of Pan-STARRS and apply difference imaging on close *Gaia* pairs with quasar-like infrared colours in WISE; they present spectroscopic follow-up with only 25% contamination from stellar systems.

We note that such a search could be applied to any lensed transient in which the image separation is greater than the resolving ability of the imaging survey, for example with the discovery of lensed supernovae in the optical, however, this technique would typically be complemented with other methods to ensure earlier detection. Finally, we note that as long as the time delay between images is much shorter than the survey length, each image should have approximately similar variability (if ignoring microlensing). Indeed a non-zero time delay is not necessary for this technique to work, in contrast to the discovery methods outlined in Sect. 4.5.

4.4 Multiple-Redshift Selection

In this section, we discuss the techniques that identify lenses from spectra exhibiting features at multiple redshifts. The Einstein cross was the first spectroscopic lens discovery (Huchra et al. 1985), thanks to the serendipitous detection of broad emission lines in a galaxy spectrum during a redshift survey of nearby galaxies. Despite no clear signs of the multiple images, the presence of a bright $z = 1.7$ quasar in the spectrum of a $z = 0.0396$ bright spiral galaxy led to the likely conclusion of strong gravitational lensing. In most cases, fibre spectra have led to the identification of lens candidates in exactly the same way: identifying a higher redshift galaxy in the spectra of lower redshift galaxies. This technique can be extended to resolved spectroscopy at a smaller scale, i.e. Integral Field Units (IFUs), while overcoming the drawbacks of fibre spectra.

4.4.1 Fibre Spectra

In the early 2000s, the arrival of the first SDSS data releases provided hundreds of thousands of high-quality spectra for targeted searches of background source features. Bolton et al. (2004) searched 50,000 LRG spectra for higher redshift oxygen and hydrogen nebular emission lines, finding 50 high-quality candidates which became the basis of the Sloan Lens ACS Survey (SLACS) – an *HST* snapshot imaging survey for strongly lensed galaxies. Figure 9 shows an example of such a lens discovered by SLACS. The efficiency of requiring multiple emission lines, well-fit LRG spectra, and larger Einstein radii led to a final sample of 70 definite strong lens systems (Bolton et al. 2008). This selection has since been repeated many times, with new datasets (Holwerda et al. 2015; Talbot et al. 2021) or machine learning selection (Li et al. 2019), and targeting specific lenses and/or source types: edge-on late-type lenses (Treu et al. 2011), higher-redshift (Brownstein et al. 2012) and lower-mass lenses (Shu et al. 2015), and higher-redshift sources (Shu et al. 2016a; Cao et al. 2020) or early-type sources (Auger et al. 2011; Oldham et al. 2017). Quasars acting as lenses have been found in a similar way (Courbin et al. 2012; Meyer et al. 2019), and examples of lensed quasars with galaxy-quasar blended SDSS spectra have been found (Lemon et al. 2019).

Due to increased fibre numbers, telescope sizes, and instrumentation improvements, next-generation spectroscopic surveys, such as the Dark Energy Spectroscopic Instrument

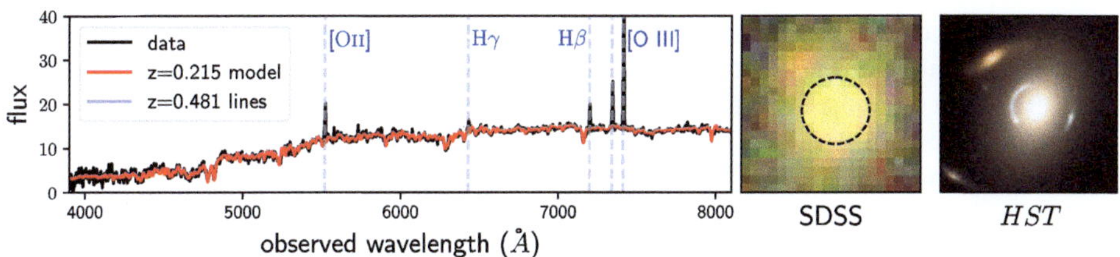

Fig. 9 SDSSJ1205+4910 discovered by Bolton et al. (2008). *Left*: an SDSS spectrum of a $z = 0.215$ LRG shows additional emission lines not fit by the model template, but that of a higher-redshift source at $z = 0.481$; *middle*: SDSS *grz* imaging with fibre overlaid; *right*: *HST* ACS F555W, F814W colour composite shows clear signs of lensing

(DESI), will obtain spectra for tens of millions of galaxies, allowing for the discovery of lensed galaxies and quasars with fainter, higher-redshift lensing galaxies.

4.4.2 IFU Spectra

While fibre spectra can efficiently provide spectra of many galaxies, they provide only integrated light within the typically 3″-wide fibres. This leads to two major drawbacks for lens searches: (i) if the lensed images lie outside of the fibre, the object will not be selected, and (ii) high-resolution follow-up imaging is almost always required to rule out a wide separation projection without evidence of strong lensing. On the other hand, imaging-only searches (Sect. 4.3) can immediately confirm the lensing nature, but follow-up spectroscopy is always required to convert lensing values into physical quantities. IFUs overcome these drawbacks, by simultaneously providing both spatial and spectral information. It is unsurprising that IFUs with wide fields of view have thus been used for targeted lens searches.

Targeted IFU observations have been used to locate the rare low-redshift lenses – particularly powerful probes of measuring the stellar mass-to-light ratio and hence the Initial Mass Function (IMF) due to probing the lens at smaller radii (where dark matter can be neglected, and thus lensing probes only the stars). The SINFONI Nearby Elliptical Lens Locator Survey (SNELLS, Smith et al. 2015) surveyed 27 galaxies selected to have the largest velocity dispersions (and thus largest lensing probability), with observations in the near-infrared to optimise the chances of observing lensed Hα, discovering two new lenses. This technique has been repeated with SDSS-MANGA (Smith 2017; Talbot et al. 2018), and MUSE with new targets and archival observations (Collier et al. 2018, 2020). Even those systems with only singly imaged nearby background objects provide competitive constraints on the IMF of nearby galaxies. Future surveys, such as Hector, will provide targeted narrow-field IFU observations of over 50,000 galaxies, and will thus offer a chance of creating larger low-redshift lens samples. The ever-increasing area of archival wide-field IFU observations also promises further lens discoveries of various source and lens types (e.g., Galbany et al. 2018).

4.5 Temporal Selection

A unique feature of gravitational lenses is their associated time delay. When the source is variable, this time delay can be measured. However, it can also be used to search for lenses with variable sources. While there is overlap with the idea of using variability of nearby objects (as outlined in Sect. 4.3.5), we focus on lens searches that require a non-zero time delay. For the following techniques, the length of the dataset must be at least the length of the time delay, and the exact success rates will depend strongly on the sampling, signal-to-noise, and intrinsic source lightcurve shape. Typically, variations with large absolute second time derivatives are most useful for detecting a time delay.

4.5.1 Lightcurves

Pindor (2005) showed that resolved lightcurves can significantly improve a search for lensed quasars by targeting systems with well-detected time delays through the use of a dispersion statistic between pairs of lightcurves. Even earlier, Geiger and Schneider (1996) proposed using autocorrelations of blended lightcurves to measure time delays in radio monitoring of lensed quasars, bypassing the need for expensive interferometric monitoring to resolve the multiple images. While their aim was not discovery, the principle has been recently revisited as a way to find lensed quasars in blended optical lightcurves, fueled by the promise of many hundreds of thousands of well-sampled quasar lightcurves in upcoming large imaging surveys. After subtracting a best-fit damped random walk (DRW) model, Shu et al. (2021) found robust features in the autocorrelation function at the expected true time delay, with true positive rates of up to 50% while keeping the false positive rate below 10%. Application on real COSMOGRAIL lightcurves returned $\sim$ 20-25% of the systems. Bag et al. (2022) propose a method of reconstructing lightcurves and identifying lenses as those with minimum fluctuations at non-zero time delays. Denissenya et al. (2022) have applied a similar technique to blended lensed supernovae using freeform templates. By considering the best-fit models for one, two, or four images, and using the Akaike Information Criterion, they find they can always recover the number of images for systems with modest magnification ratios and time delays longer than 10 days. Krone-Martins et al. (2019) note that the blended lightcurve is less stochastic than that of an unlensed quasar, and thus has lower entropy. Applying this low-entropy selection to quasar lightcurves in the Catalina Real-Time Transient Survey with multiple *Gaia* detections led to the successful discovery of new doubly imaged lensed quasars.

4.5.2 Transients

Observations of GWs, GRBs, and FRBs often provide uniquely well-sampled time baselines that would allow for an exact time delay measurement if a source were strongly lensed into multiple images. The principle for detection of a lensed event is straightforward: for each image, all the parameters should be identical (except for the arrival time, luminosity distance, and coalescence phase). Therefore, a search for signals with similar parameters may highlight pairs of lensed images from the same source. For GWs, Abbott et al. (2021) explore a ranking statistic for such a posterior overlap, and a joint posterior estimation between pairs of events. In most of the (circular, non-spinning) binary systems, we expect lensing to only change the amplitude of the signal and introduce a constant phase shift between a pair of lensed events. Thus, some methods simply search for similarity in the shape of the GW signals in the time and/or frequency dimensions as a means of identifying lensed counterparts. Accounting for overlap of sky localisation regions, population priors, selection effects, and the odds against lensing, no convincing lens pairs have yet been found. We point the reader to the Gravitational Wave Chapter for a full description of lensed GW detection methods.

For GRBs, 6 pairs have been identified as possible image pairs of a single event. A summary of relevant properties is given in Table 2. A common feature of these candidate events is their short time delays. This has led the studies' authors to posit that their lenses are black holes in the range $10^4\ M_\odot < M < 10^6\ M_\odot$.

Table 2 We list the main properties of paired GRBs: times between the pairs of events in seconds, the observed flux ratios (namely the measured luminosity of the first pulse divided by the second pulse), total magnification (summing both images) for a point mass lensing model, and the inferred lensing mass (including a factor of $1 + z$, where z is the unknown lens redshift)

Name	Δt (s)	μ_1/μ_2	μ_{macro}	$M_{\text{lens}}/M_\odot$	Reference
GRB 950830	0.4	1.8	3.5	$10^{4.4}$	Paynter et al. (2021)
GRB 200716C	1.9	1.5	5	$10^{5.3}$	Wang et al. (2021)
GRB 200716C	"	"	"	"	Yang et al. (2021)
GRB 210812A	33	4.3	1.6	10^6	Veres et al. (2021)
GRB 081126A	31	1.35	6.7	$10^{6.7}$	Lin et al. (2022b)
GRB 090717A	42	1.69	3.9	$10^{6.6}$	Lin et al. (2022b)

FRB searches are again similar, but are constrained to detect time delays below the duration of the data dumped by detection telescopes (see Sect. 5), corresponding to lens masses of $\sim 10^4 M_\odot$. Incoherent FRB lensing searches use detailed studies of individual lightcurve morphologies to distinguish between a doubly-imaged FRB and the more-common scenario of repeat emission from the same source (Liao et al. 2020; Krochek and Kovetz 2022; Zhou et al. 2022) – approximately $5 - 10\%$ of FRB sources emit repeat bursts. These searches probe minimum time delays comparable to the pulse width. Another class of searches overcomes the degeneracy between the pulse morphology and gravitational lensing by relying on the compact emission region of the FRB and detecting the interference pattern between the two images in the time domain (Kader et al. 2022; Leung et al. 2022). These coherent searches are limited by the fundamental time resolution of the telescope, though radio scattering by inhomogeneous gas in foreground haloes remains an obstacle to detection. These complementary search methods probe different timescales and therefore different lens mass ranges, and upper limits on the fraction of dark matter comprised of compact lenses have been published, though detections of lensing have not yet been claimed.

4.6 Machine Learning

Machine learning is the field of artificial intelligence devoted to the development and study of methods that are able to 'learn'. From a user-provided sample of training examples, these algorithms either:

- build arbitrarily complex relations between input attributes and numerical or discrete output values that maximise a given score measure in *supervised* learning methods; these models are then used on observations to perform predictions
- group observations according to the similarity of their input attributes according to a given similarity measurement in *unsupervised* learning methods.

Machine learning dates back to the 1950s. Examples include teaching machines to play the game of Checkers (Samuel 1959) and Rosenblatt's *perceptron* – a linear neural network capable of recognising letters. In the search for strong gravitational lenses, supervised machine learning methods are commonly used as two-class classification problems (lens vs non-lens) with output values representing the probability of the input being a lens. These classification techniques can be viewed as automatically drawing frontiers in a multidimensional input space of attributes in order to isolate objects of two intrinsically different types. Machine learning's first successful applications for lens finding were as late as 2017 (Petrillo et al. 2017; Jacobs et al. 2017; Lanusse et al. 2018), partly due to the advent of wide-field, deep surveys, but also to the requirement of large computational resources associated with the processing of astronomical images. A comparison of lens-finding algorithms, including conventional methods (e.g., arc finders and visual inspection), and machine learning methods, such as support vector machines (SVM, e.g., Hartley et al. 2017) and convolutional neural networks (CNNs), was presented in the 'strong gravitational lens-finding challenge' (Metcalf et al. 2019). Here, we will focus on the main machine learning techniques that have been applied to lens searches: CNNs and decision trees, however the field is quickly evolving, with new techniques constantly being developed and applied to new datasets. We direct the reader to Bishop (2006) for a more in-depth introduction to machine learning, and to Huertas-Company and Lanusse (2023) for an application of machine learning to astronomical survey data (which includes an overview of its application to lens finding and lens modelling).

4.6.1 CNNs

Since the early efforts by Dieleman et al. (2015) and Huertas-Company et al. (2015), deep learning algorithms have become state-of-the-art methods for galaxy morphological classifications based on pixels. These techniques similarly offer a solution to extract the pixel-level information contained in strong gravitational lenses, especially for systems with extended, spatially-resolved lensed features. The machine-learning algorithms best optimised for image analysis are supervised CNNs (LeCun et al. 1998). CNNs are able to capture the morphological patterns in imaging data by learning the coefficients of convolutional kernels and creating a range of two-dimensional feature maps. In practice, between the input layer of astronomical images, and the output layer of corresponding labels, CNN architectures comprise a range of convolutional, pooling, and fully connected hidden layers. To differentiate the network from a simple linear regression, non-linear activation functions, such as the hyperbolic tangent function, are inserted between each one of these layers.

Like all supervised learning methods, CNNs rely on a ground-truth dataset comprising images and their associated labels. During training, CNNs adjust their convolutional kernel weights by minimising a loss function, which encodes the difference between the ground truth and predicted labels. After each pass through the entire training set (refereed to as an epoch), the model is evaluated on a validation set to evaluate its performance on independent data. The evolution of training and validation losses are monitored, and the best CNN model with good generalisation to the validation set corresponds to the epoch with lowest validation loss. Information learned by the CNN is eventually stored in the feature maps. The size of labeled datasets required for training and validating CNNs for image classification vary between 10^4 to 10^6 images depending on the number of classes, image size and complexity, as well as network depth. Since the number of known strong lenses is smaller, it is necessary to create mock training sets.

The success of pioneering searches (e.g., Jacobs et al. 2017; Petrillo et al. 2017) and the results from the strong-lens classification challenge (Metcalf et al. 2019) have demonstrated the efficiency of supervised CNNs trained on strong-lens simulations to identify galaxy-galaxy configurations. While improving the CNN architectures only leads to minor gains (e.g., Schaefer et al. 2018), generating highly-realistic lens simulations and training sets that account for the complexity of the actual survey datasets is the main ingredient to reach optimal classification performances (e.g., Lanusse et al. 2018). Building on these early studies, CNNs have been applied to a large range of single and multiband imaging surveys in the optical and near-infrared to select strong-lens candidates, including DES (Jacobs et al. 2019a,b;

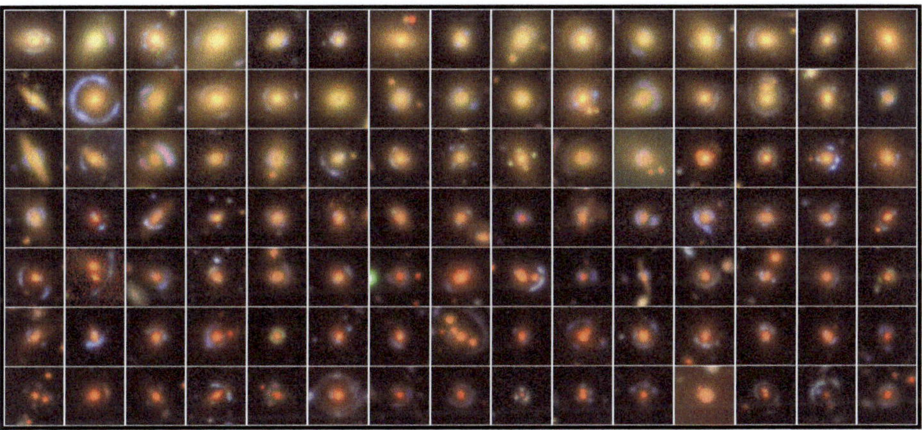

Fig. 10 Colour composite images ($10'' \times 10''$) of 105 high-quality strong lens candidates identified from the Hyper Suprime-Cam Subaru Strategic Program by Shu et al. (2022) using CNNs

Rojas et al. 2021), KiDS (Petrillo et al. 2019; Li et al. 2020, 2021), PanSTARRS (Cañameras et al. 2020), HSC (see Fig. 10; Cañameras et al. 2021; Shu et al. 2022), the DESI Legacy Imaging Surveys (Huang et al. 2020, 2021; Storfer et al. 2022), UNIONS-CFIS (Savary et al. 2022), and *HST* fields (Pourrahmani et al. 2018). In all cases, a visual inspection stage appears to be necessary for validating the CNN candidates and for increasing the sample purity by removing non-lens contaminants such as groups, spirals, and especially ring galaxies (see Rojas et al. 2021).

Each of these searches has found tens to hundreds of high-quality candidates with apparent strong lensing features and, while the grading schemes vary between different teams[2], $\sim 10^4$ new probable or definite galaxy-galaxy strong-lens candidates have been discovered with CNNs. Because confirming these candidates requires spectroscopic follow-up, ideally in combination with high-resolution imaging, the process remains highly time-consuming. Among the recent spectroscopic confirmation campaigns, Tran et al. (2022) have conducted a systematic follow-up of high-quality candidates from DES and DeCALS. The 88% success rate they obtain provides evidence for the high purity of CNN-selected galaxy-scale strong-lens candidate samples. The use of CNNs for finding group- and cluster-scale lenses has been much more limited, mainly due to the higher difficulty in simulating the diversity of this population.

CNN-based strong lens classification studies follow the supervised approach, with binary classification (lens or not lens). Multi-class models have been shown not to offer a significant benefit in lens classification (Teimoorinia et al. 2020). To the zero-th order, a supervised classifier can be regarded as a similarity checker. As a result, if the training set is biased in any specific way, decisions from such a classifier will be biased in the same way. For example, Shu et al. (2022) trained two strong-lens classifiers constructed from the same architecture with separate training sets that have different distributions in the lens galaxy redshift. Applied to the same set of strong-lens systems, it was found that, for both classifiers, the performance in lens galaxy redshift ranges that were under-represented in the training set was significantly worse than in lens galaxy redshift ranges that were well sampled. As

[2] Visual inspection by humans, including experts in strong lensing, also remains subjective (see for instance Appendix A in Shu et al. 2022) and Rojas et al. (2023).

illustrated with this example, the complex selection functions from supervised learning algorithms can fortunately be quantified with extensive tests based on real survey data (see also Cañameras et al. 2021). On the one hand, biases resulting from the design of training sets complicate the development of a self-consistent CNN classifier able to recover the variety of strong-lens morphologies, including the rare configurations (Wilde et al. 2022), while minimising confusion with non-lens contaminants. On the other hand, this sample bias can actually be exploited to design classifiers that target specific types of strong-lens systems. Jacobs (2022) quantified the sensitivity of lens-finding CNNs to various input properties in the training sample, finding the most important variables to include realistically are colour and PSF. Herle et al. (2023) report that CNNs preferentially select systems with both larger Einstein radii and larger, more concentrated sources.

Overall, these limitations are encouraging the development of complementary data-driven approaches. Among these efforts, Cheng et al. (2020) have conducted unsupervised deep learning classification of *Euclid*-like images, by combining a convolutional autoencoder with a Gaussian mixture model to cluster the extracted morphological features without the need for a training set, and Stein et al. (2021) have explored the capabilities of self-supervised representation learning using ground-based images. Keerthi Vasan et al. (2023) have demonstrated the use of semi-supervised models in which the training sets are augmented with a Generative Adversarial Network (GAN), discovering $\sim$20 promising lens candidates in the deep, small-area survey, the Deep Lens Survey. Thuruthipilly et al. (2022) demonstrated that self-attention-based encoders – namely input-dependent weighting of features – can outperform conventional CNNs in finding lenses in mock space- and ground-based imaging. These novel methods are promising, particularly for identifying exotic lens configurations that are difficult to include in training sets.

Perhaps surprisingly, CNNs have very rarely been used to discover lensed quasars. This is likely due to current samples having faint lensing galaxies, leading to confusion between quasar-star projections and double lensed quasars when using optical imaging data alone. However, quadruply imaged quasars are uniquely identifiable when their image separation is large enough. Indeed Akhazhanov et al. (2022) have shown that CNNs can reach 80% classifcation performance on mixtures of mocks and known systems, and recently such a CNN-based search has discovered a new lens in Pan-STARRS *gri* data alone (2023 a). Andika et al. (2023) find reduced contamination in real data when combining CNNs and vision transformers, and discover new candidate lensed quasars in HSC, as well as lensed galaxies as a fortuitous by-product.

4.6.2 Decision Trees

Decision trees are one of the oldest structures for building supervised learning models. They can be thought of as a sequence of if-else conditions, that visually take the form of a tree. An input passes through the tree and finishes with a classification. While passing through the tree, various attributes of the input decide which subsequent branches to take. The main advantage of decision trees is that they are easy to interpret, though they often provide poor performance given their sensitivity to the training set that is used. They are said to have a high *variance*. In order to mitigate against this effect, one can instead build N training sets, compute N trees and average all their predictions. This approach, known as Bootstrap aggregation, is used in practice in the Random Forest method along with a random selection of the input attributes. Another variant includes Extremely Randomised Trees (ERT), in which both the input attributes and observation splits are randomly selected, and many such trees are combined to provide predictions.

As such, decision tree algorithms perform well in classifying extragalactic sources, in particular quasars (e.g. Ball et al. 2006). Working at the catalogue level, they allow the user to explore large datasets with little human intervention and affordable computing time, thus selecting candidates with less stringent pre-selection criteria, maximising the precision (recovery rate), and minimising the stellar contamination. This was used as a way to remove stellar contamination from the lensed quasar search of Khramtsov et al. (2019).

As already mentioned in Sect. 4.3.3, quadruply-imaged quasars occupy specific regions of image-position and flux-ratio space. Large astrometric catalogues of individual quasar positions thus constitute an opportunity to detect lensed quasars through the use of decision trees, i.e. without having to rely on complexities of processing astronomical images. *Gaia* provides an all-sky high-resolution catalogue of quasars, ideally suited to the search of strongly lensed quasars. Using an ERT on *Gaia* DR2 data, Delchambre et al. (2019) achieved an identification rate of quadruply-imaged quasars that is better than 90% along with a misclassification rate of stellar asterisms below 1%. Nevertheless, it should be noted that the miss of a single image leads to a misidentification rate between 10 and 20%, depending on the lensed image that is unobserved, due to the dimensionality reduction and associated loss of constraints.

5 Selection Biases

Having described various lens searches, we now turn to the possible biases from these searches. Lenses are often used to probe the source or lens population, and require mass modelling. The priors that apply to these models should not be uniform because of the way that the lens system was selected. For example, magnitude-limited lens samples are biased towards higher source magnifications, and thus the sources do not lie uniformly in the source plane, but rather they preferentially lie closer to the high-magnification caustics. This might not strongly bias results when the data are strongly constraining, however population level studies can fall foul to neglecting selection effects. It is important for anyone using lenses to keep in mind the various biases of existing samples of lenses. Furthermore, understanding the limitations of previous selections might initiate novel searches aimed at removing any harmful biases from these samples.

Mandelbaum et al. (2009) provide an in-depth exploration of the selection effects of lensed point sources using a mock generation pipeline with mass models composed of both stellar and dark matter components. Sonnenfeld et al. (2023) extend this investigation to extended sources paying particular attention to stellar and dark matter masses, lens half-mass radius, and source size and surface brightness. Below, we briefly summarise several biases that exist in strong lens samples and explain their origin.

- **Magnification bias:** For a magnitude-limited sample, all of the intrinsically faint sources in the sample must have large magnifications. This leads to biases towards sources near caustics (by definition the highest magnification regions), and thus with image configurations corresponding to cusps and folds. For the same reason, lenses with larger caustics are more prevalent, which are those with more elliptical potentials or with multiple lensing galaxies (see also the quad ellipticity-shear bias this causes, described later).
- **Redshift bias:** Spectroscopic searches are limited to requiring multiple emission lines within the spectral range. The majority of spectroscopic searches are performed in the optical, and thus are limited to only one strong emission line above $z \approx 1$. Furthermore, photometric selection of sources, such as quasars, has a redshift-dependent completeness. Above $z = 2.7$, the UV-excess used to identify quasars is no longer present due to the

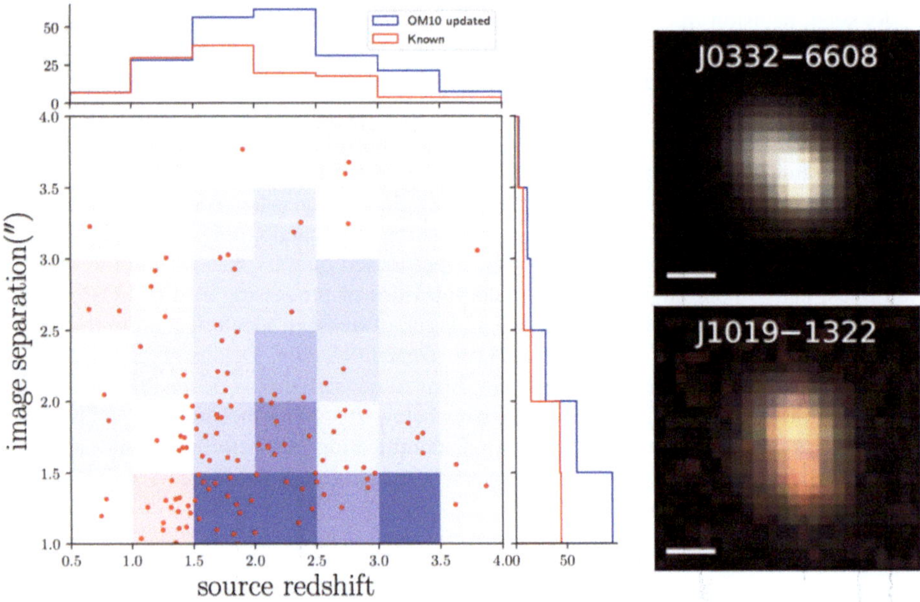

Fig. 11 *Left*: Distribution of image separations against source redshifts for known lensed quasars (red) satisfying certain 'discoverable' criteria, overlaid on an updated mock catalogue of Oguri and Marshall (2010). There is a clear lack of known lenses (regions shaded blue) with smaller separations and higher redshifts. *Right*: examples of quasar pairs with similar spectra from Lemon et al. (2023) that may represent these missing systems, lacking only a lens galaxy detection

u-band dropout of Lyman alpha, so more stringent constraints are used to remove stellar contamination, leading to less complete quasar discovery. A redshift bias can also manifest through a bias towards brighter lensing galaxies: as lensed quasar candidates with brighter lensing galaxies (as seen in existing optical imaging) are preferentially selected for limited spectroscopic confirmation, those systems with higher redshift (and thus fainter) lenses are not confirmed. Lower redshift sources naturally have lower redshift lensing galaxies, and are thus brighter and more likely to be confirmed. This is demonstrated in the known lensed quasar sample in Fig. 11.

- **Einstein radius bias:** Many search techniques rely on resolving the multiple images of a background source, however if these images are unresolved at the catalogue and/or pixel level then the lens cannot be selected in the original parent sample. This leads to a lower limit on the lensing galaxy Einstein radii. Conversely, fibre-selection of lens candidates (see Sect. 4.4.1) can miss those systems in which the brightest images lie outside the fibres, i.e. when the Einstein radii are larger than the fibre radius.
- **Model bias:** The mass models that are used to create mock lenses do not capture the complexity of real lens systems, and thus both humans and machines trained on such mocks are at risk of missing real lenses with non-standard mass distributions.
- **Microlensing bias:** Lens masses with Einstein radii comparable to the source size, such as individual stars and quasar accretion disks, can cause considerable magnifications and demagnifications of the background source. This effect depends on the convergence, internal shear, and stellar mass density at the image positions, and is more dramatic for saddle points (Schechter and Wambsganss 2002). These discrepant flux ratios can man-

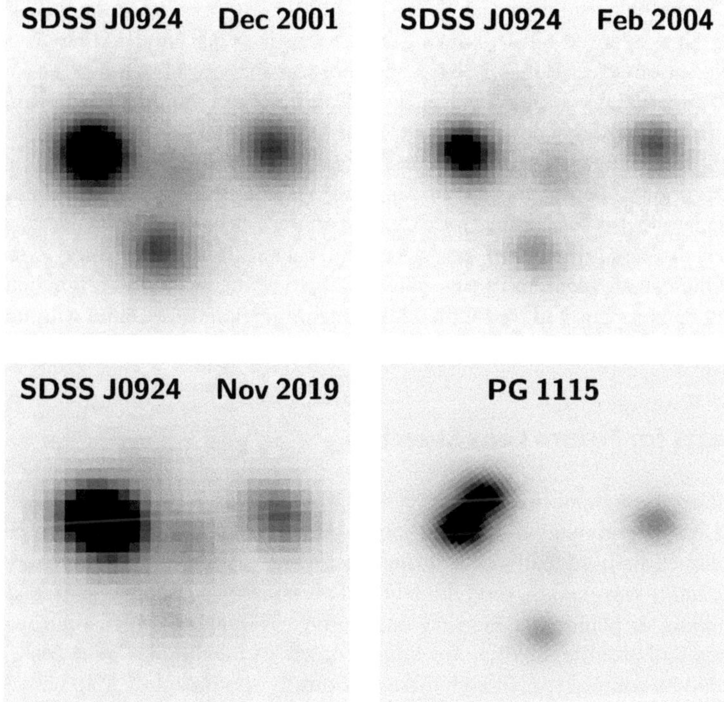

Fig. 12 i-band images of two quadruply imaged lensed quasars: SDSSJ0924+0219 shows one image being severely demagnified by microlensing over two decades, and PG1115+080 during an epoch without strong microlensing. The top two images were taken with MagIC, and the bottom two with IMACS. We note that PG1115+080 has a larger Einstein radius, and thus a larger cutout

ifest as 'missing' images in shallow imaging (see Fig. 12), which can lead to systems being missed by various magnitude-dependent selection techniques.

- **Variability bias:** Lemon et al. (2020) uncovered a *variability bias* associated with finding lensed quasars through variability alone. This bias is due to a well-known correlation between variability amplitude and quasar luminosity (see, e.g., MacLeod et al. 2012), resulting in the lenses with the faintest sources having the most variable images. Quads have, on average, higher magnifications than doubles, and so for the image brightness, quads will vary more since they probe fainter sources.
- **Survey length bias:** Oguri et al. (2003) pointed out that cadenced surveys for detecting lensed supernovae will miss images and lens systems as the duration of the survey approaches the time delay, leading to biases towards both more symmetric and smaller Einstein radius systems.
- **Time-delay selection function** Similar to the Survey length bias above, constraints on observations – both spatially or temporally – prohibit the discovery of lensed transients found in the time domain through the reappearance of extra images. For example, FRBs are of millisecond duration, and many widefield instruments (e.g., CHIME/FRB) are incapable of slewing. This imposes a "time-delay selection function" (Connor and Ravi 2022), which selects for strongly-lensed FRB systems whose time delays are either (i) less than the amount of time spent by one sky position within the beam pattern, or (ii) an integer multiple of the sidereal day.

- **Caustic area bias:** Baldwin and Schechter (2021) explore the Malmquist-like bias of quadruple image lensed quasars having larger caustic areas. Models that do not account for this selection effect typically have caustic areas that are biased low, and as such the predicted time delays are also biased low, leading to an important bias in the Hubble constant from time-delay cosmography of the order of $\sim 1\%$.
- **Quad ellipticity-shear bias:** Quadruple image lenses are biased towards more flattened lensing potentials, due to more elliptical lensing galaxy masses and/or larger external shears. This effect is less strong for extended sources than point sources, as sources can often overlap a large fraction of the background caustics. Conversely, very circular lenses have smaller caustic areas and are thus less likely to be discovered as bright quads, though this might be compensated for by the higher magnifications associated with near-circular lenses.

6 Prospects for Future Lens Searches

As emphasised throughout this Chapter, strong gravitational lensing is a rare phenomenon that spans a very broad range in lens and source properties. It is thus important that very different techniques are used to discover strong lenses. Our playground for this exercise resides in the many multi-wavelength, time-domain and even multi-messenger datasets and surveys that are ongoing or planned in the very near future. We outline which future surveys and observatories will provide the drive for finding orders of magnitudes more lenses. The section is divided by source type, though there is naturally significant overlap between surveys providing the relevant data for finding each type of lens.

6.1 Lensed Galaxies

The ESA-NASA *Euclid* survey will image 15,000 square degrees of extragalactic sky in one wide optical filter (550–920 nm) and three near-IR filters: Y, J, and H (Laureijs et al. 2011; Euclid Collaboration et al. 2022). The first strength of *Euclid* will be its spatial resolution of $0.18''$, giving access to the most compact systems with Einstein radii down to 0.3-$0.5''$. The sharp *Euclid* PSF increases the contrast between the smooth lens light and the thin arc-like images of the background source. While the broad band-pass of the *Euclid* optical filter will optimise the detection of lensed sources with strong continuum emission, it may be less effective at detecting lensed sources with flux dominated by strong emission lines. Other future space-based surveys like the Roman space telescope or the Chinese Space Station Telescope (CSST) will complement the capabilities of *Euclid* through increased depth and number of filters.

Collett (2015) create mock *Euclid* lens observations to estimate the number of discoverable systems, assuming at least partially resolved sources, magnifications above 3, and sufficient signal-to-noise. If the lens can be well subtracted, 170,000 lenses should exist in *Euclid* with a peak Einstein radius around $0.5''$. For comparison, an existing survey like DES is predicted to contain 2400 lenses with peak Einstein radii around $1.25''$.

The Rubin-LSST survey will image the whole Southern sky in 6 optical filters every few days. This adds the time and colour dimensions to *Euclid* and will surpass the depth of *Euclid* after a few months of observation but, as a ground-based survey, will be limited by atmospheric turbulence. It will play a major role in the colour pre-selection of targets, providing photometric redshifts, and in the lens-source deblending using machine learning or morpho-spectral techniques (e.g. Joseph et al. 2016; Melchior et al. 2018). Collett (2015) also predict the number of lenses to be discovered in LSST data alone: 120,000 in the optimal stack, and if the lensing galaxy must be removed through a blue-red band difference imaging technique, then 62,000. The peak of the Einstein radii is around $1''$, and source redshifts around $z = 2$.

The depth and coverage of both LSST and Euclid will expand galaxy cluster catalogues to higher redshifts, through simple conventional searches of overdensities in catalogued nearby galaxies with similar colours and magnitudes (e.g., Gladders and Yee 2005). The ongoing eROSITA X-ray survey will also play a role in finding $\sim$ 100,000 X-ray-bright clusters out to $z > 1$. The discovery of giant arcs in many of these clusters will be possible with the discovery datasets themselves.

At sub-mm wavelengths, the models that correctly reproduce the dominance of lensed systems in sub-mm galaxies (SMGs) at flux densities above 100 mJy, predict that $> 1\%$ of systems down to 1 mJy should be lensed – 10 lensed SMGs per square degree – in agreement with the number of lenses found in a common JWST-SCUBA-2 imaging field (Pearson et al. 2023). Discovery and confirmation of these systems is not possible at optical wavelengths, but could be possible in the wide-field infrared imaging of *Euclid.* JWST, ALMA, and the ngVLA (next-generation Very Large Array, to be fully operational by 2034, Selina et al. 2018) will be key to demonstrating the efficiency of these searches, and providing higher quality datasets to study the source and lens populations.

6.2 Lensed Quasars

Lensed quasar searches have relied predominantly on source selection, i.e. starting from a catalogue of known quasars and high-confidence quasar candidates. These have come from spectroscopic datasets, such as the 2dF Quasar Redshift Survey (3×10^4), SDSS/BOSS (5×10^5), as well as photometric selections due to UV-excess or typical red colours in the infrared (e.g. 10^6 from Secrest et al. 2015). These previous searches will soon be extended to fainter magnitudes, as an order of magnitude more quasars are expected from upcoming spectroscopic surveys in the next 5 years, such as the Dark Energy Spectroscopic Instrument (DESI; 4×10^6) and the 4-metre Multi-Object Spectroscopic Telescope (4MOST) extragalactic surveys (3×10^6). Pushing to even deeper magnitude limits, the Maunakea Spectroscopic Explorer (MSE) is planned on an upgraded version of the Canada France Hawaii Telescope (CFHT) with a 11 m mirror, and the capability of simultaneously obtaining 4000 spectra. Photometric selection of $z < 2.7$ quasars will also be extended thanks to the u-band imaging of the Rubin-LSST survey in the South.

However, at fainter magnitudes, lensed quasar searches can no longer be source-selected *at optical or infrared wavelengths.* As demonstrated in Fig. 13, for lensed quasars with total image brightnesses fainter than 21, the galaxy light begins to dominate. In such cases, the selection must rely on galaxies, as spectroscopic and photometric quasar catalogues will miss lensed quasars due to the lens light contamination. Lens-selected searches are already the standard for lensed galaxy searches, and should readily be applicable to quasar sources as well; for example, a SLACS-style selection to search for quasar emission lines in galaxy spectra (Sect. 4.4.1). In the X-ray and radio, there is often no contamination from the lens, offering a promising initial source-selected catalogue containing lensed quasars. The ongoing eROSITA X-ray survey will provide a deep, full-sky map at 2-10 keV at $15''$ resolution, expected to detect 3×10^6 AGN. Cross-matching optical catalogues to X-ray sources has seen success for lensed quasar discovery (e.g., between HSC and *Chandra,* Jaelani et al. 2021), and even already between eROSITA and *Gaia* (Tubín-Arenas et al. 2023).

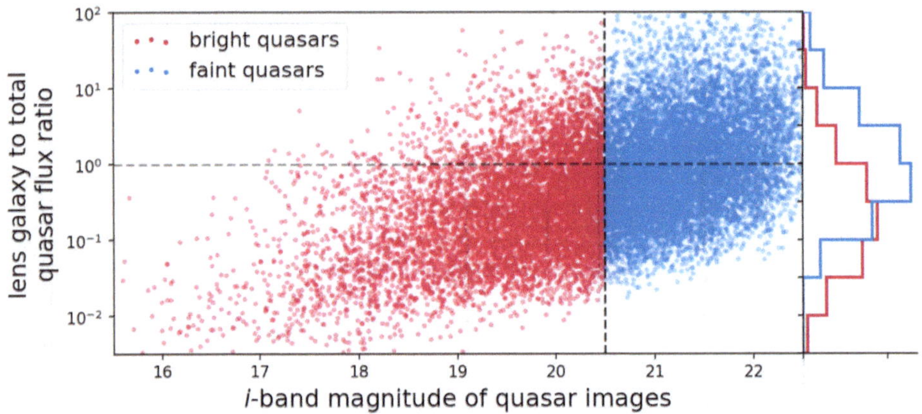

Fig. 13 Distribution of the lens-to-quasar flux ratio against total quasar magnitude for the mock lensed quasars of Oguri and Marshall (2010). The lensing galaxy population is naturally the same at all source brightnesses, so the galaxy flux (and hence colours) will start to dominate the total system flux at the fainter quasar source magnitudes expected to be discovered in future surveys like LSST and *Euclid*

Another promising selection uncontaminated by the brightness of the lensing galaxy is through extended variability (see Sect. 4.3.5). Briefly, through multiple epochs of the same field in the same filter, all lensed quasars and quasar pairs may be found through *difference imaging*. LSST will provide the ideal dataset for this search: an approximate cadence of 3 days over 20,000 square degrees in 6 filters. By stacking epochs over the 10 year nominal lifetime, variability should be detectable down to $r \sim 26.5$, or deeper when using multiple bands for detection. Taak and Treu (2023) predict ~1000 lensed quasars to have variability detectable within LSST.

We can also expect a push towards the discovery of smaller separation lensed quasars. High-resolution imaging of a handful of radio-loud or optically bright quasars has led to the smallest separation known lenses: only four systems confirmed below $0.4''$. While *Gaia* offers a $0.2''$-resolution map of point sources across the whole sky, cataloguing issues in existing data releases at small separations and in crowded local fields leads to the loss of several detections. If such issues are resolved in future releases, or the 1D single-epoch windows are provided, 100s of predicted, bright small-separation lenses might be able to be confirmed within the survey data themselves. At fainter magnitudes, *Euclid*'s space-based resolution and wide area (see Sect. 6.1) will yield many new lens systems not only from analysis of known quasars and quasar candidates, but will also detect the lensing galaxies of many existing unclassified quasar pairs, for which ground-based imaging is too shallow to detect any lenses.

At radio wavelengths, the International LOFAR Telescope (ILT) will observe 15,000 square degrees at ~150 MHz with $0.35''$ resolution and a 1σ sensitivity of 90μJy. Rezaei et al. (2022) show that CNNs can recover 95% of lensed radio sources with Einstein radii above $0.5''$ and flux densities above 2 mJy, while removing nearly all contamination from unlensed double-lobed sources. At similar resolution, a deeper all-sky survey with the Square Kilometre Array (SKA) could uncover 10,000 new radio-loud lenses, with sources mixed between lensed radio jets and lensed starbursts, with source redshifts potentially at $z > 10$ (McKean et al. 2015). Though the exact surveys of SKA are not yet confirmed, observations are planned to begin around 2028/2029.

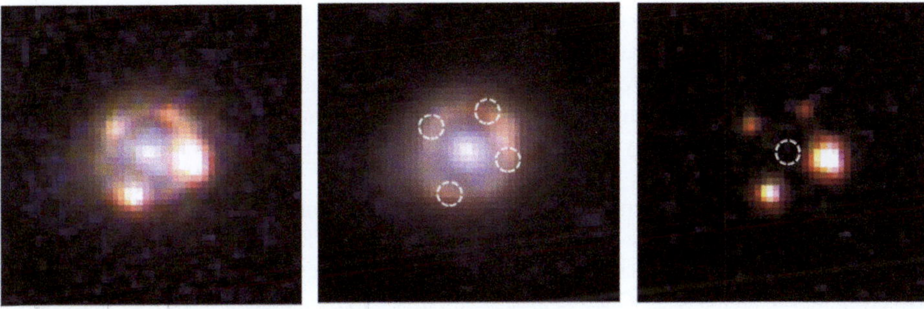

Fig. 14 Left: *HST* observation of the lensed supernova iPTF16geu; *middle*: observation after the supernova has faded; *right*: subtraction of the constant components, showing just the supernova images. Reproduced from Dhawan et al. (2020)

6.3 Lensed Supernovae

The planned Legacy Survey of Space and Time (LSST), to be carried out with the Simonyi Telescope of the Vera C. Rubin Observatory, promises an unprecedented dataset for finding lensed supernovae, thanks to its high-cadence, wide-field, and deep imaging. As discussed in Sect. 4 there are both catalogue-based schemes and pixel-based schemes for discovering lensed supernovae. The simplest of catalogue-based schemes are based on the assumption that two or more images of the lensed supernovae are recorded as transients. Given a conservative estimate of a detection limit of 22.6 mag in the i-band, an effective survey length of 2.5 years, and image separations above $0.5''$, Oguri and Marshall (2010) predict a total of 130 galaxy-scale lensed supernovae from LSST. Unresolved catalogue schemes based on strong magnifications of type Ia supernovae from measuring redshifts either photometrically or spectroscopically (see Sect. 4.2.2) are also possible, and Goldstein et al. (2019) predict 380 lensed supernovae in LSST (10 years) with such a method, broadly agreeing with the 180 lensed type Ia supernova estimate of Sainz de Murieta et al. (2023). Wojtak et al. (2019) consider both discovery methods, finding a much higher rate of 340 lensed supernovae *per year*, with $\sim$2/3 of these discoverable through their magnification alone. Combining multiple epoch observations and a variety of selection methods, such as CNN classification (e.g., Kodi Ramanah et al. 2022), will be key to discovery of these systems. The delivered image quality, cadence, observing strategy, and access to follow-up facilities will all be important factors for the true discovery rate (Huber et al. 2019a).

As LSST will also provide deep existing imaging of any candidate, and, in the absence of lens confirmation through spectroscopy, deconvolution schemes (e.g., Akhaury et al. 2022) may be used to detect a lensed host galaxy before possible further images arrive. In a similar fashion one might look for a leading image that was too faint to catalogue – "pre-covery" – as the source transient might fall near a fold caustic or due to extrinsic microlensing, similar to the case of iPTF16geu (see Fig. 14).

The number of known lensed galaxies, detected through large spectroscopic surveys or through arc detection in imaging (see Sect. 6.1), will increase by an order of magnitude before the end of the lifetime of LSST. Cross-matching known transients to lens candidates (Magee et al. 2023; Sheu et al. 2023) and monitoring of known lenses (Craig et al. 2021), will become a fruitful method for lensed supernova discovery, as it already has been for monitoring of cluster lenses (e.g., Kelly et al. 2015; Rodney et al. 2021). JWST will become a key tool for lensed supernova discovery in galaxy clusters given its unprecedented depth and resolution and thus ability to reach higher redshift fainter supernovae. Indeed, Frye et al. (2023) already report three images of a new lensed supernova in galaxy cluster G165. Based on star formation rates of source galaxies, they suggest as many as 1 more lensed supernova to be expected from monitoring of that one cluster alone. Shu et al. (2018) predict over 10 lensed supernovae occur per year in a sample of 128 known lensed galaxies from various SDSS lensed galaxy searches.

Such monitoring of clusters with the *HST* has also recently revealed individual lensed stars and star clusters (Miralda-Escude 1991; Kelly et al. 2018) thanks to the magnifications (both macro and micro) near critical curves of factors of several thousand. *JWST* is already delivering new systems (e.g., Chen et al. 2022a), and will be the most powerful observatory for the discovery of such systems for the foreseen future.

6.4 Lensed GWs

As LIGO–Virgo–Kagra (LVK) currently enter their fourth observing run (O4) at design sensitivity, detecting lensed GWs from stellar-mass black hole (BH) mergers becomes a possibility, and very likely in the fifth observing run (O5) sensitivity (estimated to start in 2027). Predictions for lensed GWs suggest rates around $0.1-5$ lensed BH-BH mergers per year in O4, and $1-50$ per year in O5, with the range caused by our uncertainty on underlying compact merger rates (e.g., Li et al. 2018; Ng et al. 2018). Lensed NS-NS events are unlikely to be detected, even during O5 (Smith et al. 2023). Most time delays will be shorter than one month, and $\sim$30% of lensed events are expected to be quadruply imaged (Li et al. 2018).

Several next-generation GW observatories (so-called third-generation) are currently planned and will be sensitive to lensed GWs of various origins. Many of these observatories will also have superior angular resolutions of $1''$or better, and extend the detection volume out to high-redshift, offering the chances of detecting optical counterparts of lensed NS-NS mergers. The Einstein telescope will have sensitivity to NS-NS events out to $z=2$, and thus even higher redshift for lensed events. Piórkowska et al. (2013) predict up to several tens of detections per year, and an order of magnitude more lensed BH-BH mergers (Biesiada et al. 2014).

The Laser Interferometer Space Antenna (LISA) is a planned (2030s) space-based set of spacecraft that will be separated by 8 light-seconds, and will provide sensitivity to larger volumes, at lower frequencies, and accordingly higher total merger masses. Optimistic models suggest up to 4 well-detected strongly lensed events within the 5-year nominal mission (Sereno et al. 2010). Between the LISA-LIGO frequency gap, DECIGO, a planned space-based mission, will operate at deci-hertz frequencies. The principle mission is to detect primordial GWs expected from inflation, however it will also detect Intermediate Mass Black Hole mergers and their lensed counterparts (Piórkowska-Kurpas et al. 2021).

At even lower frequencies, we expect to see GW signals from supermassive BH mergers. Pulsar Timing Arrays (PTAs) have recently reported strong evidence for a stochastic background consistent with such a population of BH mergers (Nanograv Collaboration et al. 2023). Khusid et al. (2022) predict 10-30 lensed binaries might be detected using PTAs with the Square Kilometre Array. Tantalising evidence at optical wavelengths for such a system already exists in a known lensed quasar (Millon et al. 2022). Figure 15 summarises the frequencies and astrophysical sources of next-generation GW observatories. The sky localisation uncertainties depend on the signal-to-noise of the signal but typical localisation areas for the observatories (in square degrees) are $\sim$1 (LISA, Kocsis et al. 2008), $\sim$10 (advanced LIGO), and $\sim$40 (SKA and iPTA, Sesana and Vecchio 2010; Finn and Lommen 2010).

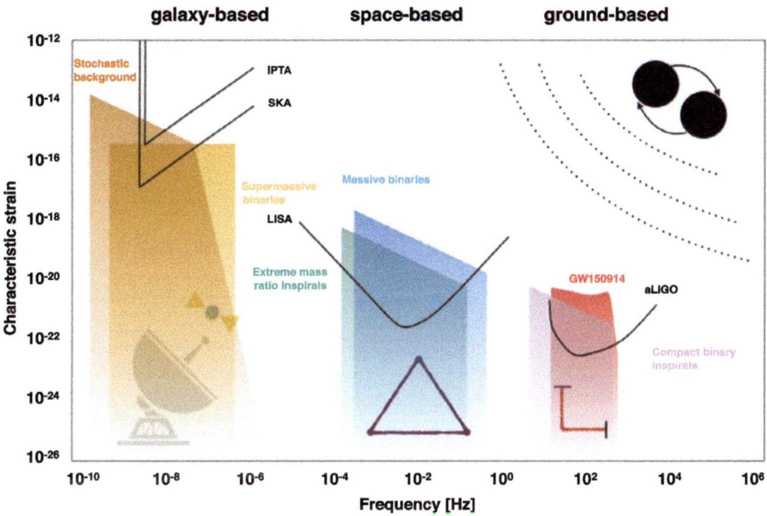

Fig. 15 Gravitational wave spectrum. It shows the frequencies at which various GW observatories (current and upcoming) are going to operate and correspondingly astrophysical sources. Figure is reproduced from K. Islo's Thesis. Credit: C. Moore, R. Cole, C. Berry and C. F. Mingarelli

6.5 Lensed FRBs

Perhaps the most promising path to detecting galaxy- or cluster-lensed FRBs involves a combination of microlensing searches and high-precision FRB localisations to their host galaxies. In this scenario, optical follow-up can confirm the lensed host galaxy. This will be enabled by widefield FRB surveys such as CHIME/FRB Outriggers (Leung et al. 2021; Mena-Parra et al. 2022), CHORD, the DSA-2000, or future coherent all-sky monitors such as BURSTT (Lin et al. 2022a), which will overcome the time-delay selection bias by extending the detection field-of-view down to the horizon.

Acknowledgements We thank the International Space Science Institute in Bern (ISSI) for their hospitality and the conveners for organizing the stimulating workshop on "Strong Gravitational Lensing".

Funding Open access funding provided by EPFL Lausanne.

Declarations

Competing Interests The authors have no relevant financial or non-financial interests to disclose. The authors have no conflicts of interest to declare that are relevant to the content of this article. All authors certify that they have no affiliations with or involvement in any organization or entity with any financial interest or non-financial interest in the subject matter or materials discussed in this manuscript. The authors have no financial or proprietary interests in any material discussed in this article.

Open Access This article is licensed under a Creative Commons Attribution 4.0 International License, which permits use, sharing, adaptation, distribution and reproduction in any medium or format, as long as you give appropriate credit to the original author(s) and the source, provide a link to the Creative Commons licence, and indicate if changes were made. The images or other third party material in this article are included in the article's Creative Commons licence, unless indicated otherwise in a credit line to the material. If material is not included in the article's Creative Commons licence and your intended use is not permitted by statutory regulation or exceeds the permitted use, you will need to obtain permission directly from the copyright holder. To view a copy of this licence, visit http://creativecommons.org/licenses/by/4.0/.

References

Abbott R, Abbott TD, Abraham S et al (2021) Astrophys J 923:14. https://doi.org/10.3847/1538-4357/ac23db
Acebron A, Grillo C, Bergamini P et al (2022) Astron Astrophys 668:A142. https://doi.org/10.1051/0004-6361/202244836
Agnello A, Treu T, Ostrovski F et al (2015) Mon Not R Astron Soc 454:1260. https://doi.org/10.1093/mnras/stv2171
Agnello A, Lin H, Kuropatkin N et al (2018) Mon Not R Astron Soc 479:4345. https://doi.org/10.1093/mnras/sty1419
Akhaury U, Starck J-L, Jablonka P, Courbin F, Michalewicz K (2022) Front Astron Space Sci 9:357. https://doi.org/10.3389/fspas.2022.1001043
Akhazhanov A, More A, Amini A et al (2022) Mon Not R Astron Soc 513:2407. https://doi.org/10.1093/mnras/stac925
Alard C (2006) arXiv e-prints astro-ph/0606757. https://doi.org/10.48550/arXiv.astro-ph/0606757
Allam SS, Tucker DL, Lin H et al (2007) Astrophys J Lett 662:L51. https://doi.org/10.1086/519520
Andika IT, Suyu SH, Cañameras R et al (2023) Astron Astrophys 678:A103. https://doi.org/10.1051/0004-6361/202347332
Anguita T, Barrientos LF, Gladders MD et al (2012) Astrophys J 748:129. https://doi.org/10.1088/0004-637X/748/2/129
Anguita T, Schechter PL, Kuropatkin N et al (2018) Mon Not R Astron Soc 480:5017. https://doi.org/10.1093/mnras/sty2172
Auger MW, Treu T, Bolton AS et al (2010) Astrophys J 724:511. https://doi.org/10.1088/0004-637X/724/1/511
Auger MW, Treu T, Brewer BJ, Marshall PJ (2011) Mon Not R Astron Soc 411:L6. https://doi.org/10.1111/j.1745-3933.2010.00980.x
Bag S, Shafieloo A, Liao K, Treu T (2022) Astrophys J 927:191. https://doi.org/10.3847/1538-4357/ac51cb
Bahcall JN, Maoz D, Doxsey R et al (1992) Astrophys J 387:56. https://doi.org/10.1086/171060
Baldwin D, Schechter PL (2021) arXiv e-prints, arXiv:2110.06378. https://doi.org/10.48550/arXiv.2110.06378
Ball NM, Brunner RJ, Myers AD, Tcheng D (2006) Astrophys J 650:497. https://doi.org/10.1086/507440
Barkana R, Blandford R, Hogg DW (1999) Astrophys J Lett 513:L91. https://doi.org/10.1086/311924
Barnabè M, Dutton AA, Marshall PJ et al (2012) Mon Not R Astron Soc 423:1073. https://doi.org/10.1111/j.1365-2966.2012.20934.x
Beckmann V, Shrader CR (2012) Active Galactic Nuclei. Wiley-VCH, Weinheim. https://doi.org/10.1002/9783527666829
Belokurov V, Evans NW, Moiseev A et al (2007) Astrophys J Lett 671:L9. https://doi.org/10.1086/524948
Bergamini P, Acebron A, Grillo C et al (2023) Astron Astrophys 670:A60. https://doi.org/10.1051/0004-6361/202244575
Bettinelli M, Simioni M, Aparicio A et al (2016) Mon Not R Astron Soc 461:L67. https://doi.org/10.1093/mnrasl/slw097
Biesiada M, Ding X, Piórkowska A, Zhu Z-H (2014) J Cosmol Astropart Phys 2014:080. https://doi.org/10.1088/1475-7516/2014/10/080
Bishop CM (2006) Pattern recognition and machine learning. Springer, Berlin
Blackburne JA, Wisotzki L, Schechter PL (2008) Astron J 135:374. https://doi.org/10.1088/0004-6256/135/1/374
Bolton AS, Burles S, Schlegel DJ, Eisenstein DJ, Brinkmann J (2004) Astron J 127:1860. https://doi.org/10.1086/382714
Bolton AS, Burles S, Koopmans LVE et al (2008) Astrophys J 682:964. https://doi.org/10.1086/589327
Browne IWA, Wilkinson PN, Jackson NJF et al (2003) Mon Not R Astron Soc 341:13. https://doi.org/10.1046/j.1365-8711.2003.06257.x
Brownstein JR, Bolton AS, Schlegel DJ et al (2012) Astrophys J 744:41. https://doi.org/10.1088/0004-637X/744/1/41
Burud I, Courbin F, Lidman C et al (1998) Astrophys J Lett 501:L5. https://doi.org/10.1086/311450
Cañameras R, Nesvadba NPH, Guery D et al (2015) Astron Astrophys 581:A105. https://doi.org/10.1051/0004-6361/201425128
Cañameras R, Schuldt S, Suyu SH et al (2020) Astron Astrophys 644:A163. https://doi.org/10.1051/0004-6361/202038219
Cañameras R, Schuldt S, Shu Y et al (2021) Astron Astrophys 653:L6. https://doi.org/10.1051/0004-6361/202141758
Cao X, Li R, Shu Y et al (2020) Mon Not R Astron Soc 499:3610. https://doi.org/10.1093/mnras/staa3058
Casey CM, Narayanan D, Cooray A (2014) Phys Rep 541:45. https://doi.org/10.1016/j.physrep.2014.02.009

Castander FJ, Treister E, Maza J, Gawiser E (2006) Astrophys J 652:955. https://doi.org/10.1086/508148
Chan JHH, Suyu SH, Chiueh T et al (2015) Astrophys J 807:138. https://doi.org/10.1088/0004-637X/807/2/138
Chan JHH, Suyu SH, Sonnenfeld A et al (2020) Astron Astrophys 636:A87. https://doi.org/10.1051/0004-6361/201937030
Chan JHH, Lemon C, Courbin F et al (2022) Astron Astrophys 659:A140. https://doi.org/10.1051/0004-6361/202142389
Chao DCY, Chan JHH, Suyu SH et al (2021) Astron Astrophys 655:A114. https://doi.org/10.1051/0004-6361/202039376
Chen Y-C, Hwang H-C, Shen Y et al (2022b) Astrophys J 925:162. https://doi.org/10.3847/1538-4357/ac401b
Chen W, Kelly PL, Treu T et al (2022a) Astrophys J Lett 940:L54. https://doi.org/10.3847/2041-8213/ac9585
Cheng T-Y, Li N, Conselice CJ et al (2020) Mon Not R Astron Soc 494:3750. https://doi.org/10.1093/mnras/staa1015
Chwolson O (1924) Astron Nachr 221:329. https://doi.org/10.1002/asna.19242212003
Coe D, Salmon B, Bradač M et al (2019) Astrophys J 884:85. https://doi.org/10.3847/1538-4357/ab412b
Collett TE (2015) Astrophys J 811:20. https://doi.org/10.1088/0004-637X/811/1/20
Collett TE, Smith RJ (2020) Mon Not R Astron Soc 497:1654. https://doi.org/10.1093/mnras/staa1804
Collier WP, Smith RJ, Lucey JR (2018) Mon Not R Astron Soc 478:1595. https://doi.org/10.1093/mnras/sty1188
Collier WP, Smith RJ, Lucey JR (2020) Mon Not R Astron Soc 494:271. https://doi.org/10.1093/mnras/staa602
Connor L, Ravi V (2022) Mon Not R Astron Soc 521:4024. https://doi.org/10.1093/mnras/stad667. arXiv:2206.14310
Connor T, Stern D, Bañados E, Mazzucchelli C (2021) Astrophys J Lett 922:L24. https://doi.org/10.3847/2041-8213/ac37b5
Coppin K, Chapin EL, Mortier AMJ et al (2006) Mon Not R Astron Soc 372:1621. https://doi.org/10.1111/j.1365-2966.2006.10961.x
Cordes JM, Chatterjee S (2019) Astron Astrophys 57:417. https://doi.org/10.1146/annurev-astro-091918-104501
Courbin F, Lidman C, Magain P (1998) Astron Astrophys 330:57. https://arxiv.org/abs/astro-ph/9707183
Courbin F, Faure C, Djorgovski SG et al (2012) Astron Astrophys 540:A36. https://doi.org/10.1051/0004-6361/201118015
Craig P, O'Connor K, Chakrabarti S et al (2021) arXiv e-prints, arXiv:2111.01680. https://doi.org/10.48550/arXiv.2111.01680
Davies FB, Wang F, Eilers A-C, Hennawi JF (2020) Astrophys J Lett 904:L32. https://doi.org/10.3847/2041-8213/abc61f
Dawes C, Storfer C, Huang X et al (2023) Astrophys J Suppl 269:61. https://doi.org/10.3847/1538-4365/ad015a. arXiv:2208.06356
Deguchi S, Watson WD (1986) Astrophys J 307:30. https://doi.org/10.1086/164389
Delchambre L, Krone-Martins A, Wertz O et al (2019) Astron Astrophys 622:A165. https://doi.org/10.1051/0004-6361/201833802
Denissenya M, Bag S, Kim AG, Linder EV, Shafieloo A (2022) Mon Not R Astron Soc 511:1210. https://doi.org/10.1093/mnras/stac143
Desira C, Shu Y, Auger MW et al (2022) Mon Not R Astron Soc 509:738. https://doi.org/10.1093/mnras/stab2960
Dhawan S, Johansson J, Goobar A et al (2020) Mon Not R Astron Soc 491:2639. https://doi.org/10.1093/mnras/stz2965
Diehl HT, Allam SS, Annis J et al (2009) Astrophys J 707:686. https://doi.org/10.1088/0004-637X/707/1/686
Diehl HT, Buckley-Geer EJ, Lindgren KA et al (2017) Astrophys J Suppl 232:15. https://doi.org/10.3847/1538-4365/aa8667
Dieleman S, Willett KW, Dambre J (2015) Mon Not R Astron Soc 450:1441. https://doi.org/10.1093/mnras/stv632
Dutton AA, Treu T (2014) Mon Not R Astron Soc 438:3594. https://doi.org/10.1093/mnras/stt2489
Dux F, Lemon C, Courbin F et al (2023a) arXiv e-prints, arXiv:2310.04494. https://doi.org/10.48550/arXiv.2310.04494
Dux F, Lemon C, Courbin F et al (2023b) arXiv e-prints, arXiv:2307.13729. https://doi.org/10.48550/arXiv.2307.13729
Dyson FW, Eddington AS, Davidson C (1920) Philos Trans R Soc Lond Ser A 220:291. https://doi.org/10.1098/rsta.1920.0009

Ebeling H, Edge AC, Henry JP (2001) Astrophys J 553:668. https://doi.org/10.1086/320958
Eigenbrod A, Courbin F, Meylan G (2007) Astron Astrophys 465:51. https://doi.org/10.1051/0004-6361:20066939
Einstein A (1936) Science 84:506. https://doi.org/10.1126/science.84.2188.506
Estrada J, Annis J, Diehl HT et al (2007) Astrophys J 660:1176. https://doi.org/10.1086/512599
Euclid Collaboration, Scaramella R, Amiaux J et al (2022) Astron Astrophys 662:A112. https://doi.org/10.1051/0004-6361/202141938
Falor C, Schechter PL (2022) Astron J 164:120. https://doi.org/10.3847/1538-3881/ac80bc
Fan X, Wang F, Yang J et al (2019) Astrophys J Lett 870:L11. https://doi.org/10.3847/2041-8213/aaeffe
Fassnacht CD, Moustakas LA, Casertano S et al (2004) Astrophys J Lett 600:L155. https://doi.org/10.1086/379004
Fassnacht CD, McKean JP, Koopmans LVE et al (2006) Astrophys J 651:667. https://doi.org/10.1086/507623
Faure C, Kneib J-P, Covone G et al (2008) Astrophys J Suppl 176:19. https://doi.org/10.1086/526426
Finn LS, Lommen AN (2010) Astrophys J 718:1400. https://doi.org/10.1088/0004-637X/718/2/1400
Flesch EW (2021) arXiv e-prints, arXiv:2105.12985. https://doi.org/10.48550/arXiv.2105.12985
Frye BL, Pascale M, Pierel J et al (2023) arXiv e-prints, arXiv:2309.07326. https://doi.org/10.48550/arXiv.2309.07326
Gaia Collaboration, Prusti T, de Bruijne JHJ et al (2016) Astron Astrophys 595:A1. https://doi.org/10.1051/0004-6361/201629272
Galbany L, Collett TE, Méndez-Abreu J et al (2018) Mon Not R Astron Soc 479:262. https://doi.org/10.1093/mnras/sty1448
Garvin EO, Kruk S, Cornen C et al (2022) Astron Astrophys 667:A141. https://doi.org/10.1051/0004-6361/202243745
Gavazzi R, Marshall PJ, Treu T, Sonnenfeld A (2014) Astrophys J 785:144. https://doi.org/10.1088/0004-637X/785/2/144
Geach JE, More A, Verma A et al (2015) Mon Not R Astron Soc 452:502. https://doi.org/10.1093/mnras/stv1243
Geiger B, Schneider P (1996) Mon Not R Astron Soc 282:530. https://doi.org/10.1093/mnras/282.2.530
Gladders MD, Yee HKC (2005) Astrophys J Suppl 157:1. https://doi.org/10.1086/427327
Goldstein DA, Nugent PE (2017) Astrophys J Lett 834:L5. https://doi.org/10.3847/2041-8213/834/1/L5
Goldstein DA, Nugent PE, Goobar A (2019) Astrophys J Suppl 243:6. https://doi.org/10.3847/1538-4365/ab1fe0
Goobar A, Leibundgut B (2011) Annu Rev Nucl Part Sci 61:251. https://doi.org/10.1146/annurev-nucl-102010-130434
Goobar A, Amanullah R, Kulkarni SR et al (2017) Science 356:291. https://doi.org/10.1126/science.aal2729
Goobar A, Johansson J, Schulze S et al (2023) Nat Astron 7:1098. https://doi.org/10.1038/s41550-023-01981-3. arXiv:2211.00656
Hartley P, Flamary R, Jackson N, Tagore AS, Metcalf RB (2017) Mon Not R Astron Soc 471:3378. https://doi.org/10.1093/mnras/stx1733
He Z, Li N, Cao X et al (2023) Astron Astrophys 672:A123. https://doi.org/10.1051/0004-6361/202245484
Hege EK, Hubbard EN, Strittmatter PA, Worden SP (1981) Astrophys J Lett 248:L1. https://doi.org/10.1086/183610
Hennawi JF, Gladders MD, Oguri M et al (2008) Astron J 135:664. https://doi.org/10.1088/0004-6256/135/2/664
Herle A, O'Riordan CM, Vegetti S (2023) arXiv e-prints, arXiv:2307.10355. https://doi.org/10.48550/arXiv.2307.10355
Hewitt JN, Turner EL, Schneider DP, Burke BF, Langston GI (1988) Nature 333:537. https://doi.org/10.1038/333537a0
Hogg DW, Blandford R, Kundic T, Fassnacht CD, Malhotra S (1996) Astrophys J Lett 467:L73. https://doi.org/10.1086/310213
Holwerda BW, Baldry IK, Alpaslan M et al (2015) Mon Not R Astron Soc 449:4277. https://doi.org/10.1093/mnras/stv589
Horesh A, Ofek EO, Maoz D et al (2005) Astrophys J 633:768. https://doi.org/10.1086/466519
Horesh A, Maoz D, Ebeling H, Seidel G, Bartelmann M (2010) Mon Not R Astron Soc 406:1318. https://doi.org/10.1111/j.1365-2966.2010.16763.x
Huang X, Storfer C, Ravi V et al (2020) Astrophys J 894:78. https://doi.org/10.3847/1538-4357/ab7ffb
Huang X, Storfer C, Gu A et al (2021) Astrophys J 909:27. https://doi.org/10.3847/1538-4357/abd62b
Huber S, Suyu SH, Noebauer UM et al (2019a) Astron Astrophys 631:A161. https://doi.org/10.1051/0004-6361/201935370
Huchra J, Gorenstein M, Kent S et al (1985) Astron J 90:691. https://doi.org/10.1086/113777
Huertas-Company M, Lanusse F (2023) Publ Astron Soc Aust 40:e001. https://doi.org/10.1017/pasa.2022.55

Huertas-Company M, Gravet R, Cabrera-Vives G et al (2015) Astrophys J Suppl 221:8. https://doi.org/10.1088/0067-0049/221/1/8
Inada N, Oguri M, Shin M-S et al (2012) Astron J 143:119. https://doi.org/10.1088/0004-6256/143/5/119
Irwin MJ, Ibata RA, Lewis GF, Totten EJ (1998) Astrophys J 505:529. https://doi.org/10.1086/306213
Jackson N (2008) Mon Not R Astron Soc 389:1311. https://doi.org/10.1111/j.1365-2966.2008.13629.x
Jackson N, Browne IWA (2007) Mon Not R Astron Soc 374:168. https://doi.org/10.1111/j.1365-2966.2006.11126.x
Jackson N, Rampadarath H, Ofek EO, Oguri M, Shin M-S (2012) Mon Not R Astron Soc 419:2014. https://doi.org/10.1111/j.1365-2966.2011.19857.x
Jacobs C (2022) In: Ruiz JE, Pierfedereci F, Teuben P (eds) Astronomical Data Analysis Software and Systems XXX. ASP Conf Ser, vol 532, p 187
Jacobs C, Glazebrook K, Collett T, More A, McCarthy C (2017) Mon Not R Astron Soc 471:167. https://doi.org/10.1093/mnras/stx1492
Jacobs C, Collett T, Glazebrook K et al (2019a) Astrophys J Suppl 243:17. https://doi.org/10.3847/1538-4365/ab26b6
Jacobs C, Collett T, Glazebrook K et al (2019b) Mon Not R Astron Soc 484:5330. https://doi.org/10.1093/mnras/stz272
Jaelani AT, More A, Oguri M et al (2020) Mon Not R Astron Soc 495:1291. https://doi.org/10.1093/mnras/staa1062
Jaelani AT, Rusu CE, Kayo I et al (2021) Mon Not R Astron Soc 502:1487. https://doi.org/10.1093/mnras/stab145
James BL, Auger M, Pettini M et al (2018) Mon Not R Astron Soc 476:1726. https://doi.org/10.1093/mnras/sty315
Joseph R, Courbin F, Metcalf RB et al (2014) Astron Astrophys 566:A63. https://doi.org/10.1051/0004-6361/201423365
Joseph R, Courbin F, Starck JL (2016) Astron Astrophys 589:A2. https://doi.org/10.1051/0004-6361/201527923
Kader Z, Leung C, Dobbs M et al (2022) Phys Rev D 106:043016. https://doi.org/10.1103/PhysRevD.106.043016
Keerthi Vasan GC, Sheng S, Jones T, Choi CP, Sharpnack J (2023) Mon Not R Astron Soc 524:5368. https://doi.org/10.1093/mnras/stad1709. arXiv:2211.00047
Keeton CRI (1998) PhD thesis, Harvard University, Massachusetts
Keeton CR, Kochanek CS (1998) Astrophys J 495:157. https://doi.org/10.1086/305272
Keeton CR, Kochanek CS, Seljak U (1997) Astrophys J 482:604. https://doi.org/10.1086/304172
Kelly PL, Rodney SA, Treu T et al (2015) Science 347:1123. https://doi.org/10.1126/science.aaa3350
Kelly PL, Diego JM, Rodney S et al (2018) Nat Astron 2:334. https://doi.org/10.1038/s41550-018-0430-3
Khramtsov V, Sergeyev A, Spiniello C et al (2019) Astron Astrophys 632:A56. https://doi.org/10.1051/0004-6361/201936006
Khusid NM, Mingarelli CMF, Natarajan P, Casey-Clyde JA, Barnacka A (2022) arXiv e-prints, arXiv:2210.00014. https://doi.org/10.48550/arXiv.2210.00014
King LJ, Browne IWA, Marlow DR, Patnaik AR, Wilkinson PN (1999) Mon Not R Astron Soc 307:225. https://doi.org/10.1046/j.1365-8711.1999.02328.x
Kochanek CS (1996) Astrophys J 466:638. https://doi.org/10.1086/177538
Kochanek CS, White M (2001) Astrophys J 559:531. https://doi.org/10.1086/322379
Kochanek CS, Mochejska B, Morgan ND, Stanek KZ (2006) Astrophys J Lett 637:L73. https://doi.org/10.1086/500559
Kocsis B, Haiman Z, Menou K (2008) Astrophys J 684:870. https://doi.org/10.1086/590230
Kodi Ramanah D, Arendse N, Wojtak R (2022) Mon Not R Astron Soc 512:5404. https://doi.org/10.1093/mnras/stac838
Kormendy J, Fisher DB, Cornell ME, Bender R (2009) Astrophys J Suppl 182:216. https://doi.org/10.1088/0067-0049/182/1/216
Kostrzewa-Rutkowska Z, Kozłowski S, Lemon C et al (2018) Mon Not R Astron Soc 476:663. https://doi.org/10.1093/mnras/sty259
Kouveliotou C, Meegan CA, Fishman GJ et al (1993) Astrophys J Lett 413:L101. https://doi.org/10.1086/186969
Krochek K, Kovetz ED (2022) Phys Rev D 105:103528. https://doi.org/10.1103/PhysRevD.105.103528
Krone-Martins A, Graham MJ, Stern D et al (2019) arXiv e-prints, arXiv:1912.08977. https://doi.org/10.48550/arXiv.1912.08977
Kubo JM, Allam SS, Annis J et al (2009) Astrophys J Lett 696:L61. https://doi.org/10.1088/0004-637X/696/1/L61

Lacki BC, Kochanek CS, Stanek KZ, Inada N, Oguri M (2009) Astrophys J 698:428. https://doi.org/10.1088/0004-637X/698/1/428

Lanusse F, Ma Q, Li N et al (2018) Mon Not R Astron Soc 473:3895. https://doi.org/10.1093/mnras/stx1665

Laureijs R, Amiaux J, Arduini S et al (2011) arXiv e-prints, arXiv:1110.3193. https://arxiv.org/abs/1110.3193

Lawrence CR, Schneider DP, Schmidt M et al (1984) Science 223:46. https://doi.org/10.1126/science.223.4631.46

LeCun Y, Bottou L, Bengio Y, Haffner P (1998) In: Proceedings of the IEEE, pp 2278–2324

Lee C-H (2017) Publ Astron Soc Aust 34:e014. https://doi.org/10.1017/pasa.2017.7

Lemon CA, Auger MW, McMahon RG, Koposov SE (2017) Mon Not R Astron Soc 472:5023. https://doi.org/10.1093/mnras/stx2094

Lemon CA, Auger MW, McMahon RG, Ostrovski F (2018) Mon Not R Astron Soc 479:5060. https://doi.org/10.1093/mnras/sty911

Lemon CA, Auger MW, McMahon RG (2019) Mon Not R Astron Soc 483:4242. https://doi.org/10.1093/mnras/sty3366

Lemon C, Auger MW, McMahon R et al (2020) Mon Not R Astron Soc 494:3491. https://doi.org/10.1093/mnras/staa652

Lemon C, Anguita T, Auger-Williams MW et al (2023) Mon Not R Astron Soc 520:3305. https://doi.org/10.1093/mnras/stac3721

Lenzen F, Schindler S, Scherzer O (2004) Astron Astrophys 416:391. https://doi.org/10.1051/0004-6361:20034619

Leung C, Mena-Parra J, Masui K et al (2021) Astron J 161:81. https://doi.org/10.3847/1538-3881/abd174

Leung C, Kader Z, Masui KW et al (2022) Phys Rev D 106:043017. https://doi.org/10.1103/PhysRevD.106.043017

Li S-S, Mao S, Zhao Y, Lu Y (2018) Mon Not R Astron Soc 476:2220. https://doi.org/10.1093/mnras/sty411

Li R, Shu Y, Su J et al (2019) Mon Not R Astron Soc 482:313. https://doi.org/10.1093/mnras/sty2708

Li R, Napolitano NR, Tortora C et al (2020) Astrophys J 899:30. https://doi.org/10.3847/1538-4357/ab9dfa

Li R, Napolitano NR, Spiniello C et al (2021) Astrophys J 923:16. https://doi.org/10.3847/1538-4357/ac2df0

Liao K, Fan X-L, Ding X, Biesiada M, Zhu Z-H (2017) Nat Commun 8:1148. https://doi.org/10.1038/s41467-017-01152-9

Liao K, Zhang SB, Li Z, Gao H (2020) Astrophys J Lett 896:L11. https://doi.org/10.3847/2041-8213/ab963e

Liao K, Biesiada M, Zhu Z-H (2022) Chin Phys Lett 39:119801. https://doi.org/10.1088/0256-307X/39/11/119801

Lin S-J, Li A, Gao H et al (2022b) Astrophys J 931:4. https://doi.org/10.3847/1538-4357/ac6505

Lin H-H, Lin K-y, Li C-T et al (2022a) Publ Astron Soc Pac 134:094106. https://doi.org/10.1088/1538-3873/ac8f71

Lintott CJ, Schawinski K, Slosar A et al (2008) Mon Not R Astron Soc 389:1179. https://doi.org/10.1111/j.1365-2966.2008.13689.x

Lotz JM, Koekemoer A, Coe D et al (2017) Astrophys J 837:97. https://doi.org/10.3847/1538-4357/837/1/97

Lucey JR, Schechter PL, Smith RJ, Anguita T (2018) Mon Not R Astron Soc 476:927. https://doi.org/10.1093/mnras/sty243

Lynds R, Petrosian V (1986) In: Bulletin of the American Astronomical Society, vol 18, p 1014

Lynds R, Petrosian V (1989) Astrophys J 336:1. https://doi.org/10.1086/166989

MacLeod CL, Ivezić Ž, Kochanek CS et al (2010) Astrophys J 721:1014. https://doi.org/10.1088/0004-637X/721/2/1014

MacLeod CL, Ivezić Ž, Sesar B et al (2012) Astrophys J 753:106. https://doi.org/10.1088/0004-637X/753/2/106

Magain P, Surdej J, Swings JP, Borgeest U, Kayser R (1988) Nature 334:325. https://doi.org/10.1038/334325a0

Magare S, Kapadia SJ, More A et al (2023) Astrophys J Lett 955:L31. https://doi.org/10.3847/2041-8213/acf668. arXiv:2302.02916

Magee MR, Sainz de Murieta A, Collett TE, Enzi W (2023) Mon Not R Astron Soc 525:542. https://doi.org/10.1093/mnras/stad2263. arXiv:2303.15439

Mandelbaum R, van de Ven G, Keeton CR (2009) Mon Not R Astron Soc 398:635. https://doi.org/10.1111/j.1365-2966.2009.15166.x

Mannucci F, Pancino E, Belfiore F et al (2022) Nat Astron 6:1185. https://doi.org/10.1038/s41550-022-01761-5

Maoz D, Bahcall JN, Doxsey R et al (1992) Astrophys J 394:51. https://doi.org/10.1086/171558

Marshall PJ, Hogg DW, Moustakas LA et al (2009) Astrophys J 694:924. https://doi.org/10.1088/0004-637X/694/2/924

Marshall PJ, Verma A, More A et al (2016) Mon Not R Astron Soc 455:1171. https://doi.org/10.1093/mnras/stv2009
Maturi M, Mizera S, Seidel G (2014) Astron Astrophys 567:A111. https://doi.org/10.1051/0004-6361/201321634
McGreer I (2013) A candidate gravitationally lensed quasar at z = 6.09. HST Proposal
McKean J, Jackson N, Vegetti S et al (2015) In: Advancing Astrophysics with the Square Kilometre Array (AASKA14), p 84
Melchior P, Moolekamp F, Jerdee M et al (2018) Astron Comput 24:129. https://doi.org/10.1016/j.ascom.2018.07.001
Mena-Parra J, Leung C, Cary S et al (2022) Astron J 163:48. https://doi.org/10.3847/1538-3881/ac397a
Metcalf RB, Meneghetti M, Avestruz C et al (2019) Astron Astrophys 625:A119. https://doi.org/10.1051/0004-6361/201832797
Meyer RA, Delubac T, Kneib J-P, Courbin F (2019) Astron Astrophys 625:A56. https://doi.org/10.1051/0004-6361/201834978
Millon M, Dalang C, Lemon C et al (2022) Astron Astrophys 668:A77. https://doi.org/10.1051/0004-6361/202244440
Miralda-Escude J (1991) Astrophys J 379:94. https://doi.org/10.1086/170486
More A, Jahnke K, More S et al (2011) Astrophys J 734:69. https://doi.org/10.1088/0004-637X/734/1/69
More A, Cabanac R, More S et al (2012) Astrophys J 749:38. https://doi.org/10.1088/0004-637X/749/1/38
More A, Oguri M, Kayo I et al (2016b) Mon Not R Astron Soc 456:1595. https://doi.org/10.1093/mnras/stv2813
More A, Verma A, Marshall PJ et al (2016a) Mon Not R Astron Soc 455:1191. https://doi.org/10.1093/mnras/stv1965
More A, Suyu SH, Oguri M, More S, Lee C-H (2017) Astrophys J Lett 835:L25. https://doi.org/10.3847/2041-8213/835/2/L25
Moustakas LA, Marshall P, Newman JA et al (2007) Astrophys J Lett 660:L31. https://doi.org/10.1086/517930
Myers ST, Jackson NJ, Browne IWA et al (2003) Mon Not R Astron Soc 341:1. https://doi.org/10.1046/j.1365-8711.2003.06256.x
Nakamura TT (1998) Phys Rev Lett 80:1138. https://doi.org/10.1103/PhysRevLett.80.1138
Nanograv Collaboration, Agazie G, Anumarlapudi A et al (2023) Astrophys J Lett 951:L8. https://doi.org/10.3847/2041-8213/acdac6
Narasimha D, Subramanian K, Chitre SM (1984) Astrophys J 283:512. https://doi.org/10.1086/162333
Negrello M, Hopwood R, De Zotti G et al (2010) Science 330:800. https://doi.org/10.1126/science.1193420
Newman AB, Treu T, Ellis RS et al (2013) Astrophys J 765:24. https://doi.org/10.1088/0004-637X/765/1/24
Newton ER, Marshall PJ, Treu T (2009) Astrophys J 696:1125. https://doi.org/10.1088/0004-637X/696/2/1125
Ng KKY, Wong KWK, Broadhurst T, Li TGF (2018) Phys Rev D 97:023012. https://doi.org/10.1103/PhysRevD.97.023012
Nightingale JW, Massey RJ, Harvey DR et al (2019) Mon Not R Astron Soc 489:2049. https://doi.org/10.1093/mnras/stz2220
O'Donnell JH, Wilkinson RD, Diehl HT et al (2022) Astrophys J Suppl 259:27. https://doi.org/10.3847/1538-4365/ac470b
Ofek EO, Oguri M, Jackson N, Inada N, Kayo I (2007) Mon Not R Astron Soc 382:412. https://doi.org/10.1111/j.1365-2966.2007.12389.x
Oguri M (2019a) Rep Prog Phys 82:126901. https://doi.org/10.1088/1361-6633/ab4fc5
Oguri M (2019b) Rep Prog Phys 82:126901. https://doi.org/10.1088/1361-6633/ab4fc5
Oguri M, Keeton CR (2004) Astrophys J 610:663. https://doi.org/10.1086/421870
Oguri M, Marshall PJ (2010) Mon Not R Astron Soc 405:2579. https://doi.org/10.1111/j.1365-2966.2010.16639.x
Oguri M, Suto Y, Turner EL (2003) Astrophys J 583:584. https://doi.org/10.1086/345431
Oguri M, Inada N, Pindor B et al (2006) Astron J 132:999. https://doi.org/10.1086/506019
Oguri M, Inada N, Strauss MA et al (2012) Astron J 143:120. https://doi.org/10.1088/0004-6256/143/5/120
Oldham L, Auger MW, Fassnacht CD et al (2017) Mon Not R Astron Soc 465:3185. https://doi.org/10.1093/mnras/stw2832
Orban de Xivry G, Marshall P (2009) Mon Not R Astron Soc 399:2. https://doi.org/10.1111/j.1365-2966.2009.14925.x
Ostrovski F, McMahon RG, Connolly AJ et al (2017) Mon Not R Astron Soc 465:4325. https://doi.org/10.1093/mnras/stw2958
Padovani P, Alexander DM, Assef RJ et al (2017) Astron Astrophys Rev 25:2. https://doi.org/10.1007/s00159-017-0102-9

Paraficz D, Courbin F, Tramacere A et al (2016) Astron Astrophys 592:A75. https://doi.org/10.1051/0004-6361/201527971
Paynter J, Webster R, Thrane E (2021) Nat Astron 5:560. https://doi.org/10.1038/s41550-021-01307-1
Pearson J, Serjeant S, Wang W-H et al (2023) arXiv e-prints, arXiv:2309.00888. https://doi.org/10.48550/arXiv.2309.00888
Petrillo CE, Tortora C, Chatterjee S et al (2017) Mon Not R Astron Soc 472:1129. https://doi.org/10.1093/mnras/stx2052
Petrillo CE, Tortora C, Vernardos G et al (2019) Mon Not R Astron Soc 484:3879. https://doi.org/10.1093/mnras/stz189
Petroff E, Hessels JWT, Lorimer DR (2019) Astron Astrophys Rev 27:4. https://doi.org/10.1007/s00159-019-0116-6
Pindor B (2005) Astrophys J 626:649. https://doi.org/10.1086/430048
Pindor B, Eisenstein DJ, Inada N et al (2004) Astron J 127:1318. https://doi.org/10.1086/381904
Piórkowska A, Biesiada M, Zhu Z-H (2013) J Cosmol Astropart Phys 2013:022. https://doi.org/10.1088/1475-7516/2013/10/022
Piórkowska-Kurpas A, Hou S, Biesiada M et al (2021) Astrophys J 908:196. https://doi.org/10.3847/1538-4357/abd482
Pourrahmani M, Nayyeri H, Cooray A (2018) Astrophys J 856:68. https://doi.org/10.3847/1538-4357/aaae6a
Powell DM, Vegetti S, McKean JP et al (2022) Mon Not R Astron Soc 516:1808. https://doi.org/10.1093/mnras/stac2350
Quimby RM, Oguri M, More A et al (2014a) Science 344:396. https://doi.org/10.1126/science.1250903
Refsdal S (1964) Mon Not R Astron Soc 128:307. https://doi.org/10.1093/mnras/128.4.307
Rezaei S, McKean JP, Biehl M, de Roo W, Lafontaine A (2022) Mon Not R Astron Soc 517:1156. https://doi.org/10.1093/mnras/stac2078
Richards GT, Myers AD, Gray AG et al (2009) Astrophys J Suppl 180:67. https://doi.org/10.1088/0067-0049/180/1/67
Ritondale E, Auger MW, Vegetti S, McKean JP (2019) Mon Not R Astron Soc 482:4744. https://doi.org/10.1093/mnras/sty2833
Robertson A, Smith GP, Massey R et al (2020) Mon Not R Astron Soc 495:3727. https://doi.org/10.1093/mnras/staa1429
Rodney SA, Brammer GB, Pierel JDR et al (2021) Nat Astron 5:1118. https://doi.org/10.1038/s41550-021-01450-9
Rojas K, Savary E, Clément B et al (2021) arXiv e-prints, arXiv:2109.00014. https://arxiv.org/abs/2109.00014
Rojas K, Collett TE, Ballard D et al (2023) arXiv e-prints, arXiv:2301.03670. https://doi.org/10.48550/arXiv.2301.03670
Rusu CE, Berghea CT, Fassnacht CD et al (2019) Mon Not R Astron Soc 486:4987. https://doi.org/10.1093/mnras/stz1142
Ryder SD, Bannister KW, Bhandari S et al (2022) arXiv e-prints, arXiv:2210.04680. https://arxiv.org/abs/2210.04680
Sainz de Murieta A, Collett TE, Magee MR et al (2023) arXiv e-prints, arXiv:2307.12881. https://doi.org/10.48550/arXiv.2307.12881
Samuel AL (1959) IBM J Res Dev 71
Sand DJ, Treu T, Ellis RS, Smith GP (2005) Astrophys J 627:32. https://doi.org/10.1086/430298
Sauer T (2008) Arch Hist Exact Sci 62:1. https://doi.org/10.1007/s00407-007-0008-4
Savary E, Rojas K, Maus M et al (2022) Astron Astrophys 666:A1. https://doi.org/10.1051/0004-6361/202142505
Schaefer C, Geiger M, Kuntzer T, Kneib JP (2018) Astron Astrophys 611:A2. https://doi.org/10.1051/0004-6361/201731201
Schechter PL, Wambsganss J (2002) Astrophys J 580:685. https://doi.org/10.1086/343856
Schechter PL, Wynne RA (2019) Astrophys J 876:9. https://doi.org/10.3847/1538-4357/ab1258
Schechter PL, Morgan ND, Chehade B et al (2017) Astron J 153:219. https://doi.org/10.3847/1538-3881/aa6899
Schmidt RW, Allen SW (2007) Mon Not R Astron Soc 379:209. https://doi.org/10.1111/j.1365-2966.2007.11928.x
Schmidt T, Treu T, Birrer S et al (2022) arXiv e-prints, arXiv:2206.04696. https://arxiv.org/abs/2206.04696
Schneider P (1984) Astron Astrophys 140:119
Secrest NJ, Dudik RP, Dorland BN et al (2015) Astrophys J Suppl 221:12. https://doi.org/10.1088/0067-0049/221/1/12
Seidel G, Bartelmann M (2007) Astron Astrophys 472:341. https://doi.org/10.1051/0004-6361:20066097

Selina RJ, Murphy EJ, McKinnon M et al (2018) Science with a next generation very large array. In: Murphy E (ed) Astronomical society of the Pacific conference series, vol 517, p 15
Sereno M, Sesana A, Bleuler A et al (2010) Phys Rev Lett 105:251101. https://doi.org/10.1103/PhysRevLett.105.251101
Sesana A, Vecchio A (2010) Class Quantum Gravity 27:084016. https://doi.org/10.1088/0264-9381/27/8/084016
Shajib AJ, Birrer S, Treu T et al (2019) Mon Not R Astron Soc 483:5649. https://doi.org/10.1093/mnras/sty3397
Shajib AJ, Treu T, Birrer S, Sonnenfeld A (2021a) Mon Not R Astron Soc 503:2380. https://doi.org/10.1093/mnras/stab536
Sharon K, Bayliss MB, Dahle H et al (2020) Astrophys J Suppl 247:12. https://doi.org/10.3847/1538-4365/ab5f13
Shen Y, Chen Y-C, Hwang H-C et al (2021) Nat Astron 5:569. https://doi.org/10.1038/s41550-021-01323-1
Sheu W, Huang X, Cikota A et al (2023) arXiv e-prints, arXiv:2301.03578. https://doi.org/10.48550/arXiv.2301.03578
Shin M-S, Strauss MA, Oguri M et al (2008) Astron J 136:44. https://doi.org/10.1088/0004-6256/136/1/44
Shin K, Masui KW, Bhardwaj M et al (2022) arXiv e-prints, arXiv:2207.14316. https://arxiv.org/abs/2207.14316
Shu Y, Bolton AS, Schlegel DJ et al (2012) Astron J 143:90. https://doi.org/10.1088/0004-6256/143/4/90
Shu Y, Bolton AS, Brownstein JR et al (2015) Astrophys J 803:71. https://doi.org/10.1088/0004-637X/803/2/71
Shu Y, Bolton AS, Kochanek CS et al (2016a) Astrophys J 824:86. https://doi.org/10.3847/0004-637X/824/2/86
Shu Y, Bolton AS, Mao S et al (2018) Astrophys J 864:91. https://doi.org/10.3847/1538-4357/aad5ea
Shu Y, Koposov SE, Evans NW et al (2019) Mon Not R Astron Soc 489:4741. https://doi.org/10.1093/mnras/stz2487
Shu Y, Belokurov V, Evans NW (2021) Mon Not R Astron Soc 502:2912. https://doi.org/10.1093/mnras/stab241
Shu Y, Cañameras R, Schuldt S et al (2022) Astron Astrophys 662:A4. https://doi.org/10.1051/0004-6361/202243203
Simpson R, Page KR, De Roure D (2014) In: Proceedings of the 23rd International Conference on World Wide Web, WWW '14 companion. Association for Computing Machinery, New York, pp 1049–1054. https://doi.org/10.1145/2567948.2579215
Sluse D, Hutsemékers D, Lamy H, Cabanac R, Quintana H (2005) Astron Astrophys 433:757. https://doi.org/10.1051/0004-6361:20042163
Smith RJ (2017) Mon Not R Astron Soc 464:L46. https://doi.org/10.1093/mnrasl/slw174
Smith RJ, Lucey JR, Conroy C (2015) Mon Not R Astron Soc 449:3441. https://doi.org/10.1093/mnras/stv518
Smith GP, Robertson A, Mahler G et al (2023) Mon Not R Astron Soc 520:702. https://doi.org/10.1093/mnras/stad140
Sonnenfeld A, Treu T, Marshall PJ et al (2015) Astrophys J 800:94. https://doi.org/10.1088/0004-637X/800/2/94
Sonnenfeld A, Chan JHH, Shu Y et al (2018) Publ Astron Soc Jpn 70:S29. https://doi.org/10.1093/pasj/psx062
Sonnenfeld A, Verma A, More A et al (2020) Astron Astrophys 642:A148. https://doi.org/10.1051/0004-6361/202038067
Sonnenfeld A, Li S-S, Despali G, Shajib AJ, Taylor EN (2023) arXiv e-prints, arXiv:2301.13230. https://doi.org/10.48550/arXiv.2301.13230
Soucail G, Fort B, Mellier Y, Picat JP (1987) Astron Astrophys 172:L14
Soucail G, Mellier Y, Fort B, Mathez G, Cailloux M (1988) Astron Astrophys 191:L19
Spingola C, McKean JP, Lee M, Deller A, Moldon J (2019) Mon Not R Astron Soc 483:2125. https://doi.org/10.1093/mnras/sty3189
Spiniello C, Agnello A, Napolitano NR et al (2018) Mon Not R Astron Soc 480:1163. https://doi.org/10.1093/mnras/sty1923
Stark DP, Auger M, Belokurov V et al (2013) Mon Not R Astron Soc 436:1040. https://doi.org/10.1093/mnras/stt1624
Stein G, Blaum J, Harrington P, Medan T, Lukic Z (2021) arXiv e-prints, arXiv:2110.00023. https://arxiv.org/abs/2110.00023
Stern D, Walton DJ (2020) Astrophys J Lett 895:L38. https://doi.org/10.3847/2041-8213/ab922c
Stern D, Assef RJ, Benford DJ et al (2012a) Astrophys J 753:30. https://doi.org/10.1088/0004-637X/753/1/30

Stern D, Djorgovski SG, Krone-Martins A et al (2021) Astrophys J 921:42. https://doi.org/10.3847/1538-4357/ac0f04
Storfer C, Huang X, Gu A et al (2022) arXiv e-prints, arXiv:2206.02764. https://arxiv.org/abs/2206.02764
Surdej J, Magain P, Swings JP et al (1987) Nature 329:695. https://doi.org/10.1038/329695a0
Taak YC, Treu T (2023) https://doi.org/10.48550/arXiv.2304.02784. arXiv e-prints, arXiv:2304.02784
Talbot MS, Brownstein JR, Bolton AS et al (2018) Mon Not R Astron Soc 477:195. https://doi.org/10.1093/mnras/sty653
Talbot MS, Brownstein JR, Dawson KS, Kneib J-P, Bautista J (2021) Mon Not R Astron Soc 502:4617. https://doi.org/10.1093/mnras/stab267
Tanaka M, Wong KC, More A et al (2016) Astrophys J Lett 826:L19. https://doi.org/10.3847/2041-8205/826/2/L19
Teimoorinia H, Toyonaga RD, Fabbro S, Bottrell C (2020) Publ Astron Soc Pac 132:044501. https://doi.org/10.1088/1538-3873/ab747b
Thuruthipilly H, Zadrozny A, Pollo A, Biesiada M (2022) Astron Astrophys 664:A4. https://doi.org/10.1051/0004-6361/202142463
Tran K-VH, Harshan A, Glazebrook K et al (2022) Astron J 164:148. https://doi.org/10.3847/1538-3881/ac7da2
Treu T, Koopmans LVE (2004) Astrophys J 611:739. https://doi.org/10.1086/422245
Treu T, Dutton AA, Auger MW et al (2011) Mon Not R Astron Soc 417:1601. https://doi.org/10.1111/j.1365-2966.2011.19378.x
Tubín-Arenas D, Lamer G, Krumpe M et al (2023) Astron Astrophys 672:L9. https://doi.org/10.1051/0004-6361/202346316
Türler M, Paltani S, Courvoisier TJ-L et al (1999) A&AS 134:89. https://doi.org/10.1051/aas:1999125
Turner EL, Ostriker JP, Gott JR, III (1984) Astrophys J 284:1. https://doi.org/10.1086/162379
van Albada TS, Sancisi R (1986) Philos Trans R Soc Lond Ser A 320:447. https://doi.org/10.1098/rsta.1986.0128
Veres P, Bhat N, Fraija N, Lesage S (2021) Astrophys J Lett 921:L30. https://doi.org/10.3847/2041-8213/ac2ee6
Vieira JD, Marrone DP, Chapman SC et al (2013) Nature 495:344. https://doi.org/10.1038/nature12001
Walsh D, Carswell RF, Weymann RJ (1979) Nature 279:381. https://doi.org/10.1038/279381a0
Wang Y, Jiang L-Y, Li C-K et al (2021) Astrophys J Lett 918:L34. https://doi.org/10.3847/2041-8213/ac1ff9
Wardlow JL, Cooray A, De Bernardis F et al (2013) Astrophys J 762:59. https://doi.org/10.1088/0004-637X/762/1/59
Weiss A, De Breuck C, Marrone DP et al (2013) Astrophys J 767:88. https://doi.org/10.1088/0004-637X/767/1/88
Weymann RJ, Latham D, Angel JRP et al (1980) Nature 285:641. https://doi.org/10.1038/285641a0
Wilde J, Serjeant S, Bromley JM et al (2022) Mon Not R Astron Soc 512:3464. https://doi.org/10.1093/mnras/stac562
Williams PR, Agnello A, Treu T et al (2018) Mon Not R Astron Soc 477:L70. https://doi.org/10.1093/mnrasl/sly043
Winn JN, Kochanek CS, McLeod BA et al (2002) Astrophys J 575:103. https://doi.org/10.1086/341265
Wisotzki L, Koehler T, Lopez S, Reimers D (1996) Astron Astrophys 315:405
Witt HJ (1996) Astrophys J Lett 472:L1. https://doi.org/10.1086/310358
Wojtak R, Hjorth J, Gall C (2019) Mon Not R Astron Soc 487:3342. https://doi.org/10.1093/mnras/stz1516
Wong KC, Chan JHH, Chao DCY et al (2022) Publ Astron Soc Jpn 74:1209. https://doi.org/10.1093/pasj/psac065
Wright EL, Eisenhardt PRM, Mainzer AK et al (2010) Astron J 140:1868. https://doi.org/10.1088/0004-6256/140/6/1868
Wynne RA, Schechter PL (2018) arXiv e-prints, arXiv:1808.06151. https://doi.org/10.48550/arXiv.1808.06151
Xu H, Chen S, Guo Y et al (2023) Res Astron Astrophys 23:075024. https://doi.org/10.1088/1674-4527/acdfa5
Yang X, Lü H-J, Yuan H-Y et al (2021) Astrophys J Lett 921:L29. https://doi.org/10.3847/2041-8213/ac2f39
Yantovski-Barth MJ, Newman JA, Dey B et al (2023) arXiv e-prints, arXiv:2307.10426. https://doi.org/10.48550/arXiv.2307.10426
Yee HKC (1988) Astron J 95:1331. https://doi.org/10.1086/114729
Young P, Gunn JE, Kristian J, Oke JB, Westphal JA (1980) Astrophys J 241:507. https://doi.org/10.1086/158365
Yue M, Fan X, Yang J, Wang F (2023) Astron J 165:191. https://doi.org/10.3847/1538-3881/acc2be. arXiv:2303.04357
Zhou H, Li Z, Liao K et al (2022) Astrophys J 928:124. https://doi.org/10.3847/1538-4357/ac510d
Zwicky F (1937) Phys Rev 51:679. https://doi.org/10.1103/PhysRev.51.679

Publisher's Note Springer Nature remains neutral with regard to jurisdictional claims in published maps and institutional affiliations.

Authors and Affiliations

Cameron Lemon[1] · Frédéric Courbin[1] · Anupreeta More[2,3] · Paul Schechter[4] · Raoul Cañameras[5,6] · Ludovic Delchambre[7] · Calvin Leung[4] · Yiping Shu[5,8] · Chiara Spiniello[9,10] · Yashar Hezaveh[11,12] · Jonas Klüter[13] · Richard McMahon[14,15]

✉ C. Lemon
cameron.lemon@fysik.su.se

1 Institute of Physics, Laboratory of Astrophysics, Ecole Polytechnique Fédérale de Lausanne (EPFL), Observatoire de Sauverny, 1290 Versoix, Switzerland

2 The Inter-University Centre for Astronomy and Astrophysics (IUCAA), Post Bag 4, Ganeshkhind, Pune 411007, India

3 Kavli Institute for the Physics and Mathematics of the Universe (IPMU), 5-1-5 Kashiwanoha, Kashiwa-shi, Chiba 277-8583, Japan

4 MIT Kavli Institute 37-635, 77 Massachusetts Avenue, Cambridge, MA, 02138-4307, USA

5 Max-Planck-Institut fur Astrophysik, Karl-Schwarzschild-Str. 1, 85748 Garching, Germany

6 TUM School of Natural Sciences, Department of Physics, Technical University of Munich, James-Franck-Straße 1, 85748, Garching, Germany

7 Institut d'Astrophysique et de Géophysique, Université de Liège, 19c, Allée du 6 Août, 4000, Liège, Belgium

8 Purple Mountain Observatory, Chinese Academy of Sciences, 3R8H+9RC, Tianwentai Rd, Xuanwu, Nanjing, 210023, P.R. China

9 INAF – Osservatorio Astronomico di Capodimonte, Via Moiariello 16, I-80131 Naples, Italy

10 Sub-Department of Astrophysics, Department of Physics, University of Oxford, Denys Wilkinson Building, Keble Road, Oxford OX1 3RH, UK

11 Département de Physique, Université de Montréal, Montreal, QC H3T 1J4, Canada

12 Center for Computational Astrophysics, Flatiron Institute, 162 Fifth Avenue, New York, NY 10010, USA

13 Department of Physics and Astronomy, Louisiana State University, 202 Nicholson Hall, Baton Rouge, LA 70803, USA

14 Institute of Astronomy, University of Cambridge, Madingley Road, Cambridge CB3 0HA, UK

15 Kavli Institute for Cosmology, University of Cambridge, Madingley Road, Cambridge CB3 0HA, UK

Space Science Reviews (2024) 220:48
https://doi.org/10.1007/s11214-024-01079-w

Time-Delay Cosmography: Measuring the Hubble Constant and Other Cosmological Parameters with Strong Gravitational Lensing

S. Birrer[1,2] · M. Millon[2,3] · D. Sluse[4] · A.J. Shajib[5,6] · F. Courbin[3,7,8] · S. Erickson[2] · L.V.E. Koopmans[9] · S.H. Suyu[10,11,12] · T. Treu[13]

Received: 3 April 2023 / Accepted: 8 May 2024 / Published online: 17 June 2024
© The Author(s) 2024

Abstract
Multiply lensed images of a same source experience a relative time delay in the arrival of photons due to the path length difference and the different gravitational potentials the photons travel through. This effect can be used to measure absolute distances and the Hubble constant (H_0) and is known as time-delay cosmography. The method is independent of the local distance ladder and early-universe physics and provides a precise and competitive measurement of H_0. With upcoming observatories, time-delay cosmography can provide a 1% precision measurement of H_0 and can decisively shed light on the current reported 'Hubble tension'. This manuscript details the general methodology developed over the past decades in time-delay cosmography, discusses recent advances and results, and, foremost, provides a foundation and outlook for the next decade in providing accurate and ever more precise measurements with increased sample size and improved observational techniques.

Keywords Gravitational lensing · Cosmology · Hubble constant · Dark energy

1 Introduction

The relative arrival times of multiply lensed sources can be used to measure an absolute distance of the Universe. The method, known to date as time-delay cosmography, was originally proposed over half a century ago, prior to the discovery of the first extragalactic gravitational lens, by Refsdal (1964). Time-delay cosmography provides a one-step measurement of the Hubble constant (H_0), independent of the local distance ladder or probes anchored with sound horizon physics, such as the cosmic microwave background (CMB).

Almost a century after its first measurement, the Hubble constant H_0 still remains arguably the most debated number in cosmology. In the past few years, a tension has emerged between a number of local measurements, and inferences from early-Universe probes such as the cosmic microwave background (CMB) and Big Bang Nucleosynthesis, under the assumption of flat Λ cold dark matter (ΛCDM) cosmology (see, e.g., Verde et al. 2019; Shah et al. 2021; Abdalla et al. 2022, for recent summaries). If this tension is real, and not due to unknown systematic uncertainties in multiple measurements and their analyses, it implies

A.J. Shajib is NHFP Einstein Fellow

Extended author information available on the last page of the article

that the standard ΛCDM model is not sufficient and new physical ingredients beyond this model are required. From a theoretical standpoint, a number of possible solutions – for example, involving early dark energy – have been proposed (e.g., Knox and Millea 2020; Di Valentino et al. 2021; Schöneberg et al. 2022, and references therein), often requiring fine-tuning of free parameters not to violate other observational constraints. From an observational standpoint, besides improving the precision of the measurements, significant attention has turned to the systematic investigation of unknown systematic uncertainties (e.g., Riess et al. 2019; Freedman et al. 2019, 2020; Riess et al. 2021; Mörtsell et al. 2022; Dainotti et al. 2021; Riess et al. 2023).

This article details the general methodology developed over the past decades in time-delay cosmography, discusses recent advances and results, and, foremost, provides a foundation and outlook for the next decade in providing accurate and ever more precise measurements with increased sample size and improved observational techniques. We will refer throughout this text to other articles of the same collection covering a wide range of strong lensing theory and applications (e.g., Saha et al. 2024; Shajib et al. 2024; Suyu et al. 2024; Vernardos et al. 2024; Lemon et al. 2024). We refer to, e.g., Treu and Marshall (2016), Suyu et al. (2018) to provide more in-depth historical perspectives on the early years of the field, Treu et al. (2022), Treu and Shajib (2023) to a broader and less technical perspective on the opportunities of time-delay cosmography in this decade, and Moresco et al. (2022) for a compact overview embedded within other cosmological probes.

We discuss primarily the methodology around lensed quasars as source objects to perform time-delay measurements because quasars are currently the most established sources with the most relevant current results. The use of lensed supernovae for cosmological and astrophysical studies is reviewed in detail by Suyu et al. (2024) in the same series; interested readers can also find descriptions of lensed supernova cosmographic results until 2023 in Suyu et al. (2024). For additional discussions on lensed supernovae and particularly on other types of lensed transients, such as gamma ray bursts and fast radio bursts, we refer to Oguri (2019) and Liao et al. (2022).

This manuscript is organized as follows: In Sect. 2 we provide the general concept and physics to turn relative time delays between multiple images of the same source into distance measurements. Section 3 provides an overview of the required analysis ingredients for individual lenses. Subsequent sections go into the details of these ingredients, the time-delay measurement (Sect. 4), determination of the lensing potential (Sect. 5), and the study of the line of sight (LOS) of the lenses (Sect. 6). Section 7 describes the cosmographic inferences and how to utilize a sample of lenses to perform an H_0 inference. In Sect. 8 we present an overview of the cosmographic method applied for galaxy clusters as the deflectors. We summarize the current status and results obtained in Sect. 9. Lastly, in Sect. 10, we look in the future and discuss the potential and challenges lying ahead for the community.

2 Time Delays and the Time-Delay Distance

2.1 Lensing Formalism and the Fermat Potential

The deflection of light due to mass over- or under-density in the Universe lead to an angular displacement between the angle of the arriving photons, where we see the image, and the angle to the originating source ignoring lensing effects.

The lens equation formally describes the lensing distortions

$$\boldsymbol{\beta} = \boldsymbol{\theta} - \boldsymbol{\alpha}(\boldsymbol{\theta}), \tag{1}$$

where $\boldsymbol{\beta}$ is the angular position of the source without the lensing effect, $\boldsymbol{\theta}$ is the corresponding angular coordinate on the sky as seen when lensed, and $\boldsymbol{\alpha}$ is the deflection angle that maps the image position to the source position in angular coordinates as seen from the observer. We refer to e.g., Saha et al. (2024) for further details into the theory, including that this formula is only valid for small angles.

There exists a scalar potential, the lensing potential ϕ, such that the gradients correspond to the deflection field

$$\nabla\phi(\boldsymbol{\theta}) = \boldsymbol{\alpha}(\boldsymbol{\theta}). \tag{2}$$

The convergence of the potential ϕ, κ, is half the Laplacian

$$\kappa(\boldsymbol{\theta}) = \frac{1}{2}\nabla^2\phi(\boldsymbol{\theta}) \tag{3}$$

and is given in the thin lens approximation for small angles as

$$\kappa(\boldsymbol{\theta}) = \frac{\Sigma(\boldsymbol{\theta})}{\Sigma_{\text{crit}}} \tag{4}$$

with $\Sigma(\boldsymbol{\theta})$ as the projected mass over- or under-density with respect to the mean background density and Σ_{crit} the critical surface density[1]

$$\Sigma_{\text{crit}} = \frac{c^2 D_{\text{s}}}{4\pi G D_{\text{ds}} D_{\text{d}}}, \tag{5}$$

where c is the speed of light and G the gravitational constant. D_{d} is the angular diameter distance to the lens, D_{s} is the angular diameter distance to the source, and D_{ds} is the angular diameter distance between the lens and the source, respectively.

When the luminosity of a strongly lensed background source varies over time, such as an active galactic nucleus (AGN), the variability pattern manifests in each of the multiple images and is delayed in time due to the different light paths of the different images (see Fig. 1). The arrival-time difference between two images $\boldsymbol{\theta}_{\text{A}}$ and $\boldsymbol{\theta}_{\text{B}}$ that originated from the same source $\boldsymbol{\beta}$, Δt_{AB}, is

$$\Delta t_{\text{AB}} = \frac{D_{\Delta t}}{c}\left[\tau(\boldsymbol{\theta}_{\text{A}}, \boldsymbol{\beta}) - \tau(\boldsymbol{\theta}_{\text{B}}, \boldsymbol{\beta})\right], \tag{6}$$

where

$$\tau(\boldsymbol{\theta}, \boldsymbol{\beta}) \equiv \left[\frac{(\boldsymbol{\theta} - \boldsymbol{\beta})^2}{2} - \phi(\boldsymbol{\theta})\right] \tag{7}$$

is the Fermat potential (Schneider 1985; Blandford and Narayan 1986), and

$$D_{\Delta t} \equiv (1 + z_{\text{d}})\frac{D_{\text{d}} D_{\text{s}}}{D_{\text{ds}}} \tag{8}$$

is the time-delay distance (Refsdal 1964; Schneider et al. 1992; Suyu et al. 2010). The Fermat potential (Eq. (7)) consists of two terms, a geometric term reflecting the geometric

[1] not to be confused with the critical density of the universe.

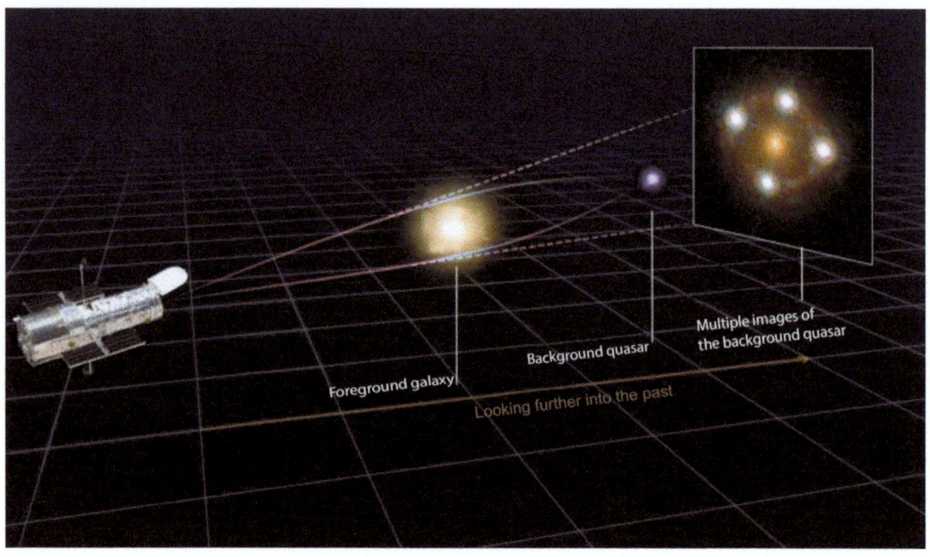

Fig. 1 Illustration of the light path of the quadruply imaged lensed quasar HE0435-1223. The different light paths result in different arrival times. The relative time delays between the images is directly proportional to the overall physical distances from the observer to the lens and source. Measuring the time delays and reconstructing the lensing effect allow one to measure an absolute scale in the universe. Graphics from: Martin Millon, Image from Hubble Space Telescope (Wong et al. 2017)

path difference, and a potential term, capturing the difference in the local spacetime dilation, known as the Shapiro delay. The optical terms stated (such as the Fermat potential) are only valid under small-angle and thin-lens assumptions. We refer to Saha et al. (2024, this collection) for these assumptions and the more general multi-plane formalism.

2.2 Angular Diameter Distances and Cosmology

The angular diameter distance between two redshifts z_1 and z_2 in an Friedmann–Lemaitre–Robertson–Walker (FLRW) metric is

$$D(z_1, z_2) = \frac{1}{1+z_2} f_K[\chi(z_1, z_2)] \tag{9}$$

where

$$\chi(z_1, z_2) = \frac{c}{H_0} \int_{z_1}^{z_2} \frac{dz'}{E(z')} \tag{10}$$

is the comoving distance with $E(z) \equiv H(z)/H_0$ as the dimensionless Friedman equation and

$$f_K(\chi) = \begin{cases} K^{-1/2} \sin\left(K^{1/2}\chi\right) & \text{for } K > 0 \\ \chi & \text{for } K = 0 \\ (-K)^{-1/2} \sinh\left[(-K)^{1/2}\chi\right] & \text{for } K < 0 \end{cases} \tag{11}$$

is the spatial curvature of the background metric.

In the ΛCDM cosmology with density parameters Ω_m for matter, Ω_k for spatial curvature, and Ω_Λ for dark energy described by the cosmological constant Λ, the dimensionless Friedman equation ($E(z)$, Eq. (10)) is given by

$$E(z) = \left(\Omega_\mathrm{m}(1+z)^3 + \Omega_k(1+z)^2 + \Omega_\Lambda\right)^{1/2} \tag{12}$$

and the spatial curvature is $K = -\Omega_k H_0^2/c^2$.

Constraints on the Fermat potential difference $\Delta\tau_\mathrm{AB}$ and a measured relative time delay Δt_AB between to images of the same source can be turned into constraints of the time-delay distance $D_{\Delta t}$. $D_{\Delta t}$ is an absolute quantity that has units of distance and anchors the scale of the Universe within the lensing configuration. The Hubble constant, the local expansion rate of the cosmological background metric, sets the locally linear relationship between relative recession velocities and physical separation of two objects. In the frame of an observer, such as on Earth, for a fixed relative velocity or redshift, the Hubble constant is inversely proportional to the absolute physical distance to the object. The Hubble constant is inversely proportional to the absolute scales of objects in the Universe for which redshifts are measured (see e.g., Eq. (10)) and thus scales with $D_{\Delta t}$ as

$$H_0 \propto D_{\Delta t}^{-1}. \tag{13}$$

The direct (inverse) proportionality of the time-delay cosmography measurable quantity $D_{\Delta t}$ and H_0 makes H_0 the primary cosmological parameter time-delay cosmography can constrain. [2] Time-delay cosmography provides primarily an *absolute* distance anchor and hence can provide valuable information to shed light on the current tension in cosmology..

There is a secondary mild dependence of the measured Hubble constant when inferred from time-delay cosmography, namely on the relative expansion history from current time ($z = 0$) to the redshift of the deflector and the source (dependence on $E(z)$ in Eq. (12)). The mild dependency on other cosmological parameters beyond H_0 can be compensated with other cosmological probes that are sensitive to the *relative* expansion history (such as SNIa luminosity distances, e.g., Taubenberger et al. 2019, Arendse et al. 2019, Liao et al. 2019, 2020), or with a large set of gravitational lenses at different lens and source redshifts (e.g., Li et al. 2024).

2.3 Observables and Degeneracies

The time delay between two images Δt_AB can be measured from light curves and is hence a direct observable (see Sect. 4). The relative Fermat potential $\Delta\tau_\mathrm{AB}$, however, is not a direct observable. The primary observations used to infer $\Delta\tau_\mathrm{AB}$ are positional constraints of multiple imaged sources and their extended distortions in the lensed arcs from the lensing effect. However, there are degeneracies inherent in gravitational lensing that limit the amount of information accessible by positional and distortion effects as observed in imaging data (e.g., Falco et al. 1985; Gorenstein et al. 1988; Kochanek 2002; Saha and Williams 2006; Schneider and Sluse 2013, 2014; Birrer et al. 2016; Unruh et al. 2017; Birrer 2021) and Saha et al. (2024).

The mass-sheet degeneracy (MSD; Falco et al. 1985) is the most prominent lensing degeneracy impacting the prediction of the Fermat potential and hence time-delay cosmography. The MSD stems formally from a multiplicative transform of the lens equation (Eq. (1))

[2] We note that the time-delay distance $D_{\Delta t}$ is not a measurement of H_0 at a specific redshift (z_lens or z_source).

which preserves image positions under a linear source displacement $\boldsymbol{\beta} \rightarrow \lambda\boldsymbol{\beta}$ combined with a transformation of the convergence field

$$\kappa_{\lambda}(\boldsymbol{\theta}) = \lambda\kappa(\boldsymbol{\theta}) + (1 - \lambda) . \quad (14)$$

Equation (14) above is known as the mass sheet transform (MST) and is a *mathematical transformation* where the term $(1 - \lambda) \equiv \kappa_{\rm MST}$ is equivalent to a uniform-surface density sheet of convergence (or mass) that extends to infinite angular scales, and hence the name mass sheet transform and the name of the experienced degeneracy, the mass sheet degeneracy. $\kappa_{\rm MST}$ can be positive or negative, since it is defined relative to the average positive density of the universe. The MST, by means of preserving image positions and being linear, also preserves any higher order *relative* differentials of the lens equation. Absolute lensing quantities, such as absolute magnification or size of structure, however, are not preserved by the MST. Only observables related to either the unlensed apparent source size ($\boldsymbol{\beta}$ vs. $\lambda\boldsymbol{\beta}$), such as the unlensed apparent brightness, or the lensing potential are able to break the MSD. For example, the same relative lensing observables can be predicted if the mass profile is scaled by the factor λ with the addition of a sheet of convergence (or mass) of $\kappa_{\rm MST}(\boldsymbol{\theta}) = (1 - \lambda)$ and re-sizing of the source by a factor λ.

The Fermat potential difference between a pair of images A& B (Eq. (7)) scales with λ as

$$\Delta\tau_{{\rm AB},\lambda} = \lambda\Delta\tau_{\rm AB}, \quad (15)$$

and so does the relative time delay as

$$\Delta t_{{\rm AB},\lambda} = \lambda\Delta t_{\rm AB}. \quad (16)$$

When transforming a lens model with an MST, the inference of the time-delay distance (Eq. (8)) from a measured time delay and previously inferred Fermat potential transforms as

$$D_{\Delta t,\lambda} = \lambda^{-1} D_{\Delta t}. \quad (17)$$

In turn, the Hubble constant, when inferred from the time-delay distance $D_{\Delta t}$, transforms as (from Eq. (13))

$$H_{0,\lambda} = \lambda H_0. \quad (18)$$

An MSD effect relative to a specified deflector model might be associated with the mass distribution of the main deflector, referred as *internal* MSD with $\lambda_{\rm int}$, or with inhomogenities along the line of sight (LOS) of the strong lens system, referred as *external* MSD.

Mass over- or under-densities relative to the mean background density along the LOS of the strong lensing system cause, to first order, shear and convergence lensing perturbations (e.g., Dalal et al. 2005; Hilbert et al. 2007; Puchwein and Hilbert 2009). Reduced shear distortions do have a measurable imprint on the azimuthal structure of the strong lensing arcs (see e.g., Birrer 2021; Hogg et al. 2023). In contrast, the convergence component of the LOS, denoted as $\kappa_{\rm ext}$, describes the focusing or de-focusing of the light rays and is equivalent to an MST, $\kappa_{\rm ext} \equiv (1 - \lambda)$, and hence not directly measurable from imaging data.

Equivalent to describing the (de-) focusing along specific line-of-sights by convergence terms, we can alter the specific angular diameter distance relative to the homogeneous background metric. In our notation, $D^{\rm lens}$ is the angular diameter distance along a specific line of sight, including all structure such as LOS. D^{bkg} is the angular diameter distance just from the homogeneous background metric without any perturbative contribution. The relation between D^{lens} and D^{bkg} are given by the convergence terms as

$$\begin{aligned} D_{\mathrm{d}}^{\mathrm{lens}} &= (1-\kappa_{\mathrm{d}})D_{\mathrm{d}}^{\mathrm{bkg}} \\ D_{\mathrm{s}}^{\mathrm{lens}} &= (1-\kappa_{\mathrm{s}})D_{\mathrm{s}}^{\mathrm{bkg}} \\ D_{\mathrm{ds}}^{\mathrm{lens}} &= (1-\kappa_{\mathrm{ds}})D_{\mathrm{ds}}^{\mathrm{bkg}}, \end{aligned} \tag{19}$$

where κ_{d} is the external convergence from the observer to the deflector, κ_{s} from the observer to the source, and κ_{ds} from the deflector to the source, respectively (Birrer et al. 2020). The individual convergence terms can be calculated in the Born approximation along undeflected light paths independent of the strong lensing deflector.

The notation of perturbed angular diameter distances allow us also to directly calculate the impact of line-of-sight structure on the cosmographic inference, and in particular the measurement of the Hubble constant. The time delay can be described as the product of three different angular diameter distances entering $D_{\Delta t}$ in Eq. (8) (Birrer et al. 2020; Fleury et al. 2021a), and hence the effective external convergence κ_{ext}impacting the time delay and time-delay distance is

$$1-\kappa_{\mathrm{ext}} = \frac{(1-\kappa_{\mathrm{d}})(1-\kappa_{\mathrm{s}})}{1-\kappa_{\mathrm{ds}}}. \tag{20}$$

We note that, also directly visible from the equation above, the lensing efficiency (see Saha et al. 2024, this collection) impacting the linear distortions for both shear and κ_{ext} is different from the standard weak lensing efficiency in the absence of a strong lensing deflector (McCully et al. 2014, 2017; Birrer et al. 2017, 2020; Fleury et al. 2021b).

Uncertainties or biases related to the MSD may also arise in regards to assumptions made in the radial density profile of the main deflector galaxy (see e.g. Kochanek 2002; Saha and Williams 2006; Read et al. 2007; Schneider and Sluse 2013; Coles et al. 2014; Xu et al. 2016; Birrer et al. 2016; Unruh et al. 2017; Sonnenfeld 2018; Kochanek 2020; Blum et al. 2020; Birrer et al. 2020; Kochanek 2021).

We refer to the MSD attributed to uncertainties in the radial density profile as the *internal* MSD, and describe its effect with the MST parameter λ_{int} relative to an assumed mass profile. We will further discuss this aspect in Sect. 5.3.

The total MST, the relevant transform to constrain for an accurate Fermat potential determination and H_0 measurement, is the product of the internal and external MST (e.g., Schneider and Sluse 2013; Birrer et al. 2016, 2020)

$$\lambda = (1-\kappa_{\mathrm{ext}}) \times \lambda_{\mathrm{int}}. \tag{21}$$

To summarize, the prediction of the time delay (Eq. (6) can be generalized to

$$\Delta t_{\mathrm{AB}} = (1-\kappa_{\mathrm{ext}})\lambda_{\mathrm{int}} \frac{D_{\Delta t}}{c} \Delta\tau_{\mathrm{AB}}. \tag{22}$$

The existence of the MST and its generalizations imply that one has to rely either on (1) non-lensing information that can specifically constrain the shape of the radial mass density profile or (2) assumptions and priors about the functional form of the radial mass density profile that limit the degrees of freedom in the direction of the MST to constrain the lens model with a sufficient level of precision for time-delay cosmography.

The line-of-sight lensing contribution, $\kappa_{\rm ext}$, can be estimated by tracers of the large-scale structure using galaxy number counts (e.g., Greene et al. 2013; Rusu et al. 2017) or weak-lensing measurements (Tihhonova et al. 2018, 2020). Galaxy number counts, paired with a cosmological model including a galaxy–halo connection, are able to constrain the probability distribution of $\kappa_{\rm ext}$ to a few per cent per sight line. The main uncertainty in this approach arise from the uncertainties of luminous matter tracing the more dominant dark matter structure.

To break the total MSD λ with observations, we require observations that are sensitive to the total MSD. Stellar kinematics is the most prominent and commonly used one to break the total MSD. The collective motion of stars is a direct tracer of the three-dimensional gravitational potential and hence provides an independent mass estimate. Joint lensing + dynamics constraints have been used to provide measurements of galaxy mass profiles (e.g., Grogin and Narayan 1996; Romanowsky and Kochanek 1999; Treu and Koopmans 2002; Koopmans 2004; Barnabè et al. 2011, 2012). The modelling of the kinematic observables in lensing galaxies range in complexity from spherical Jeans modeling (Binney and Tremaine 2008) to Schwarzschild (Schwarzschild 1979) methods.

The prediction of the LOS velocity dispersion $\sigma_{\rm v}$ from any model, regardless of the approach, can be decomposed into a cosmology-dependent and cosmology-independent part, as (see e.g., Birrer et al. 2016, 2019)

$$\sigma_{\rm v}^2 = \frac{1-\kappa_{\rm s}}{1-\kappa_{\rm ds}} \frac{D_{\rm s}}{D_{\rm ds}} c^2 J(\boldsymbol{\xi}_{\rm lens}, \boldsymbol{\beta}_{\rm ani}, \lambda_{\rm int}), \tag{23}$$

where J is a dimensionless quantity dependent on the deflector model parameters ($\boldsymbol{\xi}_{\rm lens}$), c is the speed of light, and $\boldsymbol{\beta}_{\rm ani}$ the stellar anisotropy distribution. The dimensionless factor J incorporates also the observational conditions and luminosity-weighting within the aperture of the dispersion measurement being taken (e.g., Binney and Mamon 1982; Treu and Koopmans 2004; Suyu et al. 2010). The internal component $\lambda_{\rm int}$ should be physically interpretable as a three-dimensional mass profile and incorporated into the kinematics modeling term J, in particular when there are multiple aperture measurements available (Teodori et al. 2022). In the approximate case of a very extended sheet-like perturbation, we can approximate

$$J(\boldsymbol{\xi}_{\rm lens}, \boldsymbol{\beta}_{\rm ani}, \lambda_{\rm int}) \approx \lambda_{\rm int} J(\boldsymbol{\xi}_{\rm lens}, \boldsymbol{\beta}_{\rm ani}). \tag{24}$$

Combined lensing + dynamics constraints are sensitive to the combination of terms present in Eq. (23). Only the combination of terms, i.e., $\lambda_{\rm int}$), $D_{\rm s}$, $D_{\rm ds}$, $\boldsymbol{\beta}_{\rm ani}$, $\kappa_{\rm s}$, $\kappa_{\rm ds}$, is constrained and hence assumptions or priors on parts of the terms are required to provide a precise statement other terms. For example, when assuming the relative expansion history through the involved angular diameter distance ratio $D_{\rm s}/D_{\rm ds}$, and the LOS contributions $\kappa_{\rm s}$ and $\kappa_{\rm ds}$, an inference on $\lambda_{\rm int}$ is possible. On the other hand, when assuming $\lambda_{\rm int}$ and the convergence terms, an inference on the angular diameter distance ratio, and hence the relative expansion history, is possible.

When combining time delays with lensing + dynamics, the observations of the time delay and kinematics need to be simultaneously be described by Eq. (22) and Eq. (23) in addition to the imaging data. These two independent equations can be arbitrarily algebraically combined in two-dimensional angular diameter distance constraints (Birrer et al. 2016, 2019). A convenient transform of those constraints is in the basis of

$$D_{\Delta t} = \frac{1}{(1-\kappa_{\rm ext})\lambda_{\rm int}} \frac{c\Delta t_{\rm AB}}{\Delta\tau_{\rm AB}} \tag{25}$$

and

$$D_{\mathrm{d}} = \frac{1}{1 - z_{\mathrm{d}}} \frac{1}{1 - \kappa_{\mathrm{d}}} \frac{c\Delta t_{\mathrm{AB}}}{\Delta \tau_{\mathrm{AB}}} \frac{c^2 J(\boldsymbol{\xi}_{\mathrm{lens}}, \boldsymbol{\beta}_{\mathrm{ani}}, \lambda_{\mathrm{int}})}{\lambda_{\mathrm{int}} \sigma_{\mathrm{v}}^2}. \tag{26}$$

When mapped into the $D_{\Delta t}$–D_{d} plane as outlined above, the projection on constraints in D_{d} is invariant under any pure external MSD parameter κ_{ext} (Paraficz and Hjorth 2009; Jee et al. 2015; Birrer et al. 2019; Yıldırım et al. 2023).[3] If the approximation of Eq. (24) holds, D_{d} becomes even independent of λ_{int}, and is overall less susceptible to the internal MSD.

3 Overview of Analysis Ingredients

To measure the cosmographic distances, in particular the time-delay distance $D_{\Delta t}$ (Eq. (8)), or the more general $D_{\Delta t}$–D_{d} (Eqs (25)-(26)) combination, from a strong lensing system with a time-variable source, the following data products are required:

1. discovery of a lens with a time-variable source that is multiply imaged,
2. spectroscopic redshifts of the source, z_{s}, and lens, z_{d},
3. measured time delays between at least one multiple image pair,
4. lens mass model to determine the Fermat potential between the multiple images from sufficiently high resolution imaging to resolve the positions of the multiply-imaged quasars,
5. lens environment studies to constrain external lensing effects.

A complete analysis for an individual lensing system requires the coordination of multiple independent observations. The analysis can be severely limited in its precision and reliability due to a single missing ingredient. For example, without measurements of a time delay, no constraints on absolute distances involved can be achieved, and thus, regardless of the approach or external priors chosen, no direct constraints on the Hubble constant can be made.

The spectroscopic redshifts of the quasar sources, z_{s}, are often obtained using the frequent emission lines in quasars. The redshift of the lens, z_{d}, can be challenging to measure since massive elliptical galaxies lack prominent and sharp absorption or emission lines and the bright quasar images can outshine the lens galaxy. Measuring z_{d} of a lensed quasar systems often require high signal-to-noise ratio spectra taken under good seeing conditions, to deblend the lensing galaxy from the quasar. Technically, the redshifts involved in the lensing system are not directly required for the distance measurement. However, for the cosmological interpretation of the obtained distances, the redshifts are of crucial importance.

We describe in the next sections the remaining three ingredients; time delays (Sect. 4), lensing potential (galaxy scale and cluster) (Sect. 5), and line-of-sight perturbations (Sect. 6).

4 Measuring Time Delays

4.1 Monitoring of Lensed Quasars

Lensed quasars are variable on short timescale, making the time-delay measurements possible, and sufficiently bright to be observed at cosmological distances. They were hence

[3] D_{d} is still dependent on the LOS between observer and lens, κ_{d} (Eqn. (19)).

quickly identified as excellent sources for time-delay cosmography. Lensed quasars are also currently much more common than lensed supernovae as around 300 lensed quasars have been discovered at the time of writing compared to only four lensed supernovae (see Suyu et al. 2024, this collection). Lensed quasars are typically found in the redshift range $z_s \sim$ 1-3, with massive early-type galaxies acting as the lenses located around redshift $z_d \sim 0.2 - 0.8$ (e.g., Lemon et al. 2023, and Lemon et al. 2024). This lensing configuration typically produces multiple images separated by a few arcseconds, which is sufficient to be resolved with small ground-based telescope, Typical time delays in this configuration are of the order of days to a year. The monitoring of lensed quasars and the measurement of their individual brightness fluctuations is thus challenging but possible with 1-m or 2-m class telescopes, provided that a regular and long-term access is guaranteed (see e.g. the COSMOGRAIL collaboration Eigenbrod et al. 2005; Courbin et al. 2011; Tewes et al. 2013a). The relative error on the time delays, which is directly propagated to H_0, depends on the absolute errors divided by the time-delay itself. Therefore, long-delay lenses are more valuable for cosmography as they yield smaller H_0 uncertainties from this component. The achievable precision on time-delay measurements is limited by several astrophysical, observational and instrumental factors, that are listed below.

Photometric Accuracy: In the optical, most quasars are variable on a timescale of weeks to years, and the longest variations also have the largest amplitude. This means that either long-duration light curves or high photometric accuracy are required to measure the delay reliably. In one visibility season, variations of the order of 0.2 mag are often observed, which requires a photometric accuracy of a few milli-magnitudes to identify precisely the inflection points. These inflection points are essential features in the light curves since it is not possible to measure a time delay if the quasars does not display any variations, or if the first derivative remains always constant. Reaching a photometric accuracy of only a few milli-magnitude is challenging as the quasar images are often blended with extended sources such as gravitational arcs or the lens galaxy. Consequently, the reconstruction of the Point Spread Function (PSF) and proper treatment of the contaminating light from these extended sources are usually the key to reduce the noise in the light curves.

Monitoring Cadence and Duration of the Monitoring: A fast and precise temporal sampling of the light curve is necessary if one targets the fast variations of small amplitudes of the quasar. The monitoring cadence then needs to be commensurate with the timescale of the targeted variations. The total time span of the monitoring campaign also needs to be sufficient to cover the lensing time delays and to ensure that enough variations of the quasar are recorded for multiple images at relative delayed times. To obtain light curves to such specifications requires continuous access to the telescope for at least one visibility season, which is typically 6 to 8 months.

Windowing Effects and Correlated Noise: Seasonal gaps are often unavoidable in optical light curves since only circumpolar targets are observable all year-long. The fact that data are missing every year can introduce some windowing effects, which should be accounted for when using cross-correlation techniques to measure time delays. The missing data introduces a periodic signal that must be carefully removed or taken into account before attempting to measure the time delays. Additionally, great care should be taken in the presence of correlated noise, which is often present in the light curves due to uncertainties in the assignment of flux coming from different quasar images. If no evident variations can be matched unambiguously in both light curves, it is unlikely that any statistical methods will robustly measure a time delay.

Extrinsic Variability: Extrinsic variations are often observed in the light curves. They are caused mainly by the microlensing of the quasar images, and also a variety of other astrophysical effects (see e.g. Schechter et al. 2003; Blackburne and Kochanek 2010; Dexter and Agol 2011; Sluse and Tewes 2014; Vernardos et al. 2024). Microlensing is caused by the stars in the lensing galaxy, which add some extra time-variable "micro-magnification" on top of the static "macro-magnification" produced by the lensing galaxy. As described in Vernardos et al. (2024), the modulation of the micro-magnification due to the relative motion between the quasar, the lens and the observer introduces some extrinsic variations on top of the quasar intrinsic variations. For this reason, the light curves, even shifted in time and magnitude, rarely match perfectly. These extrinsic variations can severely bias time-delay measurements if not appropriately modelled and marginalized over.

In the past two decades, these difficulties have been progressively dealt with. The advances in photometric instrumentation in the late 1990s allowed us to acquire accurate and well-sampled light curves, which yielded the first robust time-delay measurements from optical monitoring (e.g., Kundić et al. 1997a; Schechter et al. 1997; Burud et al. 2000, 2002; Hjorth et al. 2002; Colley et al. 2003; Kochanek et al. 2006) and in radio monitoring (Fassnacht et al. 1999, 2002; Biggs et al. 1999; Koopmans et al. 2000). Although some of these measurements already reached an excellent precision of a few percents, the majority had $\sim$10-15% errors, hence limiting the measurement of H_0 to the same precision. These first encouraging results led to a systematic attempt to monitor a sample of lensed quasars in both hemispheres by the COSMOGRAIL program, which started in 2003 (Courbin et al. 2005). The observing strategy then was to follow a dozen of lenses at bi-weekly cadence until the time delays can be measured to a few percent precision. This strategy yielded precise measurements for the brightest and most variable objects in about five years (Vuissoz et al. 2007, 2008; Courbin et al. 2011; Tewes et al. 2013b; Eulaers et al. 2013; Rathna Kumar et al. 2013) but required more than a decade of monitoring to obtain the time delays for most of the less variable and fainter targets (Millon et al. 2020a). An example of a light curves acquired by the COSMOGRAIL program over the past 15 years is shown in Fig. 2. Thanks to this long-term observing effort and other monitoring campaigns (e.g. Poindexter et al. 2007; Goicoechea and Shalyapin 2016; Giannini et al. 2017; Shalyapin and Goicoechea 2019), about 40 lensed quasars have now known time delays, although with variable precision, but the sample starts to be sufficiently large to vastly reduce the random uncertainties and to enable a statistical study of the time-delay lenses. In addition, three cluster-scale lensed quasars have measured time delays (Fohlmeister et al. 2007, 2008, 2013; Dahle et al. 2015; Muñoz et al. 2022) but their modelling is much more complex than galaxy-scale lenses (see Sect. 8.2 for detail).

As time-delay cosmography is now entering a new regime with an increasing number of lensed quasars being discovered every year, the time delays now need to be measured rapidly to turn these newly discovered systems into cosmological constraints. Courbin et al. (2018) demonstrated that it is possible to obtain accurate time-delay measurements in only one monitoring season thanks to the small amplitude variations of the quasars, of the order of 10 to 50 millimag, that happen on a timescale of weeks or months. These variations are faster than the microlensing variability, which varies on a typical timescale of several months or years. If detected at a sufficient signal-to-noise ratio (SNR), these features reduce the need for long light curves as it allows us to disentangle the intrinsic and microlensing variability more easily. However, this change of strategy requires almost a daily cadence to obtain a sufficient sampling of these small features in the light curves. Their amplitude is of the order of 10 mmag, which requires 2-m class telescopes to obtain a sufficient SNR in 30 minutes of exposure at magnitude as faint as $\sim$ 20. This is illustrated in Fig. 3 in the case of

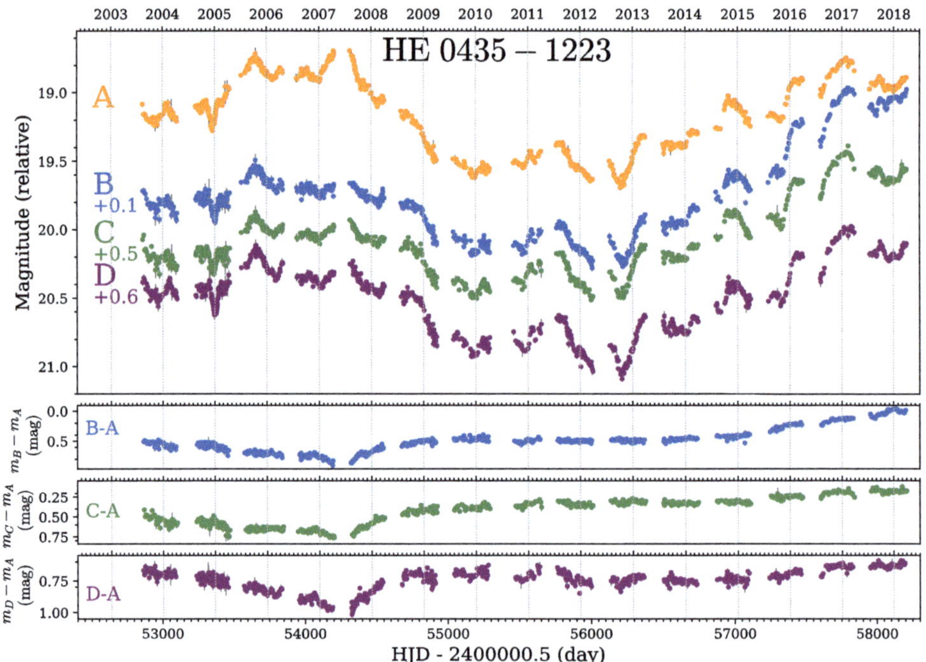

Fig. 2 R-band light curves of the lensed quasar HE0435-1223, obtained by the COSMOGRAIL program from the Euler 1.2 m Swiss Telescope, the 1.5 m telescope at the Maidanak Observatory, the Mercator 1.2 m telescope, and the SMARTS 1.3 m telescope. The bottom panels corresponds to the difference between pairs of light curves, corrected by the time delays, highlighting the microlensing variability. Figure reproduced from Millon et al. (2020a)

the bright quadruple quasar WFI 2033-4723 (Bonvin et al. 2019). The technique enabled to measure six new time delays in one single season (Millon et al. 2020c), with more to follow.

In the future, the Vera Rubin Observatory will obtain high-SNR multiband data for all southern lensed quasars, opening the possibility of building a sample of a few hundreds lensed quasars with known time delays. The cadence will however be limited to one point every few days in each band, which might not be sufficient to obtain the time delays at a few percent precision for the most interesting targets. Complementary observations from a 2-m class telescope at a daily cadence might still be useful to refine the time-delay measurements of the most promising objects.

4.2 Time-Delay Measurements Techniques

Once well-sampled light-curves have been acquired, the next step consists of identifying time-variable features that can be matched in all light curves from the individual images. The shifting of the light curve to match the features leads to a measurement of the time delays.

This step is significantly complicated by the presence of extrinsic variations due to microlensing in the light curves on top of the quasar intrinsic variations. The signature of micro-lensing can be seen in most lensed quasar light curves. In some cases, it manifests itself by a sharp rise of the luminosity in one of the multiple images, which happens when the source approaches or crosses a micro-caustic. This probably happened, for example, in 2007

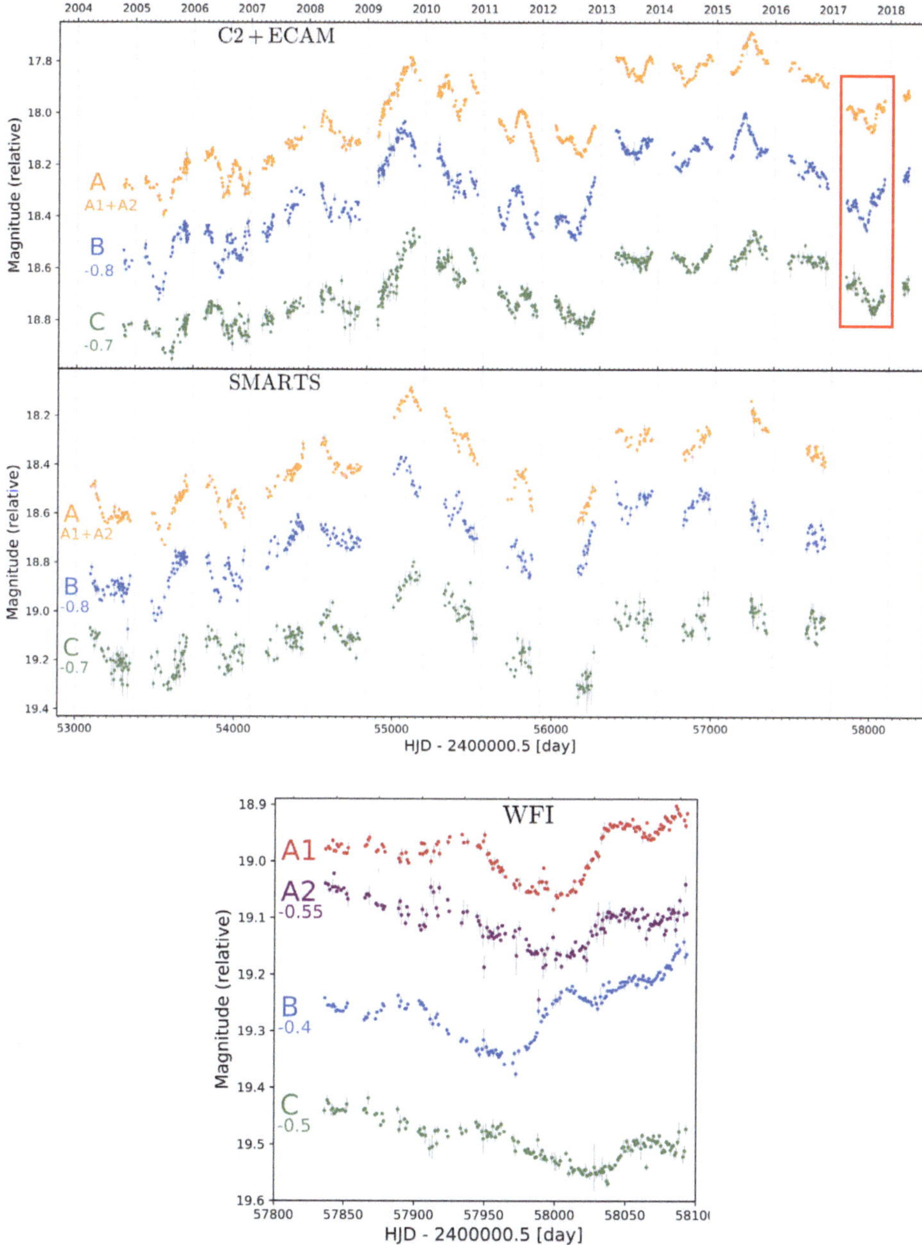

Fig. 3 Comparison of two observing cadences. On the two top panels are shown 14-year-long light curves of WFI 2033-4723, observed with a 4-day cadence at the 1.2 m Euler telescope at La Silla and at the SMARTS telescope at Las Campanas. During the season indicated with a red rectangle, the object has also been observed daily with the MPIA 2.2 m at La Silla (bottom panel), unveiling exquisite small-scale structures that vary faster than microlensing. Such observations allow to measure time delays in 1 single season, with similar accuracy and precision than the lower cadence data over 14 years (Bonvin et al. 2019; Millon et al. 2020a)

for image A of HE0435–1223. In Fig. 2, we show the light curves of HE0435–1223 as well as the differential curves between the images, highlighting the microlensing signal. Caustic crossing events are not the only signature of microlensing visible in the data. Most quasar light curves also exhibit slow variations of the microlensing over several years. An example of this phenomenon is image B of HE0435–1223, which slowly raised by ~ 0.5 mag between 2013 and 2018. It typically happens when the stellar density is high and the quasar is located in regions where the microcaustics overlap. The net effect is a smooth variation of the microlensing magnification as the quasar moves through these crowded regions. The extrinsic variations introduced by microlensing contains valuable information about the quasar accretion disk structure but is a real source of nuisance when measuring the time delays from optical monitoring. However, radio light curves are generally less influenced by microlensing compared to their optical counterparts. This is because the region from which radio waves are emitted is usually significantly larger than the microcaustics, so the impact of microlensing tends to be averaged out and less noticeable. In the optical range, the accretion disk, responsible for most of the UV and optical emission, has a comparable size to the microcaustics. Consequently, optical light curves are more affected by microlensing, leading to important extrinsic variations (for a detailed description of how microlensing differently impacts various emission regions, refer to Vernardos et al. 2024, this collection).

To deal with this issue, several curve-shifting algorithms have been proposed over the years, which can be classified into two categories. On one hand, some methods are based on the light-curve cross-correlations (e.g. Pelt et al. 1996), sometimes without attempting to subtract the microlensing variability (e.g. the smoothing and cross-correlation method by Aghamousa and Shafieloo 2015). On the other hand, several techniques rely on the analytical modelling of the intrinsic variability of the quasars and/or microlensing variations with, for example, splines (Tewes et al. 2013a) or Gaussian Processes (e.g. Hojjati et al. 2013). When explicitly modelled, the microlensing variations are removed from the light curves before attempting to find the optimal time delays. Due to the broad band nature of the monitored signal, mixing flux arising from multiple emission regions, microlensing is rarely perfectly removed but this is shown to have in general a negligible impact on the delay (Sluse and Tewes 2014). One can also mention the recent work by Tak et al. (2016), Donnan et al. (2021) and Meyer et al. (2023), aiming to infer the time delays in a Bayesian framework, including an explicit modelling of the microlensing variations.

These methods were tested in the "Time Delay Challenge" (TDC; Dobler et al. 2015), a blind data challenge aiming at assessing the precision and accuracy of the curve shifting algorithm on simulated but realistic data, which includes the microlensing variability. The results and conclusions of the challenge are presented in Liao et al. (2015) as well as in individual papers (Hojjati and Linder 2014; Bonvin et al. 2016). The problem is in fact more complicated than it sounds since a large fraction of the participating teams did not meet the requirements in term of precision and accuracy on the first and simplest rung of the challenge. Among the qualified teams to participate to the more advanced rungs of the TDC, the different proposed techniques showed overall good performance given the actual quality of the data. Several teams reached an accuracy of $\lesssim 1\%$ on the most variable light curves. However, it remains to be checked if this level of performance holds if more realistic accretion disk emission mechanisms and source-size effects are included in the simulations.

Among these source-size effects, microlensing time delay, which is described in details in Tie and Kochanek (2018), may be a more subtle manifestation of microlensing acting as a nuisance to measure the time delay. Although it has never been detected so far in lensed quasar light curves directly, this effect may arise when different emission regions of the accretion disc are differentially magnified. A simple model to explain the UV and optical variability of the quasars is the "lamp post" model (e.g. Cackett et al. 2007; Starkey et al. 2017), where the luminosity fluctuations originate close from the supermassive black hole and then illuminate the rest of the disk. This triggers temperature fluctuations in the disk, which result into delayed UV and optical emission due to the light travel time from the center. In the absence of differential magnification caused by microlensing, this time lag cancels out between the multiple images and only the "cosmological" time delay is observed. However, if one of the multiple image is affected by microlensing, the time lag originating from a particular region of the disk might be amplified by microlensing and a net excess of microlensing time delay can add to the "cosmological" time-delay. This effect could reach a few hours to a couple of days, which is negligible for most of the systems with long time delays but it can significantly increase the uncertainties for systems with short time delays. This effect can however be mitigated with multi-band light curves (Chan et al. 2021; Liao 2020, 2021) or a proper Bayesian treatment of this effect as a source of nuisance (Chen et al. 2018).

Future developments of curve shifting algorithms might also include time-delay measurements from unresolved light curves (e.g. Hirv et al. 2007; Shu et al. 2021; Biggio et al. 2021; Springer and Ofek 2021; Bag et al. 2023), which will open the possibility to monitor small-separation ($< 1''$) lensed quasars. While precise delays from unresolved lightcurves have already been measured in the gamma-ray range (Barnacka et al. 2011; Cheung et al. 2014), they cannot be used for cosmography as the location of the gamma-ray emission w.r.t. the central AGN remains unknown. In the optical range, space-based large sky surveys, such as Euclid, are expected to discover thousands of small-separation systems, which will not be fully resolved from the ground with our current follow-up facilities. To turn this large sample of small-separation lensed quasars into a useful cosmological probe will require to develop these new techniques.

5 Determining Lensing Potential

Determining the lensing potential ϕ is a crucial ingredient in time-delay cosmography as it directly enters the time-delay prediction through the Fermat potential (Eq. (7)). The Fermat potential is generally dominated by a massive elliptical galaxy acting as the main deflector and intervening line-of-sight over- and under-densities. To achieve a precise and accurate cosmographic inference, knowledge of both the line-of-sight structure and the mass distribution within the main deflector need to be known. One of the major limitations in a precise determination of the Fermat potential is the MSD, and the inability with imaging data to constrain the Fermat potential. Thus, either physical assumptions based on what we know from other modelled galaxies, e.g. in the nearby universe, or external data, such as stellar kinematics, is required to constrain the Fermat potential. The measurement of the Hubble constant and constraining the galaxy density profiles are tightly connected and most of the questions asked and techniques being used in Shajib et al. (2024, this collection) are of the same relevance and applicability for time-delay cosmography.

We discuss observables and inferences, first, for historical context on positional constraints alone in Sect. 5.1, and then from imaging data in Sect. 5.2. We then discuss assumptions on mass profiles in Sect. 5.3 and what external information provide necessary constraints in Sect. 5.4.

5.1 Conjugate Point Analysis

Historically, the first lens models constructed for time-delay cosmography were based on positional constraints of the quasar images alone, and in combination with the time delays, and to some extent the image magnifications. Valid models must satisfy the lens equation (Eq. (1)) such that all images must map back to the same source position. In the case of a quadruply lensed system, the four image positions result in constraints of five relative distortion angles (plus translation and rotation of the system). With such limited information, investigators had to assume a very simple form for the lens mass distribution, such as a singular isothermal sphere (Koopmans and Fassnacht 1999), for meaningful constraints on the model. Different assumptions on the mass profiles lead to vastly different results (see e.g., Kochanek and Schechter 2004).

5.2 Inference from Imaging Data

High-resolution imaging of gravitational lenses with constraints from hundreds/thousands of high signal-to-noise surface brightness pixels is able to measure accurate astrometry of multiple images and capture the detailed distorted images of extended source structure. This information is crucial to capture the relative lensing deflection and to achieve a precise determination of the relative Fermat potential between multiple images of the time-variable source. Modeling the imaging data on the pixel level has become the standard over the last two decades. In this section, we discuss the necessary aspects of the mass distribution that imaging data can constrain, apart from the remaining degeneracies.

To derive constraints on the mass of the gravitational lens and its deflection field from imaging data, models of the imaging data with different deflection fields are compared to the data in a Bayesian way on the likelihood level of individual pixels. Besides a description of the deflection field, all light emission components have to be described, containing the light emission of the source and the deflector. All components that affect the imaging data need to be modeled and accounted for, in particular around the region impacted by lensing features. Required modeled components include, but are not limited to, the extended source component of host galaxy of the time-variable source, the image positions of the time-variable source and its resulting approximate point-like flux emission, the surface brightness of the deflector galaxy, differential dust extinction caused by the deflector galaxy on the background source, and any other sources of surface brightness, such as satellite galaxies. In addition, instrument effects, such as the point spread function (PSF), instrumental noise and shot noise, and pixelization, as well as potential data reduction artifacts need to be accurately taken into account in the modeling and comparison with the data.

The lensing effect distorting an extended surface brightness $S(\boldsymbol{\beta})$, such as from the extended host galaxy, can be computed with a method known as 'backwards ray-tracing'. Making use of the fact that surface brightness is conserved through lensing, the surface brightness at a position in the image $I(\boldsymbol{\theta})$ can be computed by the surface brightness in the source plane associated with the corresponding coordinate $\boldsymbol{\beta}(\boldsymbol{\theta})$ as $I(\boldsymbol{\theta}) = S(\boldsymbol{\beta}(\boldsymbol{\theta}))$, where $\boldsymbol{\beta}(\boldsymbol{\theta}) = \boldsymbol{\theta} - \boldsymbol{\alpha}(\boldsymbol{\theta})$ is the 'ray tracing' term given by the lens equation (Eq. (1)).

For unresolved point-like images, the backwards ray-tracing is numerically inefficient. To guarantee that multiple images precisely come from the same location in the source plane within the astrometric requirements for an accurate time-delay prediction, the lens equation has to be solved for the point source constraints within the astrometric precision, or alternatively, solutions not satisfying the astrometric requirement (e.g., Birrer and Treu 2019) need to be discarded.

Given a lens model with parameters $\boldsymbol{\xi}_{\text{mass}}$ (which includes all aspects of the deflection field, including line-of-sight structure and nearby perturbers) and surface brightness model with parameters $\boldsymbol{\xi}_{\text{light}}$, a model of the imaging data can be constructed, $\boldsymbol{d}_{\text{model}}$. The full process of simulating a modeled image can be cast as a consecutive application of operators as follows: starting with the surface brightness operator $\mathcal{S}$, the lensing operator $\mathcal{L}$ is applied on the lensed source, followed by a PSF convolution operation $\mathcal{C}$, and finally an operator $\mathcal{P}$ matching the pixel resolution of the data, formally an integral of the convolved surface brightness over the size of a pixel. With this notation and $\odot$ denoting the consecutive application of operators from left to right, we can write the generation of modeled data as

$$\boldsymbol{d}_{\text{model}} = \mathcal{P} \odot \mathcal{C} \odot \left[\mathcal{L}(\boldsymbol{\xi}_{\text{lens}}) \odot \mathcal{S}_{\text{source}}(\boldsymbol{\xi}_{\text{light}}) + \mathcal{S}_{\text{lens}}(\boldsymbol{\xi}_{\text{light}})\right]. \tag{27}$$

The Bayesian analysis to constrain the lens model is performed on the pixel-level likelihood of the imaging data. The likelihood is computed at the individual pixel level accounting for the noise properties from background and other noise properties, such as read-out, as well as the Poisson contribution from the sources. In the Gaussian limit the imaging likelihood is given by

$$p(\mathcal{D}_{\text{img}} \mid \boldsymbol{\xi}_{\text{mass}}, \boldsymbol{\xi}_{\text{light}}) = \frac{\exp\left[-\frac{1}{2}(\boldsymbol{d}_{\text{data}} - \boldsymbol{d}_{\text{model}})^{\text{T}} \boldsymbol{\Sigma}_{\text{pixel}}^{-1} (\boldsymbol{d}_{\text{data}} - \boldsymbol{d}_{\text{model}})\right]}{\sqrt{(2\pi)^k \det(\boldsymbol{\Sigma}_{\text{pixel}})}}, \tag{28}$$

where k is the number of pixels used in the likelihood and $\boldsymbol{\Sigma}_{\text{pixel}}$ is the error covariance matrix. We also note that for the flux noise, the error covariance matrix $\boldsymbol{\Sigma}_{\text{pixel}}$ is a function of the brightness of the model $\boldsymbol{d}_{\text{model}}$ and hence not independent of the model prediction. Current analyses assume uncorrelated noise properties in the individual pixels and the covariance matrix becomes diagonal.

The primary target of an imaging analysis is to retrieve the lens model parameter posteriors marginalized over other model parameters, in particular the surface brightness and regularization parameters as

$$p(\boldsymbol{\xi}_{\text{mass}} \mid \mathcal{D}_{\text{img}}) = \int p(\mathcal{D}_{\text{img}} \mid \boldsymbol{\xi}_{\text{mass}}, \boldsymbol{\xi}_{\text{light}}) p(\boldsymbol{\xi}_{\text{mass}}, \boldsymbol{\xi}_{\text{light}}) d\boldsymbol{\xi}_{\text{light}}, \tag{29}$$

where $p(\boldsymbol{\xi}_{\text{mass}}, \boldsymbol{\xi}_{\text{light}})$ denotes the prior on the lens and light model parameters.

To jointly marginalize over an unknown yet possibly complex source morphology, different techniques have been developed. Such techniques include regularized pixelated source reconstruction (e.g., Warren and Dye 2003; Treu and Koopmans 2004; Koopmans 2005; Suyu et al. 2006, 2009), a set of basis functions, such as shapelets (e.g., Birrer et al. 2015; Birrer and Amara 2018) or wavelets (Joseph et al. 2019; Galan et al. 2021), or simply parameterized surface brightness profiles, such as Sérsic profiles. The methods mentioned above have in common that their surface brightness amplitude components create all a linear response on the pixel values of the data. The optimization of the often numerous linear coefficients to provide the maximum likelihood of the data given a proposed model for the other parameters (such as surface brightness shape parameters and lensing parameters) can be cast as a linear problem with a solution obtained by a weighted least square minimization. The Gaussian covariance matrix of the linear weighted least square minimization can be used to analytically marginalize over the prior of the linear coefficients (e.g., Suyu et al. 2006; Vegetti and Koopmans 2009; Birrer et al. 2015).[4]

The joint sampling of lens and light model parameters to infer the lens model posterior distribution is then a semi-linear process. While the amplitudes of the light model coefficients can be solved linearly, the remaining parameters, including those pertaining to the

[4]This is not a statement about the validity of the form of the prior of the linear surface brightness coefficients.

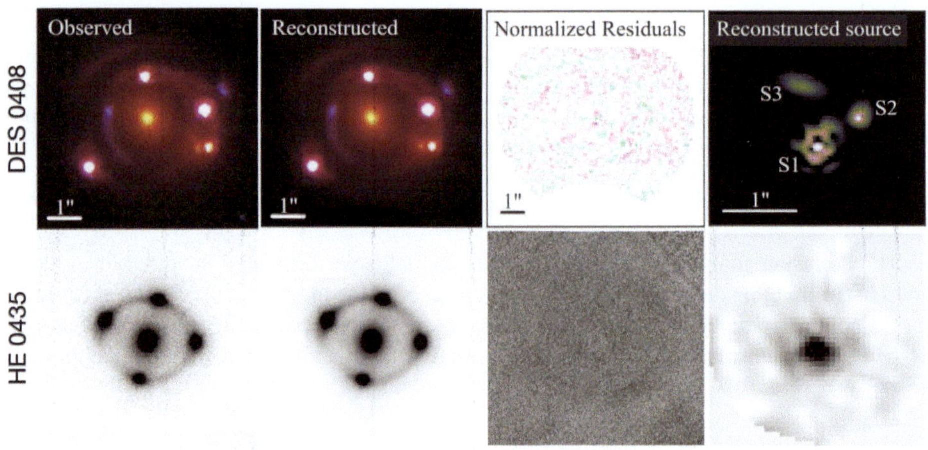

Fig. 4 Illustration of imaging modeling for two lenses. From left to right: Imaging data, the reconstructed model, the reduced residuals $(\boldsymbol{d}_{\text{data}} - \boldsymbol{d}_{\text{model}})/\sigma$, reconstructed source. Top row: HST data and model for DES J0408–5354 in three bands, from Shajib et al. (2020), with shapelet and parameterized source reconstruction using the modeling software LENSTRONOMY. Bottom row: Keck adaptive optics imaging and modeling of HE0435–1223, from Chen et al. (2019), with pixelated source reconstruction using the modeling software GLEE

lens mass model, and other shape-related surface brightness parameters have to be sampled non-linearly.

Often it is not clear at the beginning of an investigation what the level of complexity in the model is required to describe the data and to guarantee an accurate modeling. Current procedures are to start with a simple model and subsequently increase the complexity in the different model components until a satisfactory fit is achieved. Current criteria for a goodness of fit in use are the Bayesian Information Criteria (BIC) (Birrer et al. 2019) and the Bayesian Evidence (Shajib et al. 2020).

Imaging modeling is primarily performed on high resolution space based *Hubble Space Telescope* (*HST*) or ground-based adaptive optics (AO; Chen et al. 2016, 2019, 2021b) imaging. Figure 4 illustrates, as an example, the imaging data and models for two quadruply imaged lensed quasars, originally presented by Shajib et al. (2020) and Chen et al. (2019). Figure 5 presents the key lens model posteriors from the imaging modeling fit of the lens HE0435–1223 by Chen et al. (2019) for both, *HST* and AO imaging, for a power-law elliptical mass distribution with external shear contribution. To enhance the signal in the data set, and to distinguish deflector and source light components, the lens modeling is often performed simultaneously with multiple filters and combined on the likelihood level. With multiple filters, a better differentiation between source and deflector light can be drawn, with the deflector often being bright in the infrared channels, and the lensed source often dominant in the optical and ultraviolet channels. The infrared channels are often brighter, and hence contain more signal, while the optical channels can resolve smaller angular scales with more prominant source morphologies. The modeling across bands is performed with an identical lens model, while the surface brightness solutions are flexible to change (i.e., independent linear optimization or even different surface brightness components). Very high relative astrometric solutions of order $\sim$ 1mas across bands are required for the modeling. Current modeling fits the relative astrometric solutions across bands in the modeling process (e.g., Shajib et al. 2020).

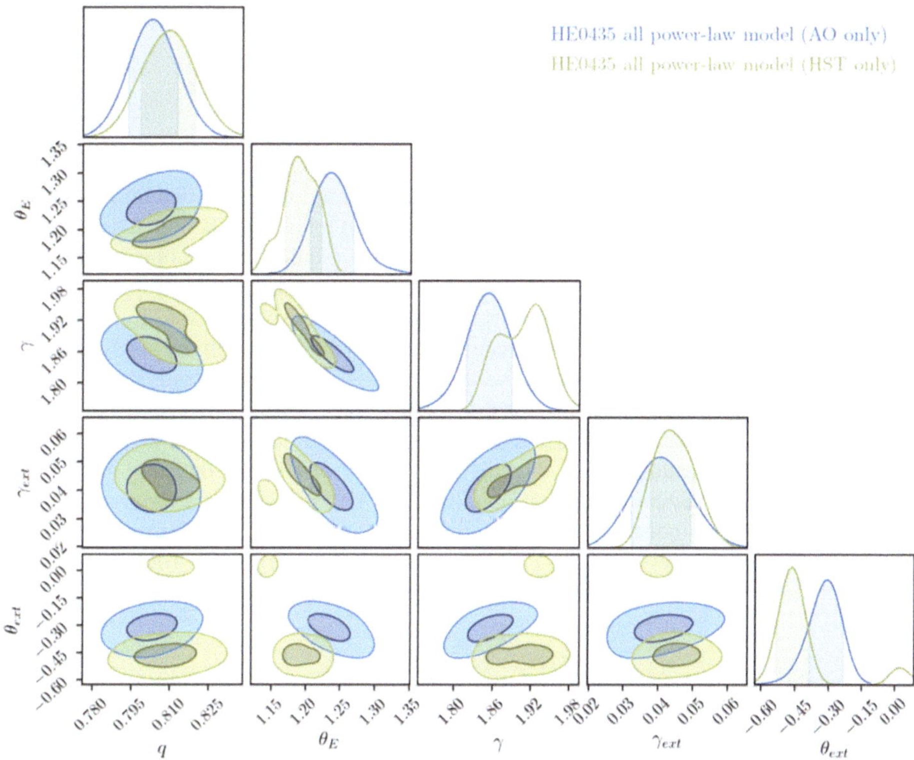

Fig. 5 Key lens model parameter posterior from the fit to imaging data (*HST* and Keck adaptive optics). q is the semi-minor to semi-major axis ratio in the projected mass density profile, θ_E is the Einstein radius, γ the three-dimensional radial power-law density slope, γ_{ext} is the external shear strength, and θ_{ext} is the shear angle. Figure adopted from Chen et al. (2019)

The PSF needs to be characterized very accurately, both to provide a high astrometric precision of the images of the time-variable sources (Birrer and Treu 2019; Chen et al. 2021a) and for the detailed modeling of the extended source structure without spurious signal of bright quasar images. Current methods to obtain a precise PSF model contain an iterative procedure during the model fitting process to extract improved constraints of the PSF from the data itself (e.g., Chen et al. 2016; Birrer et al. 2016).

The lens model and Fermat potential posteriors are marginalized over a set of systematic effects and modifications in the choice of the source reconstruction and other modeling choices.

5.3 Mass Profile Assumptions of the Main Deflector

Resolved multiply imaged structure is an exquisite data product to provide constraints on the relative deflection field (see Sect. 5.2). Imaging data, in the absence of the knowledge of the unlensed apparent source size or brightness, is not able to break the MST and its generalization, the source position transformation (SPT; Schneider and Sluse 2014). The quantity that is invariant under the MST in the radial direction and hence can be constrained by imaging data is (Kochanek 2002; Sonnenfeld 2018; Kochanek 2020; Birrer 2021)

$$\xi_{\rm rad} \equiv \frac{\theta_{\rm E}\alpha''_{\rm E}}{1-\alpha'_{\rm E}} \propto \frac{\theta_{\rm E}\alpha''_{\rm E}}{1-\kappa_{\rm E}}, \tag{30}$$

where $\alpha'_{\rm E}$ is the derivative and $\alpha''_{\rm E}$ is the second derivative of the deflection angle at the Einstein radius $\theta_{\rm E}$, respectively, and $\kappa_{\rm E}$ is the convergence at θ_E. On azimuthal invariances and observable lensing features, we refer to Birrer (2021) and references therein. Constraining a global mass density profile based on imaging data alone requires assumptions on the radial profile. For example, when imposing the assumption that the mass density profile follows a single power law, the power-law slope $\gamma_{\rm pl}$ has a direct correspondence to $\xi_{\rm rad}$ (Eqn. (30)) with $\xi_{\rm rad} = \gamma_{\rm pl} - 2$ and a precise (few percent uncertainty) inference on the Fermat potential is possible. Relaxing the assumption on the shape of the density profile leads to significantly widened constraints. When allowing for a full mass-sheet degree of freedom, the Fermat potential is fully degenerate (i.e. $\xi_{\rm rad}$ remains unchanged by a MST). A pure mass sheet is unphysical as no localized three-dimensional density profile can describe it. However, approximate parameterizations can be found that can be expressed as a three-dimensional density profile and are indistinguishable based on imaging data (Schneider and Sluse 2013; Blum et al. 2020; Birrer et al. 2020; Yıldırım et al. 2023). For example, Blum et al. (2020) introduced the cored density profile which has a three-dimensional density distribution

$$\rho_c(r) = \frac{2}{\pi}\Sigma_c R_c^3 \left(R_c^2 + r^2\right)^{-2} \tag{31}$$

with the convergence profile as

$$\kappa_c(\theta) = \left(1 + \frac{\theta^2}{\theta_c^2}\right)^{-3/2}. \tag{32}$$

The approximate MST can then be written as

$$\kappa_{\lambda_c}(\theta) = \lambda_c \kappa(\theta) + (1-\lambda_c)\kappa_c(\theta). \tag{33}$$

Figure 6 illustrates an example of how an approximate MST can be physically interpreted when applied to a composite profile described with a stellar and a dark matter component.

There are also possibilities in deviations in the mass density profiles that do not directly mimic an MST. Any radial mass profile that satisfies the same constraints on $\xi_{\rm rad}$ (Eqn. (30)) provides an equally good fit to the data.[5] Azimuthal assumptions of the mass density profile do also matter in the interpretation of the radial components (Birrer 2021; Kochanek 2021) and assumptions on disky and boxiness, ellipticity gradients and isodensity twists of the density profile may also impact the Fermat potential differences (Van de Vyvere et al. 2022b,a; Gomer and Williams 2020, 2021).

From a physics point of view, the matter distribution of the main deflector is made of stellar mass, gas, and dark matter, where the stellar mass is dominating the inner-most parts. The dark matter fraction within the Einstein radius is about ∼10–60% (e.g., Auger et al. 2010; Ferreras et al. 2005). Invisible substructure in the lens and along the LOS can also perturb the Fermat potential (e.g., Oguri 2007; Keeton and Moustakas 2009). Gilman et al.

[5] $\xi_{\rm rad}$ is the first-order Taylor expansion term affecting the observed radial distortions. The quantity is well defined for an azimuthally symmetric lens. The generalization of the relevant radial quantity for elliptical mass models needs to be investigated.

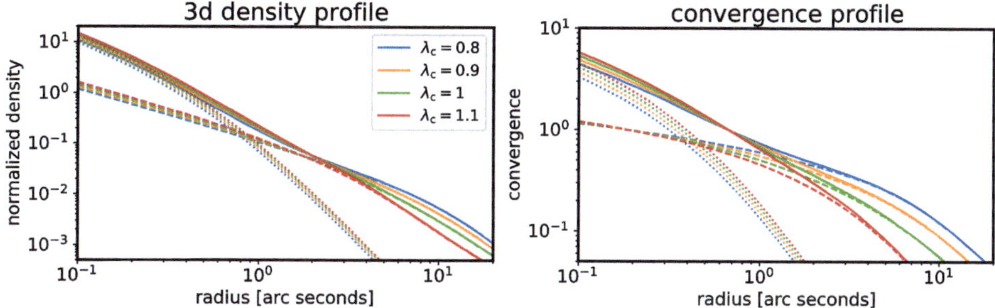

Fig. 6 Illustration of a composite profile consisting of a stellar component (Hernquist profile, dotted lines) and a dark matter component (NFW + cored component with λ_c acting as an approximate MST (from Blum et al. 2020, Eqns. (32), (33)), dashed lines) which transform according to an approximate MST (joint as solid lines). The stellar component gets rescaled by the MST while the cored component transforms the dark matter component. Physically, the profiles of each color differ by a 10% different mass-to-light ratio combined with a slightly more extended or contracted dark matter profile also on the 10% level. **Left:** profile components in three dimensions. **Right:** profile components in projection. Each profile provides a 10% difference in the predicted time delay, and hence H_0 inference. The transforms presented here cannot be distinguished by imaging data alone and require i.e., stellar kinematics constraints. Figure from Birrer et al. (2020)

(2020) showed that omitting dark substructure does not bias inferences of H_0. However, perturbations from substructure contribute an additional source of random uncertainty in the inferred value of H_0 ranging from 0.7–2.4% depending on the redshift and image configuration. We also highlight that the lensing mass and convergence only accounts for the mass over-density in regard to the cosmological background density.

Different approaches running with different assumptions have been taken in the literature to describe the deflector lensing potential. Among the assumptions being used are single power-law mass profiles, composite models involving a mass-follows-light component with a separate component describing the dark matter profile, free-form pixelated mass profiles (e.g., Saha and Williams 2004; Coles et al. 2014; Denzel et al. 2021), pixelated lensing potential corrections (e.g., Suyu et al. 2009), or an explicit internal mass profile MST component (Blum et al. 2020; Birrer et al. 2020).

There are multiple considerations in the specific choice of an investigation. On one side, there are physical considerations. What basic assumptions in the modeling are justified? What priors to chose in the Bayesian modeling? Then there are also practical considerations. What aspects of the model can be constrained by the data? Is it feasible to perform a posterior inference in a finite amount of time with given resources?

Among the simplest models employed is the single power-law profile. It has a one-to-one relation to the radial quantity described in Eq. (30) and breaks the MST. A power-law elliptical mass profile is an efficient parameterization to describe the first order radial and azimuthal features in strong lensing imaging data. Composite models do relate to certain physical assumptions of mass follows light and assert a stiffness in the profile that implicitly also break the MST .[6]

An explicit parameterization of the MST in the model denies any prior or assumptions to break the MST and is maximally agnostic to the MST with minimal added parameter degrees.

[6]The breaking of the MST for composite models is dependent on imposed mass and concentration priors of the dark matter profile, as well as mass-to-light gradients or the absence of it.

On the high-complexity end of lens models are free-form methods, such as pixelated mass distributions (Saha and Williams 2004; Coles et al. 2014). Free-form models come with very few restrictions on the lens mass distribution and offer a different modeling strategy compared to the simply parametrized approaches. The ensemble of models allowed by the data can be interpreted as the model posterior distribution, with the regularization scheme proposing models without data constraints being the prior.

Increased flexibility in the parameterization better guarantees that the underlying truth in the mass distribution, and in particular the prediction of the Fermat potential entering the time delays, can be represented by the model. On the other hand, increased flexibility in the model at fixed data constraining power increases the uncertainty in the posterior-predictive model. Less constraining posteriors put inevitably more weight and reliance on the priors applied, whether they are explicitly in a parameterized form, or implicit within an over-parameterized, free-form approach. No matter what choices are being made in the modeling of lenses, mitigating the dependence on the explicit or implicit priors becomes important when combining a set of multiple lenses, as we will discuss in Sect. 7.2.

We discuss additional data sets that can constrain the lens mass profile in Sect. 5.4.

5.4 Non-lensing Observables

The currently used primary observation to break the MSD is stellar kinematics from the deflector galaxy (Treu and Koopmans 2002; Koopmans et al. 2003; Koopmans 2004). The kinematics of stars, in particular their velocity dispersion, is a direct and lensing-independent tracer of the three-dimensional gravitational potential. The kinematic measurement is performed by targeting stellar absorption lines and measuring their width with spectrographs. Figure 7 shows a Keck/LRIS spectrum of HE0435–1223.

The line-of-sight stellar velocity dispersion is an integrated quantity of the radial and tangential components of the velocity dispersion projected along the line of sight. The orbital anisotropy, i.e., the ratio of the radial and tangential velocity dispersion components, is unknown *a priori* and thus introduces a degeneracy in the predicted line-of-sight velocity dispersion corresponding to the same 3D mass profile. This degeneracy is known as the mass–anisotropy degeneracy. Typically, a prior on the anisotropy profile, e.g., isotropic or Osipkov–Merritt (Osipkov 1979; Merritt 1985a,b), is assumed. The Osipkov–Merritt profile allows the anisotropy to be isotropic near the center and gradually more radial farther away from the center, which is motivated by the observed properties of the stellar orbits in local elliptical galaxies. The isotropic profile is thus a special case of the Osipkov–Merritt profile. The transition scale radius r_{ani} from isotropic to radial orbits in the Osipkov–Merritt profile is *a priori* unknown, which directly manifests in the mass–anisotropy degeneracy. Thus, the prior on r_{ani} has significant impact on the kinematics prediction (e.g., Shajib et al. 2018; Birrer et al. 2020). We note that there are many other forms of the radial anisotropy distribution and the specific choice of functional model used might impact the results, as well as what priors are adopted. One way to mitigate this degeneracy is to obtain spatially resolved measurement of the velocity dispersion instead of an unresolved (or, integrated) velocity dispersion (Shajib et al. 2018; Yıldırım et al. 2020, 2023; Birrer and Treu 2021).

Other proposed observations and analyses methods that can break the MST and provide the necessary constraints on the mass density profile are standardizable magnifications (e.g., Kolatt and Bartelmann 1998; Oguri and Kawano 2003; Foxley-Marrable et al. 2018; Birrer et al. 2022), lens population statistics of appearances and asymmetry in the multiple images (e.g., Sonnenfeld and Cautun 2021; Sonnenfeld 2021), and galaxy-galaxy weak gravitational lensing (Khadka et al. 2024).

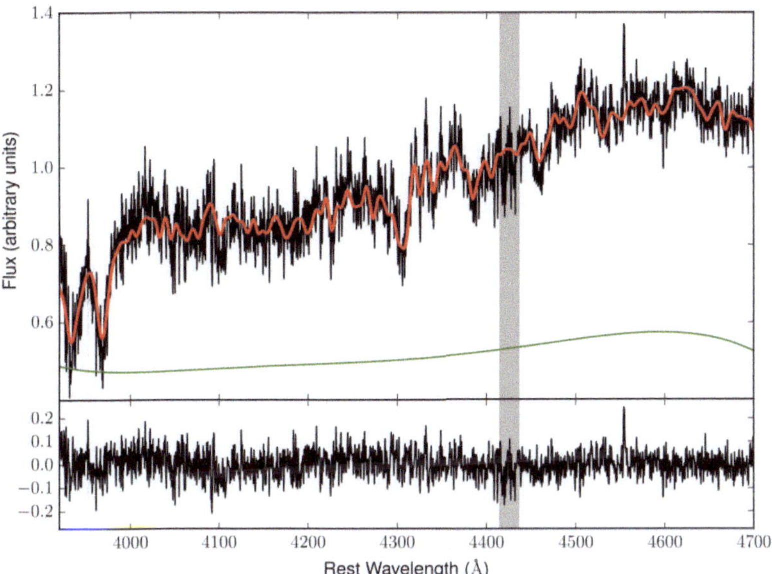

Fig. 7 **Top**: Keck/LRIS spectrum of HE0435−1223 with the best-fitting model overplotted in red and a polynomial continuum, which accounts for contamination from the lensed QSO images and template mismatch, shown in green. The measurement results in an integrated velocity dispersion $\sigma_{\rm V} = 222 \pm 15\ \mathrm{km\ s^{-1}}$, including systematic uncertainties due to the templates used, the region of the spectrum that was fitted, and the order of the polynomial continuum. The grey vertical band represents a wavelength range that is excluded from the fit due to the presence of a strong Mg II absorption system. Bottom: Residuals from the best fit. Figure adopted from Wong et al. (2017)

We emphasize that these non-lensing observations are primarily sensitive to the total MST, the combination of LOS and internal mass density profile degeneracies (Eq. (20)). Decoupling of the different projected effects in the lensing potential is not necessary to perform an accurate cosmographic inference since the time-delay prediction only requires the combined product. However, when combining different lenses with potentially different selections, the priors and assumptions imposed in either of the two components impacting the MST can become important.

6 Estimating Line-of-Sight Contributions

Strong lensing requires a high projected mass density. Strong lenses are hence biased toward more massive galaxies, which are biased toward overdense environments. The contribution of the mass density fluctuations along the line of sight to the lensed source is generally of order few percent, and commonly lower than 10% of the total convergence of the lens. While this may appear to be small, it is not negligible when it comes to estimating the Hubble parameter to percent accuracy. A constant effective contribution of a few percents caused by the line-of-sight is equivalent to an external mass-sheet $\kappa_{\rm ext}$ (Sect. 2).

The exact impact of the line-of-sight objects depends on whether the dominant-lens approximation is valid, in which case the critical density of the line-of-sight objects is very small compared to the main deflector critical density, and on whether the tidal regime is applicable, which happens when the perturber's gravitational field is small compared to the changes of the deflection $\alpha(\theta)$ (e.g. McCully et al. 2014; Birrer et al. 2017; Fleury et al. 2021b). When one of those approximation is invalid, an explicit treatment is needed, requiring potentially to solve the multi-plane lens equation (see e.g., McCully et al. 2014; Wong et al. 2020; Shajib et al. 2020; Li et al. 2021). We refer to Saha et al. (2024) for a detailed discussion on the multi-plane gravitational lensing formalism. Conversely, when line-of-sight objects can be treated as small perturbations that only introduce convergence that is constant over the extent of the lensed system, a statistical treatment is sufficient. In practice, a hybrid scheme needs to be followed most of the time, including explicitly modelling those objects that modify differently the Fermat potential for each lensed images, and calculating the statistical contribution of the other objects that shift the Fermat potential in linear order.

From an information perspective, there is only limited direct data available of the total matter distribution on the universe at the scales relevant ton constrain $\kappa_{\rm ext}$. Hence, any method relies on some assumptions on how mass traces light. These assumptions are well motivated by large scale structure probes, but are only validated in a statistical way.

The following subsections present the various methods that have been considered to estimate $\kappa_{\rm ext}$. Section 6.1 presents a direct modeling, Sect. 6.2 presents galaxy number counts statistics, Sect. 6.3 weak lensing measurements, and Sect. 6.4 a hybrid approach.

6.1 Direct Modeling

The most direct approach is to collect the positions, redshifts, stellar masses and potentially even velocity dispersion measurement of the galaxies located in the field of view towards the lens and explicitly model the matter distribution of all relevant objects. A complete direct reconstruction is near-impossible. A simple approach that has been developed to estimate $\kappa_{\rm ext}$ consists in identifying which galaxies form massive galaxy groups that contribute the most significant impact along the line of sight (e.g. Fassnacht and Lubin 2002; Momcheva et al. 2006; Fassnacht et al. 2006; Sluse et al. 2017). An estimate of $\kappa_{\rm ext}$ may then be derived by fitting analytical mass density profiles on those groups (Auger et al. 2007; Wilson et al. 2017). This approach generally yields estimates of $\kappa_{\rm ext}$ typically uncertain to a factor 2-4 depending on the specific assumptions one may reasonably make on the group mass density (e.g., halo associated to each individual galaxies, a common halo for all the systems), but also sometimes with low precision due to the uncertainties associated to the group identification (fields of view are never complete and group finders have their own biases). To overcome this problem, Collett et al. (2013) have proposed a simple halo model prescription to reproduce the mass along the line-of-sight from a photometric catalogue of galaxies. The convergence $\kappa_{\rm h}$ from each halo has then been calibrated against $\kappa_{\rm ext}$ derived from ray-tracing estimate through numerical simulations. This method does not account explicitly for the convergence due to dark structures and divergence due to voids, but those effects are included statistically owing to the calibration of $\kappa_{\rm h}$ against $\kappa_{\rm ext}$.

6.2 Number Counts

An alternative to the direct modeling of the line of sight consists of measuring the galaxy number density in the vicinity of the lens, describe it in a summary statistic, and comparing it to reference fields. This procedure will determine whether the LOS is over- or under-dense compared to average (Suyu et al. 2010; Fassnacht et al. 2011; Greene et al. 2013; Rusu et al. 2017; Wells et al. 2023). First, a detailed characterization of the line of sight towards the lens is required, using deep imaging data, complemented by spectroscopic data (see Fig. 8 for an illustration). Similar LOS are then searched for in large volume, and high resolution

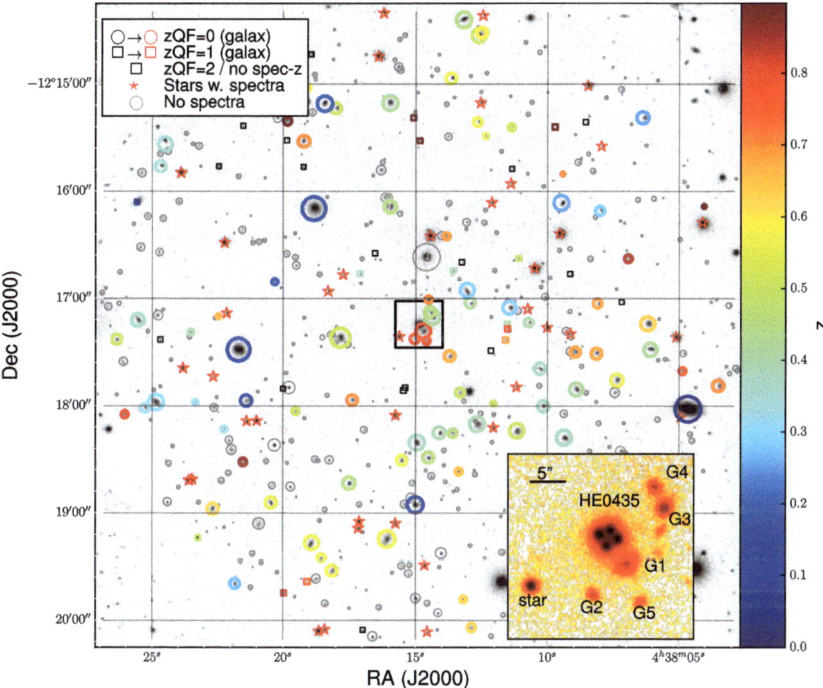

Fig. 8 Example of the result of the spectroscopic characterization of the field of view of the gravitational lens HE0435–1223. The redshifts of the objects measured in the field of view are used to identify which objects (or galaxy groups) needs to be explicitly included in the macro-model. The number counts analysis uses the galaxies located at projected distances of $<45''$ and $<120''$ from the lens to estimate $P(\kappa_{\rm ext})$. Courtesy of Sluse et al. (2017)

numerical simulations. The surface mass density of matter along these line of sight being calculated using a ray-tracing technique (Hilbert et al. 2007, 2008, 2009), it is then possible to derive a probability distribution of external convergence compatible with the observed lens matching the summary statistics.

In practice, summary statistics that deviate from pure number counts can be a better informed statistics of the underlying over- or under- density. For example, if $N_{\rm gal}$ galaxies are observed in the field of view of a lens, one can calculate a weighted number counts $W_q = \sum_{i=1}^{N_{\rm gal}} q_i$ with q being a particular type of weight, such as the inverse of the distance to the lens, i.e. $1/r$. To calculate a weighted density of galaxies, ζ_q, it is necessary to perform the same measurement over an ensemble of control fields, such that for each control field (CF, j) one derives a density $\zeta_q^j = W_q / W_q^{{\rm CF},j}$. To avoid introducing any bias through this normalisation procedure, it is important to choose enough control fields but also ensure that those fields have characteristics that match closely those of the imaging data of the observed lens system. This allows one to account for sample variance and to assess that galaxy detection biases are similar for the lens and for the control fields. In particular, one should ascertain that the lens and control fields have similar depth, seeing, and pixel scale, the latter quantities being critical in the framework of source deblending and identification. It happens that some regions of the control fields and potentially the data around the lens target are masked due to saturation of stars, cosmic rays, or camera defects. To guarantee unbiased estimates of ζ_q, it is important to apply the same mask to the weighted count of the lens field and the control field. A large variety of weighting schemes has been explored, some of them involving a proxy on some of the galaxy properties such as the redshift and the stellar mass. Those quantities are derived using a photometric redshift code, such as LEPHARE (Ilbert et al. 2006). This implies the availability of deep multi-band photometric data. Having a similar depth for the lens and comparison field is important to matching galaxy properties in photometric surveys. Also, similar set-up for magnitude measurements are required to minimize systematic errors caused by aperture and/or object deblending uncertainties. When spectroscopic redshift are also available, they may generally be preferred to the photometric ones.

The conversion of the probability density distribution $p(\zeta_q)$ into $p(\kappa_{\rm ext})$ requires the use of numerical simulations for which $\kappa_{\rm ext}$ has been derived through ray-tracing. Large simulation volumes are required to minimise the impact of sample variance and cosmic variance. The Millennium simulation (Springel et al. 2005) being dark matter only, galaxy photometric properties are inpainted using physically motivated prescriptions. A common choice has been to use the semi-analytic model of De Lucia and Blaizot (2007). Densities ζ_q can be estimated in those catalogues in similar way as for real data except that the reference fields is now the simulation catalog itself. As explained in Rusu et al. (2017), the probability $p(\kappa_{\rm ext} \mid \mathbf{d})$, where $\mathbf{d}$ stands for the available data, is given by:

$$p(\kappa_{\rm ext} \mid \mathbf{d}) = \int d\zeta_q \frac{p(\kappa_{\rm ext}, \zeta_q) p(\zeta_q, \mathbf{d})}{p(\zeta_q) p(\mathbf{d})} = \int d\zeta_q p(\kappa_{\rm ext} \mid \zeta_q) p(\zeta_q \mid \mathbf{d}). \tag{34}$$

Greene et al. (2013) has shown that the precision on $\kappa_{\rm ext}$ is improved by a factor 2 when using a combination of weights, the two main ones being the standard number counts ($q = 1$), and as the inverse of the projected distance to the lens ($q = 1/r$). The addition of a weight based on the modeled amplitude of the shear, $\gamma_{\rm ext}$ is also often considered, such that the general expression of $p(\kappa_{\rm ext} \mid \mathbf{d})$ becomes:

$$p(\kappa_{\rm ext} \mid \mathbf{d}) = \int d\zeta_1 d\zeta_{1/r} d\zeta_{q \neq 1,1/r} d\zeta_{\gamma_{\rm ext}} p_{\rm MS}(\kappa_{\rm ext} \mid \zeta_1, \zeta_{1/r}, \ldots$$

$$\ldots \zeta_{q \neq 1,1/r}, \zeta_{\gamma_{\rm ext}}) p(\zeta_1, \zeta_{1/r}, \zeta_{q \neq 1,1/r}, \zeta_{\gamma_{\rm ext}} \mid \mathbf{d}) \,. \tag{35}$$

The addition of a fourth weight, $q \neq 1, 1/r$ allows one to evaluate systematic errors involved by the specific choice of equally motivated weighting schemes and explore which combination of weight yields the best precision on $\kappa_{\rm ext}$. The first application of this technique in the framework of time-delay cosmography has been presented in Suyu et al. (2010). Subsequent time-delay cosmography analyses from H0LICOW, STRIDES and TDCOSMO have used an approach that broadly follows the strategy outlined above (see Greene et al. 2013; Rusu et al. 2017, for a more in-depth description of the method), but proposing small variations and tests in terms of weighting schemes and choices of comparison fields (Birrer et al. 2019; Sluse et al. 2019; Buckley-Geer et al. 2020). Figure 9 displays $p(\kappa_{\rm ext})$ as derived with different weighting schemes for the lens system HE0435–1223.

A novel method for estimating $\kappa_{\rm ext}$ has recently been proposed by Park et al. (2021). They replace the weighted number count scheme by a machine learning approach. Specifically, they have trained a Bayesian Graph Neural Network on LSST DESC DC2 sky survey (LSST Dark Energy Science Collaboration (LSST DESC) 2021) in order to derive a distribution of κ_{ext} for arbitrary gravitational lens sight-line.

The reliance on cosmological simulations and their cosmological assumptions poses a slight circularity in the inference when the very goal of time-delay cosmography is to test

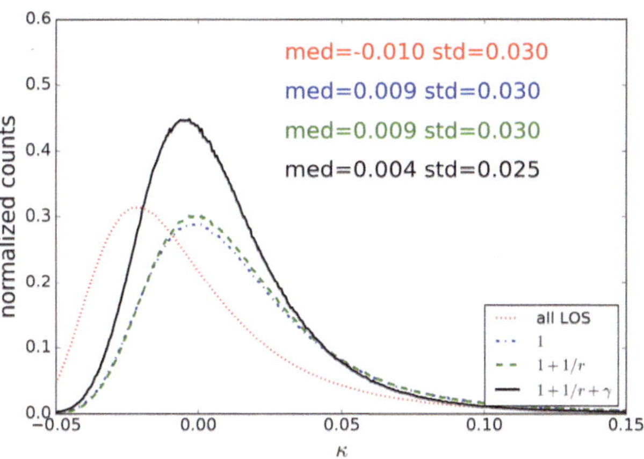

Fig. 9 Probability distribution $P(\kappa_{\text{ext}})$ for HE0435–1223 and an aperture radius of 45″. The different curves correspond to all lines of sights (dotted red), considering only lines of sights with the same overdensity as the data (dash-dotted blue), using a weighting inversely proportional to the distance (dashed green) and the additional constraint from the shear (solid black). Courtesy of Rusu et al. (2017)

and challenge cosmological models. However, the line-of-sight correction term that is constrained relying on cosmological assumptions is perturbative, i.e., even if the actual cosmology resulted in a $\sim 10\%$ relative difference in the line of sight characteristics, it would be a sub 1% change to the distance measurements since the expected effect is only a few percent.

6.3 Weak Lensing

Weak lensing, the linear shape distortion of background galaxies due to foreground structure, is a direct probe of the LOS structure. On linear scales, the cosmic shear measurements can be translated to convergence in a unique mapping (Kaiser and Squires 1993). Hence, this technique does neither rely on priors from numerical simulations nor of a galaxy-halo connection. However, there are also several drawbacks. The angular scale of a weak lensing measurement is limited by the number density of lensed sources, and a high S/N measurement can only be achieved at scales of arc minutes. Thus, weak lensing is an excellent observable to quantify large scale cosmic density distributions but other smaller scale density perturbations down to the scales of arc seconds are not well captured. Another limitation is that the weak lensing source population is not at the same redshift as the strongly lensed source. One needs to translate the weak lensing convergence map to a different lensing kernel, which comes with additional statistical uncertainties (e.g., Kuhn et al. 2021).

In the strong lensing context, for example, Fischer and Tyson (1997), Nakajima et al. (2009), Fadely et al. (2010) relied on the weak lensing effect produced by massive structures in the vicinity of the deflector. They constrained the external convergence by integrating the tangential weak gravitational shear in the area around the lens. More recently, Tihhonova et al. (2018, 2020) applied the weak lensing techniques to the quadruply lensed quasar systems HE0435–1223 and B1608+656 and performed a convergence map reconstruction based on HST imaging. Kuhn et al. (2021) performed a convergence map reconstruction of the COSMOS field at the position of discovered strong lenses.

6.4 Hybrid Framework

Given the strengths and weaknesses of the direct modeling and summary statistics approaches, as well as weak lensing measurements, a hybrid approach can leverage the complementary methodologies. Summary statistics can be most effectively employed for objects that mostly cumulatively affect the lensing convergence while explicit modeling of LOS objects makes a difference for massive or very close by objects. The specific decision of where to split the analysis between a statistical approach and explicit modeling is primarily impacted by two factors. The first is the accuracy in the deflection properties, both in terms of higher-order lensing distortions and the need for a multi-plane lensing approach. The second is the available information and the handling of priors in the absence of sufficient information.

A method to account accurately for the line-of-sight has been proposed by McCully et al. (2014, 2017). It consists in a multi-plane lens equation where only the planes associated to important perturbing groups/clusters/galaxies are included. The other perturbers along the LOS are treated under the tidal approximation. In order to identify those objects, McCully et al. (2017) proposes to use a threshold based on the value of the flexion-shift, i.e. $\Delta_3 x$ whose expression is given by:

$$\Delta_3 x = f(\beta) \times \frac{(\theta_{\mathrm{E}}\, \theta_{\mathrm{E,p}})^2}{\theta^3}, \tag{36}$$

where θ_{E} and $\theta_{\mathrm{E,p}}$ are the Einstein radius of the main lens and of the perturber, and θ is the angular separation on the sky between the lens and the perturber. The function $f(\beta) = (1-\beta)^2$ if the perturber is behind the main lens, and $f(\beta) = 1$ if the galaxy is in the foreground. In that expression, β is the pre-factor of the lens deflection in the multiplane lens equation:

$$\beta = \frac{D_{\mathrm{dp}} D_{\mathrm{os}}}{D_{\mathrm{op}} D_{\mathrm{ds}}}, \tag{37}$$

where $D_{ij} = D(z_i, z_j)$ are angular diameter distances between redshifts z_i and z_j, corresponding to the observer (o), deflector (d), perturber (p) and source (s). Missing to accounting for a foreground perturber may have a stronger impact on the models than missing a background one. The reason is that the background perturber will have a multiplicative effect on the source position, while the deflection from the foreground pertuber enters the lens equation inside the argument of the deflection of the main lens galaxy. In other words, the foreground perturber modifies the coordinates of the lensed images positions compared to the main lens case. These non-linear effects require a multi-plane treatment to be properly accounted for. From a set of simulation of time-delay lens systems resembling real ones, and their subsequent modeling based on point-source image positions, McCully et al. (2017) suggests that a value $\Delta_3 x < 10^{-4}$ arcsec yields to a bias on H_0 of less than a percent. Since Sluse et al. (2017), this prescription is used by the H0LICOW and TDCOSMO collaborations to select the objects that they explicitly include in the lens mass modelling.

Birrer et al. (2017), Kuhn et al. (2021) combined the study of the environment using the halo-rendering approach, i.e. linking the galaxy stellar masses to the underlying mass distribution, with the external shear measurements of the strong lens system. Their combined approach yielded tighter constraints on the inferred external convergence compared to a halo-rendering approach only.

7 Cosmographic Inference

Having established the necessary observations and analyses components in the previous sections, in this section we discuss how an end-to-end combined analysis leads to constraints on H_0 and other relevant cosmological parameters. First we discuss the analysis for a single lens (Sect. 7.1) and then state the analysis for a set of multiple lenses (Sect. 7.2).

7.1 Single Lens Cosmography

For each individual strong lens, there are preferrably four data sets available: (i) imaging data of the strong lensing features and the deflector galaxy, $\mathcal{D}_{\text{img}}$; (2) time-delay measurements between the multiple images, $\mathcal{D}_{\text{td}}$; (3) stellar kinematics measurement of the main deflector galaxy, $\mathcal{D}_{\text{spec}}$; (4) line-of-sight galaxy count and weak lensing statistics, $\mathcal{D}_{\text{los}}$. These data sets are independent and so are their likelihoods in a joint cosmographic inference. Hence, we can write the likelihood of the joint set of the data

$$\mathcal{D} = \{\mathcal{D}_{\text{img}}, \mathcal{D}_{\text{td}}, \mathcal{D}_{\text{spec}}, \mathcal{D}_{\text{los}}\} \tag{38}$$

given the cosmographic parameters $\{D_{\text{d}}, D_{\text{s}}, D_{\text{ds}}\} \equiv D_{\text{d,s,ds}}$ as

$$\begin{aligned}\mathcal{L}(\mathcal{D}|D_{\text{d,s,ds}}) = &\int \mathcal{L}(\mathcal{D}_{\text{img}}|\xi_{\text{mass}}, \xi_{\text{light}}) \\ &\times \mathcal{L}(\mathcal{D}_{\text{td}}|\xi_{\text{mass}}, \xi_{\text{light}}, \lambda, D_{\Delta t}) \\ &\times \mathcal{L}(\mathcal{D}_{\text{spec}}|\xi_{\text{mass}}, \xi_{\text{light}}, \beta_{\text{ani}}, \lambda, D_{\text{s}}/D_{\text{ds}})\mathcal{L}(\mathcal{D}_{\text{los}}|\kappa_{\text{ext}}) \\ &\times p(\xi_{\text{mass}}, \xi_{\text{light}}, \lambda_{\text{int}}, \kappa_{\text{ext}}, \beta_{\text{ani}})d\xi_{\text{mass}}d\xi_{\text{light}}d\lambda_{\text{int}}d\kappa_{\text{ext}}d\beta_{\text{ani}}.\end{aligned} \tag{39}$$

In the expression above we only included the relevant model components of the individual likelihoods. ξ_{light} formally includes the source and lens light surface brightness.[7]

The sampling of the cosmographic posterior from the joint likelihood of Eq. (39) can be split in parts to simplify the problem. For example, we can first perform the imaging analysis providing constraints on ξ_{lens} and ξ_{light} without sampling the cosmological or distance parameters. In turn, simple sampling the ξ_{lens} and ξ_{light} posteriors in post-processing when evaluating the time-delay likelihood and stellar kinematic likelihood can translate the posteriors into distance posteriors in $D_{\Delta t} - D_{\text{d}}$ space. Marginalization over different modeling choices can also be done in the $D_{\Delta t} - D_{\text{d}}$ posterior space. Figure 10 provides an example of $D_{\Delta t} - D_{\text{d}}$ for a set of different modeling choices.

Weighting the posteriors of different models can be done with Bayesian model comparison. This weighting also allows one to combine models in a single posterior, which then includes systematics considerations. Discrete and finite choices made in the models and scatter in the sampling and BIC calculation can lead to over-constraint model selections. Procedures to take noise and finite model selection in the BIC estimate into account have been developed (Birrer et al. 2019).

7.2 Population Level Analysis

The overarching goal of time-delay cosmography is to provide a robust inference of cosmological parameters, $\boldsymbol{\pi}$, and in particular the absolute distance scale, the Hubble constant H_0, and possibly other parameters describing the expansion history of the Universe (such as Ω_Λ or Ω_{m}), from a sample of gravitational lenses with measured time delays.

In Bayesian language, we want to calculate the probability of the cosmological parameters, $\boldsymbol{\pi}$, given the strong lensing data set, $p(\boldsymbol{\pi}|\{\mathcal{D}_i\}_N)$, where $\mathcal{D}_i$ is the data set of an individual lens (including imaging data, time-delay measurements, kinematic observations and line-of-sight galaxy properties) and N the total number of lenses in the sample.

[7] To evaluate the time-delay likelihood, we require the time-variable image positions from the set of ξ_{light} parameters.

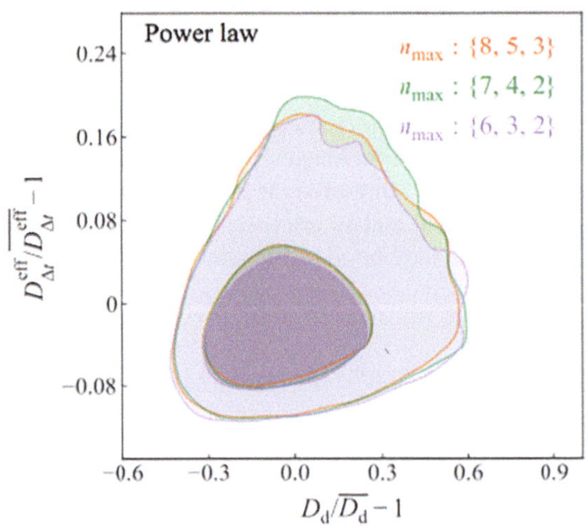

Fig. 10 Median-subtracted and mean-divided relative angular diameter distance posteriors, $D_{\Delta t}$ and D_{d} for a power-law mass density profile with three different source reconstruction settings. $n_{\max}$ in this figure refers to the polynomial order in the shapelets with the set of three numbers corresponding to three different sources being simultaneously modelled in the image. Figure adopted from Shajib et al. (2020)

In addition to $\boldsymbol{\pi}$, we denote $\boldsymbol{\xi}$ all the model parameters part of either a single lens analysis (Sect. 7.1) or present on the population level. Using Bayes rule and considering that the data of each individual lens $\mathcal{D}_i$ is independent, we can write, following Birrer et al. (2020):

$$p(\boldsymbol{\pi} \mid \{\mathcal{D}_i\}_N) \propto \mathcal{L}(\{\mathcal{D}_i\}_N|\boldsymbol{\pi})p(\boldsymbol{\pi}) = \int \mathcal{L}(\{\mathcal{D}_i\}_N \mid \boldsymbol{\pi}, \boldsymbol{\xi})p(\boldsymbol{\pi}, \boldsymbol{\xi})d\boldsymbol{\xi}$$
$$= \int \prod_i^N \mathcal{L}(\mathcal{D}_i|\boldsymbol{\pi}, \boldsymbol{\xi})p(\boldsymbol{\pi}, \boldsymbol{\xi})d\boldsymbol{\xi}. \quad (40)$$

In the following, we divide the nuisance parameter, $\boldsymbol{\xi}$, into a subset of parameters that we constrain independently per lens, $\boldsymbol{\xi}_i$, and a set of parameters that require to be sampled across the lens sample population globally, $\boldsymbol{\xi}_{\text{pop}}$. The parameters of each individual lens, $\boldsymbol{\xi}_i$, include the lens model, source and lens light surface brightness and any other relevant parameter of the model to predict the data. Hence, we can express the hierarchical inference (Eqn. (40)) as

$$p(\boldsymbol{\pi} \mid \{\mathcal{D}_i\}_N) \propto \int \prod_i \left[\mathcal{L}(\mathcal{D}_i \mid D_{\text{d,s,ds}}(\boldsymbol{\pi}), \boldsymbol{\xi}_i, \boldsymbol{\xi}_{\text{pop}})p(\boldsymbol{\xi}_i)\right]$$
$$\times \frac{p(\boldsymbol{\pi}, \{\boldsymbol{\xi}_i\}_N, \boldsymbol{\xi}_{\text{pop}})}{\prod_i p(\boldsymbol{\xi}_i)} d\boldsymbol{\xi}_{\{i\}} d\boldsymbol{\xi}_{\text{pop}} \quad (41)$$

where $\{\boldsymbol{\xi}_i\}_N = \{\boldsymbol{\xi}_1, \boldsymbol{\xi}_2, \ldots, \boldsymbol{\xi}_N\}$ is the set of the parameters applied to the individual lenses and $p(\boldsymbol{\xi}_i)$ are the interim priors on the model parameters in the inference of an individual lens. The cosmological parameters $\boldsymbol{\pi}$ are fully encompassed in the set of angular diameter distances, $\{D_{\text{d}}, D_{\text{s}}, D_{\text{ds}}\} \equiv D_{\text{d,s,ds}}$, and thus, instead of stating $\boldsymbol{\pi}$ in Eq. (41), we now state $D_{\text{d,s,ds}}(\boldsymbol{\pi})$. Up to this point, no approximation was applied to the full hierarchical expression (Eqn. (40)).

Key differences among different inferences of H_0 from a set of lenses involve, beyond the assumptions on individual lenses, assumptions on the covariant nature and the prior on the population level of the governing hyper-parameters. For example, Wong et al. (2020)

assumes full independence of the nuisance priors from one lens to another. Formally, within Bayes Theorem, this approach assumes perfect knowledge of the governing population hyper-parameter distribution prior (Eq. (41)). In this approach, the distance posteriors of individual lenses can be interpreted as measurements and the cosmographic analysis can be done in solely operating in the $D_{\Delta t} - D_{\rm d}$ space with a direct independent and easy accessible likelihood description.

Millon et al. (2020b) performed an analysis exploring the difference between two different radial mass density profile families, assuming that either all lenses are of one type or another, effectively treating one modeling choice as a covariant nuisance parameter in their inference while keeping all other priors independent with an assumed population.

Denzel et al. (2021) is using a free-form approach in the modeling of individual lenses. The ensemble of models allowed by the data for an individual lens is providing the model posterior distribution. The underlying regularization scheme is the implicit prior applied on individual lenses. The identical regularization scheme is applied to all lenses assuming independence in the priors without covariances in the choice of the regularization scheme between lenses. Denzel et al. (2021) did not use any external information to break the MST. Hence, the specific choice of the regularization scheme with their underlying physical and regularization priors is responsible for the breaking of the MST on the population level.

Birrer et al. (2020) introduced the hierarchical analysis framework into time delay cosmography and identified few key parameters, that on a per lens basis are not sufficiently well constrained and thus the population prior can significantly affect the outcome of the analysis. The parameters hierarchically sampled, beyond the cosmological ones, were the MST population $\lambda_{\rm int}$ (Eq. (21)), and the stellar anisotropy distribution (see Sect. 5.4).

Park et al. (2023) implemented a Bayesian hierarchical framework to determine the external convergence distribution on the population level for a full sample of lens systems used for time-delay cosmography and demonstrated how to correct for a selection bias in the population of lenses when there is limited information on an individual lens basis.

The required population-level description of priors, in particular of parameters that can not be constrained to high precision (overcoming the prior in the analysis) do also need to take accurately into account potential differences among subsets of the population. For example, different lens discovery channels might preferentially select a different lens and line-of-sight population.

8 Cosmography with Galaxy Clusters

In this section, we discuss current and past application of cosmograpy with galaxy clusters. We first discuss relative expansion history constraints from multiple source redshifts (Sect. 8.1) and then discuss time-delay cosmography applications (Sect. 8.2). This section is aimed to provide a brief overview over these aspects and we refer to the specific literature referenced in this section for further details.

8.1 Relative Expansion History with Galaxy Clusters

For a given deflector, changing the source redshift alters the angular diameter distances in Eq. (5), whilst the other terms in Eqs (1)-(4) are unchanged. Hence, for two photons passing through the same point in the lens plane, but originating on different source planes, the ratio of scaled deflection angles, α_1, α_2 is given by the cosmological scaling factor, β,

$$\frac{\alpha_1}{\alpha_2} = \frac{D_{ls1} D_{s2}}{D_{s1} D_{ls2}} \equiv \beta. \tag{42}$$

It has been realized that lenses with multiple source planes can additionally provide constraints on cosmological distance ratios sensitive to the relative expansion history and geometry of the Universe (e.g., Paczynski and Gorski 1981; Link and Pierce 1998; Cooray 1999; Golse et al. 2002; Sereno 2002; Soucail et al. 2004; Gavazzi et al. 2008; Gilmore and Natarajan 2009). Link and Pierce (1998) showed that the cosmological sensitivity of the angular size-redshift relation could be exploited using sources at distinct redshifts and developed a methodology to simultaneously invert the lens and derive cosmological constraints.

In particular, galaxy clusters with a large strong lensing cross section do have multiple sources at different redshifts and are exquisite objects to study the geometrical effect between different source redshifts. In fact, early studies of galaxy clusters already indicated the presence of a cosmological constant (e.g., Paczynski and Gorski 1981; Sereno 2002; Soucail et al. 2004).

The method to probe relative angular diameter distances, in addition to multiple sources at different redshifts, also requires a complete understanding of the lens density profile and other perturbing masses along the line of sight.

With more exquisite deep multi-colour imaging and spectroscopy for a small subset of galaxy clusters, such as with the Hubble Frontier Fields program (Lotz et al. 2017) has led to the discovery of hundreds of multiple images and thus to a significant improvement of cluster mass estimates (Jauzac et al. 2014; Diego et al. 2016; Lagattuta et al. 2017; Monna et al. 2017). Cluster lenses are complex and most of the efforts have been spent in accurately reconstruct cluster lensing profiles (see e.g., Jullo et al. 2010; D'Aloisio and Natarajan 2011; Magaña et al. 2015; Caminha et al. 2016; Acebron et al. 2017). The mass modelling of strong lensing clusters can be carried out in different manners: parametric and non-parametric methods are equally used; the primary distinction between them being that parametric modelling assumes that luminous cluster galaxies trace the cluster mass whereas non-parametric does not.

The method of using sources at multiple redshifts can also be applied to galaxy-scale lenses, thought double source plane lenses are more rare Biesiada (2006), Gavazzi et al. (2008), Collett et al. (2012). The Einstein radius is a function of the lens mass and the cosmological distances. The ratio of Einstein radii in a lens with sources at two or more redshifts is independent of the deflector mass (e.g., Gavazzi et al. 2008; Collett et al. 2012).

In both cases, galaxy and cluster scales, the method also requires a complete understanding of the lens density profile and additional lensing by the source galaxies and other perturbing masses along the line of sight.

8.2 H_0 with Galaxy Clusters

To date, galaxy-scale lenses have dominated the literature on H_0 determination in the number of measurements and precision. We have recently witnessed competitive constraints from galaxy clusters in measuring H_0 (Kelly et al. 2023a; Napier et al. 2023; Liu et al. 2023; Pascale et al. 2024) and other cosmological quantities (Grillo et al. 2024). Massive clusters are rich in multiple images and have definite advantages over individual galaxies. The main one is that sources at multiple redshifts break the mass-sheet or steepness degeneracy (e.g., Bradač et al. 2004), which is the main degeneracy and hence source of uncertainty affecting galaxy-scale determination of H_0 (see also Sect. 5). Having a much larger image separation in clusters compared to galaxy-scale lenses resulting in overall longer time delays of order months to years. Those time delays are relatively easily determined to a few percent precision, rivalling time-delay determinations from quasar sources in galaxy-scale lenses. However, the longer time-delay implies years of monitoring to obtain lightcurves with sufficient overlap in the case of lensed quasars (Fohlmeister et al. 2008, 2013; Muñoz et al.

2022, e.g.) or dedicated HST follow-up at the time of reappearance in the case of lensed supernovae (Kelly et al. 2023b). The drawback of clusters is that their mass distributions are more complex: they are dynamically younger than galaxies, and their multiple image regions sample a much larger fraction of the clusters' virial radius than in galaxies. Therefore the multiple image region of clusters is expected to be more abundant in substructure, and hence harder to model. These difficulties can be circumvented if there are a few tens or hundreds of multiple images, then H_0 can be estimated to a 1-few % precision (Ghosh et al. 2020). At present, in a cluster lens like MACS 1149, one can estimate H_0 to 6%, assuming a conservative 3% uncertainty on the observed time delay (Grillo et al. 2018).

The first cluster lens to produce a precise estimate of H_0 was MACS 1149, where the first confirmed multiply imaged supernova was observed a few years ago (Kelly et al. 2015). The long time delay before the reappearance of the last arriving image—saddle in the arrival time surface of the cluster—allowed the lensing community to make model predictions for the time of the reappearance. Most models agreed reasonably well on 250-350 day delay (Kelly et al. 2016; Treu et al. 2016). Very recently, Frye et al. (2024) discovered another lensed supernovae in a cluster, this time of type Ia, which even allows for a standardization of the lensing magnification (Pierel et al. 2024a), and (Pierel et al. 2024b) discovered for the first time a second supernovae in the same host galaxy, with the initial supernovae discovered in archival data by (Rodney et al. 2021), further demonstrating the future prospects using cluster-scale lenses.

9 Current Status and Results

We go in length through recent results using lensed quasars and also present the recent results using lensed SNe. Figure 11 summarizes a selection of current measurements and comparison with other probes.

9.1 Recent Results from Quasars

The independent analysis of six lensed quasar systems (Suyu et al. 2010, 2013; Wong et al. 2017; Bonvin et al. 2017; Birrer et al. 2019; Chen et al. 2019; Rusu et al. 2020) by the H0LiCOW collaboration (Suyu et al. 2017) inferred a Hubble constant value of $H_0 = 73.3^{+1.7}_{-1.8}$ km s^{-1}Mpc^{-1}. This measurement uses parameteric forms of the mass density profile of the deflector, either described as a power-law or stars (constant mass-to-light ratio) plus dark matter halos following an NFW (Navarro et al. 1997) profile with priors on the mass and concentration of the halo reflecting the population of haloes in N-body simulations (Wong et al. 2020). The H0LiCOW result is a 2% precision measurement on H_0, in excellent agreement with the local distance ladder measurement by the SH0ES team (Riess et al. 2019, 2021). Moreover, the H0LiCOW measurement is not more than 3σ statistical tension with early-Universe probes (e.g., Planck Collaboration 2020; Aiola et al. 2020). An additional lens analyzed by the STRIDES collaboration with the same mass profile assumptions as the H0LiCOW collaboration further provided the most precise single-lens measurement of $H_0 = 74.2^{+2.7}_{-3.0}$ km s^{-1}Mpc^{-1} (Shajib et al. 2020). In summary, if the mass density profiles of the H0LiCOW and STRIDES lenses are well described by a power-law or a baryonic component with a constant mass-to-light ratio plus dark matter profiles from standard N-body dark matter only simulations, and under the assumption that the covariances are negligible, the tension is significant from the strong lensing measurements alone, and corroborating other measurements (e.g., Riess et al. 2021).

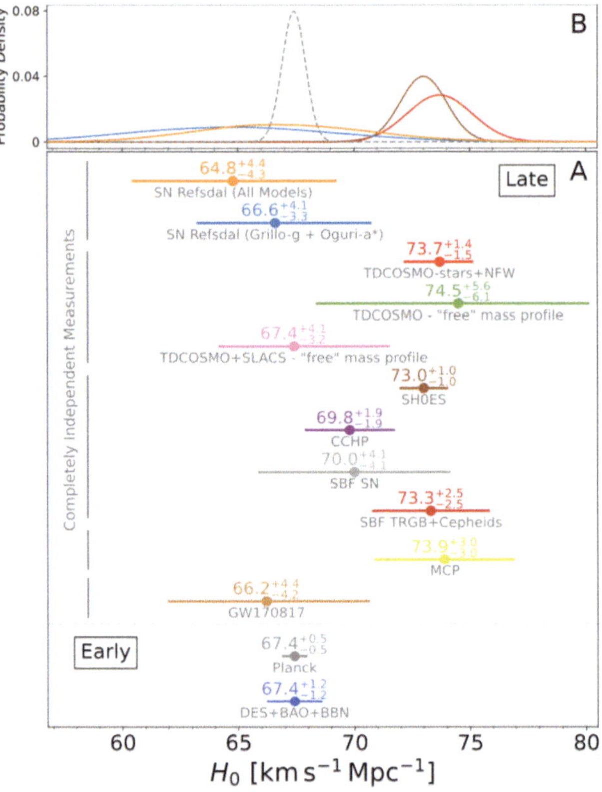

Fig. 11 Comparison of recent H_0 measurements in the literature. Presented are the time-delay cosmography constraints from Kelly et al. (2023a) of SN Refsdal, for the full set of eight models (orange) and the subset of the two best models (blue), from the TDCOSMO collaboration (red) of six TDCOSMO time-delay lenses (five H0LiCOW lenses (Wong et al. 2020) and one STRIDES lens by Shajib et al. (2020) assuming parameteric forms of the mass density profile of the deflector, either described as a power-law or stars (constant mass-to-light ratio) plus dark matter halos (Millon et al. 2020b), from the TDCOSMO collaboration of the same lenses as in Millon et al. (2020b) with virtually no assumption on the radial mass density profile of the lens galaxy, and taken into account the covariance between the lenses (green) (Birrer et al. 2020). The TDCOSMO + SLACS measurement comes from the joint analysis of the TDCOSMO sample and 33 SLACS lenses with SDSS spectroscopy. The "free" mass profile assumptions of the two measurements by Birrer et al. (2020) are constrained only by the stellar kinematics and fully accounts for the uncertainty related to the mass sheet transformation (MST). Aside from time-delay studies, the local measurements by SH0ES + Gaia (Riess et al. 2022), the Carnegie-Chicago Hubble Program (CCHP) (Freedman et al. 2019), surface brightness fluctuations (SBF) SN (Khetan et al. 2021), SBF Tip of the Red Giant Branch (TRGB) + Cepheids (Blakeslee et al. 2021), Megamaser Cosmology Project (MCP) (Pesce et al. 2020), gravitational wave (GW) event 170817 (Dietrich et al. 2020), Planck (Planck Collaboration 2020, ; dashed grey), and Dark Energy Survey (DES) + Baryon Acoustic Oscillation (BAO) + Big Bang nucleosynthesis (BBN) (Abbott et al. 2018) are presented. Error bars in panel (B) show the 16th, 50th, and 84th percentile confidence levels. Dashed horizontal line separates measurements from observations of the universe early in its evolution from those late in its evolution. H_0 measurements bracketed by different vertical gray bars are entirely independent of each other. Figure from Kelly et al. (2023a) which was generated using a previous comparison (Bonvin and Millon 2020)

Given that incompatibilities between the local value of H_0 and the ΛCDM model extrapolated H_0 inference from the CMB or other early-universe physics anchored inherently break the standard model of cosmology and likely may require new physics, several groups, including H0LiCOW, STRIDES and SHARP, now TDCOSMO collaboration, are investigating potential systematics in the H_0 measurements.

The TDCOSMO collaboration found, combining six lenses from H0LiCOW, SHARP and STRIDES, that the results when assuming that all lenses are either of one or the other previously assumed forms of the mass density profile are in good agreement with each other when measuring H_0. The good agreement in the H_0 results between power-law and composite profiles was interpreted by Millon et al. (2020b) as a consequence of the 'bulge-halo conspiracy' that the combined baryonic and dark matter density components form a power-law profile (e.g., Koopmans et al. 2006, 2009; van de Ven et al. 2009). Denzel et al. (2021) analyzed 8 quadruply imaged quasars with a free-form modeling approach and obtained $H_0 = 71.8^{+3.9}_{-3.3}$ km s^{-1}Mpc^{-1}. Gilman et al. (2020) investigated the effect of unaccounted for subhalos and small undetected line-of-sight halos in the uncertainty budget and found insignificant residual uncertainties to mitigate the tension of the measurements with the CMB and large scale structure probes. Van de Vyvere et al. (2022b,a) showed that a variety of expected azimuthal structures in the mass distribution (i.e. multipoles, twists and ellipticity gradients) should leave H_0 unaffected at the population level unless there are specific selection effect in the galaxy population.

The attention further turned to assessing and relaxing the radial profile assumption (see Sect. 5.3), as well as the introduction of population priors for parameters that cannot be constrained on a lens-by-lens basis for a covariant treatment of their uncertainties. Birrer et al. (2020) addressed the radial profile assumption by choosing a parametrization of the radial mass density profile that is maximally degenerate with H_0, via the MST. This is the most explicit and direct way addressing the MST effect on the time-delay cosmographic analysis. With this more flexible parametrization, H_0 is only constrained if the measured time delays and imaging data are supplemented by stellar kinematics. Applying this extremely conservative choice to the TDCOSMO sample of 7 lenses increases the uncertainty on H_0 from 2% to 8% resulting in $H_0 = 74.5^{+5.6}_{-6.1}$ km s^{-1}Mpc^{-1}, without changing the mean inferred value significantly.

Birrer et al. (2020) further introduced a hierarchical framework (see also Sect. 7.2) in which external datasets can be combined with the time-delay lenses to improve the precision, in particular on the MST parameter of the population, and hence on H_0. A secondary required parameter that must be constrained when using stellar kinematics is the stellar anisotropy, due to the mass-anisotropy degeneracy. External data sets with spatially resolved kinematics measurements can aid breaking this degeneracy to constrain the MST parameter. Birrer et al. (2020) achieved a 5% precision measurement on H_0 by combining the TDCOSMO lenses with imaging modeling and stellar kinematic measurements of a sample of lenses from the Sloan Lens ACS (SLACS) survey with no time-delay information (Bolton et al. 2008; Auger et al. 2009; Shajib et al. 2021) and measured $H_0 = 67.4^{+4.1}_{-3.2}$ km s^{-1}Mpc^{-1}. The mean of the TDCOSMO + SLACS measurement is offset with respect to the TDCOSMO-only value, effectively matching the CMB inferred value, although still statistically consistent with previous assumptions given the uncertainties in the measurement. The Birrer et al. (2020) measurements with and without the SLACS dataset added are in statistical agreement with each other and with the earlier H0LiCOW/SHARP/STRIDES measurements based on the radial mass profile assumptions. The result by Birrer et al. (2020) is also consistent, with the work by Shajib et al. (2021) studying more flexible mass density profiles and mass-to-light gradients. The studies by Birrer et al. (2020) and Shajib et al. (2021) share the same measurements for the SLACS lenses and the consistency is implied by construction. Shajib et al. (2021) concluded that NFW + stars, when using wider priors on mass and concentration than earlier H0LiCOW/SHARP/STRIDES measurements, is a sufficiently accurate description of the mass density profile of the SLACS lenses. However, a larger flexibility in the mass-concentration relation on the population level and small departures from those radial forms are allowed by the data, resulting in the uncertainties reflected by the Birrer et al. (2020) analysis. The shift in the mean in H_0 when adding the SLACS lenses could be real or it could be due to an intrinsic difference between the deflector population in the TDCOSMO and SLACS samples. Differences in the deflectors might arise from unequal selection effects. For example, the two samples are well matched in stellar velocity dispersion (total mass), but they differ in redshift. Potentially unaccounted for evolutionary trends in the mass profiles could bias the results when adding samples of lenses at different redshifts. Another example is that the TDCOSMO sample is source selected, meaning the main characteristics for the data set to be discovered and selected are properties of the source as seen when lensed, and composed mostly of quadruply imaged quasars, while the SLACS sample is deflector selected, meaning the primary criteria for the selection arises from properties of the deflector irrespective of the source and its geometric lensing effect, and dominated by doubly imaged galaxies.

More recently, Shajib et al. (2023) performed the first analysis with spatially resolved stellar kinematics measurement with the same conservative assumptions as Birrer et al. (2020) and achieved a $\sim 9\%$ measurement from a single quadruply lensed quasar, finding results in agreement with the previous analysis based on power law and composite models Suyu et al. (2014). This work demonstrates the constraining power of kinematic data in the absence of priors on the shape of the mass density profile.

9.2 Recent Results from Lensed Supernovae

In addition to lensed quasars, the discovery of the first multiply-imaged supernovae (SN) (Kelly et al. 2015) in the cluster MACS 1149 allowed to measure H_0 with lensed SNe (Vega-Ferrero et al. 2018; Kelly et al. 2023a). Kelly et al. (2023a) presents results of a combination of several models from different independent modeling teams which were done truly blind, before the measured time delays were known. A Bayesian model selection was done based on the precision and accuracy of the predicted image positions of the reappearance of SN Refsdal, and the predicted magnification ratio (which are independent of H_0). Combining these Bayesian weights with the weighted uncertainties of all the eight individual models, they found $H_0 = 64.8^{+4.4}_{-4.3}$ km s^{-1}Mpc^{-1}. Kelly et al. (2023a) found that models that assign dark-matter halos to individual galaxies and the overall cluster best reproduce the observations. When combining the two best performing models, both consistent within their uncertainties with each other, Kelly et al. (2023a) found $H_0 = 66.6^{+4.1}_{-3.3}$ km s^{-1}Mpc^{-1}. Very recently, a H_0 measurement was made by a second lensed SN, supernova "H0pe" (Frye et al. 2024). A combination of a spectroscopic and photometric time-delay measurement (Chen et al. 2024; Pierel et al. 2024a) were compared to the predictions of seven independently constructed cluster lens models to measure a value for the Hubble constant. In combination with the standardizable magnification of the type Ia nature of the SN, this resulted $H_0 = 75.4^{+8.1}_{-5.5}$ km s^{-1}Mpc^{-1}. These are very encouraging and precise measurements. For further discussions of lensed SNe, we refer to Suyu et al. (2024, this collection).

10 Outlook in the (Near) Future

The goal of time-delay cosmography is to provide a robust measurement of the Hubble constant to 1% precision to decisively tell the outcome of the currently observed tension between late and early time measurements of H_0. In the previous Sect. 9 we presented current results. In this section, we discuss the potential of the time-delay method in the near future. To do so, we first provide some details on the error budget of current analyses which allows us to assess and scale to the expected future samples and data quality (Sect. 10.1). Second, we describe the data and instrumentation which enable us to push ahead (Sect. 10.2). Third, we will highlight avenues where continuing work is required in assessing the methodology to maintain accuracy while increasing the precision of the measurements (Sect. 10.3). Finally, we leave some concluding remarks about the prospects of time-delay cosmography in Sect. 10.4.

10.1 Error Budget

Table 1 presents an overview of the error budget of time-delay cosmography divided into different aspects of the analysis for individual lenses, the current work of 7 lenses, and a forecast of a future analysis of 40 lenses with improved data. The error budget is split between the three different components of time-delay cosmography: The time-delay measurement (Sect. 4), the main deflector profile for the prediction of Fermat potential (Sect. 5), and the line-of-sight contribution (Sect. 6). For the deflector error budget, we provide an uncertainty on the lens model excluding the potential systematics of the MST, which can be achieved with high-resolution imaging, and a total error budget including the MST, which depends on additional data (such as currently stellar kinematics of the deflector). For the analysis of individual lenses, we report uncertainties from Wong et al. (2020), Millon et al. (2020b) excluding the MST, and from Birrer et al. (2020) including the MST in the deflector profile. The errors between the time-delay measurement, deflector model and line of sight are effectively uncorrelated, and hence the total error budget adds in quadrature. Excluding the MST-related uncertainties, the three components are roughly on equal footing, and the best single lenses alone can provide an H_0 uncertainty of $< 4\%$ (e.g., Shajib et al. 2020). With stringent priors on the deflector potential ignoring additional MST-related uncertainties, the current sample of 7 lenses assuming uncorrelated uncertainties result in a $\sim 2\%$ uncertainty on H_0 (Wong et al. 2020; Millon et al. 2020b).

The MST-related uncertainty for an individual lens depends on the imposed priors and can be constrained to a $\sim 10-20\%$ level with current pre-JWST kinematic measurements.

Table 1 Approximate error budget of time-delay cosmography divided into different aspects of the analysis for individual lenses, the current work by Birrer et al. (2020) of 7 lenses, and a forecast of a future analysis of 40 lenses with improved data. The error budget is split between the three different components of time-delay cosmography: The time-delay measurement (see Sect. 4), the main deflector profile for the prediction of Fermat potential (see Sect. 5, and the line-of-sight contribution (see Sect. 6. For the deflector error budget, we provide an uncertainty on the lens model excluding the potential systematics of the MST (ex MST), which can be achieved with high-resolution imaging, and a total error budget including the MST, which depends on additional data (such as currently stellar kinematics of the deflector). The forecast for 40 lenses is based on the same mix of quality in the data as for the current 7 lens constraints

	Individual lens	7 lens sample	40 lens sample
Time delay	$1-20\%$	$\lesssim 0.5\%$	$\lesssim 0.1\%$
Deflector model (ex MST)	$1-10\%$	$\sim 0.8\%$	$\lesssim 0.3\%$
Deflector model	$10-20\%$	7%	$\lesssim 1\%$
Line of sight	$1-5\%$	$\lesssim 0.5\%$	$\lesssim 0.2\%$
Total (ex MST)	$3-20\%$	2.3%	$\lesssim 0.5\%$
Total	$10-25\%$	8%	$\sim 1\%$

The combination of 7 lenses by Birrer et al. (2020) lead to a $\sim 8\%$ uncertainty on H_0, dominated by a $\sim 7\%$ uncertainty on the MST. Given the uncertainties and their expected scaling ($\Delta H_0/H_0 \propto \sqrt{N}$ for N number of lenses), for 40 lenses, the MST-related uncertainty is the dominant and single most relevant uncertainty to achieve a 1% measurement of H_0. The required data and methodology improvements are laid out in the following subsections.

The secondary uncertainties of the relative expansion history on H_0 is $< 1\%$, not mentioned in Table 1, but included in the total error budget on the population and can be mitigated by either an informed prior on the relative expansion history, or by a combination with probes sensitive to the relative expansion history.

10.2 Future Data

We expect there to be several 10,000 galaxy-galaxy lenses, several hundred quadruply lensed quasars and more than a thousand doubly lensed quasars on the full sky (e.g., Oguri and Marshall 2010; Collett 2015). With the upcoming large area (wide) and sensitive to faint objects (deep) ground- and space-based surveys, such as the Vera C. Rubin Observatory, Roman Observatory, and Euclid, we expect many of those lenses to be discovered within a decade. Compared to the current analyses conducted on few lenses (e.g., 7 lenses in case of current TDCOSMO results), these are several e-foldings of the number of lenses possibly suitable for time-delay analyses. The sheer number of lenses will transform the measurements and new approaches are going to be required in the domain of time-delay cosmography to efficiently and accurately make use of all the data available.

The first step in utilizing these lenses present on the sky is to discover them in the large data sets. We refer to Lemon et al. (2024, this collection) for an extensive review of techniques, recent successes and an outlook in the searches and discovery of strong lenses. The next step is to acquire all the necessary follow-up data products to conduct accurate and precise cosmographic analyses (see Sect. 3 and subsequent sections). The data products range from monitoring data for a time-delay measurements, high-resolution imaging, to spectroscopic information about the source and lens redshift as well as velocity dispersion of the deflector. This step is very resource expensive and there are going to be challenges in how to allocate these limited resources. Decisions will have to be made to decide which lenses are followed-up. We comment in Sect. 10.3 about developments of methodology that can deal with less constraining or incomplete data for a larger lensing data set. Some lenses might require less substantial monitoring follow-up in case where LSST light curves are good enough for a time-delay measurement (Liao et al. 2015). Some lenses may also automatically obtain high-resolution and sufficiently high signal-to-noise ratio imaging data from wide field space surveys, such as Euclid or Roman (Meng et al. 2015). Understanding to what extent the acquired data products impact the precision on H_0 is key to assess the need for allocating follow-up resources and on which lenses to spend it. Besides the limited resources, follow-up decisions are currently also impacted by the limited access to adaptive optics (AO) instrument on ground-based large-diameter telescopes. With the next-generation AO instrumentation and their commissioning on both hemispheres, we expect a full sky accessibility that allows the community to target every single gravitational lens on the sky.

The dominant uncertainty in the current measurement of the Hubble constant with time-delay cosmography is attributed to uncertainties in the mass profiles of the main deflector galaxies (see e.g., Sect. 5.3). There are multiple independent avenues available in the near future to approach a 1% measurement of H_0 with different data sets. We will focus on these pathways with improved instrumentation and increased data sets in this section.

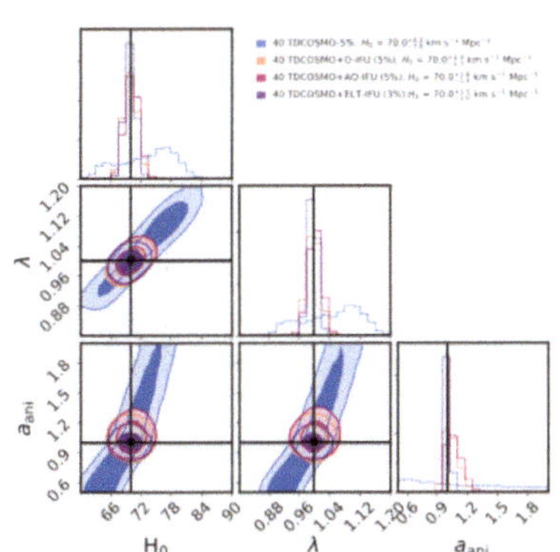

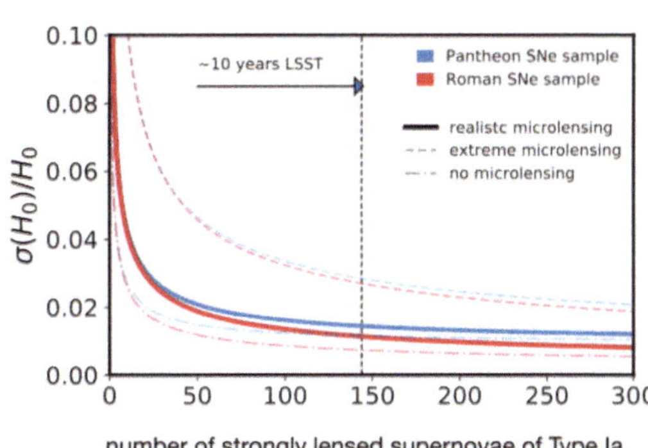

Fig. 12 Forecast for H_0 measurements in the near future with the upcoming facilities. Left: Spatially resolved kinematics measurements of a sample of 40 time-delay lenses enable a precision on H_0 of 1.5% JWST (Figure adopted from Birrer and Treu 2021). Right: Lensed supernovae with standardizable magnification measurements. An expected yield of $\sim$ 144 gravitationally lensed supernovae over the span of the 10 years LSST survey enable a precision on H_0 of 1.5% (Figure adopted from Birrer et al. 2022). Both approaches, stellar kinematics and standardized magnifications, do provide independent observational constrains on the MST with different systematics

Spatially resolved stellar kinematics of the deflector galaxy (see Sect. 5.4 for details on methodology) with the next generation space (James Webb Space Telescope; JWST) and ground-based (extremely large telescopes; E-ELT, GMT, TMT) instruments provide precise measurements of the kinematics of stars. Such two-dimensional observations of the kinematics, paired with the lensing measurements, have the ability to break the mass-anisotropy degeneracy, a currently limiting systematic when interpreting and de-projecting integrated kinematic measurements to measure the three-dimensional gravitational potential. Birrer and Treu (2021) forecasts, based on the methods and assumptions used by Birrer et al. (2020) without relying on mass-density profile assumptions to break the MST, that with 40 time-delay lenses with exquisite spatially resolved kinematics and otherwise similar measurements as the 7 lens TDCOSMO sample, a 1.5% precision on H_0 can be achieved (Fig. 12 left graphic), and see also e.g. Yıldırım et al. (2020, 2023). Such a strategy with exquisite data on the sample of time-delay lenses is one way to make progress. Another approach is to infer the mass density profile properties from a larger set of non-time-delay lenses and apply the constraints on the mass density profile and stellar anisotropy distribution on the time-delay lenses (Birrer et al. 2020; Birrer and Treu 2021; Gomer et al. 2022). In particular, resolved spectroscopy can also be employed on non-time delay lenses without bright and contaminating quasar images, either as prior constraints or by directly incorporating into a hierarchical analysis, to further improve the kinematic measurement precision.

Standardizable magnifications with gravitationally lensed supernovae (glSNe) provide another promising avenue to constrain the mass density profiles and open up an avenue for a percent measurement of H_0 in the near future with the onset of LSST. Standardizable magnifications are able to constrain the absolute lensing magnification and hence constrain the density profile (incl. the MST) (e.g., Kolatt and Bartelmann 1998; Oguri and Kawano 2003; Foxley-Marrable et al. 2018; Birrer et al. 2022). Birrer et al. (2022) (Fig. 12 right panel) provides a forecast with glSNe in constraining H_0 independently of stellar kinematics. They conclude that the standardizable nature of glSNe of type Ia enables a 1.5% H_0 measurement with a 10 years LSST survey. This forecast is contingent to a near-optimal discovery and follow-up effort of glSNe. We refer to Suyu et al. (2024) for a detailed review and in-depth discussion on the discovery, expected number of glSNe, the challenges of following them up and the caveats of micro-lensing.

Another method is to make use of the statistical distribution of images under the assumption of knowing the distribution of sources in the source plane with a statistical combination of a large sample of time-delay lenses, relying purely on strong lensing data (Sonnenfeld 2021).

Yet another method to constrain the radial density profile is to use galaxy-galaxy weak gravitational lensing for a large sample of deflectors analogue to the strong gravitational lenses (Khadka et al. 2024).

Overall, the trade-offs of analysing all (or most) of the lenses, with most of them with limited data, or to focus on a few of the best (e.g. "golden") lenses, has yet to be seen and explored in detailed. Different approaches have advantages and inconveniences in regard to precision and accuracy on the H_0 measurements.

10.3 Methodology Improvements

With the expected wealth of data and the increase in the number of time-delay and non-time-delay lenses, the prospect of measuring H_0 to 1% precision can become a reality. The employed methodology and assumptions must keep up to provide the accuracy requirement. In the following we discuss methodology improvements and validations in the domain of galaxy density profiles (Sect. 10.3.1), assumption in the interpretation of non-lensing constraints (Sect. 10.3.2), selection effects (Sect. 10.3.3), automatization (Sect. 10.3.4) and general aspects of methodology verification (Sect. 10.3.5). These sections are not meant to be complete but to provide guidance in the near future on where focused effort is required.

10.3.1 Galaxy Density Profiles

The currently employed model mitigating the MST effect by Birrer et al. (2020) is parameterized with a pure MST parameter λ. This parameterization is foremost of mathematical nature and leaves the physical interpretation (e.g., Blum et al. 2020) ambiguous. A pure MST parameterization may in certain regimes even become unphysical, e.g. resulting in total mass profiles with negative density in the outskirts.[8] Such a one-parameter extension to previously considered more simple and rigid mass profiles may also not encompass the necessary flexibility beyond the pure MST that can affect kinematics observations (e.g., Birrer et al. 2020; Yıldırım et al. 2023), or to deal with more generalized forms of lensing degeneracies, such as the SPT. To make progress, the full degeneracy inherent in gravitational lensing needs to be folded into flexible, but physically motivated, mass profile parameters. Such an approach was explored by (Shajib et al. 2021) constraining the extended mass density profiles of the SLACS galaxy-galaxy lenses, but has not yet been employed for time-delay cosmography. Quasar microlensing studies might also help to constrain the stellar mass to light ratio in massive elliptical galaxies. Ambitious measurements below the 10% level might additionally help to constrain the mass density profiles and would allow the focus on the dark matter portion of the profile. We refer to Vernardos et al. (2024) for techniques and prospects of this methodology.

[8] Projected lensing convergence can come slightly negative to the extent that the physical mass drops below the mean background density.

10.3.2 Non-lensing Constraints

Significant constraints on the MST, and in general mass density profiles, are expected to come from non-lensing observables. These measurements, as well as their model interpretation, need to be tested to the percent level. For example, for the kinematic measurements, the impact of stellar template fitting needs to be further assessed and validated. For the interpretation of the measurements, de-projection assumptions, rotational structure, as well as stellar anisotropy need to be rigorously tested and assessed for covariant systematics on the population level. For upcoming magnification measurements with glSNe, micro- and milli-lensing effects need to be assessed and incorporated into the model self-consistently. Furthermore, more knowledge about the structure and size of the variable quasar accretion disc are required to determine the strength of the micro-lensing time delay effect.

10.3.3 Selection Effects

The phenomena of strong gravitational lensing is inherently a very specific selection effect of an otherwise mostly weak lensing field. Quantifying the selection effect of where and in what form strong lensing phenomena occur is going to be crucial requirement to maintain accuracy when increasing precision on H_0 in the years to come. The strong lensing phenomena is impacted by both, the nature of line-of-sight structure, and the main deflector. For the line-of-sight structure, the convergence either raises or lowers the lensing efficiency of an equal mass galaxy to act as a strong lensing deflector, and the cosmic shear changes the geometry of the caustic structure, making it more likely to have quadruply imaged sources. Similarly, for the main deflector, more concentrated mass distribution, or favorable projections along the line of sight, lead to higher lensing efficiencies, and more elliptical mass profiles (also in projection) lead to a more extended inner caustic region. Including the differential selection effects among different samples of lenses is required when combining information coming from differently selected populations. For example, quadruply lensed quasars are visible only when the source quasar lies within the diamond caustic of the lensing galaxy. This condition creates a Malmquist-like selection effect in the population of observed quadruply lensed quasars, increasing the true caustic area (Baldwin and Schechter 2024).

Many of these effects are hard or near-impossible to quantify on a lens-by-lens basis. These selection effects need to be modeled and inferred on the population level, with the focus of making sure that relative selection effects between different sub-populations are being understood.

There are two distinct and complementary approaches to understand and mitigate selection effects in the analysis. First, one can attempt to understand the selection from first principles which then can be used to explicitly account for in the analysis procedure. This approach requires extensive simulations including all relevant aspects, starting from the full sky a prior abundance and population of phenomena and a reproducible selection function of the discovery channel and follow-up decision being made to either include a lens in the sample or not. For example, Collett and Cunnington (2016) simulated a sample of double- and quadruple-image systems and when assuming reasonable thresholds on image separation and flux, based on current lens monitoring campaigns, they found that the typical density profile slopes of monitorable lenses are significantly shallower than the input ensemble. Second, one can empirically determine a relative selection function by comparing a set of observables of a sample of lenses compared to random galaxies or sight lines on the sky. Deviations on the set of observables on the population level indicate then the level of selection bias in the sample. Observables may include, but are not restricted to, central velocity dispersion, stellar mass, size and morphology of the deflector, number of subhaloes and line-of-sight projected galaxies nearby, redshift of the deflector, among others. Deviations from established scaling relations among the galaxy properties are then indications of selection biases We refer to Sect. 6 for data and approaches to quantify line of sight effects. We also stress that these techniques rely on underlying priors and model assumption on the population bias and an explicit de-biasing is required to constrain hierarchically unknown selection effects (see e.g., Park et al. 2023). Currently, neither of the two approaches have been successfully applied.

With the expected large number of lenses in the next few year of the mid 2020s, and the more uniform data set of large and deep surveys, both, the theoretical forward modeling and the empirical hierarchical modeling, will become feasible. We also advocate for analyses to take into account the specific discovery channel of the lenses when performing population level inferences. Understanding the selection function may or may not imply to effectively re-discover the lenses in the analysis to guarantee a uniform and reproducible selection and analysis.

10.3.4 Automatization

Current state-of-the-art analyses of single lens systems takes up more than a year of work, with the involvement of many people, as well as several hundred of thousands of CPU hours of computational cost. To utilize the upcoming larger lens samples and to achieve a high-precision H_0 measurement, the time to analyse a single system has to be reduced significantly. Automated decision-making and model choices (e.g., Schmidt et al. 2023; Ertl et al. 2023), as well as GPU assisted computations (e.g., Gu et al. 2022) hold promises in these regards. Moreover, analyses have to be able to be repeated with modifications to test for assumptions covariant among all lenses multiple times. The faster the entire analysis runs, the more explorations of potential systematics in the choices can be executed. The challenge in finding uniform analyses choices are that every lens is different from another and particularities have been notices that needed special attention for lenses on the individual basis. The analyses conducted need to be uniform in their choices and approaches such that impacts on assumptions can be tested on the ensemble level. Uniformity of analyses can also reduce human errors and sets the analyses on quantifiable priors.

There is currently an effort in homogenizing the analysis procedure, for both time-delay lenses (Shajib et al. 2019; Schmidt et al. 2023; Ertl et al. 2023) and non-time delay lenses (Shajib et al. 2021) and further effort is underway. In parallel, alternative methodology in the modeling and posterior inference are being explored with machine learning techniques, which have the potential to speed up the analysis by orders of magnitude (e.g., Park et al. 2021).

10.3.5 Methodology Verification

Guaranteeing accuracy with ever more precise measurements is a challenge throughout the cosmological community. High-precision measurements of quantities to relevance of fundamental physics is a relatively new field and we dedicate a separate subsection highlighting different strategies to verify the methodology and to perform to the necessary quality standard to maintain accuracy.

- Realistic simulations offer a validation of a methodology on a known truth (see e.g., Xu et al. 2016; Tagore et al. 2018). It is important that the complexity in the simulations are realistic to explore avenues of potential systematics and gain a deep understanding of what data products are able to constrain what aspects of the model. Simulations eventually need to encompass all aspects of the analysis, including the selection effect and the entire line-of-sight structure within the full cosmological and astrophysical context.

- Data modeling challenges, such as the Time-Delay Challenge (Liao et al. 2015) and the Time-Delay Lens Modeling Challenge (Ding et al. 2021) offer platforms to validate currently employed methodology on mock data sets, explore new ways of analyzing the data and can provide a transparent overview of the current state of the field.
- Blind analyses prevent experimenter bias. The analysis should be guided by the assessment of uncertainties regardless of the anticipated result. Blind analyses have regularly been performed by the H0LiCOW and TDCOSMO collaborations.
- Open source accessibility of the raw data, processed data product, analysis software and entire end-to-end analysis pipelines can best guarantee reproducibility, form community trust and provides access to the community to alter and improve existing methodology.

10.4 Concluding Remarks

Time-delay cosmography has an exciting time ahead. The method has come along way since its original proposal by Refsdal (1964). Current measurements of the Hubble constant with time-delay cosmography are at the few percent level, enabled by detailed analyses and precise measurements of different aspects of the analysis. With the expected increase in the lensing sample and the advances in instrumentation, the path towards a percent precision measurement of H_0 becomes in reach.

Measuring the Hubble constant to percent level precision is a challenging endeavor, regardless of the cosmological probe. In this manuscript, we aimed to provide a detailed account of the methodology and measurements to provide guidance to achieve a precise and accurate measurement of H_0 at the one-percent level. We emphasized the challenges and systematics in the different components of the analysis and strategies to mitigate them. Above all, in Carl Sagans words: "Extraordinary claims require extraordinary evidence".

Acknowledgements We thank the International Space Science Institute in Bern (ISSI) for their hospitality and the conveners for organizing the stimulating workshop on "Strong Gravitational Lensing". We thank Liliya Williams for contributions to Sect. 8 and useful comments, and Paul Schechter for useful comments in the preparation of the manuscript. SB is partially supported by the Kavli Foundation and Stony Brook University. MM acknowledges support by the SNSF through grant P500PT_203114 and funding from the European Research Council (ERC) under the European Union's Horizon 2020 research and innovation programme (COSMICLENS: grant agreement No 787886). AJS is supported by NASA through the NASA Hubble Fellowship grant HST-HF2-51492 by the Space Telescope Science Institute, which is operated by the Association of Universities for Research in Astronomy, Inc., for NASA, under contract NAS5-26555. FC and DS acknowledge funding from the European Research Council (ERC) under the European Union's Horizon 2020 research and innovation programme (COSMICLENS: grant agreement No 787886). SHS thanks the Max Planck Society for support through the Max Planck Research Group. This research is supported in part by the Excellence Cluster ORIGINS which is funded by the Deutsche Forschungsgemeinschaft (DFG, German Research Foundation) under Germany's Excellence Strategy – EXC-2094 – 390783311. TT acknowledges support by the National Science Foundation through grants AST-1906976 and AST-15436, by NASA through grants HST-GO-15652, and by the Gordon and Betty Moore Foundation through grant 8548.

Funding Open Access funding provided thanks to the CRUE-CSIC agreement with Springer Nature.

Declarations

Competing Interests The authors declare no competing interests.

Open Access This article is licensed under a Creative Commons Attribution 4.0 International License, which permits use, sharing, adaptation, distribution and reproduction in any medium or format, as long as you give appropriate credit to the original author(s) and the source, provide a link to the Creative Commons licence, and indicate if changes were made. The images or other third party material in this article are included in the article's Creative Commons licence, unless indicated otherwise in a credit line to the material. If material is not included in the article's Creative Commons licence and your intended use is not permitted by statutory regulation or exceeds the permitted use, you will need to obtain permission directly from the copyright holder. To view a copy of this licence, visit http://creativecommons.org/licenses/by/4.0/.

References

Abbott TMC, Abdalla FB, Annis J et al (2018) Dark energy survey year 1 results: a precise H_0 estimate from DES Y1, BAO, and D/H data. Mon Not R Astron Soc 480(3):3879–3888. https://doi.org/10.1093/mnras/sty1939

Abdalla E, Abellán GF, Aboubrahim A et al (2022) Cosmology intertwined: a review of the particle physics, astrophysics, and cosmology associated with the cosmological tensions and anomalies. J High Energy Astrophys 34:49–211. https://doi.org/10.1016/j.jheap.2022.04.002

Acebron A, Jullo E, Limousin M, Tilquin A, Giocoli C, Jauzac M, Mahler G, Richard J (2017) Hubble Frontier Fields: systematic errors in strong lensing models of galaxy clusters - implications for cosmography. Mon Not R Astron Soc 470(2):1809–1825. https://doi.org/10.1093/mnras/stx1330

Aghamousa A, Shafieloo A (2015) Fast and reliable time delay estimation of strong lens systems using the smoothing and cross-correlation methods. Astrophys J 804(1):39. https://doi.org/10.1088/0004-637X/804/1/39

Aiola S, Calabrese E, Maurin L et al (2020) The Atacama cosmology telescope: DR4 maps and cosmological parameters. J Cosmol Astropart Phys 2020(12):047. https://doi.org/10.1088/1475-7516/2020/12/047

Arendse N, Agnello A, Wojtak RJ (2019) Low-redshift measurement of the sound horizon through gravitational time-delays. Astron Astrophys 632:91. https://doi.org/10.1051/0004-6361/201935972

Auger MW, Fassnacht CD, Abrahamse AL, Lubin LM, Squires GK (2007) The gravitational lens-galaxy group connection. II. Groups associated with B2319 + 051 and B1600 + 434. Astron J 134(2):668–679. https://doi.org/10.1086/519238

Auger MW, Treu T, Bolton AS, Gavazzi R, Koopmans LVE, Marshall PJ, Bundy K, Moustakas LA (2009) The Sloan Lens ACS Survey. IX. Colors, lensing, and stellar masses of early-type galaxies. Astrophys J 705(2):1099–1115. https://doi.org/10.1088/0004-637X/705/2/1099

Auger MW, Treu T, Bolton AS, Gavazzi R, Koopmans LVE, Marshall PJ, Moustakas LA, Burles S (2010) The Sloan Lens ACS Survey. X. Stellar, dynamical, and total mass correlations of massive early-type galaxies. Astrophys J 724:511–525. https://doi.org/10.1088/0004-637X/724/1/511

Bag S, Sohn W, Shafieloo A, Liao K (2023) Harnessing unresolved lensed quasars: the mathematical foundation of the fluctuation curve. Astrophys J 950(2):158. https://doi.org/10.3847/1538-4357/accea3

Baldwin D, Schechter PL (2024) A Malmquist-like bias in the inferred areas of diamond caustics and the resulting bias in inferred time delays for gravitationally lensed quasars. Astrophys J 962:108. https://doi.org/10.3847/1538-4357/ad175a

Barnabè M, Czoske O, Koopmans LVE, Treu T, Bolton AS (2011) Two-dimensional kinematics of SLACS lenses - III. Mass structure and dynamics of early-type lens galaxies beyond z = 0.1. Mon Not R Astron Soc 415(3):2215–2232. https://doi.org/10.1111/j.1365-2966.2011.18842.x

Barnabè M, Dutton AA, Marshall PJ, Auger MW, Brewer BJ, Treu T, Bolton AS, Koo DC, Koopmans LVE (2012) The SWELLS survey - IV. Precision measurements of the stellar and dark matter distributions in a spiral lens galaxy. Mon Not R Astron Soc 423(2):1073–1088. https://doi.org/10.1111/j.1365-2966.2012.20934.x

Barnacka A, Glicenstein J-F, Moudden Y (2011) First evidence of a gravitational lensing-induced echo in gamma rays with Fermi LAT. Astron Astrophys 528:3. https://doi.org/10.1051/0004-6361/201016175

Biesiada M (2006) Strong lensing systems as a probe of dark energy in the universe. Phys Rev D 73(2):023006. https://doi.org/10.1103/PhysRevD.73.023006

Biggio L, Domi A, Tosi S, Vernardos G, Ricci D, Paganin L, Bracco G (2021) Time delay estimation in unresolved lensed quasars. arXiv e-prints, arXiv:2110.01012

Biggs AD, Browne IWA, Helbig P, Koopmans LVE, Wilkinson PN, Perley RA (1999) Time delay for the gravitational lens system B0218 + 357. Mon Not R Astron Soc 304(2):349–358. https://doi.org/10.1046/j.1365-8711.1999.02309.x

Binney J, Mamon GA (1982) M/L and velocity anisotropy from observations of spherical galaxies, of must M 87 have a massive black hole? Mon Not R Astron Soc 200:361–375. https://doi.org/10.1093/mnras/200.2.361

Binney J, Tremaine S (2008) Galactic dynamics, 2nd edn. Princeton University Press
Birrer S (2021) Gravitational lensing formalism in a curved arc basis: a continuous description of observables and degeneracies from the weak to the strong lensing regime. Astrophys J 919(1):38. https://doi.org/10.3847/1538-4357/ac1108
Birrer S, Amara A (2018) Lenstronomy: multi-purpose gravitational lens modelling software package. Phys Dark Universe 22:189–201. https://doi.org/10.1016/j.dark.2018.11.002
Birrer S, Treu T (2019) Astrometric requirements for strong lensing time-delay cosmography. Mon Not R Astron Soc 489(2):2097–2103. https://doi.org/10.1093/mnras/stz2254
Birrer S, Treu T (2021) TDCOSMO. V. Strategies for precise and accurate measurements of the Hubble constant with strong lensing. Astron Astrophys 649:61. https://doi.org/10.1051/0004-6361/202039179
Birrer S, Amara A, Refregier A (2015) Gravitational lens modeling with basis sets. Astrophys J 813(2):102. https://doi.org/10.1088/0004-637X/813/2/102
Birrer S, Amara A, Refregier A (2016) The mass-sheet degeneracy and time-delay cosmography: analysis of the strong lens RXJ1131-1231. J Cosmol Astropart Phys 2016(8):020. https://doi.org/10.1088/1475-7516/2016/08/020
Birrer S, Welschen C, Amara A, Refregier A (2017) Line-of-sight effects in strong lensing: putting theory into practice. J Cosmol Astropart Phys 2017(4):049. https://doi.org/10.1088/1475-7516/2017/04/049
Birrer S, Treu T, Rusu CE, Bonvin V, Fassnacht CD, Chan JHH, Agnello A, Shajib AJ, Chen GC-F, Auger M, Courbin F, Hilbert S, Sluse D, Suyu SH, Wong KC, Marshall P, Lemaux BC, Meylan G (2019) H0LiCOW - IX. Cosmographic analysis of the doubly imaged quasar SDSS 1206 + 4332 and a new measurement of the Hubble constant. Mon Not R Astron Soc 484(4):4726–4753. https://doi.org/10.1093/mnras/stz200
Birrer S, Shajib AJ, Galan A, Millon M, Treu T, Agnello A, Auger M, Chen GC-F, Christensen L, Collett T, Courbin F, Fassnacht CD, Koopmans LVE, Marshall PJ, Park J-W, Rusu CE, Sluse D, Spiniello C, Suyu SH, Wagner-Carena S, Wong KC, Barnabè M, Bolton AS, Czoske O, Ding X, Frieman JA, Van de Vyvere L (2020) TDCOSMO. IV. Hierarchical time-delay cosmography – joint inference of the Hubble constant and galaxy density profiles. Astron Astrophys 643:165. https://doi.org/10.1051/0004-6361/202038861
Birrer S, Dhawan S, Shajib AJ (2022) The Hubble constant from strongly lensed supernovae with standardizable magnifications. Astrophys J 924:2. https://doi.org/10.3847/1538-4357/ac323a
Blackburne JA, Kochanek CS (2010) The effect of a time-varying accretion disk size on quasar microlensing light curves. Astrophys J 718(2):1079–1084. https://doi.org/10.1088/0004-637X/718/2/1079
Blakeslee JP, Jensen JB, Ma C-P, Milne PA, Greene JE (2021) The Hubble constant from infrared surface brightness fluctuation distances. Astrophys J 911(1):65. https://doi.org/10.3847/1538-4357/abe86a
Blandford R, Narayan R (1986) Fermat's principle, caustics, and the classification of gravitational lens images. Astrophys J 310:568. https://doi.org/10.1086/164709
Blum K, Castorina E, Simonović M (2020) Could quasar lensing time delays hint to a core component in halos, instead of H_0 tension? Astrophys J Lett 892(2):27. https://doi.org/10.3847/2041-8213/ab8012
Bolton AS, Burles S, Koopmans LVE, Treu T, Gavazzi R, Moustakas LA, Wayth R, Schlegel DJ (2008) The Sloan Lens ACS Survey. V. The full ACS strong-lens sample. Astrophys J 682(2):964–984. https://doi.org/10.1086/589327
Bonvin V, Millon M (2020) H0LiCOW H0 tension plotting notebook. https://doi.org/10.5281/zenodo.3635517
Bonvin V, Tewes M, Courbin F, Kuntzer T, Sluse D, Meylan G (2016) COSMOGRAIL: the COSmological MOnitoring of GRAvItational lenses. XV. Assessing the achievability and precision of time-delay measurements. Astron Astrophys 585:88. https://doi.org/10.1051/0004-6361/201526704
Bonvin V, Courbin F, Suyu SH, Marshall PJ, Rusu CE, Sluse D, Tewes M, Wong KC, Collett T, Fassnacht CD, Treu T, Auger MW, Hilbert S, Koopmans LVE, Meylan G, Rumbaugh N, Sonnenfeld A, Spiniello C (2017) H0LiCOW - V. New COSMOGRAIL time delays of HE 0435-1223: H_0 to 3.8 per cent precision from strong lensing in a flat ΛCDM model. Mon Not R Astron Soc 465(4):4914–4930. https://doi.org/10.1093/mnras/stw3006
Bonvin V, Millon M, Chan JH-H, Courbin F, Rusu CE, Sluse D, Suyu SH, Wong KC, Fassnacht CD, Marshall PJ, Treu T, Buckley-Geer E, Frieman J, Hempel A, Kim S, Lachaume R, Rabus M, Chao DC-Y, Chijani M, Gilman D, Gilmore K, Rojas K, Williams P, Anguita T, Kochanek CS, Morgan C, Motta V, Tewes M, Meylan G (2019) COSMOGRAIL. XVIII. Time delays of the quadruply lensed quasar WFI2033-4723. Astron Astrophys 629:97. https://doi.org/10.1051/0004-6361/201935921
Bradač M, Lombardi M, Schneider P (2004) Mass-sheet degeneracy: fundamental limit on the cluster mass reconstruction from statistical (weak) lensing. Astron Astrophys 424:13–22. https://doi.org/10.1051/0004-6361:20035744

Buckley-Geer EJ, Lin H, Rusu CE et al (2020) STRIDES: spectroscopic and photometric characterization of the environment and effects of mass along the line of sight to the gravitational lenses DES J0408-5354 and WGD 2038-4008. Mon Not R Astron Soc 498(3):3241–3274. https://doi.org/10.1093/mnras/staa2563

Burud I, Hjorth J, Jaunsen AO, Andersen MI, Korhonen H, Clasen JW, Pelt J, Pijpers FP, Magain P, Østensen R (2000) An optical time delay estimate for the double gravitational lens system B1600 + 434. Astrophys J 544(1):117–122. https://doi.org/10.1086/317213

Burud I, Hjorth J, Courbin F, Cohen JG, Magain P, Jaunsen AO, Kaas AA, Faure C, Letawe G (2002) Time delay and lens redshift for the doubly imaged BAL quasar SBS 1520 + 530. Astron Astrophys 391:481–486. https://doi.org/10.1051/0004-6361:20020856

Cackett EM, Horne K, Winkler H (2007) Testing thermal reprocessing in active galactic nuclei accretion discs. Mon Not R Astron Soc 380(2):669–682. https://doi.org/10.1111/j.1365-2966.2007.12098.x

Caminha GB, Grillo C, Rosati P, Balestra I, Karman W, Lombardi M, Mercurio A, Nonino M, Tozzi P, Zitrin A, Biviano A, Girardi M, Koekemoer AM, Melchior P, Meneghetti M, Munari E, Suyu SH, Umetsu K, Annunziatella M, Borgani S, Broadhurst T, Caputi KI, Coe D, Delgado-Correal C, Ettori S, Fritz A, Frye B, Gobat R, Maier C, Monna A, Postman M, Sartoris B, Seitz S, Vanzella E, Ziegler B (2016) CLASH-VLT: a highly precise strong lensing model of the galaxy cluster RXC J2248.7-4431 (Abell S1063) and prospects for cosmography. Astron Astrophys 587:80. https://doi.org/10.1051/0004-6361/201527670

Chan JHH, Rojas K, Millon M, Courbin F, Bonvin V, Jauffret G (2021) Measuring accretion disk sizes of lensed quasars with microlensing time delay in multi-band light curves. Astron Astrophys 647:115. https://doi.org/10.1051/0004-6361/202038971

Chen GC-F, Suyu SH, Wong KC, Fassnacht CD, Chiueh T, Halkola A, Hu IS, Auger MW, Koopmans LVE, Lagattuta DJ, McKean JP, Vegetti S (2016) SHARP - III. First use of adaptive-optics imaging to constrain cosmology with gravitational lens time delays. Mon Not R Astron Soc 462(4):3457–3475. https://doi.org/10.1093/mnras/stw991

Chen GC-F, Chan JHH, Bonvin V, Fassnacht CD, Rojas K, Millon M, Courbin F, Suyu SH, Wong KC, Sluse D, Treu T, Shajib AJ, Hsueh J-W, Lagattuta DJ, Koopmans LVE, Vegetti S, McKean JP (2018) Constraining the microlensing effect on time delays with a new time-delay prediction model in H_0 measurements. Mon Not R Astron Soc 481(1):1115–1125. https://doi.org/10.1093/mnras/sty2350

Chen GC-F, Fassnacht CD, Suyu SH, Rusu CE, Chan JHH, Wong KC, Auger MW, Hilbert S, Bonvin V, Birrer S, Millon M, Koopmans LVE, Lagattuta DJ, McKean JP, Vegetti S, Courbin F, Ding X, Halkola A, Jee I, Shajib AJ, Sluse D, Sonnenfeld A, Treu T (2019) A SHARP view of H0LiCOW: H_0 from three time-delay gravitational lens systems with adaptive optics imaging. Mon Not R Astron Soc 490(2):1743–1773. https://doi.org/10.1093/mnras/stz2547

Chen GC-F, Fassnacht CD, Suyu SH, Koopmans LVE, Lagattuta DJ, McKean JP, Auger MW, Vegetti S, Treu T (2021b) SHARP VIII: J0924 + 0219 lens mass distribution and time-delay prediction through adaptive-optics imaging. arXiv e-prints, arXiv:2107.10304

Chen GC-F, Treu T, Fassnacht CD, Ragland S, Schmidt T, Suyu SH (2021a) Point-spread function reconstruction of adaptive-optics imaging: Meeting the astrometric requirements for time-delay cosmography. arXiv e-prints, arXiv:2106.11060

Chen W, Kelly PL, Frye BL, Pierel J, Willner SP, Pascale M, Cohen SH, Conselice CJ, Engesser M, Furtak LJ, Gilman D, Grogin NA, Huber S, Jha SW, Johansson J, Koekemoer AM, Larison C, Meena AK, Siebert MR, Windhorst RA, Yan H, Zitrin A (2024) JWST Spectroscopy of SN H0pe: Classification and Time Delays of a Triply-imaged Type Ia Supernova at $z = 1.78$. arXiv e-prints, arXiv:2403.19029. https://doi.org/10.48550/arXiv.2403.19029

Cheung CC, Larsson S, Scargle JD, Amin MA, Blandford RD, Bulmash D, Chiang J, Ciprini S, Corbet RHD, Falco EE, Marshall PJ, Wood DL, Ajello M, Bastieri D, Chekhtman A, D'Ammando F, Giroletti M, Grove JE, Lott B, Ojha R, Orienti M, Perkins JS, Razzano M, Smith AW, Thompson DJ, Wood KS (2014) Fermi large area telescope detection of gravitational lens delayed γ-ray flares from Blazar B0218 + 357. Astrophys J Lett 782(2):14. https://doi.org/10.1088/2041-8205/782/2/L14

Coles JP, Read JI, Saha P (2014) Gravitational lens recovery with GLASS: measuring the mass profile and shape of a lens. Mon Not R Astron Soc 445(3):2181–2197. https://doi.org/10.1093/mnras/stu1781

Collett TE (2015) The population of galaxy-galaxy strong lenses in forthcoming optical imaging surveys. Astrophys J 811(1):20. https://doi.org/10.1088/0004-637X/811/1/20

Collett TE, Cunnington SD (2016) Observational selection biases in time-delay strong lensing and their impact on cosmography. Mon Not R Astron Soc 462(3):3255–3264. https://doi.org/10.1093/mnras/stw1856

Collett TE, Auger MW, Belokurov V, Marshall PJ, Hall AC (2012) Constraining the dark energy equation of state with double-source plane strong lenses. Mon Not R Astron Soc 424(4):2864–2875. https://doi.org/10.1111/j.1365-2966.2012.21424.x

Collett TE, Marshall PJ, Auger MW, Hilbert S, Suyu SH, Greene Z, Treu T, Fassnacht CD, Koopmans LVE, Bradač M, Blandford RD (2013) Reconstructing the lensing mass in the universe from photometric catalogue data. Mon Not R Astron Soc 432(1):679–692. https://doi.org/10.1093/mnras/stt504

Colley WN, Schild RE, Abajas C, Alcalde D, Aslan Z, Bikmaev I, Chavushyan V, Chinarro L, Cournoyer J-P, Crowe R, Dudinov V, Evans AKD, Jeon Y-B, Goicoechea LJ, Golbasi O, Khamitov I, Kjernsmo K, Lee HJ, Lee J, Lee KW, Lee MG, Lopez-Cruz O, Mediavilla E, Moffat AFJ, Mujica R, Ullan A, Muñoz J, Oscoz A, Park M-G, Purves N, Saanum O, Sakhibullin N, Serra-Ricart M, Sinelnikov I, Stabell R, Stockton A, Teuber J, Thompson R, Woo H-S, Zheleznyak A (2003) Around-the-clock observations of the Q0957 + 561A,B gravitationally lensed quasar. II. Results for the second observing season. Astrophys J 587(1):71–79. https://doi.org/10.1086/368076

Cooray AR (1999) An upper limit on Ω_m using lensed arcs. Astrophys J 524(2):504–509. https://doi.org/10.1086/307861

Courbin F, Eigenbrod A, Vuissoz C, Meylan G, Magain P (2005) COSMOGRAIL: the COSmological MOnitoring of GRAvItational lenses. In: Mellier Y, Meylan G (eds) Gravitational lensing impact on cosmology, Proc IAU Symposium, vol 225. Cambridge University Press, pp 297–303. https://doi.org/10.1017/S1743921305002097

Courbin F, Chantry V, Revaz Y, Sluse D, Faure C, Tewes M, Eulaers E, Koleva M, Asfandiyarov I, Dye S, Magain P, van Winckel H, Coles J, Saha P, Ibrahimov M, Meylan G (2011) COSMOGRAIL: the COSmological MOnitoring of GRAvItational lenses. IX. Time delays, lens dynamics and baryonic fraction in HE 0435-1223. Astron Astrophys 536:53. https://doi.org/10.1051/0004-6361/201015709

Courbin F, Bonvin V, Buckley-Geer E et al (2018) COSMOGRAIL: the COSmological MOnitoring of GRAvItational lenses. XVI. Time delays for the quadruply imaged quasar DES J0408-5354 with high-cadence photometric monitoring. Astron Astrophys 609:71. https://doi.org/10.1051/0004-6361/201731461

Dahle H, Gladders MD, Sharon K, Bayliss MB, Rigby JR (2015) Time delay measurements for the cluster-lensed sextuple quasar SDSS J2222 + 2745. Astrophys J 813(1):67. https://doi.org/10.1088/0004-637X/813/1/67

Dainotti MG, De Simone B, Schiavone T, Montani G, Rinaldi E, Lambiase G (2021) On the Hubble constant tension in the SNe Ia pantheon sample. Astrophys J 912(2):150. https://doi.org/10.3847/1538-4357/abeb73

Dalal N, Hennawi JF, Bode P (2005) Noise in strong lensing cosmography. Astrophys J 622(1):99–105. https://doi.org/10.1086/427323

D'Aloisio A, Natarajan P (2011) Cosmography with cluster strong lenses: the influence of substructure and line-of-sight haloes. Mon Not R Astron Soc 411(3):1628–1640. https://doi.org/10.1111/j.1365-2966.2010.17795.x

De Lucia G, Blaizot J (2007) The hierarchical formation of the brightest cluster galaxies. Mon Not R Astron Soc 375(1):2–14. https://doi.org/10.1111/j.1365-2966.2006.11287.x

Denzel P, Coles JP, Saha P, Williams LLR (2021) The Hubble constant from eight time-delay galaxy lenses. Mon Not R Astron Soc 501(1):784–801. https://doi.org/10.1093/mnras/staa3603

Dexter J, Agol E (2011) Quasar accretion disks are strongly inhomogeneous. Astrophys J Lett 727(1):24. https://doi.org/10.1088/2041-8205/727/1/L24

Di Valentino E, Mena O, Pan S, Visinelli L, Yang W, Melchiorri A, Mota DF, Riess AG, Silk J (2021) In the realm of the Hubble tension-a review of solutions. Class Quantum Gravity 38(15):153001. https://doi.org/10.1088/1361-6382/ac086d

Diego JM, Broadhurst T, Wong J, Silk J, Lim J, Zheng W, Lam D, Ford H (2016) A free-form mass model of the Hubble Frontier Fields cluster AS1063 (RXC J2248.7-4431) with over one hundred constraints. Mon Not R Astron Soc 459(4):3447–3459. https://doi.org/10.1093/mnras/stw865

Dietrich T, Coughlin MW, Pang PTH, Bulla M, Heinzel J, Issa L, Tews I, Antier S (2020) Multimessenger constraints on the neutron-star equation of state and the Hubble constant. Science 370(6523):1450–1453. https://doi.org/10.1126/science.abb4317

Ding X, Treu T, Birrer S, Chen GC-F, Coles J, Denzel P, Frigo M, Galan A, Marshall PJ, Millon M, More A, Shajib AJ, Sluse D, Tak H, Xu D, Auger MW, Bonvin V, Chand H, Courbin F, Despali G, Fassnacht CD, Gilman D, Hilbert S, Kumar SR, Lin JY-Y, Park JW, Saha P, Vegetti S, Van de Vyvere L, Williams LLR (2021) Time delay lens modelling challenge. Mon Not R Astron Soc 503(1):1096–1123. https://doi.org/10.1093/mnras/stab484

Dobler G, Fassnacht CD, Treu T, Marshall P, Liao K, Hojjati A, Linder E, Rumbaugh N (2015) Strong lens time delay challenge. I. Experimental design. Astrophys J 799(2):168. https://doi.org/10.1088/0004-637X/799/2/168

Donnan FR, Horne K, Hernández Santisteban JV (2021) Bayesian analysis of quasar lightcurves with a running optimal average: new time delay measurements of COSMOGRAIL gravitationally lensed quasars. Mon Not R Astron Soc 508(4):5449–5467. https://doi.org/10.1093/mnras/stab2832

Eigenbrod A, Courbin F, Vuissoz C, Meylan G, Saha P, Dye S (2005) COSMOGRAIL: the COSmological MOnitoring of GRAvItational lenses. I. How to sample the light curves of gravitationally lensed quasars to measure accurate time delays. Astron Astrophys 436(1):25–35. https://doi.org/10.1051/0004-6361:20042422

Ertl S, Schuldt S, Suyu SH, Schmidt T, Treu T, Birrer S, Shajib AJ, Sluse D (2023) TDCOSMO. X. Automated modeling of nine strongly lensed quasars and comparison between lens-modeling software. Astron Astrophys 672:2. https://doi.org/10.1051/0004-6361/202244909

Eulaers E, Tewes M, Magain P, Courbin F, Asfandiyarov I, Ehgamberdiev S, Rathna Kumar S, Stalin CS, Prabhu TP, Meylan G, Van Winckel H (2013) COSMOGRAIL: the COSmological MOnitoring of GRAvItational lenses. XII. Time delays of the doubly lensed quasars SDSS J1206 + 4332 and HS 2209 + 1914. Astron Astrophys 553:121. https://doi.org/10.1051/0004-6361/201321140

Fadely R, Keeton CR, Nakajima R, Bernstein GM (2010) Improved constraints on the gravitational lens Q0957 + 561. II. Strong lensing. Astrophys J 711(1):246–267. https://doi.org/10.1088/0004-637X/711/1/246

Falco EE, Gorenstein MV, Shapiro II (1985) On model-dependent bounds on H_0 from gravitational images: application to Q 0957 + 561 A, B. Astrophys J 289:1–4. https://doi.org/10.1086/184422

Fassnacht CD, Lubin LM (2002) The gravitational lens-galaxy group connection. I. Discovery of a group coincident with CLASS B0712 + 472. Astron J 123(2):627–636. https://doi.org/10.1086/338648

Fassnacht CD, Pearson TJ, Readhead ACS, Browne IWA, Koopmans LVE, Myers ST, Wilkinson PN (1999) A determination of H_0 with the CLASS gravitational lens B1608 + 656. I. Time delay measurements with the VLA. Astrophys J 527(2):498–512. https://doi.org/10.1086/308118

Fassnacht CD, Xanthopoulos E, Koopmans LVE, Rusin D (2002) A determination of H_0 with the CLASS gravitational lens B1608 + 656. III. A significant improvement in the precision of the time delay measurements. Astrophys J 581(2):823–835. https://doi.org/10.1086/344368

Fassnacht CD, Gal RR, Lubin LM, McKean JP, Squires GK, Readhead ACS (2006) Mass along the line of sight to the gravitational lens B1608 + 656: galaxy groups and implications for H_0. Astrophys J 642(1):30–38. https://doi.org/10.1086/500927

Fassnacht CD, Koopmans LVE, Wong KC (2011) Galaxy number counts and implications for strong lensing. Mon Not R Astron Soc 410(4):2167–2179. https://doi.org/10.1111/j.1365-2966.2010.17591.x

Ferreras I, Saha P, Williams LLR (2005) Stellar and total mass in early-type lensing galaxies. Astrophys J Lett 623(1):5–8. https://doi.org/10.1086/429995

Fischer P, Tyson JA (1997) The mass distribution of the most luminous X-ray cluster RXJ 1347.5-1145 from gravitational lensing. Astron J 114:14–24. https://doi.org/10.1086/118447

Fleury P, Larena J, Uzan J-P (2021a) Gravitational lenses in arbitrary space-times. Class Quantum Gravity 38(8):085002. https://doi.org/10.1088/1361-6382/abea2d

Fleury P, Larena J, Uzan J-P (2021b) Line-of-sight effects in strong gravitational lensing. J Cosmol Astropart Phys 2021(08):024. https://doi.org/10.1088/1475-7516/2021/08/024

Fohlmeister J, Kochanek CS, Falco EE, Wambsganss J, Morgan N, Morgan CW, Ofek EO, Maoz D, Keeton CR, Barentine JC, Dalton G, Dembicky J, Ketzeback W, McMillan R, Peters CS (2007) A time delay for the cluster-lensed quasar SDSS J1004 + 4112. Astrophys J 662(1):62–71. https://doi.org/10.1086/518018

Fohlmeister J, Kochanek CS, Falco EE, Morgan CW, Wambsganss J (2008) The rewards of patience: an 822 day time delay in the gravitational lens SDSS J1004 + 4112. Astrophys J 676(2):761–766. https://doi.org/10.1086/528789

Fohlmeister J, Kochanek CS, Falco EE, Wambsganss J, Oguri M, Dai X (2013) A two-year time delay for the lensed quasar SDSS J1029 + 2623. Astrophys J 764(2):186. https://doi.org/10.1088/0004-637X/764/2/186

Foxley-Marrable M, Collett TE, Vernardos G, Goldstein DA, Bacon D (2018) The impact of microlensing on the standardization of strongly lensed Type Ia supernovae. Mon Not R Astron Soc 478(4):5081–5090. https://doi.org/10.1093/mnras/sty1346

Freedman WL, Madore BF, Hatt D, Hoyt TJ, Jang IS, Beaton RL, Burns CR, Lee MG, Monson AJ, Neeley JR, Phillips MM, Rich JA, Seibert M (2019) The carnegie-Chicago Hubble program. VIII. An independent determination of the Hubble constant based on the tip of the red giant branch. Astrophys J 882(1):34. https://doi.org/10.3847/1538-4357/ab2f73

Freedman WL, Madore BF, Hoyt T, Jang IS, Beaton R, Lee MG, Monson A, Neeley J, Rich J (2020) Calibration of the tip of the red giant branch. Astrophys J 891(1):57. https://doi.org/10.3847/1538-4357/ab7339

Frye BL, Pascale M, Pierel J, Chen W, Foo N, Leimbach R, Garuda N, Cohen SH, Kamieneski PS, Windhorst RA, Koekemoer AM, Kelly P, Summers J, Engesser M, Liu D, Furtak LJ, Polletta MdC, Harrington KC, Willner SP, Diego JM, Jansen RA, Coe D, Conselice CJ, Dai L, Dole H, D'Silva JCJ, Driver SP, Grogin NA, Marshall MA, Meena AK, Nonino M, Ortiz R, Pirzkal N, Robotham A, Ryan RE, Strolger L,

Tompkins S, Willmer CNA, Yan H, Yun MS, Zitrin A (2024) The JWST discovery of the triply imaged type Ia "Supernova H0pe" and observations of the galaxy cluster PLCK G165.7 + 67.0. Astrophys J 961(2):171. https://doi.org/10.3847/1538-4357/ad1034

Galan A, Peel A, Joseph R, Courbin F, Starck J-L (2021) SLITRONOMY: towards a fully wavelet-based strong lensing inversion technique. Astron Astrophys 647:176. https://doi.org/10.1051/0004-6361/202039363

Gavazzi R, Treu T, Koopmans LVE, Bolton AS, Moustakas LA, Burles S, Marshall PJ (2008) The Sloan Lens ACS Survey. VI. Discovery and analysis of a double Einstein ring. Astrophys J 677(2):1046–1059. https://doi.org/10.1086/529541

Ghosh A, Williams LLR, Liesenborgs J (2020) Free-form grale lens inversion of galaxy clusters with up to 1000 multiple images. Mon Not R Astron Soc 494(3):3998–4014. https://doi.org/10.1093/mnras/staa962

Giannini E, Schmidt RW, Wambsganss J, Alsubai K, Andersen JM, Anguita T, Bozza V, Bramich DM, Browne P, Calchi Novati S, Damerdji Y, Diehl C, Dodds P, Dominik M, Elyiv A, Fang X, Figuera Jaimes R, Finet F, Gerner T, Gu S, Hardis S, Harpsøe K, Hinse TC, Hornstrup A, Hundertmark M, Jessen-Hansen J, Jørgensen UG, Juncher D, Kains N, Kerins E, Korhonen H, Liebig C, Lund MN, Lundkvist MS, Maier G, Mancini L, Masi G, Mathiasen M, Penny M, Proft S, Rabus M, Rahvar S, Ricci D, Scarpetta G, Sahu K, Schäfer S, Schönebeck F, Skottfelt J, Snodgrass C, Southworth J, Surdej J, Tregloan-Reed J, Vilela C, Wertz O, Zimmer F (2017) MiNDSTEp differential photometry of the gravitationally lensed quasars WFI 2033-4723 and HE 0047-1756: microlensing and a new time delay. Astron Astrophys 597:49. https://doi.org/10.1051/0004-6361/201527422

Gilman D, Birrer S, Treu T (2020) TDCOSMO. III. Dark matter substructure meets dark energy. The effects of (sub)halos on strong-lensing measurements of H_0. Astron Astrophys 642:194. https://doi.org/10.1051/0004-6361/202038829

Gilmore J, Natarajan P (2009) Cosmography with cluster strong lensing. Mon Not R Astron Soc 396(1):354–364. https://doi.org/10.1111/j.1365-2966.2009.14612.x

Goicoechea LJ, Shalyapin VN (2016) Gravitational lens system SDSS J1339 + 1310: microlensing factory and time delay. Astron Astrophys 596:77. https://doi.org/10.1051/0004-6361/201628790

Golse G, Kneib J-P, Soucail G (2002) Constraining the cosmological parameters using strong lensing. Astron Astrophys 387:788–803. https://doi.org/10.1051/0004-6361:20020448

Gomer M, Williams LLR (2020) Galaxy-lens determination of H_0: constraining density slope in the context of the mass sheet degeneracy. J Cosmol Astropart Phys 2020(11):045. https://doi.org/10.1088/1475-7516/2020/11/045

Gomer MR, Williams LLR (2021) Galaxy-lens determination of H_0: the effect of the ellipse + shear modelling assumption. Mon Not R Astron Soc 504(1):1340–1354. https://doi.org/10.1093/mnras/stab930

Gomer MR, Sluse D, van de Vyvere L, Birrer S, Courbin F (2022) TDCOSMO VIII. A key test of systematics in the hierarchical method of time-delay cosmography. Astron Astrophys 667:A86. https://doi.org/10.1051/0004-6361/202244324

Gorenstein MV, Falco EE, Shapiro II (1988) Degeneracies in parameter estimates for models of gravitational lens systems. Astrophys J 327:693. https://doi.org/10.1086/166226

Greene ZS, Suyu SH, Treu T, Hilbert S, Auger MW, Collett TE, Marshall PJ, Fassnacht CD, Blandford RD, Bradač M, Koopmans LVE (2013) Improving the precision of time-delay cosmography with observations of galaxies along the line of sight. Astrophys J 768(1):39. https://doi.org/10.1088/0004-637X/768/1/39

Grillo C, Rosati P, Suyu SH, Balestra I, Caminha GB, Halkola A, Kelly PL, Lombardi M, Mercurio A, Rodney SA, Treu T (2018) Measuring the value of the Hubble constant "à la Refsdal". Astrophys J 860(2):94. https://doi.org/10.3847/1538-4357/aac2c9

Grillo C, Pagano L, Rosati P, Suyu SH (2024) Cosmography with Supernova Refsdal through time-delay cluster lensing: independent measurements of the Hubble constant and geometry of the Universe. Astron Astrophys 684:L23. https://doi.org/10.1051/0004-6361/202449278

Grogin NA, Narayan R (1996) A new model of the gravitational lens 0957 + 561 and a limit on the Hubble constant. Astrophys J 464:92. https://doi.org/10.1086/177302

Gu A, Huang X, Sheu W, Aldering G, Bolton AS, Boone K, Dey A, Filipp A, Jullo E, Perlmutter S, Rubin D, Schlafly EF, Schlegel DJ, Shu Y, Suyu SH (2022) GIGA-lens: fast Bayesian inference for strong gravitational lens modeling. Astrophys J 935(1):49. https://doi.org/10.3847/1538-4357/ac6de4

Hilbert S, White SDM, Hartlap J, Schneider P (2007) Strong lensing optical depths in a ΛCDM universe. Mon Not R Astron Soc 382(1):121–132. https://doi.org/10.1111/j.1365-2966.2007.12391.x

Hilbert S, White SDM, Hartlap J, Schneider P (2008) Strong-lensing optical depths in a ΛCDM universe - II. The influence of the stellar mass in galaxies. Mon Not R Astron Soc 386(4):1845–1854. https://doi.org/10.1111/j.1365-2966.2008.13190.x

Hilbert S, Hartlap J, White SDM, Schneider P (2009) Ray-tracing through the millennium simulation: born corrections and lens-lens coupling in cosmic shear and galaxy-galaxy lensing. Astron Astrophys 499(1):31–43. https://doi.org/10.1051/0004-6361/200811054

Hirv A, Eenmäe T, Liivamägi LJ, Pelt J (2007) Estimation of time delays from two blended light curves of gravitational lenses. Balt Astron 16:241–250

Hjorth J, Burud I, Jaunsen AO, Schechter PL, Kneib J-P, Andersen MI, Korhonen H, Clasen JW, Kaas AA, Østensen R, Pelt J, Pijpers FP (2002) The time delay of the quadruple quasar RX J0911.4 + 0551. Astrophys J Lett 572(1):11–14. https://doi.org/10.1086/341603

Hogg NB, Fleury P, Larena J, Martinelli M (2023) Measuring line-of-sight shear with Einstein rings: a proof of concept. Mon Not R Astron Soci 520(4):5982–6000. https://doi.org/10.1093/mnras/stad512

Hojjati A, Linder EV (2014) Next generation strong lensing time delay estimation with Gaussian processes. Phys Rev D 90(12):123501. https://doi.org/10.1103/PhysRevD.90.123501

Hojjati A, Kim AG, Linder EV (2013) Robust strong lensing time delay estimation. Phys Rev D 87(12):123512. https://doi.org/10.1103/PhysRevD.87.123512

Ilbert O, Arnouts S, McCracken HJ et al (2006) Accurate photometric redshifts for the CFHT legacy survey calibrated using the VIMOS VLT deep survey. Astron Astrophys 457(3):841–856. https://doi.org/10.1051/0004-6361:20065138

Jauzac M, Clément B, Limousin M, Richard J, Jullo E, Ebeling H, Atek H, Kneib J-P, Knowles K, Natarajan P, Eckert D, Egami E, Massey R, Rexroth M (2014) Hubble Frontier Fields: a high-precision strong-lensing analysis of galaxy cluster MACSJ0416.1-2403 using ~200 multiple images. Mon Not R Astron Soc 443(2):1549–1554. https://doi.org/10.1093/mnras/stu1355

Jee I, Komatsu E, Suyu SH (2015) Measuring angular diameter distances of strong gravitational lenses. J Cosmol Astropart Phys 2015(11):033. https://doi.org/10.1088/1475-7516/2015/11/033

Joseph R, Courbin F, Starck J-L, Birrer S (2019) Sparse Lens Inversion Technique (SLIT): lens and source separability from linear inversion of the source reconstruction problem. Astron Astrophys 623:14. https://doi.org/10.1051/0004-6361/201731042

Jullo E, Natarajan P, Kneib J-P, D'Aloisio A, Limousin M, Richard J, Schimd C (2010) Cosmological constraints from strong gravitational lensing in clusters of galaxies. Science 329:924–927. https://doi.org/10.1126/science.1185759

Kaiser N, Squires G (1993) Mapping the dark matter with weak gravitational lensing. Astrophys J 404:441. https://doi.org/10.1086/172297

Keeton CR, Moustakas LA (2009) A new channel for detecting dark matter substructure in galaxies: gravitational lens time delays. Astrophys J 699(2):1720–1731. https://doi.org/10.1088/0004-637X/699/2/1720

Kelly PL, Rodney SA, Treu T, Foley RJ, Brammer G, Schmidt KB, Zitrin A, Sonnenfeld A, Strolger L-G, Graur O, Filippenko AV, Jha SW, Riess AG, Bradac M, Weiner BJ, Scolnic D, Malkan MA, von der Linden A, Trenti M, Hjorth J, Gavazzi R, Fontana A, Merten JC, McCully C, Jones T, Postman M, Dressler A, Patel B, Cenko SB, Graham ML, Tucker BE (2015) Multiple images of a highly magnified supernova formed by an early-type cluster galaxy lens. Science 347(6226):1123–1126. https://doi.org/10.1126/science.aaa3350

Kelly PL, Rodney SA, Treu T, Strolger L-G, Foley RJ, Jha SW, Selsing J, Brammer G, Bradač M, Cenko SB, Graur O, Filippenko AV, Hjorth J, McCully C, Molino A, Nonino M, Riess AG, Schmidt KB, Tucker B, von der Linden A, Weiner BJ, Zitrin A (2016) Deja vu all over again: the reappearance of supernova Refsdal. Astrophys J Lett 819(1):8. https://doi.org/10.3847/2041-8205/819/1/L8

Kelly PL, Rodney S, Treu T, Birrer S, Bonvin V, Dessart L, Foley RJ, Filippenko AV, Gilman D, Jha S, Hjorth J, Mandel K, Millon M, Pierel J, Thorp S, Zitrin A, Broadhurst T, Chen W, Diego JM, Dressler A, Graur O, Jauzac M, Malkan MA, McCully C, Oguri M, Postman M, Schmidt KB, Sharon K, Tucker BE, von der Linden A, Wambsganss J (2023b) The magnificent five images of supernova Refsdal: time delay and magnification measurements. Astrophys J 948(2):93. https://doi.org/10.3847/1538-4357/ac4ccb

Kelly PL, Rodney S, Treu T, Oguri M, Chen W, Zitrin A, Birrer S, Bonvin V, Dessart L, Diego JM, Filippenko AV, Foley RJ, Gilman D, Hjorth J, Jauzac M, Mandel K, Millon M, Pierel J, Sharon K, Thorp S, Williams L, Broadhurst T, Dressler A, Graur O, Jha S, McCully C, Postman M, Schmidt KB, Tucker BE, von der Linden A (2023a) Constraints on the Hubble constant from supernova Refsdal's reappearance. Science 380(6649):1322. https://doi.org/10.1126/science.abh1322

Khadka N, Birrer S, Leauthaud A, Nix H (2024) Breaking the mass-sheet degeneracy in strong lensing mass modeling with weak lensing observations arXiv e-prints, arXiv:2404.01513. https://doi.org/10.48550/arXiv.2404.01513

Khetan N, Izzo L, Branchesi M, Wojtak R, Cantiello M, Murugeshan C, Agnello A, Cappellaro E, Della Valle M, Gall C, Hjorth J, Benetti S, Brocato E, Burke J, Hiramatsu D, Howell DA, Tomasella L, Valenti S (2021) A new measurement of the Hubble constant using Type Ia supernovae calibrated with surface brightness fluctuations. Astron Astrophys 647:72. https://doi.org/10.1051/0004-6361/202039196

Knox L, Millea M (2020) Hubble constant hunter's guide. Phys Rev D 101(4):043533. https://doi.org/10.1103/PhysRevD.101.043533

Kochanek CS (2002) What do gravitational lens time delays measure? Astrophys J 578(1):25–32. https://doi.org/10.1086/342476

Kochanek CS (2020) Overconstrained gravitational lens models and the Hubble constant. Mon Not R Astron Soc 493(2):1725–1735. https://doi.org/10.1093/mnras/staa344

Kochanek CS (2021) Overconstrained models of time delay lenses redux: how the angular tail wags the radial dog. Mon Not R Astron Soc 501(4):5021–5028. https://doi.org/10.1093/mnras/staa4033

Kochanek CS, Schechter PL (2004) The Hubble constant from gravitational lens time delays. In: Freedman WL (ed) Measuring and modeling the universe. Carnegie Observatories Astrophysics Series, vol 2. Cambridge University Pres, pp 117–137. arXiv:astro-ph/0306040

Kochanek CS, Morgan ND, Falco EE, McLeod BA, Winn JN, Dembicky J, Ketzeback B (2006) The time delays of gravitational lens HE 0435-1223: an early-type galaxy with a rising rotation curve. Astrophys J 640(1):47–61. https://doi.org/10.1086/499766

Kolatt TS, Bartelmann M (1998) Gravitational lensing of type IA supernovae by galaxy clusters. Mon Not R Astron Soc 296(3):763–772. https://doi.org/10.1046/j.1365-8711.1998.01466.x

Koopmans LVE (2004) Gravitational lensing & stellar dynamics: dark-matter and baryons in early-type galaxies to $z = 1$. PoS BDMH2004:066. https://doi.org/10.22323/1.014.0066

Koopmans LVE (2005) Gravitational imaging of cold dark matter substructures. Mon Not R Astron Soc 363(4):1136–1144. https://doi.org/10.1111/j.1365-2966.2005.09523.x

Koopmans LVE, Fassnacht CD (1999) A determination of H_0 with the CLASS gravitational lens B1608 + 656. II. Mass models and the Hubble constant from lensing. Astrophys J 527(2):513–524. https://doi.org/10.1086/308120

Koopmans LVE, de Bruyn AG, Xanthopoulos E, Fassnacht CD (2000) A time-delay determination from VLA light curves of the CLASS gravitational lens B1600 + 434. Astron Astrophys 356:391–402

Koopmans LVE, Treu T, Fassnacht CD, Blandford RD, Surpi G (2003) The Hubble constant from the gravitational lens B1608 + 656. Astrophys J 599(1):70–85. https://doi.org/10.1086/379226

Koopmans LVE, Treu T, Bolton AS, Burles S, Moustakas LA (2006) The Sloan Lens ACS Survey. III. The structure and formation of early-type galaxies and their evolution since $z \approx 1$. Astrophys J 649(2):599–615. https://doi.org/10.1086/505696

Koopmans LVE, Bolton A, Treu T, Czoske O, Auger MW, Barnabè M, Vegetti S, Gavazzi R, Moustakas LA, Burles S (2009) The structure and dynamics of massive early-type galaxies: on homology, isothermality, and isotropy inside one effective radius. Astrophys J Lett 703(1):51–54. https://doi.org/10.1088/0004-637X/703/1/L51

Kuhn FA, Birrer S, Bruderer C, Amara A, Refregier A (2021) Combining strong and weak lensing estimates in the Cosmos field. J Cosmol Astropart Phys 2021(4):010. https://doi.org/10.1088/1475-7516/2021/04/010

Kundić T, Turner EL, Colley WN, Gott I, Richard J, Rhoads JE, Wang Y, Bergeron LE, Gloria KA, Long DC, Malhotra S, Wambsganss J (1997a) A robust determination of the time delay in 0957 + 561A, B and a measurement of the global value of Hubble's constant. Astrophys J 482(1):75–82. https://doi.org/10.1086/304147

Lagattuta DJ, Richard J, Clément B, Mahler G, Patrício V, Pelló R, Soucail G, Schmidt KB, Wisotzki L, Martinez J, Bina D (2017) Lens modelling Abell 370: crowning the final frontier field with MUSE. Mon Not R Astron Soc 469(4):3946–3964. https://doi.org/10.1093/mnras/stx1079

Lemon C, Anguita T, Auger M, Courbin F, Galan A, McMahon R, Neira F, Oguri M, Schechter P, Shajib A, Treu T (2023) Gravitationally lensed quasars in Gaia – IV. 150 new lenses, quasar pairs, and projected quasars. Mon Not R Astron Soc 520(3):3305–3328. https://doi.org/10.1093/mnras/stac3721

Lemon C, Courbin F, More A, Schechter P, Cañameras R, Delchambre L, Leung C, Shu Y, Spiniello C, Hezaveh Y, Klüter J, McMahon R (2024) Searching for strong gravitational lenses. Space Sci Rev 220:23. https://doi.org/10.1007/s11214-024-01042-9

Li N, Becker C, Dye S (2021) The impact of line-of-sight structures on measuring H_0 with strong lensing time delays. Mon Not R Astron Soc 504(2):2224–2234. https://doi.org/10.1093/mnras/stab984

Li T, Collett TE, Krawczyk CM, Enzi W (2024) Cosmology from large populations of galaxy-galaxy strong gravitational lenses. Mon Not R Astron Soc 527(3):5311–5323. https://doi.org/10.1093/mnras/stad3514

Liao K (2020) Chromatic microlensing time delays. Astrophys J Lett 899(2):33. https://doi.org/10.3847/2041-8213/abadfd

Liao K (2021) Impact of microlensing on observational strategies for gravitational time-delay measurements. Astrophys J 906(1):26. https://doi.org/10.3847/1538-4357/abc876

Liao K, Treu T, Marshall P, Fassnacht CD, Rumbaugh N, Dobler G, Aghamousa A, Bonvin V, Courbin F, Hojjati A, Jackson N, Kashyap V, Rathna Kumar S, Linder E, Mandel K, Meng X-L, Meylan G, Moustakas LA, Prabhu TP, Romero-Wolf A, Shafieloo A, Siemiginowska A, Stalin CS, Tak H, Tewes

M, van Dyk D (2015) Strong lens time delay challenge. II. Results of TDC1. Astrophys J 800(1):11. https://doi.org/10.1088/0004-637X/800/1/11

Liao K, Shafieloo A, Keeley RE, Linder EV (2019) A model-independent determination of the Hubble constant from lensed quasars and supernovae using Gaussian process regression. Astrophys J Lett 886(1):23. https://doi.org/10.3847/2041-8213/ab5308

Liao K, Shafieloo A, Keeley RE, Linder EV (2020) Determining model-independent H_0 and consistency tests. Astrophys J Lett 895(2):29. https://doi.org/10.3847/2041-8213/ab8dbb

Liao K, Biesiada M, Zhu Z-H (2022) Strongly lensed transient sources: a review. Chin Phys Lett 39(11):119801. https://doi.org/10.1088/0256-307X/39/11/119801

Link R, Pierce MJ (1998) Cosmological parameters from multiple-arc gravitational lensing systems. I. Smooth lensing potentials. Astrophys J 502(1):63–74. https://doi.org/10.1086/305892

Liu Y, Oguri M, Cao S (2023) Hubble constant from the cluster-lensed quasar system SDSS J 1004 + 4112: investigation of the lens model dependence. Phys Rev D 108(8):083532. https://doi.org/10.1103/PhysRevD.108.083532

Lotz JM, Koekemoer A, Coe D, Grogin N, Capak P, Mack J, Anderson J, Avila R, Barker EA, Borncamp D, Brammer G, Durbin M, Gunning H, Hilbert B, Jenkner H, Khandrika H, Levay Z, Lucas RA, MacKenty J, Ogaz S, Porterfield B, Reid N, Robberto M, Royle P, Smith LJ, Storrie-Lombardi LJ, Sunnquist B, Surace J, Taylor DC, Williams R, Bullock J, Dickinson M, Finkelstein S, Natarajan P, Richard J, Robertson B, Tumlinson J, Zitrin A, Flanagan K, Sembach K, Soifer BT, Mountain M (2017) The Frontier Fields: survey design and initial results. Astrophys J 837(1):97. https://doi.org/10.3847/1538-4357/837/1/97

LSST Dark Energy Science Collaboration (LSST DESC), Abolfathi B et al (2021) The LSST DESC DC2 simulated sky survey. Astrophys J Suppl Ser 253(1):31. https://doi.org/10.3847/1538-4365/abd62c

Magaña J, Motta V, Cárdenas VH, Verdugo T, Jullo E (2015) A magnified glance into the dark sector: probing cosmological models with strong lensing in A1689. Astrophys J 813(1):69. https://doi.org/10.1088/0004-637X/813/1/69

McCully C, Keeton CR, Wong KC, Zabludoff AI (2014) A new hybrid framework to efficiently model lines of sight to gravitational lenses. Mon Not R Astron Soc 443(4):3631–3642. https://doi.org/10.1093/mnras/stu1316

McCully C, Keeton CR, Wong KC, Zabludoff AI (2017) Quantifying environmental and line-of-sight effects in models of strong gravitational lens systems. Astrophys J 836(1):141. https://doi.org/10.3847/1538-4357/836/1/141

Meng X-L, Treu T, Agnello A, Auger MW, Liao K, Marshall PJ (2015) Precision cosmology with time delay lenses: high resolution imaging requirements. J Cosmol Astropart Phys 2015(9):059. https://doi.org/10.1088/1475-7516/2015/09/059

Merritt D (1985a) Distribution functions for spherical galaxies. Mon Not R Astron Soc 214:25–28. https://doi.org/10.1093/mnras/214.1.25P

Merritt D (1985b) Spherical stellar systems with spheroidal velocity distributions. Astron J 90:1027–1037. https://doi.org/10.1086/113810

Meyer AD, van Dyk DA, Tak H, Siemiginowska A (2023) TD-CARMA: Painless, accurate, and scalable estimates of gravitational-lens time delays with flexible CARMA processes. Astrophys J 950:37. https://doi.org/10.3847/1538-4357/acbea1

Millon M, Courbin F, Bonvin V, Buckley-Geer E, Fassnacht CD, Frieman J, Marshall PJ, Suyu SH, Treu T, Anguita T, Motta V, Agnello A, Chan JHH, Chao DC-Y, Chijani M, Gilman D, Gilmore K, Lemon C, Lucey JR, Melo A, Paic E, Rojas K, Sluse D, Williams PR, Hempel A, Kim S, Lachaume R, Rabus M (2020c) TDCOSMO. II. Six new time delays in lensed quasars from high-cadence monitoring at the MPIA 2.2 m telescope. Astron Astrophys 642:193. https://doi.org/10.1051/0004-6361/202038698

Millon M, Courbin F, Bonvin V, Paic E, Meylan G, Tewes M, Sluse D, Magain P, Chan JHH, Galan A, Joseph R, Lemon C, Tihhonova O, Anderson RI, Marmier M, Chazelas B, Lendl M, Triaud AHMJ, Wyttenbach A (2020a) COSMOGRAIL. XIX. Time delays in 18 strongly lensed quasars from 15 years of optical monitoring. Astron Astrophys 640:105. https://doi.org/10.1051/0004-6361/202037740

Millon M, Galan A, Courbin F, Treu T, Suyu SH, Ding X, Birrer S, Chen GC-F, Shajib AJ, Sluse D, Wong KC, Agnello A, Auger MW, Buckley-Geer EJ, Chan JHH, Collett T, Fassnacht CD, Hilbert S, Koopmans LVE, Motta V, Mukherjee S, Rusu CE, Sonnenfeld A, Spiniello C, Van de Vyvere L (2020b) TDCOSMO. I. An exploration of systematic uncertainties in the inference of H_0 from time-delay cosmography. Astron Astrophys 639:101. https://doi.org/10.1051/0004-6361/201937351

Momcheva I, Williams K, Keeton C, Zabludoff A (2006) A spectroscopic study of the environments of gravitational lens galaxies. Astrophys J 641(1):169–189. https://doi.org/10.1086/500382

Monna A, Seitz S, Balestra I, Rosati P, Grillo C, Halkola A, Suyu SH, Coe D, Caminha GB, Frye B, Koekemoer A, Mercurio A, Nonino M, Postman M, Zitrin A (2017) Precise strong lensing mass profile of the CLASH galaxy cluster MACS 2129. Mon Not R Astron Soc 466(4):4094–4106. https://doi.org/10.1093/mnras/stx015

Moresco M, Amati L, Amendola L, Birrer S, Blakeslee JP, Cantiello M, Cimatti A, Darling J, Della Valle M, Fishbach M, Grillo C, Hamaus N, Holz D, Izzo L, Jimenez R, Lusso E, Meneghetti M, Piedipalumbo E, Pisani A, Pourtsidou A, Pozzetti L, Quartin M, Risaliti G, Rosati P, Verde L (2022) Unveiling the universe with emerging cosmological probes. Living Rev Relativ 25(6). https://doi.org/10.1007/s41114-022-00040-z

Mörtsell E, Goobar A, Johansson J, Dhawan S (2022) Sensitivity of the Hubble constant determination to Cepheid calibration. Astrophys J 933:212. https://doi.org/10.3847/1538-4357/ac756e

Muñoz JA, Kochanek CS, Fohlmeister J, Wambsganss J, Falco E, Forés-Toribio R (2022) The longest delay: a 14.5 yr campaign to determine the third time delay in the lensing cluster SDSS J1004 + 4112. Astrophys J 937(1):34. https://doi.org/10.3847/1538-4357/ac8877

Nakajima R, Bernstein GM, Fadely R, Keeton CR, Schrabback T (2009) Improved constraints on the gravitational lens Q0957 + 561. I. Weak lensing. Astrophys J 697(2):1793–1804. https://doi.org/10.1088/0004-637X/697/2/1793

Napier K, Sharon K, Dahle H, Bayliss M, Gladders MD, Mahler G, Rigby JR, Florian M (2023) Hubble constant measurement from three large-separation quasars strongly lensed by galaxy clusters. Astrophys J 959(2):134. https://doi.org/10.3847/1538-4357/ad045a

Navarro JF, Frenk CS, White SDM (1997) A universal density profile from hierarchical clustering. Astrophys J 490(2):493–508. https://doi.org/10.1086/304888

Oguri M (2007) Gravitational lens time delays: a statistical assessment of lens model dependences and implications for the global Hubble constant. Astrophys J 660(1):1–15. https://doi.org/10.1086/513093

Oguri M (2019) Strong gravitational lensing of explosive transients. Rep Prog Phys 82(12):126901. https://doi.org/10.1088/1361-6633/ab4fc5

Oguri M, Kawano Y (2003) Gravitational lens time delays for distant supernovae: breaking the degeneracy between radial mass profiles and the Hubble constant. Mon Not R Astron Soc 338(4):25–29. https://doi.org/10.1046/j.1365-8711.2003.06290.x

Oguri M, Marshall PJ (2010) Gravitationally lensed quasars and supernovae in future wide-field optical imaging surveys. Mon Not R Astron Soc 405(4):2579–2593. https://doi.org/10.1111/j.1365-2966.2010.16639.x

Osipkov LP (1979) Spherical systems of gravitating bodies with an ellipsoidal velocity distribution. Sov Astron Lett 5:42

Paczynski B, Gorski K (1981) Another possible case of a gravitational lens. Astrophys J Lett 248:101–104. https://doi.org/10.1086/183634

Paraficz D, Hjorth J (2009) Gravitational lenses as cosmic rulers: Ω_m, Ω_Λ from time delays and velocity dispersions. Astron Astrophys 507(3):49–52. https://doi.org/10.1051/0004-6361/200913307

Park JW, Wagner-Carena S, Birrer S, Marshall PJ, Lin JY-Y, Roodman A (LSST Dark Energy Science Collaboration) (2021) Large-scale gravitational lens modeling with Bayesian neural networks for accurate and precise inference of the Hubble constant. Astrophys J 910(1):39. https://doi.org/10.3847/1538-4357/abdfc4

Park JW, Birrer S, Ueland M, Cranmer M, Agnello A, Wagner-Carena S, Marshall PJ, Roodman A (2023) Hierarchical inference of the lensing convergence from photometric catalogs with Bayesian graph neural networks. Astrophys J 953:178. https://doi.org/10.3847/1538-4357/acdc25

Pascale M, Frye BL, Pierel JDR, Chen W, Kelly PL, Cohen SH, Windhorst RA, Riess AG, Kamieneski PS, Diego JM, Meena AK, Cha S, Oguri M, Zitrin A, Jee MJ, Foo N, Leimbach R, Koekemoer AM, Conselice CJ, Dai L, Goobar A, Siebert MR, Strolger L, Willner SP (2024) SN H0pe: the first measurement of H_0 from a multiply-imaged type Ia Supernova. Discovered by JWST. arXiv e-prints, arXiv:2403.18902. https://doi.org/10.48550/arXiv.2403.18902

Pelt J, Kayser R, Refsdal S, Schramm T (1996) The light curve and the time delay of QSO 0957 + 561. Astron Astrophys 305:97

Pesce DW, Braatz JA, Reid MJ, Riess AG, Scolnic D, Condon JJ, Gao F, Henkel C, Impellizzeri CMV, Kuo CY, Lo KY (2020) The megamaser cosmology project. XIII. Combined Hubble constant constraints. Astrophys J Lett 891(1):1. https://doi.org/10.3847/2041-8213/ab75f0

Pierel JDR, Frye BL, Pascale M et al (2024a) JWST photometric time-delay and magnification measurements for the triply-imaged type Ia “Supernova H0pe” at z = 1.78. Astrophys J 967:50. https://doi.org/10.3847/1538-4357/ad3c43

Pierel JDR, Newman AB, Dhawan S et al (2024b) Lensed type Ia Supernova “Encore” at z = 2: the first instance of two multiply-imaged Supernovae in the same host galaxy. https://doi.org/10.48550/arXiv.2404.02139. arXiv e-prints, arXiv:2404.02139

Planck Collaboration, Aghanim N et al (2020) Planck 2018 results. VI. Cosmological parameters. Astron Astrophys 641:6. https://doi.org/10.1051/0004-6361/201833910

Poindexter S, Morgan N, Kochanek CS, Falco EE (2007) Mid-IR observations and a revised time delay for the gravitational lens system quasar HE 1104-1805. Astrophys J 660(1):146–151. https://doi.org/10.1086/512773

Puchwein E, Hilbert S (2009) Cluster strong lensing in the millennium simulation: the effect of galaxies and structures along the line-of-sight. Mon Not R Astron Soc 398(3):1298–1308. https://doi.org/10.1111/j.1365-2966.2009.15227.x

Rathna Kumar S, Tewes M, Stalin CS, Courbin F, Asfandiyarov I, Meylan G, Eulaers E, Prabhu TP, Magain P, Van Winckel H, Ehgamberdiev S (2013) COSMOGRAIL: the COSmological MOnitoring of GRAvItational Lenses. XIV. Time delay of the doubly lensed quasar SDSS J1001 + 5027. Astron Astrophys 557:44. https://doi.org/10.1051/0004-6361/201322116

Read JI, Saha P, Macciò AV (2007) Radial density profiles of time-delay lensing galaxies. Astrophys J 667(2):645–654. https://doi.org/10.1086/520714

Refsdal S (1964) On the possibility of determining Hubble's parameter and the masses of galaxies from the gravitational lens effect. Mon Not R Astron Soc 128:307. https://doi.org/10.1093/mnras/128.4.307

Riess AG, Casertano S, Yuan W, Macri LM, Scolnic D (2019) Large Magellanic Cloud Cepheid standards provide a 1% foundation for the determination of the Hubble constant and stronger evidence for physics beyond ΛCDM. Astrophys J 876(1):85. https://doi.org/10.3847/1538-4357/ab1422

Riess AG, Casertano S, Yuan W, Bowers JB, Macri L, Zinn JC, Scolnic D (2021) Cosmic distances calibrated to 1% precision with Gaia EDR3 parallaxes and Hubble Space Telescope photometry of 75 Milky Way cepheids confirm tension with ΛCDM. Astrophys J Lett 908(1), 6. https://doi.org/10.3847/2041-8213/abdbaf

Riess AG, Yuan W, Macri LM, Scolnic D, Brout D, Casertano S, Jones DO, Murakami Y, Anand GS, Breuval L, Brink TG, Filippenko AV, Hoffmann S, Jha SW, D'arcy Kenworthy W, Mackenty J, Stahl BE, Zheng W (2022) A comprehensive measurement of the local value of the Hubble constant with 1 km s^{-1} mpc^{-1} uncertainty from the Hubble Space Telescope and the SH0ES team. Astrophys J Lett 934(1):7. https://doi.org/10.3847/2041-8213/ac5c5b

Riess AG, Anand GS, Yuan W, Casertano S, Dolphin A, Macri LM, Breuval L, Scolnic D, Perrin M, Anderson RI (2023) Crowded no more: the accuracy of the Hubble constant tested with high-resolution observations of cepheids by JWST. Astrophys J Lett 956(1):18. https://doi.org/10.3847/2041-8213/acf769

Rodney SA, Brammer GB, Pierel JDR, Richard J, Toft S, O'Connor KF, Akhshik M, Whitaker KE (2021) A gravitationally lensed supernova with an observable two-decade time delay. Nat Astron 5:1118–1125. https://doi.org/10.1038/s41550-021-01450-9

Romanowsky AJ, Kochanek CS (1999) Constraints on H_0 from the central velocity dispersions of lens galaxies. Astrophys J 516(1):18–26. https://doi.org/10.1086/307077

Rusu CE, Fassnacht CD, Sluse D, Hilbert S, Wong KC, Huang K-H, Suyu SH, Collett TE, Marshall PJ, Treu T, Koopmans LVE (2017) H0LiCOW - III. Quantifying the effect of mass along the line of sight to the gravitational lens HE 0435-1223 through weighted galaxy counts. Mon Not R Astron Soc 467(4):4220–4242. https://doi.org/10.1093/mnras/stx285

Rusu CE, Wong KC, Bonvin V, Sluse D, Suyu SH, Fassnacht CD, Chan JHH, Hilbert S, Auger MW, Sonnenfeld A, Birrer S, Courbin F, Treu T, Chen GC-F, Halkola A, Koopmans LVE, Marshall PJ, Shajib AJ (2020) H0LiCOW XII. Lens mass model of WFI2033-4723 and blind measurement of its time-delay distance and H_0. Mon Not R Astron Soc 498(1):1440–1468. https://doi.org/10.1093/mnras/stz3451

Saha P, Williams LLR (2004) A portable modeler of lensed quasars. Astron J 127(5):2604–2616. https://doi.org/10.1086/383544

Saha P, Williams LLR (2006) Gravitational lensing model degeneracies: is steepness all-important? Astrophys J 653(2):936–941. https://doi.org/10.1086/508798

Saha P, Sluse D, Wagner J, Williams LLR (2024) Essentials of strong gravitational lensing. Space Sci Rev 220:12. https://doi.org/10.1007/s11214-024-01041-w

Schechter PL, Bailyn CD, Barr R, Barvainis R, Becker CM, Bernstein GM, Blakeslee JP, Bus SJ, Dressler A, Falco EE, Fesen RA, Fischer P, Gebhardt K, Harmer D, Hewitt JN, Hjorth J, Hurt T, Jaunsen AO, Mateo M, Mehlert D, Richstone DO, Sparke LS, Thorstensen JR, Tonry JL, Wegner G, Willmarth DW, Worthey G (1997) The quadruple gravitational lens PG 1115 + 080: time delays and models. Astrophys J Lett 475(2):85–88. https://doi.org/10.1086/310478

Schechter PL, Udalski A, Szymański M, Kubiak M, Pietrzyński G, Soszyński I, Woźniak P, Żebruń K, Szewczyk O, Wyrzykowski Ł (2003) Microlensing of relativistic knots in the quasar HE 1104-1805 AB. Astrophys J 584(2):657–663. https://doi.org/10.1086/345716

Schmidt T, Treu T, Birrer S et al (2023) STRIDES: automated uniform models for 30 quadruply imaged quasars. Mon Not R Astron Soc 518(1):1260–1300. https://doi.org/10.1093/mnras/stac2235

Schneider P (1985) A new formulation of gravitational lens theory, time-delay, and Fermat's principle. Astron Astrophys 143(2):413–420

Schneider P, Sluse D (2013) Mass-sheet degeneracy, power-law models and external convergence: impact on the determination of the Hubble constant from gravitational lensing. Astron Astrophys 559:37. https://doi.org/10.1051/0004-6361/201321882

Schneider P, Sluse D (2014) Source-position transformation: an approximate invariance in strong gravitational lensing. Astron Astrophys 564:103. https://doi.org/10.1051/0004-6361/201322106

Schneider P, Ehlers J, Falco EE (1992) Gravitational lenses. https://doi.org/10.1007/978-3-662-03758-4

Schöneberg N, Abellán GF, Pérez Sánchez A, Witte SJ, Poulin V, Lesgourgues J (2022) The H_0 Olympics: a fair ranking of proposed models. Phys Rep 984:1. https://doi.org/10.1016/j.physrep.2022.07.001

Schwarzschild M (1979) A numerical model for a triaxial stellar system in dynamical equilibrium. Astrophys J 232:236–247. https://doi.org/10.1086/157282

Sereno M (2002) Probing the dark energy with strong lensing by clusters of galaxies. Astron Astrophys 393:757–764. https://doi.org/10.1051/0004-6361:20021048

Shah P, Lemos P, Lahav O (2021) A buyer's guide to the Hubble constant. Astron Astrophys Rev 29(9). https://doi.org/10.1007/s00159-021-00137-4

Shajib AJ, Treu T, Agnello A (2018) Improving time-delay cosmography with spatially resolved kinematics. Mon Not R Astron Soc 473:210–226. https://doi.org/10.1093/mnras/stx2302

Shajib AJ, Birrer S, Treu T et al (2019) Is every strong lens model unhappy in its own way? Uniform modelling of a sample of 13 quadruply + imaged quasars. Mon Not R Astron Soc 483(4):5649–5671. https://doi.org/10.1093/mnras/sty3397

Shajib AJ, Birrer S, Treu T et al (2020) STRIDES: a 3.9 per cent measurement of the Hubble constant from the strong lens system DES J0408-5354. Mon Not R Astron Soc 494(4):6072–6102. https://doi.org/10.1093/mnras/staa828

Shajib AJ, Treu T, Birrer S, Sonnenfeld A (2021) Dark matter haloes of massive elliptical galaxies at z $\sim$ 0.2 are well described by the Navarro-Frenk-White profile. Mon Not R Astron Soc 503(2):2380–2405. https://doi.org/10.1093/mnras/stab536

Shajib AJ, Vernardos G, Collett TE, Motta V, Sluse D, Williams LLR, Saha P, Birrer S, Spiniello C, Treu T (2024) Strong lensing by galaxies. Space Sci Rev 220

Shajib AJ, Mozumdar P, Chen GC-F, Treu T, Cappellari M, Knabel S, Suyu SH, Bennert VN, Frieman JA, Sluse D, Birrer S, Courbin F, Fassnacht CD, Villafaña L, Williams PR (2023) TDCOSMO. XII. Improved Hubble constant measurement from lensing time delays using spatially resolved stellar kinematics of the lens galaxy. Astron Astrophys 673:A9. https://doi.org/10.1051/0004-6361/202345878

Shalyapin VN, Goicoechea LJ (2019) Gravitationally lensed quasar SDSS J1442 + 4055: redshifts of lensing galaxies, time delay, microlensing variability, and intervening metal system at z $\sim$ 2. Astrophys J 873(2):117. https://doi.org/10.3847/1538-4357/ab08f0

Shu Y, Belokurov V, Evans NW (2021) Discovering strongly lensed QSOs from unresolved light curves. Mon Not R Astron Soc 502(2):2912–2921. https://doi.org/10.1093/mnras/stab241

Sluse D, Tewes M (2014) Imprints of the quasar structure in time-delay light curves: microlensing-aided reverberation mapping. Astron Astrophys 571:60. https://doi.org/10.1051/0004-6361/201424776

Sluse D, Sonnenfeld A, Rumbaugh N, Rusu CE, Fassnacht CD, Treu T, Suyu SH, Wong KC, Auger MW, Bonvin V, Collett T, Courbin F, Hilbert S, Koopmans LVE, Marshall PJ, Meylan G, Spiniello C, Tewes M (2017) H0LiCOW - II. Spectroscopic survey and galaxy-group identification of the strong gravitational lens system HE 0435-1223. Mon Not R Astron Soc 470(4):4838–4857. https://doi.org/10.1093/mnras/stx1484

Sluse D, Rusu CE, Fassnacht CD et al (2019) H0LiCOW - X. Spectroscopic/imaging survey and galaxy-group identification around the strong gravitational lens system WFI 2033-4723. Mon Not R Astron Soc 490(1):613–633. https://doi.org/10.1093/mnras/stz2483

Sonnenfeld A (2018) On the choice of lens density profile in time delay cosmography. Mon Not R Astron Soc 474(4):4648–4659. https://doi.org/10.1093/mnras/stx3105

Sonnenfeld A (2021) Statistical strong lensing. II. Cosmology and galaxy structure with time-delay lenses. Astron Astrophys 656:A153. https://doi.org/10.1051/0004-6361/202142062

Sonnenfeld A, Cautun M (2021) Statistical strong lensing. I. Constraints on the inner structure of galaxies from samples of a thousand lenses. Astron Astrophys 651:18. https://doi.org/10.1051/0004-6361/202140549

Soucail G, Kneib J-P, Golse G (2004) Multiple-images in the cluster lens Abell 2218: constraining the geometry of the universe? Astron Astrophys 417:33–37. https://doi.org/10.1051/0004-6361:20040077

Springel V, White SDM, Jenkins A, Frenk CS, Yoshida N, Gao L, Navarro J, Thacker R, Croton D, Helly J, Peacock JA, Cole S, Thomas P, Couchman H, Evrard A, Colberg J, Pearce F (2005) Simulations of the formation, evolution and clustering of galaxies and quasars. Nature 435(7042):629–636. https://doi.org/10.1038/nature03597

Springer OM, Ofek EO (2021) Measuring time delays - I. Using a flux time series that is a linear combination of time-shifted light curves. Mon Not R Astron Soc 506(1):864–876. https://doi.org/10.1093/mnras/stab1600

Starkey D, Horne K, Fausnaugh MM et al (2017) Space telescope and optical reverberation mapping project.VI. Reverberating disk models for NGC 5548. Astrophys J 835(1):65. https://doi.org/10.3847/1538-4357/835/1/65

Suyu SH, Marshall PJ, Hobson MP, Blandford RD (2006) A Bayesian analysis of regularized source inversions in gravitational lensing. Mon Not R Astron Soc 371(2):983–998. https://doi.org/10.1111/j.1365-2966.2006.10733.x

Suyu SH, Marshall PJ, Blandford RD, Fassnacht CD, Koopmans LVE, McKean JP, Treu T (2009) Dissecting the gravitational lens B1608 + 656. I. Lens potential reconstruction. Astrophys J 691(1):277–298. https://doi.org/10.1088/0004-637X/691/1/277

Suyu SH, Marshall PJ, Auger MW, Hilbert S, Blandford RD, Koopmans LVE, Fassnacht CD, Treu T (2010) Dissecting the gravitational lens B1608 + 656. II. Precision measurements of the Hubble constant, spatial curvature, and the dark energy equation of state. Astrophys J 711(1):201–221. https://doi.org/10.1088/0004-637X/711/1/201

Suyu SH, Auger MW, Hilbert S, Marshall PJ, Tewes M, Treu T, Fassnacht CD, Koopmans LVE, Sluse D, Blandford RD, Courbin F, Meylan G (2013) Two accurate time-delay distances from strong lensing: implications for cosmology. Astrophys J 766(2):70. https://doi.org/10.1088/0004-637X/766/2/70

Suyu SH, Treu T, Hilbert S, Sonnenfeld A, Auger MW, Blandford RD, Collett T, Courbin F, Fassnacht CD, Koopmans LVE, Marshall PJ, Meylan G, Spiniello C, Tewes M (2014) Cosmology from gravitational lens time delays and Planck data. Astrophys J Lett 788(2):35. https://doi.org/10.1088/2041-8205/788/2/L35

Suyu SH, Bonvin V, Courbin F, Fassnacht CD, Rusu CE, Sluse D, Treu T, Wong KC, Auger MW, Ding X, Hilbert S, Marshall PJ, Rumbaugh N, Sonnenfeld A, Tewes M, Tihhonova O, Agnello A, Blandford RD, Chen GC-F, Collett T, Koopmans LVE, Liao K, Meylan G, Spiniello C (2017) H0LiCOW - I. H_0 lenses in COSMOGRAIL's wellspring: program overview. Mon Not R Astron Soc 468(3):2590–2604. https://doi.org/10.1093/mnras/stx483

Suyu SH, Chang T-C, Courbin F, Okumura T (2018) Cosmological distance indicators. Space Sci Rev 214(5):91. https://doi.org/10.1007/s11214-018-0524-3

Suyu SH, Goobar A, Collett T, More A, Vernardos G (2024) Strong gravitational lensing and microlensing of supernovae. Space Sci Rev 220:13. https://doi.org/10.1007/s11214-024-01044-7

Tagore AS, Barnes DJ, Jackson N, Kay ST, Schaller M, Schaye J, Theuns T (2018) Reducing biases on H_0 measurements using strong lensing and galaxy dynamics: results from the EAGLE simulation. Mon Not R Astron Soc 474(3):3403–3422. https://doi.org/10.1093/mnras/stx2965

Tak H, Mandel K, van Dyk DA, Kashyap VL, Meng X-L, Siemiginowska A (2016) Bayesian Estimates of Astronomical Time Delays between Gravitationally Lensed Stochastic Light Curves. arXiv:1602.01462

Taubenberger S, Suyu SH, Komatsu E, Jee I, Birrer S, Bonvin V, Courbin F, Rusu CE, Shajib AJ, Wong KC (2019) The Hubble constant determined through an inverse distance ladder including quasar time delays and Type Ia supernovae. Astron Astrophys 628:7. https://doi.org/10.1051/0004-6361/201935980

Teodori L, Blum K, Castorina E, Simonović M, Soreq Y (2022) Comments on the mass sheet degeneracy in cosmography analyses. J Cosmol Astropart Phys 2022(07):027. https://doi.org/10.1088/1475-7516/2022/07/027

Tewes M, Courbin F, Meylan G (2013a) COSMOGRAIL: the COSmological MOnitoring of GRAvItational lenses. XI. Techniques for time delay measurement in presence of microlensing. Astron Astrophys 553:120. https://doi.org/10.1051/0004-6361/201220123

Tewes M, Courbin F, Meylan G, Kochanek CS, Eulaers E, Cantale N, Mosquera AM, Magain P, Van Winckel H, Sluse D, Cataldi G, Vörös D, Dye S (2013b) COSMOGRAIL: the COSmological MOnitoring of GRAvItational lenses. XIII. Time delays and 9-yr optical monitoring of the lensed quasar RX J1131-1231. Astron Astrophys 556:22. https://doi.org/10.1051/0004-6361/201220352

Tie SS, Kochanek CS (2018) Microlensing makes lensed quasar time delays significantly time variable. Mon Not R Astron Soc 473(1):80–90. https://doi.org/10.1093/mnras/stx2348

Tihhonova O, Courbin F, Harvey D, Hilbert S, Rusu CE, Fassnacht CD, Bonvin V, Marshall PJ, Meylan G, Sluse D, Suyu SH, Treu T, Wong KC (2018) H0LiCOW VIII. A weak-lensing measurement of the external convergence in the field of the lensed quasar HE 0435-1223. Mon Not R Astron Soc 477(4):5657–5669. https://doi.org/10.1093/mnras/sty1040

Tihhonova O, Courbin F, Harvey D, Hilbert S, Peel A, Rusu CE, Fassnacht CD, Bonvin V, Marshall PJ, Meylan G, Sluse D, Suyu SH, Treu T, Wong KC (2020) H0LiCOW - XI. A weak lensing measurement of the external convergence in the field of the lensed quasar B1608 + 656 using HST and Subaru deep imaging. Mon Not R Astron Soc 498(1):1406–1419. https://doi.org/10.1093/mnras/staa1436

Treu T, Koopmans LVE (2002) The internal structure of the lens PG1115 + 080: breaking degeneracies in the value of the Hubble constant. Mon Not R Astron Soc 337(2):6–10. https://doi.org/10.1046/j.1365-8711.2002.06107.x

Treu T, Koopmans LVE (2004) Massive dark matter halos and evolution of early-type galaxies to $z \approx 1$. Astrophys J 611(2):739–760. https://doi.org/10.1086/422245

Treu T, Marshall PJ (2016) Time delay cosmography. Astron Astrophys Rev 24(1):11. https://doi.org/10.1007/s00159-016-0096-8

Treu T, Shajib AJ (2023) Strong Lensing and H_0. arXiv e-prints, arXiv:2307.05714. https://doi.org/10.48550/arXiv.2307.05714

Treu T, Brammer G, Diego JM, Grillo C, Kelly PL, Oguri M, Rodney SA, Rosati P, Sharon K, Zitrin A, Balestra I, Bradač M, Broadhurst T, Caminha GB, Halkola A, Hoag A, Ishigaki M, Johnson TL, Karman W, Kawamata R, Mercurio A, Schmidt KB, Strolger L-G, Suyu SH, Filippenko AV, Foley RJ, Jha SW, Patel B (2016) "Refsdal" meets popper: comparing predictions of the re-appearance of the multiply imaged supernova behind MACSJ1149.5 + 2223. Astrophys J 817(1):60. https://doi.org/10.3847/0004-637X/817/1/60

Treu T, Suyu SH, Marshall PJ (2022) Strong lensing time-delay cosmography in the 2020s. Astron Astrophys Rev 30(1):8. https://doi.org/10.1007/s00159-022-00145-y

Unruh S, Schneider P, Sluse D (2017) Ambiguities in gravitational lens models: the density field from the source position transformation. Astron Astrophys 601:77. https://doi.org/10.1051/0004-6361/201629048

van de Ven G, Mandelbaum R, Keeton CR (2009) Galaxy density profiles and shapes - I. Simulation pipeline for lensing by realistic galaxy models. Mon Not R Astron Soc 398(2):607–634. https://doi.org/10.1111/j.1365-2966.2009.15167.x

Van de Vyvere L, Gomer MR, Sluse D, Xu D, Birrer S, Galan A, Vernardos G (2022b) TDCOSMO. VII. Boxyness/discyness in lensing galaxies: detectability and impact on H_0. Astron Astrophys 659:127. https://doi.org/10.1051/0004-6361/202141551

Van de Vyvere L, Sluse D, Gomer MR, Mukherjee S (2022a) Consequences of the lack of azimuthal freedom in the modeling of lensing galaxies. Astron Astrophys 663:A179. https://doi.org/10.1051/0004-6361/202243382

Vega-Ferrero J, Diego JM, Miranda V, Bernstein GM (2018) The Hubble constant from SN Refsdal. Astrophys J Lett 853(2):31. https://doi.org/10.3847/2041-8213/aaa95f

Vegetti S, Koopmans LVE (2009) Bayesian strong gravitational-lens modelling on adaptive grids: objective detection of mass substructure in Galaxies. Mon Not R Astron Soc 392(3):945–963. https://doi.org/10.1111/j.1365-2966.2008.14005.x

Verde L, Treu T, Riess AG (2019) Tensions between the early and late universe. Nat Astron 3:891–895. https://doi.org/10.1038/s41550-019-0902-0

Vernardos G, Sluse D, Pooley D, Schmidt RW, Millon M, Weisenbach L, Motta V, Anguita T, Saha P, O'Dowd M, Peel A, Schechter PL (2024) Microlensing of strongly lensed quasars. Space Sci Rev 220(1):14. https://doi.org/10.1007/s11214-024-01043-8

Vuissoz C, Courbin F, Sluse D, Meylan G, Ibrahimov M, Asfandiyarov I, Stoops E, Eigenbrod A, Le Guillou L, van Winckel H, Magain P (2007) COSMOGRAIL: the COSmological MOnitoring of GRAvItational lenses. V. The time delay in SDSS J1650 + 4251. Astron Astrophys 464(3):845–851. https://doi.org/10.1051/0004-6361:20065823

Vuissoz C, Courbin F, Sluse D, Meylan G, Chantry V, Eulaers E, Morgan C, Eyler ME, Kochanek CS, Coles J, Saha P, Magain P, Falco EE (2008) COSMOGRAIL: the COSmological MOnitoring of GRAv-Itational lenses. VII. Time delays and the Hubble constant from WFI J2033-4723. Astron Astrophys 488(2):481–490. https://doi.org/10.1051/0004-6361:200809866

Warren SJ, Dye S (2003) Semilinear gravitational lens inversion. Astrophys J 590(2):673–682. https://doi.org/10.1086/375132

Wells P, Fassnacht CD, Rusu CE (2023) TDCOSMO. XIV. Practical techniques for estimating external convergence of strong gravitational lens systems and applications to the SDSS J0924 + 0219 system. Astron Astrophys 676:95. https://doi.org/10.1051/0004-6361/202346093

Wilson ML, Zabludoff AI, Keeton CR, Wong KC, Williams KA, French KD, Momcheva IG (2017) A spectroscopic survey of the fields of 28 strong gravitational lenses: implications for H_0. Astrophys J 850(1):94. https://doi.org/10.3847/1538-4357/aa9653

Wong KC, Suyu SH, Auger MW, Bonvin V, Courbin F, Fassnacht CD, Halkola A, Rusu CE, Sluse D, Sonnenfeld A, Treu T, Collett TE, Hilbert S, Koopmans LVE, Marshall PJ, Rumbaugh N (2017) H0LiCOW - IV. Lens mass model of HE 0435-1223 and blind measurement of its time-delay distance for cosmology. Mon Not R Astron Soc 465(4):4895–4913. https://doi.org/10.1093/mnras/stw3077

Wong KC, Suyu SH, Chen GC-F, Rusu CE, Millon M, Sluse D, Bonvin V, Fassnacht CD, Taubenberger S, Auger MW, Birrer S, Chan JHH, Courbin F, Hilbert S, Tihhonova O, Treu T, Agnello A, Ding X, Jee I, Komatsu E, Shajib AJ, Sonnenfeld A, Blandford RD, Koopmans LVE, Marshall PJ, Meylan G (2020) H0LiCOW - XIII. A 2.4 per cent measurement of H_0 from lensed quasars: 5.3σ tension between early- and late-universe probes. Mon Not R Astron Soc 498(1):1420–1439. https://doi.org/10.1093/mnras/stz3094

Xu D, Sluse D, Schneider P, Springel V, Vogelsberger M, Nelson D, Hernquist L (2016) Lens galaxies in the Illustris simulation: power-law models and the bias of the Hubble constant from time delays. Mon Not R Astron Soc 456(1):739–755. https://doi.org/10.1093/mnras/stv2708

Yıldırım A, Suyu SH, Halkola A (2020) Time-delay cosmographic forecasts with strong lensing and JWST stellar kinematics. Mon Not R Astron Soc 493(4):4783–4807. https://doi.org/10.1093/mnras/staa498

Yıldırım A, Suyu SH, Chen GC-F, Komatsu E (2023) TDCOSMO. XIII. Cosmological distance measurements in light of the mass-sheet degeneracy: Forecasts from strong lensing and integral field unit stellar kinematics. Astron Astrophys 675:A21. https://doi.org/10.1051/0004-6361/202142318

Publisher's Note Springer Nature remains neutral with regard to jurisdictional claims in published maps and institutional affiliations.

Authors and Affiliations

S. Birrer[1,2] · M. Millon[2,3] · D. Sluse[4] · A.J. Shajib[5,6] · F. Courbin[3,7,8] · S. Erickson[2] · L.V.E. Koopmans[9] · S.H. Suyu[10,11,12] · T. Treu[13]

✉ S. Birrer
simon.birrer@stonybrook.edu

✉ F. Courbin
frederic.courbin@epfl.ch

1 Department of Physics and Astronomy, Stony Brook University, Stony Brook, NY 11794, USA

2 Kavli Institute for Particle Astrophysics and Cosmology and Department of Physics, Stanford University, Stanford, CA 94305, USA

3 Institute of Physics, Laboratory of Astrophysics, Ecole Polytechnique Fédérale de Lausanne (EPFL) Observatoire de Sauverny, 1290 Versoix, Switzerland

4 STAR Institute, Quartier Agora - Allée du six Août, 19c B-4000 Liège, Belgium

5 Department of Astronomy and Astrophysics, University of Chicago, Chicago, IL 60637, USA

6 Kavli Institute for Cosmological Physics, University of Chicago, Chicago, IL 60637, USA

7 ICC-UB Institut de Ciències del Cosmos, University of Barcelona, Martí Franquès, 1, E-08028 Barcelona, Spain

8 ICREA, Pg. Lluís Companys 23, Barcelona, E-08010, Spain

9 Kapteyn Astronomical Institute, University of Groningen, P.O. Box 800, 9700AV Groningen, The Netherlands

10 Max-Planck-Institut für Astrophysik, Karl-Schwarzschild-Str. 1, 85748 Garching, Germany

11 TUM School of Natural Sciences, Physics Department, Technical University of Munich, James-Franck-Straße 1, 85748 Garching, Germany

12 Academia Sinica Institute of Astronomy and Astrophysics (ASIAA), 11F of ASMAB, No.1, Sect. 4, Roosevelt Road, Taipei 10617, Taiwan

13 Department of Physics and Astronomy, University of California, Los Angeles, CA 90095, USA

Space Science Reviews (2024) 220:58
https://doi.org/10.1007/s11214-024-01087-w

Strong Gravitational Lensing as a Probe of Dark Matter

S. Vegetti[1] · S. Birrer[2] · G. Despali[3,4] · C.D. Fassnacht[5] · D. Gilman[6] · Y. Hezaveh[7,8,9,10,11] · L. Perreault Levasseur[7,8,9,10,11] · J.P. McKean[12,13,14] · D.M. Powell[1] · C.M. O'Riordan[1] · G. Vernardos[15]

Received: 17 June 2023 / Accepted: 1 July 2024 / Published online: 30 July 2024
© The Author(s) 2024

Abstract
Dark matter structures within strong gravitational lens galaxies and along their lines of sight leave a gravitational imprint on the multiple images of lensed sources. Strong gravitational lensing provides, therefore, a key test of different dark matter models. In this article, we describe how galaxy-scale strong gravitational lensing observations are sensitive to the physical nature of dark matter. We provide an historical perspective of the field, and review its current status. We discuss the challenges and advances in terms of data, treatment of systematic errors and theoretical predictions, that will enable one to deliver a stringent and robust test of different dark matter models in the next decade. With the advent of the next generation of sky surveys, the number of known strong gravitational lens systems is expected to increase by several orders of magnitude. Coupled with high-resolution follow-up observations, these data will provide a key opportunity to constrain the properties of dark matter with strong gravitational lensing.

Keywords Gravitational lensing: strong · Haloes: number · Structure · Dark matter

1 Introduction

In the standard cosmological model, 85 per cent of the total amount of matter in the Universe is made of non-baryonic dark matter (Planck Collaboration et al. 2020b). Evidence of the existence of dark matter spans a wide range of independent astronomical observations: the cosmic micro-wave background (Planck Collaboration et al. 2020a), the rotation curves of disk galaxies (e.g. van Albada et al. 1985; Rubin 1991), the motion of galaxies within clusters (e.g. Zwicky 1933), gravitational lensing by galaxies and galaxy clusters (e.g Treu and Koopmans 2004; Mandelbaum et al. 2006; Clowe et al. 2006; Auger et al. 2009; Barnabè et al. 2011; Sonnenfeld et al. 2022; Shajib et al. 2024), hot gas in galaxy clusters (e.g. Ettori et al. 2013), weak lensing and galaxy clustering (e.g. Alam et al. 2017; Heymans et al. 2021; Abbott et al. 2022). To this day, the physical nature of dark matter remains an unsolved problem. There is no dark matter particle within the standard model of particle physics. So far, several candidates, spanning 90 orders of magnitude in mass, have been proposed (Bertone and Tait 2018). In the 1980s, weakly interactive massive particles (WIMPs) emerged as the favourite candidate for cold dark matter. The success of WIMPs stems from the fact that their existence is predicted by Supersymmetry theories (Fox 2018, and references therein),

Extended author information available on the last page of the article

according to which they are thermally produced with a self-interaction cross-section that leads to the correct amount of dark matter at the present day. However, decades of particle physics experiments have failed to produce a WIMP detection (Arcadi et al. 2018).

Other viable dark matter candidates include warm (WDM), self-interacting (SIDM) and fuzzy (FDM) dark matter. We refer the reader to Sect. 2 for a more detailed description of the various models. What makes these alternative models interesting is that they predict a different distribution of dark matter on subgalactic scales.

Strong gravitational lensing is sensitive to the distribution of matter and dark matter between the observer and the source. It is a purely gravitational probe and does not rely on the presence and distribution of baryons. It therefore provides a channel to observationally constrain the physical nature of dark matter. Other probes include the Lyman-α forest (e.g. Villasenor et al. 2023), the satellite galaxies of the Milky Way (e.g. Nadler et al. 2021) and stellar streams in the Local Group (e.g. Erkal et al. 2017).

This article focuses on galaxy-scale strong gravitational lensing as a probe of dark matter and is organized as follows. In Sect. 2, we give a description of the dark matter models that have so far been tested with strong gravitational lensing observations. We expect this list to increase in the future as theoretical predictions for structure formation in more dark matter models become available. In Sect. 3, we discuss how lensing observables (i.e. image positions, fluxes, and time delays) are affected by the distribution of dark matter on small scales. In Sect. 4, we describe the process of lens modelling and how one can constrain the properties of dark matter from strong gravitational lensing observations as an inference problem. Another important aspect is the role played by degeneracies, systematic errors and unknowns, all of which are discussed in Sect. 5. The article then continues with an historical perspective of the field (Sect. 6), its current status (Sect. 7) and how it is likely to evolve in the near future (Sect. 8). We recognize that the last two sections may soon be obsolete. Therefore, we encourage the interested reader to complement this text with the latest relevant publications. Finally, we present our concluding remarks in Sect. 9.

2 Dark Matter Models

We begin this section by introducing theoretical predictions on the properties of low-mass haloes in the CDM model as derived from numerical simulations. Alternative dark matter models can affect both the number and the structural properties of these objects and, as a consequence, the lensing signal that they produce. In the rest of this section, we describe the most studied models and how they differ from CDM. In Sect. 5.6, we discuss current theoretical uncertainties (e.g. the effect of baryons) relative to the subhalo mass function and mass density profile, and how these affect the robustness of dark matter constraints from strong gravitational lensing observations.

2.1 Cold Dark Matter

Dark matter models where the particle is non-relativistic are described as cold. Possible candidates include WIMPs and axions (e.g. Feng 2010). At the same time, dark matter does not have to be a particle and primordial black holes have also been proposed as a CDM candidate (e.g. Green and Kavanagh 2021). Thanks to its several successes at reproducing many observations (especially on large scales), CDM is, at present, the prevailing cosmological model. However, it remains largely untested on subgalactic scales, where strong gravitational lensing represents a key opportunity to further assess its validity. One of the fundamental predictions of CDM is the fact that dark matter structures form hierarchically and bottom-up: low-mass haloes form first and subsequently merge into structures of increasing mass and size. The structure and evolution of CDM haloes has been extensively studied and precisely characterised using numerical simulations (see Zavala and Frenk 2019, for a review). The collisionless nature of CDM and hierarchical structure formation result in a population of haloes with well defined properties.

Halo mass function: at the low-mass end, the number density n of haloes is well-described, to first approximation, by a power-law distribution:

$$\frac{dn}{dM} \propto M^{-\alpha}. \tag{1}$$

For halo masses smaller than a mass of $M \sim 10^{11} M_{\odot}$ the slope is $\alpha = 1.9$, and the number of structures increases with decreasing halo mass. The shape of the halo mass function is well understood from statistical arguments. These are based on the properties of the initial density field of fluctuations and the gravitational collapse process that leads to the formation of virialised haloes. This is at the basis of the formalism first introduced by Press and Schechter (1974), and subsequently extended and improved by Sheth and Tormen (1999), Sheth et al. (2001), Tinker et al. (2008) and Despali et al. (2016).

Subhalo mass function: the number density of haloes that have been accreted onto larger ones is also thought to be a power-law with a normalisation that depends on the host halo mass and redshift (e.g. Gao et al. 2004; Springel et al. 2008; Giocoli et al. 2008; Angulo et al. 2009; Despali and Vegetti 2017). Due to the interaction between the subhaloes and the host, the former are tidally stripped and sometimes destroyed (e.g. Angulo et al. 2009; Berezinsky et al. 2014; Delos 2019; Green and van den Bosch 2019; Green et al. 2021). As a consequence of this process, the total number of subhaloes in hydrodynamical simulations is reduced by between 20 and 50 per cent, depending on the galaxy formation model, relative to dark-matter-only simulations (Sawala et al. 2017; Despali and Vegetti 2017; Chua et al. 2017). The number density of subhaloes also changes as a function of distance from the halo centre according to an Einasto profile (Springel et al. 2008). However, at fixed subhalo mass, the projected number density distribution is constant with radius (Xu et al. 2015; Despali and Vegetti 2017).

Halo mean mass density profile: the mass density $\rho(r)$ of a dark matter halo as a function of radius r is described by the Navarro, Frenk and White profile (NFW, Navarro et al. 1996):

$$\rho(r) = \frac{\rho_s}{\frac{r}{r_s}\left(1 + \frac{r}{r_s}\right)^2}. \tag{2}$$

Here, r_s is the scale radius and ρ_s is the density normalization. The NFW profile can also be defined in terms of the halo virial mass $M_{\rm vir}$ (i.e. the mass within the radius, $r_{\rm vir}$, that encloses a virial overdensity $\Delta_{\rm vir}$, defined following Bryan and Norman 1998), and the virial concentration $c_{\rm vir} = r_{\rm vir}/r_s$.

Subhalo mean mass density profile: due to the tidal interactions between the subhaloes and the host halo, the mass density profiles of the former tend to deviate from a standard NFW profile and are significantly more concentrated than isolated haloes of the same mass (e.g. Moliné et al. 2017). For this reason, their properties are better described in terms of the peak circular velocity $V_{\rm max}$ and the corresponding $R_{\rm max}$ radius.

2.2 Warm Dark Matter

Dark matter particles with (close to) relativistic free-streaming velocities in the early Universe, for example, light neutrinos, are commonly described as Hot Dark Matter candidates

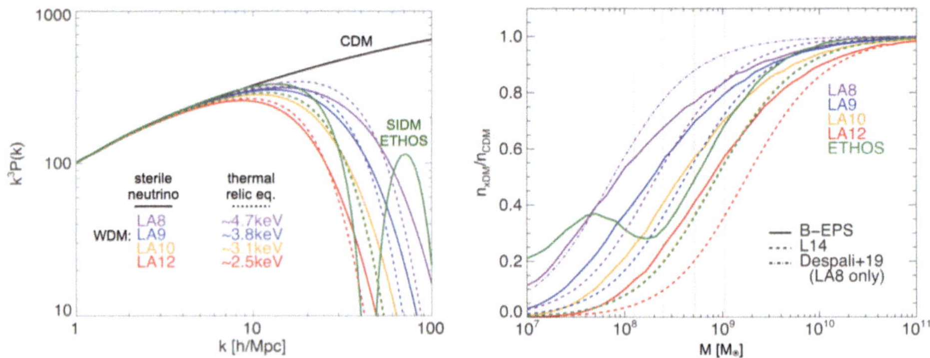

Fig. 1 Power spectra (left) and mass functions (right) in warm and self-interacting dark matter models. In the left panel, the CDM $P(k)$ is shown in black, while different colours represent four WDM models and one SIDM model (ETHOS, see Vogelsberger et al. 2016) in which the power-spectrum is modified. In the WDM case, the suppression is shown both for sterile neutrino models (solid) and their thermal-relic closest counterpart (dashed). The right panel shows instead the suppression in the halo mass function with respect to CDM, calculated with the extended Press-Schechter (EPS) formalism, or measured in simulations (Lovell et al. 2014; Despali et al. 2018). The figures are adapted from Fig. 1 and 2 in Lovell (2020), respectively

(Doroshkevich et al. 1981). A universe predominantly made of HDM has already been ruled out by observations of the clustering of galaxies on large scales (White et al. 1983). Between CDM and HDM lies a class of dark matter models known as warm dark matter (WDM), whose candidates include the gravitino and sterile neutrinos (e.g. Boyarsky et al. 2009, 2019; Paduroiu 2022). They are non-relativistic, but have a non-negligibile free-streaming velocity at early times. This property leads to the suppression of the power-spectrum of the mass-density fluctuations on scales smaller than the half-mode scale $\lambda_{\rm hm}$ (Viel et al. 2012; Iršič et al. 2017) - see Fig. 1. The corresponding suppression in the number density of low-mass haloes relative to CDM can be expressed as follows (Schneider et al. 2012; Lovell et al. 2012; Bose et al. 2016; Lovell 2020):

$$\frac{n_{\rm WDM}}{n_{\rm CDM}} = \left(1 + \gamma \frac{M_{\rm hm}}{M}\right)^{\beta} . \tag{3}$$

$M_{\rm hm}$ is the *half-mode mass* and is defined as the mass-scale at which the square root of the WDM linear matter power-spectrum is 50 per cent smaller than in CDM. In practice, the half-mode mass is inversely proportional to the dark matter particle mass m_{DM}: the lighter the particle candidate, the stronger the suppression in the mass function at small scales.

As a result of the increased particle velocity in WDM, structure formation is delayed to a time when the Universe is less dense. As a consequence, haloes of a given mass are not only less numerous, but also less concentrated (e.g. Ludlow et al. 2016). By suppressing both halo abundance and halo concentration, the latter quantity determining the lensing efficiency of haloes, WDM models predict less small-scale perturbations to the strong gravitationally lensed images than one would expect from CDM (Despali et al. 2018; Li et al. 2017; Despali et al. 2020).

2.3 Self-Interacting Dark Matter

Self-interacting dark matter (SIDM) postulates that dark matter particles are not collisionless, but have non-gravitational interactions in which they exchange energy and momentum.

The term SIDM refers to a variety of models that include elastic or inelastic scattering and a constant or velocity dependent interaction cross-section (Vogelsberger et al. 2012; Rocha et al. 2013; Kaplinghat et al. 2014; Vogelsberger et al. 2016; Sameie et al. 2018; Robertson et al. 2018; Lovell et al. 2019; Kaplinghat et al. 2019; Vogelsberger et al. 2019; Despali et al. 2019; Sameie et al. 2020; Kaplinghat et al. 2020; Robertson et al. 2021; Correa 2021).

The distinctive signature of SIDM is a modification of the central mass density profile of haloes and subhaloes. The interactions between dark matter particles in high-density regions lead to a transfer of heat and the formation of a density core with a depth and size that is related to the strength of the self-interaction cross-section σ. In some cases, however, the halo and subhalo subsequently undergo a runaway contraction, also known as gravo-thermal catastrophe or core collapse. This phenomenon results in the formation of a cuspy density profile. It is accelerated by the presence of baryons (e.g. Feng et al. 2021) and, in the case of subhaloes, by tidal stripping (Nishikawa et al. 2020). From a strong gravitational lensing perspective, these diverse changes to the central mass density distribution imply that certain (sub-)haloes will be more efficient lenses than others (Despali et al. 2019; Robertson et al. 2019; Gilman et al. 2021, 2022; Amorisco et al. 2022).

In addition to altering the internal structure of haloes, SIDM can also suppress the abundance of subhaloes relative to CDM through ram-pressure stripping of the subhaloes while they are accreted by the host galaxy (see Fig. 1) and Nadler et al. 2020; Zeng et al. 2022). As ram-pressure stripping occurs due to self-interactions between dark matter particles (as distinct from baryonic ram-pressure stripping) bound to the subhaloes and those bound to the host, the efficiency of this mechanism depends on the amplitude of the self-interaction cross-section. For example, in velocity-independent SIDM models with elastic scattering, a very high cross-section ($\sigma \sim 10$ cm^2 g^{-1}) is required. These models are currently ruled out by observations of galaxy clusters (Sagunski et al. 2021; Andrade et al. 2022).

Strong gravitational lensing, being sensitive to the amount and concentration of low-mass haloes (i.e. with a central velocity dispersion from less than ~ 10 km s^{-1} to ~ 50 km s^{-1}), provides, therefore, an independent avenue to constrain SIDM models. It allows one to constrain the self-interaction cross-section at low velocities and complements the constraints derived from galaxies and galaxy clusters (e.g. Loudas et al. 2022).

2.4 Fuzzy Dark Matter

Fuzzy dark matter is a particular form of dark matter made of *ultra-light* bosons, i.e., $m_{\rm DM}c^2 \sim 10^{-22}$ eV. This particle mass is orders of magnitude smaller than that of WIMPs and WDM models. As a result, the de Broglie wavelength is larger than the inter-particle separation and waves better describe the behaviour of the FDM field. This effect leads to a series of distinctive phenomenologies with respect to the other dark matter models so far considered (see Hui 2021, for a review). For example, numerical simulations, which model the full non-linear evolution (Schive et al. 2014a; May and Springel 2021), show that the wave-like behaviour (e.g. interference effects) of FDM leads to the formation of a soliton core at the centre of haloes and density granules on scales smaller than a kpc.

Similarly to WDM, there is a cut-off in the FDM transfer function at small scales, though via a different mechanism that is dependent on the de Broglie wavelength rather than a free-streaming length.[1] As a consequence, from a strong gravitational lensing perspective, FDM models are expected to lead to fewer perturbations of the lensed images compared to CDM.

[1] Note, however, that the wave-like nature of ultra-light dark matter particles results in the FDM power-spectrum to briefly exceeds that in CDM on $\mathcal{O}$(kpc) scales that correspond to the mean de Broglie wavelength.

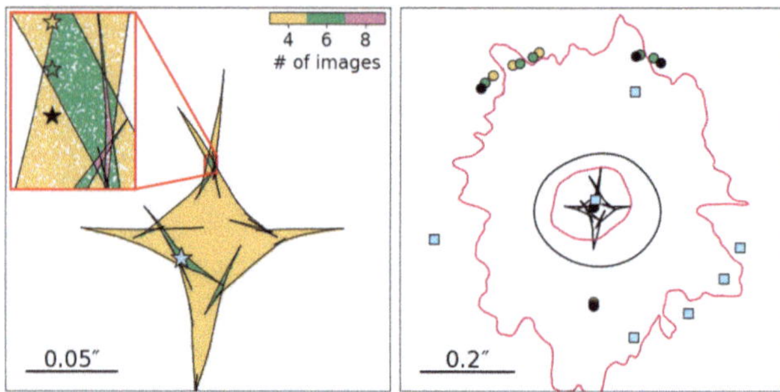

Fig. 2 Lensed image multiplicity in FDM. The left panel shows the caustic structure of a galaxy-scale lens in FDM. The colour indicates the number of lensed images for a point source located at different positions on the source plane. The right panel shows the corresponding lensed image position on the observed plane. Rare hexad and octad images become more likely in FDM models. Figure reproduced with permission from Chan et al. (2020), copyright by APS

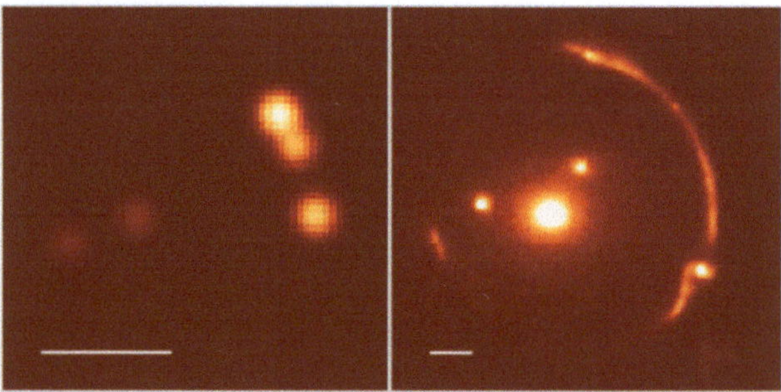

Fig. 3 Two examples of galaxy-scale lenses with a unresolved (left) and resolved (right) background source. On the left panel, the gravitational lens system B2045+265 shows a strong flux-ratio anomaly which is inconsistent with a smooth lens mass distribution (Fassnacht et al. 1999). On the right, the gravitational lens system SDSSJ120602.09+514229.5, also known as the Clone, displays a distorted arc due to the presence of a small luminous satellite galaxy (Vegetti et al. 2010)

However, the granular structure in the halo density profile of lens galaxies has been shown to lead to a distinct new source of small-scale perturbations of the lensed images, which is unique to FDM models (see Fig. 2 and Chan et al. 2020; Laroche et al. 2022; Powell et al. 2023; Amruth et al. 2023).

3 Strong Lensing as a Probe of Dark Matter

Changes to the matter distribution on subgalactic scales within lens galaxies and along their lines of sight leave a gravitational imprint on the strong lensing data. Here, we briefly describe the nature and strength of this effect. In particular, we discuss changes induced on the lensing potential, its first and second derivative, and how they affect the observed time-delay, image positions and magnifications, respectively. For an historical perspective of the field, we refer the reader to Sect. 6.

Image magnification: The largest effect produced by local fluctuations in the lensing mass density distribution (either in the form of subhaloes, field haloes or FDM granules)

is a change in the relative magnification of unresolved lensed images (see Fig. 3 for an example), and the creation of so called flux-ratio anomalies. These anomalies could also be related to micro-lensing by stars (see Chap. 5) or propagation effects such as free-free absorption at long wavelength (Mittal et al. 2007) and dust extinction (Elíasdóttir et al. 2006) in the lens galaxy. The observed ratios are also affected by intrinsic and extrinsic (such as microlensing or interstellar scattering) variability of the background source (Koopmans et al. 2003; Biggs 2023). As such, they can only be used as a probe of dark matter with observations at wavelengths where the angular size of the lensed object is larger than the scale of the micro-lenses, and at which dust and free electron absorption and/or scattering is low. Since flux-ratio anomalies are related to a local change of the second derivative of the lensing potential, they are an effective tool to detect some of the lowest mass perturbations. Lenses with small image separation (relative to the scale of the macro-lens) in the fold and cusp configurations (i.e. where the source lies on the cusp or fold of the caustic curves) are the most sensitive to the perturbations of the lensing potential and least sensitive to intrinsic variability. In these cases, the following R_{fold} and R_{cusp} relations have been sometimes used to quantify the strengths of the anomaly:

$$R_{\text{fold}} = \frac{\mu_A + \mu_B}{|\mu_A| + |\mu_B|}, \tag{4}$$

and

$$R_{\text{cusp}} = \frac{\mu_B + \mu_A + \mu_c}{|\mu_A| + |\mu_B| + |\mu_C|}. \tag{5}$$

Here, $\mu_{A,B}$ in R_{fold} and $\mu_{A,B,C}$ in R_{cusp} are the magnifications of the merging double and triplet images, respectively. As the opening angles between the images A and C ($\Delta\phi$) in the cusp configuration and A and B (ϕ_1) in the fold configuration approach zero, so do R_{fold} and R_{cusp}, when the lens mass distribution is smooth and the background source is a point-source. However, it is worth noting that both relations suffer from a few limitations. First, astrophysical sources of emission, whether they be the accretion disk, the narrow emission line region or the warm (> 50 K) dust torus around a black hole or the relativistic jet that they produce are not point-like, but have some angular scale. Second, the fold relation is only reliable when the image separation is smaller than the distance at which the image with negative parity would have a less than unity magnification. While the cusp relation breaks down unless the source is very close to the bisector.

Image positions: The second largest effect is a local change of the relative positions of the multiple lensed images and the creation of so called astrometric perturbations. These anomalies are related to a local change of the first derivative of the lensing potential. Therefore, they can only be produced by a gravitational perturbation and cannot be caused by micro-lensing. In the case of resolved sources, astrometric anomalies appear as perturbations to the surface brightness distribution of highly magnified arcs and Einstein rings, and are sometimes referred to as surface-brightness anomalies (see Fig. 3 for an example). The level of perturbations that can be detected using a surface-brightness anomaly is set by the quality of the data and structure of the source surface brightness distribution (see Sect. 7 and Despali et al. 2022). For example, a subhalo with a mass of $10^6 M_\odot$ will produce distortions on angular scales of a couple of milli-arcsecond.

Image time delay: The weakest effect is a change of the lensing potential itself. In the case of multiply imaged quasars with a flux-varying source, this phenomenon is observed as a perturbation to the time delay (see Chap. 7) between the multiple images (Keeton and

Moustakas 2009; Cyr-Racine et al. 2016; Gilman et al. 2020). Unlike flux-ratio anomalies, time-delay anomalies are not affected by dust extinction in the lens galaxy. However, they suffer from stellar micro-lensing which induces time-delay changes of the order of days (Tie and Kochanek 2018). Time-delay anomalies due to subhaloes are typically of the order of a fraction of a day. At present, this is smaller than the typical time-delay uncertainty (equal or larger than a day, e.g., Fassnacht et al. 2002). To be an effective probe of the nature of dark matter, time-delay anomalies require, therefore, sensitive measurements with a high observing cadence.

Image polarisation angle: The presence of a field of Axion-Like Particles (ALPs) and its interaction with photons leads to the so-called phenomenon of birefringence whereby the polarisation angle of a linearly polarised source is rotated. If the field oscillates over time, so does the change in the polarisation angle in a way that is related to the mass of the particles. Due to gravitational time delay, the multiple images of strongly lensed sources experience a different level of rotation, leading to differential birefringence, which can be used to probe the particle mass and its coupling with photons (Basu et al. 2021).

4 Lens Modelling

The process of constraining the properties of dark matter with strong gravitational lensing is best understood as a (hierarchical) Bayesian inference problem with the following unknowns: the intrinsic properties of the source $\vec{s}$, the parameters of the main lens(es) mass distribution $\vec{\eta}$, and the amount and properties of low-mass haloes or fuzzy dark matter granules $\vec{\eta}_{\rm pert}$. These have to be simultaneously inferred from the observed data $\vec{d}$ to which they are related in a statistical sense via the following posterior distribution:

$$P(\vec{s}, \vec{\eta}, \vec{\eta}_{\rm pert}|\vec{d}) = \frac{P(\vec{d}|\vec{s}, \vec{\eta}, \vec{\eta}_{\rm pert})P(\vec{s}, \vec{\eta}, \vec{\eta}_{\rm pert})}{P(\vec{d}|M)} . \quad (6)$$

In the numerator, the first term is the Likelihood function and the second is the prior on the parameters of the model M. The term in the denominator is the marginalised Likelihood (also known as Bayesian evidence). A detailed description of each ingredient is provided in the following sections. The equation above is an exact expression of the inference problem at hand, and it encodes how the different components are related to each other. However, depending on the modelling approach (see Sect. 4.4), one may not make direct use of this posterior probability and instead adopt simplifying assumptions to make the inference problem more tractable.

4.1 The Likelihood

The Likelihood function returns the probability of observing the data, given a choice of model. In the context of strongly lensed unresolved sources, the data $\vec{d}$ is the set of image fluxes and positions. For resolved sources, the data is the surface brightness distribution in each image pixel in optical observations and a set of visibilities for interferometric ones. In the case of radio polarised emission, the data is given by the visibilities of the coherency vector. In studies that combine multiple data-sets of the same lens system at different wavelengths, the data can be thought of as a concatenation of the different observations, which originate from the same lensing potential, but with a different surface brightness distribution of the source. In many cases, the Likelihood function of a model image being correct given the data is well approximated by a Gaussian distribution with uncorrelated noise.[2]

4.2 The Source

Sources in gravitational lens systems can be of a large variety. In the following, we consider two idealised cases: sources that appear as unresolved and those that are instead resolved by the observations. OPtically faint AGN and QSOs are examples of sources that tend to be unresolved by most observations. Resolved sources include galaxies and radio jets.

Unresolved sources: Traditionally, for gravitationally lensed sources that appear unresolved, either a point or a Gaussian source is assumed. Free parameters of the model are therefore the source position and flux, and in the latter case also the source size. Assumptions on the source size and shape can introduce systematic errors in the dark matter inference that are discussed in more detail below.

Resolved sources: Typically, one of the following methods is used to describe the surface brightness distribution on the source plane: an analytical (e.g. a Sérsic) profile, a pixellated model, a basis function regression (e.g. starlets and shapelets) or a deep generative model. Here, we briefly discuss the main strengths and weaknesses of each approach.

The advantage of analytical models is that the number of free parameters is small, and the modelling procedure is fast. The main drawback is that such simplistic sources are an unlikely representation of lensed star-forming galaxies, which are often clumpy and irregular (e.g. Belokurov et al. 2009; Borsato et al. 2024; Dye et al. 2018; Ritondale et al. 2019b).

Pixellated sources have enough freedom to fit complicated light distributions well. However, as the lensing problem is poorely constrained, they require a regularising prior (e.g. Suyu et al. 2006; Birrer et al. 2015; Vernardos and Koopmans 2022; Galan et al. 2022). The choice of prior is non-trivial and may not necessarily be physically motivated. A possible solution is the introduction of hyper-priors (Rizzo et al. 2018; Vernardos and Koopmans 2022), which have the advantage of retaining the freedom of a pixellated source while imposing physically-meaningful constraints. Another limitation of free-form models is that they are challenging from a computational perspective, especially for high-resolution interferometric data (Hezaveh et al. 2016; Powell et al. 2021). One more aspect to consider is that these models require constructing a regular or adaptive grid on the source plane. We refer the reader to Tagore and Keeton (2014) and Nightingale and Dye (2015) for a detailed discussion of biases related to different choices of source discretisation and regularisation schemes.

Recently, machine learning techniques have been introduced to model the source galaxies (e.g. Chianese et al. 2020; Adam et al. 2022; Karchev et al. 2022). These approaches can overcome some of the above limitations. However, their performance is sensitive to the choice of training data. It is currently unclear how to generate large training samples of realistic lensed galaxies at high angular resolution. It is, for example, unlikely that nearby galaxies are a good description of the high-redshift population. Similarly, lensed and unlensed galaxies at the same redshift are not observed on the same angular scales. Holzschuh et al. (2022), have shown how generative models can be used to create arbitrarily large samples of source galaxies from hydrodynamical simulations. However, it is unclear how well these simulated galaxies represent the population of observed lensed sources. As for pixellated sources, deep learning techniques also suffer from biases related to the data discretisation.

[2]However, optical images may have correlated noise as a result of the pixel re-sampling using, for example, the drizzle image processing.

Independently of the chosen model, some level of degeneracy always exists between structures in the source and structures in the lensing potential. Therefore, any assumption on the source light has important consequences for the inference of dark matter with strong gravitational lensing. We provide a more detailed discussion of this issue in Sect. 5.

4.3 The Lensing Potential

In this article, we focus our attention on galaxy-scale lenses. While galaxy clusters could also provide constraints on small-scale fluctuations in the dark matter distribution, most studies so far have focused on smaller scale deflectors. This choice is related to the challenge of modelling the more complicated mass distribution of galaxy clusters with enough precision.

4.3.1 Main Lens

For galaxy scale lenses, the main deflector is typically a single galaxy, most commonly but not limited to a massive early-type. For many years, the most common parametrization for its mass density distribution was a single power law (SPL) or a singular isothermal (SIE) elliptical profile, plus external shear. This choice was motivated by the analysis of relatively large samples of lens galaxies (e.g. Koopmans et al. 2009; Barnabè et al. 2011). However, it is worth noting that no existing sample of strong lenses is representative of the actual distribution of observable lenses in the Universe. That is, the sample of lens systems used so far for dark matter studies (e.g. SLACS, BELLS and CLASS) have very specific selection functions. *Euclid*, on the other hand, is expected to identify a single homogeneous sample that is much closer to the underlying distribution found in nature - e.g. much smaller Einstein radii and lenses at larger redshift, on average (e.g. Sonnenfeld 2022).

An important question is the level of complexity in the mass distribution of lens galaxies and how it affects the constraints on dark matter. For example, deep Near-Infrared (NIR) observations have revealed the presence of a disk component in several lenses that were previously assumed to be purely elliptical. It was also shown that these disks have a non-negligible lensing effect and account for most of the observed flux ratio anomaly (Hsueh et al. 2016, 2017). Similarly, Spingola et al. (2018) and Powell et al. (2022) have shown with VLBI observations that an SPL is a good description of the lens mass distribution of MG J0751+2716 only down to scales of a few milli-arcseconds. Below these scales more complex angular and radial structures become important, and lens models that are smooth on milli-arcsecond scales are not able to fully reproduce the data. In light of these and other results (see Sect. 5), more recent analyses have included the effect of multipole mass moments, such as boxiness and diskiness, as well as that of nearby satellite galaxies. The degeneracy between different forms of complexity in the lensing potential (e.g. subhaloes versus disks or multipoles) is a source of systematic error in the inference of dark matter. We discuss this problem and possible solutions in Sect. 5.

4.3.2 Low-Mass Haloes

In the following, we refer to subhaloes and field haloes collectively as low-mass haloes. The parameters describing this population are: the number of objects, their masses, positions, redshifts and mass density profiles. Different assumptions regarding these quantities can be found in the literature. Here, we provide a description of those that are most frequently used.

Analytical: most commonly, low-mass haloes are modelled as spherical systems with a Pseudo-Jaffe (PJ, e.g. Dalal and Kochanek 2002) or a (truncated) NFW (e.g. Gilman et al. 2019; Hsueh et al. 2020) profile. The mass-concentration relation of the NFW depends on the dark matter properties and is either taken from numerical simulations (e.g. Gilman et al. 2019) or is a free parameter of the model (e.g. Gilman et al. 2020a). In fully forward models, the number of objects as a function of their redshift and mass are drawn from a Poisson distribution with an expectation value derived from the halo and subhalo mass functions. The latter are set by the dark matter model and are taken from numerical simulations (see Sect. 2). On each redshift plane, the projected positions of low-mass haloes are generally assumed to be uniformly distributed, as motivated by numerical simulations (Xu et al. 2015).

Pixellated: low-mass haloes are described as linear local corrections to the lensing potential (Koopmans 2005; Vegetti and Koopmans 2009). Individually detected objects are identified as positive convergence corrections of the otherwise smooth lens mass distribution. As for pixellated sources, a regularising prior for the potential corrections must be defined (e.g. Vernardos and Koopmans 2022). Due to its free-form nature, this approach does not make a priori assumptions on the number and mass density profile of the low-mass haloes. Typically, the connection to specific dark matter models is done a posteriori as described in Sect. 4.4.

Gaussian Random Field: low-mass haloes are represented by a Gaussian Random Field (GRF, e.g. Hezaveh et al. 2016; Diaz Rivero et al. 2018; Díaz Rivero et al. 2018; Chatterjee and Koopmans 2018; Cyr-Racine et al. 2019; Bayer et al. 2023a,b). The corresponding power-spectrum, which is by construction well represented by a power-law, then carries information on the low-mass haloes abundance, mass function and density profile. Hence, these quantities do not have to be assumed a priori. As the assumption of Gaussianity only holds for the very low mass haloes that appear in great number, the more rare and massive objects have to be individually detected and separately treated. The connection to the properties of dark matter is done a posteriori and in terms of the power-spectrum (see Sect. 4.4).

4.3.3 Fuzzy Dark Matter Granules and Subhaloes

In FDM cosmologies, the small-scale structure of galaxy-scale haloes and their subhalo populations are markedly different from their CDM or WDM analogues. The main difference is that the main dark matter halo exhibits wave intereference on $\sim$kpc scales due to the ultra-low mass of the dark matter particle; this gives rise to $\mathcal{O}(1)$ fluctuations in the halo density, which are commonly termed “granules”. FDM haloes that are self-consistent with regard to the governing Schrödinger-Poisson equations can be obtained via numerical simulation (Schive et al. 2014b; May and Springel 2023) or direct construction of wave-function eigen-modes (Yavetz et al. 2022). However, for the practical purpose of gravitational lens modeling, a faster analytic prescription is often preferred. To this end, Chan et al. (2020) and Kawai et al. (2022) derive statistical properties of FDM granules that can be used to quickly generate perturbations to a smooth lensing potential, which are consistent with an FDM halo. Laroche et al. (2022) implement this approach by randomly placing a large population of Gaussian density profiles in the lens, while Powell et al. (2023) use a Fourier-space approach to achieve a similar result.

4.4 Modelling Approaches

4.4.1 Semi-Linear

Warren and Dye (2003) presented a lens modelling approach for the analysis of data with a resolved source in which the latter is pixellated (see Sect. 4.2). The most probable a posteriori (MAP) source and main lens parameters are inferred from the posterior probability in Eq. (6). In the case of quadratic prior distributions, the MAP source is obtained by solving a linear system (hence the term semi-linear - see Birrer et al. 2017; Galan et al. 2022, for examples of a semi-linear techinique with a basis function regression model for the source). The Bayesian evidence (denominator in Eq. (6)) is used to compare different models for the lensing potential and different choices of regularizations.

Koopmans (2005) introduced the so-called gravitational imaging technique for the detection of low-mass haloes with galaxy-galaxy lensing. In this approach, the source is pixellated and the low-mass haloes are described as linear, pixellated corrections to the analytical main lensing potential (see Sect. 4.3.2). The methodology was fully embedded in the framework of Bayesian statistics with an adaptive source by Vegetti and Koopmans (2009) and extended to the 3D (one frequency and two spatial dimensions) and interferometric domain respectively by Rizzo et al. (2018) and Powell et al. (2021). Recently, Vernardos and Koopmans (2022) have further extended the original method by Koopmans (2005) to include physically motivated priors. Due to the pixellated nature of the potential corrections, the gravitational imaging methodology does not require any assumption on the number, mass and position of low-mass haloes. Indeed, Vegetti and Koopmans (2009) have explicitly shown that more than one subhalo can be identified (provided that they have an effect on the lensed images), and Dhanasingham et al. (2022) have introduced a formalism to differentiate between subhaloes and field haloes based on the two-point function of the effective deflection angle field. Moreover, the potential corrections are not limited to capturing the effect of low-mass haloes. As shown by Barnabè et al. (2009), Ritondale et al. (2019a) and Galan et al. (2022) they can be used to identify components in the lensing potential that are not captured by the main parametric lens model. Indeed, the freedom allowed to the lensing potential is one of the main advantages of this approach as one can directly identify and differentiate different forms of complexity. One disadvantage is that the method is not fully forward, and two more steps are required to derive constraints on the properties of the dark matter: assessing the statistical relevance of detections and non-detections, and the interpretation of these in the context of theoretical predictions.

Detections: Vegetti et al. (2010), Vegetti et al. (2012) and Ritondale et al. (2019a) have introduced the following criteria to define the detection of a low-mass halo as statistically robust. (i) The mass and position of the pixellated convergence corrections have to be consistent with those inferred from an analytical description of the low-mass halo. The latter is inferred from the posterior distribution with a non-informative prior on the object mass and position, and a given choice of mass density profile. (ii) The model that includes low-mass haloes is preferred over the smooth one with a Bayes factor of at least 50. Under the assumption of statistical Gaussian errors, this difference in Bayesian evidence corresponds roughly to a $\geq$10-σ detection. While this may sound overly conservative, this choice is made due to the presence of unaccounted-for systematic errors that can result in false positives at lower significance levels (see Ritondale et al. 2019a, for a study on false positive detections).

Recently, this issue was more systematically quantified by Nightingale et al. (2022). We discuss their work in more detail in Sect. 5. It should be noted that false detections are not limited to this lens modelling approach. They are intrinsic to the general problem of detecting low-mass haloes with lensing and their degeneracy with other aspects of the lens inference problem. If anything, requiring that the free-form and the analytical model are consistent with each other can significantly mitigate some of these issues and reduce the incidence of false positives.

Non detections: Vegetti et al. (2014) and Ritondale et al. (2019a) quantify the statistical relevance of the non-detections using the so called sensitivity function. For each pixel on the image plane the sensitivity function returns the lowest subhalo mass that could have been detected with a Bayes factor of 50. One of the main draw-backs of the sensitivity function is that, depending on the size of the data, it can be computationally expensive to evaluate. For each pixel on the image plane one has to calculate the Bayesian evidence of the model with a subhalo of a given mass. Hence marginalizing over redshift, number, mass and mass density profile could be practically unfeasible. O'Riordan et al. (2023) have recently shown that, in principle, one can successfully overcome these computational limitations by calculating the sensitivity function with machine-learning approaches. In Sect. 7 we will discuss in more detail what sets the sensitivity of a given data set to the presence of low-mass haloes of a given mass.

Dark matter constraints: From the sensitivity function, one can then interpret detections and non-detections within a given dark matter model. This is obtained by calculating the posterior probability of the dark matter particle mass assuming that the number of low-mass haloes has a Poisson distribution with expectation value given by the halo and subhalo mass function from that specific dark matter model (Vegetti et al. 2014, 2018; Ritondale et al. 2019a; Enzi et al. 2021).

Hezaveh et al. (2016) have introduced a new formalism in which the presence of subhaloes in the lens galaxy is modelled as a GRF (see Sect. 4.3.2). From the lensing observations, one constrains the power spectrum of projected density fluctuations, which amplitude and shape can then be a-posteriori compared to predictions from different dark matter models. The advantage of this approach is that no a priori assumption is made on the properties of the low-mass haloes as these can be inferred from the analysis itself. As should be the case, Diaz Rivero et al. (2018) have shown that the amplitude and shape of the power-spectrum are sensitive to the abundance, mass density profile, and concentration of subhaloes. The main disadvantage is that the assumption of a GRF only holds for the lowest-mass objects $M < 5 \times 10^7 M_\odot$ (Hezaveh et al. 2016). Hence, the larger ones have to be first individually identified and then explicitly included in the mass model. Moreover, in its current implementation, this approach does not allow one to identify other forms of complexity in the lens mass distribution (these are unlikely to be well described by a GRF), which then introduce a systematic bias on the dark matter inference (e.g. Bayer et al. 2023b).

4.4.2 ABC

Approximate Bayesian Computation (ABC, Rubin 1984; Diggle and Gratton 2018; Tavaré et al. 1997; Turner and Van Zandt 2012; Liepe et al. 2014) is an algorithm rooted in Bayesian statistics to circumvent the direct calculation of intractable Likelihood functions. As such, it enables inference analysis based on simulated data sets computed in a forward fashion. ABC methods follow a common general process: (i) emulate the data many times with different underlying target parameters as well as noise realizations; (ii) compress the difference between the simulation and the data in a set of summary statistics to provide a metric distance between the simulation and the data; (iii) accept the proposed simulations and their underlying parameters if the distance metric is within a certain threshold ϵ and (iv) the accepted samples can be interpreted as posterior distributions with the prior being the draws of the simulator. If the acceptance criteria converges to the identical matching of the simulation and data, $\epsilon \rightarrow 0$, the accepted sample is identical to the posterior from the exact Likelihood expression applied on the summary statistic.

An accurate ABC inference requires three main components. The first one is accurate simulations including all relevant aspects affecting the data (or more specifically the summary statistic). The second ingredient is a summary statistic that captures significant aspects of the signal of interest. Finally, one needs a sufficient number of simulations such that a narrow acceptance criteria (small ϵ) can be chosen that leads to convergence and an accurate posterior prediction. One of the main advantages of an ABC approach is that, in contrast to some machine-learning-based methods (see Sect. 4.4.4 for more details), the summary statistic is explicit and is a means of understanding the impact of possible sources of systematic errors. However, the choice of summary statistic is arbitrary, potentially leading to a significant loss of information and constraining power. Specifically, the choice of summary statistic and level of data compression determine the number of simulations needed for the ABC process to converge, potentially resulting in a loss of computational efficiency relative to machine learning approaches.

In the context of inferring the properties of dark matter with strong gravitational lensing, Birrer et al. (2017) were the first to propose an ABC method. In particular, they made use of elaborate summary statistics including of a substructure scanning approach and the map of relative Likelihood (Sect. 4.1) values in the reconstruction with and without a substructure. This scanning mechanism was meant to filter signal and not absorb other spurious effects between the simulations and the data. The methodology was designed for strong gravitational lens systems with a resolved source, which was modelled with a basis function regression method in the form of shapelets. We refer the reader to Enzi et al. (2020), He et al. (2023), Bayer et al. (2023a) and Bayer et al. (2023b) for other examples of summary statistics commonly used when modelling strongly lensed galaxies. Gilman et al. (2018), Gilman et al. (2019) and Gilman et al. (2020b) have made use of an ABC approach to analyse the fluxes and positions of strongly lensed quasars. In this case, as the size of the data $\vec{d}$ was small, no compression was required and the summary statistics were set to the difference in the flux ratio of the multiple images between the simulation and the data.

As the ABC is a forward method, subhaloes and field haloes are typically described by analytical mass profiles and have properties statistically drawn as described in Sect. 4.3.2.

4.4.3 Trans-Dimensional

One of the challenges in constraining the properties of dark matter with strong gravitational lensing is that the number of low-mass haloes in any given lens system is unknown. As a consequence, the model describing the low-mass halo population and its properties needs to have a variable and a-priori unknown number of free parameters. An additional challenge is that most of the low-mass haloes will be at or below the detection limit. Brewer et al. (2016) and Daylan et al. (2018) have proposed the use of trans-dimensional Bayesian inference approaches (Green 1995, 2003) to overcome these problems.

These methods apply probabilistic cataloging to images of strongly lensed systems (Hogg and Lang 2010; Brewer et al. 2013, 2016; Daylan et al. 2017; Portillo et al. 2017) and output an ensemble of probability-weighted (sub-)halo catalogs providing a good fit to data. The prior distribution for the properties of the low-mass haloes is specified hierarchically, so that their mass function is a natural output of the method. Sampling of the posterior can be done with reversible jump Markov Chain Monte Carlo with differences in the explicit sampling methods, such as Diffusive Nested Sampling (DSN, Brewer et al. 2016) or PCAT (Daylan et al. 2018). One can also evaluate the marginal Likelihood of the model, including over the unknown number of low-mass haloes, and the source and lens properties. In most applications, both the source and the lens are described by analytical models. However, there is, in principle, no limitation to couple a trans-dimensional treatment of low-mass haloes with, for example, pixellated sources. As for Machine-Learning based techniques (see following section), the trans-dimensional approach has been applied so far only to simulated observations.

4.4.4 Machine Learning

In recent years, it has been shown that neural networks can efficiently estimate the parameters of strong lens models directly from observations (Hezaveh et al. 2017; Perreault Levasseur et al. 2017; Schuldt et al. 2021). Motivated by these results, a number of works have explored machine learning methods for identifying subhaloes in mock observations (Ostdiek et al. 2020; Yao-Yu Lin et al. 2020; Diaz Rivero and Dvorkin 2020). Others have shown that subhalo summary statistics can be directly estimated from mock observations (Brehmer et al. 2019; Alexander et al. 2020a,b; Varma et al. 2020; Alexander et al. 2021; Vattis et al. 2021; Ostdiek et al. 2022). However, these works do not provide any rigorous uncertainty estimates.

To address this issue, Brehmer et al. (2019) proposed an approach relying on Neural Ratio Estimators (NREs). NREs transform the problem of inferring the value of a continuous variable into a classification problem between two sets. Using this, it is possible to calculate the posterior of a parameter as the ratio of the Likelihood, multiplied by the prior, and the evidence. This allows for better uncertainty estimates. However, the approach of Brehmer et al. (2019) is extremely data-hungry, making its scaling to realistic data non-trivial. A number of variants of NREs have been proposed to solve this problem (Cranmer et al. 2020; Coogan et al. 2020; Montel et al. 2022).

Uncertainties on the subhalo population can also be obtained similarly to those of the macro-model as proposed by Perreault Levasseur et al. (2017). Wagner-Carena et al. (2023) showed that it is possible to infer the parameters of the subhalo mass function over a population of gravitational lenses, in a Bayesian hierarchical formalism. They trained a neural network that directly predicts a parameterized distribution approximating the posterior of the low-dimensional macro-model parameters and subhalo mass function normalization (see also Vernardos et al. 2020). One of the existing challenges is how to consistently incorporate realistic sources of biases (e.g. complex selection functions from lens-finding neural networks) in the inference process. A possible solution to this issue has been proposed by Legin et al. (2022).

Together, these works have shown that neural networks can extract low dimensional information about the density fields, that is, the parameters of the mass function or m_{WDM}. The main limitation of these methods, however, is that they are only tractable for the inference of low-dimensional representations, implicitly marginalizing over all nuisance parameters that are not explicitly estimated. At first glance, implicit marginalization appears appealing because the marginal posterior of the dark matter model parameters is the desired outcome. However, it hinders reproducibility and confirmation of the results by alternative, more traditional approaches. Machine learning models do not predict, at least for now, the actual distribution of density in the foreground, or the surface brightness of the background source. One cannot therefore compare their predictions with the data by ray-tracing through the model favoured by the neural network.

If, for example, excess power is detected in the mass function, it would be challenging to identify the specific feature in the data that accounts for it. Investigating whether a neural network's preference for a given dark matter model comes from a massive subhalo or a population of low mass subhaloes would be intractable in most machine learning frameworks developed to date. The lack of explicit predictions of physical features like density or brightness distributions, makes it difficult to verify measurements of low-dimensional substructure statistics by neural networks using more traditional methods.

One of the central issues in modelling the physical features of strong gravitational lenses, either with traditional analysis methods or with deep learning, has been that of defining priors over such high dimensional spaces. Recent advances in denoising diffusion-based models are now making this previously-intractable problem possible (Song et al. 2020). Further, Adam et al. (2022) have shown that a neural network can learn the score of the prior over background source images, learned from unlensed high-resolution images of galaxies. By adding the Likelihood score to this learnt prior score and using a reverse-time stochastic differential equation solver, the authors can obtain samples from the posterior over background source pixels. Generalizing the framework of denoising score-based models to solve highly non-linear problems such as the reconstruction of density maps in strong gravitational lenses is an active area of research.

5 Degeneracies and Systematic Errors

When assessing the robustness of strong gravitational lensing constraints on dark matter it is important to consider the effect of degeneracies (intended here as a single degeneracy, represented by an n-dimensional subspace of the model parameters, and related to the inadequacy of the available data to constrain the model under consideration) and systematic errors. Some of these are related to the measurement at hand or unsolved theoretical questions and affect both unresolved and resolved sources. Others are lens modelling and data dependent. All of them are discussed in this section.

5.1 Degeneracy with Complex Macro-Models

For any analytical model of the main deflector, a degeneracy exists between the macro-model parameters (e.g. Einstein radius and external shear) and the presence and properties of low-mass haloes. Its strength depends on the data signal-to-noise ratio (SNR) and angular resolution, and the properties of the low-mass haloes themselves. For low-mass haloes that can be individually detected, the more concentrated their density profile and the closer to the lensed emission their position, the smaller the degeneracy with the lens properties (Despali et al. 2022).

At the same time, real galaxies are unlikely to be simple elliptical objects and are expected to have complex radial and angular structures (e.g. Spingola et al. 2018; Powell et al. 2022; Van de Vyvere et al. 2022). Unmodelled components in the lensing potential are likely to lead to an overestimation of the amount of low-mass haloes or FDM granules, resulting in a biased inference on the properties of dark matter. For example, Xu et al. (2015) have shown that the population of CDM subhaloes from the Aquarius (Springel et al. 2008) and Phoenix simulations (Gao et al. 2012) alone cannot reproduce the level of flux-ratio anomalies in eight strongly lensed quasars from the CLASS survey (Myers et al. 2003; Browne et al. 2003), under the assumption that those lenses are well represented by an SIE plus external shear model. As field haloes from the Millenium-II simulation (Springel et al. 2005) alleviate, but do not solve this discrepancy (Xu et al. 2012), they concluded that other forms of departure from a power-law macro-model may be contributing to the observed signal. Thanks to deep NIR observations of the CLASS gravitational lens systems B1555+375 and B0712+472, Hsueh et al. (2016) and Hsueh et al. (2017) were able to show that previously undetected stellar disks are responsible for most of the observed flux-ratio anomaly in these systems. Indeed, using mock simulated data, Gilman et al. (2017) and Hsueh et al. (2018) have reported that baryonic structures in the main lens can lead to an increase of the probability of large flux-ratio anomalies of between 10 and 20 per cent, depending on the lens galaxy morphology (see Fig. 4). Similarly, from the analysis of the Cosmic Horseshoe lens

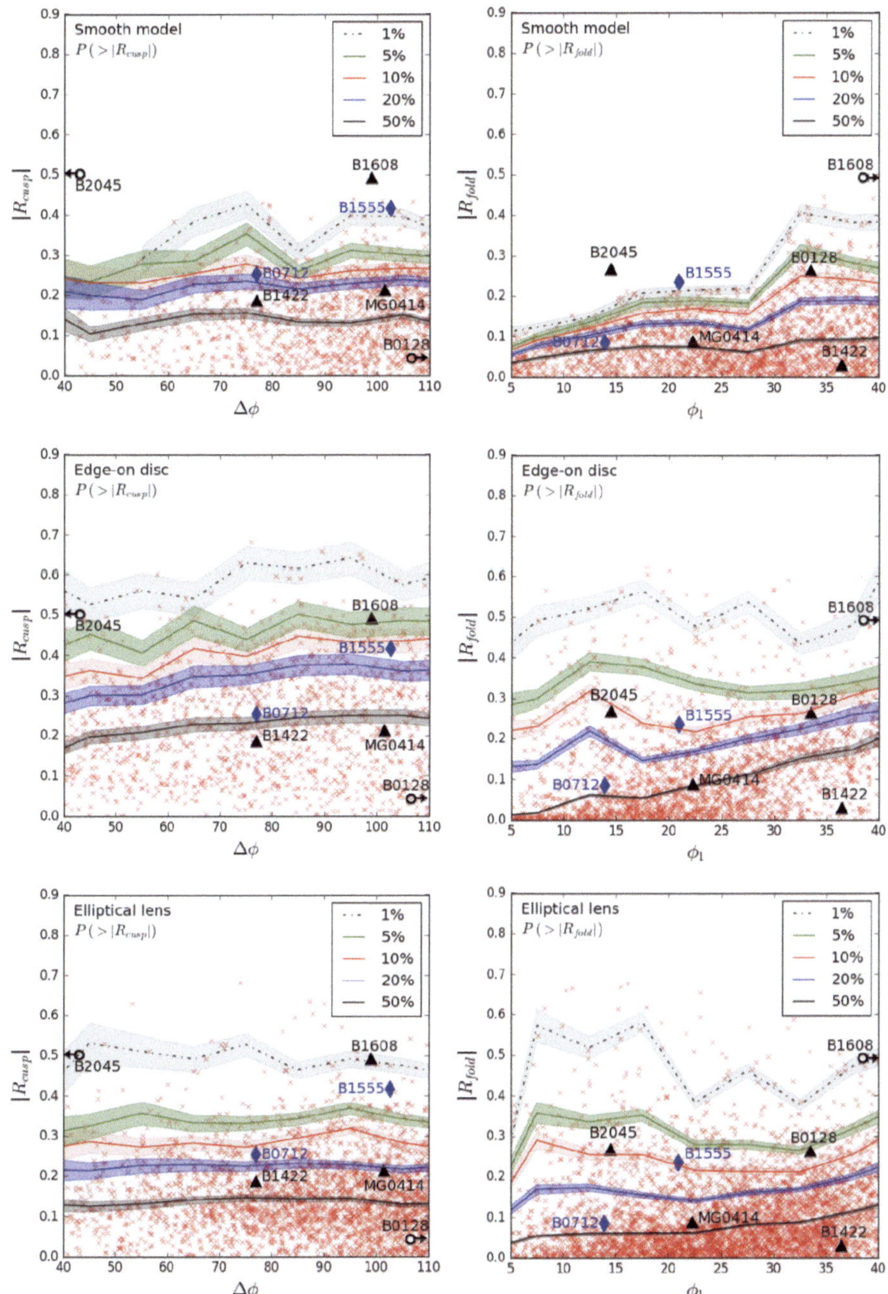

Fig. 4 Strength of the flux-ratio anomaly for a cusp (left) and fold (right) configuration arising from different mass distributions: an elliptical smooth power-law (top), an edge-on stellar disk (middle) and an early-type galaxy (bottom) selected from the Illustris simulation (Vogelsberger et al. 2014b). The strength of the anomaly is expressed in terms of the $|R_{\rm fold}|$ and $|R_{\rm cusp}|$ values, which are related to the observed flux-ratio anomalies via the equations (4) and (5), respectively. The curves represent 1, 5, 10, 20, and 50 per cent probabilities to find values of $|R_{\rm fold}|$ and $|R_{\rm cusp}|$ larger than a given value for fixed opening angles ϕ_1 (∘) and $\Delta\phi$ (∘). The panels are taken from Figs. 5, 6 and 7 in Hsueh et al. (2017)

system, Brewer et al. (2016) concluded that the inferred substructure population might have partially mimicked larger scale components not included in the primary lens model. He et al. (2023) have found that deviations from an elliptical shape that are not explicitly included in the macro-model lead to a biased characterisation of correctly identified subhaloes as well as false positive detections. Similarly, O'Riordan and Vegetti (2024) showed that multipole perturbations to the angular structure of the lensing galaxy are a significant source of false positive subhalo detections at larger masses. They concluded that these perturbations must be included in the model for detections to be reliable.

Nightingale et al. (2022) have performed a systematic study of the degeneracies between the macro-model and isolated subhaloes. From the analysis of a sample of fifty-four gravitationally lensed galaxies from the SLACS and BELLS GALLERY surveys (Bolton et al. 2008; Shu et al. 2016), they initially reported thirty-four subhalo detections at low significance. Upon further analysis, it is found that sixteen of these thirty-four are due to a degeneracy with the lens light (see Sect. 5.3) and five are due to insufficient source plane resolution. The remaining thirteen false positives are macro-model dependent with eight due to an overly simple lens mass model: a subhalo is favoured because it can replicate the effect of the missing complexity, instead when radial or angular structure in the lens is included, the addition of a subhalo is penalized by the Bayesian evidence. These false positives can be accounted for by including in the lens mass models more radial (e.g. the broken power-law model) and angular (e.g. multipoles) structure in two and six cases, respectively.

In general, the analysis by Nightingale et al. (2022) shows that more complicated mass distributions can both remove and add a subhalo candidate depending on the lens system and the type of model considered (e.g. multipoles versus composite mass distribution). As a consequence of this effect, results based on one lens system cannot be easily generalised to other objects. The situation is further confounded by the fact that the nature of the macro-model is unknown a priori and likely dependent on each lens galaxy formation history. While stellar disks may be visually identified with appropriate observations (Hsueh et al. 2016, 2017; Nierenberg et al. 2020), this is not necessarily the case for other forms of radial and angular structure. Moreover, depending on the spatial scales affected, the latter may be identified and characterized only with very-high angular-resolution data (Powell et al. 2022). These results highlight the importance of allowing for models other than a smooth SPL and modelling each lens in as much detail as the data will support, before proceeding with the detection of low-mass haloes.

While this degeneracy is intrinsic to the measurement itself and affects any type of lens modelling technique and data, it is particularly thorny for unresolved sources and models that solely rely on analytical macro-models.[3] In the first case, the data (at most, eight positions and four flux measurements) provide only limited information on the lens model. In the second case, for each lens system, one needs to marginalise over a wide range of macro-model assumptions. Resolved sources, where the potential is modelled with a free-form approach (Sect. 4.4) have in this respect an advantage: the data has more constraining power and the freedom allowed to the potential makes it possible to detect and potentially differentiate various mass components from each other (e.g. Galan et al. 2022).

In reality, the degeneracy between the lens macro-model and dark matter fluctuations, does not act in isolation. Its net effect is the result of its interplay with another degeneracy, that between the lensing potential and the source light distribution. We discuss the latter in the following section.

[3]These also include the case where low-mass haloes are described by a GRF, as the latter is an unlikely characterisation of (baryonic) structures in the lens mass distribution.

5.2 Degeneracy with the Source

5.2.1 Unresolved Sources

The angular size of the background source relative to that of a low-mass halo of a given mass sets the level of perturbation to the lensed images magnification. This effect introduces a degeneracy between the amount of low-mass haloes, micro-lensing by stars in the main deflector and the unknown size of the source. Micro-lensing by stars (see Chap. 5) becomes problematic when the image flux is gathered from a region surrounding the background quasar that is less than ~ 0.1 pc in diameter. It is responsible for an overestimation of the amount of low-mass haloes and hence a bias in favour of colder dark matter models. One can mitigate or even remove its effect with observations at wavelengths that are known to originate from an extended region around the source or by monitoring the system for a sufficiently large amount of time.

Provided that the source is large enough to avoid micro-lensing, uncertainties in the intrinsic size of the emission region translates into uncertainties in the amount of low-mass haloes. An overestimation (underestimation) of the former leads to an underestimation of the latter and a bias in favour of warmer (colder) dark matter models. Typically, one tries to account for this degeneracy by adopting observationally motivated priors (e.g. Koopmans et al. 2003; Chiba et al. 2005; Müller-Sánchez et al. 2011) and marginalising over the size of the source.

Another source of systematic error is related to flux variations in the emission region over time. Flux changes appear with a time delay between different positions on the image plane. As a result, single-epoch measurements sample the intrinsic light curve of the source at different times for the different lensed images. This effect leads to flux measurement errors that can be as large as 20 per cent (Dalal and Kochanek 2002; Koopmans et al. 2003).[4] It can be mitigated by monitoring and averaging the observed flux over a long period of time, or correcting for the time-delay.

At radio wavelengths, propagation effects, such as free-free absorption and scatter broadening can alter the measured properties of the different lensed images. As these effects have a strong wavelength dependence, they can be identified and corrected for with multi-wavelength observations (e.g. Winn et al. 2004; Biggs et al. 2003; Mittal et al. 2007).

5.2.2 Resolved Sources

The detection of low-mass haloes with resolved sources is based on the idea that the many pixels on the image plane provide redundant information about the source surface brightness distribution, allowing one to separate structures in the background object (which appear two or four times in the data) and structures in the lensing potential (which produce a relatively localised effect on the lensed images). In practice, however, due to the presence of noise in the data and the smoothing effect of the telescope point spread function, a degeneracy exists.

The extent of this degeneracy is strongly dependent on how the source and lensing potential are modelled. Vernardos and Koopmans (2022) have studied in detail the case of a

[4]It is possible for radio sources to show intrinsic variability of 10 per cent or more on the order of less than 100 days (Fassnacht et al. 2002). This may be because some of the emission is coming from much smaller regions, such as shocks in the jet. On the other hand, Koopmans and de Bruyn (2000) and Koopmans et al. (2003) detected changes to the fluxes of the individual lensed images of the order of 20 per cent. While they found these to be rare, at least some of them, were related to calibration issues, for others microlensing could also be an explanation..

pixellated source and potential corrections. They concluded that the two can partly absorb each other's complex structures in a way that depends on the form of the regularisation, the pixellatisation resolution and the actual complexity in the data (see also Bayer et al. 2023a). For example, when modelling a lens system, which is well described by a smooth analytical lens mass distribution and a complex source, the potential corrections can partly absorb fluctuations in the latter that are not well captured by the grid resolution and regularisation. Similarly, the source can adapt to compensate for the lack of structure in the model for the lensing potential. Interestingly, Vernardos and Koopmans (2022) find that the degeneracy between complexity in the source and in the potential mainly results in a biased lens macro-model and source, while the statistical properties of the potential corrections are well recovered. This suggests that, at least in the free-form approach, this degeneracy should not result in biased constraints on the properties of dark matter. However, a detailed investigation of this issue as a function of data type (e.g. optical or interferometric) and quality (e.g. angular resolution, signal-to-noise ratio and uv-coverage) is still lacking.

At the other end of the spectrum, models with analytical sources and potentials are likely to result in dark-matter constraints that are biased towards models that are colder or have a lower FDM particle mass. As the only complexity allowed in the analysis is in the form of low-mass haloes or FDM granules, these are then likely to absorb structures in the source as well as in the macro-model. Identifying suitable priors for the lensing potential and source light is therefore a key ingredient to infer the properties of dark matter with strong gravitational lensing.

5.3 Degeneracy with the Lens Light

Optical data includes light emission from the lens galaxy. This contribution is either pre-subtracted from the data (e.g. Vegetti et al. 2010) or inferred during the lens modelling analysis (e.g. Ritondale et al. 2019b,a) and sometimes used as an extra constraint on the lens mass distribution (e.g. Nightingale et al. 2022). Typically, it is described via a two dimensional B-spline function (e.g. Vegetti et al. 2014) or an analytical profile such as a (combination of) Sérsic model (e.g. Ritondale et al. 2019a; Nightingale et al. 2022). As for the lens mass distribution, simplistic models for the lens light may fail to reproduce complex radial and angular structures, as for example, boxy or disky isophotes and dust lanes. As a result, false positive detections of low-mass haloes may be obtained (Nightingale et al. 2022). It is therefore important to test candidate detections against different assumptions for the lens light (e.g. Vegetti et al. 2012; Nightingale et al. 2022) or, if possible, with observations at different frequencies.[5] For example, one can use observations in two or more wavelengths and the information carried by the multiple lensed images to correct for the effect of dust in the lens galaxy (e.g., Suyu et al. 2009).

5.4 Some Considerations on the Low-Mass Halo Density Profile

Low-mass haloes with a more concentrated mass density profile are more efficient lenses and are, therefore, more likely to be detected (e.g. Despali et al. 2018). This effect opens up the possibility of extracting further information on the properties of dark matter (e.g. Amorisco et al. 2022; Gilman et al. 2020a, 2021). At the same time, however, it introduces two related challenges in the interpretation of the observations within a given dark matter model. The first challenge is related to the robustness of the theoretical predictions and how they are

[5] Ideally at frequencies at which the lens does not emit any light.

affected by baryonic physics (Heinze et al. 2023) and the resolution of the simulations (see Sect. 5.6 for further discussions). The second one is related to the fact that the detected low-mass haloes are likely to be a biased sub-sample of the general population, an effect that needs to be taken into account in the inference process.

5.5 Some Considerations on Interferometric Data

At present, radio interferometetry (e.g. Smirnov 2011) provides the highest angular resolution available for strong gravitational lensing observations. In principle, this makes it, (for resolved sources), the most sensitive probe of the halo and subhalo mass functions, with very long baseline interferometry (VLBI) arrays capable of individual halo detections down to $10^6 M_\odot$ (McKean et al. 2015). However, modeling radio interferometric observations of strong gravitational lenses is challenging given the very large data sizes, and is an area of active research.

It is preferable to model radio interferometric observations directly in the native visibility space (a visibility is a Fourier component of the sky brightness as measured by two antennas at a given time and frequencey interval). In principle, it is possible to model a gravitational lens observation by first imaging the data using some established deconvolution technique (e.g. CLEAN, Hogbom 1974; or maximum-entropy, Cornwell and Evans 1985), then applying standard lens modelling techniques to the deconvolved image. However, on the space of the CLEANed images, the noise is correlated and not well characterised, and the CLEANing process may introduce artefacts in the surface brightness distribution that mimic the effect of low-mass haloes. Source-plane deconvolution algorithms for radio observations of gravitational lenses (such as LensClean; Kochanek and Narayan 1992; Ellithorpe et al. 1996; Wucknitz 2004) were the first to be explored as a solution to this issue. Direct χ^2 fitting in the visibility plane was then applied by several authors (Bussmann et al. 2012, 2013; Hezaveh et al. 2013) to observations of lensed dusty star-forming galaxies (DSFGs) taken with the Sub-Millimetre Array (SMA) and the Atacama Large Millimetre Array (ALMA). A fully Bayesian treatment for fitting interferometric observations, which included prior information on the source surface brightness, was finally realized by Rybak et al. (2015) and Hezaveh et al. (2016).

Amplitude and phase calibration errors (e.g. Pearson and Readhead 1984) in interferometric data can potentially masquerade as false-positive detections of low-mass dark matter haloes. This was investigated using ALMA observations of the DSFG SDP.81 by Hezaveh et al. (2016), with a test of the data sensitivity to the subhalo detection both with and without a treatment for phase errors. Hezaveh et al. (2013) and Hezaveh et al. (2016) addressed this issue by including a single antenna-based phase correction as a free parameter in the lens modeling, self-consistently incorporating uncertainties due to this systematic effect into the model posterior. It is expected that the problem of calibration errors is less relevant for cm-wavelength VLBI observations as the atmosphere is more stable over time and the antenna receivers are more sensitive. Strongly lensed sources observed with VLBI are typically radio-bright jets containing both extended and compact features, which help to provide robust phase calibration solutions prior to the lens modeling step. This is in contrast to mm-wavelength observations of DSFGs, which feature rather diffuse, low-surface-brightness emission; in this case modeling phase errors as part of the lens modeling pipeline likely yields more robust calibration solutions than *a priori* self-calibration. While it is clear that residual phase calibration errors have an effect on the rate of false-positive detections and sensitivity to low-mass dark matter haloes, there is yet to be a systematic study of this effect.

The size of the data is important when it comes to modeling antenna phases directly; the ALMA observation to which this phase-correction model was applied contains a small enough number of visibilities (5×10^5 after some additional binning) that a standard linear-algebra solver framework could be applied to the source inversion step (Warren and Dye 2003; Suyu et al. 2006). For cm-wavelength VLBI data, for which averaging can introduce additional systematics that degrades sensitivity to low-mass dark haloes, an FFT-based iterative solver is required in order to be computationally tractable (Powell et al. 2021, 2022).

5.6 Theoretical Unknowns

5.6.1 Resolution, Halo and Subhalo Mass Functions

Theoretical predictions for the abundance and structure of CDM haloes and subhaloes initially came from analytical models, such as perturbation theory (Zel'dovich 1970), the Press-Schechter model (Press et al. 1992), the statistics of peaks in Gaussian random fields (Bardeen et al. 1986) and the excursion set approach (Bond et al. 1991; Sheth and Lemson 1999). Predictions from these theoretical models have then been compared to the results from N-body (i.e. dark-matter-only) numerical simulations. This comparison has led to precise fitting functions for the halo mass function that take into account the non-linear evolution of structure formation (Sheth et al. 2001; Giocoli et al. 2008; Tinker et al. 2008; Despali et al. 2016). Similarly, the functional form of the NFW profile (Navarro et al. 1996) reflects the expectations for the density profile of a dark-matter-dominated structure that forms via hierarchical accretion. CDM N-body simulations (i.e. without baryonic physics) created with different codes from the same initial conditions agree very well with each other: the impact of numerical effects and that of different (sub)halo identification methods have been widely studied and are well understood, resulting in (sub)halo mass functions in agreement within a few per cent (Knebe et al. 2011; Onions et al. 2012; Okabe et al. 2013).

However, some uncertainties remain. These are mainly related to the resolution limit of the numerical simulations and the consequent fact that analytical predictions have only been tested down to a finite scale. Numerical effects are especially problematic for the subhalo population. For example, the artificial disruption of subhaloes related to the limited spatial resolution of the simulations (Green and van den Bosch 2019; Green et al. 2021) can lead to an underestimation of the number of such objects on scales of the resolution limit of a factor of 10 to 20 per cent. Moreover, while the density profile of isolated haloes is well understood, subhaloes are affected by tidal disruption and stripping inside the main halo and thus show a larger variety of profiles (Sawala et al. 2017; Moliné et al. 2017), which are not all well described by the same functional form.

WDM N-body simulations, in which the power spectrum cutoff is resolved, are known to undergo artificial fragmentation in filaments producing spurious clumps that, close to the resolution limit, can outnumber real structures. One challenge is thus to correctly identify and remove them from the (sub)halo catalogues: this can be achieved by studying the shapes of the Lagrangian initial regions that correspond to the final structures, and eliminating those that are very elongated (Lovell et al. 2012). Alternatively, Stücker et al. (2020) have developed a method to smooth the density field using phase-sheet methods and a high-resolution force calculation, in order to completely circumvent the issue and avoid the formation of spurious subhaloes.

FDM theories predict differences in the subhalo population of a lens, with a suppression of the halo mass function at low masses due to the large de Broglie wavelength of the dark matter particle. Characterizing the low-mass halo population in FDM cosmologies is an area of active research. Analytic approximations to the FDM halo mass function in FDM have been attempted (Marsh and Silk 2014; Kulkarni and Ostriker 2022), but at this time only mass functions obtained from numerical simulations (Schive et al. 2016; May and Springel 2023) have been applied in a lens modeling context (Laroche et al. 2022). The mass-concentration relation for FDM haloes is also highly uncertain; Laroche et al. (2022) modelled it using an extended Press-Schechter formalism (Schneider 2015) assuming some correspondence between the gravitational collapse time-scales of FDM and CDM haloes. Additionally, FDM haloes contain a characteristic soliton core that alters their density profiles (Schive et al. 2014b; Chan et al. 2022). An accurate model for subhaloes around a lens galaxy is important to consider for inferences based on flux-ratio anomalies of unresolved sources, as subhaloes and granules can produce similar observational signatures. However, for FDM particle masses lower than $\sim 5 \times 10^{-21}$ eV, subhaloes are too few and too diffuse to impart a small-scale signature on the observed source morphology, and can be absorbed into a sufficiently complex macro-model. This leaves the presence of granules as the main source of constraint on the particle mass.

5.6.2 The Effect of Baryons

In the past few years, numerical simulations, and especially those with a CDM cosmology, have made significant progress. Large-scale structure simulations are now able to reproduce realistic galaxy morphologies and the observed scaling relations (Vogelsberger et al. 2014b; Schaye et al. 2015; Pillepich et al. 2018; Dubois et al. 2021), reducing some of the tensions between CDM and observations (Brooks et al. 2013). Despite these successes, uncertainties, which may affect the interpretation of strong gravitational lensing studies within a given model, persist. Feedback processes, that cause a loss of baryonic mass, alter the total halo mass in a non-trivial way that depends on the halo mass and the galaxy formation model (Sawala et al. 2015; Despali and Vegetti 2017; Pillepich et al. 2018; Garrison-Kimmel et al. 2019). As a result, the number density of low-mass haloes ($10^7 M_\odot < M < 10^{10} M_\odot$) is suppressed in hydrodynamical simulations by a factor between 10 and 40 per cent. This suppression is smaller than the one seen in most WDM models in the same mass range, as the inclusion of baryons further suppresses the halo and subhalo mass function (Lovell et al. 2019; Despali and Vegetti 2017). However, due to the lack of hydrodynamical simulations with high-enough mass resolution, it is at present unclear how these mass functions are affected by the baryon and dark matter physics below the probed mass limit.

Baryonic physics also affects the inner mass density profile of galaxies in a way that depends on the feedback model (e.g. Mukherjee et al. 2021) and its interplay with the dark matter model. For example, Robertson et al. (2019), Despali et al. (2019) and Shen et al. (2022) have found that, in hydro-simulations with elastic SIDM, haloes have a larger variety of density profiles (with respect to the non-elastic SIDM case) and that the shape of haloes is much closer to the CDM hydrodynamical case, than initially inferred from dark-matter-only simulations. Mocz et al. (2019) studied the interplay between baryons and fuzzy dark matter, finding that the first stars can form in filamentary structures along the cosmic web, instead of only in the collapsed haloes. These results, could have interesting implication for the analysis of strong gravitational lens galaxies. However, the limited spatial resolution available in simulations at the scale of massive galaxies prevents us from a robust comparison between observations and simulations. For example, simulated galaxies display a central density core, which is large enough to produce strongly lensed image configurations that are not observed. While these uncertainties affect simulations in all dark models, they are currently more problematic for alternative non-CDM models. The vast majority of simulations where dark matter is different from CDM does not include baryonic physics (Lovell et al. 2012; Vogelsberger et al. 2014a) and is not yet at the level of the CDM case, in terms of resolution, number of objects and volume (Adhikari et al. 2022). Moreover, even when baryons are included, the sub-grid physics processes that describe their behaviour often remain calibrated on CDM simulations. Hence, it is unclear whether non-CDM hydrodynamical simulations provide a correct description of the interplay between dark-matter and baryons. Obtaining large samples of numerical simulations with different galaxy formation and dark matter physics, as well as the volume and resolution required for a meaningful comparison with strong gravitational lensing observations is a fundamental step for future works.

6 Historical Perspective

6.1 Unresolved Sources

The use of strong gravitational lensing as a probe of dark matter was first proposed by Mao and Schneider (1998). They showed that anomalies in the flux ratios (i.e. ratios that deviated from those predicted by smooth mass models) of lensed quasars could be accounted for by subhaloes in the lens galaxy. In particular, they focused on the two and three brightest images in a fold and cusp configuration, respectively. For a smooth lens mass distribution, the fluxes of these lensed quasar images should satisfy the asymptotic relations in Eqs. (4) and (5). Discrepancies between the observed flux ratios and the generic predictions are therefore an indication of some type of small-scale structure in the mass distribution. Mao and Schneider (1998) also pointed out that flux-ratio anomalies observed in optical data could easily be due to micro-lensing by stars in the lensing galaxy. Focusing on radio observations was, therefore, a better way to investigate the presence of subhaloes. At these frequencies, the lensed emission had a large enough angular extent as to be insensitive to micro-lensing by stars (although see Koopmans and de Bruyn 2000). The paper investigated both compact mass distributions (globular clusters with masses of $\sim 10^6 M_\odot$) and smoother fluctuations such as spiral arms in the lensing galaxy.

Mao and Schneider (1998) thus truly set the stage for subsequent investigations of perturbations by dark matter haloes by: (1) considering perturbations by larger-scale structures than stars and, (2) pointing out that the observations had to be conducted using sources that had angular scales that were large compared to the Einstein radii of the stars in the lensing galaxy. This second requirement ruled out using emission from the accretion disks and broad-line regions associated with the lensed AGN, which are sensitive to micro-lensing. However, observations at radio or mid-IR wavelengths could be used to investigate dark matter since the regions in the background objects that produce emission at these wavelengths are large enough to be mostly unaffected by micro-lensing by stars in the lensing galaxy, but small enough to be sensitive to perturbation by relatively low-mass dark matter haloes. Unfortunately, it was difficult to obtain large samples at these wavelengths, since only $\sim$10 per cent of AGN are radio loud, and ground-based observations at mid-IR wavelengths are extraordinarily difficult due to thermal emission from the Earth's atmosphere. Thus, for many years the sample of lenses that were useful for dark matter investigations was on the order of 10 systems, primarily discovered in the MG (Bennett et al. 1986; Langston et al. 1990; Griffith et al. 1990, 1991), JVAS (Patnaik et al. 1992; Browne et al. 1998; Wilkinson et al. 1998), and CLASS (Myers et al. 2003; Browne et al. 2003) radio surveys. Initial investigations focused on individual lenses, particularly radio-loud four-image systems that strongly violated the standard relationships for merging images (Fassnacht et al. 1999; Marlow et al. 1999; Trotter et al. 2000; Biggs et al. 2004), and included some work that incorporated numerical simulations to understand subhaloes (Bradač et al. 2002).

The seminal paper by Dalal and Kochanek (2002) was the first analysis of flux-ratio measurements in a statistically significant sample, with a goal of testing the ΛCDM model. They examined the flux ratios in a sample of seven lens systems and found broad consistency with CDM. Although this analysis had a high impact for many years, it did have several shortcomings in the context of current approaches to using flux-ratio statistics to draw inferences on the nature of dark matter. These include using CDM-only simulations, using fairly simple lens models, only considering subhaloes within the halo of the primary lensing galaxy while not including line-of-sight haloes, using PJ mass profiles for the subhaloes, and being restricted to using the somewhat uncertain flux-ratio measurements that were available at the time.

For nearly two decades following this analysis, no new lens systems that had high-sensitivity radio or mid-IR flux ratio measurements were discovered, so subsequent investigations had to focus on extending or improving the analysis rather than working with larger samples. One improvement came from a monitoring program to look for extrinsic variability in radio-loud lenses, which had the additional benefit of providing high-precision flux-ratio measurements after correcting for any time delays in the systems (Koopmans et al. 2003). Other work investigated astrometric shifts and parity dependence as methods for determining the abundance of subhaloes (Chen et al. 2007; Chen 2009).

Another major effort addressed the problem by conducting ray-tracing analyses through high-resolution numerical simulations, which allowed the inclusion of lower subhalo masses and, thus, more thoroughly to explore the lensing effects of dark-matter subhaloes and other possible perturbers (Mao et al. 2004; Xu et al. 2009, 2010; Richardson et al. 2022). These studies incorporated not only the number of anomalies in the observed radio lenses, but also how much the observed flux ratios deviated from the predictions of smooth mass models.

Interestingly, several of these studies found that CDM subhaloes were not sufficient to explain the observations fully (Xu et al. 2009, 2015), even when including line-of-sight haloes (Xu et al. 2012), suggesting that more complex macro models were needed. Earlier work had reached a related conclusion from simulated data, finding that an edge-on stellar disk could cause violations of the standard cusp relation for the lensed image magnifications (Bradač et al. 2004). Subsequent work discovered exactly this type of situation in observed lens systems. In some systems with the most extreme flux-ratio anomalies, high-resolution imaging from ground-based adaptive optics and HST data revealed edge-on disk components of the lensing galaxies. These additional baryonic components could explain the observed flux-ratio anomalies without needing to resort to either subhaloes or line-of-sight haloes (Hsueh et al. 2016, 2017). Further investigations used simulated galaxies to confirm the importance of baryonic structures in the lensing galaxies (Gilman et al. 2017; Hsueh et al. 2018).

In parallel with the work on extending and improving the analyses was an effort to increase the sample size of observed four-image lenses by pushing past the traditional radio and mid-IR observations that had provided the primary samples for the analyses above. In particular, Moustakas and Metcalf (2003) proposed a spectroscopic technique based on the differential magnification of several different emitting regions in the lensed quasar, i.e., the continuum, broad-line region, and narrow-line region (NLR). This approach allowed them to separate contributions from the smooth lens model, micro-lensing, and lensing by subhaloes. This method was applied to a single very low-redshift lens that was observed with an integral field unit (IFU) spectrograph, and seemed promising (Metcalf et al. 2004). It is difficult to apply this technique to higher redshift systems using the seeing-limited ground-based observations used in Metcalf et al. (2004), due to angular resolution issues. However, a similar approach has been successful when using either ground-based adaptive optics IFUs (Nierenberg et al. 2014) or HST grism observations (Nierenberg et al. 2017). This technique has led to the first major increase in the sample size of four-image systems with measurements that are useful for flux-ratio investigations (Nierenberg et al. 2020).

6.2 Resolved Sources

Koopmans (2005) was the first to propose the use of galaxy-galaxy lensing observations to detect subhaloes within the lens using the semi-linear pixellated approach described in Sect. 4.4. A few years later, Vegetti and Koopmans (2009) extended the original idea to be fully embedded within the context of Bayesian statistics.

Then, Vegetti and Koopmans (2009) proposed a statistical framework for the interpretation of both detections and non detections within the context of specific theoretical predictions. They also showed how the achievable level of constraints on the subhalo mass function parameters is related to the number of lenses in the sample and the sensitivity of the data. Recently, Despali et al. (2022) have provided a detailed quantification of how the latter depends on the data properties. We discuss their findings and relative implications in Sect. 7.

Vegetti et al. (2010) and Vegetti et al. (2012) were the first one to apply the pixellated gravitational imaging technique to real data taken with the HST and Keck-AO. They reported a 16-σ and 12-σ detection of two individual subhaloes in the gravitational lens systems SDSSJ0946+1006 (also known as the double ring or the Jackpot) and B1938+666, respectively. These detections, obtained in a pixellated fashion, were then modelled with a PJ profile. From the latter, a total mass of $\sim 10^9 M_\odot$ and $\sim 10^8 M_\odot$ was inferred for each system, respectively. Both detections have been independently confirmed (Minor et al. 2021; Sengül et al. 2022). However, Despali et al. (2018) and Sengül et al. (2022) have found that the detection in the system B1938+666 is more likely a field halo. Whether the concentration of these two objects is consistent with CDM predictions is currently under investigation (e.g. Minor et al. 2021; Şengül and Dvorkin 2022). From the analysis of the ALMA long baseline campaign data for the lens system SDP.81, Hezaveh et al. (2016) reported the detection at the 5-σ level of a subhalo with a total PJ mass of $\sim 10^9 M_\odot$. Inoue et al. (2016) also found a detection in this lens system. However, the inferred subhalo position and lens macro-model are inconsistent between the two analyses. Further investigations are required to understand the origin of this discrepancy.

Focusing, for the first time, on a larger number of strong gravitational lens systems, Vegetti et al. (2014) found no additional subhaloes in a sample of ten SLACS lenses, in the mass regime probed by the data. Recently, Nightingale et al. (2022) have searched for the presence of subhaloes in fifty-four lens systems (the largest number considered so far) from the SLACS and BELLS gallery samples. They reported two candidate detections, one of which matches the one by Vegetti et al. (2010) in the Jackpot lens.

While early studies considered the subhalo population only, Li et al. (2017), Despali et al. (2018) and Amorisco et al. (2022) have shown the contribution from field haloes to be important and in some cases dominant. This is a significant result. Unlike for subhaloes the properties and number of field haloes are better understood from a theoretical perspective (see Sect. 5.6), and the resulting increase in the number of detectable objects per lens allows for stronger constraints on dark matter with fewer systems. Allowing for the contributions of both populations, Ritondale et al. (2019a) concluded that their lack of detection in twenty-one lens systems from the HST BELLS gallery sample is consistent with the CDM model.

Dark-matter constraints from strongly lensed resolved sources have been mainly limited by the amount of available data with enough angular resolution. In recent years, efforts have been made to increase the number of known strong gravitational lens systems (e.g. Lanusse et al. 2018; Huang et al. 2021; Petrillo et al. 2019; Cañameras et al. 2021; Rezaei et al. 2022), as well as the number of observations with improved angular resolution with, for example, Keck-AO (Lagattuta et al. 2012) in the NIR, ALMA in the sub-mm (Spilker et al. 2016) and VLBI at cm-wavelenght (Spingola et al. 2019). At same time, several new lens modelling approaches have been developed (see Sect. 4.4 and references therein).

Existing data-sets as well as simulated ones, have been used to improve our knowledge of systematic errors and degeneracies (see Sect. 5 and references therein), and our understanding of the signal under study. For example, Amorisco et al. (2022) showed that an improvement on the level of dark matter constraints can be obtained by taking into account the low-mass haloes mass-concentration relation (and how it changes with the dark-matter model) and its scatter. Similarly, ray-tracing through numerical simulations has been instrumental to quantify the lensing effect arising from smaller- and larger-scale structures (e.g. Enzi et al. 2020; He et al. 2023) in different dark matter models (Despali et al. 2019, 2020).

7 Dark Matter Constraints

7.1 Unresolved Sources

Since 2019, four new analyses (Hsueh et al. 2020; Gilman et al. 2020b,a; Laroche et al. 2022) have characterized the properties of low-mass haloes using a sample of quadruply-imaged quasars. These works interpret strong lensing data in the context of a variety of scenarios, including classes of dark matter such as warm dark matter, ultra-light dark matter, and self-interacting dark matter, as well as early Universe physics that alters the power spectrum of primordial density fluctuations.

Hsueh et al. (2020) and Gilman et al. (2020b) analyzed a sample of strong lens systems in the context of warm dark matter. Their inference places an upper limit on the free-streaming length of dark matter equivalent to that of a 5 to 6 keV thermal relic dark matter particle, and constrains a variety of sterile neutrino models with varying production mechanisms (Zelko et al. 2022). Gilman et al. (2020a) used the sensitivity of the relative magnifications among the lensed images to infer the concentrations of CDM haloes, an analysis that was later generalized by Gilman et al. (2022) to make a direct connection between the abundance and internal structure of dark matter haloes and the primordial matter power spectrum on small scales $\left(k > 10\,\mathrm{Mpc}^{-1}\right)$. The resulting inference on the concentration-mass relation and power spectrum agreed with the CDM prediction. Laroche et al. (2022) interpreted the same sample of eleven four-image lenses as analysed by Gilman et al. (2022) in the context of ultra-light dark matter, showing that the granular structure of the host halo density profile that arises from quantum wave interference effects can impact image flux ratios in a similar manner to dark matter haloes. Finally, building on work by Gilman et al. (2021), Gilman et al. (2022) interpreted the same sample of eleven lenses in the context of self-interacting dark matter and showed that existing data disfavors SIDM models with large amplitudes at low speeds, such as those that can arise from resonances in the self-interaction cross section, assuming the large amplitude of self-interaction cross section drives low-mass haloes to core-collapse. By combining observations of the Milky Way satellites with the sample of strongly lensed unresolved sourses from Gilman et al. (2022), Nadler et al. (2021) have derived a limit on the half-mode mass of $M_{\rm hm} < 10^7 M_\odot$ (i.e., $m_{\rm WDM} > 9.7$ keV) at the 95 per cent confidence level (see Fig. 5).

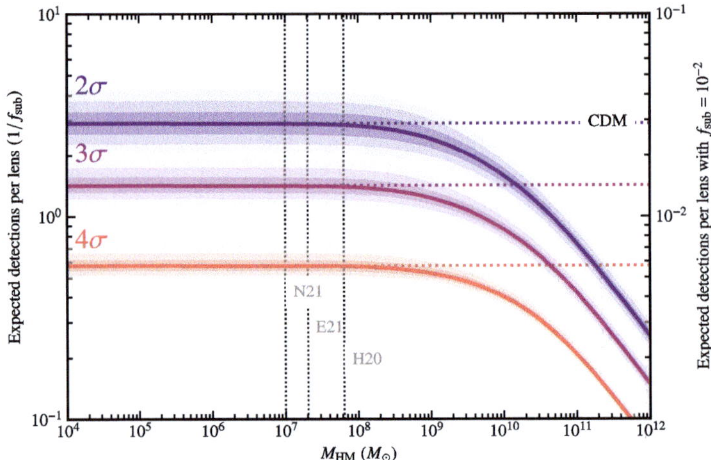

Fig. 5 Expected number of subhaloes per lens as a function of the half-mode mass from Euclid-like observations. Each curve (with 64 per cent, 95 per cent, and 99 per cent confidence areas) is for a different level of significance of subhalo detection. The horizontal dotted lines display the expected number of detectable subhaloes in CDM (see O'Riordan et al. 2023, for more details). The vertical dotted lines show the current 95 per cent upper limits on the half-mode mass from Nadler et al. (2021) and Hsueh et al. (2020) together with the 1/20th of the maximum Likelihood from Enzi et al. (2021)

These analyses have incorporated the latest theoretical understanding of structure formation in CDM and alternative dark matter models, such as the halo mass function and concentation-mass relation in warm dark matter, the process of core-collapse that is expected to occur in self-interacting dark matter, and the phenomenon of wave-interference unique to ultra-light dark matter. At the same time, Basu et al. (2021) have introduced a novel approach to constrain Axion-Like Particles dark matter models from the differential birefringence effect imparted on the strongly lensed images. Using broad-band polarisation observations of the lens system B1152+199 from the Karl G. Jansky Very Large Array (VLA), they derived an upper bound on the ALP-photon coupling between $g_{a\gamma} \leq 9.2 \times 10^{-11}$ and 7.7×10^{-8} eV at the 95 per cent confidence limit for an ALP mass between $m_a = 3.6 \times 10^{-21}$ eV and 4.6×10^{-18} eV.

Two recent innovations have expanded the scope of strong lensing of unresolved sources as a probe of fundamental dark matter physics. First, the sample size of lenses suitable for a subhalo inference doubled with measurements of relative image fluxes from narrow-line emission around the background quasar. Nuclear narrow-line emission, which emanates from an area in the source plane with a typical size between 10 to 100 pc, subtends angular scales on the sky much larger than a micro-arcsecond. This renders the relative brightness of lensed images immune to micro-lensing, while retaining sensitivity to milli-lensing by dark matter haloes. Second, open-source software to model gravitational lens systems and to generate populations of dark matter haloes for lensing computations have enabled ray-tracing simulations to be performed in parallel on computing clusters with realistic background sources and full populations of subhaloes. At the same time, recent analyses (from 2021 onwards) have made some efforts to mitigate potential systematic uncertainties associated with the lens macro-model by including the contribution of multipole mass moments.

7.2 Resolved Sources

Most studies based on resolved lensed sources are so far consistent with the CDM model. For example, Vegetti et al. (2014) and Hezaveh et al. (2016) have inferred a dark matter fraction in subhaloes (i.e. the normalisation of the subhalo mass function) that is consistent with the subhalo population in CDM simulations (Springel et al. 2008; Xu et al. 2015; Despali and Vegetti 2017). From an analysis (which only include the contribution of subhaloes) of the lens system RXJ1131−1231, Birrer et al. (2017) have derived a 2-σ lower limit on the particle of a thermal relic dark matter model of $m_{\rm th} > 2$ keV. On the other hand, Bayer et al. (2023b) have found an upper limit on the power-spectrum of mass density fluctuations in the lens system SDSS J0252+0039 that exceeds the value expected from CDM. Their result is likely driven by their choice of macro-model (SPL) and the lack of other form complexity beyond subhaloes.

Allowing for the contribution of field haloes, Vegetti et al. (2018) have derived constraints on sterile neutrinos that show a preference for colder dark matter models: $\log M_{\rm hm}[M_\odot] < 12.0$ at the 2-σ level. This result excludes sterile neutrino models with neutrino masses $m_{\rm s} < 0.8$ keV at any value of the lepton asymmetry L_6. Ritondale et al. (2019a) reported zero detection of low-mass haloes in a sample of twenty-one strong gravitational lens systems from the BELLS survery, in agreement with CDM given the quality of the data. Recently, Enzi et al. (2021) have derived constraints on thermal relic and sterile neutrino dark matter models by combining the lensing results by Vegetti et al. (2018) and Ritondale et al. (2019a) with observations of the Lyman-α forest and the Milky Way satellite galaxies. They derived a joint limit on the thermal relic mass of $m_{\rm th} > 6.048$ keV (i.e. $M_{\rm hm} < 3 \times 10^7 M_\odot h^{-1}$) at the 95 per cent confidence level (see Fig. 5). This result is mainly set by the Milky Way and the Lyman-α forest. The lensing measurements alone lead to a 95 per cent confidence level lower limit of $m_{\rm th} > 0.6$ keV and $m_{\rm th} > 0.1$ keV, for the SLACS and BELLS samples, respectively.

While most published analyses of resolved sources show a preference for CDM, they do not rule out alternative and still viable dark matter models yet. This lack of constraints is partly due to the low number of lens systems available and the relatively limited quality of the data: signal-to-noise ratio and in particular angular resolution. Using realistic mock observations of varying data quality, Despali et al. (2022) have recently quantified how the sensitivity to subhaloes depends on the properties of the lens system (e.g. source structure and position relative to the caustics) and the data (i.e. signal-to-noise ratio and angular resolution). They concluded that the lowest detectable subhalo mass decreases linearly with signal-to-noise ratio, and more strongly with the angular resolution. An increase in the latter by a factor of two leads on average to an increase in mass sensitivity of a similar factor (for a fixed signal-to-noise ratio, lensing configuration and source properties). For example, existing HST observations ($<\mathrm{SNR}> \sim 3.5$ and $\mathrm{PSF}_{\rm FWHM} = 0.09$ mas) have a maximum sensitivity of $M_{\rm vir}^{\rm NFW} = 10^9 M_\odot$, leading to an expected number of detectable subhaloes per lens in CDM consistent with the current number of objects detected so far (Vegetti et al. 2014; Ritondale et al. 2019a; Nightingale et al. 2022). These effects also explain why the constraints are, at present, less stringent (though possibly more robust) than those obtained with unresolved sources: predictions from different dark matter models mostly differ at halo masses smaller than currently probed.

Observations at very high-angular resolution are therefore key to probe the nature of dark matter with resolved sources. For example, from observations of the gravitational lens system MG J0751+2716 taken with the Global Very Long Baseline Interferometry array, Powell et al. (2023) have ruled out scalar FDM models with a particle mass $m_\chi \leq 4.4 \times 10^{-21}$

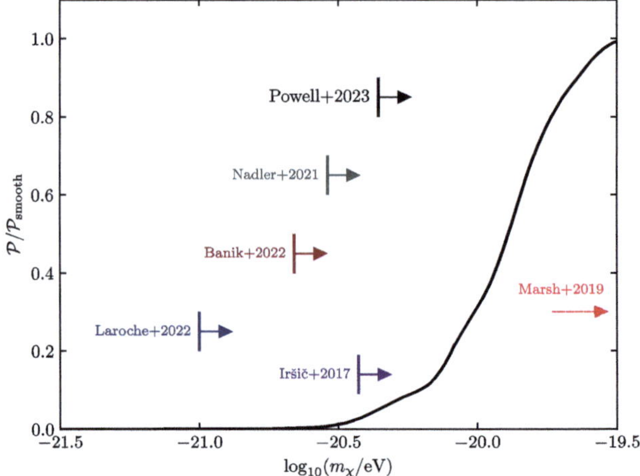

Fig. 6 Bounds on the FDM particle mass from different observational probes. The solid black curve and arrow show the fully marginalized posterior odds ratio and relative lower bound from Powell et al. (2023), respectively. These were derived from the analysis of mas-resolution observations of the lens system MG J0751+2716 taken with the Global Very Long Baseline Interferometry array. The results by Laroche et al. (2022, blue arrow) were obtained from a sample of eleven strongly lensed QSOs. Also shown are the constraints from the Lyman-α forest (Iršič et al. 2017, purple arrow), the Milky Way satellites (Nadler et al. 2021, green arrow), stellar streams in the Milky Way (Banik et al. 2021, dark red arrow). The limit by Marsh and Niemeyer (2019, red dashed arrow) lie beyond the limit of the plot, and were derived from observations of star clusters within the Milky Way Ultra-faint Dwarf satellite galaxy Eridanus II

eV. Their constraints, which were obtained from a single lens system with milli-arcsecond angular resolution, are more stringent than previously published results from larger samples of lensed quasars (see Fig. 6). Their work, together with the results by Despali et al. (2022) and O'Riordan et al. (2023), clearly demonstrates how a smaller number of high-angular resolution observations can be more effective at constraining the properties of dark matter than many lens systems with lower quality data. We will discuss this further in the context of future surveys in the following section.

8 Future Prospects

At present, the level of constraints that can be obtained on the properties of dark matter from both resolved and unresolved sources has mainly been limited by the amount of available data. Strong gravitational lensing is a rare phenomenon and the number of known systems with suitable data (e.g. high enough signal-to-noise ratio and angular resolution for resolved sources and the right frequency coverage for unresolved sources) is small (of the order of a few tens).

Thanks to ongoing wide-sky surveys the number of known strong gravitational lens system has significantly increased in recent years (e.g. Storfer et al. 2022). With the advent of the next generation of sky surveys with, e.g., Euclid, the Low-Frequency Array (LOFAR), the Square Kilometre Array (SKA) and the Vera Rubin Observatory, this number is expected to increase even further (e.g. Oguri and Marshall 2010; McKean et al. 2015). The size of these samples is unprecedented. Coupled with follow-up observations they will provide robust and meaningful constraints on dark matter.

8.1 Unresolved Sources

Increasing the sample size for four-image systems will reduce the statistical uncertainties of measurements made with these types of system. Provided that all sources of systematic error can be accounted for, this increased constraining power can lead to stronger constraints over various dark matter models, such as warm and self-interacting dark matter (Gilman et al. 2019, 2021; Hsueh et al. 2020). The existing analysis techniques applied to interpret data from multiply-imaged quasars can scale to match the increased sample size. Thus, computational costs will not limit the scientific output achievable with forthcoming data.

Follow-up observations will be required to precisely measure the positions and relative magnifications among the unresolved images of the lensed source. These observations can be performed with spaced-based observatories, such as HST and JWST (while they remain operational), but most likely with ground-based facilities with adaptive optics for optical and infrared-bright quasars (Nierenberg et al. 2014), and with the Very Large Array (VLA), the enhanced Multi Element Remotely Linked Interferometer Network (e-MERLIN) and VLBI for unresolved lensed radio sources. Upcoming observational facilities, such as the Keck All sky Precision Adaptive optics program (Wizinowich et al. 2022) and thirty-metre class telescopes, like the European Extremely Large Telescope (ELT), will provide exquisite astrometry and sensitivity with which to measure the relative image fluxes and positions. Likewise, the SKA will provide high resolution imaging at a resolution of 30 to 70 mas, which, when combined with VLBI, can provide astrometric measurements at sub-mas-arcsecond precision (e.g. Spingola and Barnacka 2020).

The JWST will soon deliver precise measurements of relative image fluxes in the mid-infrared at >20 μm (Nierenberg et al. 2021), emission that emanates from a more compact region (1 to 10 pc) around the background source than the nuclear narrow-line emission (10 to 100 pc) that has been measured with the HST and Keck-AO (Nierenberg et al. 2020). The spatial extent of the mid-IR emission renders these data immune to stellar micro-lensing and variability of the background quasar, systematic effects that one could conflate with the perturbation by low-mass haloes .[6] The more compact mid-IR emission that is measurable with the JWST increases sensitivity to perturbations by low-mass haloes relative to existing data because the minimum deflection angle that can affect the data scales with the angular size of the lensed source (Dobler and Keeton 2006). Mid-IR flux ratios measured with the JWST have anticipated sensitivity to populations of haloes less massive than 10^7 $M_\odot$ (Nierenberg et al. 2021). Moreover, the compact background sources measurable with JWST will increase sensitivity to the internal structure of low-mass haloes. This improved sensitivity, in combination with the expanded sample size of strong lens systems, will enable searches for populations of core-collapsed haloes, such as structures with logarithmic central density profile slopes steeper than -2 (Balberg et al. 2002; Turner et al. 2021). If detected, the existence of these objects would suggest dark matter has a velocity-dependent cross section with strong self-interactions at low speeds that triggers core collapse (Gilman et al. 2021; Turner et al. 2021; Yang et al. 2023; Gilman et al. 2022). Increased sensitivity to the internal structure of low-mass haloes will also enable stronger constraints on the concentration-mass relation of low-mass haloes (Gilman et al. 2020a). This type of measurement can be interpreted in the context of the primordial matter power spectrum on scales $k > 10$ Mpc^{-1} (Zentner and Bullock 2003; Gilman et al. 2022). However, as the background sources will

[6] Scattering of light into our field of view by the spatially-extended nuclear narrow-line and mid-IR emission regions acts as a low-pass filter that washes out variability in the quasar light curves on timescales less than the light crossing time.

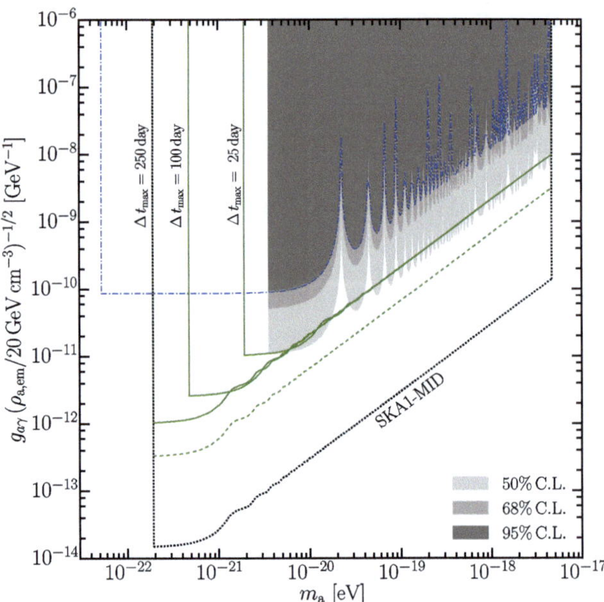

Fig. 7 Limits on the ALP-photon coupling from strong gravitational lensing. The dashed area shows current bounds from the gravitational lens system B1152+199. The dash-dotted blue line shows the expected constraint that can be obtained from monitoring B1152+199 over the course of ~ 5 year. The solid and dashed green lines show the predicted bounds for a sample of 100 and 1000 gravitational lens systems for different values of the maximum time delay ($\Delta\,t_{\rm max}$) on the observed plane. The dotted black line shows the parameter space that can be probed with the SKA. The figure is taken from Basu et al. (2021)

remain unresolved, their unknown surface brightness distribution will still stand as a potential source of systematic error. Moreover, in those cases without any extended emission from the quasar-host galaxy, the limited number of observational constraints provided by the data may lead to a potential systematic error in the macro-model.

At radio wavelengths (between around 4 and 18 GHz), large fractional bandwidths with, for example, the ngVLA and the SKA, will detect the emission from pc-scale radio sources, which are expected to be immune from micro-lensing (but see, e.g., Koopmans and de Bruyn 2000; Biggs 2023). The orders of magnitude improvement in sensitivity will allow monitoring of the lensed radio sources for cosmology (in the case of any intrinsic variability) or for dark-matter studies, when the radio source is found not to vary. While the large fractional bandwidths will allow radio propagation effects, such as free-free absorption or scattering, to be identified, which can either corrected for or used to create samples of lensed radio sources that do not show such issues. However, the main contribution that radio-selected samples can provide is our ability to characterise the source structure on mas-scales through observations with VLBI (e.g McKean et al. 2007; Spingola and Barnacka 2020). In general, it is worth noting that combining the data from multiple emission regions (mid-IR-bright torus, NRL, radio-jets) will allow various systematics associated with the source model to be quantified and possibly corrected for. This may include better constraints to the macro-model via large scale source properties, such as the optical/infrared emission from the stellar component of the quasar host galaxy or through high resolution imaging of the kpc-scale dust emission with ALMA (e.g. Stacey et al. 2021).

Finally, high-sensitivity observations with the LOFAR2.0 and SKA-MID of large samples of polarised and strongly lensed sources are expected to improve upon current bounds on the ALP-photon coupling by one to two orders of magnitude, depending on the ALP mass range (see Basu et al. 2021, and Fig. 7). Providing, thereby, competitive constraints on dark-matter models made of ALPs.

8.2 Resolved Sources

The larger number of strong gravitational lens systems with a resolved source that will be discovered by wide-sky surveys will represent a unique opportunity. However, follow-up observations at high angular resolution will be necessary to obtain meaningful constraints on the properties of dark matter. Recently, O'Riordan et al. (2023) calculated the sensitivity function for 16000 Euclid-VIS-like mock observations (see Fig. 5). They found most of the lenses in the sample to be completely insensitive to the presence of subhaloes with a mass lower than $M_{\rm max}^{\rm NFW} = 10^{11} M_{\odot}$. The most sensitive pixels yield a lowest detectable mass of $M_{\rm max}^{\rm NFW} = 10^{8.8\pm0.2} M_{\odot}$. Assuming CDM and a dark matter fraction in subhaloes of 0.01, this sensitivity leads to one detectable object in every seventy lenses. These results are in line with Despali et al. (2022) and the current number of detected objects (see previous section). Hence, Euclid will allow one to detect a number of subhaloes that is significantly larger than what has been possible so far, and potentially deliver tighter constraints on the amount of dark matter in subhaloes (i.e. the normalisation of the subhalo mass function) than currently possible. However, most of these objects will be relatively massive, and will probe (thermal relic) dark matter models with a half-mode mass of $M_{\rm hm} > 10^8 M_{\odot}$. These models have already been ruled out by other observations, including strongly lensed quasars. Similarly, if not slightly worse results are expected from the Vera Rubin Observatory, given its slightly worse angular resolution.

While the sensitivity to the presence of low-mass haloes increases linearly with decreasing signal-to-noise ratio, the angular resolution of the observation sets a hard limit on the lowest detectable halo and the spatial scales on which the dark matter distribution can be probed (Despali et al. 2022; O'Riordan et al. 2023). For this reason, follow-up observations with Keck-AO, HST and ALMA are likely to lead to an increase in sensitivity (relatively to Euclid) between two to three orders in magnitude in halo mass. With an angular resolution of a few milli-arcseconds, Global VLBI observations at 1.6 to 15 GHz, will allow one to detect masses as low as $10^6 M_{\odot}$ (McKean et al. 2015), and probe the general spatial distribution of dark matter on sub-kpc scales (e.g. Powell et al. 2023). Moreover, searches for milli-arcsecond scale separation images with VLBI will potentially reveal the presence of super-critical low-mass haloes, which existence is predicted by certain SIDM models (Casadio et al. 2021; Loudas et al. 2022). In the future, 30-metre class optical telescopes such as the ELT are likely to provide a similar level of sensitivity, while interferometric observations in the 10 to 100 GHz range with SKA-VLBI or the ngVLA will lead to an even stronger increase. This significant improvement in data quality will also result in better constraints on the properties of the lens macro-model. This effect will significantly reduce the role of one of the main sources of systematic uncertainty and will potentially have interesting implications for other scientific applications of strong gravitational lensing. A relatively small number of high-resolution observations is, therefore, expected to potentially deliver stronger constraints on the properties of dark matter than the several orders of magnitude larger sample of Euclid and Vera Rubin lenses. Note, however, that these facilities and also LOFAR and the SKA will be vital for finding the lenses needed for high-resolution follow-up observations.

9 Summary and Conclusions

Strong gravitational lensing provides a unique channel to constrain the dark matter distribution on subgalactic scales, and hence provide a key test of several dark matter models. In this article, we described the process of inferring the properties of dark matter from strong lensing observations in terms of a Bayesian inference problem, and showed how it can be approached from different perspectives (Sect. 4). For each possible prior choice and lens modelling technique, we then discussed their advantages and limitations (Sect. 4.4). In Sect. 5 we provided an extensive discussion of the degeneracies and theoretical unknowns. In particular, we highlighted how better numerical simulations are needed to answer important and still open theoretical questions (Sect. 5.6).

While we only focused on galaxy-scale lenses, all of the approaches discussed above can potentially be extended to cluster-scale lenses. These type of studies would allows one to probe larger cosmological volumes and hence potentially require fewer observations to reach an equal level of constraints on dark matter. However, the lensing potential in galaxy clusters is significantly more complex than those of galaxies and it remains yet to be demonstrated that one can constrain their mass distribution with enough precision to avoid contamination by complex macro-models (see Sect. 5.1 for a discussion).

Currently, constraints on dark matter from lensing are based on relative small samples of systems (about ten for unresolved sources and about 50 for resolved ones). Ongoing and upcoming wide-sky surveys, with for example Euclid, the Vera Rubin Telescope, LOFAR and the SKA are expected to deliver orders of magnitudes more lensed galaxies and quasars. On the other hand, high-resolution follow-up with, for example, ALMA, the Global VLBI Network and the ELT will likely provide the data quality necessary to probe the dark matter distribution on small scales with resolved sources as well as obtain better constraints on the macro-model for unresolved ones. We can expect, therefore, that in the next five to ten years, the field of strong gravitational lensing will provide robust and statistically meaningful constraints on the nature of dark matter.

Funding Open Access funding enabled and organized by Projekt DEAL. S. V. thanks the Max Planck Society for support through a Max Planck Lise Meitner Group and funding from the European Research Council (ERC) under the European Union's Horizon 2020 research and innovation programme (LEDA: grant agreement No 758853). G. V. has received funding from the European Union's Horizon 2020 research and innovation programme under the Marie Sklodovska-Curie grant agreement No 897124. This research was made possible by the generosity of Eric and Wendy Schmidt by recommendation of the Schmidt Futures program. G. D. was supported by a Gliese Fellowship. C. D. F acknowledges support for this work from the National Science Foundation under Grant No. AST-1715611. D. G. was partially supported by a HQP grant from the McDonald Institute (reference number HQP 2019-4-2). Y. H. and L. P. are supported by a generous donation by Eric and Wendy Schmidt with the recommendation of the Schmidt Futures Foundation, as well as the National Sciences and Engineering Council of Canada, the Fonds de recherche du Québec, and the Canada Research Chairs Program. J. P. M. acknowledges support from the Netherlands Organization for Scientific Research (NWO, project number 629.001.023) and the Chinese Academy of Sciences (CAS, project number 114A11KYSB20170054). This work is based on the research supported in part by the National Research Foundation of South Africa (Grant Number: 128943).

Declarations

Competing Interests The authors declare no competing interests.

Open Access This article is licensed under a Creative Commons Attribution 4.0 International License, which permits use, sharing, adaptation, distribution and reproduction in any medium or format, as long as you give appropriate credit to the original author(s) and the source, provide a link to the Creative Commons licence, and indicate if changes were made. The images or other third party material in this article are included in the article's Creative Commons licence, unless indicated otherwise in a credit line to the material. If material is not included in the article's Creative Commons licence and your intended use is not permitted by statutory regulation or exceeds the permitted use, you will need to obtain permission directly from the copyright holder. To view a copy of this licence, visit http://creativecommons.org/licenses/by/4.0/.

References

Abbott TMC et al [DES Collaboration] (eds) (2022) Dark Energy Survey Year 3 results: cosmological constraints from galaxy clustering and weak lensing. Phys Rev D 105(2):023520. https://doi.org/10.1103/PhysRevD.105.023520

Adam A, Coogan A, Malkin N, Legin R, Perreault-Levasseur L, Hezaveh Y, Bengio Y (2022) Posterior samples of source galaxies in strong gravitational lenses with score-based priors. ArXiv e-prints, arXiv:2211.03812 [astro-ph.IM]. https://doi.org/10.48550/arXiv.2211.03812

Adhikari S, Banerjee A, Boddy KK, Cyr-Racine F-Y, Desmond H, Dvorkin C, Jain B, Kahlhoefer F, Kaplinghat M, Nierenberg A, Peter AHG, Robertson A, Sakstein J, Zavala J (2022) Astrophysical Tests of Dark Matter Self-Interactions. ArXiv e-prints, arXiv:2207.10638 [astro-ph.CO]. https://doi.org/10.48550/arXiv.2207.10638

Alam S, Ata M, Bailey S, Beutler F, Bizyaev D, Blazek JA, Bolton AS, Brownstein JR, Burden A, Chuang C-H et al (2017) The clustering of galaxies in the completed SDSS-III Baryon Oscillation Spectroscopic Survey: cosmological analysis of the DR12 galaxy sample. Mon Not R Astron Soc 470(3):2617–2652

Alexander S, Gleyzer S, McDonough E, Toomey MW, Usai E (2020b) Deep learning the morphology of dark matter substructure. Astrophys J 893(1):15. https://doi.org/10.3847/1538-4357/ab7925

Alexander S, Gleyzer S, Parul H, Reddy P, Toomey MW, Usai E, Von Klar R (2020a) Decoding Dark Matter Substructure without Supervision. ArXiv e-prints, arXiv:2008.12731 [astro-ph.CO]. https://doi.org/10.48550/arXiv.2008.12731

Alexander S, Gleyzer S, Reddy P, Tidball M, Toomey MW (2021) Domain Adaptation for Simulation-Based Dark Matter Searches Using Strong Gravitational Lensing. ArXiv e-prints, arXiv:2112.12121 [astro-ph.CO]. https://doi.org/10.48550/arXiv.2112.12121

Amorisco NC, Nightingale J, He Q, Amvrosiadis A, Cao X, Cole S, Etherington A, Frenk CS, Li R, Massey R, Robertson A (2022) Halo concentration strengthens dark matter constraints in galaxy-galaxy strong lensing analyses. Mon Not R Astron Soc 510(2):2464–2479. https://doi.org/10.1093/mnras/stab3527

Amruth A, Broadhurst T, Lim J, Oguri M, Smoot GF, Diego JM, Leung E, Emami R, Li J, Chiueh T, Schive H-Y, Yeung MCH, Li Kei S (2023) Anomalies in Gravitational-Lensed Images revealing Einstein rings modulated by wavelike dark matter. Nat Astron 7:736. https://doi.org/10.1038/s41550-023-01943-9. arXiv:2304.09895 [astro-ph.CO]

Andrade KE, Fuson J, Gad-Nasr S, Kong D, Minor Q, Roberts MG, Kaplinghat M (2022) A stringent upper limit on dark matter self-interaction cross-section from cluster strong lensing. Mon Not R Astron Soc 510(1):54–81. https://doi.org/10.1093/mnras/stab3241

Angulo RE, Lacey CG, Baugh CM, Frenk CS (2009) The fate of substructures in cold dark matter haloes. Mon Not R Astron Soc 399(2):983–995. https://doi.org/10.1111/j.1365-2966.2009.15333.x

Arcadi G, Dutra M, Ghosh P, Lindner M, Mambrini Y, Pierre M, Profumo S, Queiroz FS (2018) The waning of the WIMP? A review of models, searches, and constraints. Eur Phys J C 78(3):203. https://doi.org/10.1140/epjc/s10052-018-5662-y

Auger MW, Treu T, Bolton AS, Gavazzi R, Koopmans LVE, Marshall PJ, Bundy K, Moustakas LA (2009) The Sloan Lens ACS Survey. IX. Colors, lensing, and stellar masses of early-type galaxies. Astrophys J 705(2):1099–1115. https://doi.org/10.1088/0004-637X/705/2/1099

Balberg S, Shapiro SL, Inagaki S (2002) Self-interacting dark matter halos and the gravothermal catastrophe. Astrophys J 568(2):475–487. https://doi.org/10.1086/339038

Banik N, Bovy J, Bertone G, Erkal D, de Boer TJL (2021) Novel constraints on the particle nature of dark matter from stellar streams. J Cosmol Astropart Phys 2021(10):043. https://doi.org/10.1088/1475-7516/2021/10/043

Bardeen JM, Bond JR, Kaiser N, Szalay AS (1986) The statistics of peaks of Gaussian random fields. Astrophys J 304:15–61. https://doi.org/10.1086/164143

Barnabè M, Nipoti C, Koopmans LVE, Vegetti S, Ciotti L (2009) Crash-testing the CAULDRON code for joint lensing and dynamics analysis of early-type galaxies. Mon Not R Astron Soc 393(4):1114–1126. https://doi.org/10.1111/j.1365-2966.2008.14208.x

Barnabè M, Czoske O, Koopmans LVE, Treu T, Bolton AS (2011) Two-dimensional kinematics of SLACS lenses - III. Mass structure and dynamics of early-type lens galaxies beyond $z \sim 0.1$. Mon Not R Astron Soc 415(3):2215–2232. https://doi.org/10.1111/j.1365-2966.2011.18842.x

Basu A, Goswami J, Schwarz DJ, Urakawa Y (2021) Searching for axionlike particles under strong gravitational lenses. Phys Rev Lett 126(19):191102. https://doi.org/10.1103/PhysRevLett.126.191102

Bayer D, Chatterjee S, Koopmans LVE, Vegetti S, McKean JP, Treu T, Fassnacht CD, Glazebrook K (2023b) Probing sub-galactic mass structure with the power spectrum of surface-brightness anomalies in high-resolution observations of galaxy-galaxy strong gravitational lenses. II. Observational constraints on the subgalactic matter power spectrum. Mon Not R Astron Soc 523(1):1310–1325. https://doi.org/10.1093/mnras/stad1402

Bayer D, Koopmans LVE, McKean JP, Vegetti S, Treu T, Fassnacht CD, Glazebrook K (2023a) Probing sub-galactic mass structure with the power spectrum of surface-brightness anomalies in high-resolution observations of galaxy-galaxy strong gravitational lenses - I. Power-spectrum measurement and feasibility study. Mon Not R Astron Soc 523(1):1326–1345. https://doi.org/10.1093/mnras/stad1403

Belokurov V, Evans NW, Hewett PC, Moiseev A, McMahon RG, Sanchez SF, King LJ (2009) Two new large-separation gravitational lenses from SDSS. Mon Not R Astron Soc 392(1):104–112. https://doi.org/10.1111/j.1365-2966.2008.14075.x

Bennett CL, Lawrence CR, Burke BF, Hewitt JN, Mahoney J (1986) The MIT–Green Bank (MG) 5 GHz survey. Astrophys J Suppl Ser 61:1. https://doi.org/10.1086/191108

Berezinsky VS, Dokuchaev VI, Eroshenko YN (2014) Small-scale clumps of dark matter. Phys Usp 57(1):1–36. https://doi.org/10.3367/UFNe.0184.201401a.0003

Bertone G, Tait TMP (2018) A new era in the search for dark matter. Nature 562(7725):51–56. https://doi.org/10.1038/s41586-018-0542-z

Biggs AD (2023) A VLA monitoring study of JVAS B1422+231: investigation of time delays and detection of extrinsic variability. Mon Not R Astron Soc 522(1):426–437. https://doi.org/10.1093/mnras/stad870

Biggs AD, Browne IWA, Jackson NJ, York T, Norbury MA, McKean JP, Phillips PM (2004) Radio, optical and infrared observations of CLASS B0128+437. Mon Not R Astron Soc 350(3):949–961. https://doi.org/10.1111/j.1365-2966.2004.07701.x

Birrer S, Amara A, Refregier A (2015) Gravitational lens modeling with basis sets. Astrophys J 813(2):102. https://doi.org/10.1088/0004-637X/813/2/102

Birrer S, Amara A, Refregier A (2017) Lensing substructure quantification in RXJ1131-1231: a 2 keV lower bound on dark matter thermal relic mass. J Cosmol Astropart Phys 2017(5):037. https://doi.org/10.1088/1475-7516/2017/05/037

Bolton AS, Burles S, Koopmans LVE, Treu T, Gavazzi R, Moustakas LA, Wayth R, Schlegel DJ (2008) The Sloan Lens ACS Survey. V. The full ACS strong-lens sample. Astrophys J 682(2):964–984. https://doi.org/10.1086/589327

Bond JR, Cole S, Efstathiou G, Kaiser N (1991) Excursion set mass functions for hierarchical Gaussian fluctuations. Astrophys J 379:440–460. https://doi.org/10.1086/170520

Borsato E, Marchetti L, Negrello M, Corsini EM, Wake D, Amvrosiadis A, Baker AJ, Bakx TJLC, Beelen A, Berta S, Beyer A, Clements DL, Cooray A, Cox P, Dannerbauer H, de Zotti G, Dye S, Eales SA, Enia A, Farrah D, Gonzalez-Nuevo J, Hughes DH, Ismail D, Jin S, Lapi A, Lehnert MD, Neri R, Pérez-Fournon I, Riechers DA, Rodighiero G, Scott D, Serjeant S, Stanley F, Urquhart S, van der Werf P, Vaccari M, Wang L, Yang C, Young A (2024) Characterization of Herschel-selected strong lens candidates through HST and sub-mm/mm observations. Mon Not R Astron Soc 528(4):6222–6279. https://doi.org/10.1093/mnras/stad3381

Bose S, Hellwing WA, Frenk CS, Jenkins A, Lovell MR, Helly JC, Li B (2016) The Copernicus complexio: statistical properties of warm dark matter haloes. Mon Not R Astron Soc 455(1):318–333. https://doi.org/10.1093/mnras/stv2294

Boyarsky A, Ruchayskiy O, Shaposhnikov M (2009) The role of sterile neutrinos in cosmology and astrophysics. Annu Rev Nucl Part Sci 59(1):191–214. https://doi.org/10.1146/annurev.nucl.010909.083654

Boyarsky A, Drewes M, Lasserre T, Mertens S, Ruchayskiy O (2019) Sterile neutrino dark matter. Prog Part Nucl Phys 104:1–45. https://doi.org/10.1016/j.ppnp.2018.07.004

Bradač M, Schneider P, Steinmetz M, Lombardi M, King LJ, Porcas R (2002) B1422+231: the influence of mass substructure on strong lensing. Astron Astrophys 388:373–382. https://doi.org/10.1051/0004-6361:20020559

Bradač M, Schneider P, Lombardi M, Steinmetz M, Koopmans LVE, Navarro JF (2004) The signature of substructure on gravitational lensing in the ΛCDM cosmological model. Astron Astrophys 423:797–809. https://doi.org/10.1051/0004-6361:20040168

Brehmer J, Mishra-Sharma S, Hermans J, Louppe G, Cranmer K (2019) Mining for dark matter substructure: inferring subhalo population properties from strong lenses with machine learning. Astrophys J 886(1):49. https://doi.org/10.3847/1538-4357/ab4c41

Brewer BJ, Foreman-Mackey D, Hogg DW (2013) Probabilistic catalogs for crowded stellar fields. Astron J 146(1):7. https://doi.org/10.1088/0004-6256/146/1/7

Brewer BJ, Huijser D, Lewis GF (2016) Trans-dimensional Bayesian inference for gravitational lens substructures. Mon Not R Astron Soc 455(2):1819–1829. https://doi.org/10.1093/mnras/stv2370

Brooks AM, Kuhlen M, Zolotov A, Hooper D (2013) A baryonic solution to the missing satellites problem. Astrophys J 765(1):22. https://doi.org/10.1088/0004-637X/765/1/22

Browne IWA, Wilkinson PN, Patnaik AR, Wrobel JM (1998) Interferometer phase calibration sources - II. The region $0^\circ \leq \delta_{B1950} \leq +20^\circ$. Mon Not R Astron Soc 293(3):257–287. https://doi.org/10.1046/j.1365-8711.1998.01072.x

Browne IWA et al (2003) The Cosmic Lens All-Sky Survey - II. Gravitational lens candidate selection and follow-up. Mon Not R Astron Soc 341:13–32. https://doi.org/10.1046/j.1365-8711.2003.06257.x

Bryan GL, Norman ML (1998) Statistical properties of X-ray clusters: analytic and numerical comparisons. Astrophys J 495:80. https://doi.org/10.1086/305262

Bussmann RS, Gurwell MA, Fu H, Smith DJB, Dye S, Auld R, Baes M, Baker AJ, Bonfield D, Cava A, Clements DL, Cooray A, Coppin K, Dannerbauer H, Dariush A, De Zotti G, Dunne L, Eales S, Fritz J, Hopwood R, Ibar E, Ivison RJ, Jarvis MJ, Kim S, Leeuw LL, Maddox S, Michałowski MJ, Negrello M, Pascale E, Pohlen M, Riechers DA, Rigby E, Scott D, Temi P, Van der Werf PP, Wardlow J, Wilner D, Verma A (2012) A detailed gravitational lens model based on submillimeter array and Keck adaptive optics imaging of a Herschel-ATLAS submillimeter galaxy at $z = 4.243$. Astrophys J 756(2):134. https://doi.org/10.1088/0004-637X/756/2/134

Bussmann RS, Pérez-Fournon I, Amber S, Calanog J, Gurwell MA, Dannerbauer H, De Bernardis F, Fu H, Harris AI, Krips M, Lapi A, Maiolino R, Omont A, Riechers D, Wardlow J, Baker AJ, Birkinshaw M, Bock J, Bourne N, Clements DL, Cooray A, De Zotti G, Dunne L, Dye S, Eales S, Farrah D, Gavazzi R, González Nuevo J, Hopwood R, Ibar E, Ivison RJ, Laporte N, Maddox S, Martínez-Navajas P, Michalowski M, Negrello M, Oliver SJ, Roseboom IG, Scott D, Serjeant S, Smith AJ, Smith M, Streblyanska A, Valiante E, van der Werf P, Verma A, Vieira JD, Wang L, Wilner D (2013) Gravitational lens models based on submillimeter array imaging of Herschel-selected strongly lensed sub-millimeter galaxies at $z > 1.5$. Astrophys J 779(1):25. https://doi.org/10.1088/0004-637X/779/1/25

Cañameras R, Schuldt S, Shu Y, Suyu SH, Taubenberger S, Meinhardt T, Leal-Taixé L, Chao DC-Y, Inoue KT, Jaelani AT, More A (2021) HOLISMOKES. VI. New galaxy-scale strong lens candidates from the HSC-SSP imaging survey. Astron Astrophys 653:6. https://doi.org/10.1051/0004-6361/202141758

Casadio C, Blinov D, Readhead ACS, Browne IWA, Wilkinson PN, Hovatta T, Mandarakas N, Pavlidou V, Tassis K, Vedantham HK, Zensus JA, Diamantopoulos V, Dolapsaki KE, Gkimisi K, Kalaitzidakis G, Mastorakis M, Nikolaou K, Ntormousi E, Pelgrims V, Psarras K (2021) SMILE: search for MIlli-LEnses. Mon Not R Astron Soc 507(1):6–10. https://doi.org/10.1093/mnrasl/slab082

Chan JHH, Schive H-Y, Wong S-K, Chiueh T, Broadhurst T (2020) Multiple images and flux ratio anomaly of fuzzy gravitational lenses. Phys Rev Lett 125(11):111102. https://doi.org/10.1103/PhysRevLett.125.111102

Chan HYJ, Ferreira EGM, May S, Hayashi K, Chiba M (2022) The diversity of core-halo structure in the fuzzy dark matter model. Mon Not R Astron Soc 511(1):943–952. https://doi.org/10.1093/mnras/stac063

Chatterjee S, Koopmans LVE (2018) The inner mass power spectrum of galaxies using strong gravitational lensing: beyond linear approximation. Mon Not R Astron Soc 474(2):1762–1772. https://doi.org/10.1093/mnras/stx2674

Chen J (2009) Parity dependence in strong lens systems as a probe of dark matter substructure. Astron Astrophys 498(1):49–60. https://doi.org/10.1051/0004-6361/200811134

Chen J, Rozo E, Dalal N, Taylor JE (2007) Astrometric perturbations in substructure lensing. Astrophys J 659(1):52–68. https://doi.org/10.1086/512002

Chianese M, Coogan A, Hofma P, Otten S, Weniger C (2020) Differentiable strong lensing: uniting gravity and neural nets through differentiable probabilistic programming. Mon Not R Astron Soc 496(1):381–393. https://doi.org/10.1093/mnras/staa1477

Chiba M, Minezaki T, Kashikawa N, Kataza H, Inoue KT (2005) Subaru mid-infrared imaging of the quadruple lenses PG 1115+080 and B1422+231: limits on substructure lensing. Astrophys J 627(1):53–61. https://doi.org/10.1086/430403

Chua KTE, Pillepich A, Rodriguez-Gomez V, Vogelsberger M, Bird S, Hernquist L (2017) Subhalo demographics in the illustris simulation: effects of baryons and halo-to-halo variation. Mon Not R Astron Soc 472(4):4343–4360. https://doi.org/10.1093/mnras/stx2238

Clowe D, Bradač M, Gonzalez AH, Markevitch M, Randall SW, Jones C, Zaritsky D (2006) A direct empirical proof of the existence of dark matter. Astrophys J Lett 648(2):109–113. https://doi.org/10.1086/508162

Coogan A, Karchev K, Weniger C (2020) Targeted Likelihood-Free Inference of Dark Matter Substructure in Strongly-Lensed Galaxies. ArXiv e-prints, arXiv:2010.07032 [astro-ph.CO]

Cornwell TJ, Evans KF (1985) A simple maximum entropy deconvolution algorithm. Astron Astrophys 143(1):77–83

Correa CA (2021) Constraining velocity-dependent self-interacting dark matter with the Milky Way's dwarf spheroidal galaxies. Mon Not R Astron Soc 503(1):920–937. https://doi.org/10.1093/mnras/stab506

Cranmer K, Brehmer J, Louppe G (2020) The frontier of simulation-based inference. Proc Natl Acad Sci 117(48):30055–30062. https://doi.org/10.1073/pnas.1912789117

Cyr-Racine F-Y, Moustakas LA, Keeton CR, Sigurdson K, Gilman DA (2016) Dark census: statistically detecting the satellite populations of distant galaxies. Phys Rev D 94(4):043505. https://doi.org/10.1103/PhysRevD.94.043505

Cyr-Racine F-Y, Keeton CR, Moustakas LA (2019) Beyond subhalos: probing the collective effect of the universe's small-scale structure with gravitational lensing. Phys Rev D 100(2):023013. https://doi.org/10.1103/PhysRevD.100.023013

Dalal N, Kochanek CS (2002) Direct detection of cold dark matter substructure. Astrophys J 572(1):25–33. https://doi.org/10.1086/340303

Daylan T, Portillo SKN, Finkbeiner DP (2017) Inference of unresolved point sources at high galactic latitudes using probabilistic catalogs. Astrophys J 839(1):4. https://doi.org/10.3847/1538-4357/aa679e

Daylan T, Cyr-Racine F-Y, Diaz Rivero A, Dvorkin C, Finkbeiner DP (2018) Probing the small-scale structure in strongly lensed systems via transdimensional inference. Astrophys J 854(2):141. https://doi.org/10.3847/1538-4357/aaaa1e

Delos MS (2019) Evolution of dark matter microhalos through stellar encounters. Phys Rev D 100(8):083529. https://doi.org/10.1103/PhysRevD.100.083529

Despali G, Vegetti S (2017) The impact of baryonic physics on the subhalo mass function and implications for gravitational lensing. Mon Not R Astron Soc 469:1997–2010. https://doi.org/10.1093/mnras/stx966

Despali G, Giocoli C, Angulo RE, Tormen G, Sheth RK, Baso G, Moscardini L (2016) The universality of the virial halo mass function and models for non-universality of other halo definitions. Mon Not R Astron Soc 456:2486–2504. https://doi.org/10.1093/mnras/stv2842

Despali G, Vegetti S, White SDM, Giocoli C, van den Bosch FC (2018) Modelling the line-of-sight contribution in substructure lensing. Mon Not R Astron Soc 475(4):5424–5442. https://doi.org/10.1093/mnras/sty159

Despali G, Sparre M, Vegetti S, Vogelsberger M, Zavala J, Marinacci F (2019) The interplay of self-interacting dark matter and baryons in shaping the halo evolution. Mon Not R Astron Soc 484:4563–4573. https://doi.org/10.1093/mnras/stz273

Despali G, Lovell M, Vegetti S, Crain RA, Oppenheimer BD (2020) The lensing properties of subhaloes in massive elliptical galaxies in sterile neutrino cosmologies. Mon Not R Astron Soc 491(1):1295–1310. https://doi.org/10.1093/mnras/stz3068

Despali G, Vegetti S, White SDM, Powell DM, Stacey HR, Fassnacht CD, Rizzo F, Enzi W (2022) Detecting low-mass haloes with strong gravitational lensing I: the effect of data quality and lensing configuration. Mon Not R Astron Soc 510(2):2480–2494. https://doi.org/10.1093/mnras/stab3537

Dhanasingham B, Cyr-Racine F-Y, Peter AHG, Benson A, Gilman D (2022) Interlopers speak out: studying the dark universe using small-scale lensing anisotropies. Mon Not R Astron Soc. https://doi.org/10.1093/mnras/stac2993

Diaz Rivero A, Dvorkin C (2020) Direct detection of dark matter substructure in strong lens images with convolutional neural networks. Phys Rev D 101(2):023515. https://doi.org/10.1103/PhysRevD.101.023515

Diaz Rivero A, Cyr-Racine F-Y, Dvorkin C (2018) Power spectrum of dark matter substructure in strong gravitational lenses. Phys Rev D 97(2):023001. https://doi.org/10.1103/PhysRevD.97.023001

Díaz Rivero A, Dvorkin C, Cyr-Racine F-Y, Zavala J, Vogelsberger M (2018) Gravitational lensing and the power spectrum of dark matter substructure: insights from the ETHOS N-body simulations. Phys Rev D 98(10):103517. https://doi.org/10.1103/PhysRevD.98.103517

Diggle PJ, Gratton RJ (2018) Monte Carlo methods of inference for implicit statistical models. J R Stat Soc, Ser B, Methodol 46(2):193–212. https://doi.org/10.1111/j.2517-6161.1984.tb01290.x

Dobler G, Keeton CR (2006) Finite source effects in strong lensing: implications for the substructure mass scale. Mon Not R Astron Soc 365(4):1243–1262. https://doi.org/10.1111/j.1365-2966.2005.09809.x

Doroshkevich AG, Khlopov MI, Sunyaev RA, Szalay AS, Zeldovich IB (1981) Cosmological impact of the neutrino rest mass. Ann NY Acad Sci 375:32–42. https://doi.org/10.1111/j.1749-6632.1981.tb33688.x

Dubois Y, Beckmann R, Bournaud F, Choi H, Devriendt J, Jackson R, Kaviraj S, Kimm T, Kraljic K, Laigle C, Martin G, Park M-J, Peirani S, Pichon C, Volonteri M, Yi SK (2021) Introducing the NEWHORIZON simulation: galaxy properties with resolved internal dynamics across cosmic time. Astron Astrophys 651:109. https://doi.org/10.1051/0004-6361/202039429

Dye S, Furlanetto C, Dunne L, Eales SA, Negrello M, Nayyeri H, van der Werf PP, Serjeant S, Farrah D, Michałowski MJ, Baes M, Marchetti L, Cooray A, Riechers DA, Amvrosiadis A (2018) Modelling high-resolution ALMA observations of strongly lensed highly star-forming galaxies detected by Herschel. Mon Not R Astron Soc 476(4):4383–4394. https://doi.org/10.1093/mnras/sty513

Elíasdóttir Á, Hjorth J, Toft S, Burud I, Paraficz D (2006) Extinction curves of lensing galaxies out to $z = 1$. Astrophys J Suppl Ser 166(2):443–469. https://doi.org/10.1086/507131

Ellithorpe JD, Kochanek CS, Hewitt JN (1996) Visibility LensClean and the reliability of deconvolved radio images. Astrophys J 464:556. https://doi.org/10.1086/177346

Enzi W, Vegetti S, Despali G, Hsueh J-W, Metcalf RB (2020) Systematic errors in strong gravitational lensing reconstructions, a numerical simulation perspective. Mon Not R Astron Soc 496(2):1718–1729. https://doi.org/10.1093/mnras/staa1224

Enzi W, Murgia R, Newton O, Vegetti S, Frenk C, Viel M, Cautun M, Fassnacht CD, Auger M, Despali G, McKean J, Koopmans LVE, Lovell M (2021) Joint constraints on thermal relic dark matter from strong gravitational lensing, the Ly α forest, and Milky Way satellites. Mon Not R Astron Soc 506(4):5848–5862. https://doi.org/10.1093/mnras/stab1960

Erkal D, Koposov SE, Belokurov V (2017) A sharper view of Pal 5's tails: discovery of stream perturbations with a novel non-parametric technique. Mon Not R Astron Soc 470(1):60–84. https://doi.org/10.1093/mnras/stx1208

Ettori S, Donnarumma A, Pointecouteau E, Reiprich TH, Giodini S, Lovisari L, Schmidt RW (2013) Mass profiles of galaxy clusters from X-ray analysis. Space Sci Rev 177(1–4):119–154. https://doi.org/10.1007/s11214-013-9976-7

Fassnacht CD, Blandford RD, Cohen JG, Matthews K, Pearson TJ, Readhead ACS, Womble DS, Myers ST, Browne IWA, Jackson NJ, Marlow DR, Wilkinson PN, Koopmans LVE, de Bruyn AG, Schilizzi RT, Bremer M, Miley G (1999) B2045+265: a new four-image gravitational lens from CLASS. Astron J 117(2):658–670. https://doi.org/10.1086/300724

Fassnacht CD, Xanthopoulos E, Koopmans LVE, Rusin D (2002) A determination of H_0 with the CLASS gravitational lens B1608+656. III. A significant improvement in the precision of the time delay measurements. Astrophys J 581(2):823–835. https://doi.org/10.1086/344368

Feng JL (2010) Dark matter candidates from particle physics and methods of detection. Annu Rev Astron Astrophys 48:495–545. https://doi.org/10.1146/annurev-astro-082708-101659

Feng W-X, Yu H-B, Zhong Y-M (2021) Seeding supermassive black holes with self-interacting dark matter: a unified scenario with baryons. Astrophys J Lett 914(2):26. https://doi.org/10.3847/2041-8213/ac04b0

Fox PJ (2018) TASI Lectures on WIMPs and Supersymmetry. PoS(TASI2028)005. https://doi.org/10.22323/1.333.0005

Galan A, Vernardos G, Peel A, Courbin F, Starck J-L (2022) Using wavelets to capture deviations from smoothness in galaxy-scale strong lenses. Astron Astrophys 668:155. https://doi.org/10.1051/0004-6361/202244464

Gao L, White SDM, Jenkins A, Stoehr F, Springel V (2004) The subhalo populations of ΛCDM dark haloes. Mon Not R Astron Soc 355(3):819–834. https://doi.org/10.1111/j.1365-2966.2004.08360.x

Gao L, Navarro JF, Frenk CS, Jenkins A, Springel V, White SDM (2012) The Phoenix project: the dark side of rich galaxy clusters. Mon Not R Astron Soc 425(3):2169–2186. https://doi.org/10.1111/j.1365-2966.2012.21564.x

Garrison-Kimmel S, Hopkins PF, Wetzel A, Bullock JS, Boylan-Kolchin M, Kereš D, Faucher-Giguère C-A, El-Badry K, Lamberts A, Quataert E, Sanderson R (2019) The local group on FIRE: dwarf galaxy populations across a suite of hydrodynamic simulations. Mon Not R Astron Soc 487(1):1380–1399. https://doi.org/10.1093/mnras/stz1317

Gilman D, Agnello A, Treu T, Keeton CR, Nierenberg AM (2017) Strong lensing signatures of luminous structure and substructure in early-type galaxies. Mon Not R Astron Soc 467(4):3970–3992. https://doi.org/10.1093/mnras/stx158

Gilman D, Birrer S, Treu T, Keeton CR, Nierenberg A (2018) Probing the nature of dark matter by forward modelling flux ratios in strong gravitational lenses. Mon Not R Astron Soc 481(1):819–834. https://doi.org/10.1093/mnras/sty2261

Gilman D, Birrer S, Treu T, Nierenberg A, Benson A (2019) Probing dark matter structure down to 10^7 solar masses: flux ratio statistics in gravitational lenses with line-of-sight haloes. Mon Not R Astron Soc 487(4):5721–5738. https://doi.org/10.1093/mnras/stz1593

Gilman D, Birrer S, Nierenberg A, Treu T, Du X, Benson A (2020b) Warm dark matter chills out: constraints on the halo mass function and the free-streaming length of dark matter with eight quadruple-image strong gravitational lenses. Mon Not R Astron Soc 491(4):6077–6101. https://doi.org/10.1093/mnras/stz3480

Gilman D, Birrer S, Treu T (2020) TDCOSMO. III. Dark matter substructure meets dark energy. The effects of (sub)halos on strong-lensing measurements of H_0. Astron Astrophys 642:194. https://doi.org/10.1051/0004-6361/202038829

Gilman D, Du X, Benson A, Birrer S, Nierenberg A, Treu T (2020a) Constraints on the mass-concentration relation of cold dark matter halos with 11 strong gravitational lenses. Mon Not R Astron Soc 492(1):12–16. https://doi.org/10.1093/mnrasl/slz173

Gilman D, Bovy J, Treu T, Nierenberg A, Birrer S, Benson A, Sameie O (2021) Strong lensing signatures of self-interacting dark matter in low-mass haloes. Mon Not R Astron Soc 507(2):2432–2447. https://doi.org/10.1093/mnras/stab2335

Gilman D, Benson A, Bovy J, Birrer S, Treu T, Nierenberg A (2022) The primordial matter power spectrum on sub-galactic scales. Mon Not R Astron Soc 512(3):3163–3188. https://doi.org/10.1093/mnras/stac670

Gilman D, Zhong Y-M, Bovy J (2022) Constraining resonant dark matter self-interactions with strong gravitational lenses. ArXiv e-prints, 2207–13111. https://doi.org/10.48550/arXiv.2207.13111

Giocoli C, Pieri L, Tormen G (2008) Analytical approach to subhalo population in dark matter haloes. Mon Not R Astron Soc 387:689–697. https://doi.org/10.1111/j.1365-2966.2008.13283.x

Green PJ (1995) Reversible jump Markov chain Monte Carlo computation and Bayesian model determination. Biometrika 82(4):711–732. https://doi.org/10.1093/biomet/82.4.711

Green PJ (2003) Trans-dimensional Markov chain Monte Carlo. In: Green PJ, Hjort NL, Richardson S (eds) Highly structured stochastic systems. Oxford University Press, Oxford, pp 179–198. https://doi.org/10.1093/oso/9780198510550.003.0017

Green AM, Kavanagh BJ (2021) Primordial black holes as a dark matter candidate. J Phys G, Nucl Phys 48(4):043001. https://doi.org/10.1088/1361-6471/abc534

Green SB, van den Bosch FC (2019) The tidal evolution of dark matter substructure - I. subhalo density profiles. Mon Not R Astron Soc 490(2):2091–2101. https://doi.org/10.1093/mnras/stz2767

Green SB, van den Bosch FC, Jiang F (2021) The tidal evolution of dark matter substructure - II. The impact of artificial disruption on subhalo mass functions and radial profiles. Mon Not R Astron Soc 503(3):4075–4091. https://doi.org/10.1093/mnras/stab696

Griffith M, Langston G, Heflin M, Conner S, Lehar J, Burke B (1990) The third MIT–Green Bank 5 GHz survey. Astrophys J Suppl Ser 74:129. https://doi.org/10.1086/191495

Griffith M, Langston G, Heflin M, Conner S, Burke B (1991) The fourth MIT–Green Bank 5 GHz survey. Astrophys J Suppl Ser 75:801. https://doi.org/10.1086/191549

He Q, Nightingale J, Robertson A, Amvrosiadis A, Cole S, Frenk CS, Massey R, Li R, Amorisco NC, Metcalf RB, Cao X, Etherington A (2023) Testing strong lensing subhalo detection with a cosmological simulation. Mon Not R Astron Soc 518(1):220–239. https://doi.org/10.1093/mnras/stac2779

Heinze FM, Despali G, Klessen RS (2023) Not all subhaloes are created equal: modelling the diversity of subhalo density profiles in TNG50. Mon Not R Astron Soc 527(4):11996–12015. https://doi.org/10.1093/mnras/stad3894

Heymans C, Tröster T, Asgari M, Blake C, Hildebrandt H, Joachimi B, Kuijken K, Lin C-A, Sánchez AG, van den Busch JL, Wright AH, Amon A, Bilicki M, de Jong J, Crocce M, Dvornik A, Erben T, Fortuna MC, Getman F, Giblin B, Glazebrook K, Hoekstra H, Joudaki S, Kannawadi A, Köhlinger F, Lidman C, Miller L, Napolitano NR, Parkinson D, Schneider P, Shan H, Valentijn EA, Verdoes Kleijn G, Wolf C (2021) KiDS-1000 cosmology: multi-probe weak gravitational lensing and spectroscopic galaxy clustering constraints. Astron Astrophys 646:140. https://doi.org/10.1051/0004-6361/202039063

Hezaveh Y, Dalal N, Holder G, Kuhlen M, Marrone D, Murray N, Vieira J (2013) Dark matter substructure detection using spatially resolved spectroscopy of lensed dusty galaxies. Astrophys J 767(1):9. https://doi.org/10.1088/0004-637X/767/1/9

Hezaveh YD, Marrone DP, Fassnacht CD, Spilker JS, Vieira JD, Aguirre JE, Aird KA, Aravena M, Ashby MLN, Bayliss M, Benson BA, Bleem LE, Bothwell M, Brodwin M, Carlstrom JE, Chang CL, Chapman SC, Crawford TM, Crites AT, De Breuck C, de Haan T, Dobbs MA, Fomalont EB, George EM, Gladders MD, Gonzalez AH, Greve TR, Halverson NW, High FW, Holder GP, Holzapfel WL, Hoover S, Hrubes JD, Husband K, Hunter TR, Keisler R, Lee AT, Leitch EM, Lueker M, Luong-Van D, Malkan M, McIntyre V, McMahon JJ, Mehl J, Menten KM, Meyer SS, Mocanu LM, Murphy EJ, Natoli T, Padin S, Plagge T, Reichardt CL, Rest A, Ruel J, Ruhl JE, Sharon K, Schaffer KK, Shaw L, Shirokoff E, Stalder B, Staniszewski Z, Stark AA, Story K, Vanderlinde K, Weiß A, Welikala N, Williamson R (2013) ALMA observations of SPT-discovered, strongly lensed, dusty, star-forming galaxies. Astrophys J 767(2):132. https://doi.org/10.1088/0004-637X/767/2/132

Hezaveh Y, Dalal N, Holder G, Kisner T, Kuhlen M, Perreault Levasseur L (2016) Measuring the power spectrum of dark matter substructure using strong gravitational lensing. J Cosmol Astropart Phys 2016(11):048. https://doi.org/10.1088/1475-7516/2016/11/048

Hezaveh YD, Dalal N, Marrone DP, Mao Y-Y, Morningstar W, Wen D, Blandford RD, Carlstrom JE, Fassnacht CD, Holder GP, Kemball A, Marshall PJ, Murray N, Perreault Levasseur L, Vieira JD, Wechsler RH (2016) Detection of lensing substructure using ALMA observations of the dusty galaxy SDP.81. Astrophys J 823:37. https://doi.org/10.3847/0004-637X/823/1/37

Hezaveh YD, Perreault Levasseur L, Marshall PJ (2017) Fast automated analysis of strong gravitational lenses with convolutional neural networks. Nature 548(7669):555–557. https://doi.org/10.1038/nature23463

Högbom JA (1974) Aperture synthesis with a non-regular distribution of interferometer baselines. Astron Astrophys Suppl Ser 15:417

Hogg DW, Lang D (2010) Telescopes don't make catalogues! In: Turon C, Meynadier F, Arenou F (eds). GAIA: at the frontiers of astrometry. EAS Publications Series, vol 45, pp 351–358. https://doi.org/10.1051/eas/1045059

Holzschuh BJ, O'Riordan CM, Vegetti S, Rodriguez-Gomez V, Thuerey N (2022) Realistic galaxy images and improved robustness in machine learning tasks from generative modelling. Mon Not R Astron Soc 515(1):652–677. https://doi.org/10.1093/mnras/stac1188

Hsueh J-W, Fassnacht CD, Vegetti S, McKean JP, Spingola C, Auger MW, Koopmans LVE, Lagattuta DJ (2016) SHARP - II. Mass structure in strong lenses is not necessarily dark matter substructure: a flux ratio anomaly from an edge-on disc in B1555+375. Mon Not R Astron Soc 463(1):51–55. https://doi.org/10.1093/mnrasl/slw146

Hsueh J-W, Oldham L, Spingola C, Vegetti S, Fassnacht CD, Auger MW, Koopmans LVE, McKean JP, Lagattuta DJ (2017) SHARP - IV. An apparent flux-ratio anomaly resolved by the edge-on disc in B0712+472. Mon Not R Astron Soc 469(3):3713–3721. https://doi.org/10.1093/mnras/stx1082

Hsueh J-W, Despali G, Vegetti S, Xu D, Fassnacht CD, Metcalf RB (2018) Flux-ratio anomalies from discs and other baryonic structures in the illustris simulation. Mon Not R Astron Soc 475(2):2438–2451. https://doi.org/10.1093/mnras/stx3320

Hsueh J-W, Enzi W, Vegetti S, Auger MW, Fassnacht CD, Despali G, Koopmans LVE, McKean JP (2020) SHARP - VII. New constraints on the dark matter free-streaming properties and substructure abundance from gravitationally lensed quasars. Mon Not R Astron Soc 492(2):3047–3059. https://doi.org/10.1093/mnras/stz3177

Huang X, Storfer C, Gu A, Ravi V, Pilon A, Sheu W, Venguswamy R, Banka S, Dey A, Landriau M, Lang D, Meisner A, Moustakas J, Myers AD, Sajith R, Schlafly EF, Schlegel DJ (2021) Discovering new strong gravitational lenses in the DESI Legacy Imaging Surveys. Astrophys J 909(1):27. https://doi.org/10.3847/1538-4357/abd62b

Hui L (2021) Wave dark matter. Annu Rev Astron Astrophys 59:247–289. https://doi.org/10.1146/annurev-astro-120920-010024

Inoue KT, Minezaki T, Matsushita S, Chiba M (2016) ALMA imprint of intergalactic dark structures in the gravitational lens SDP.81. Mon Not R Astron Soc 457(3):2936–2950. https://doi.org/10.1093/mnras/stw168

Iršič V, Viel M, Haehnelt MG, Bolton JS, Becker GD (2017) First constraints on fuzzy dark matter from Lyman-α forest data and hydrodynamical simulations. Phys Rev Lett 119(3):031302. https://doi.org/10.1103/PhysRevLett.119.031302

Kaplinghat M, Keeley RE, Linden T, Yu H-B (2014) Tying dark matter to baryons with self-interactions. Phys Rev Lett 113(2):021302. https://doi.org/10.1103/PhysRevLett.113.021302

Kaplinghat M, Valli M, Yu H-B (2019) Too big to fail in light of Gaia. Mon Not R Astron Soc 490(1):231–242. https://doi.org/10.1093/mnras/stz2511

Kaplinghat M, Ren T, Yu H-B (2020) Dark matter cores and cusps in spiral galaxies and their explanations. J Cosmol Astropart Phys 2020(6):027. https://doi.org/10.1088/1475-7516/2020/06/027

Karchev K, Coogan A, Weniger C (2022) Strong-lensing source reconstruction with variationally optimized Gaussian processes. Mon Not R Astron Soc 512(1):661–685. https://doi.org/10.1093/mnras/stac311

Kawai H, Oguri M, Amruth A, Broadhurst T, Lim J (2022) An analytic model for the subgalactic matter power spectrum in fuzzy dark matter halos. Astrophys J 925(1):61. https://doi.org/10.3847/1538-4357/ac39a2

Keeton CR, Moustakas LA (2009) A new channel for detecting dark matter substructure in galaxies: gravitational lens time delays. Astrophys J 699(2):1720–1731. https://doi.org/10.1088/0004-637X/699/2/1720

Knebe A, Knollmann SR, Muldrew SI, Pearce FR, Aragon-Calvo MA, Ascasibar Y, Behroozi PS, Ceverino D et al (2011) Haloes gone MAD: the halo-finder comparison project. Mon Not R Astron Soc 415:2293–2318. https://doi.org/10.1111/j.1365-2966.2011.18858.x

Knebe A, Pearce FR, Lux H, Ascasibar Y, Behroozi P, Casado J, Moran CC, Diemand J et al (2013) Structure finding in cosmological simulations: the state of affairs. Mon Not R Astron Soc 435:1618–1658. https://doi.org/10.1093/mnras/stt1403

Kochanek CS, Narayan R (1992) LensClean: an algorithm for inverting extended, gravitationally lensed images with application to the radio ring lens PKS 1830-211. Astrophys J 401:461. https://doi.org/10.1086/172078

Koopmans LVE (2005) Gravitational imaging of cold dark matter substructures. Mon Not R Astron Soc 363(4):1136–1144. https://doi.org/10.1111/j.1365-2966.2005.09523.x

Koopmans LVE, de Bruyn AG (2000) Microlensing of multiply-imaged compact radio sources. Evidence for compact halo objects in the disk galaxy of B1600+434. Astron Astrophys 358:793–811. https://doi.org/10.48550/arXiv.astro-ph/0004112

Koopmans LVE, Biggs A, Blandford RD, Browne IWA, Jackson NJ, Mao S, Wilkinson PN, de Bruyn AG, Wambsganss J (2003) Extrinsic radio variability of JVAS/CLASS gravitational lenses. Astrophys J 595(2):712–718. https://doi.org/10.1086/377434

Koopmans LVE, Bolton A, Treu T, Czoske O, Auger MW, Barnabè M, Vegetti S, Gavazzi R, Moustakas LA, Burles S (2009) The structure and dynamics of massive early-type galaxies: on homology, isothermality, and isotropy inside one effective radius. Astrophys J Lett 703(1):51–54. https://doi.org/10.1088/0004-637X/703/1/L51

Kulkarni M, Ostriker JP (2022) What is the halo mass function in a fuzzy dark matter cosmology? Mon Not R Astron Soc 510(1):1425–1430. https://doi.org/10.1093/mnras/stab3520

Lagattuta DJ, Vegetti S, Fassnacht CD, Auger MW, Koopmans LVE, McKean JP (2012) SHARP - I. A high-resolution multiband view of the infrared Einstein ring of JVAS B1938+666. Mon Not R Astron Soc 424(4):2800–2810. https://doi.org/10.1111/j.1365-2966.2012.21406.x

Langston GI, Heflin MB, Conner SR, Lehar J, Carilli CL, Burke BF (1990) The second MIT–green bank 5 GHz survey. Astrophys J Suppl Ser 72:621. https://doi.org/10.1086/191427

Lanusse F, Ma Q, Li N, Collett TE, Li C-L, Ravanbakhsh S, Mandelbaum R, Póczos B (2018) CMU DeepLens: deep learning for automatic image-based galaxy-galaxy strong lens finding. Mon Not R Astron Soc 473(3):3895–3906. https://doi.org/10.1093/mnras/stx1665

Laroche A, Gilman D, Li X, Bovy J, Du X (2022) Quantum fluctuations masquerade as haloes: bounds on ultra-light dark matter from quadruply imaged quasars. Mon Not R Astron Soc 517(2):1867–1883. https://doi.org/10.1093/mnras/stac2677

Legin R, Stone C, Hezaveh Y, Perreault-Levasseur L (2022) Population-Level Inference of Strong Gravitational Lenses with Neural Network-Based Selection Correction. ArXiv e-prints, arXiv:2207.04123 [astro-ph.IM]. https://doi.org/10.48550/arXiv.2207.04123

Li R, Frenk CS, Cole S, Wang Q, Gao L (2017) Projection effects in the strong lensing study of subhaloes. Mon Not R Astron Soc 468(2):1426–1432. https://doi.org/10.1093/mnras/stx554

Liepe J, Kirk P, Filippi S, Toni T, Barnes CP, Stumpf MPH (2014) A framework for parameter estimation and model selection from experimental data in systems biology using approximate Bayesian computation. Nat Protoc 9:439–456. https://doi.org/10.1038/nprot.2014.025

Loudas N, Pavlidou V, Casadio C, Tassis K (2022) Discriminating power of milli-lensing observations for dark matter models. Astron Astrophys 668:166. https://doi.org/10.1051/0004-6361/202244978

Lovell MR (2020) The halo mass function in alternative dark matter models. Mon Not R Astron Soc 493(1):11–15. https://doi.org/10.1093/mnrasl/slaa005

Lovell MR (2020) Toward a general parameterization of the warm dark matter halo mass function. Astrophys J 897(2):147. https://doi.org/10.3847/1538-4357/ab982a

Lovell MR, Eke V, Frenk CS, Gao L, Jenkins A, Theuns T, Wang J, White SDM, Boyarsky A, Ruchayskiy O (2012) The haloes of bright satellite galaxies in a warm dark matter universe. Mon Not R Astron Soc 420:2318–2324. https://doi.org/10.1111/j.1365-2966.2011.20200.x

Lovell MR, Frenk CS, Eke VR, Jenkins A, Gao L, Theuns T (2014) The properties of warm dark matter haloes. Mon Not R Astron Soc 439(1):300–317. https://doi.org/10.1093/mnras/stt2431

Lovell MR, Zavala J, Vogelsberger M (2019) ETHOS - an effective theory of structure formation: formation of the first haloes and their stars. Mon Not R Astron Soc 485(4):5474–5489. https://doi.org/10.1093/mnras/stz766

Ludlow AD, Bose S, Angulo RE, Wang L, Hellwing WA, Navarro JF, Cole S, Frenk CS (2016) The mass-concentration-redshift relation of cold and warm dark matter haloes. Mon Not R Astron Soc 460:1214–1232. https://doi.org/10.1093/mnras/stw1046

Mandelbaum R, Seljak U, Kauffmann G, Hirata CM, Brinkmann J (2006) Galaxy halo masses and satellite fractions from galaxy-galaxy lensing in the Sloan Digital Sky Survey: stellar mass, luminosity, morphology and environment dependencies. Mon Not R Astron Soc 368(2):715–731. https://doi.org/10.1111/j.1365-2966.2006.10156.x

Mao S, Schneider P (1998) Evidence for substructure in lens galaxies? Mon Not R Astron Soc 295(3):587–594. https://doi.org/10.1046/j.1365-8711.1998.01319.x

Mao S, Jing Y, Ostriker JP, Weller J (2004) Anomalous flux ratios in gravitational lenses: for or against cold dark matter? Astrophys J Lett 604(1):5–8. https://doi.org/10.1086/383413

Marlow DR, Myers ST, Rusin D, Jackson N, Browne IWA, Wilkinson PN, Muxlow T, Fassnacht CD, Lubin L, Kundić T, Blandford RD, Pearson TJ, Readhead ACS, Koopmans L, de Bruyn AG (1999) CLASS B1555+375: a new four-image gravitational lens system. Astron J 118(2):654–658. https://doi.org/10.1086/300987

Marsh DJE, Niemeyer JC (2019) Strong constraints on fuzzy dark matter from ultrafaint dwarf galaxy eridanus II. Phys Rev Lett 123(5):051103. https://doi.org/10.1103/PhysRevLett.123.051103

Marsh DJE, Silk J (2014) A model for halo formation with axion mixed dark matter. Mon Not R Astron Soc 437(3):2652–2663. https://doi.org/10.1093/mnras/stt2079

May S, Springel V (2021) Structure formation in large-volume cosmological simulations of fuzzy dark matter: impact of the non-linear dynamics. Mon Not R Astron Soc 506(2):2603–2618. https://doi.org/10.1093/mnras/stab1764

May S, Springel V (2023) The halo mass function and filaments in full cosmological simulations with fuzzy dark matter. Mon Not R Astron Soc 524(3):4256–4274. https://doi.org/10.1093/mnras/stad2031

McKean JP, Koopmans LVE, Flack CE, Fassnacht CD, Thompson D, Matthews K, Blandford RD, Readhead ACS, Soifer BT (2007) High-resolution imaging of the anomalous flux ratio gravitational lens system

CLASS B2045+265: dark or luminous satellites? Mon Not R Astron Soc 378(1):109–118. https://doi.org/10.1111/j.1365-2966.2007.11744.x
McKean J, Jackson N, Vegetti S, Rybak M, Serjeant S, Koopmans LVE, Metcalf RB, Fassnacht C, Marshall PJ, Pandey-Pommier M (2015) Strong gravitational lensing with the SKA. In: Advancing Astrophysics with the Square Kilometre Array (AASKA14), p 84
Metcalf RB, Moustakas LA, Bunker AJ, Parry IR (2004) Spectroscopic gravitational lensing and limits on the dark matter substructure in Q2237+0305. Astrophys J 607(1):43–59. https://doi.org/10.1086/383243
Minor Q, Gad-Nasr S, Kaplinghat M, Vegetti S (2021) An unexpected high concentration for the dark substructure in the gravitational lens SDSSJ0946+1006. Mon Not R Astron Soc 507(2):1662–1683. https://doi.org/10.1093/mnras/stab2247
Mittal R, Porcas R, Wucknitz O (2007) Free-free absorption in the gravitational lens JVAS B0218+357. Astron Astrophys 465(2):405–415. https://doi.org/10.1051/0004-6361:20066127
Mocz P, Fialkov A, Vogelsberger M, Becerra F, Amin MA, Bose S, Boylan-Kolchin M, Chavanis P-H, Hernquist L, Lancaster L, Marinacci F, Robles VH, Zavala J (2019) First star-forming structures in fuzzy cosmic filaments. Phys Rev Lett 123(14):141301. https://doi.org/10.1103/PhysRevLett.123.141301
Moliné Á, Sánchez-Conde MA, Palomares-Ruiz S, Prada F (2017) Characterization of subhalo structural properties and implications for dark matter annihilation signals. Mon Not R Astron Soc 466:4974–4990. https://doi.org/10.1093/mnras/stx026
Montel NA, Coogan A, Correa C, Karchev K, Weniger C (2022) Estimating the warm dark matter mass from strong lensing images with truncated marginal neural ratio estimation. Mon Not R Astron Soc. https://doi.org/10.1093/mnras/stac3215
Moustakas LA, Metcalf RB (2003) Detecting dark matter substructure spectroscopically in strong gravitational lenses. Mon Not R Astron Soc 339(3):607–615. https://doi.org/10.1046/j.1365-8711.2003.06055.x
Mukherjee S, Koopmans LVE, Metcalf RB, Tortora C, Schaller M, Schaye J, Vernardos G, Bellagamba F (2021) SEAGLE - II. Constraints on feedback models in galaxy formation from massive early-type strong-lens galaxies. Mon Not R Astron Soc 504(3):3455–3477. https://doi.org/10.1093/mnras/stab693
Müller-Sánchez F, Prieto MA, Hicks EKS, Vives-Arias H, Davies RI, Malkan M, Tacconi LJ, Genzel R (2011) Outflows from active galactic nuclei: kinematics of the narrow-line and coronal-line regions in seyfert galaxies. Astrophys J 739(2):69. https://doi.org/10.1088/0004-637X/739/2/69
Myers ST, Jackson NJ, Browne IWA, de Bruyn AG, Pearson TJ, Readhead ACS, Wilkinson PN, Biggs AD, Blandford RD, Fassnacht CD, Koopmans LVE, Marlow DR, McKean JP, Norbury MA, Phillips PM, Rusin D, Shepherd MC, Sykes CM (2003) The cosmic lens all-sky survey - I. Source selection and observations. Mon Not R Astron Soc 341:1–12. https://doi.org/10.1046/j.1365-8711.2003.06256.x
Nadler EO, Banerjee A, Adhikari S, Mao Y-Y, Wechsler RH (2020) Signatures of velocity-dependent dark matter self-interactions in Milky Way-mass halos. Astrophys J 896(2):112. https://doi.org/10.3847/1538-4357/ab94b0
Nadler EO, Birrer S, Gilman D, Wechsler RH, Du X, Benson A, Nierenberg AM, Treu T (2021) Dark matter constraints from a unified analysis of strong gravitational lenses and Milky Way satellite galaxies. Astrophys J 917(1):7. https://doi.org/10.3847/1538-4357/abf9a3
Nadler EO, Drlica-Wagner A, Bechtol K, Mau S, Wechsler RH, Gluscevic V, Boddy K, Pace AB, Li TS, McNanna M, Riley AH, García-Bellido J, Mao Y-Y, Green G, Burke DL, Peter A, Jain B, Abbott TMC, Aguena M, Allam S, Annis J, Avila S, Brooks D, Carrasco Kind M, Carretero J, Costanzi M, da Costa LN, De Vicente J, Desai S, Diehl HT, Doel P, Everett S, Evrard AE, Flaugher B, Frieman J, Gerdes DW, Gruen D, Gruendl RA, Gschwend J, Gutierrez G, Hinton SR, Honscheid K, Huterer D, James DJ, Krause E, Kuehn K, Kuropatkin N, Lahav O, Maia MAG, Marshall JL, Menanteau F, Miquel R, Palmese A, Paz-Chinchón F, Plazas AA, Romer AK, Sanchez E, Scarpine V, Serrano S, Sevilla-Noarbe I, Smith M, Soares-Santos M, Suchyta E, Swanson MEC, Tarle G, Tucker DL, Walker AR, Wester W (2021) Constraints on dark matter properties from observations of Milky Way satellite galaxies. Phys Rev Lett 126:091101. https://doi.org/10.1103/PhysRevLett.126.091101
Navarro JF, Frenk CS, White SDM (1996) The structure of cold dark matter halos. Astrophys J 462:563. https://doi.org/10.1086/177173
Nierenberg AM, Treu T, Wright SA, Fassnacht CD, Auger MW (2014) Detection of substructure with adaptive optics integral field spectroscopy of the gravitational lens B1422+231. Mon Not R Astron Soc 442(3):2434–2445. https://doi.org/10.1093/mnras/stu862
Nierenberg AM, Treu T, Brammer G, Peter AHG, Fassnacht CD, Keeton CR, Kochanek CS, Schmidt KB, Sluse D, Wright SA (2017) Probing dark matter substructure in the gravitational lens HE 0435-1223 with the WFC3 grism. Mon Not R Astron Soc 471(2):2224–2236. https://doi.org/10.1093/mnras/stx1400
Nierenberg AM, Gilman D, Treu T, Brammer G, Birrer S, Moustakas L, Agnello A, Anguita T, Fassnacht CD, Motta V, Peter AHG, Sluse D (2020) Double dark matter vision: twice the number of compact-source lenses with narrow-line lensing and the WFC3 grism. Mon Not R Astron Soc 492(4):5314–5335. https://doi.org/10.1093/mnras/stz358810.48550/arXiv.1908.06344

Nierenberg A, Bennert VN, Benson A, Birrer S, Djorgovski SG, Du X, Fassnacht C, Gilman D, Hoenig SF, Kusenko A, Malkan MA, Motta V, Moustakas LA, Sluse D, Stern DK, Treu TL (2021) A definitive test of the dark matter paradigm on small scales. JWST Proposal Cycle 1:#2046

Nightingale JW, Dye S (2015) Adaptive semi-linear inversion of strong gravitational lens imaging. Mon Not R Astron Soc 452(3):2940–2959. https://doi.org/10.1093/mnras/stv1455

Nightingale JW, He Q, Cao X, Amvrosiadis A, Etherington A, Frenk CS, Hayes RG, Robertson A, Cole S, Lange S, Li R, Massey R (2022) Scanning for Dark Matter Subhalos in Hubble Space Telescope Imaging of 54 Strong Lenses. ArXiv e-prints, arXiv:2209.10566 [astro-ph.CO]

Nishikawa H, Boddy KK, Kaplinghat M (2020) Accelerated core collapse in tidally stripped self-interacting dark matter halos. Phys Rev D 101(6):063009

Oguri M, Marshall PJ (2010) Gravitationally lensed quasars and supernovae in future wide-field optical imaging surveys. Mon Not R Astron Soc 405(4):2579–2593. https://doi.org/10.1111/j.1365-2966.2010.16639.x

Okabe N, Smith GP, Umetsu K, Takada M, Futamase T (2013) LoCuSS: the mass density profile of massive galaxy clusters at $z = 0.2$. Astrophys J Lett 769:35. https://doi.org/10.1088/2041-8205/769/2/L35

Onions J, Knebe A, Pearce FR, Muldrew SI, Lux H, Knollmann SR, Ascasibar Y, Behroozi P, Elahi P, Han J, Maciejewski M, Merchán ME, Neyrinck M, Ruiz AN, Sgró MA, Springel V, Tweed D (2012) Subhaloes going notts: the subhalo-finder comparison project. Mon Not R Astron Soc 423:1200–1214. https://doi.org/10.1111/j.1365-2966.2012.20947.x

O'Riordan CM, Vegetti S (2024) Angular complexity in strong lens substructure detection. Mon Not R Astron Soc 528(2):1757–1768. https://doi.org/10.1093/mnras/stae153

O'Riordan CM, Despali G, Vegetti S, Lovell MR, Moliné Á (2023) Sensitivity of strong lensing observations to dark matter substructure: a case study with Euclid. Mon Not R Astron Soc 521(2):2342–2356. https://doi.org/10.1093/mnras/stad650

Ostdiek B, Diaz Rivero A, Dvorkin C (2020) Extracting the Subhalo Mass Function from Strong Lens Images with Image Segmentation. Astrophys J 927:83. https://doi.org/10.3847/1538-4357/ac2d8d

Ostdiek B, Diaz Rivero A, Dvorkin C (2022) Image segmentation for analyzing galaxy-galaxy strong lensing systems. Astron Astrophys 657:14. https://doi.org/10.1051/0004-6361/202142030

Paduroiu S (2022) Warm dark matter in simulations. Universe 8(2):76. https://doi.org/10.3390/universe8020076

Patnaik AR, Browne IWA, Wilkinson PN, Wrobel JM (1992) Interferometer phase calibration sources - I. The region $35° \leq \delta \leq 75°$. Mon Not R Astron Soc 254:655. https://doi.org/10.1093/mnras/254.4.655

Pearson TJ, Readhead ACS (1984) Image formation by self-calibration in radio astronomy. Annu Rev Astron Astrophys 22:97–130. https://doi.org/10.1146/annurev.aa.22.090184.000525

Perreault Levasseur L, Hezaveh YD, Wechsler RH (2017) Uncertainties in parameters estimated with neural networks: application to strong gravitational lensing. Astrophys J Lett 850(1):7. https://doi.org/10.3847/2041-8213/aa9704

Petrillo CE, Tortora C, Vernardos G, Koopmans LVE, Verdoes Kleijn G, Bilicki M, Napolitano NR, Chatterjee S, Covone G, Dvornik A, Erben T, Getman F, Giblin B, Heymans C, de Jong JTA, Kuijken K, Schneider P, Shan H, Spiniello C, Wright AH (2019) LinKS: discovering galaxy-scale strong lenses in the kilo-degree survey using convolutional neural networks. Mon Not R Astron Soc 484(3):3879–3896. https://doi.org/10.1093/mnras/stz189

Pillepich A, Springel V, Nelson D, Genel S, Naiman J, Pakmor R, Hernquist L, Torrey P, Vogelsberger M, Weinberger R, Marinacci F (2018) Simulating galaxy formation with the IllustrisTNG model. Mon Not R Astron Soc 473(3):4077–4106. https://doi.org/10.1093/mnras/stx2656

Planck Collaboration, Aghanim N et al (2020a) Planck 2018 results. I. Overview and the cosmological legacy of Planck. Astron Astrophys 641:1. https://doi.org/10.1051/0004-6361/201833880

Planck Collaboration, Aghanim N et al (2020b) Planck 2018 results. VI. Cosmological parameters. Astron Astrophys 641:6. https://doi.org/10.1051/0004-6361/201833910

Portillo SKN, Lee BCG, Daylan T, Finkbeiner DP (2017) Improved point-source detection in crowded fields using probabilistic cataloging. Astron J 154(4):132. https://doi.org/10.3847/1538-3881/aa8565

Powell D, Vegetti S, McKean JP, Spingola C, Rizzo F, Stacey HR (2021) A novel approach to visibility-space modelling of interferometric gravitational lens observations at high angular resolution. Mon Not R Astron Soc 501(1):515–530. https://doi.org/10.1093/mnras/staa2740

Powell DM, Vegetti S, McKean JP, Spingola C, Stacey HR, Fassnacht CD (2022) A lensed radio jet at milliarcsecond resolution I: Bayesian comparison of parametric lens models. Mon Not R Astron Soc 516(2):1808–1828. https://doi.org/10.1093/mnras/stac2350

Powell DM, Vegetti S, McKean JP, White SDM, Ferreira EGM, May S, Spingola C (2023) A lensed radio jet at milli-arcsecond resolution II: Constraints on fuzzy dark matter from an extended gravitational arc. ArXiv e-prints, 2302–10941 https://doi.org/10.48550/arXiv.2302.10941

Press WH, Schechter P (1974) Formation of galaxies and clusters of galaxies by self-similar gravitational condensation. Astrophys J 187:425–438. https://doi.org/10.1086/152650

Press WH, Teukolsky SA, Vetterling WT, Flannery BP (1992) Numerical recipes in C. The art of scientific computing

Rezaei S, McKean JP, Biehl M, de Roo W, Lafontaine A (2022) A machine learning based approach to gravitational lens identification with the international LOFAR telescope. Mon Not R Astron Soc 517(1):1156–1170. https://doi.org/10.1093/mnras/stac2078

Richardson TRG, Stücker J, Angulo RE, Hahn O (2022) Non-halo structures and their effects on gravitational lensing. Mon Not R Astron Soc 511(4):6019–6032. https://doi.org/10.1093/mnras/stac493

Ritondale E, Auger MW, Vegetti S, McKean JP (2019b) Resolving on 100 pc scales the UV-continuum in Lyman-α emitters between redshift 2 and 3 with gravitational lensing. Mon Not R Astron Soc 482(4):4744–4762. https://doi.org/10.1093/mnras/sty2833

Ritondale E, Vegetti S, Despali G, Auger MW, Koopmans LVE, McKean JP (2019a) Low-mass halo perturbations in strong gravitational lenses at redshift z $\sim$ 0.5 are consistent with CDM. Mon Not R Astron Soc 485(2):2179–2193. https://doi.org/10.1093/mnras/stz464

Rizzo F, Vegetti S, Fraternali F, Di Teodoro E (2018) A novel 3D technique to study the kinematics of lensed galaxies. Mon Not R Astron Soc 481(4):5606–5629. https://doi.org/10.1093/mnras/sty2594

Robertson A, Massey R, Eke V, Tulin S, Yu H-B, Bahé Y, Barnes DJ, Bower RG, Crain RA, Dalla Vecchia C, Kay ST, Schaller M, Schaye J (2018) The diverse density profiles of galaxy clusters with self-interacting dark matter plus baryons. Mon Not R Astron Soc 476:20–24. https://doi.org/10.1093/mnrasl/sly024

Robertson A, Harvey D, Massey R, Eke V, McCarthy IG, Jauzac M, Li B, Schaye J (2019) Observable tests of self-interacting dark matter in galaxy clusters: cosmological simulations with SIDM and baryons. Mon Not R Astron Soc 488(3):3646–3662. https://doi.org/10.1093/mnras/stz1815

Robertson A, Massey R, Eke V, Schaye J, Theuns T (2021) The surprising accuracy of isothermal jeans modelling of self-interacting dark matter density profiles. Mon Not R Astron Soc 501(3):4610–4634. https://doi.org/10.1093/mnras/staa3954

Rocha M, Peter AHG, Bullock JS, Kaplinghat M, Garrison-Kimmel S, Oñorbe J, Moustakas LA (2013) Cosmological simulations with self-interacting dark matter - I. Constant-density cores and substructure. Mon Not R Astron Soc 430:81–104. https://doi.org/10.1093/mnras/sts514

Rubin DB (1984) Bayesianly justifiable and relevant frequency calculations for the applied statistician. Ann Stat 12. https://doi.org/10.1214/aos/1176346785

Rubin VC (1991) Evidence for dark matter from rotation curves: ten years later. In: Holt SS, Bennett CL, Trimble V (eds) After the first three minutes. AIP Conference Proceedings, vol 222, pp 371–380. https://doi.org/10.1063/1.40398

Rybak M, McKean JP, Vegetti S, Andreani P, White SDM (2015) ALMA imaging of SDP.81 - I. A pixelated reconstruction of the far-infrared continuum emission. Mon Not R Astron Soc 451:40–44. https://doi.org/10.1093/mnrasl/slv058

Sagunski L, Gad-Nasr S, Colquhoun B, Robertson A, Tulin S (2021) Velocity-dependent self-interacting dark matter from groups and clusters of galaxies. J Cosmol Astropart Phys 2021(1):024. https://doi.org/10.1088/1475-7516/2021/01/024

Sameie O, Creasey P, Yu H-B, Sales LV, Vogelsberger M, Zavala J (2018) The impact of baryonic discs on the shapes and profiles of self-interacting dark matter haloes. Mon Not R Astron Soc 479:359–367. https://doi.org/10.1093/mnras/sty1516

Sameie O, Yu H-B, Sales LV, Vogelsberger M, Zavala J (2020) Self-interacting dark matter subhalos in the Milky Way's tides. Phys Rev Lett 124(14):141102. https://doi.org/10.1103/PhysRevLett.124.141102

Sawala T, Frenk CS, Fattahi A, Navarro JF, Bower RG, Crain RA, Dalla Vecchia C, Furlong M, Jenkins A, McCarthy IG, Qu Y, Schaller M, Schaye J, Theuns T (2015) Bent by baryons: the low-mass galaxy-halo relation. Mon Not R Astron Soc 448:2941–2947. https://doi.org/10.1093/mnras/stu2753

Sawala T, Pihajoki P, Johansson PH, Frenk CS, Navarro JF, Oman KA, White SDM (2017) Shaken and stirred: the Milky Way's dark substructures. Mon Not R Astron Soc 467:4383–4400. https://doi.org/10.1093/mnras/stx360

Schaye J, Crain RA, Bower RG, Furlong M, Schaller M, Theuns T, Dalla Vecchia C, Frenk CS et al (2015) The EAGLE project: simulating the evolution and assembly of galaxies and their environments. Mon Not R Astron Soc 446:521–554. https://doi.org/10.1093/mnras/stu2058

Schive H-Y, Chiueh T, Broadhurst T (2014a) Cosmic structure as the quantum interference of a coherent dark wave. Nat Phys 10(7):496–499. https://doi.org/10.1038/nphys2996

Schive H-Y, Liao M-H, Woo T-P, Wong S-K, Chiueh T, Broadhurst T, Hwang W-YP (2014b) Understanding the core-halo relation of quantum wave dark matter from 3D simulations. Phys Rev Lett 113(26):261302. https://doi.org/10.1103/PhysRevLett.113.261302

Schive H-Y, Chiueh T, Broadhurst T, Huang K-W (2016) Contrasting galaxy formation from quantum wave dark matter, ψDM, with ΛCDM, using Planck and Hubble data. Astrophys J 818(1):89. https://doi.org/10.3847/0004-637X/818/1/89

Schneider A (2015) Structure formation with suppressed small-scale perturbations. Mon Not R Astron Soc 451(3):3117–3130. https://doi.org/10.1093/mnras/stv1169

Schneider A, Smith RE, Macciò AV, Moore B (2012) Non-linear evolution of cosmological structures in warm dark matter models. Mon Not R Astron Soc 424:684–698. https://doi.org/10.1111/j.1365-2966.2012.21252.x

Schuldt S, Suyu SH, Meinhardt T, Leal-Taixé L, Cañameras R, Taubenberger S, Halkola A (2021) HOLISMOKES. IV. Efficient mass modeling of strong lenses through deep learning. Astron Astrophys 646:126. https://doi.org/10.1051/0004-6361/202039574

Şengül AÇ, Dvorkin C (2022) Probing dark matter with strong gravitational lensing through an effective density slope. Mon Not R Astron Soc 516(1):336–357. https://doi.org/10.1093/mnras/stac2256

Sengül AÇ, Dvorkin C, Ostdiek B, Tsang A (2022) Substructure detection reanalysed: dark perturber shown to be a line-of-sight halo. Mon Not R Astron Soc 515(3):4391–4401. https://doi.org/10.1093/mnras/stac1967

Shajib AJ, Vernardos G, Collett TE, Motta V, Sluse D, Williams LLR, Saha P, Birrer S, Spiniello C, Treu T (2024) Space Sci Rev 220

Shen X, Brinckmann T, Rapetti D, Vogelsberger M, Mantz A, Zavala J, Allen SW (2022) X-ray morphology of cluster-mass haloes in self-interacting dark matter. Mon Not R Astron Soc 516(1):1302–1319. https://doi.org/10.1093/mnras/stac2376

Sheth RK, Lemson G (1999) The forest of merger history trees associated with the formation of dark matter haloes. Mon Not R Astron Soc 305(4):946–956. https://doi.org/10.1046/j.1365-8711.1999.02477.x

Sheth RK, Tormen G (1999) Large-scale bias and the peak background split. Mon Not R Astron Soc 308(1):119–126. https://doi.org/10.1046/j.1365-8711.1999.02692.x

Sheth RK, Mo HJ, Tormen G (2001) Ellipsoidal collapse and an improved model for the number and spatial distribution of dark matter haloes. Mon Not R Astron Soc 323:1–12. https://doi.org/10.1046/j.1365-8711.2001.04006.x

Shu Y, Bolton AS, Mao S, Kochanek CS, Pérez-Fournon I, Oguri M, Montero-Dorta AD, Cornachione MA, Marques-Chaves R, Zheng Z, Brownstein JR, Ménard B (2016) The BOSS emission-line lens survey. IV. Smooth lens models for the BELLS GALLERY sample. Astrophys J 833(2):264. https://doi.org/10.3847/1538-4357/833/2/264

Smirnov OM (2011) Revisiting the radio interferometer measurement equation. I. A full-sky Jones formalism. Astron Astrophys 527:106. https://doi.org/10.1051/0004-6361/201016082

Song Y, Sohl-Dickstein J, Kingma DP, Kumar A, Ermon S, Poole B (2020) Score-Based Generative Modeling through Stochastic Differential Equations. ArXiv e-prints, arXiv:2011.13456 [cs.LG]. https://doi.org/10.48550/arXiv.2011.13456

Sonnenfeld A (2022) Statistical strong lensing. III. Inferences with complete samples of lenses. Astron Astrophys 659:132. https://doi.org/10.1051/0004-6361/202142301

Sonnenfeld A, Tortora C, Hoekstra H, Asgari M, Bilicki M, Heymans C, Hildebrandt H, Kuijken K, Napolitano NR, Roy N, Valentijn E, Wright AH (2022) The dark matter halo masses of elliptical galaxies as a function of observationally robust quantities. Astron Astrophys 662:55. https://doi.org/10.1051/0004-6361/202142511

Spilker JS, Marrone DP, Aravena M, Béthermin M, Bothwell MS, Carlstrom JE, Chapman SC, Crawford TM, de Breuck C, Fassnacht CD, Gonzalez AH, Greve TR, Hezaveh Y, Litke K, Ma J, Malkan M, Rotermund KM, Strandet M, Vieira JD, Weiss A, Welikala N (2016) ALMA imaging and gravitational lens models of south pole telescope—selected dusty, star-forming galaxies at high redshifts. Astrophys J 826(2):112. https://doi.org/10.3847/0004-637X/826/2/112

Spingola C, Barnacka A (2020) Constraining VLBI-optical offsets in high redshift galaxies using strong gravitational lensing. Mon Not R Astron Soc 494(2):2312–2326. https://doi.org/10.1093/mnras/staa870

Spingola C, McKean JP, Auger MW, Fassnacht CD, Koopmans LVE, Lagattuta DJ, Vegetti S (2018) SHARP - V. Modelling gravitationally lensed radio arcs imaged with global VLBI observations. Mon Not R Astron Soc 478(4):4816–4829. https://doi.org/10.1093/mnras/sty1326

Spingola C, McKean JP, Lee M, Deller A, Moldon J (2019) A novel search for gravitationally lensed radio sources in wide-field VLBI imaging from the mJIVE-20 survey. Mon Not R Astron Soc 483(2):2125–2153. https://doi.org/10.1093/mnras/sty3189

Springel V, White SDM, Jenkins A, Frenk CS, Yoshida N, Gao L, Navarro J, Thacker R, Croton D, Helly J, Peacock JA, Cole S, Thomas P, Couchman H, Evrard A, Colberg J, Pearce F (2005) Simulations of the formation, evolution and clustering of galaxies and quasars. Nature 435(7042):629–636. https://doi.org/10.1038/nature03597

Springel V, Wang J, Vogelsberger M, Ludlow A, Jenkins A, Helmi A, Navarro JF, Frenk CS, White SDM (2008) The aquarius project: the subhaloes of galactic haloes. Mon Not R Astron Soc 391(4):1685–1711. https://doi.org/10.1111/j.1365-2966.2008.14066.x

Stacey HR, McKean JP, Powell DM, Vegetti S, Rizzo F, Spingola C, Auger MW, Ivison RJ, van der Werf PP (2021) The rocky road to quiescence: compaction and quenching of quasar host galaxies at z $\sim$ 2. Mon Not R Astron Soc 500(3):3667–3688. https://doi.org/10.1093/mnras/staa3433

Storfer C, Huang X, Gu A, Sheu W, Banka S, Dey A, Jain A, Kwon J, Lang D, Lee V, Meisner A, Moustakas J, Myers AD, Tabares-Tarquinio S, Schlafly EF, Schlegel DJ (2022) New Strong Gravitational Lenses from the DESI Legacy Imaging Surveys Data Release 9. ArXiv e-prints, arXiv:2206.02764 [astro-ph.CO]. https://doi.org/10.48550/arXiv.2206.02764

Stücker J, Hahn O, Angulo RE, White SDM (2020) Simulating the complexity of the dark matter sheet I: numerical algorithms. Mon Not R Astron Soc 495(4):4943–4964. https://doi.org/10.1093/mnras/staa1468

Suyu SH, Marshall PJ, Hobson MP, Blandford RD (2006) A Bayesian analysis of regularized source inversions in gravitational lensing. Mon Not R Astron Soc 371(2):983–998. https://doi.org/10.1111/j.1365-2966.2006.10733.x

Suyu SH, Marshall PJ, Blandford RD, Fassnacht CD, Koopmans LVE, McKean JP, Treu T (2009) Dissecting the gravitational lens B1608+656. I. Lens potential reconstruction. Astrophys J 691(1):277–298. https://doi.org/10.1088/0004-637X/691/1/277

Tagore AS, Keeton CR (2014) Statistical and systematic uncertainties in pixel-based source reconstruction algorithms for gravitational lensing. Mon Not R Astron Soc 445(1):694–710. https://doi.org/10.1093/mnras/stu1671

Tavaré S, Balding DJ, Griffiths RC, Donnelly P (1997) Inferring coalescence times from DNA sequence data. Genetics 145(2):505–518. https://doi.org/10.1093/genetics/145.2.505

Tie SS, Kochanek CS (2018) Microlensing makes lensed quasar time delays significantly time variable. Mon Not R Astron Soc 473(1):80–90. https://doi.org/10.1093/mnras/stx2348

Tinker J, Kravtsov AV, Klypin A, Abazajian K, Warren M, Yepes G, Gottlöber S, Holz DE (2008) Toward a halo mass function for precision cosmology: the limits of universality. Astrophys J 688:709–728. https://doi.org/10.1086/591439

Treu T, Koopmans LVE (2004) Massive dark matter halos and evolution of early-type galaxies to $z \approx 1$. Astrophys J 611(2):739–760. https://doi.org/10.1086/422245

Trotter CS, Winn JN, Hewitt JN (2000) A multipole-Taylor expansion for the potential of the gravitational lens MG J0414+0534. Astrophys J 535(2):671–691. https://doi.org/10.1086/308861

Turner B, Van Zandt T (2012) A tutorial on approximate Bayesian computation. J Math Psychol 56:69–85. https://doi.org/10.1016/j.jmp.2012.02.005

Turner HC, Lovell MR, Zavala J, Vogelsberger M (2021) The onset of gravothermal core collapse in velocity-dependent self-interacting dark matter subhaloes. Mon Not R Astron Soc 505(4):5327–5339. https://doi.org/10.1093/mnras/stab1725

van Albada TS, Bahcall JN, Begeman K, Sancisi R (1985) Distribution of dark matter in the spiral galaxy NGC 3198. Astrophys J 295:305–313. https://doi.org/10.1086/163375

Van de Vyvere L, Gomer MR, Sluse D, Xu D, Birrer S, Galan A, Vernardos G (2022) TDCOSMO. VII. Boxyness/discyness in lensing galaxies: detectability and impact on H_0. Astron Astrophys 659:127. https://doi.org/10.1051/0004-6361/202141551

Varma S, Fairbairn M, Figueroa J (2020) Dark matter subhalos, strong lensing and machine learning. ArXiv e-prints, arXiv:2005.05353 [astro-ph.CO]. https://doi.org/10.48550/arXiv.2005.05353

Vattis K, Toomey MW, Koushiappas SM (2021) Deep learning the astrometric signature of dark matter substructure. Phys Rev D 104(12):123541. https://doi.org/10.1103/PhysRevD.104.123541

Vegetti S, Koopmans LVE (2009) Bayesian strong gravitational-lens modelling on adaptive grids: objective detection of mass substructure in galaxies. Mon Not R Astron Soc 392(3):945–963. https://doi.org/10.1111/j.1365-2966.2008.14005.x

Vegetti S, Koopmans LVE (2009) Statistics of mass substructure from strong gravitational lensing: quantifying the mass fraction and mass function. Mon Not R Astron Soc 400(3):1583–1592. https://doi.org/10.1111/j.1365-2966.2009.15559.x

Vegetti S, Czoske O, Koopmans LVE (2010) Quantifying dwarf satellites through gravitational imaging: the case of SDSSJ120602.09+514229.5. Mon Not R Astron Soc 407(1):225–231. https://doi.org/10.1111/j.1365-2966.2010.16952.x

Vegetti S, Koopmans LVE, Bolton A, Treu T, Gavazzi R (2010) Detection of a dark substructure through gravitational imaging. Mon Not R Astron Soc 408(4):1969–1981. https://doi.org/10.1111/j.1365-2966.2010.16865.x

Vegetti S, Lagattuta DJ, McKean JP, Auger MW, Fassnacht CD, Koopmans LVE (2012) Gravitational detection of a low-mass dark satellite galaxy at cosmological distance. Nature 481(7381):341–343. https://doi.org/10.1038/nature10669

Vegetti S, Koopmans LVE, Auger MW, Treu T, Bolton AS (2014) Inference of the cold dark matter substructure mass function at z = 0.2 using strong gravitational lenses. Mon Not R Astron Soc 442(3):2017–2035. https://doi.org/10.1093/mnras/stu943

Vegetti S, Despali G, Lovell MR, Enzi W (2018) Constraining sterile neutrino cosmologies with strong gravitational lensing observations at redshift z $\sim$ 0.2. Mon Not R Astron Soc 481(3):3661–3669. https://doi.org/10.1093/mnras/sty2393

Vernardos G, Koopmans LVE (2022) The very knotty lenser: exploring the role of regularization in source and potential reconstructions using Gaussian process regression. Mon Not R Astron Soc 516(1):1347–1372. https://doi.org/10.1093/mnras/stac1924

Vernardos G, Tsagkatakis G, Pantazis Y (2020) Quantifying the structure of strong gravitational lens potentials with uncertainty-aware deep neural networks. Mon Not R Astron Soc 499(4):5641–5652. https://doi.org/10.1093/mnras/staa3201

Viel M, Markovič K, Baldi M, Weller J (2012) The non-linear matter power spectrum in warm dark matter cosmologies. Mon Not R Astron Soc 421(1):50–62. https://doi.org/10.1111/j.1365-2966.2011.19910.x

Villasenor B, Robertson B, Madau P, Schneider E (2023) New Constraints on Warm Dark Matter from the Lyman-α Forest Power Spectrum. Phys Rev D 108:023502. https://doi.org/10.1103/PhysRevD.108.023502

Vogelsberger M, Zavala J, Loeb A (2012) Subhaloes in self-interacting galactic dark matter haloes. Mon Not R Astron Soc 423:3740–3752. https://doi.org/10.1111/j.1365-2966.2012.21182.x

Vogelsberger M, Genel S, Springel V, Torrey P, Sijacki D, Xu D, Snyder G, Nelson D, Hernquist L (2014b) Introducing the illustris project: simulating the coevolution of dark and visible matter in the universe. Mon Not R Astron Soc 444:1518–1547. https://doi.org/10.1093/mnras/stu1536

Vogelsberger M, Zavala J, Simpson C, Jenkins A (2014a) Dwarf galaxies in CDM and SIDM with baryons: observational probes of the nature of dark matter. Mon Not R Astron Soc 444:3684–3698. https://doi.org/10.1093/mnras/stu1713

Vogelsberger M, Zavala J, Cyr-Racine F-Y, Pfrommer C, Bringmann T, Sigurdson K (2016) ETHOS - an effective theory of structure formation: dark matter physics as a possible explanation of the small-scale CDM problems. Mon Not R Astron Soc 460:1399–1416. https://doi.org/10.1093/mnras/stw1076

Vogelsberger M, Zavala J, Schutz K, Slatyer TR (2019) Evaporating the Milky Way halo and its satellites with inelastic self-interacting dark matter. Mon Not R Astron Soc 484(4):5437–5452. https://doi.org/10.1093/mnras/stz340

Wagner-Carena S, Aalbers J, Birrer S, Nadler EO, Darragh-Ford E, Marshall PJ, Wechsler RH (2023) From images to dark matter: end-to-end inference of substructure from hundreds of strong gravitational lenses. Astrophys J 942(2):75. https://doi.org/10.3847/1538-4357/aca525

Warren SJ, Dye S (2003) Semilinear gravitational lens inversion. Astrophys J 590(2):673–682. https://doi.org/10.1086/375132

White SDM, Frenk CS, Davis M (1983) Clustering in a neutrino-dominated universe. Astrophys J Lett 274:1–5. https://doi.org/10.1086/184139

Wilkinson PN, Browne IWA, Patnaik AR, Wrobel JM, Sorathia B (1998) Interferometer phase calibration sources - III. The regions $+20 \leq \delta_{\rm B1950} \leq +35^{\circ}$ and $+75 \leq \delta_{\rm B1950} \leq +90^{\circ}$. Mon Not R Astron Soc 300(3):790–816. https://doi.org/10.1046/j.1365-8711.1998.01941.x

Wizinowich P, Lu JR, Cetre S, Chin J, Correia C, Delorme J-R, Gers L, Lilley S, Lyke J, Marin E, Ragland S, Richards P, Surendran A, Wetherell E, Chen C-F, Chu D, Do T, Fassnacht C, Freeman M, Gautam A, Ghez A, Hunter L, Jones T, Liu MC, Mawet D, Max C, Morris M, Phillips M, Ruffio J-B, Rundquist N-E, Sabhlok S, Terry S, Treu T, Wright S (2022) Keck all sky precision adaptive optics program overview. In: Schreiber L, Schmidt D, Vernet E (eds) Adaptive optics systems VIII. Society of Photo-Optical Instrumentation Engineers (SPIE) conference series, vol 12185, p 121850. https://doi.org/10.1117/12.2628275

Wucknitz O (2004) LENSCLEAN revisited. Mon Not R Astron Soc 349(1):1–13. https://doi.org/10.1111/j.1365-2966.2004.07513.x

Xu DD, Mao S, Wang J, Springel V, Gao L, White SDM, Frenk CS, Jenkins A, Li G, Navarro JF (2009) Effects of dark matter substructures on gravitational lensing: results from the Aquarius simulations. Mon Not R Astron Soc 398(3):1235–1253. https://doi.org/10.1111/j.1365-2966.2009.15230.x

Xu DD, Mao S, Cooper AP, Wang J, Gao L, Frenk CS, Springel V (2010) Substructure lensing: effects of galaxies, globular clusters and satellite streams. Mon Not R Astron Soc 408(3):1721–1729. https://doi.org/10.1111/j.1365-2966.2010.17235.x

Xu DD, Mao S, Cooper AP, Gao L, Frenk CS, Angulo RE, Helly J (2012) On the effects of line-of-sight structures on lensing flux-ratio anomalies in a ΛCDM universe. Mon Not R Astron Soc 421(3):2553–2567. https://doi.org/10.1111/j.1365-2966.2012.20484.x

Xu D, Sluse D, Gao L, Wang J, Frenk C, Mao S, Schneider P, Springel V (2015) How well can cold dark matter substructures account for the observed radio flux-ratio anomalies. Mon Not R Astron Soc 447(4):3189–3206. https://doi.org/10.1093/mnras/stu2673

Yang D, Nadler EO, Yu H-B (2023) Strong dark matter self-interactions diversify halo populations within and surrounding the Milky Way. Astrophys J 949:67. https://doi.org/10.3847/1538-4357/acc73e

Yao-Yu Lin J, Yu H, Morningstar W, Peng J, Holder G (2020) Hunting for Dark Matter Subhalos in Strong Gravitational Lensing with Neural Networks. ArXiv e-prints, arXiv:2010.12960 [astro-ph.CO]. https://doi.org/10.48550/arXiv.2010.12960

Yavetz TD, Li X, Hui L (2022) Construction of wave dark matter halos: numerical algorithm and analytical constraints. Phys Rev D 105(2):023512. https://doi.org/10.1103/PhysRevD.105.023512

Zavala J, Frenk CS (2019) Dark matter haloes and subhaloes. Galaxies 7(4):81. https://doi.org/10.3390/galaxies7040081

Zel'dovich YB (1970) Gravitational instability: an approximate theory for large density perturbations. Astron Astrophys 5:84–89

Zelko IA, Treu T, Abazajian KN, Gilman D, Benson AJ, Birrer S, Nierenberg AM, Kusenko A (2022) Constraints on sterile neutrino models from strong gravitational lensing, Milky Way satellites, and the Lyman-α forest. Phys Rev Lett 129(19):191301. https://doi.org/10.1103/PhysRevLett.129.191301

Zeng ZC, Peter AHG, Du X, Benson A, Kim S, Jiang F, Cyr-Racine F-Y, Vogelsberger M (2022) Core-collapse, evaporation, and tidal effects: the life story of a self-interacting dark matter subhalo. Mon Not R Astron Soc 513(4):4845–4868. https://doi.org/10.1093/mnras/stac1094

Zentner AR, Bullock JS (2003) Halo substructure and the power spectrum. Astrophys J 598(1):49–72. https://doi.org/10.1086/378797

Zwicky F (1933) Die Rotverschiebung von extragalaktischen Nebeln. Helv Phys Acta 6:110–127

Publisher's Note Springer Nature remains neutral with regard to jurisdictional claims in published maps and institutional affiliations.

Authors and Affiliations

S. Vegetti[1] · S. Birrer[2] · G. Despali[3,4] · C.D. Fassnacht[5] · D. Gilman[6] · Y. Hezaveh[7,8,9,10,11] · L. Perreault Levasseur[7,8,9,10,11] · J.P. McKean[12,13,14] · D.M. Powell[1] · C.M. O'Riordan[1] · G. Vernardos[15]

✉ S. Vegetti
svegetti@mpa-garching.mpg.de

1 Max Planck Institut für Astrophysik, Karl-Schwarzschild-Strasse 1, D-85748 Garching bei München, Germany

2 Department of Physics and Astronomy, Stony Brook University, Stony Brook, NY 11794, USA

3 Dipartimento di Fisica e Astronomia "A. Righi", Alma Mater Studiorum Università di Bologna, Via Piero Gobetti 93/2, I-40129 Bologna, Italy

4 Institut für Theoretische Astrophysik, Zentrum für Astronomie, Heidelberg Universität, Albert-Ueberle-Str. 2, 69120, Heidelberg, Germany

5 Dept. of Physics and Astronomy, University of California, Davis, 1 Shields Ave. Davis, CA 95616, USA

6 Department of Astronomy and Astrophysics, University of Toronto, Toronto, ON, M5S 3H4, Canada

7 Department of Physics, Université de Montréal, Montréal, Canada

8 Mila - Quebec Artificial Intelligence Institute, Montréal, Canada

9 Ciela - Montreal Institute for Astrophysical Data Analysis and Machine Learning, Montréal, Canada

10 Center for Computational Astrophysics, Flatiron Institute, 162 5th Avenue, 10010, New York, NY, USA

11 Perimeter Institute for Theoretical Physics, Waterloo, Ontario, Canada, N2L 2Y5

[12] Kapteyn Astronomical Institute, University of Groningen, PO Box 800, NL-9700 AV Groningen, The Netherlands

[13] South African Radio Astronomy Observatory (SARAO), P.O. Box 443, Krugersdorp 1740, South Africa

[14] Department of Physics, University of Pretoria, Lynnwood Road, Hatfield, Pretoria, 0083, South Africa

[15] Institute of Physics, Laboratory of Astrophysique, École Polytechnique Fédérale de Lausanne (EPFL), Observatoire de Sauverny, 1290 Versoix, Switzerland

Space Science Reviews (2024) 220:14
https://doi.org/10.1007/s11214-024-01043-8

Microlensing of Strongly Lensed Quasars

G. Vernardos[1,2,3] · D. Sluse[4] · D. Pooley[5,6] · R.W. Schmidt[7] · M. Millon[1,8] · L. Weisenbach[9] · V. Motta[10] · T. Anguita[11,12] · P. Saha[13] · M. O'Dowd[2,3,14] · A. Peel[1] · P.L. Schechter[15,16]

Received: 27 April 2023 / Accepted: 30 December 2023 / Published online: 7 February 2024
© The Author(s) 2024

Abstract
Strong gravitational lensing of quasars has the potential to unlock the poorly understood physics of these fascinating objects, as well as serve as a probe of the lensing mass distribution and of cosmological parameters. In particular, gravitational microlensing by compact bodies in the lensing galaxy can enable mapping of quasar structure to $< 10^{-6}$ arcsec scales. Some of this potential has been realized over the past few decades, however the upcoming era of large sky surveys promises to bring this promise to full fruition. In this article, we review the theoretical framework of this field, describe the prominent current methods for parameter inference from quasar microlensing data across different observing modalities, and discuss the constraints so far derived on the geometry and physics of quasar inner structure. We also review the application of strong lensing and microlensing to constraining the granularity of the lens potential, i.e. the contribution of the baryonic and dark matter components, and the local mass distribution in the lens, i.e. the stellar mass function. Finally, we discuss the future of the field, including the new possibilities that will be opened by the next generation of large surveys and by new analysis methods now being developed.

Keywords Gravitational lensing: strong · Gravitational lensing: micro

1 Introduction

Microlensing is a phenomenon that allows us to probe very small spatial and mass scales in the Universe that are otherwise inaccessible by current and next-generation instruments. The underlying physical principles of gravitational light deflection are the same as in the case of strong lensing by galaxies and clusters, however, the main difference is that these deflections happen in such small angular scales ($\sim 10^{-6}$ arcsec) that the resulting multiple images of a source are unresolved. The only remaining quantity with an observable effect is the total magnification, i.e. the total flux received from all the unresolved microimages (the centroid of the images can be affected too, but this requires higher sensitivity to detect). A demonstration of the phenomenon is shown in Fig. 1. Given our solid theoretical understanding of microlensing, we can use it as a tool that is, in many ways, more powerful than any telescope we could build in the next several decades. With this tool, we can learn both about the background source and the lens itself (the massive galaxy along the line of sight). In this case, the type of sources that we examine are quasars, although microlensing

Extended author information available on the last page of the article

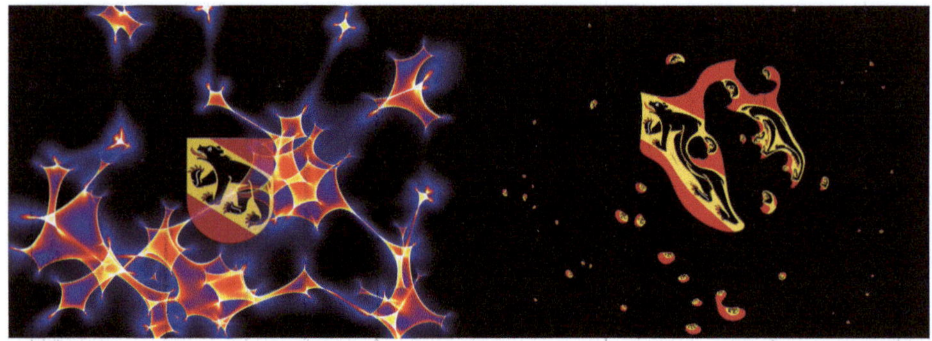

Fig. 1 Illustration of the effect of microlensing on a source, in this case, the coat-of-arms of Bern, Switzerland, the seat of the International Space Science Institute, where the "Strong Gravitational Lensing" workshop was held between 18-22 of July 2022 that lead to the writing of this review. *Right*: A collection of microlenses on the image plane (the same as the one shown in Fig. 4) produces magnified and distorted micro-images of the source. *Left*: The corresponding magnification pattern is shown on the source plane (the same as the one shown in Fig. 10). For an interactive version of this figure visit: https://austinpeel.github.io/gladius-microlensing/

can also affect lensed supernovae (see Suyu et al. 2024) and other more 'exotic' sources like Fast Radio Bursts (Lewis 2020), Gamma-Ray Bursts (Mao 1993), and gravitational waves (Diego et al. 2019). In this review, we present the state-of-the-art of the field of quasar microlensing and expand on previous works by Schneider et al. (1992), Schneider et al. (2006), and Schmidt and Wambsganss (2010).

The first discovery of a gravitational lens, the doubly imaged quasar Q 0957+561 (Walsh et al. 1979), was followed by the pioneering works of Chang and Refsdal (1979), Gott (1981), Young (1981), and Chang and Refsdal (1984) who suggested that light from the multiple images could be further affected by the presence of stellar mass objects near the line of sight. Indeed, microlensing was confirmed a decade later in the quadruply lensed Q 2237+0305 (Irwin et al. 1989), also known as Huchra's lens (Huchra et al. 1985) or "The Einstein Cross", shown in Fig. 2. These, and the studies of Paczynski (1986) and Kayser et al. (1986), set the foundations of the field of quasar microlensing.

Although throughout this review we refer to the "quasar" as if it was a single, well-defined object, a quasar, or Active Galactic Nucleus (AGN, we use both terms indistinctly), is a composite of several different structures around a Super-Massive Black Hole (SMBH). Each of these regions produces its own signature radiation and together they sum up to give an astrophysical object that is bright across the whole electromagnetic spectrum. Through decades of dedicated work (Netzer 2015; Padovani et al. 2017), a general picture of the various quasar regions has been inferred. However, given the small physical scales around the black hole ($\sim$ ld) and the large distances to these objects ($z \gtrsim 0.5$), the dimensions and exact geometry of these regions remain largely unresolved by any current or planned telescope. With microlensing, we can *measure* and place direct constraints on the sizes and locations of the emitting regions of a quasar at a cosmological distance on scales from micro- to nano-arcsec. In Fig. 3 we summarize our current understanding of quasar structure and highlight the microlensing effect on its different components.

What we can learn about the composition of galaxies from microlensing is as important as it is inaccessible by any other means. While every astronomer knows that a galaxy is "a gravitationally bound system of stars, stellar remnants, interstellar gas, dust, and dark matter", establishing the relative proportions of each component is one of the great challenges

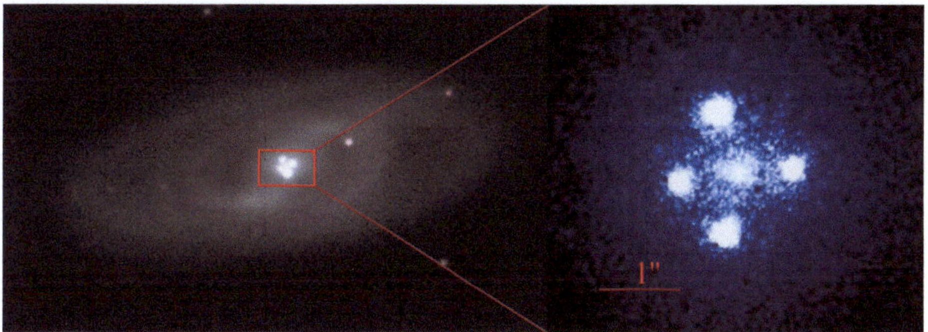

Fig. 2 Multiply imaged quasar Q 2237+0305. *Left*: the lensed quasar seen through the central regions of a foreground spiral galaxy-lens (credits: J. Rhoads, S. Malhotra, I. Dell'Antonio, NOAO/WIYN/NSF). *Right*: zoom on the bulge of the galaxy-lens seen in the centre, surrounded by multiple images of the background quasar produced by gravitational lensing (credits: NASA, ESA, and STScI)

of extragalactic astronomy. In particular, the contribution of either dark matter and stellar mass is subject to much disagreement and uncertainty (see Sect. 4 in Shajib et al. 2024). For example, there are important works that measure stellar mass-to-light ratios for elliptical galaxies (see Cappellari 2016, and references therein), but buried within these papers the authors invariably caution that uncertainties in the faint end of the stellar mass function, where the stars are practically invisible, render their results uncertain by a factor of two. Because microlensing is sensitive only to mass, it can be used to determine the amount of mass in individual stars, including stellar remnants, brown dwarfs, and red dwarfs that are too faint to produce any photometric or spectroscopic signatures. In other words, microlensing is an independent and direct method to determine the graininess of the gravitational potential.

In order to prepare the unfamiliar reader with the data and methods used in microlensing studies, in the remaining of this section we summarise some general theoretical and observational properties. In Sect. 2 we set the foundations of the theory of microlensing as well as our best current theoretical understanding of the structure of quasar emission regions. Section 3 presents the different approaches to analyze how microlensing manifests across time and wavelength. The application of these methods to real data leads to measurements of quasar structure and lens galaxy mass, which are reviewed in Sects. 4 and 5 respectively. We conclude in Sect. 6 with the promising future of quasar microlensing, which will be revolutionized by the upcoming all-sky surveys of this decade, and the challenges that need to be addressed in order to deliver groundbreaking scientific results.

1.1 General Theoretical Properties and Facts

Microlensing is caused by compact astrophysical objects with dimensions much smaller than their Einstein radii,[1] θ_E (i.e. stars and stellar remnants), as opposed to smoothly varying mass distributions over large scales (e.g. galaxies and clusters). In fact, this radius, which is typically of the order of 10^{-6} arcsec for lensed quasars (hence the prefix "micro", see also Kayser 1992), is critical because it determines the scale length of the phenomenon. The

[1] Here we assume that the reader is familiar with basic concepts of gravitational lensing, like the Einstein radius, critical lines, and caustics – see Saha et al. (2024) for an overview.

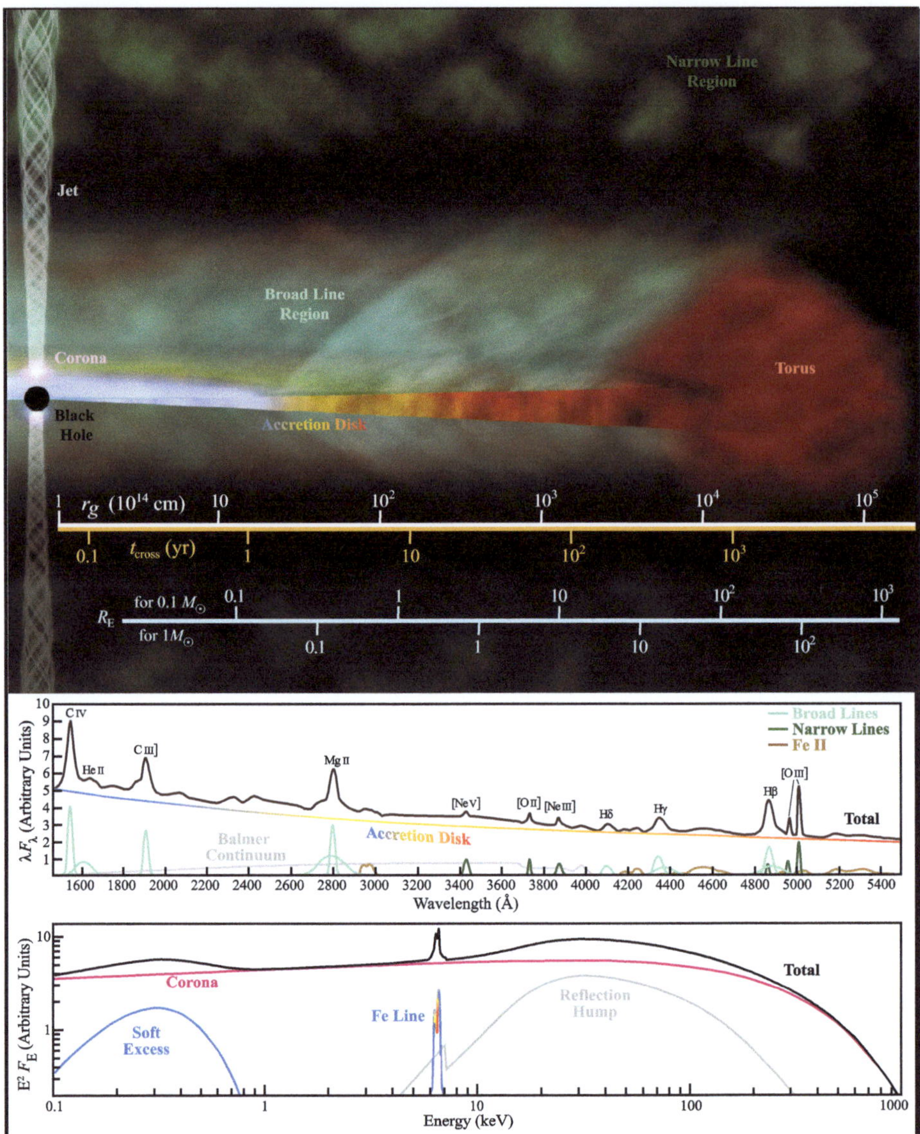

Fig. 3 *Top*: Schematic view of the structure of a quasar with the main regions that can be probed by microlensing color-coded as follows, starting from the smallest/innermost (left to right): X-ray corona (purple), accretion disc (blue to red), Broad Line Region (lighter green), dust torus (red), and Narrow Line Emission (darker green). For completeness, the black hole and the jet are also shown. Different characteristic lengths are shown in logarithmic scale in the bottom of this panel: the gravitational radius, r_g, for a 7×10^8 $M_\odot$ black hole (see Eq. (35)), the corresponding time it takes to cross this length, t_{cross}, for an effective velocity of 500 km/s (see Sect. 3.4.1), and the Einstein radius on the source plane, R_E, for lens and quasar located at redshifts of 0.5 and 2 respectively (see Eq. (7)). The latter is given for two different masses of the microlenses. Inspired from Fig. 1 of Moustakas et al. (2019). *Middle*: Composite UV-optical spectrum of an AGN with a few main emission lines identified. Its different components have been color-coded to match the corresponding regions shown in the top panel. *Bottom*: Same as the middle panel but for the X-rays

rule to remember is that the smaller (larger) a quasar emitting region with respect to θ_E, the stronger (weaker) the amplitude of microlensing.

The optical depth to microlensing is dramatically increased along the line of sight to a quasar that is already strongly lensed by a foreground galaxy (see Fig. 2). Hence, we almost always expect a population of compact objects acting as microlenses and in very few cases single, isolated point masses. Such populations create caustic features of very high magnification, often superimposed in a non-linear fashion, but because flux has to be conserved (e.g. compared to a smooth mass sheet with the same total mass) there need to be large swathes of de-magnified regions on the source plane as well. Depending on where the source lies within this caustic pattern it can be magnified or de-magnified.

Due to the relative velocity of observer, lens, microlenses, and quasar, which is of the order of 1 θ_E per decade (Mosquera and Kochanek 2011), we expect to see significant variations ($\sim$1 mag) of the microlensing magnification as the quasar emitting regions cross caustics and areas of demagnification. Moreover, because the quasar region(s) emitting at any given wavelength can vary in size, there is a chromatic dependence of this (de)magnification. It is important to note that quasars are intrinsically variable objects, therefore, in order to obtain the microlensing signal (in a single epoch or as a function of time) one needs to cleanly subtract this intrinsic variability after taking into account the time delays per multiple image due to the presence of the lensing galaxy (see Birrer et al. 2024).

1.2 Observational Considerations

From an observer's point of view, one may wonder which kind of data are needed in order to use microlensing for one of the above mentioned science applications. The size of the source with respect to θ_E, together with the dynamic nature of the phenomenon may guide the answer to this question. The wavelength range to consider covers almost the whole electromagnetic spectrum, from the innermost X-ray corona and accretion disc to the Broad Line Region (BLR, see Fig. 3), while the time scales when large variations are expected are in general shorter for the most compact regions. When the quasar is crossing a caustic, the magnification increases rapidly with time and therefore nightly cadence is required to capture in detail how the different quasar regions respond. Under a "normally" microlensed quasar, i.e. not undergoing a high magnification event, the daily cadence requirement can be relaxed and weekly observations should suffice for long-term microlensing signals. In practice, the existing observational resources are quite restricted both in wavelength coverage and cadence (e.g. season gaps).

Over the years, widely different observations have been performed to capture the various manifestations of microlensing variability: from decade-long monitoring to multiwavelength snapshots and spectra. The limited resources have resulted in two main strategies, each with its own advantages and drawbacks: either long-term weekly monitoring but, with very few exceptions, in a single band, or multi-wavelength snapshots at a given moment in time. For example, light curves provide more data points to constrain quasar structure and lens mass but require an additional model component (the effective velocity) and a good measurement of the time delay, while snapshots can constrain the accretion disc temperature profile but are prone to other systematic biases and limitations, like contamination from massive substructures (e.g. anomalous flux ratios, see Vegetti et al. 2024) and application to image pairs with negligible, or known, time delays. Although very information-rich on quasar structure, data for high magnification events have so far been scarce due to their rarity and quick evolution. Requiring almost daily cadence, these have been almost exclusively observed in a single system (the Einstein cross).

2 Background and Theory

In this section, we present the theoretical background that is specific to microlensing, as opposed to strong lensing in general. For an overview of the basic principles and theory of lensing we refer the reader to Saha et al. (2024).

2.1 Light Deflection

To study the lensing effect of an ensemble of compact objects in the lens plane on a background quasar, we start again by setting up the lens equation (Saha et al. 2024), but this time for many objects:

$$\boldsymbol{\beta} = \boldsymbol{\theta} - \frac{D_{\rm ds}}{D_{\rm d} D_{\rm s}} \sum_{i=1}^{N} \hat{\boldsymbol{\alpha}}_i(\boldsymbol{\theta}), \tag{1}$$

where we are summing over the deflection angles $\hat{\boldsymbol{\alpha}}_i$ due to N compact objects with masses M_i situated at $\boldsymbol{\theta}_i$ in the lens plane.

In addition to the surface mass density at the quasar image position due to compact objects within a radius θ:

$$\kappa_* = \Sigma_*(<\theta)/\Sigma_{\rm cr}, \tag{2}$$

with $\Sigma_{\rm cr} = \frac{c^2}{4\pi G}\frac{D_{\rm s}}{D_{\rm ds} D_{\rm d}}$ (Saha et al. 2024), we need to include the effect of a constant surface density of matter $\kappa_{\rm c}$ in the lens plane to account for a smooth dark matter component (also expressed in units of $\Sigma_{\rm cr}$). The total surface mass density is then:

$$\kappa = \kappa_* + \kappa_{\rm c}. \tag{3}$$

Finally, we need to consider the effect of shear due to the tidal field of the lensing galaxy at the position of the quasar image (Saha et al. 2024). Using Einstein's formula for the deflection (Saha et al. 2024) the lens equation due to N compact masses M_i in the presence of a mass sheet $\kappa_{\rm c}$ and shear $\gamma = \sqrt{\gamma_1^2 + \gamma_2^2}$ becomes:

$$\boldsymbol{\beta} = \begin{pmatrix} 1 - \kappa_{\rm c} - \gamma_1 & -\gamma_2 \\ -\gamma_2 & 1 - \kappa_{\rm c} + \gamma_1 \end{pmatrix} \boldsymbol{\theta} + \frac{4G}{c^2}\frac{D_{\rm ds}}{D_{\rm d} D_{\rm s}} \sum_{i=1}^{N} M_i \frac{\boldsymbol{\theta}_i - \boldsymbol{\theta}}{|\boldsymbol{\theta}_i - \boldsymbol{\theta}|^2}. \tag{4}$$

For a homogeneous mass distribution with surface density κ_*, the second term becomes $\propto -\kappa_* \boldsymbol{\theta}$. This equation can also be written as:

$$\boldsymbol{\beta} = \begin{pmatrix} 1 - \kappa_{\rm c} - \gamma_1 & -\gamma_2 \\ -\gamma_2 & 1 - \kappa_{\rm c} + \gamma_1 \end{pmatrix} \boldsymbol{\theta} + \sum_{i=1}^{N} \theta_{{\rm E}i}^2 \frac{\boldsymbol{\theta}_i - \boldsymbol{\theta}}{|\boldsymbol{\theta}_i - \boldsymbol{\theta}|^2}, \tag{5}$$

where we used the Einstein radii $\theta_{{\rm E}i}$ of the compact objects in the lens plane (Saha et al. 2024). This is a fundamental quantity for microlensing as it defines the scale length of the phenomenon. Thus, we define it here again in the image plane, as in the equation above:

$$\theta_{\rm E} = \sqrt{\frac{4GM}{c^2}\frac{D_{ds}}{D_d D_s}} \tag{6}$$

and its projection on the source plane:

$$R_{\rm E} = \sqrt{\frac{4GM}{c^2}\frac{D_s D_{ds}}{D_d}}, \tag{7}$$

in units of length.

How can we distribute N masses such that a certain κ_* is achieved? From the homogeneous mass case $\boldsymbol{\beta} \propto -\kappa_* \boldsymbol{\theta}$ (Eq. (4)) and $\boldsymbol{\beta} \propto -(\sum_{i=1}^{N} \theta_{{\rm E}_i}^2/\theta^2)\,\boldsymbol{\theta}$ in Eq. (5), it can be derived that the normalized surface density in compact objects κ_* equals the summed "Einstein circles" divided by the area in which the point masses are distributed:

$$\kappa_* = \frac{\pi {\theta_{\rm E}}^2}{\pi \theta^2} \sum_{i=1}^{N} m_i = \frac{\pi \sum_{i=1}^{N} m_i}{A_{enc}} \tag{8}$$

with $m_i = M_i/{\rm M}_\odot$. In the last term, the area A_{enc}, in which the N point masses are distributed, has been normalized by the Einstein radius for one solar mass $\theta_{\rm E}^2$ squared. To create a mass ensemble of compact objects with surface density κ_*, one needs to distribute point masses with summed masses m_i:

$$\sum_{i=1}^{N} m_i = \frac{\kappa_* A_{enc}}{\pi} \tag{9}$$

in the area A_{enc}.

It is customary to also rewrite the microlensing mapping from Eq. (5) in a normalized way by dividing by $(1-\kappa_{\rm c})/\sqrt{|1-\kappa_{\rm c}|}$ (for $\kappa_{\rm c} \neq 1$). Then, one can define the normalized lens-plane and source-plane coordinates:

$$\boldsymbol{z} = \frac{\boldsymbol{\theta}}{\theta_{\rm E}/\sqrt{|1-\kappa_{\rm c}|}} \quad \text{and} \quad \boldsymbol{\zeta} = \frac{1}{1-\kappa_{\rm c}} \frac{\boldsymbol{\beta}}{\theta_{\rm E}/\sqrt{|1-\kappa_{\rm c}|}}. \tag{10}$$

The normalized lens equation follows as (Paczynski 1986; Kayser et al. 1986):

$$\boldsymbol{\zeta} = \begin{pmatrix} 1-g_1 & -g_2 \\ -g_2 & 1+g_1 \end{pmatrix} \boldsymbol{z} + sign\left(\frac{\kappa_*}{1-\kappa_{\rm c}}\right) \sum_{i=1}^{N} m_i \frac{\boldsymbol{z}_i - \boldsymbol{z}}{|\boldsymbol{z}_i - \boldsymbol{z}|^2}, \tag{11}$$

where g_1 and g_2 are called the components of the reduced shear (see Saha et al. 2024). Employing the signum function is due to Paczynski (1986), who used it because it shows that only the normalized surface mass density $\kappa_*/(1-\kappa_{\rm c})$ and the reduced shear g are needed to describe a quasar microlensing situation (note that $sign(\kappa_*) > 0$ always). We note that these two quantities, normalized surface mass density and reduced shear, are also known as "effective" convergence and shear and are further discussed in Sect. 2.8.

Often in the literature, the compact-object surface density κ_* and the smooth surface density $\kappa_{\rm c}$ are both given in addition to the shear γ. It should be noted, however, that the above described degeneracy between the three quantities holds (e.g. Saha 2000). In the case of $\kappa_{\rm c} > 1$ (sometimes called over-focussing) the deflection due to the individual masses is formally counted as repulsive. However, as Paczynski put it this is just a "trick" to make equation Eq. (11) simpler. In the following we shall assume $\kappa_{\rm c} < 1$ for simplicity because this is the most common scenario for observations (but see also Dobler et al. 2007).

2.2 Magnification for Ensembles of Point Masses

The magnification of a microlensed image is denoted by the symbol μ. An image can be magnified, $|\mu| > 1$, or demagnified, $|\mu| < 1$. In addition the image can be mirror-inverted, which corresponds to negative magnification $\mu < 0$. The sign of the magnification μ is also called parity (see also Saha et al. 2024).

For the lens mapping described by the lens equation introduced above, the magnification of a point source is given by the inverse of the Jacobi determinant (see Saha et al. 2024):

$$A = \det J = \det \left(\frac{\partial \zeta}{\partial z} \right). \tag{12}$$

The total magnification can be found by taking the sum:

$$\mu_{\rm tot} = \sum_{j=1}^{M} \frac{1}{|A|_{z=z_j}}, \tag{13}$$

over all M images at locations $\boldsymbol{z}_j$, or $\boldsymbol{\theta}_j$, on the lens plane. Calculating the magnification for extended sources, such as a quasar accretion disc, can proceed by splitting up the source into sub-sources (e.g. Witt and Mao 1994). However, in Sect. 2.7 it is shown how to calculate the magnification using the ray-shooting method. This latter technique is the most widely used approach.

As an example, the magnification μ for a constant sheet of matter $\kappa_{\rm c} > 0$ with shear γ is given by (see also Saha et al. 2024):

$$\mu = \frac{1}{(1-\kappa_{\rm c})^2 - \gamma^2}. \tag{14}$$

We note that this is often referred to as the "macro-magnification" because $\kappa_{\rm c}$ and γ can be attributed to the lensing galaxy's macroscopic potential at the location of the multiple images. It can be seen from this equation that for positive parity images the magnification is always $|\mu| > 1$. For negative parity the image can be magnified, $|\mu| > 1$, or demagnified, $|\mu| < 1$.

2.3 Complex Notation for Microlensing

The normalized lens equation Eq. (11) describes a mapping from the lens plane $\boldsymbol{z}$ to the source plane $\boldsymbol{\zeta}$. Starting with Bourassa et al. (1973), Bourassa and Kantowski (1975), complex numbers were used to describe this mapping. While they partly treated real and imaginary parts separately, Witt (1990) introduced a consequent complex notation.

Normalized positions in the lens plane are represented by $z = x + iy$ and in the source plane by $\zeta = \xi + i\eta$. The shear can be described as a complex quantity a well, $g = g_1 + ig_2 = |g| \exp^{2i\varphi}$. Since for complex numbers $z/|z|^2 = 1/\bar{z}$ (the bar indicates the complex conjugate), the normalized lens equation Eq. (11) can be re-written as:

$$\zeta = z - g\bar{z} + \sum_{i=1}^{N} \frac{m_i}{\bar{z}_i - \bar{z}}. \tag{15}$$

The complex formulation has the advantage that it can be well treated by a computer language that can deal with complex numbers (e.g. Fortran, C, python). For example, the magnification of an image at z can be calculated using the inverse of the complex version of Eq. (12) (Witt 1990):

$$\det J = \left(\frac{\partial \zeta}{\partial z}\right)^2 - \frac{\partial \zeta}{\partial \bar{z}}\overline{\frac{\partial \zeta}{\partial \bar{z}}}$$

$$= 1 - \left(-g + \sum_{i=1}^{N} \frac{m_i}{(\bar{z}_i - \bar{z})^2}\right)\left(-\bar{g} + \sum_{i=1}^{N} \frac{m_i}{(z_i - z)^2}\right). \tag{16}$$

In the last step, the corresponding derivatives of Eq. (15) were calculated with respect to z, $\bar{z}$ rather than the more usual pair x, y (i.e. the complex plane) are used. They have the same properties as "normal" derivatives (i.e. linearity, product rule, chain rule, conjugation), but can be used more efficiently with the complex lens equation.

2.4 Finding the Micro-Images

The micro-images of a point source at a given position produced by a large number of point mass lenses can be found using explicit search algorithms in the lens plane (e.g. Paczynski 1986; Saha and Williams 2011). This can be done very efficiently using the complex notation.

The maximal number of micro-images that can be observed of a point source at ζ has been studied thoroughly (Witt 1990, 1991; Petters 1992; Petters et al. 2001). Witt (1990) has shown that all images of a point source at ζ can be calculated by complex conjugation of Eq. (15), by multiplying with $\prod_{i=1}^{N}(z_i - z)$, and by re-inserting Eq. (15) for z. The result is a large polynomial of degree $(N+1)^2$ that only depends on $\bar{z}$. Taking the conjugate of the complex roots of this polynomial yields all possible $(N+1)^2$ solutions – for $g = 0$ the maximal number of images is $N^2 + 1$. Not all of those solutions, however, also satisfy the real-valued lens-equation Eq. (11) and a check needs to be performed. This polynomial-technique to determine the solutions of the normalized lens equation is very successfully used in the field of planetary microlensing, where N is only a few (binary, triple system etc, e.g. Bozza 2010).

Another elegant procedure to find all images corresponding to a point source is due to Witt (1993) (see also Lewis et al. 1993, for a different implementation of the same idea). Witt (1993) shows that all images of a point source can be found by following a straight source track from far away from the ensemble of point masses to the position of interest ζ:

- For a source position ζ_0 very far away from the ensemble of N point masses there exists one image close to the source and N images close to the masses m_i.
- To find these images one first needs to solve the lens equation iteratively to find the far image (except for $|g| > 1$, see Witt 1993):

$$z_{[k+1]} = \zeta_0 + g\,\bar{z}_{[k]} - \sum_{i=1}^{N} \frac{m_i}{\bar{z}_i - \bar{z}_{[k]}}, \tag{17}$$

where k is the number of iteration.

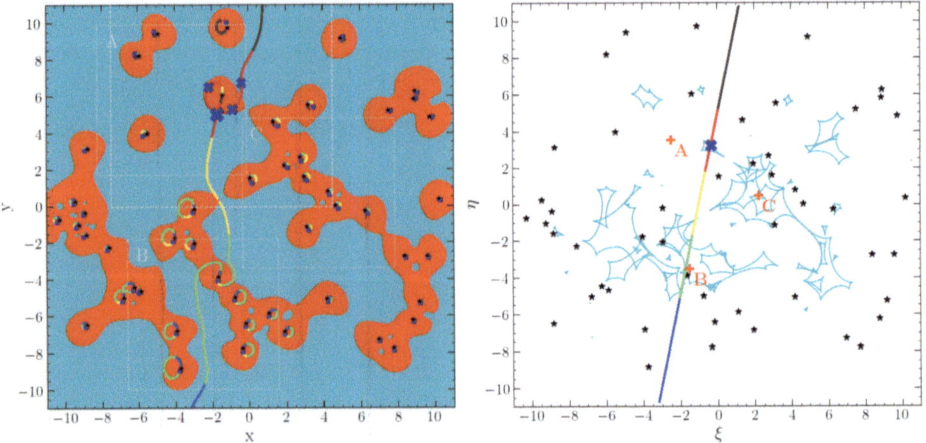

Fig. 4 *Left*: Example images of a straight line lensed by an ensemble of 50 point masses (star symbols) distributed randomly within a circle of radius 11.2 θ_E without shear. *Right*: corresponding source plane and caustic curves. Each image of the line can consist of multiply imaged parts, as indicated by the colored segments. The main image line passes through 2 stars. The image plane is covered by regions of negative (orange) and positive (blue) parity separated by the critical lines. There is at least one image close to every star, but those images with negative parity (micro-saddles) are usually highly demagnified and can get arbitrarily close to the point mass positions when the source is far away. The small gaps in the image lines are numerical artifacts near the critical lines (see also Sect. 2.5). The micro-images for the particular position $\zeta = (-0.35, 3.20)$ on the source line (right panel) are marked with blue crosses. Approx. 98 per cent of the total magnification of the source, $\mu = 6.7$, is due to the images on the main track (see also Fig. 6). For these images, the symbols are scaled to the corresponding magnification (the other symbol sizes remain fixed for clarity). Coordinates in both panels are in units of the Einstein radius (the dashed rectangle corresponds to Fig. 6 and the labelled ones - A,B,C - as well as the corresponding source plane locations on the right panel, to Fig. 7)

- The other images near the stars are found from:

$$z \approx z_i - m_i / \left[\bar{\zeta}_0 - \bar{z}_i + g\, z_i - \sum_{j=1, j\neq i}^{N} \frac{m_j}{z_j - z_i} \right]. \tag{18}$$

- After all the initial image positions are found, the paper goes on to show how to find all images of a straight line, and thus effectively all images for all source positions ζ on this line.

An example is given in the left panel of Fig. 4, where a straight line on the source plane is lensed into a main image of a “wiggly” line and many small “loops” around the microlenses. All images consist of single or multiply imaged parts of the straight line on the source plane. Depending on the configuration, loops can sometimes combine to form bigger loops and the main line can also be connected to stars, as can be seen on the figure.

The total magnification is actually dominated by only a few images (e.g. see Rauch et al. 1992; Wambsganss 1992; Schechter and Wambsganss 2002; Granot et al. 2003; Saha and Williams 2011). This is apparent in the left panel of Fig. 4 and illustrated further in Fig. 6. The overwhelming majority of the micro-images are faint negative-parity images (saddle-points of the arrival time) near each star. There are formally also images coincident with the stars; these are maxima of the arrival time, and are so demagnified that they are negligible.

The important images are a small number of micro-minima and equal number of micro-saddle points, whose properties we describe in Sect. 2.6.

2.5 Caustics of the Deflectors

In the case of quasar microlensing, the combination of many point masses and a shear creates a complicated pattern of critical curves. The critical curves can be determined as the places in the lens plane where the determinant Eq. (12) vanishes. A small area in the lens plane would be mapped to a point, identifying a locus of formally infinite magnification in the source plane.

Using the complex formalism above, Witt (1990) has shown that Eq. (16) can be used to work out the location for all critical curves in the lens plane. Because the determinant vanishes, the expression in brackets can be written as:

$$-g + \sum_{i=1}^{N} \frac{m_i}{(\bar{z}_i - \bar{z})^2} = e^{i\varphi}. \tag{19}$$

Similar to Sect. 2.4, multiplying by $\prod_{i=1}^{N}(\bar{z}_i - \bar{z})^2$ yields a polynomial of degree $2N$. All critical curves can be found by solving for all the roots $\varphi = 0$ and then tracing all critical curves for $0 <= \varphi <= 2\pi$. This can be easily achieved using a simple root-finder like the Newton-Raphson method.

In the left panel of Fig. 4 the critical lines for the $N = 50$ stars were calculated in this way by starting at the $2\,N = 100$ complex roots and following the critical curves for φ between 0 and 2π. The corresponding caustic lines in the source plane can be found by mapping the critical lines using the microlensing equation Eq. (15) (or Eq. (4) and Eq. (5)) and are shown in the right panel of Fig. 4. The caustic lines separate regions of differing image multiplicity; whenever a source crosses a caustic, two images either appear or disappear. The source position denoted by the cross symbol in Fig. 4 is inside the slightly elongated astroid caustic, so that two additional images appear, as seen in the left panel of the same figure and in Fig. 6.

The total magnification of a point source moving along the straight line in the right panel of Fig. 4 can be calculated as a function of position or time. Characteristic spikes of extreme magnification are expected whenever the source is crossing a caustic. Such a plot of magnification against time, also known as a "light curve", is shown in Fig. 15 and is examined in detail in Sect. 3.4 as it is a very important microlensing observable.

2.6 Properties of the "Swarm" of Microimages

Finding all the microimages, albeit is feasible as described in Sect. 2.4, is quite costly. However, there is a number of key properties of the "swarm" of microimages, such as their number, distribution, and individual magnifications, that can be obtained more easily. It is helpful to begin our examination of these properties with the time delay surface:

$$\tau = \frac{1}{2}(\boldsymbol{\theta} - \boldsymbol{\beta})^2 - \psi(\boldsymbol{\theta}), \tag{20}$$

where ψ is some lensing potential (see Saha et al. 2024). In the absence of any gravitational lensing, this surface is a circular paraboloid with a single extremum and one image is seen at the position of the source with magnification $\mu = 1$. Under the influence of some background gravitational potential $\psi_B(\boldsymbol{\theta})$ with constant second derivatives (i.e. constant κ and

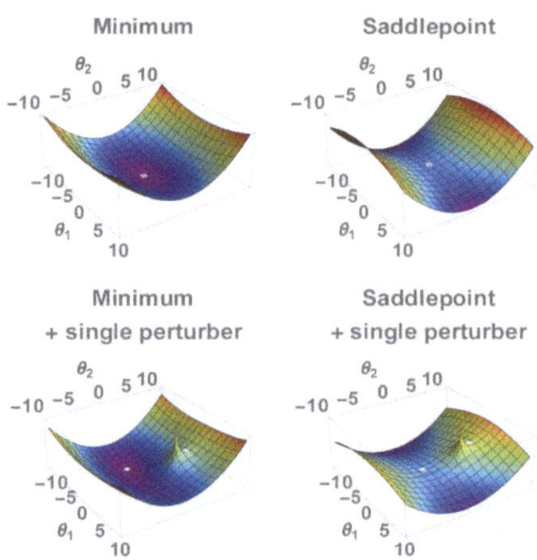

Fig. 5 *Top*: The time delay surface for a minimum (left) and saddle point image (right). *Bottom*: Same but with the addition of a point-mass perturber. The white circles mark the locations of the micro-images

γ), the time delay surface transforms into an elliptical or hyperbolic paraboloid. Whether the image, which corresponds to the observed macro-image, is a minimum or a saddle point then depends on the curvature of the time delay surface.

Point masses introduce logarithmic terms into the gravitational potential (see Saha et al. 2024). If the point masses have some convergence κ_* the time delay surface becomes:

$$\tau = \frac{1}{2}(\boldsymbol{\theta} - \boldsymbol{\beta})^2 - \psi_B(\boldsymbol{\theta}) - \theta_E^2 \sum_{i=1}^{N} m_i \ln|\boldsymbol{\theta} - \boldsymbol{\theta}_i| - \frac{1}{2}(-\kappa_*)\boldsymbol{\theta}^2, \tag{21}$$

where a term due to a negative constant surface mass density equal to the mass in stars is included to conserve the total convergence. In the limit of few deflecting masses, the time delay is largely unperturbed, as shown in Fig. 5. There are scattered logarithmic spikes, but the likelihood of one lying close to the line-of-sight to the source is low.

The locations of the microimages satisfy the lens equation, which we re-write here as:

$$\boldsymbol{\beta} = \boldsymbol{\theta} - \boldsymbol{\alpha}_B(\boldsymbol{\theta}) - \boldsymbol{\alpha}_*(\boldsymbol{\theta}), \tag{22}$$

where $\boldsymbol{\alpha}_{\rm B}(\boldsymbol{\theta})$ and $\boldsymbol{\alpha}_*(\boldsymbol{\theta})$ are the deflection angles due to the potential $\psi_{\rm B}$ and the microlenses (with the negative constant convergence correction) respectively. For constant second derivatives of the potential $\psi_{\rm B}$ (constant κ and γ), the first two terms of the lens equation become proportional to $\boldsymbol{\theta}$. For microimages far away from the source position, $\boldsymbol{\beta}$, these terms become very large. Consequently, the sum of the deflections due to the microlenses (see Eq. (1)) must also become very large. Because point mass deflections are proportional to $1/|\boldsymbol{\theta} - \boldsymbol{\theta}_i|$, the location of the microimage is required to be very close to some microlens so that a single term in the sum dominates. It can be shown that the magnifications of these far away micro-images behave as $\mu \propto |\boldsymbol{\theta}|^{-4}$ (Paczynski 1986; Schneider and Weiss 1987), and they are therefore faint saddle points.

For larger values of κ_*, microlenses are more likely to lie closer to the macro-image and affect it. So long as there is external shear, a single point mass close to the macro-image position can split it into four, instead of just two, extra images (counting the infinitely de-magnified micro-maximum) if the source lies within the star's caustic. As κ_* increases, there is no longer a single micro-image that can be associated with the macro-image, which is split into an increasingly more complex "swarm" of microimages. Figure 6 shows the microimages along with contours of the light travel time that pass through micro-saddle-points.

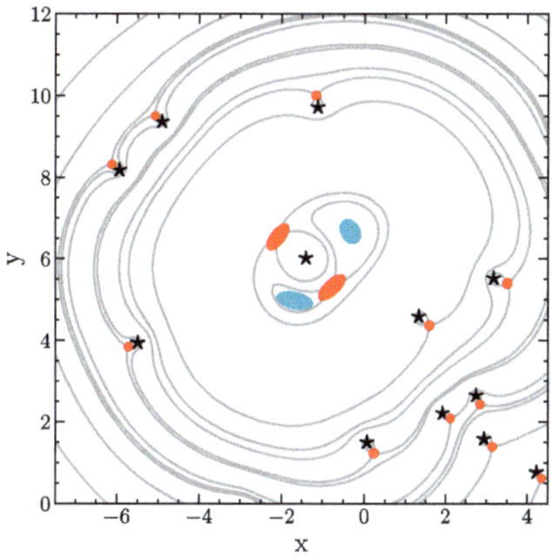

Fig. 6 Zoomed-in region around the brightest images in the left panel of Fig. 4. The gray curves are saddle-point contours of the arrival time, at whose self-intersection points saddle-point micro-images form (red ellipses). The micro-minima (blue ellipses) and positions of the microlenses (black stars) are also shown. The latter coincide with the locations of zero-brightness micro-maxima. The size and shapes of the ellipses are indicative of the magnification tensor

Close to the macro-image position, the micro-images are not restricted to lie very close to an associated microlens anymore; the deflection term of the macro-potential is small enough to be counterbalanced by the sum of the deflections due to all the microlenses without any specific term dominating. Instead, there is an effective region within which they tend to lie that determines the size of the micro-image swarm.

2.6.1 The Size of the Micro-Image Swarm

The magnification of a point source located at $\boldsymbol{\beta}$ is (Venumadhav et al. 2017):

$$\mu(\boldsymbol{\beta}) = \int \delta(\boldsymbol{\theta} - \boldsymbol{\alpha}_B(\boldsymbol{\theta}) - \boldsymbol{\alpha}_*(\boldsymbol{\theta}) - \boldsymbol{\beta})\, \mathrm{d}^2\boldsymbol{\theta}. \tag{23}$$

This can be seen by utilizing properties of the Dirac delta function; Eq. (23) can be rewritten as a sum:

$$\mu(\boldsymbol{\beta}) = \int \sum_i \mu(\boldsymbol{\theta}_i)\delta(\boldsymbol{\theta} - \boldsymbol{\theta}_i)\, \mathrm{d}^2\boldsymbol{\theta}, \tag{24}$$

where the $\boldsymbol{\theta}_i$ are the locations of the microimages. The deflection angle $\boldsymbol{\alpha}_*(\boldsymbol{\theta})$ due to the point masses in Eq. (23) can be viewed as a random variable that changes based on different realizations of the random point mass positions. Averaging over all such realizations, we get:

$$\langle \mu(\boldsymbol{\beta}) \rangle = \int \delta(\boldsymbol{\theta} - \boldsymbol{\alpha}_B(\boldsymbol{\theta}) - \boldsymbol{\alpha}_* - \boldsymbol{\beta}) p(\boldsymbol{\alpha}_*)\, \mathrm{d}^2\boldsymbol{\alpha}_*\, \mathrm{d}^2\boldsymbol{\theta}, \tag{25}$$

where $p(\boldsymbol{\alpha}_*)$ is the probability density function of the deflection angle. By transforming coordinates from the image plane to the source plane with the change of variable $\boldsymbol{\beta}' = \boldsymbol{\theta} - \boldsymbol{\alpha}_B(\boldsymbol{\theta}) - \boldsymbol{\beta}$, we arrive at:

$$\langle \mu(\boldsymbol{\beta}) \rangle = \int \delta(\boldsymbol{\beta}' - \alpha_*) p(\boldsymbol{\alpha}_*) \mu_B(\boldsymbol{\beta}' + \boldsymbol{\beta})\, \mathrm{d}^2\boldsymbol{\alpha}_*\, \mathrm{d}^2\boldsymbol{\beta}'. \tag{26}$$

Performing the integral over $\boldsymbol{\alpha}_*$, this simplifies to:

$$\langle \mu(\boldsymbol{\beta}) \rangle = \int p(\boldsymbol{\beta}') \mu_B(\boldsymbol{\beta}' + \boldsymbol{\beta})\, \mathrm{d}^2\boldsymbol{\beta}'. \tag{27}$$

This resulting expression is a convolution of the microlens deflection probability density with the background magnification model. What was once a point source has, on average, been "smeared out" into a source with a profile that looks like $p(\boldsymbol{\alpha}_*)$.

The form of the probability density function of $\boldsymbol{\alpha}_*$ was first worked out by Katz et al. (1986), while Schneider et al. (1992) provide it in a slightly different form and Petters et al. (2009a) present a more thorough mathematical treatment. This function is isotropic and depends only on the magnitude of the deflection angle, not the direction. It is a combination of a bivariate normal distribution, whose width depends on the number of point masses N as:

$$\sigma_* = \theta_{\rm E}\kappa_*^{1/2}\left[\ln(2e^{1-\gamma_E}N^{1/2})\right]^{1/2}, \tag{28}$$

where $\gamma_E \approx 0.577$ is the Euler-Mascheroni constant, with a tail that behaves as:

$$p(\boldsymbol{\alpha}_*) = \frac{\theta_{\rm E}^2\kappa_*}{\pi\,|\boldsymbol{\alpha}_*|^4}, \tag{29}$$

for large values of $\boldsymbol{\alpha}_*$.

To determine the size of the image swarm for a point source, one can integrate $p(\boldsymbol{\alpha}_*)$ over a portion of the source plane that contains a large fraction of the flux. If we only integrate over a region with width σ_*, the probability density function behaves like a normal distribution and the resulting isophote in the image plane contains microimages that on average constitute only 39% of the flux; in order to include 90% of the flux (on average), integrating over a region with width $2.15\sigma_*$ is required. We can transform Eq. (28) for σ_*, which depends on the number of point masses N, into a function of the macro-parameters only. By considering the tail of the probability density function, one finds that 99% of the average flux is contained within a circle of radius $10\kappa_*^{1/2}\theta_{\rm E}$. The resulting size of the image swarm is then given by transforming this circle to the image plane, resulting in an ellipse with axes given by:

$$r_\pm = \frac{10\kappa_*^{1/2}\theta_{\rm E}}{|1-\kappa\pm\gamma|}. \tag{30}$$

Figure 7 shows various micro-image configurations for the three source positions indicated in Fig. 4. The magnification lying within the marked contours is slightly different than the expected average due to sample variance of the source and microlens positions.

The size of the image swarm determines the allowable area where bright microimages can appear. This in turn affects the possible astrometric shifts from microlensing. Large astrometric shifts can come about only due to caustic crossings and the creation or annihilation of pairs of microimages, as otherwise the microimages only change positions smoothly. Treyer and Wambsganss (2004) found that strong astrometric microlensing shifts are seen in systems with $\kappa=\gamma=0.4$ and $\kappa=\gamma=0.6$, while weak shifts are seen in systems with $\kappa=\gamma=0.2$ and $\kappa=\gamma=0.8$. This is easily explained by the axis ratio of the image swarm ellipse for each set of parameters, which additionally provides the reason for a dependence on the direction of the external shear noted in Treyer and Wambsganss (2004). The maximum allowable astrometric shift is roughly given by Eq. (30) and is therefore on the order of 10 s of micro-arcseconds. In more extreme cases of magnification from, e.g., cluster critical curves, the astrometric shifts from microlensing may approach more observable milliarcsecond levels.

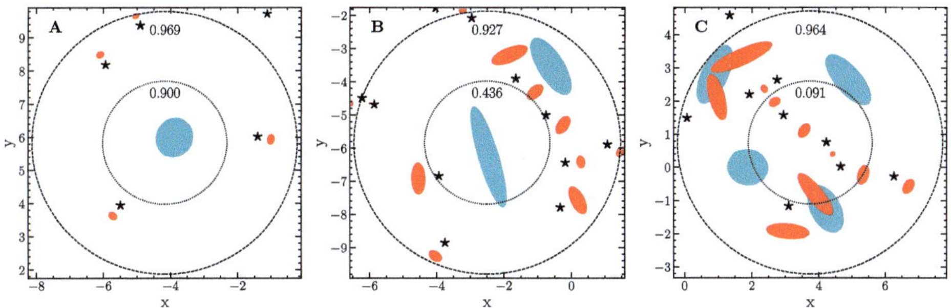

Fig. 7 Microimages corresponding to a point source located at the three source positions indicated in the right panel of Fig. 4. The blue (orange) ellipses represent the micro-minima (micro-saddles). The sizes of the ellipses scale logarithmically with magnification in order to show a greater dynamic range, while their orientation shows the micro-shear from the stars. The dashed and dotted ellipses (circles in this case because there is no macro-shear) denote the isophotes of 90 per cent and $1\sigma_*$ respectively, while the numbers indicate the actual fraction of the total magnification that lies within them

For an extended source with some profile $I(\boldsymbol{\beta})$, the microimages in general simply result in a broadening of the source profile. The width of $p(\boldsymbol{\alpha}_*)$, which according to the previous discussion is of order $\kappa_*^{1/2}\theta_{\mathrm{E}}$, is added in quadrature to the width of the source σ_{I} (Dai and Pascale 2021), resulting in an effective width:

$$\sigma_{\mathrm{eff}} \approx \sqrt{\sigma_{\mathrm{I}}^2 + \kappa_* \theta_{\mathrm{E}}^2}. \tag{31}$$

2.6.2 The Number of Micro-Images

While every point mass has an associated saddle point image, the total number of images for a field of N point masses with external shear cannot exceed $5N$ (Khavinson and Neumann 2004). This implies there are at most $2N-1$ micro-minima and $3N-1$ micro-saddles, plus one additional micro-image of either parity corresponding to the macro-minimum or saddle (Petters et al. 2009b). In practice, for macro-images that are not near a macro-critical curve (i.e. outside the extreme high magnification regime), the number of expected micro-minima (extra image pairs) is fairly low, with some simulations suggesting an approximately Poisson distribution (Granot et al. 2003).

The micro-critical curves split the image plane into regions of positive and negative parity, as shown in Fig. 4. The multiplicity with which regions of positive parity overlap when mapped back to the source plane gives the mean number of positive parity images $\langle n \rangle$, for which analytic formulas are given in Wambsganss (1992), Granot et al. (2003). Figure 8 shows the caustic pattern of Fig. 7, color-coded to show the number of micro-minima. The histogram of the resulting distribution of the number of microminima is shown in Fig. 9.

Finally, by extending arguments found in Wambsganss (1992), Granot et al. (2003), it can be shown that the average magnification of a single micro-minimum is:

$$\langle \mu_{\mathrm{mm}} \rangle = \frac{\mu_{\mathrm{macro}}}{\langle n \rangle} P_{\mu>0}(x_{img}), \tag{32}$$

where $\mu_{\mathrm{macro}}/\langle n \rangle$ is the macro-magnification from Eq. (14) divided by the expected number of micro-minima, and P is the probability that location x_{mm} is a minimum and not a saddle-point.

Fig. 8 Caustics (black lines, same as in the right panel of Fig. 4) and corresponding number of micro-minima from each region they enclose. The region of the source plane shown here is the same as in Fig. 10

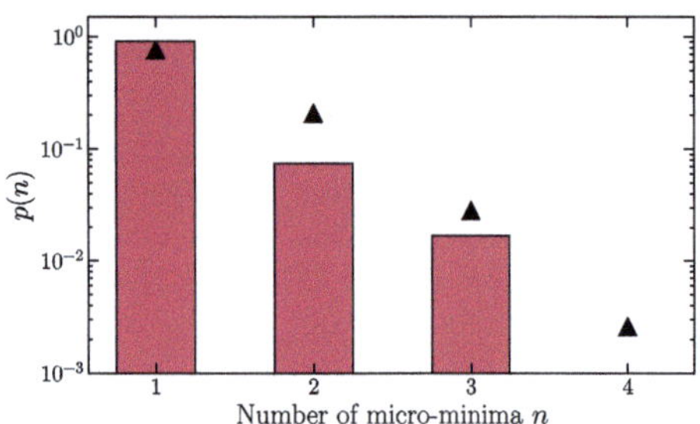

Fig. 9 Histogram showing the distribution of the number of micro-minima. The black triangles denote the expected probabilities for a Poisson distribution with a mean $\langle n \rangle - 1$ (as the image is a macro-minimum)

2.7 Microlensing Maps

A very useful tool for microlensing studies is the magnification field on the source plane induced by an ensemble of microlenses. This can be used together with a model for the source to match simulated light curves and flux ratios with observations, as will be discussed in Sect. 3. To calculate this magnification, we have to transform Eq. (13), the sum of magnifications due to all the microlenses, from the lens to the source plane. This is hardly practical, and becomes quickly intractable as the number of microlenses increases. However, there has been a number of techniques developed specifically to optimize this task, which we review here.

Our starting point is Eq. (5), where the image positions $\boldsymbol{\theta}$ map uniquely to a position on the source plane, $\boldsymbol{\beta}$ – a one-to-one mapping. As we have seen, the inverse is not true and a source can have multiple images. Finding these solutions to the lens equation is a complex problem, if not impossible (see Schneider et al. 1992, for analytical solutions in a few simple cases). Therefore, an alternative approach is to proceed in the inverse manner and calculate the total magnification as the sum of the intensity of all the microimages in a finite region of the source plane, as shown in Fig. 10. This technique, called inverse ray-shooting, was introduced by Kayser et al. (1986) and comprises propagating a grid of rays backwards from the observer through the lens plane, where each ray is deflected using Eq. (5), to the source plane, where its final position is mapped. By dividing the source plane into pixels, counting the number of rays reaching each pixel, N_{ij}, and comparing it to the number of rays that would have reached that pixel if there was no lensing taking place, N_{rays}, we obtain an estimate for the magnification, i.e. a pixelated magnification map:

$$\mu_{ij} = N_{ij}/N_{\text{rays}}. \tag{33}$$

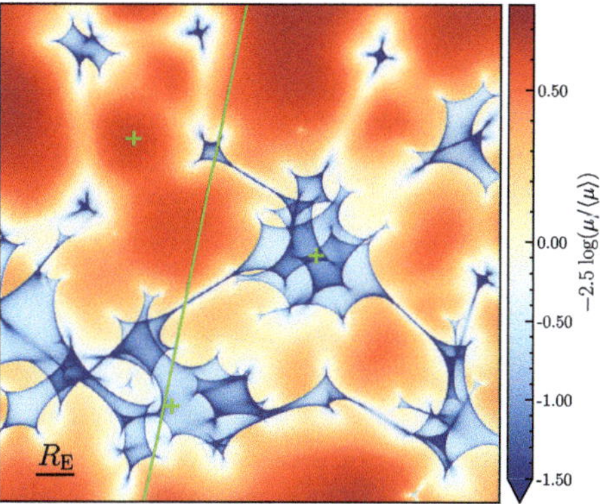

Fig. 10 Magnification map corresponding to the microlens configuration shown in Fig. 4, generated by the inverse ray-shooting code GPUD Thompson et al. (2010, 2014). The scale is in magnitudes over the mean magnification, $\langle\mu\rangle$, which approaches the macroscopic one given by Eq. (14) (a function of the local convergence and shear) as the number of rays increases. The field of view is $13 \times 13\,\theta_E$ on the source plane and includes the track from the right panel of Fig. 4 (green line) and the locations used to produce Fig. 7 (green crosses). Corresponding light curves for different accretion discs are shown in Fig. 15

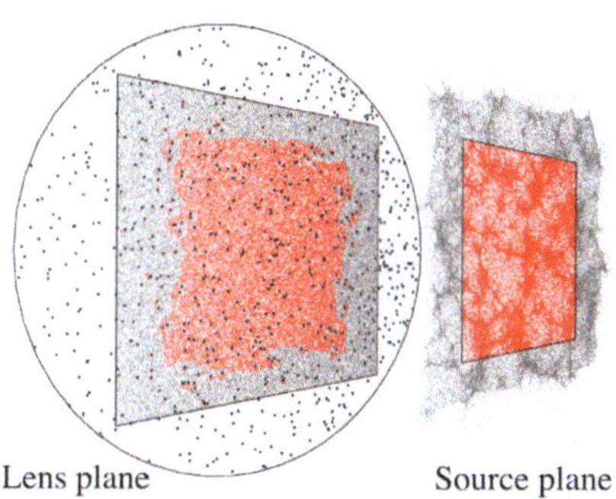

Fig. 11 Schematic representation of ray-shooting. Microlenses (black dots) are distributed in a circle of area A on the lens plane. A grid of light-rays (grey and red dots) is projected backwards from the observer through the lens plane, where the deflection by each lens is calculated for each ray using Eq. (5), and mapped onto the source plane. A rectangular area on the source plane is selected (red points) away from edge effects, divided into pixels, and used to calculate the magnification from Eq. (33). Taken from Vernardos and Fluke (2014a)

This relation is an approximation because it does not take into account light rays from outside the defined grid that could be deflected inside the region of the source plane under examination. In practice, by choosing a grid of rays large enough with respect to the size of the source plane, the effect of such extreme deflections can be neglected. The above outlined procedure is schematically shown in Fig. 11.

Performing inverse ray-shooting directly is a computationally expensive procedure, with a forbidding cost for large scale applications (see Bate and Fluke 2012; Vernardos and Fluke 2014a, for some benchmarks). Hence, efficient approximate implementations have been developed, as well as direct approaches taking advantage of hardware acceleration, which we outline below.

2.7.1 The Tree Code

One way of reducing the calculations required to produce magnification maps is by approximating the sum of all the deflections due to the microlenses. The key idea here is that the deflection angle of a given light ray due to a given microlens is inversely proportional to the distance between the two, in an analogous way to gravitational N-body problems. In the hierarchical tree code approach developed by Wambsganss (1990, 1999), microlenses that are further away from a light ray are grouped together in cells, or pseudo-particles, whose size and total mass depends on the distance to the light ray. This approximation greatly reduces the final number of deflections that need to be calculated per light ray and, subsequently, increases the speed of the calculations, at the cost of a slight computational overhead and a loss in accuracy. This method has been used extensively in the past to constrain properties of the background quasar and the lens (Rauch and Blandford 1991; Keeton et al. 2006; Bate et al. 2008; Pooley et al. 2012; Dai and Guerras 2018; Hutsemekers et al. 2019). Due to its efficiency and the fact that it was the only widely available code for producing magnification maps for a long period of time, this approach can be considered as the "industry standard" in quasar microlensing.

Although the tree code has led to a tremendous speed-up of the calculations compared to direct ray-shooting, there are a few disadvantages. The required accuracy of the code – the level at which a number of microlenses at a given distance from a light ray are grouped into a pseudo-particle or treated individually – is something to be determined empirically by running the code multiple times until the results are acceptable. Hence, a certain amount of expert knowledge is required to implement and use it. The memory overhead of the tree data structure used to group the microlenses into cells may grow too large with the number of microlenses for a single processor to handle. For this reason, a parallel version has been developed (Garsden and Lewis 2010) that overcomes this by distributing the computations across many nodes on a computer cluster. Using this approach, magnification patterns of up to a billion microlenses can be computed in realistic timescales.

2.7.2 Polygon Mapping

Another approach to speed up the solution of Eq. (33) is to reduce the number of rays shot, N_{rays}, as does the inverse polygon mapping technique developed by Mediavilla et al. (2006, 2011), Shalyapin et al. (2021). Here, the lens plane is divided into cells that are mapped onto the source plane using the lens equation. By counting the portions of the area of the cells that are within a given source plane pixel, as opposed to the standard inverse ray-shooting that counts rays, an estimate of the magnification is obtained. The main advantage of this method is the increased computational efficiency, i.e. there are no unused rays that got deflected outside the source plane region of interest, while there is an improved resolution around caustics as well. However, it has its own computational overhead due to finding the correct tiling of the lens plane and the large number of cells required near critical lines. Another advantage of this method is that the error on its estimate of the number of rays per pixel follows $N_{ij}^{-3/4}$, which is smaller than the Poisson error $N_{ij}^{-1/2}$ of randomly shot rays. In fact, the former is a direct result of using a regular grid of rays (instead of random positions Kayser et al. 1986). Selected examples of studies employing the inverse polygon mapping technique are Guerras et al. (2013a), Jiménez-Vicente et al. (2014), Rojas et al. (2014), Esteban-Gutiérrez et al. (2020).

2.7.3 Direct Approaches

Numerous scientific problems, whose solution has so far been possible only through complex approximations, if available, can now be revisited due to the advent of Graphics Processing Units (GPU) and their design focused on massively parallel processing (see the general brute-force and algorithm analysis techniques described in Barsdell et al. 2010; Fluke et al. 2011). Although originally developed as specialized hardware to accelerate the generation of graphics on the computer screen, particularly for the game industry, the GPU architecture together with the emergence of the notion of general purpose programming libraries, has allowed for speed-ups of $\mathcal{O}(10)$ to $\mathcal{O}(100)$ for various algorithms. This is the case for inverse ray-shooting, whose direct implementation is "embarassingly parallel": each deflection by a single microlens is independent of any other microlenses and the deflections for each light ray are independent of all the other light rays. Such an algorithm – GPUD – has been implemented by Thompson et al. (2010, 2014), with an obvious advantage being its simplicity and ease to modify and maintain. Another advantage is the benefit from Moore's law (Moore 1965) for GPUs – a doubling of the speed every 1-2 years due to hardware improvements – without any software modification (see Fig. 2 in Vernardos and Fluke 2014a), which has ceased since 2005 for CPUs. GPUD was used to carry out GERLUMPH, the largest parameter space exploration of microlensing maps discussed in Sect. 2.8, which enabled a series of previously unfeasible studies (Vernardos 2018; Foxley-Marrable et al. 2018; Vernardos 2019; Vernardos and Tsagkatakis 2019; Neira et al. 2020). Recent developments by Zheng et al. (2022) can increase the speed of GPUD by 100 in the regime where it is the slowest, i.e. large numbers of lenses and high resolution maps.

The Fourier-based approach by Kochanek (2004) separates the long- and short-range effects of the microlenses on ray deflections using a particle-particle/particle-mesh (P3M) algorithm, assumes the lens and source planes to be spatially periodic, and approximates the long-range deflections via a Fourier transform. As a consequence, the edge effects in the source plane are removed and more of the magnification map is usable (higher efficiency). However, there is an upper bound on the size of the generated magnification maps (approx. 8192 pixels on a side) due to the increasing amount of memory required by the Fourier transforms. Selected examples of studies employing the Fourier-based method are Poindexter et al. (2008), Chartas et al. (2009), Dai et al. (2010), MacLeod et al. (2015), Morgan et al. (2018), Cornachione et al. (2020a).

2.7.4 Combined Approaches

Apart from GPUD's direct approach to inverse ray-shooting, the tree and polygon based algorithms can also benefit from the massive GPU parallelization, at the cost of more complex algorithm development. A spectacular example is the Teralens[2] algorithm, which follows the same general principles as the tree code of Wambsganss (1999) but has a dramatically different algorithmic design tuned for maximum parallel efficiency. One can also envisage merging the tree-based method that approximates the lens deflections with shooting polygons instead of rays, since the two approximate different parts of the lens equation. Recently, Jiménez-Vicente and Mediavilla (2022) achieved this by combining the inverse polygon method with the fast multiple method (Greencard and Rokhlin 1987), an algorithm similar in concept to a tree or particle-mesh algorithm. Implementing this approach on the GPU would result in the fastest conceivable magnification map generating code. Finally,

[2] https://github.com/illuhad/teralens.

alternative GPU/CPU parallel architectures could provide another path forward, requiring minimal modifications for speeding up existing CPU codes (Chen et al. 2017).

2.8 How Caustic Structures Respond to Macro Parameters

The stellar-mass compact object populations that cause microlensing are unobservable – one would require to measure masses and positions of hundreds to thousands of such objects within galaxies at cosmological distances, orders of magnitude below the resolution of any conceivable telescope. Hence, we have to resort to a statistical description of such populations at the location of the observed quasar multiple images that undergo microlensing. This is based on local properties of the macroscopic mass and light distribution of the lensing galaxy. The three major macro-parameters statistically defining a microlensing population are:

- the total convergence, κ, defined in Eq. (3),
- its compact matter component, κ_*, related to the number of microlenses via Eq. (8), and
- the shear, γ.

The κ, γ are introduced and linked to the lens potential within the general formalism presented in Sect. 3 of Saha et al. (2024), while specific choices of lens models and partitioning the mass between a smooth/dark and a compact/stellar component are discussed in Sects. 2 and 3 of Shajib et al. (2024). Although the shear is a vector, its direction becomes important only when comparing to other multiple images, for example when generating light curves (e.g. see Fig. 17). Therefore, without loss of generality, we can align the shear with the x-axis, leading to $\gamma_2 = 0$ in Eq. (5) – a useful trick to maximize the efficiency in using magnification maps. This leads to all three parameters being scalar fields, functions of the position on the lens plane (an example of how these three parameters define microlensing populations which in turn produce corresponding magnification maps and microlensing observables is shown in Fig. 17).

Figure 12 demonstrates the effect of different macro-parameters on the magnification maps in the case where the whole mass is in the form of microlenses, i.e. $\kappa_c = 0$ in Eq. (5). The density of the caustics is proportional to the total magnification (and consequently the number of microlenses from Eq. (13)), whose contours are shown in the top panel of Fig. 13. The effect of the shear is also striking, stretching the caustics and producing stratified maps with elongated "filaments" of caustics interspaced with demagnification troughs. Interestingly, for super-critical regions where $\kappa > 1$ (these correspond to highly demagnified saddle-point or maximum images), the anisotropic magnification (stretching) due to the shear changes direction by $\pi/2$ due to the $1 - \kappa$ term being negative (see Eq. (76) of Saha et al. 2024). This is clearly seen as the horizontal caustic structures for $\kappa < 1$ become vertical for $\kappa > 1$.

These diverse caustic structures have an impact on the microlensing effect present in observations, i.e. through light curves and magnification probability distributions of flux ratios (see the next section). Vernardos and Fluke (2013) have explored the latter case, producing a similar tiling of the parameter space as Fig. 12 but with magnification distributions (see their Fig. 4). In summary, they found that along the critical curve (where magnification diverges), the distributions are centered around the macro-magnification, extend in a narrow area around it, $0.2 < \log(\mu/\mu_{\text{macro}}) < 5$, where μ_{macro} is given by Eq. (14), and appear symmetric (in log space). Away from the critical curve, the distributions generally become wider and asymmetric, mostly due to the appearance of a high magnification tail and/or a secondary peak.

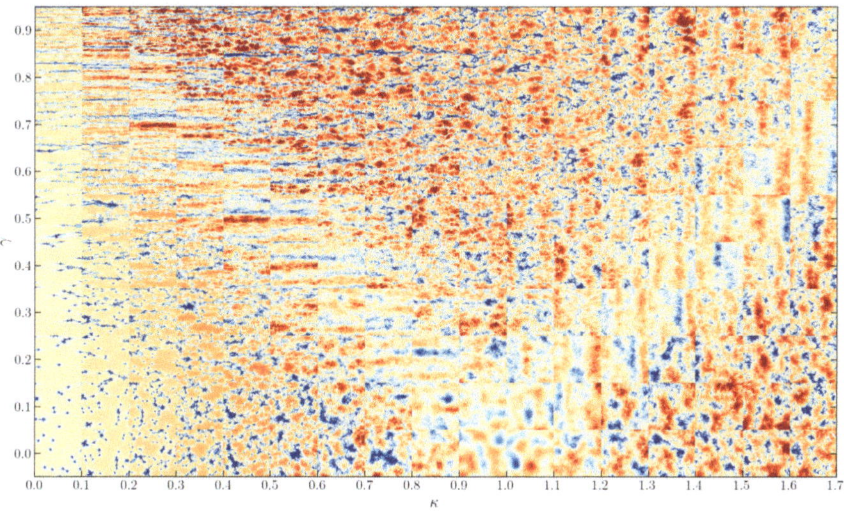

Fig. 12 Tiling of the $\kappa - \gamma$ parameter space by 17×10 magnification maps with $\kappa_c = 0$ from the (GERLUMPH) database. The central point of each tile corresponds to the $\kappa - \gamma$ values that were used to generate it. The caustics become denser as the number density of microlenses increases along the x-axis, while they appear increasingly stretched due to higher shear along the y-axis (the stretching always occurs along the x-axis of the maps themselves, see Fig. 17 for the proper alignment of the maps in a real lensed system). The color scale is the same as in Fig. 10

Including an additional convergence component due to smooth matter has fundamental implications that trace back to the mass-sheet degeneracy (Saha et al. 2024). The effect is two-fold: on one hand, the number of microlenses is reduced since now there is less mass in the form of compact objects (see Eq. (8)), while on the other hand, the resulting caustic networks, which are now less dense, are isotropically magnified by the uniform mass-sheet described by κ_c (see Eq. (5)). An example of such maps is shown in Fig. 14.

Figure 12 is actually showing one slice of a three dimensional parameter space – the third dimension being κ_*. As Paczynski (1986) has shown, the κ, γ, κ_* parameter space can be transformed to an equivalent, two-dimensional parameter space of effective convergence and shear:

$$\kappa_{\text{eff}} = \frac{\kappa_*}{1 - \kappa + \kappa_*}, \quad \gamma_{\text{eff}} = \frac{\gamma}{1 - \kappa + \kappa_*}. \tag{34}$$

In the above equation, κ_{eff} is entirely due to compact matter. The effect of the above transformation can be understood in the following way, as also illustrated in the top panel of Fig. 14. Starting from a given location κ_0, γ_0 with $\kappa_c = 0$ (locations A and D), maps with $\kappa_c > 0$ lie in the third dimension in the traditional κ, γ, κ_* parameter space. But in the effective space they lie along straight lines defined by κ_0, γ_0 and $(1, 0)$. In terms of the magnification distribution, the above mass-sheet transformation leads to statistically equivalent maps (Vernardos et al. 2014), i.e. we expect the same microlensing deviations from the macro-magnification, as shown in Fig. 14. Minor differences arise due to sample variance from using a finite field of stars to generate maps for a finite source plane region. Such differences can be minimized by, e.g., averaging over multiple maps or using a larger region in the source plane. In terms of light curves, however, there are other slight differences: we expect less peaks because there are fewer caustics to cross, but also longer intervals for the variability because caustics cover now a bigger area on the source plane.

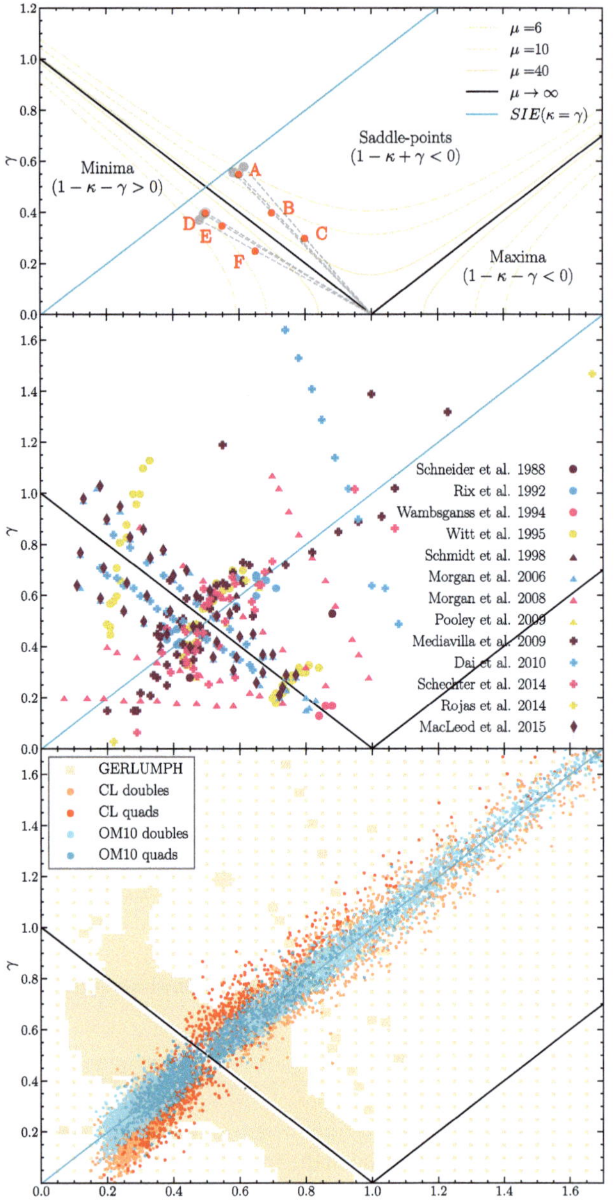

Fig. 13 Theory (top), macromodels (middle), and simulations (bottom) shown in the $\kappa - \gamma$ parameter space. The critical line, i.e. where the magnification given by Eq. (14) diverges dividing the parameter space to the minimum, saddle-point, and maximum image regions in direct correspondence to the macromodel multiple images, and the $\kappa = \gamma$ line, i.e. the locus of a Singular Isothermal Ellipsoid (SIE) macromodel, are shown in all panels for reference. *Top*: Selected locations with $\kappa_c = 0$ (orange points) have the same $\kappa_{\text{eff}}, \gamma_{\text{eff}}$ with any other location along the lines to (1,0) (grey dashed lines) with κ_c given by Eq. (34). Because locations B, C and E, F lie very close to the lines emanating from A and D respectively, they have very similar $\kappa_{\text{eff}}, \gamma_{\text{eff}}$ values (grey points) for suitable κ_c (this is illustrated in Fig. 14 in more detail). Magnification contours from Eq. (14) are also indicated. *Middle*: Selected macromodels from the literature – see https://gerlumph.swin.edu.au/macromodels/ for an interactive version of this plot. *Bottom*: Two simulated populations of lensed quasars are shown, one from Oguri and Marshall (2010) and an identical one generated with lens potential ellipticity and external shear distributions with lower and higher mean values respectively (0.15 and 0.1 as opposed to 0.3 and 0.05, based on Luhtaru et al. 2021). The footprint of the GERLUMPH database is also shown

2.9 Quasar Structure Overview

Because the microlensing signal depends on the size and structure of the lensed source, we aim in this section at providing an overview of the characteristics of the AGN emission regions over the electromagnetic spectrum, e.g. shape and size of the X-ray corona, size and temperature profile of the accretion disc, geometry of the broad line region, etc. The generally accepted structure of a quasar has been pieced together by painstaking work over many decades to account for all of the various emission and absorption features that have

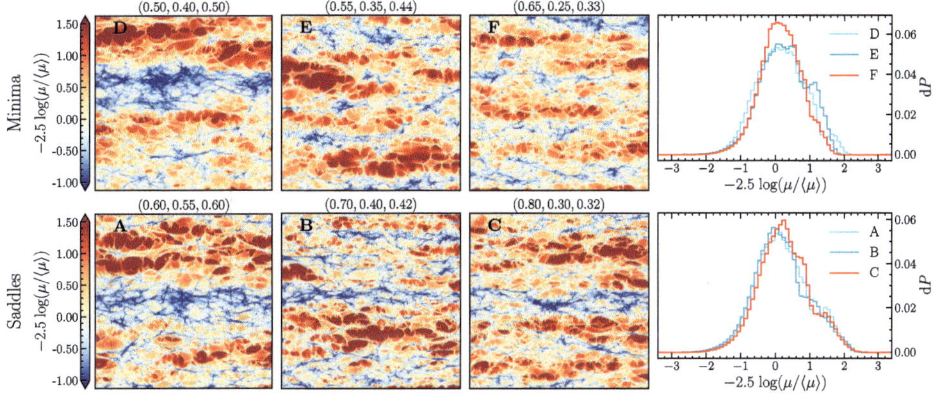

Fig. 14 Magnification maps and histograms corresponding to the locations marked on the top panel of Fig. 13. The κ, γ, κ_c values are given above each map and correspond to $\kappa_{\rm eff}, \gamma_{\rm eff} = (0.6 \pm 0.02, 0.55 \pm 0.02)$ and $(0.5 \pm 0.02, 0.4 \pm 0.02)$ for the saddle-points and minima respectively (grey points on the top panel of Fig. 13). Although the κ, γ, κ_c of these maps are very different, their magnification histograms are very similar, making them 'equivalent' maps (per image type, see Vernardos et al. 2014). The minor differences between the histograms can be attributed to the sample variance of examining a finite source plane region. The same cannot be readily said for light curves resulting from these maps

been seen across the electromagnetic spectrum. Figure 3 shows a schematic representation of the different components.

A scale commonly used as "unit length" when it comes to describing inner region of AGN is the gravitational radius r_g, also called Schwarzschild radius:

$$r_g = \frac{2GM_{\rm BH}}{c^2}, \tag{35}$$

where $M_{\rm BH}$ is the mass of the black hole, and G and c are the gravitational constant and speed of light. This radius defines the event horizon of a (non-spinning) black hole. For a typical mass of $M_{\rm BH} = 10^8\ M_\odot$, we have $r_g \sim 3 \times 10^8$ km that corresponds to ~ 2 AU, 10^{-2} lt-days, or $\sim 10^{-5}$ pc. In the perspective of microlensing applications, this should be compared to the typical Einstein radius of a microlens in the source plane (taken as the average from Mosquera and Kochanek 2011, see also Eq. (7)), $R_{\rm E} = 2.5 \times 10^{16} \sqrt{\langle M \rangle / 0.3\ {\rm M}_\odot}$ cm $\sim$ 9.7 $\sqrt{\langle M \rangle / 0.3\ {\rm M}_\odot}$ l-d $\sim 840 \sqrt{\langle M \rangle / 0.3\ {\rm M}_\odot}\ r_g$ (for $M_{\rm BH} = 10^8\ {\rm M}_\odot$). Finally, another quantity of interest from general relativity typically associated with the inner edge of the disc is the radius of the innermost stable circular orbit of a photon, $R_{\rm ISCO} = \alpha r_g$, where $\alpha = 3$ for a non-rotating black hole and 9/2 (1/2) for a maximally pro- (retro-) grade spinning one.

Our discussion will "follow the photons" from their production in the accretion disc through their encounters with other material in the quasar on their way to the observer. The aim is to help the reader associate features in the electromagnetic radiation with physical structures in the quasar. We do not aim to give a complete description of every aspect of quasar emission; instead we focus on those most relevant to microlensing.

The hot material in the accretion disc produces an abundance of optical/UV photons. Some of those encounter a population of high-energy electrons in the innermost regions near the black hole and are inverse-Compton upscattered to X-ray energies; this is the "primary" X-ray emission. Some of this energy will reach the observer directly and result in an X-ray continuum spectrum, while the rest of the emission from this X-ray corona will interact with other parts of the quasar and give rise to reflection features, the most prominent of which being the Fe Kα emission line at 6.4 keV. This line can have two components: a broad component that comes from the inner part of the accretion disc, and a narrow component that comes from material farther out (i.e., the broad-line region or torus).

The matter accretion onto the central supermassive black hole can reach a rate of a few solar masses per year and radiate with an efficiency ranging between 5.7% in the non spinning case and 43% for high spin (Thorne 1974), much larger than nuclear fusion ($\sim 0.8\%$). This energy release causes the accretion disc to heat up and emit a power-law continuum radiation from UV to optical ranges. This emission arises from the inner part of the disc that corresponds to distances from the central black hole ranging from several to tens of astronomical units. The high energy continuum from the inner disc ionizes the gas in the surrounding area, producing broad emission lines characteristic of quasar optical spectra (Fig. 3).

Based on the variety of observed line widths and ratios, AGN were originally classified into various types and subtypes. In the nineties, a unification scheme based on orientation has been proposed to explain this phenomenology in a coherent, geometric way (Antonucci 1993; Urry and Padovani 1995). The model postulates that obscuring material in the equatorial plane (assumed to be a dust torus[3]) and orientation with respect to the line of sight are causing this diversity.

There is growing evidence that the physical properties of AGN also play a role in their spectral appearance. In particular, variation in the accretion rate, which is suspected to be related to the launch of radio jets[4] (Laor and Behar 2008; White et al. 2015), also impacts the spectral appearance of AGN (Marziani et al. 2001; Shen and Ho 2014; Elitzur et al. 2014).

The following subsections describe more quantitatively our knowledge of the AGN emission regions that are the most relevant for microlensing studies, as shown in Fig. 3. As direct imaging is not possible, we briefly explain the methods used to build up our understanding of the unified model of AGN structure. Our journey into the heart of AGNs starts with the most compact regions, which are also the most susceptible to microlensing, and ends at the interface between the AGN and its host galaxy, namely the torus and the narrow line region.

2.9.1 The Most Compact Emission

After the first X-ray satellites were launched in the 1970s (Uhuru, Ariel 5, SAS-3, OSO-8), the X-ray spectral properties of quasars and (suspected) black hole X-ray binaries became known, and much work was done to understand their features. In particular, the X-ray continuum power-law spectrum seen in quasars and the high energy emission of black hole X-ray binaries was explained as the Compton upscattering (also called "inverse Compton scattering" or just "Compton scattering" in the literature) of lower energy photons by a population of hot, energetic electrons (e.g. Shapiro et al. 1976; Katz 1976; Pozdnyakov et al. 1976; Galeev et al. 1979; Sunyaev and Truemper 1979; Sunyaev and Titarchuk 1980). As more and higher quality observations of quasars and Seyfert galaxies were obtained by the HEAO I and then EXOSAT observatories, the remarkable similarity of the X-ray spectra (a single power-law continuum with spectral index of $\sim$0.7 Rothschild et al. 1983; Mushotzky 1984; Turner and Pounds 1989) suggested a common origin. After the Ginga satellite detected the Fe emission line and a broad hump reflection feature (Rothschild et al. 1983) that

[3]The validity of this paradigm, with dust being in a torus in the equatorial plane, is currently under scrutiny by the community as interferometric observations of dust in AGN are also consistent with a polar emission.

[4]Only approximately 10% of AGN are radio-loud and show synchrotron emission as evidence for a jet. Other AGN, however, are not radio-silent, but the origin of their radio emission is not yet fully established.

were predicted by Guilbert and Rees (1988) and Lightman and White (1988), Mushotzky (1984), Turner and Pounds (1989), a "two-phase" model AGN was proposed by Haardt and Maraschi (1991, 1993) in which a hot, tenuous corona exists above the accretion disc and Compton upscatters UV and optical photons from the disc to X-ray energies with a power-law continuum spectral shape. The corona is thought to be supplied with energetic particles accelerated by the magnetic fields anchored in the accretion disc (e.g. Haardt et al. 1994; Di Matteo 1998; Merloni and Fabian 2001). For decades, the geometry and extent of the corona was largely unknown, with arguments that it may be patchy, extend over parts of the inner disc, and vary with time (e.g. Gallo et al. 2015; Wilkins and Gallo 2015). However, spectral and timing studies suggest a compact, centrally located corona (e.g. Brenneman and Reynolds 2006; Fabian et al. 2009, 2013; Parker et al. 2015). Microlensing has strongly constrained the X-ray emitting corona to be compact.

2.9.2 The Accretion Disc

The powerhouse of AGN emission originates from within 10-1000 r_g (i.e. 0.1-10 lt-days for a black hole mass $M_{BH} \sim 10^8 M_\odot$) of the central supermassive black hole. This structure is thought to be a geometrically thin (in the vertical direction) but optically thick disc that is heated locally by the dissipation of gravitational binding energy through accretion of matter (the "thin-disc" model, Lynden-Bell 1969; Shakura and Sunyaev 1973). Ignoring relativistic effects, the temperature profile for such a thin-disc can be expressed as (e.g. Zdziarski et al. 2022):

$$T(R) \propto R^{-\beta} \left(1 - \sqrt{\frac{R_{\text{in}}}{R}}\right)^{1/4}, \tag{36}$$

where R_{in} is the inner edge of the disc, and $\beta = 3/4$. The inner edge effects are commonly ignored (see Zdziarski et al. 2022, regarding the impact of disc truncation and wind), and the dependence $T \propto R^{-\beta}$ is often kept as a first order generalisation of the thin-disc.

For this model, the radius at which the temperature coincides with the rest wavelength of the observations ($kT = h_p c/\lambda_{\text{rest}}$) is:

$$R_\lambda^{\text{flux}} \simeq \frac{3.4 \times 10^{15}}{\sqrt{\cos i}} \frac{D_s}{r_H} \left(\frac{\lambda}{\mu\text{m}}\right)^{3/2} \left(\frac{\text{zpt}}{3631\ \text{Jy}}\right)^{1/2} 10^{-0.2(m-19)} h^{-1}\ \text{cm}, \tag{37}$$

where r_H is the Hubble radius, i is the disc inclination angle, zpt is the zero point of the AB magnitude system,[5] and m is the intrinsic magnitude of the source (Morgan et al. 2010). A handy quantity for comparison with microlensing studies is the half-light radius of the disc. Cornachione and Morgan (2020) refers to the latter as "luminosity size", and its expression, based on the observed specific flux per wavelength (in a particular band), $F_{\lambda,\text{obs}}$, is given by:

$$\begin{aligned} R_{\lambda,1/2}^{\text{flux}} &= C(\beta)\ R_\lambda^{\text{flux}} \\ &= C(\beta) \left(\frac{\lambda_{\text{obs}}^5 D_L^2 F_{\lambda,\text{obs}}}{4\pi h c^2 \cos(i)(1+z)^4}\right)^{1/2} \times \left(\int_{u_{\text{in}}}^{\infty} \frac{u\, du}{\exp(u^\beta) - 1}\right)^{-1/2}, \end{aligned} \tag{38}$$

[5] This can be modified to use another wavelength, for instance, for HST F814W filter $\lambda = 0.814$ μm and zpt = 2409 Jy.

where D_L is the luminosity distance to the quasar, $u = R/r_\lambda$, and the factor $\cos(i)$ accounts for disc inclination. The factor $C(\beta)$ is required to transform the radius $R_\lambda^{\rm flux}$ into a half-light radius and is given in Li et al. (2019) as a numerical solution to

$$\int_{u_{\rm in}}^{C(\beta)} \frac{u\,{\rm d}u}{\exp(u^\beta) - 1} = \frac{1}{2}\int_{u_{\rm in}}^{\infty} \frac{u\,{\rm d}u}{\exp(u^\beta) - 1}. \tag{39}$$

For the standard disc, we have $C(\beta = 3/4) = 2.44$. On the other hand, knowing the mass of the central black hole, $M_{\rm BH}$, and the luminosity, L, the disc half-light radius[6] can be calculated as:

$$R_{\lambda,1/2}^{\rm BH} = 2.37 \times 10^{16} \left(\frac{\lambda_{\rm rest}}{\mu{\rm m}}\right)^{4/3} \left(\frac{M_{\rm BH}}{10^9 {\rm M}_\odot}\right)^{2/3} \left(\frac{L}{\eta L_{\rm E}}\right)^{1/3} \ {\rm cm}, \tag{40}$$

where L_E is the Eddington luminosity and η is the accretion efficiency (Marconi and Hunt 2003; Graham 2007, 2016). Typical values for these parameters are $L/L_{\rm E} \sim 1/3$ and $\eta = 0.1$ (Kollmeier et al. 2006; Shen et al. 2008; Edelson et al. 2015). Although both methods assume the same model, $R_{\lambda,1/2}^{\rm flux}$ estimations are a factor of $\sim 2-3$ smaller than $R_{\lambda,1/2}^{\rm BH}$ (Collin et al. 2002; Pooley et al. 2007; Morgan et al. 2010) and cannot be reconciled neither through uncertainties in the measured black hole mass, nor, in the case of strongly lensed quasars, through lensing magnification (e.g. Mosquera et al. 2011).

Extended versions of the standard thin-disc model have been formulated (see e.g. Abramowicz and Fragile 2013; Middleton et al. 2015; Lasota et al. 2015) that include general relativistic corrections, radiative transfer in the disc atmosphere, black hole spin, or disc winds (Novikov and Thorne 1973; Thorne 1974; Hubeny et al. 2000; Sądowski et al. 2011; Davis and Laor 2011). While the standard accretion disc remains the dominant paradigm, a growing number of observations challenge at least its universality. A non exhaustive list of alternative models that have been proposed includes: advection dominated accretion discs (Ichimaru 1977; Narayan and Yi 1994), slim discs (Abramowicz et al. 1988), inhomogeneous accretion (Dexter and Agol 2011), magnetically arrested discs (Zamaninasab et al. 2014), discs with modified viscosity laws to account for magnetization (Grzędzielski et al. 2017), torn discs (Nixon et al. 2012; Hall et al. 2014), and puffy discs Lančová et al. (2019), Wielgus et al. (2022).

Constraints on the accretion disc temperature profile have been achieved over the last decade by continuum reverberation mapping. This technique consists in measuring how the UV-optical continuum responds to variations of the X-ray emission. Measuring the time lag of this response, τ, in multiple bands can be translated into an increase in size of the emission region with wavelength. The data accumulated for a growing number of AGN indicate that the disc size varies as expected for the thin-disc model of Shakura and Sunyaev (1973). Uncertainties on the slope of the temperature profile are, however, still too large to robustly rule out alternatives. In addition, the disc size is found to be larger than predicted by the model (e.g. Edelson et al. 2015; Cackett et al. 2020; Guo et al. 2022). Diffuse continuum emission, originating from the inner part of the BLR (such as Balmer and Paschen emission; Korista et al. 1995; Korista and Goad 2001; Gardner and Done 2017) may explain the excess UV-optical size in several reverberation mapped systems, but other sources of diffuse emission (e.g. scattering, thick gas clouds, etc) may also be present in AGN (e.g. Lawrence 2012; Lawther et al. 2023). All in all, our theoretical understanding of accretion

[6] We have converted the commonly quoted disc size $R_{\lambda,1/2}^{\rm BH}$ into a half-light radius by multiplying it by 2.44 (Eq. (39)).

discs is under tension. On the one hand, theoretical developments indicate that the accretion disc should not be universally described by a thin-disc model. On the other hand, the thin-disc cannot explain all the observations: the size is found to be larger than expected but its dependence on wavelength follows the $R \propto \lambda^{4/3}$ relation from the standard model. Unfortunately, firm conclusions are hard to draw due to the blending of the disc emission with diffuse pseudo-continuum emission of debated nature.

2.9.3 Intermediate Sizes: The Broad Line Region

One of the most salient features detected in UV-optical quasar spectra are broad emission lines. They arise mostly from hydrogen and helium recombination lines, but also permitted and semi-forbidden lines, such as C IV, and C III], and also complex multiplets from Fe II. These lines, observed over 7 orders of magnitude in AGN luminosities, arise from the Broad Line Region, BLR, a region of radius ranging from ~ 4 up to 100 times larger than the accretion disc, and possibly originating in the form of a clumpy wind launched from the latter (see Fig. 3 and Elitzur et al. 2014; Czerny et al. 2017). The BLR is an important probe of the physical conditions (e.g. gas density, hydrogen column density, metallicity, ...) prevailing in the direct vicinity of the black hole. It is composed of a large number of clouds or of an inhomogeneous/clumpy wind of gas, photoionized by the continuum emission. As is the case for the accretion disc, the BLR is also too compact to be spatially resolved with current telescopes. Therefore, our knowledge of the geometry (e.g. disc-like, bi-conic, bowl-like) and kinematics (Keplerian rotation, inflow, outflow) of BLR gas is yet limited despite of more than fifty years of investigation. The physical conditions in the BLR are explored through advanced photoionization codes, such as the state-of-the-art code CLOUDY (Ferland et al. 2013). This code enables one to reproduce most of the line ratios observed in AGN and has improved our understanding of the chemical evolution of AGN host galaxies (Nagao et al. 2006). Some challenges remain, however, such as reproducing the iron multiplet or the ratio between optical and UV Fe II emission, which shows a large scatter at the population level (Ferland et al. 2009; Sarkar et al. 2021). Our inability to accurately reproduce this major component of AGN spectra is likely linked to uncertainties on the distribution and kinematics of the iron emission in the BLR (Ferland et al. 2009).

Most of our knowledge on the structure of the BLR comes from the reverberation mapping technique (Blandford and Mckee 1982). This technique has enabled the measurement of the luminosity-weighted distance, R_{BLR}, of the BLR from the continuum for the Hβ line in about 70 local AGN (Peterson et al. 1985; Horne et al. 1991; Kaspi et al. 2000; Bentz et al. 2013; Bentz and Katz 2015). This distance scales with the square root of the quasar luminosity, confirming that the BLR gas is mostly photoionized. Time-lags for UV emission lines have been harder to obtain. The Mg II lags have been measured for about 50 AGNs and the luminosity-size relation agrees with the one obtained for Hβ (Homayouni et al. 2020; Yu et al. 2023). Time lags for higher ionization lines, such as C IV or He II have been measured for about 50 systems (Peterson et al. 2005; Kaspi et al. 2007; Grier et al. 2019) and found to be systematically shorter than for Hβ, indicating a stratification of the BLR (i.e. high ionization gas is found closer to the continuum). The line broadening of the (optical) Fe II, as well as reverberation mapping data of a small sample of local Seyfert, indicate that this blend arises from a region at least as large as the Balmer BLR (e.g. Barth et al. 2013; Hu et al. 2015).

Information on the geometry and kinematics of the BLR is more difficult to achieve. Direct modeling of the lines using radiative transfer codes can successfully reproduce their shapes (Murray and Chiang 1997; Borguet and Hutsemékers 2010; Higginbottom et al.

2014). However, the same line shape can be reproduced by a variety of geometries and kinematics of the emitting region, limiting the usefulness of this approach in constraining BLR structure. Spectro-polarimetry of emission lines has been a useful complementary probe, revealing, for instance, the rotating disc-structure of Hα emission in some Seyfert galaxies (Smith et al. 2005). But such measurements are limited to significantly polarized AGN, and results arguably depend on the (poorly known) location of the scattering region at the origin of the polarization. Velocity-resolved delay maps, i.e. measurements of the reverberation time-lag for various velocity slices in the line, are another indirect probe of BLR structure (Horne et al. 1991; Peterson and Wandel 1999). Results obtained for a few dozens of systems are often difficult to interpret due to the unknown transfer function that encodes the time-delay distribution across the broad line as a function of the line-of-sight velocity (Villafaña et al. 2022, and references therein). The more direct method developed by Pancoast et al. (2011) does not require knowledge of the transfer function but has other limitations, especially in presence of complex variability features (see e.g. Li et al. 2013; Pancoast et al. 2014; Bentz et al. 2021). Overall, the data indicate a disc-like geometry for the region emitting the Hβ line (which is the best studied line) but suggest a surprising diversity of BLR kinematics, including Keplerian discs, inflows and outflows.

2.9.4 Further Out: The Torus, NLR, and the Radio Domain

Above ~ 1.5 μm (rest-frame), the contribution of the accretion disc to the continuum emission becomes subdominant compared to the emission from the dust. Observations of dozens of local AGN indicate that the near/mid-infrared dust emission arises from two regions: a "compact" region characterised by hot/warm dust temperature ($T \sim 1400$ K), and a colder component with $T \sim 300$ K that may be 100 to 1000 times more extended than the hotter emission (Kishimoto et al. 2011a). Interferometric studies of local AGN suggest that the ratio of effective areas between these two components is of the order of 400, such that cold emission dominates above typically 7 μm, while hot emission peaks around 2 μm (Kishimoto et al. 2011b). The latter component is the only one compact enough to be susceptible to microlensing. Dust reverberation mapping studies show that the K-band reverberation radius (dominated by dust emission) scales with the square root of the luminosity, establishing a solid dependence of the dust emission on AGN luminosity (e.g. Suganuma et al. 2006; Koshida et al. 2014; Yang et al. 2020). This reverberation radius can be used as the foundation of a ring-like model of the torus by fitting it as a function of luminosity. Kishimoto et al. (2007) derived the scaling of the inner radius of the torus as a function of the luminosity in the V-band, i.e. $\nu L_\nu(5500\ \text{Å})$, as:

$$R_{\text{in}} = 0.47 \left(\frac{6\nu L_\nu(5500\ \text{Å})}{10^{46}\ \text{erg/s}} \right) \text{pc}. \tag{41}$$

The remaining model components are the surface brightness and outer radius. The surface brightness is observationally unknown but theoretical arguments favour a power-law decrease (Barvainis 1987). Interferometric data suggest that $R_{\text{out}}/R_{\text{in}} \leq 2$ at 2.2 μm and reach a factor of several at longer wavelengths (Kishimoto et al. 2007).

Finally, we may stress that the geometry and clumpiness of the dust emitting region is yet debated. A torus-like structure has been favoured for decades as it provides a strong support to the orientation-based AGN manifestations within the framework of the unified model. However, a disc-like ring with dust winds launched from it is suggested in some systems instead of a torus (e.g. Pfuhl et al. 2020). Overall, a lot of open questions remain regarding the dust emitting region. Depending of the AGN luminosity and torus properties, the near-/mid-infrared AGN emission region can be small enough to be slightly affected by microlensing.

The narrow emission lines (with FWHM ≤ 800 km s^{-1}) commonly detected in quasar spectra either arise from the AGN host galaxy or from the Narrow Line Region (NLR). The gas in the NLR is photo-ionised, while narrow emission in the host can be associated, for example, with star formation and may therefore appear to be spatially offset with respect to the NLR (and possibly spatially resolved in high resolution images). The gas in the NLR is exterior to the torus, reaching tens to thousands of parsecs (depending on the luminosity), and can also be spatially resolved (Bennert et al. 2002; Dempsey and Zakamska 2018). The frequent asymmetry in the shape of narrow lines, in particular of [O III] $\lambda\lambda 4959, 5007$ Å, indicates the common presence of outflowing material that can regulate stellar activity in the host (e.g. Speranza et al. 2021). Due to these characteristics, the NLR is expected to be too large to undergo any microlensing.

AGN emission at radio wavelengths reveals a dichotomy whose origin (and even existence) is still debated. About 15% of AGN are radio-loud, while the remaining are qualified as radio-quiet. The emission from radio-loud quasars is mostly due to synchrotron emission. Interferometric radio data often reveal an unresolved core (at sub-parsec scales) and multiple components of a jet that sometimes extends over kiloparsecs. The strongest radio emission is often observed in blazars and flat spectrum systems, whose (often relativistic) jet is aligned to our line-of-sight. Quasars classified as radio-quiet can still have some energy emitted in the radio. Thermal free-free emission (i.e. bremsstrahlung) from either a stellar or AGN component is possible. Additionally, magnetic heating of the corona, thin free-free emission from winds, or synchrotron emission from the base of a small jet or particles accelerated in shocks are also plausible mechanisms of this radio emission (see e.g. Silpa et al. 2020, and reference therein). The radio emission from AGN is commonly assumed to be insensitive to microlensing, but this complex picture of radio AGN does not fully imply this. The existence of radio variability on time scales of a few days (e.g. Biggs and Browne 2018) indicates that regions compact enough to be microlensed should contribute a substantial fraction of the radio emission. Observation of microlensed radio-emission of AGN remains, however, elusive (Koopmans and Bruyn 2000; Biggs 2023).

2.10 Microlensing and the Quasar Structure

The main reason a detailed physical picture of AGN structure is not yet in place is our inability to spatially resolve their innermost regions. "Direct" imaging of the vicinity of a supermassive black-hole has been possible only for a handful of nearby systems thanks to the Event Horizon Telescope, which required a tremendous technical and observational effort to turn our planet into a giant radio interferometer. This enabled the reconstruction of an image of the shadow of the black hole in M87* and of the one lying in the center of our own Galaxy (Event Horizon Telescope Collaboration et al. 2019, 2022). While EHT has been used to observe some distant quasars ($z > 0.5$) at 20 μarcsec resolution at 230 GHz (Jorstad et al. 2023), we are still far from reaching such a spatial resolution for a sizeable sample of systems over the whole electromagnetic spectrum. Our most powerful near-infrared interferometers resolve AGN on scales of ~ 1 pc, corresponding to emission from the "dust torus" (Amorim et al. 2021), but we are yet far from getting a comprehensive view of the innermost parsec.

The fortunate match between the microlensing Einstein radius and the size of the otherwise unresolved AGN regions represents an opportunity to shed light on quasar structure.

Microlensing (de)magnification directly scales with source size (see next section), turning variable signals in time and wavelength into an astrophysical ruler that enables a sensitive measurement of the heart of quasars at multiple scales. For instance, it is the only technique that enables a measurement of the size of the X-ray continuum emission. Contrary to reverberation mapping that gets hampered by relativistic time dilation at high redshift,, microlensing does not rely on observing the intrinsic quasar variability. It is therefore particularly suitable to probe the size of distant quasar emitting regions, nicely complementing reverberation mapping measurements in the local Universe. Differential microlensing between regions of different sizes also provides a tool to study finer properties, like the temperature profile of the disc or the geometry and kinematics of the BLR. Here as well, microlensing nicely complements standard techniques (e.g. reverberation mapping, photoionization, line shape modeling) that rely on different working assumptions. At the scale of $\gtrsim 0.1 pc$, the radial structure of the hot torus may also potentially be constrained by microlensing (or its absence of). The possibility of using microlensing to zoom in the radio and sub-milimeter emission regions is less clear, but observations hint that microlensed effects in these wavelengths may not be immediately excluded. A presentation of the main results obtained with microlensing techniques is given in Sect. 4.

3 Methods

Microlensing offers two main methods to probe quasar structure and the partition of matter in the lens. The first one, known as the "single-epoch method" (described in Sect. 3.3), takes advantage of the differential microlensing occurring between regions of different sizes. For instance, our basic understanding that inner (outer) parts of accretion discs are hotter (cooler) and hence emit in bluer (redder) wavelengths (see Fig. 3 and Eqs. (37) and (40)) postulates that microlensing magnification measured at the same epoch will depend on wavelength. As we can anticipate from the structure of AGNs (Sect. 2.9), the wavelength dependence of microlensing is not restricted to the disc. The second method, known as the "light curve method" (described in Sect. 3.4), analyses the amplitude and rate of the time variability of the microlensing effect. The timescale of variability is shorter for the smaller regions, explaining why it is mostly applied to study the X-ray (corona) and optical continuum (disc) regions. Other methods, generally sharing concepts with the single-epoch and light curve methods, exist, but they are usually tailored to a particular type of data or scenario. A non-exhaustive overview of some selected such methods is given in Sect. 3.5.

All these techniques require identifying the presence of microlensing in the data and measuring its amplitude. In order to infer properties of the source or the lens, a forward modeling method is generally followed that simulates microlensing data and compares them to the observations. Section 3.1 presents important aspects that need to be considered upon designing the simulations, while Sect. 3.2 explains how the microlensing signal can be extracted from the data in the most common cases.

3.1 General Modeling Considerations

Microlensing variability depends most crucially on the size of different emitting regions of the quasar with respect to the Einstein radius of the microlenses (see Eq. (7)). In order to simulate the magnification for any size and shape of the source one simply needs to convolve a magnification map with the source's brightness profile. The magnification range produced by the stars in the lens galaxy increases as the source becomes more compact, which is

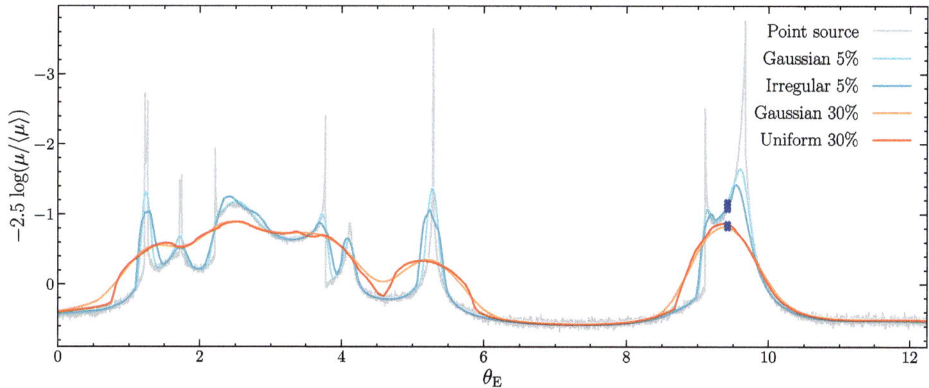

Fig. 15 Light curves for sources of different size and shape moving along the trajectory shown in the right panel of Fig. 4 and in Fig. 10 (from bottom to top). When the point source (gray curve) enters and exits the areas bordered by caustic lines, characteristic double-horned profiles can be seen. Larger sources lead to smoother curves without the extremely magnified peaks, as it can be seen by the Gaussian-shaped sources with half-light radii at 5 (blue) and 30 (red) per cent of the Einstein radius. The shape of the source plays a secondary role and only modulates small scale features in the light curves, as demonstrated by a uniform and irregular disc (the latter being simply a greyscale, black background photo of the face of one of the authors) with the same half-light radii as the Gaussian profiles. Statistical errors (noise) introduced by the inverse ray-shooting technique in the magnification map pixels can be seen for the point source, which are averaged out as soon as the source has a finite size of just a few pixels. The source position marked with the blue cross in the right panel of Fig. 4 is also indicated here

imprinted on all kinds of microlensing data, i.e. flux ratios, light curves, spectra, and high magnification events (see Table 1). An example for the case of light curves is shown in Fig. 15, where we see a clear dependence primarily on the size of the source – short dramatic changes and smooth extended variations for small and large sources respectively – and to second order on its shape. Another example is illustrated in Fig. 16, where the magnification probability distribution is shown for a "point source" ($< 0.0025\ \theta_E$, see below), and three Gaussian brightness profiles of increasing size. In fact, it has been shown (Mortonson et al. 2005; Vernardos 2019) that it is mainly the half-light radius of the source, $r_{1/2}$, that determines the extent of microlensing effects, while its detailed projected two-dimensional shape plays a secondary role (as long as $r_{1/2}$ is kept the same). For this reason, a common choice is a Gaussian luminosity profile for the source ($r_{1/2} = 1.18\sigma$). In a similar way, a common choice of the dependence of size on wavelength, particularly for the accretion disc, is the parametric model:

$$r_{1/2} = r_0 \left(\frac{\lambda_{\text{rest}}}{\lambda_0}\right)^{1/\beta}, \tag{42}$$

where λ_0 and r_0 are a reference wavelength and size, and β is the slope of the temperature profile. This allows for more flexibility, as opposed to, for example, the thin-disc model that has a fixed temperature slope of $\beta = 3/4$ (and depends on other physical parameters like the black hole mass, etc, see Eq. (40)).

A macroscopic model of the lensing galaxy defines the general properties of microlensing variability. It is the starting point of almost all of the methods used to infer physical quantities from a microlensing signal. The macromodel values of κ, γ, κ_* at the locations of the observed multiple images (see also Shajib et al. 2024) are used to obtain corresponding

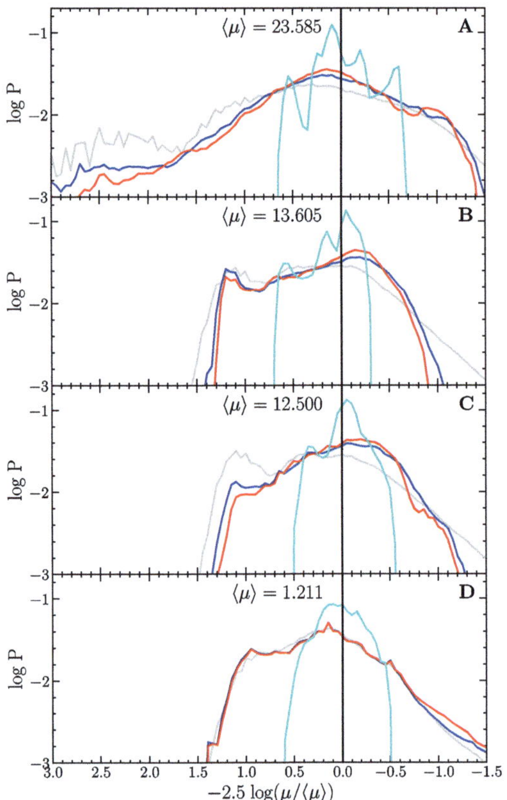

Fig. 16 Magnification probability distributions from the maps corresponding to the multiple images of RXJ 1131−1231 (shown in Fig. 17) for a point source ($< 0.0025\ R_\mathrm{E}$, grey lines) and three Gaussian brightness profiles with sizes of 0.117 (blue), 0.2 (red) and 1.6 (cyan) R_E, respectively (matching the source sizes shown in Fig. 19). The vertical lines indicate the macromagnification $\langle\mu\rangle$. We can see that the larger the source the smaller the extent of the deviations from the macromagnification due to microlensing. However, in this example we used the same 25^2-θ_E, $10,000^2$-pixel maps for all the sources, which have a sufficiently large resolution compared to the size of the smallest sources (blue and red circles on Fig. 19) but are probably not wide enough to result in unbiased magnifications for the largest source. The high resolution is not needed in this case because of the smoothness of the chosen Gaussian profiles

microlensing magnification maps, as illustrated in Fig. 17 for a quad. Such maps are the indispensable tool to model microlensing variability and are computed with one of the techniques presented in Sect. 2.7. Because the κ, γ, κ_* parameters only define the microlenses at the population level, e.g. they do not define the positions and masses of the microlenses uniquely, each map is one of many[7] random and equivalent realizations that can be used interchangeably. This is important because maps have a finite size and resolution that can be limiting depending on the application, hence using equivalent maps can help. In addition, averaging over multiple realizations of maps can help avoid the sample variance in the histogram from using a single map.

A map constitutes a finite region of the source plane and has to be large enough to be representative, i.e. cover enough "troughs" and clusters of superimposed caustics (see Fig. 12) in order not to be biased towards lower or higher magnification.[8] The higher the resolution of the map compared to θ_E, the better the pixels approximate the size of a point source. For pixel sizes of $< 0.006\ \theta_\mathrm{E}$ (practically point sources), Vernardos and Fluke (2013) found that the minimum size for unbiased maps is $24 \times 24\ \theta_\mathrm{E}$. For larger sources, there are two reasons for having an adequately sized map: avoiding convolution edge effects and covering

[7] As many as the different configurations of the microlens positions and the ways κ_*, the mass density in microlenses, can be distributed to them – by all measures, infinite.

[8] This is required for statistical studies, e.g. flux ratio or light curve analysis, but does not need to be the case for high magnification events, which are usually dominated by a single caustic crossing that has well-defined, universal properties (Fluke and Webster 1999).

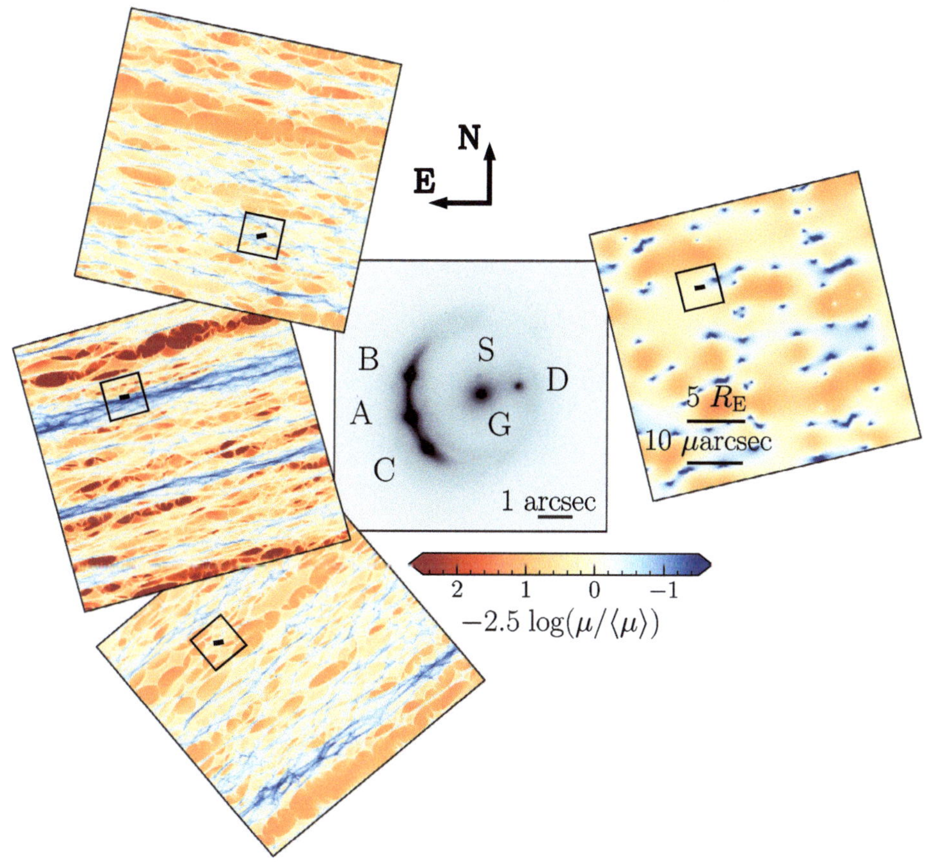

Fig. 17 Lens model of quadruply lensed quasar RXJ 1131−1231 taken from Chen et al. (2016) and corresponding magnification maps at the locations of the quasar multiple images. The maps are consistent with the mass model, i.e. have the corresponding κ, κ_*, γ (see Table 1 in Vernardos 2022) and are aligned with the local total shear vector. A trial trajectory for a quasar moving (from right to left) across the source plane is shown on each map (black line with length of $\approx 0.5\ \theta_E$; the corresponding microlensing light curves are shown in Fig. 18). Adapted from Vernardos (2022)

enough area for unbiased magnifications, i.e. mitigating the sample variance (e.g. studies of the BLR usually use maps whose size can be up to 50-200 θ_E). Statistical studies need to sample the stochastically varying magnification field in an unbiased way. It should be noted that locations that are within a source-size from each other become correlated via convolution.[9] This correlation may be desirable when studying time series, whose values are not independent from each other. The number of sources that can fit in a map of given size without overlap decreases with the source size and the more overlap between source positions the more correlated their magnification values become. The issue merely reflects the already mentioned problem of sample variance: if a few, long-range features dominate a

[9]That is, if the source is placed in these locations it will overlap with itself. This applies equally to tracks across maps. For practical purposes, the full extent of the source could be replaced by some fraction of its half-light radius.

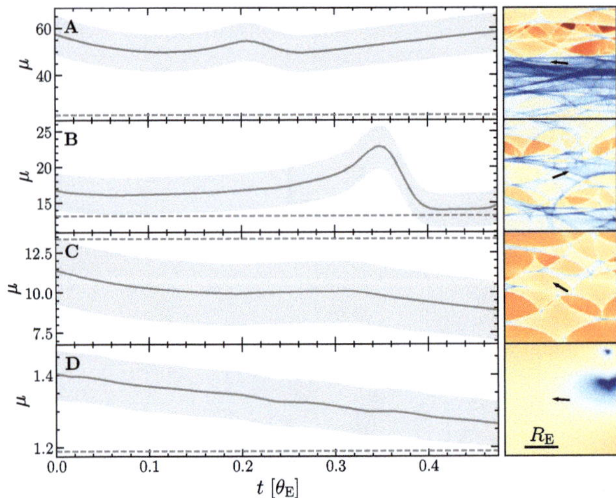

Fig. 18 Realization of magnification as a function of time for each multiple image of RXJ 1131−1231 and corresponding zoomed-in region of the magnification maps shown in Fig. 17. The source trajectory is indicated by the arrow (same as the trajectories in Fig. 17). The horizontal dashed lines correspond to the macromagnification of each image. The shaded areas indicate the Poisson error (after convolution, Vernardos et al. 2015) of the magnification resulting from the numerical inverse ray-shooting technique (Kayser et al. 1986) that was used to generate the maps

magnification map, even after smoothing via convolution with a large source, then larger or statistically equivalent maps should be used in order to obtain unbiased magnifications.

Once the map size is set, then the minimum pixel size (resolution) should be small/high enough so that the smallest source would still be covered by a few map pixels. This rule-of-thumb is to avoid extreme (de)magnifications that can appear if a source is equal to or smaller than a map pixel.[10] Similarly, because each pixel is a numerical approximation of the magnification by a finite number of rays, statistical (Poisson) fluctuations average out over a few pixels. Both of these effects can be seen in the light curves shown in Figs. 15 and 18. In fact, the pixel size should be small enough to resolve any part of the source that contains a large fraction of the total flux (e.g. a hotspot within an accretion disc). Conversely, if there are mostly small brightness gradients over the parts of the source that emit most of the flux, then there is no need for high resolution. This is the case for the magnification of the BLR, which is smoothed out due to its extent over several θ_E. In the examples shown in Fig. 16, the probability distributions for the largest source could be biased and using multi-scale, multi-resolution maps cannot be avoided, at the cost of making sure they are consistent with each other.

3.2 General Observational Considerations

At any given time, there is a high probability that microlensing takes place in at least one image of a lensed quasar due to the high optical depth for microlensing. This probability is

[10] A map pixel is simply a finite area in the source plane that indiscriminately collects rays that are shot from the image plane. If, for example, the pixel happens to contain a caustic, which is a line of infinitesimal width, the very high resulting magnification is attributed to the entire pixel and any source that may lie within it, regardless of its exact position.

Table 1 Summary of the most common types of data sets used depending on the microlensing application. Temperature and time delay are denoted as T and Δt

	Long-term variability	Snapshots	High magnification events
Observational characteristics	Decade-long monitoring Weekly cadence Single band	Fixed moment in time Multiple bands / spectrum	Month-long monitoring Daily cadence Multiple bands / spectrum
Science applications	Accretion disc size Baryonic/dark matter ratio Mean microlens mass	Accretion disc T profile Accretion disc size BLR structure+kinematics Baryonic/dark matter ratio Mean microlens mass	Detailed disc geometry and T profile BLR structure+kinematics
Advantages/ drawbacks	Additional velocity model Needs a measured Δt	Needs an independently measured Δt	Rare (1 per decade/system)† Hard to predict

†Except in the singular case of Q2237+0305 that has a very high effective velocity.

the largest for quads, where there is almost always one image that is subject to substantial microlensing (Witt et al. 1995). The presence of microlensing can be assessed by measuring fluxes, in one or more bands, or spectral ratios, at a single or at multiple epochs, as summarized in Table 1, and then comparing these measurements to the microlensing-free case. The latter serves as a "micro-lensing zero-point" or "no-microlensing baseline", and has to be determined from ancillary data or models. However, it is difficult to obtain this for single images, as well as disentangling the intrinsic variability of the quasar, without additional information (e.g. the time delay, see below). For this reason, multiple images are considered in pairs.

For an image i, the magnification from the macro-model M_i (Eq. (14)), can be used to derive the baseline magnification ratio between a pair of images (i, j), i.e. $M = M_i/M_j$. There is, however, no a priori guarantee that the macromodel captures the full complexity of the system. For instance, the presence of a nearby satellite galaxy may produce a flux ratio anomaly compared to a standard macro-model (e.g. Vegetti et al. 2024). If the latter is not accounted for, the estimated amplitude of microlensing from the data will be biased. For that reason, one can alternatively derive the amplitude of microlensing directly from the data, as explained in Sect. 3.2.1 and Sect. 3.2.2.

The requirement to compare pairs of lensed images implies that the interpretation of the microlensing signal is in general ambiguous: either one image is magnified or the other one is demagnified. It can also be that microlensing is taking place simultaneously in both images in a pair. The comparison of multiple pairs of images in quads helps resolving this ambiguity. The combination of different kinds of data, such as time-series and multi-wavelength data, may also be helpful for determining a plausible microlensing scenario in a lensed system.

One can distinguish three main strategies for carrying out observations suitable for microlensing studies (Table 1): snapshot data, long-term monitoring, and targeted monitoring triggered based on specific criteria. The single-epoch method (see Sect. 3.3) is observationally the simplest one, as it consists of one-time observations of a system (snapshots) in many wavelengths. Flux ratios are assumed to have been corrected for time delays, otherwise, it is necessary to simulate the impact of intrinsic variability, either by theoretical models (e.g. Vernardos et al. 2023, submitted) or by means of Monte-Carlo simulations. Long-term monitoring data have commonly been a by-product of campaigns aiming at measuring the time delay between lensed images (e.g. Tewes et al. 2013, and Birrer et al. 2024). Once corrected for the time delay, these light curves can be used to extract microlensing signals. Surveys observing regularly the same region of the sky, like the Rubin Observatory Legacy Survey of Space and Time (LSST), will yield an increase in the availability of such data in the next decade. Finally, there is a category of suitably timed observations targeting specific systems that are undergoing High Magnification Events (HME). These are generally defined as sharp increases of brightness (e.g. > 1 mag in the optical) in one of the images in a pair over a period of a few weeks to months or even years, owing to a very compact region of the quasar, most likely very close to the SMBH, going through a very high magnification region. These events are usually attributed to single caustic crossings, although more complex situations can produce similar events (see Neira et al. 2023, submitted, and Fig. 4). Triggering such follow-up observations is not easy due to the large number of systems that have to be monitored in order to capture the onset of an event. Although a handful of candidate HMEs have been captured by dedicated campaigns, we expect a few hundred events per year to be observed by surveys like LSST (Neira et al. 2020).

3.2.1 Measuring Microlensing from Imaging Data

Good proxies of the microlensing-free flux ratios can be measured in regions that are considered large enough to be free of microlensing, i.e. tens of θ_E (see Refsdal and Stabell 1997, for an estimate of the predicted amplitude of microlensing fluctuations for large sources). As explained in Sect. 2.9.4, a sweet-spot may be the radio and mid-infrared ranges. While those regions may be the ones less affected, it is not guaranteed that they are totally free of microlensing. For instance, microlensing is suspected to occasionally occur at radio wavelengths (Koopmans and Bruyn 2000; Biggs 2023), but also radio emission may be spatially offset from the optical disc or blend multiple sources of emission only discernible with very high resolution interferometry (e.g. Hartley et al. 2019; Pashchenko et al. 2020). In the mid-infrared range, one needs to disentangle between different emission regions to derive flux ratios totally free of microlensing (Stalevski et al. 2012; Sluse et al. 2013). Rest-frame emission with $\lambda > 11$ μm is probably the least susceptible to microlensing (Sect. 2.9.4), but is observationally out of reach for quasars at $z \gtrsim 2$. When the source size gets too large, one should also account for finite size effects arising because the macro-magnification is not constant over the source profile. In some rare situations, the source may even cross a macro-caustic introducing more dramatic changes of flux ratios but generally accompanied by spatially resolved emission.

When spectra are available, NLR flux ratios may also be used (see Sect. 3.2.2), however, they may sometimes deviate from the baseline and become spatially resolved (e.g. Sluse et al. 2007). Because of small asymmetries in the emission region, there can be a luminosity centroid that is slightly offset compared to the emission from the accretion disc. The differential magnification between the disc and the NLR due to strong lensing may yield a small bias on flux ratio measurements. Even in the case where a NLR brightness peak is in projection coincident with the location of the disc, biases can occur due to the way the flux ratio is obtained. The measurement needs to be performed over the whole resolved emission from the NLR, and a correction for source size effects associated to strong lensing has to be applied.

It is important to keep in mind that other effects can yield a deviation of monochromatic flux ratios from the baseline. The intrinsic quasar variability, as well as the differential extinction caused by different amounts of dust along the path of individual lensed images, inevitably introduce additional uncertainties. The former can be accounted for by observing a system at multiple epochs separated by the time delay, while the latter requires multi-wavelength data to be corrected for (see Sect. 3.2.3).

3.2.2 Measuring Microlensing in Spectra

The spectra of quasars combine emissions arising from regions of different sizes, which are therefore subject to a different amount of microlensing. Consequently, the shape of a microlensed spectrum will be deformed compared to the intrinsic one. In general, at a given epoch, the smallest region will be more affected than more extended regions[11] This basic rule helps to anticipate the shape of a spectral deformation.

The simplest diagnostic for the presence of microlensing in spectra consists of calculating spectral ratios between pairs of gravitationally lensed images (see also Fig. 19). Let us consider the spectra of two lensed images of an AGN, $F_1(\lambda)$ and $F_2(\lambda)$, as observed in the optical rest-frame range at two epochs separated by the time delay. In absence of microlensing, those two images are simply magnified by the macro-magnification M_1 and M_2. The spectra may contain emission from the power-law continuum,[12] which arises from the compact accretion disc with size of typically $\sim 0.1\ \theta_E$, emission from broad lines originating from the BLR with size $\sim 2\ \theta_E$, and emission from narrow emission lines that arise from regions with size $\sim 50\ \theta_E$ (Fig. 3). In a no-microlensing scenario, we should have $M(\lambda) = F_1/F_2 = M_1/M_2$ being a constant across wavelengths. If extinction due to the lensing galaxy is present, $M(\lambda)$ is not constant anymore and differs from the macro-magnification M (Sect. 3.2.3).

Figure 19c shows an example of spectral deformation expected for a pair of lensed images due to microlensing. For simplicity, this figure displays the case where F_2 is virtually free of microlensing (i.e. $\mu \sim 1$), and the continuum of F_1 is microlensed by $\mu \sim 2.15$ at the wavelength of $H\beta$. The spectral ratio F_1/F_2 (Fig. 19e) reveals two features. First, a dip is visible at the wavelengths corresponding to the emission lines. This can be easily understood in terms of the expected relative sizes of the emitting regions: since the continuum comes from the smallest region, it is on average more microlensed than the broad and narrow lines, resulting in more continuum than line flux. Second, a chromatic slope is seen because of differential magnification of the continuum itself: bluer emission arises from the inner, smaller part of the disc and is more magnified than redder. The strength of this chromatic effect depends mostly on the amplitude of microlensing and on the relative change of size of the disc as a function of wavelength. The flux ratio in the narrow lines will be the closest to M due to the large size of the NLR. It is important to realise that an accurate measurement of M based on the narrow lines (when unresolved by strong lensing) cannot be achieved by simply looking at the spectral ratios due to the presence of both continuum and line flux at the same wavelength. Instead, it should be based on the integrated flux in the lines, i.e. after continuum subtraction [see panel (e) of Fig. 19]. In order to better characterise the potential role of variability, one can also compare flux ratios of the same lensed images at multiple epochs. This method, in combination with quantitative measurements on the spectral ratio of pairs of images, has been used by Popović and Chartas (2005) to disentangle microlensing from intrinsic variability and millilensing in several systems.

The spectral ratios mostly provide a simple and quick diagnostic of the presence and amplitude of microlensing. A more quantitative criterion of differential microlensing between

[11] However, this is not strictly always true. While less frequent, one can find specific configurations of the source with respect to the microlensing network, where, for example, a more compact emitting region can be less magnified because of being located in a demagnification trough compared to a more extended emission region that can cover several caustics (e.g. Hutsemékers et al. 2023).

[12] For the sake of simplicity, we ignore here pseudo-continuum emission from, for example, the AGN host galaxy or from Fe II, which are known to be present in AGN spectra.

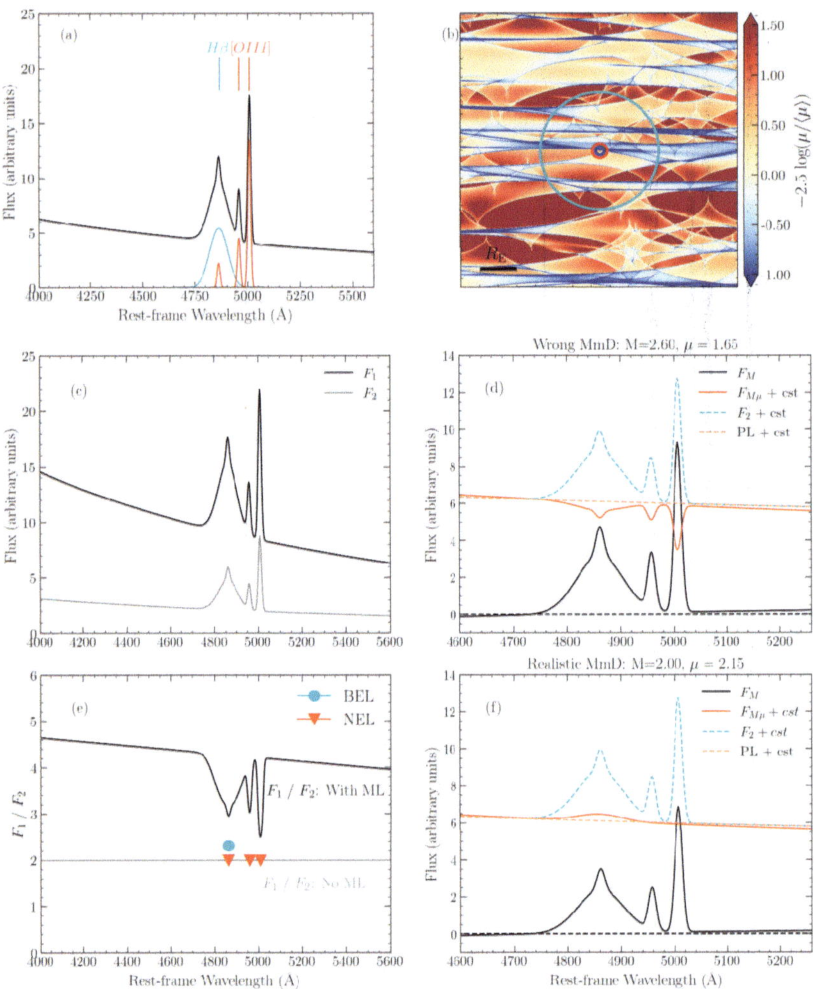

Fig. 19 Illustration of the spectral deformation introduced by microlensing on an AGN spectrum. *Panel (a)*: The rest-frame range 4000-6000 Å is shown. The emission components considered are a power-law continuum, a broad $H\beta$ λ 4864 Å emission arising from the BLR, and 3 narrow emission lines arising from the NLR and associated to $H\beta$ and [O III] λλ4959, 5007. The broad line components (i.e. after continuum subtraction) are shown with colored solid lines. *Panel (b)*: The colored circles correspond to $r_{1/2}$, of the continuum emission at 4000 Å (blue; $r_{1/2} = 0.117\ R_E$), 6000 Å (red; $r_{1/2} = 0.2\ R_E$), and the BLR (turquoise; $r_{1/2} = 1.6\ R_E$), overlaid on the magnification map of image A of RXJ 1131−1231 (Fig. 17). The NLR is much larger than the map ($r_{1/2} = 48\ R_E$) and hence not shown. The micro-magnification of the different components are $\mu_{\rm blue} = 2.46$, $\mu_{\rm red} = 1.86$, $\mu_{\rm BLR} = 1.16$. The solid black horizontal bar corresponds to 1 R_E. *Panel (c)*: The spectra of two lensed images F_1 and F_2. The macro-magnification ratio between the two images is fixed to $M = M_1/M_2 = 2$. *Panel (e)*: Spectral ratio for the microlensing situation displayed (black), and in absence of microlensing (grey). The ratios derived from the narrow lines (triangles) and from the broad lines (circles) are also shown. *Panels (d & f)*: Result of the MmD decomposition method, explained in Sect. 3.2.2. This method derives $F_{M\mu}$ (F_M), which corresponds to the fraction of the flux that is (not) microlensed. *Panel (d)* shows the decomposition when an incorrect pair (M, μ) is used. *Panel (f)* shows a realistic decomposition achieved by minimizing the appearance of the NEL emission in $F_{M\mu}$. Due to the chromatic microlensing of the continuum, we have $F_M < 0$ for λ < 4700 Å and $F_M > 0$ for λ < 4700 Å. The value $\mu = 2.15$ retrieved is $\mu_{\rm blue} > \mu > \mu_{\rm red}$. It corresponds to μ at λ ∼ 4900 Å. Detrending the chromaticity of the continuum microlensing prior to the MmD can be performed to achieve a more precise decomposition

the continuum and the lines can be obtained by measuring the line equivalent width:

$$W = \int_{\lambda_1}^{\lambda_2} \left(\frac{F_\lambda}{\mathcal{F}} - 1 \right) \mathrm{d}\lambda, \tag{43}$$

where $\mathcal{F}$ is the continuum flux at the level of the line and F_λ the total observed spectral emission at wavelength λ. Because it involves a normalisation by the continuum, it is easy to verify that W measured in two lensed images will differ only if the amplitude of microlensing is different in the line and in the continuum (e.g. Lewis and Belle 1998; Sluse et al. 2015, and Vernardos et al. 2023, submitted).

Another method to derive microlensing signals from spectral deformations consists of modeling the pairs of spectra with the same intrinsic components, such as the sum of a power law continuum and of a series of Gaussian/Lorentzian line profiles that represent the emission (and/or absorption) components, as depicted in panels (d) and (f) of Fig. 19. The comparison of the flux ratios in each of these components may reveal the presence of microlensing (Sluse et al. 2007, 2011). The drawback of this method is that there is no guarantee that each of these components effectively represents a spatially different emission region (e.g. a single-peaked Gaussian line profile could result from the superposition of a double-peaked emission from a Keplerian emission and of a Gaussian profile arising in a wind). It is also possible that the intrinsic emission profile will get deformed due to microlensing, such that a model valid for a non-microlensed line will fail for a microlensed one.

To cope with the limitation of the previous method, Sluse et al. (2007) have proposed isolating the flux arising from regions affected by microlensing from the flux arising from larger, microlensing-free regions by linearly combining pairs of spectra. This method, called Macro-micro decomposition (MmD), is similar to the one proposed by Angonin et al. (1990) and O'Dowd et al. (2015). It assumes that one spectrum is minimally microlensed (hereafter F_2) and is a good template of the intrinsic spectrum. If μ is the average microlensing magnification of the continuum in the wavelength range of the line, $M = M_1/M_2$ is the macro-magnification ratio, $F_{M\mu}$ the microlensed part of the flux, and F_M the non-microlensed part, then one can write:

$$F_1 = M \times (\mu\, F_{M\mu} + F_M), \tag{44}$$

$$F_2 = F_M + F_{M\mu}. \tag{45}$$

These two equations can easily be combined to isolate F_M and $F_{M\mu}$:

$$F_M = \frac{-A}{A - M} \left(\frac{F_1}{A} - F_2 \right), \tag{46}$$

$$F_{M\mu} = \frac{M}{A - M} \left(\frac{F_1}{M} - F_2 \right), \tag{47}$$

where we have defined $A = M \times \mu$. This method is illustrated in panels (d) and (f) of Fig. 19. In addition, any decomposition should verify $F_M > 0$ and $F_{M\mu} > 0$. In general, these two equations can be verified for a range of (positive) values of (M, μ). The procedure described below enables the derivation of the values of those parameters that minimize the amount of microlensing in the broad line. First, one determines the product, $A = M \times \mu$, by measuring the flux ratio F_1/F_2 in a region that contains only continuum emission. This implies that $F_M = 0$ in the continuum. Once A is derived, M is fined-tuned until the flux $F_{M\mu}$ mimics as closely as possible a continuum emission. $F_{M\mu}$ can deviate from a pure continuum emission if other regions are microlensed (e.g. the broad lines). Note that M can also be set a priori from the macro-model or from ancillary imaging data as explained in Sect. 3.2.1. To first order, this method also generally works when the second lensed image is also microlensed (see appendix C of Sluse et al. 2012). In that case, only the relative microlensing between the pair of images is retrieved. When more than two lensed images are observed, the comparison between multiple pairs of images generally enables one to identify which image is minimally microlensed. There is a number of situations where the above decomposition does not work. For instance, when the line flux is magnified only for some velocity bins while others are demagnified, e.g. the blue wing of the line is magnified while the red wing is demagnified. The method also fails when some velocity bins are more (de-)magnified than the underlying local continuum. A description of the method when both absorption and emission lines are present can be found in Hutseméкers et al. (2010). If applied to pairs of spectra that have not been corrected by the time delay, intrinsic variability could mimic microlensing-induced line deformations, and/or chromatic microlensing. These effects remain in principle small for time delays $\lesssim$ 40 days Yonehara et al. (2008), Sluse et al. (2012).

Guerras et al. (2013a) proposed another method to measure microlensing from spectra. It consists in deriving the amount of microlensing present in a line by calculating the difference of magnitude between the line wings and core. First, one needs to locally fit the continuum flux based on two regions on each side of the line and free of pseudo-continuum emission. From the continuum subtracted spectrum, one can measure the magnitude $m_{1,2} = -2.5\log(F^{\rm line}_{1,2})$ of the line, either in its core or in the line wings. The differential microlensing between the wing and core of the line is given by:

$$\Delta m = (m_1 - m_2)_{\rm wings} - (m_1 - m_2)_{\rm core}. \tag{48}$$

As shown by Guerras et al. (2013a), Δm is naturally corrected for differential extinction, which can be assumed to be constant over the wavelength range covered by the line. Δm exactly corresponds to the microlensing of the line only if the line core is effectively free of microlensing.

Braibant et al. (2017) have introduced several indicators to quantify the effect of differential magnification between the wings and core of a line and/or line asymmetry. They can be calculated using the ratio between the line profiles in pairs of lensed images, are independent of the shape of the line profile, and can be calculated on both simulated or observed emission lines. These indicators are useful in establishing quantitative diagnostics of line deformations (see Braibant et al. 2017), and can be calculated even for moderate signal-to-noise ratios. Calculating such indicators instead of using the full profile may however be seen as a loss / degradation of the signal. Instead, one may rather use continuum subtracted line profiles in pairs of spectra to derive microlensing as a function of velocity: $\mu(v) = F^{\rm line}_1/(M * F^{\rm line}_2)$. In the situation where the line is also contaminated by pseudo-continuum emission (such as Fe II), a model of the latter needs to be subtracted as well and uncertainties propagated to the estimate of $\mu(v)$. The use of $\mu(v)$ is only possible for spectra with a sufficient signal-to-noise ratio. Even in that case, $\mu(v)$ may need to be truncated at high velocity because the line flux becomes too low to enable a reliable estimate of μ.

3.2.3 The Role of Extinction

Extinction caused by dust along the line of sight to different lensed images can lead to a different no-microlensing baseline and disguise itself as chromatic microlensing (wavelength dependence). To mitigate this, let us consider that the intrinsic flux of an AGN lensed image at a given wavelength, λ (observer's frame), and time, t, can be expressed as $m_0(\lambda, t)$ (in mag). An observed lensed image i will be magnified by $M_i(\lambda, t)$ (the total magnification includes the macrolens and microlensing that could vary with wavelength and time), and have a time delay Δt_i, resulting in the observed spectrum (Falco et al. 1999):

$$\begin{aligned} m_i(\lambda, t) = m_0\left(\frac{\lambda}{1+z_S}, t - \Delta t_i\right) - 2.5\log\left[M_i\left(\frac{\lambda}{1+z_L}, t\right)\right] \\ + E_i R_i\left(\frac{\lambda}{1+z_L}\right) + E_{\rm Gal} R_{\rm Gal}(\lambda) + E_S R_S\left(\frac{\lambda}{1+z_S}\right), \end{aligned} \tag{49}$$

where E_i, $E_{\rm Gal}$, E_S are the image i extinctions, $E(B-V)$, produced by the lens galaxy, our Galaxy, and the source host respectively, and $R_i(\lambda)$, $R_{\rm Gal}(\lambda)$, and $R_S(\lambda)$ indicate the corresponding extinction laws. This equation can be simplified considering several assumptions. First, the terms associated with our Galaxy and the host galaxy extinction can be considered as negligible because the separation between rays at the source and at the observer is considered small and thus the amount of extinction should be the same for all images. On the contrary, the contribution of the lens galaxy should be more important because the images pass through regions separated by kiloparsecs, which are likely to contain different amounts of dust (Cardelli and Savage 1988). Finally, assuming that the magnification is independent of λ and t (i.e. no microlensing), the source spectrum is independent of t and time delays, and the lens galaxy extinction curve is the same for all images, we can estimate the magnitude difference between two images i and j as (Falco et al. 1999):

$$m_i(\lambda) - m_j(\lambda) = -2.5\log\left(\frac{M_i}{M_j}\right) + (E_i - E_j)R\left(\frac{\lambda}{1+z_L}\right), \tag{50}$$

which depends only on the constant macro-magnification ratios, M_i/M_j, the extinction difference, $E_i - E_j$, and the extinction curve in the lens rest frame, $R(\lambda/(1+z_L))$. The previous assumptions might not hold for the smallest regions of the AGN (e.g. the continuum could be affected by chromatic microlensing magnification, (see Fig. 19), thus it is better to use narrow emission lines (which are only affected by the macrolensing magnification and extinction). For example, the doubly imaged SBS0909+523 system was used to estimate the extinction curve of the lens galaxy and confirm the lens redshift by detecting the 2175 Åblue bump (Motta et al. 2002; Mediavilla et al. 2005), while also estimating chromatic microlensing in the system.

3.3 The Single-Epoch Method

Here we consider the application of the single-epoch method to photometric data. These consist of single or combined exposures in a narrow or broad band, from which flux measurements for point sources, i.e. the quasar multiple images, are extracted. In this case, the flux between a pair of macroscopically observed multiple images A and B, $\Delta m_{\rm AB}^{\rm obs,k} = m_{\rm B}^{\rm obs,k} - m_{A}^{\rm obs,k}$ where k denotes the photometric band used, is measured with respect to a baseline that is believed unaffected by microlensing, $\Delta m_{\rm AB}^{{\rm ref},k} = m_{\rm B}^{{\rm ref},k} - m_{\rm A}^{{\rm ref},k}$. The microlensing signal can then be defined as:

$$\Delta m_{\rm AB}^{k} = \Delta m_{\rm AB}^{\rm obs,k} - \Delta m_{\rm AB}^{{\rm ref},k}. \tag{51}$$

We note that the no-microlensing baseline may or may not depend on the band/wavelength, for example if there is extinction or if it is taken to be the macromodel flux ratio respectively (see Sect. 3.2).

Simulated microlensing flux ratios (or more precisely magnification ratios) can be obtained from the magnification maps corresponding to images A and B given a model of the source light profile. The former are defined mainly by the κ, γ, κ_* parameters, but can also depend on other assumptions like the mass function of the microlenses, which we can collectively denote as η_m. The latter essentially boils down to a way of determining $r_{1/2}$ of the source as a function of wavelength. This radius may be calculated based on a physical model of the source, for instance, in the case of an accretion disc one can use Eq. (40) that depends on the black hole mass, luminosity and accretion efficiency, or be purely phenomenological (for example Eq. (42)). We can collectively denote all the free parameters of the source as η_s. Simulated microlensing flux ratios as a function of wavelength are in fact a function of these free model parameters:

$$\Delta m_{\rm AB}^{\rm mod}(\lambda_{\rm rest}) \equiv \Delta m_{\rm AB}^{\rm mod}(\lambda_{\rm rest}|\eta_m, \eta_s). \tag{52}$$

To obtain the microlensing flux ratio in any photometric band one should in principle integrate over a range of wavelengths and take into account the band's response function. In practice, only the central wavelength of the band can be considered and the dependence on λ_{rest} can be replaced by the index k. Finally, to obtain the ratios between images A and B we randomly select locations on the convolved maps and divide their magnification values pairwise. We note that these locations have to be the same across the different bands considered.

The observed and simulated microlensing flux ratios can be compared through a chi-squared statistic:

$$\chi_n^2 = \sum_i \sum_{j>i} \sum_k^K \left[\frac{\Delta m_{ij}^{\rm obs,k} - \Delta m_{ij}^{{\rm ref},k} - \Delta m_{ij}^{\rm mod,k,n}(\eta_m, \eta_s)}{\sigma_{ij}^k} \right]^2, \tag{53}$$

where i, j is $1-4$ for a quadruply lensed quasar, σ_{ij}^k is the error associated with the observed flux ratios, K is the number of bands, and n denotes a given pair of locations between two magnification maps. We can thus obtain the final likelihood for a given combination of parameters η_m, η_s as the sum:

$$\mathcal{L}(\boldsymbol{d}|\eta_m, \eta_s) = \sum_n^N e^{-\chi_n^2/2}, \tag{54}$$

where $\boldsymbol{d}$ represents the flux ratio observations between all image pairs across all bands and we use N trials of simulated flux ratios (i.e. pairs of locations between maps).

The likelihood above can be used in a Bayesian setup with the probability density of the posterior given by:

$$\frac{{\rm d}P}{{\rm d}\eta_m\,{\rm d}\eta_s} \propto \mathcal{L}(\boldsymbol{d}|\eta_m, \eta_s) \frac{{\rm d}p}{{\rm d}\eta_m} \frac{{\rm d}p}{{\rm d}\eta_s}, \tag{55}$$

where p are the priors on the free parameters of the problem – common choices being uniform or logarithmic. In this way, the posterior probabilities for any parameter of the model, e.g. the dependence of size on wavelength (ν, for an accretion disc) or κ_*, can be obtained by marginalizing over all the remaining free parameters of the model (e.g. Bate et al. 2018).

Finally, one can extend this method to constrain a common underlying source model with data from multiple systems (e.g. Jiménez-Vicente et al. 2015a, who rescaled the size of the accretion disc by the black hole mass). The final joint likelihood is obtained by the product:

$$\frac{\mathrm{d}P}{\mathrm{d}\boldsymbol{\eta}_s} \propto \prod_l^L \frac{\mathrm{d}p_l}{\mathrm{d}\boldsymbol{\eta}_s} \int \mathcal{L}(\boldsymbol{d}_l|\boldsymbol{\eta}_{m,l}, \boldsymbol{\eta}_s) \frac{\mathrm{d}p_l}{\mathrm{d}\boldsymbol{\eta}_{m,l}} \mathrm{d}\boldsymbol{\eta}_{m,l} \,, \tag{56}$$

where each of the L observed lensed systems has its own map parameters $\boldsymbol{\eta}_m$ (e.g. the κ, γ for each image) and may have its individual priors p_l for the source parameters $\boldsymbol{\eta}_s$.

3.3.1 Advantages and Drawbacks

The main advantage of the single-epoch method is that both data and simulations are easy to handle. One needs snapshots of a lensed system in one or more bands, which are easy to schedule and obtain, and the method can be applied to several systems simultaneously. Creating pairs of simulated microlensing magnification ratios for different sources is straightforward. The parameter space of the model[13] is usually smooth and sampling the bulk of the probability does not rely heavily on priors. However, the resulting probability distribution can be quite extended and in some cases can lead only to upper limits for some parameters (usually the accretion disc size, see Sect. 4.3). Bate et al. (2018) have shown that this method can have poor constraining power depending on the chromaticity amplitude, i.e. by how much the flux ratio between different bands varies, and the offset with respect to the no-microlensing baseline. Other difficulties include controlling systematic uncertainties like extinction, the no-microlensing baseline (see Sect. 3.2), leakage of broad line emission in the wavelength range covered by a broad band filter (see also Sect. 4.3.3), and quasar intrinsic variability. To mitigate the latter, configurations with small time delays like crosses or close image pairs are preferred. Although one could still apply this method to any image pair with a known time delay, two snapshots would be required in this case correctly spaced in time.

3.4 The Light Curve Method

In a lensed quasar, the observer, lensing galaxy, micro-lenses, and background emission regions are not static. The combined velocity of all the different components makes quasar microlensing a dynamic phenomenon in timescales that vary from weeks to decades. Several lensed quasars have been monitored for more than a decade, mainly for time-delay cosmography applications (see Birrer et al. 2024) and mostly in a single photometric band. Once time delays are measured, it is possible to shift the observed light curves of each image by the corresponding delay and subtract them pair-wise to cancel out the intrinsic variations of the quasar. The resulting difference curves are the observed microlensing signal, i.e. the ratio of microlensing magnification between the pair of multiple images, which contains valuable information on quasar structure and lensing galaxy mass partition and IMF. An example of such data for the doubly lensed quasar Q J0158−4325 is shown in Fig. 20.

[13] In almost all cases, the free parameters consist of the source parameters and κ_* (see Sects. 4 and 5), while κ, γ are assumed to be known from the macromodel (although they can be included in the analysis, as in Vernardos 2018).

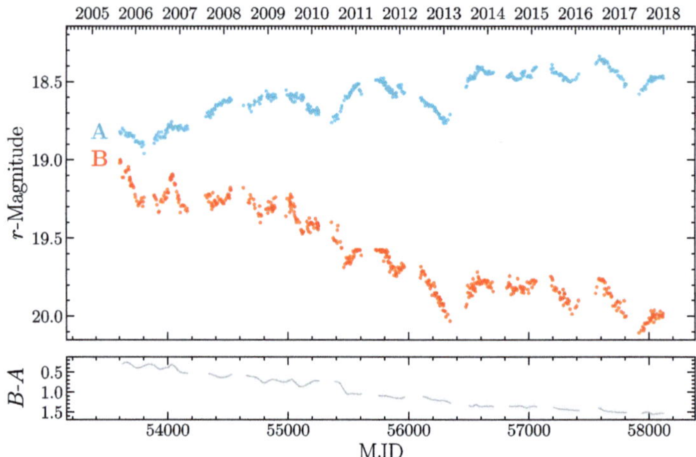

Fig. 20 Luminosity variation of the doubly lensed quasar J 0158−4325 in the r-band, from 13 years of monitoring at the Euler Swiss Telescope in La Silla Observatory. *Top*: Light curves of the two lensed images, A (minimum) and B (saddle). Measurement errors are indicated by the grey vertical lines (mostly smaller than the size of the points). *Bottom*: A spline model of the A image light curve subtracted from another spline model of B (Paic et al. 2022, see also Fig. 3 in). Importantly, the B spline model is shifted by the best time-delay estimate ($\Delta t_A B = 22.7$ d, image A leading Millon et al. 2020a). This system displays strong microlensing effects as indicated by the magnitude difference between the two images that steadily increases by > 1 mag over the 13 years of the monitoring campaign

3.4.1 The Effective Velocity Model

To describe time-varying microlensing it is necessary to define a model for the effective projected velocity, which results in the background source moving along trajectories or "tracks" on top of a magnification pattern. These tracks can then be used to extract magnification values as a function of time that translate into a light curve. In fact, this is the biggest difference and complication with respect to the single-epoch method from the simulations' point of view. The two additional degrees of freedom, the length and direction of the simulated trajectories, are critically increasing the number of simulations required (see next section).

Let us define the effective velocity of the source, $\boldsymbol{v}_e$, as the vector sum of the transverse velocities of the observer, $\boldsymbol{v}_o$, microlenses, $\boldsymbol{v}_\star$, lensing galaxy, $\boldsymbol{v}_l$, and source, $\boldsymbol{v}_s$ (see e.g. Kayser et al. 1986; Kochanek 2004; Neira et al. 2020):

$$\boldsymbol{v}_e = \frac{\boldsymbol{v}_o}{1+z_l}\frac{D_{ls}}{D_{ol}} + \frac{\boldsymbol{v}_\star + \boldsymbol{v}_l}{1+z_l}\frac{D_{os}}{D_{ol}} + \frac{\boldsymbol{v}_s}{1+z_s}, \tag{57}$$

where time is measured in the observer's rest frame and length on the source plane. These velocity components are also schematically shown in Fig. 21. The only fully measurable of these vector quantities is $\boldsymbol{v}_o$, which is determined relative to the Cosmic Microwave Background velocity dipole ($\boldsymbol{v}_{\text{CMB}}$):

$$\boldsymbol{v}_0 = \boldsymbol{v}_{\text{CMB}} - (\boldsymbol{v}_{\text{CMB}} \cdot \hat{z})\,\hat{z}, \tag{58}$$

where $\hat{z}$ is the direction of the line-of-sight. The unknown peculiar velocities of both the lensing galaxy and the background quasar can be considered as uniformly random in direc-

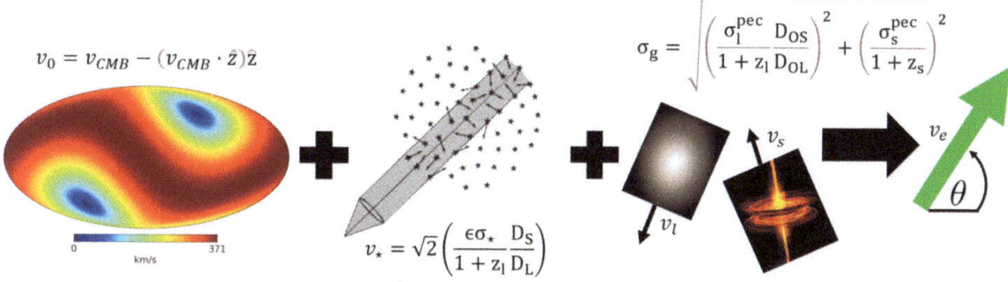

Fig. 21 Schematic representation of the components of the relative transverse velocity defined in equation (57). *Left*: the magnitude and direction of the transverse CMB dipole velocity, i.e. the velocity of the observer. *Middle*: the individual velocities (arrows) of the microlenses (star symbols) that lie along the line of sight (grey cone). *Right*: the random peculiar velocities of lensing galaxy and quasar

tion with magnitude in the source plane drawn from a normal distribution $\mathcal{N}(0, \sigma_g)$ with:

$$\sigma_g = \left[\left(\frac{\sigma_l^{\text{pec}}}{1+z_l} \frac{D_{os}}{D_{ol}} \right)^2 + \left(\frac{\sigma_s^{\text{pec}}}{1+z_s} \right)^2 \right]^{1/2}, \tag{59}$$

where σ_l^{pec} and σ_s^{pec} are the standard deviations of the peculiar velocity distributions of the lens and the source respectively. A proper treatment of the final velocity component, the individual velocities of the microlenses, would require the use of computationally expensive "moving patterns", where the individual movement of each star is taken into account (see Fig. 1). However, Kundic and Wambsganss (1993) showed that this relative movement was equivalent to an increase in the magnitude of the effective velocity. Wyithe et al. (2000a) showed that the magnitude of this effect can be approximated by a "bulk velocity" of the micro-lenses such that:

$$\upsilon_\star = \sqrt{2}\epsilon\sigma_\star, \tag{60}$$

where $\sigma_\star$ is the velocity dispersion at the lens center and ϵ is an efficiency factor that depends on local κ and γ values. Since in the general case only $\boldsymbol{\upsilon}_o$ is known, a direction and magnitude for the final effective velocity vector need to be drawn from a distribution, as defined by equation (57). By construction, this distribution is skewed towards the fixed direction of $\boldsymbol{\upsilon}_o$ and has a minimum magnitude defined by $\boldsymbol{\upsilon}_\star$. As an example, Fig. 22 shows histograms of effective velocity realizations for three well-known lensed quasar systems.

Once an effective velocity distribution and the duration of an observing period have been established, one can finally define tracks of the source traveling through the magnification pattern. These tracks have a length given by the duration of the observations times the velocity modulus and a direction defined by the vector sum of the velocity components (see Fig. 17). Extraction of the pixel values by superposition of this track on top of a magnification pattern convolved with the chosen quasar model results in a simulated light curve (see Fig. 18). Unlike static microlensing, in this case, in addition to the strength, also the steepness and the timescales of the variations shown in the light curve will depend on the specific model of the source (see Fig. 15).

3.4.2 Fitting Light Curves

The standard approach to analyse light curve data consists of fitting simulated light curves to difference curves and was introduced by Kochanek (2004). To simplify a bit the equations,

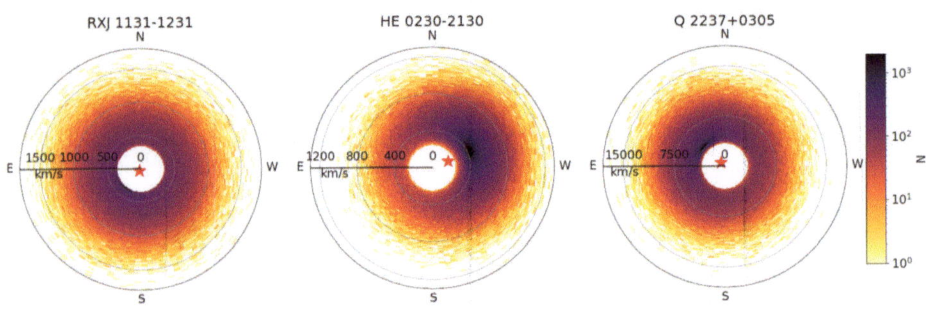

Fig. 22 Histograms of simulated effective velocities for three known lensed quasar systems. The minimum possible value for the velocity (inner radius of the ring-like structure) is defined by $v_{\star}$, which depends on the lensing galaxy's velocity dispersion. The peaks of the distributions are skewed towards the direction defined by the fixed velocity component v_l. The star symbol denotes the direction and magnitude of this component

we focus on a single photometric band light curve and a single system. In this case, we define the time series that constitute a difference light curve between a pair of macroscopically observed multiple images A and B as $\Delta m_{\mathrm{AB}}^{\mathrm{obs,t}} = m_{\mathrm{B}}^{\mathrm{obs,t}} - m_{\mathrm{A}}^{\mathrm{obs,t}}$, where t denotes one of T measurements from the difference light curve. The no-microlensing baseline is $\Delta m_{\mathrm{AB}}^{\mathrm{ref}} = m_{\mathrm{B}}^{\mathrm{ref}} - m_{\mathrm{A}}^{\mathrm{ref}}$ and does not depend on time. The microlensing signal can then be defined as:

$$\Delta m_{\mathrm{AB}}^{t} = \Delta m_{\mathrm{AB}}^{\mathrm{obs,t}} - \Delta m_{\mathrm{AB}}^{\mathrm{ref}}. \tag{61}$$

Before generating simulated microlensing light curves, a number of hyper-parameters need to be defined for the magnification maps, $\boldsymbol{\eta}_m$, and the source, $\boldsymbol{\eta}_s$, in the same way as for the single-epoch method. Here, we also need to include the hyper-parameters that are related to the effective velocity, i.e. its magnitude and direction, which we denote as $\boldsymbol{\eta}_v$. Finally, the simulated microlensing light curves that are sampled from trial trajectories on convolved maps are a function of these free model parameters:

$$\Delta m_{\mathrm{AB}}^{\mathrm{mod}}(t) \equiv \Delta m_{\mathrm{AB}}^{\mathrm{mod}}(t\,|\,\boldsymbol{\eta}_m, \boldsymbol{\eta}_s, \boldsymbol{\eta}_v). \tag{62}$$

The same notes as for the single-epoch method apply here, i.e. we can replace a broad photometric band by its central wavelength, we can randomly select starting locations for the trajectories on the convolved maps and divide their magnification values pairwise, and these locations have to remain the same if we are considering data in many bands. However, the pair of trajectories must have the same effective velocity and the orientation of each map, or more precisely the orientation of the local shear at the position of each multiple image, has to be taken into account, as illustrated in Fig. 17.

The observed and simulated microlensing light curves can be compared through a chi-squared statistic:

$$\chi_n^2 = \sum_i \sum_{j>i} \sum_t^T \left[\frac{\Delta m_{ij}^{\mathrm{obs,t}} - \Delta m_{ij}^{\mathrm{ref}} - \Delta m_{ij}^{\mathrm{mod,t,n}}(\boldsymbol{\eta}_m, \boldsymbol{\eta}_s, \boldsymbol{\eta}_v)}{\sigma_{ij}^t} \right]^2, \tag{63}$$

where i, j is $1-4$ for a quadruply lensed quasar, σ_{ij}^t is the error associated with each point in the difference light curve, and n denotes a given pair of trajectories between two magnification maps. The observant reader may have noticed the similarity between this and Eq. (53), however, we note that because of $T \gg K$ and the larger parameter space due to $\boldsymbol{\eta}_v$, inference of the target physical parameters becomes much harder in this case. We can define a likelihood function in the same way as in Eq. (54) using N pairs of simulated light curves, and cast it in a Bayesian framework to obtain:

$$\frac{\mathrm{d}P}{\mathrm{d}\boldsymbol{\eta}_m\,\mathrm{d}\boldsymbol{\eta}_s\,\mathrm{d}\boldsymbol{\eta}_v} \propto \mathcal{L}(\boldsymbol{d}|\boldsymbol{\eta}_m,\boldsymbol{\eta}_s,\boldsymbol{\eta}_v)\frac{\mathrm{d}p}{\mathrm{d}\boldsymbol{\eta}_m}\frac{\mathrm{d}p}{\mathrm{d}\boldsymbol{\eta}_s}\frac{\mathrm{d}p}{\mathrm{d}\boldsymbol{\eta}_v}, \tag{64}$$

where $\boldsymbol{d}$ is now difference light curve data, i.e. a set of measurements $\Delta m^{\mathrm{obs,t}}$ and their associated uncertainties, σ^t for one or more pairs of multiple images. The velocity model presented in Sect. 3.4.1 can serve as the prior for $\boldsymbol{\eta}_v$ here. The posterior probabilities for any parameter of the model can be obtained by marginalizing over all the other remaining free parameters. Extending Eq. (63) to include multiple wavelengths and systems (as in Eq. (56)) is left as an exercise for the reader.

3.4.3 Advantages, Drawbacks, and Other Approaches

The light curve method is quite appealing because of the larger amount of data – hundreds as opposed to a single epoch – that, in principle, should result in more constraining power over the source and lens galaxy models. That alone is a sufficiently strong argument to consider this a major advantage over other methods. Of course, the first step for the applicability of this method is the availability of the data, meaning that the required monitoring observations for periods of years should have been scheduled and taken place, a demanding and expensive task. LSST will dramatically improve on the availability of microlensing light curves by monitoring thousands of lensed quasars for at least a decade. However, the potential advantages of the light curve method come at the cost of complicated calculations, additional model degeneracies, and some "hidden" complexity in the data.

The signal contained in the difference light curves can in fact be much more complex an amplitude modulation as the source is crossing the magnification field produced by the microlenses, as shown in 15. A rapid flickering on a timescale of weeks to months is often observed (Schild 1996; Schechter et al. 2003; Millon et al. 2020b), which is too fast to be explained by the transverse velocity models described in Sect. 3.4.1, as it can be seen in Fig. 20. Several explanations have been put forth to explain these fast microlensing variations: Schild (1996) proposes that a population of planetary mass objects act as microlenses, whereas Blackburne and Kochanek (2010) attribute this flickering to a variation of the accretion disc size over time, Gould and Miralda-Escude (1997), Wyithe and Loeb (2002), Schechter et al. (2003), Dexter and Agol (2011) invoke inhomogeneities or broad absorption clouds in relativistic motion in the accretion disc being magnified by microlensing, and Millon et al. (2022) propose a secondary black hole in orbit within the disc. A different explanation simply invokes the combination of reverberation of the BLR and microlensing, as introduced by Sluse and Tewes (2014) (see also Eq. (12) in Paic et al. 2022). Because the accretion disc and the BLR have different physical sizes, they are affected differently by microlensing. In addition to this, the BLR is expected to reverberate the intrinsic variability of the quasar with some delay due to its location further away from the SMBH (and some smoothing due to its size, see Cackett et al. 2021). This leads to a difference in contrast between the reverberated and the direct signal that can introduce variations in the difference curves on the same timescale as the quasar intrinsic variations. These variations are much faster than the usual microlensing timescale and could explain the complex signals present on such shorter timescales.

When a source moves through a caustic network, whose characteristic size is described by θ_{E}, there is a three-way degeneracy that is introduced between the source size, effective velocity, and the mass of the microlenses. The relation between these parameters is not exact, but can be understood by the following examples. If we disregard any velocity component, then the source size and microlens mass are obviously degenerate through the ratio $r_{1/2}/R_E$ that determines the extent of microlensing effects. For a given effective velocity, a small source moving in a sparse caustic field (few high-mass microlens) and a large source moving through a dense field of caustics (many small-mass microlenses) will result a similar small-amplitude microlensing signal. A similar signal would result from a slow and small or a fast and large source moving through a fixed-mass caustic network. Finally, for a fixed source size, small θ_E and low velocity lead to magnification similar to having a large θ_E and high velocity. Eventually, if the observations are long enough then these degeneracies are easier to break, but this is not always the case. Therefore, priors on these three parameters are often imposed, with the velocity model presented in Sect. 3.4.1 being a commonly used one (Kochanek 2004). Other sources of systematic bias in resulting measurements have not been adequately or at all explored (see Bate et al. 2018, for a study of such biases for the single-epoch technique). For example, these could be the peak-to-peak amplitude in the light curves as a function of wavelength and the deviation of the mean magnification from the no-microlensing baseline.

A more practical, but possibly the main disadvantage of this method, is its very high computational cost, as the probability of finding a trajectory that is a good-fit to the data is decreasing with the length of the light curve. To mitigate this, authors have divided full light curves into separately-fitted segments (Poindexter and Kochanek 2010) or averaged data within a season (Morgan et al. 2012). Another common simplifying assumption is to assume that microlensing is taking place in only one of the two images and thus simulate trajectories on a single map (Kochanek 2004). With the advent of machine learning methods, Vernardos and Tsagkatakis (2019) have investigated the use of Convolutional Neural Networks to infer the source size from the microlensing features seen in the difference curves. Although promising, careful thinking needs to go into the design of the training set for such methods in order to reflect the complexity of real world scenarios as closely as possible.

3.5 Other Microlensing Data Analysis Methods

Apart from the two general methods to analyze microlensing light curve and flux ratio data described until now, there is a number of other more specialized approaches tailored to specific situations and data. Here we give a short summary of these methods and refer the reader to the corresponding papers for more details.

3.5.1 Measuring SMBH Properties from HMEs

Observing a lensed quasar at any random moment, either by obtaining flux ratios or through monitoring, can be used to constrain its overall size (the half-light radius) as a function of wavelength. However, when a microlensing HME takes place there is much more information to be extracted on the immediate vicinity of the SMBH, e.g. its event horizon, mass, and spin. The magnification during such events can often be described analytically, which can make modelling simpler (see also Sect. 3.5.6). Mediavilla et al. (2015) use three such events in Q 2237+0305 whose fine structure at the peak they attribute to the Innermost Stable Circular Orbit (ISCO) at ≈ 3 gravitational radii. This result is obtained by fitting both classical and relativistic accretion disc models convolved with an analytic function for the caustic magnification (a straight fold combined with a linear term to factor in long-range magnification gradients due to neighbouring microlenses, i.e. not responsible for this particular caustic). Their relativistic model includes the effect of beaming, Doppler shift, and gravitational redshift, which effectively improves their fit to observed data. (Best et al. 2022) on the other hand adopt a machine learning approach to measuring the ISCO, which they train with thousands of simulated microlensing light curves and accretion disc models that include relativistic beaming, Doppler shifts, and lensing from the central SMBH.

3.5.2 Modelling X-ray Line Deformation

Several authors have studied with increasing sophistication how the shape of the Fe Kα reflection line changes as a result of microlensing. The emitted 6.4 keV photons in the inner regions of the disc are affected by the gravitational redshift, bending, Doppler broadening, and relativistic beaming, resulting in a well-known two peaked profile (Fabian et al. 1989; Laor 1991). During an HME or caustic crossing, some regions of the disc producing these photons will be magnified more than others, resulting in additional peaks (Heyrovský and Loeb 1997; Popović et al. 2006; Jovanović et al. 2009) and edges (Neronov and Vovk 2016) to the profile. As pointed out by (Krawczynski and Chartas 2017), this may explain the Fe Kα line variations observed in RX J1131−1231 (Chartas et al. 2012).

The above mentioned studies used methods similar to those described below in Sect. 3.5.3. The work by Ledvina et al. (2018) first makes maps at representative inclinations of the Fe line intensity and energy shift (g-factor) around a quasar (in a $60 \times 60 r_g$ region) and then sweeps a high-magnification, narrow, linear fold caustic across those maps. At specific moments in time, the total observed (microlensed) Fe Kα line profile is constructed. One of the most prominent microlensing-induced spectral features is seen as the caustic sweeps across the inner disc: a sharp and highly magnified peak dominates the spectral profile. The location of the peak changes in time as the caustic moves as different parts of the disc get highly magnified (e.g., their Figs. 2 and 3).

3.5.3 Modeling Broad Emission Line Deformations

Because the broadening of an emission line is due to different velocity components of the emitting material that arise from spatially distinct regions, it is necessary to calculate the microlensing signal for an ensemble of luminosity profiles, each associated to a velocity slice in the BLR, as shown in Fig. 23. These luminosity profiles are subsequently convolved with microlensing maps to calculate the line deformation. This is similar to the procedure followed to model the temperature profile of the accretion disc, but instead of calculating the disc luminosity profile at different wavelengths (corresponding to different temperatures), one calculates the luminosity profile of the BLR for different velocities. A simple power-law decrease of the BLR brightness with distance to the central black hole is assumed, as well as simple geometries, such as a keplerian disc, a polar or equatorial wind, and various velocity fields able to reproduce emission line profiles (e.g. Murray et al. 1995; Murray and Chiang 1997). To simulate such profiles, one could use analytical models (e.g. Abajas et al. 2002) or radiative transfer simulations that can produce the light profile of the BLR for arbitrary velocity slice(s) (e.g. Braibant et al. 2017). For radiative transfer models, the photo-ionizing source of light is supposed to be the quasar accretion disc, assumed to be smaller and have the same inclination as the BLR. The microlensing of the continuum is calculated together with the line deformation to ensure that the model can self-consistently reproduce the data. The exact luminosity profile of the disc is not critical (see Sect. 3.1) and a uniform profile has commonly been considered. The need for the simulations to reproduce both the continuum and the BLR emission implies that the resolution of the microlensing maps should be high enough to sample the disc.

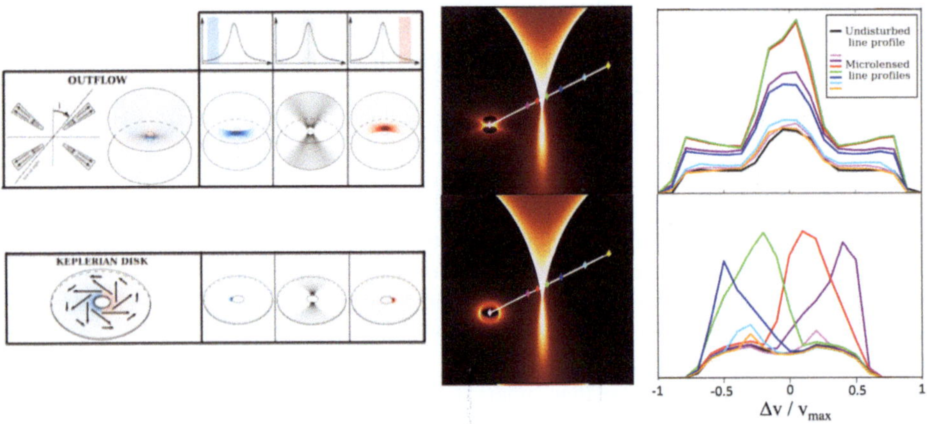

Fig. 23 Illustration of broad line deformation produced by microlensing for two different models of the broad emission line region, with $r_{1/2} \sim 0.3 R_E$ (the size chosen to be smaller than expected BLR size to highlight line deformations). Two models are sketched the polar wind (top) and the keplerian disc (bottom) The left most panel shows the two models at intermediate inclination. The blue and red colors indicate the location of the approaching and receding gas, corresponding to blue and red velocity components of the line. The right panel depicts the line deformation corresponding to different location of the BLR with respect to the caustic, as shown in the middle pannel (colored losanges corresponding to the colored line profiles on the right). For the depicted event, the microlensing of the Keplerian disc yields asymmetric red/blue line profile distortions while microlensing of the biconical outflow is characterized by symmetric wings/core distortions. Adapted from Hutsemékers et al. (2017)

Once deformed line profiles have been calculated for a variety of BLR models, they can be compared to the data in different ways. For example, O'Dowd et al. (2011) used the ratio between the microlensed and un-lensed profiles, which is proportional to the amplitude of microlensing as a function of the gas velocity (i.e. $\mu(v)$), Abajas et al. (2007) calculated relative changes of the line FWHM, while Braibant et al. (2017) integrated the flux over a fraction of the line profile to quantify the red-blue asymmetry and its wing-to-core magnification/deformation. In order to minimize the loss of information inevitable by the use of such "integrated" quantities, which may even lead to overfitting, Hutsemékers and Sluse (2021), Hutsemékers et al. (2023) used the velocity dependent microlensed amplitude, $\mu(v)$ (Sect. 3.2.2). This choice minimizes the loss of information but requires high signal-to-noise data and in the case of too simple BLR models the method may be unable to reproduce the observed signal at all.

The comparison of models to data can proceed through a standard χ^2 statistic. For instance, one can compare the amplitude of the simulated microlensing signal in the continuum, $\mu_{\rm C}^{\rm mod}$, and the profile of the line deformation, $\mu_{\rm B}^{\rm mod}(v)$, for an ensemble of velocity bins, Δv_k (see e.g. Sect. 3.2.2 for methods enabling to measure those quantities). If $\boldsymbol{\eta}_B$ and $\boldsymbol{\eta}_C$ are the ensemble of model parameters of the BLR and the disc (i.e. continuum emission), then the total χ_n^2 associated to a model realisation n can be calculated as the sum of the corresponding BLR and continuum terms:

$$\chi_n^2 = \sum_k \left[\frac{\mu_{\rm B}^{\rm obs}(\Delta v_k) - \mu_{\rm B}^{\rm mod}(\Delta v_k | \boldsymbol{\eta}_B)}{\sigma_{\rm B}^{\rm obs}(\Delta v_k)} \right]^2 + \left[\frac{\mu_{\rm C}^{\rm obs} - \mu_{\rm C}^{\rm mod}(\boldsymbol{\eta}_C)}{\sigma_{\rm C}^{\rm obs}} \right]^2. \tag{65}$$

We note that the first term in the above equation is not unique. Instead of $\mu_{\rm B}(\Delta v_k)$, one can use other measurements of line deformations like the "integrated" indicators described above. The inference on the model parameters can be cast in a Bayesian framework as described in Sects. 3.3 and 3.4.2 for single- and multi- epoch data respectively.

Existing works show that the largest constraining power arises when the amplitude of microlensing is large. For low amplitudes, the whole range of possible line deformations are very similar for any BLR model (Braibant et al. 2017). On the other hand, there are indications that more complex BLR models than the three listed above (i.e. keplerian disc, polar and equatorial wind) may be needed to accurately reproduce the line deformations observed in some systems (Hutsemekers et al. 2019; Hutsemékers and Sluse 2021).

3.5.4 Extensions to the Single-Epoch Technique

A single flux ratio measurement obtained from a snapshot of a lensing system at a given band has little constraining power; compared to the degrees of freedom of the microlensing models (source structure and mass in the lens) the problem is under-constrained. To mitigate this, flux ratios obtained at different epochs could be used, however, successful incorporation of such additional data requires attention to both the time separation between observations and the analysis method used. If enough time has passed between observations to result in the source having moved to an entirely different region of the magnification map, then the measurements can be considered independent, otherwise their covariance needs to be estimated and considered in, for example, χ^2 fits. These time intervals are shorter for smaller sources because magnification is less correlated on short distances through convolution.

In the X-rays, where the source is expected to be the smallest, Pooley et al. (2012) consider such flux ratios observed in multiple epochs as independent measurements. Guerras et al. (2020) used ≈ 10 flux ratio measurements in the X-rays for each of four systems over periods of 5-15 years. This timescale is most likely longer than the source-size crossing time in the X-rays. They did not consider these measurements as independent, but used the mean and a second-order moment of their distribution instead, effectively compressing the X-ray light curve to these two summary statistics. Their simulated flux ratio distributions where obtained along tracks on magnification maps whose length – an important parameter that is proportional to the effective velocity – was arbitrarily chosen and not fitted for.

Fian et al. (2016, 2018, 2021b) have followed a similar approach to measure the accretion disc size by collapsing observed difference light curves to flux ratio probability distributions. However, instead of measuring the mean and width of the distribution, they calculate its distance from the one obtained from magnification maps for different source sizes. While the method has been applied to very long light curve data (between 10 and 20 years), the microlensing variability observed over that period is not guaranteed to be representative of the full magnification map. Indeed, the corresponding tracks on the maps may not be longer than a few θ_E – the exact value depending on the unknown effective velocity. More work may be needed to demonstrate that this caveat does not bias the final results of this method.

3.5.5 The Microlensing Time-Delay as a Probe for Accretion Disc Size

In addition to the well known macroscopic time delay in lensed quasars, which can be used as an independent way to perform cosmography and measure the Hubble constant (see Birrer et al. 2024), Tie and Kochanek (2018) introduced the microlensing time delay that can extend or shorten the arrival times of the signal in the different multiple images by a few days in an uncorrelated way. The origin of this delay is due to the combination of the driving variability mechanism of the quasar, assumed to be a lamppost, and how microlensing can magnify different parts of the accretion disc. Based on the fact that this effect depends on the source size, which in turn depends on wavelength, Chan et al. (2021) proposed using measured differences in the time delays between different bands to constrain the accretion disc size. Their method is complementary to traditional curve shifting techniques, which consider the amplitude of microlensing only, but, unlike these methods, they take into account the distortion and delay caused by the convolution of the lamppost variability with the light profile of the accretion disc and the caustic network.

The method requires a set of measured time delays in different bands with respect to some baseline with an uncertainty of 0.1 days, which is about 10 times better than the highest precision delays achieved nowadays. Additional assumptions are that the quasar varies as a lamppost-like model and its accretion disc brightness profile shape (more precisely, its spatial derivative Tie and Kochanek 2018) is known. Forward modelling and fitting of the time delays as a function of wavelength, and consequently half-light radius of the disc, follows. The value of the micro-magnification of a multiple image can be used to drastically speed up the calculations by pre-selecting the matching locations on magnification maps. Using multi-band light curves of a fiducial quadruply imaged quasar, the authors were able to forecast that the measured accretion disc size from such data would be within a factor of 0.2 and 2 of the truth. This method provides measurements that are independent of the choice of Initial Mass Function for the microlenses in the lens galaxy (see also Sect. 5.2).

3.5.6 Reconstructing Accretion Disc Profiles

The methods to measure accretion discs discussed until now correspond to the so-called forward problem: microlensing observables are calculated and fitted to data based on disc models prescribed a priori. However, the problem can be approached in an inverse way as well, i.e. reconstructing an unknown disc profile from the measured data. An observed microlensing light curve results from the convolution of a source profile with the magnification field of the microlenses. Hence, the situation is similar to deconvolution which requires solving an ill-posed inverse problem with the use of regularization to avoid overfitting and stabilize against noise.

The first to adopt this treatment were Grieger et al. (1988, 1991), followed by Mineshige and Yonehara (1999), Agol and Krolik (1999), and (Bogdanov and Cherepashchuk 2002). These studies made two main assumptions: they described the source with a one-dimensional "strip" profile and focused on HMEs. The first assumption is justified by the fact that the observed magnification depends mostly on the brightness distribution of the source along the direction perpendicular to the caustic (see Fig. 1 in Mineshige and Yonehara 1999). On long time scales, of the order of the θ_E crossing time, the source can be affected by several microlenses that lead to superimposed caustics and a complex and non-linear magnification field (see Sect. 3.4). But in the case of short HMEs, the magnification field corresponds to that of a caustic crossing, which is simpler and can be described analytically to a good approximation (Gaudi and Petters 2002a,b). This simplifies the problem of deconvolution considerably but can lead to some problems, e.g. the caustic can be discontinuous in the case of a cusp, it can have significant curvature, and magnification can vary along its direction (Agol and Krolik 1999). It turns out that the latter two can be resolved if the source has a size of 10% of θ_E (Grieger et al. 1988).

This method has been demonstrated to be stable on simulations using basic regularization schemes (see Benning and Burger 2018, for a review), showing potential of resolving the accretion disc structure within a few gravitational radii from the central SMBH. When it comes to real observations, the best and only system to provide the necessary HMEs has been the quadruply lensed quasar Q 2237+0305. Its accretion disc was found to have a brightness profile consistent with the thin-disc model (Agol and Krolik 1999; Bogdanov and Cherepashchuk 2002; Koptelova et al. 2007).

3.5.7 Time Varying Equivalent Widths of Non-lensed Quasars

Looking for signatures of lensing in the variability of non-lensed quasars is an idea that has been explored by several authors over decades (Press and Gunn 1973; Canizares 1982; Schneider 1993; Zackrisson et al. 2003; Bruce et al. 2017; Hawkins 2022). Vernardos et al. (2023, submitted) revisit this in the context of variable emission line equivalent widths. The data, in this case, are two measured equivalent widths of the same line obtained by spectroscopic observations of the same system taken ≈ 10 years apart. The authors develop a probabilistic model for these measurements using the magnification of a single compact microlens (similar to Galactic microlensing models). Once the most promising candidates are selected from spectroscopic surveys and re-observed to confirm that lensing is indeed responsible for their varying equivalent widths, the microlens mass and effective velocity can be measured.

3.5.8 Astrometric Microlensing

The variations in flux due to microlensing are correlated with astrometric shifts due to the individual microimages (Lewis and Ibata 1998). In particular, the creation or annihilation of pairs of microimages during a caustic crossing event can give rise to astrometric shifts on the order of 10 s of micro-arcseconds (Treyer and Wambsganss 2004). While the detection of such shifts have been long predicted, they have yet to be confirmed observationally on extra-galactic scales. They may, however, contaminate studies of proper motions of lensed AGN from Gaia (Makarov and Secrest 2022). Astrometric shifts from stellar microlensing have been observed however; see e.g. McGill et al. (2023).

4 Quasar Results

4.1 Black Hole Tomography with HMEs

Due to the rarity of observations of HMEs in the currently known lensed quasars, there are few results regarding their very promising use for constraining the inner accretion disc and the black hole. HMEs have almost exclusively been observed in one system: the Einstein cross (Fig. 2). This system, due to its exceptionally low lens redshift ($z = 0.039$) that results in short variability time scales, has been the target of several intensive monitoring campaigns (Ostensen et al. 1996; Udalski et al. 2006; Goicoechea et al. 2020). A few HMEs in this object have been captured in almost 3-decades worth of data and two indicative examples are shown in Fig. 24 (see Fig. 1 of Eigenbrod et al. 2008b, for another example). Study of such HMEs has allowed to place the inner edge of the quasar accretion disc at 3 R_g (Mediavilla et al. 2015) and detect a possible warped disc geometry (Abolmasov and Shakura 2012b). Several monitoring programs have provided a rich dataset for this system, but such rapid variability is a statistical outlier.

For the vast majority of lensed quasars, a typical HME is expected in roughly every 20 years.[14] Therefore, monitoring of tens of systems for several years would be required in

[14]This estimate is provided by Mosquera and Kochanek (2011). However, as introduced in Sect. 3.2, HME is an overarching term for any variation with a high-enough magnification – see Neira et al. 2023 submitted, for more examples.

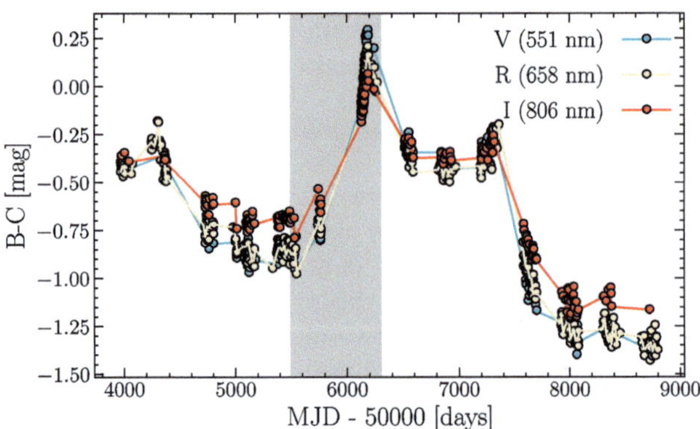

Fig. 24 Difference light curve between images B and C of Q 2237+0305 (i.e. the Einstein Cross) observed in the VRI bands. The double-peaked signal is typical of the source entering and exiting a single diamond caustic (e.g. as in Galactic microlensing events). Notice how the brightest band changes from I (red) to V (blue) within the grey-shaded area - a crucial indicator of an imminent HME. Adapted from Goicoechea et al. (2020)

order to observe a statistically meaningful sample of HMEs. Such monitoring has been performed by the COSMOGRAIL program for 30 systems over more than a decade. Although its main goal has been measuring the time delays between the macroscopic quasar images, microlensing signals can be extracted from the observed light curves as a by-product. A few potential HMEs can be spotted in the final data (Millon et al. 2020a), albeit none unambiguously due to lack of observations during its peak. The GLENDAMA archive also provides a compilation of monitoring data from various telescopes and filters for 9 systems (Gil-Merino et al. 2018), where a few events can be identified.

The majority of observations and analyses have employed single-band photometry, which is a cheaper and more viable approach for long-term monitoring. If daily observations are taken during a HME, then such single-band data should be enough to provide valuable insights to quasar structure. However, the real wealth of information on accretion disc structure, which is accessible only during the event's peak, can be extracted through multi-band, or even better, spectroscopic observations. With existing facilities, such observations are expensive and impractical for long monitoring periods, therefore, predictions and early detection of the onset of HMEs and their peaks are mandatory to trigger multi-wavelength follow-up.

4.1.1 HME Prediction and Triggers

An alert system has been used by OGLE (Woźniak et al. 2000) and a triggering mechanism for multi-band follow-up has been proposed by Wyithe et al. (2000b). However, such prediction algorithms have not been used extensively and as of yet there has not been any clear prediction and detailed follow-up of a HME as it unfolds, i.e. predicting the peak from light curves like those shown in Fig. 24 and scheduling detailed observations accordingly (spectroscopy, X-ray spectra and imaging, etc). This could also be attributed to using only just a single band as the base signal from which to predict HMEs. Every HME is preceded by a rise in magnification in the light curve, but such a rise could be due to other reasons too, e.g. improperly subtracted intrinsic variability, reverberated variability from the BLR (Sluse et al. 2013; Paic et al. 2022), or just long-term microlensing (see Sect. 3.2). Microlensing, however, has a distinct signature as a function of wavelength: larger (redder) sources are expected to get magnified sooner than smaller (bluer) ones (Young 1981; Schneider and Weiss 1987), which can be beautifully seen in the VRI band monitoring data of the Einstein cross shown in Fig. 24. Therefore, having such multi-wavelength information – even in two bands – is paramount for obtaining a clear and coherent signal that will lead to early and robust predictions of the onset of HMEs.

Table 2 X-ray Microlensing Results on the size of the X-ray corona

System	$\log(r_{1/2,X}/\text{cm})$	Reference
PG 1115+080	$15.6^{+0.6}_{-0.9}$	Morgan et al. (2008)
HE 1104−1805	< 15.3	Chartas et al. (2009)
RX J1131−1231	$14.4^{+0.2}_{-0.3}$	Dai et al. (2010)
Q J0158−4325	< 14.6	Morgan et al. (2012)
Q 2237+0305	$15.5^{+0.3}_{-0.3}$	Mosquera et al. (2013)
HE 0435−1223	< 14.8	Blackburne et al. (2014)
SDSS J0924+0219	< 15	MacLeod et al. (2015)

4.2 Corona

The first indication from microlensing that the X-ray emitting corona is fairly compact came from the study of *Chandra* observations of 10 quadruply lensed quasars by Pooley et al. (2007). They showed that, in almost every case where a flux ratio anomaly is present, it is more extreme in X-rays than in the optical. This indicates that the X-rays must arise from a region much smaller than the optical half-light radius. Pooley et al. (2007) qualitatively conclude that the X-ray emission region size is consistent with the inner edge of the accretion disc or smaller.

Quantitative results first came when the light curve method (Kochanek 2004) was applied to multiple *Chandra* and *HST* observations of PG 1115+080 by Morgan et al. (2008). They found that the X-ray half-light radius (rest frame 1.4–21.8 keV) is $\log(r_{1/2,X}/\text{cm}) = 15.6^{+0.6}_{-0.9}$. For a black hole mass of $1.2 \times 10^9 M_\odot$, this is consistent with the inner edge of the disc or smaller. As *Chandra* has observed and followed-up on more lensed quasars and the light curve method has been applied to many of them, it is clear that a compact size for the X-ray emitting region is common for all systems observed to date, typically within 10 r_g or less. A summary of the results so far is given in Table 2.

4.3 Accretion Disc

4.3.1 Inner Edge of the Disc

The Fe Kα reflection feature is one of the best probes we have of the inner edge of the accretion disc. Detection and possible microlensing of this emission feature has been reported for H1413+117 (Oshima et al. 2001; Chartas et al. 2004, 2007), MG J0414+0534 (Chartas et al. 2002), QSO 2237+0305 (Dai et al. 2003; Chen et al. 2012), SDSS J1004+4112 (Ota et al. 2006; Chen et al. 2012), RXJ 1131−1231 (Chartas et al. 2012), QJ 0158−4325, HE 0435−1223, SDSS J0924+0219, and HE 1104−1805 (Chen et al. 2012). In almost all cases, the equivalent width of the Fe Kα line is much larger for these lensed quasars than it is for comparable, non-lensed AGN. This indicates substantial microlensing and suggests a compact size for that region, comparable to or even smaller than the X-ray continuum region (i.e. the corona, e.g. Chen et al. 2012).

Chartas et al. (2017) apply their g-distribution method to the Fe lines seen in RX J1131−1231, QJ 0158−4325, and SDSS J1004+4112. In this method, the observed Fe line shifts are due to microlensing caustics near the inner edge of the disc, and the distribution of the shifts can provide estimates of the size of the innermost stable circular orbit (ISCO), spin, and inclination (i, with zero being perpendicular to the line-of-sight) of the disc. To date, results have only been obtained on RX J1131−1231, for which Chartas et al. (2017) report values of $r_{\rm ISCO} \lesssim 8.5 r_{\rm g}$ and $i \gtrsim 55°$.

Strong amplitude of microlensing has been found in the Fe III multiplet visible in rest-frame UV (i.e. range [2039-2113] Å) suggesting that emission arises in the direct vicinity of the disc and in a region almost as compact as the latter (Guerras et al. 2013b; Shalyapin and Goicoechea 2014; Fian et al. 2021a). By combining the SMBH mass inferred from the redshift and broadening of this line, with microlensing-based estimates of BLR sizes, Mediavilla et al. (2020) constrained the virial factor, f, of several low ionization lines in an ensemble of 10 quasars.

4.3.2 Size

Just a year after the first detection of a microlensing-induced variability in Q 2237+0305 (Irwin et al. 1989), Wambsganss et al. (1990) compared the observed variations to simulated light curves and predicted the Optical-UV size of its accretion disc radius to be $< 2\,10^{15}$ cm. This early result showed consistency with the expectations of the thin-disc theory. However, further analysis of this and another ∼20 systems showed that the sizes measured from microlensing are systematically larger by a factor of 3-5 than the expectations from the thin-disc model (see Cornachione et al. 2020a, for a recent compilation). These results were obtained from the analysis of the microlensing signal in optical light curves (see Sect. 3.4.2), focusing either on a single HME (Anguita et al. 2008; Eigenbrod et al. 2008b; Mediavilla et al. 2015) or using all variations in the difference light curves over a long period of time (Hainline et al. 2012, 2013; MacLeod et al. 2015; Morgan et al. 2008, 2012, 2018; Cornachione and Morgan 2020). Measurements based on single-epoch observations also generally suggest larger disc sizes (e.g. Blackburne and Kochanek 2010; Jiménez-Vicente et al. 2012). Although it seems that this method consistently results in larger sizes than predicted by the thin-disc theory, no consensus on the physical implications for quasar accretion discs has been reached.

Before discussing possible explanations for this discrepancy, it is important to remember that single-band multi-epoch analysis can constrain only a combination of the slope of the temperature profile of the disc, β, and its size, r_s (Eq. (36)). This means that this tension should be investigated in the $\beta - r_s$ plane, shown in Fig. 25. In this perspective, a possible solution could be a shallower slope than predicted by the thin-disc theory (Cornachione et al. 2020a). However, slopes are more prone to systematic uncertainties than absolute source sizes, as discussed in the next subsection (Sect. 4.3.3).

Several alternative explanations have been proposed to reconcile the disc size discrepancy. One suggestion to explain the larger sizes inferred by microlensing is the inhomogeneous accretion model proposed by Dexter and Agol (2011). However, the level of inhomogeneities required by this model to match the observations would need to be unrealistically high. Another plausible explanation could be that part of the UV-continuum emission is actually coming from a much more extended region, hence biasing the microlensing measurement. For example, Hutsemékers et al. (2015) presented spectropolarimetric observations of the BAL lensed quasar H 1413+117 that showed two distinct sources of continuum emission. Coupling the observations with microlensing simulations, they noted that these regions differ significantly in their size with a compact unpolarized emission coming directly from the disc and a much larger non-microlensed continuum coming from an extended region located along the polar axis. The fraction of total flux arising from that extended emission may however be less than ∼40 per cent. This would reduce the microlensing source size by less than a factor two (Dai et al. 2010; Sluse et al. 2015).

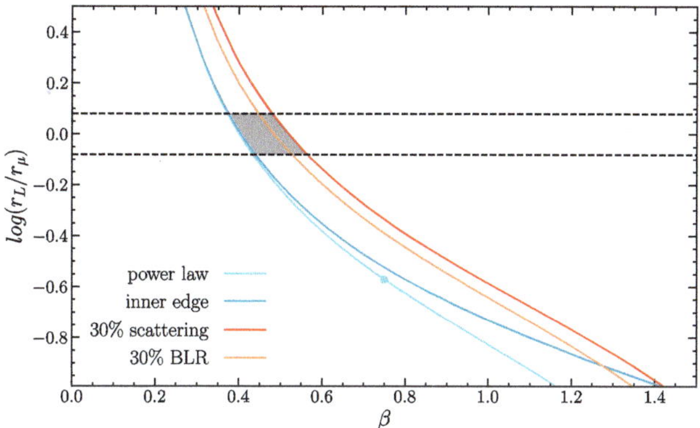

Fig. 25 Ratio between the luminosity half-light radius $r_L \equiv R^{\text{flux}}_{\lambda,1/2}$ (Eq. (38)) and the half-light radius measured from microlensing $r_{1/2}$ ($\equiv r_\mu$ on the figure). Discs described by a power-law temperature profile are represented by the light blue line whereas the dot indicates the particular case of $\beta = 3/4$, corresponding to the thin-disc model. Alternative disc models including an inner edge or contamination from the BLR or scattered light are also plotted. These alternative models, however, are insufficient to reconcile luminosity and microlensing sizes. For this to happen, measurements should fall in the shaded grey region, which means that the slope of the temperature profile should be reduced drastically compared to the thin-disc model. Figure reproduced from Morgan et al. (2010) and Cornachione and Morgan (2020)

Alternatively, Abolmasov and Shakura (2012a) proposed a super-Eddington accretion rate for these specific systems, leading to the formation of an optically thick envelope scattering the radiation emitted from the disc. This effect would make the apparent disc size larger and independent of wavelength. Similarly, low-density scattering atmospheres could produce non-black-body emission spectra (see the model by Hall et al. 2018), which would result in a flatter spectral energy distribution than a thin-disc. This could result in a broken power-law temperature profile, increasing the apparent half-light radius of the disc. A similar broken power-law profile could also appear from iron opacity (Jiang et al. 2016).

The disc wind model (e.g. Li et al. 2019) with mass outflows could also explain the discrepancy. The principle is again the same: the disc wind is flattening the radial temperature profile that increases the source size measured by microlensing. It might also be that the solution to this problem is related to our detailed understanding of the origin of X-ray emission and its interaction with the disc. (Papadakis et al. 2022) calculated that for a disc illuminated by the X-ray corona, energy is injected into the disc increasing the temperature mainly at larger radii. Therefore, X-ray illuminated discs will have an increased apparent size. Finally, we note that both changes in the disc structure and the presence of unaccounted extended emission could be at play at the same time, as proposed by Zdziarski et al. (2022). These authors showed that a combination of disc truncation at an inner radius larger than the innermost stable circular orbit, disc winds, and local color corrections to the disc black-body emission could explain larger discs.

4.3.3 Thermal Slope

Any measurement of accretion disc sizes as a function of wavelength using multi-band or spectroscopic observations (single-epoch or monitoring) allows one to infer the thermal slope of the accretion disc (because $r_\lambda \propto \lambda^{1/\beta}$ or $T \propto R^{-\beta}$; Eq. (36) and Sect. 2.9.2). In 2008, several studies using the techniques described in Sect. 3 obtained the first results for Q 2237+0305 (Anguita et al. 2008; Eigenbrod et al. 2008a; Mosquera et al. 2009) but also for a few other systems (Poindexter et al. 2008; Bate et al. 2008; Floyd et al. 2009). These results showed consistency with the expectation for a thin-disc model, i.e. $\beta = 3/4$ Sect. 2.9.2. The error bars of these measurements, were too large to rule out alternative disc models. Thanks to multi-epoch data gathered for a sample of 11 systems, Morgan et al. (2010) alerted that, even if the size of the disc increased as $M_{BH}^{2/3}$, in agreement with the thindisc model, their too large absolute sizes implied thermal slopes in conflict with that model. The results from Blackburne et al. (2011), who used single-epoch multi-band data from X-ray to NIR for 12 lensed quasar systems, are in agreement with Morgan et al. (2010). They infer a variety of thermal slope measurements, but most of them significantly steeper that expected ($\beta > 3/4$); even consistent with no size variation with wavelength at all (i.e. $1/\beta \rightarrow 0$). Several additional studies thereafter also found steeper than expected slopes ($\beta \gtrsim 1$, Muñoz et al. 2011; Motta et al. 2012; Jiménez-Vicente et al. 2014; Muñoz et al. 2016). On the other hand, a few studies using either the single-epoch (Rojas et al. 2014; Bate et al. 2018) or the light curve method (Cornachione and Morgan 2020) found slopes shallower than expected ($\beta \lesssim 0.5$).

At the time of writing, the tension between different microlensing measurements still exists. There are, however, several possible avenues being explored to explain the contradicting results. A first attempt at explaning this investigates the impact of flux not arising from within the disc. As one may notice by looking at an AGN spectrum (Fig. 3), broad band data can include some fraction of emission from the broad emission lines. The contribution to the total flux from the two regions will vary in each lensed quasar image due to differential microlensing (see Fig. 15). To minimise contamination, Jiménez-Vicente et al. (2014) have specifically used data originating from narrow band imaging, spectroscopy, and broad-band imaging free of salient emission lines. This sound observational strategy does not however correct for the contribution arising from less salient lines, like the Fe II pseudo-continuum, Balmer continuum emission or other pseudo-continua, which are present in different amounts in AGN spectra.[15] Sluse et al. (2015) measured the disc temperature profile of H1413+117 from single-epoch data and shown that not correcting for the observed extended continuum pushes the measured slope towards steeper temperature profiles. Due to the different amount of non-disc emission among quasars, but also of its possible variability as a function of wavelength (and possibly time), it is difficult to predict a generic impact on the measured slope. The latter effectively corresponds to a ratio of sizes, and may be thus prone to any systematic uncertainty affecting the reference source size. To mitigate this effect, Cornachione et al. (2020a), have chosen to explore the inconsistency between the source size and luminosity size r_L (Eq. (38), Fig. 25) as a function of β.

A second explanation is systematic biases and uncertainties in the methods. For the single-epoch technique, (Bate et al. 2018) showed that low amplitudes of chromaticity, i.e. < 0.4 mag, defined as the difference of microlensing magnification between the reddest and bluest wavelengths, may yield slopes that are steeper than expected. On the other hand, the selection of observations displaying large amplitude of microlensing may systematically favour smaller sizes. Overall, simulations suggest that selection effects need to be carefully assessed when interpreting single-epoch measurements (Bate et al. 2018; Guerras et al. 2020). Monitoring data may be less prone to biases but the impact of sampling, number of bands, and length of monitoring on the results still needs to be quantified. A related source of systematic uncertainty is the one associated to the macro-model and specifically to the exact value of the macro-magnification. This can be uncertain by a factor of a few in each multiple image (see Sect. 3.2). This may impact the theory size of the source (see Sect. 2.9.2), which

[15] The impact of this "blended" emission on slope measurements is not specific to microlensing as it also affects reverberation mapping (e.g. Cackett et al. 2020).

depends on the magnification-corrected size of the source, i.e. $R_\lambda^{\rm BH} \propto L^{1/3}$ (see e.g. Cornachione et al. 2020b). A biased estimate of the black hole mass may have an effect when comparing microlensing sizes to theory as $R_\lambda^{\rm BH} \propto M_{\rm BH}^{2/3}$ (Morgan et al. 2018).

Finally, a third explanation is that intrinsic scatter of disc properties between AGN naturally takes place. For instance, the accretion efficiency, η, directly impacts the source size (Eq. (40)), but depends on the black hole spin (e.g. Thorne 1974). However, as discussed by Morgan et al. (2010), the efficiency alone may be insufficient explaining all the microlensing data. On the other hand, a non universal disc model is also a possibility, with a slim disc or an advection dominated flow occuring for high or low Eddington accretion rates respectively (e.g. Lasota 2023). Therefore, it could be that the various disc models invoked to explain over-sized discs (see discussion in Sect. 4.3.2) are in fact the case for different subsamples of systems.

In any case, none of the above solutions may be sufficient on its own to explain all microlensing measurements. Large multi-epoch surveys of the sky, like LSST, but also targeted works through e.g. spectroscopic monitoring, in combination with a careful account for potential sources of systematic uncertainties, may enable to design the optimal experiment for revealing the properties of the accretion disc.

4.4 Broad Line Region

Microlensing of the BLR has long been thought to be a marginal effect (Schneider and Wambsganss 1990; Lewis et al. 1998). After the early measurements of the size of BLR with reverberation mapping at the end of the 1990s, several authors have suggested that BLR microlensing may be larger than previously though (Abajas et al. 2002; Lewis and Ibata 2004). This has been soon after confirmed with the first striking line deformation observed by Richards et al. (2004) in the large separation lensed quasar J1004+4112. It is now securely detected in more than 30 AGN and suspected to be visible in at least 80% of the microlensed AGN (Sluse et al. 2012; Guerras et al. 2013a). Often, only a small fraction ($\lesssim 10\%$) of the line flux is found to be microlensed, but extreme cases have been observed where almost the whole line is magnified by ML (Richards et al. 2004; Keeton et al. 2006; Sluse et al. 2007, 2011). Except in the situation where some velocities are magnified and other are demagnified (Hutsemékers et al. 2023), the amplitude of microlensing of the broad lines provides a robust estimate of the projected and luminosity weighted BLR size. Information on the structure and kinematics of the BLR is encoded in the microlensing induced deformation of the emission line profile (Fig. 23). A variety of line deformations are observed, affecting either differently each wing of the line (Richards et al. 2004; Braibant et al. 2014), or symmetrically the whole line, with however the higher velocities being more magnified than the lower ones. That finding supports a BLR kinematics where the highest velocity component takes place closer to the central engine. Overall, the diversity of broad emission line deformations is qualitatively supported by simulations. The following subsections review important insights on the BLR obtained for various emission lines.

4.4.1 High Ionization Lines

Guerras et al. (2013a) have shown, from the analysis of microlensing in image pairs in 16 strongly lensed systems that the size of high ionization lines (O VI] λλ1035, Ly α+N V λλ 1216, Si IV+O IV λλ 1400 and C IV λλ 1549) is substantially smaller than that of the lower ionization lines, providing a confirmation of the stratification of the BLR expected from photo-ionization models, and observed in reverberation mapping data. This result has recently been confirmed from an expanded sample of systems by Fian et al. (2021a). Large sample studies (Sluse et al. 2012; Guerras et al. 2013a), also reveal frequent occurrence of microlensing in one of the line wings, indicating that the high ionization BLR may not have a spherically symmetric geometry.

Detailed studies of individual systems have provided further insights on the size and structure of the BLR for specific quasars. To-date, the best studied system has been the Einstein cross. O'Dowd et al. (2011) have shown that the microlensing-induced line deformation observed in that system favour a gravitationally dominated dynamics over an accelerating outflow. Furthermore, constraints on the size and geometry of the region emitting C IV have obtained owing to the spectroscopic monitoring carried out by Eigenbrod et al. (2008a). On the one hand, the microlensing-induced size has been found to be compatible with the $R_{BLR} - L$ relation measured in local AGN (Sluse et al. 2011; Hutsemékers and Sluse 2021). On the other hand, forward models of line deformations observed for 3 geometries and kinematics models (and a range of orientations with respect to the observer) ruled out an equatorial wind model, and favoured a keplerian disc over a biconical polar wind (Hutsemékers and Sluse 2021). In addition, inclination of ~ 40 deg of BLR had been favoured by these authors, in agreement with the Type 1 properties of this system, and previous constraints on the inclination derived from microlensing of the accretion disc (Poindexter and Kochanek 2010). Another system displaying salient line deformation is the large separation quad J1004+4112. The recurrent asymmetric blue-wing enhancement of C IV observed in that system has been qualitatively reproduced by Abajas et al. (2007) using a biconical geometry. The microlensing interpretation of the atypical microlensing signal observed in this system (Lamer et al. 2006) has been challenged by Green (2006) who has proposed a non-microlensing explanation of the data, where each strongly-lensed image is crossing different regions of warm outflowing material. As shown by several authors (e.g. Lamer et al. 2006; Hutsemékers et al. 2023), the choice of normalisation of the line plays an important role in the apparent recurrence of the blue wing enhancement and in the quantification of the velocity dependant amplitude of microlensing. Using as baseline the flux ratios from radio data (Hartley et al. 2021), Hutsemékers et al. (2023) have shown that a microlensing-induced deformation of the line stable over 15 years exists, with the red wing being demagnified while the blue wing is strongly magnified. This scenario is supported by simulations that enable them to show that a keplerian disc or an equatorial wind provide a good description of the BLR.

4.4.2 Low Ionization Lines

Emission lines requiring lower ionisation energy, such as Balmer lines (in e.g. Hα λ6562.8 and Hβ λ4861.3) and Mg II $\lambda\lambda$2798, are observed at larger wavelengths than higher ionization lines. Because most lensed AGNs lie at $z > 1.5$, those lines have been less regularly observed and scrutinized for microlensing than higher ionization ones. Mg II is found to be on average less microlensed than higher ionization lines, which means that their sizes are a few times larger than high ionization lines Guerras et al. (2013a), Fian et al. (2021a). This is in agreement with the expected stratification of the BLR with increasing ionization degree.

Symmetric line deformations are the most common, but asymmetric deformation are occasionally observed. There are no statistical constraints on line deformations of Balmer lines, but both symmetric and asymmetric deformation of these lines have been reported (Sluse et al. 2007; Braibant et al. 2016, 2014). For two systems (Q 2237+0305 and HE 0435−1223), a forward modeling of the observed deformation has been performed (cf. Sect. 3.5.3), strongly favouring a keplerian disc geometry (Braibant et al. 2016; Hutsemekers et al. 2019; Hutsemékers and Sluse 2021). A half-light radius $r_{1/2}(\mathrm{H}\beta) = 47 \pm 19$ lt-days has been derived for Q 2237+0305, about an order of magnitude below the expectation from $R_{BLR} - L$ relation derived from reverberation mapping (Grier et al. 2017). Hutsemékers and Sluse (2021) have proposed that this could be related to the high accretion rate estimated for that system.

The comparison of line deformations based on data obtained at the same epoch (or pairs of data accounting for the time delay) provide further insights on the existence of a common BLR structure for various atomic species. To our knowledge, there are only a handful of systems where both Mg II and a Balmer line have been scrutinized. In HE 0435−1223, Mg II and Hα are found to be similarly microlensed (Braibant et al. 2014). For WGD2038−4008, both lines are found to be free of microlensing deformation (Melo et al. 2021). This contrasts with the results obtained for RXJ 1131−1231 where a very broad and spatially compact emission has been identified only in Mg II (Sluse et al. 2007). While differences are also observed between these two lines in SBS0909+532, this is at two very different epochs Mediavilla et al. (2005), Guerras et al. (2013a)). Therefore, the microlensing configuration could have varied substantially between the two observations, precluding to draw robust conclusions.

The C III] λ1908.73 line, sometimes classified as intermediate ionization line, is commonly present is lensed AGNs spectra but suffers from a strong blending with Al III and Si III lines, which complicates the identification of microlensing induced line deformation. Nevertheless, substantial microlensing is often identified in this line, with appearance of blue/red asymmetries similar to what is found for C IV (Sluse et al. 2012; Fian et al. 2021a). This line may therefore arise from a region smaller than the one emitting low-ionization lines. In their analysis of the spectro-photometric monitoring data of Q 2237+0305, Sluse et al. (2011) found an emitting size of C III] emission region compatible with the one of C IV, but with hints of a different structure in the two regions.

4.4.3 UV-Optical Iron Lines

The Iron emission blend covers almost the whole UV to optical range (Sect. 2.9.3). Guerras et al. (2013b) performed the only systematic study of microlensing of UV Fe II in a sizeable sample of 13 quasars. They focused on the range [2050-2650] Åand identified strong microlensing in 4 systems. They derived a size of the corresponding emitting region of several light-days, comparable to the accretion disc size. Evidence for compact UV Fe II emission has also been reported in Q 2237+0305 (Sluse et al. 2011), H1413+117 (O'Dowd et al. 2015) (in [1590-1680] Å) and in J1131-1231 (in [3080-3540] Å) by Sluse et al. (2007). The low redshift of the latter system enabled those authors to compare microlensing of the UV and optical Fe II. They found a microlensing of optical Fe II comparable to that of the Balmer lines, at the exception of the region [4630-4800] Å potentially arising from a more compact region.

4.4.4 Broad Absorption Lines

Broad absorption lines ($FWHM > 2000$ km s^{-1}) are present in 10 to 20% of quasars at $2 < z < 4$, and are predominantly observed in resonance lines of ionized species such as C IV, Si IV or N V (Allen et al. 2011). The blueshifts from 1000 km s^{-1} to up to 20% of the speed of light displayed by the absorber may be the signature of outflowing gas. At X-ray wavelengths, BALs quasars are generally X-ray weak, probably due to absorption[16]

[16]We may note that mini-BAL, characterised by a small level of UV absorption, and narrow absorption line quasars, characterised by absorption width with $FWHM < 500$ km s^{-1}, generally show different X-ray properties (Hamann et al. 2013; Chartas et al. 2021).

(Gallagher et al. 2006). The idea of using microlensing for probing the structure of the outflowing gas in BAL quasars has been proposed very early, but yet applied to a limited number of systems (e.g. Hutsemékers 1993, 1994; Lewis et al. 1998; Chelouche 2005).

The best-studied system is H1413+117, a.k.a. the cloverleaf. The main insights on the geometry of the outflow come from the decomposition of the line profiles using the MmD decomposition (see Sect. 3.2.2). The analysis of this system (Hutsemékers et al. 2010; O'Dowd et al. 2015) reveals (a) a nearly black absorption of C IV in the microlensed flux, and unveils the underlying line emission in the non microlensed component; (b) an onset velocity of the absorption of about 1500 $\mathrm{km\,s^{-1}}$ for C IV and a larger velocity onset of Si IV and Al III; (c) part of the C IV emission is reabsorbed over a small wavelength range. These characteristics support radial changes in the absorbing material, which can be interpreted as a two components outflowing wind: one component which may be co-spatial with the emission, and another one, more distant, that partially re-absorb the BEL emission (Borguet and Hutsemékers 2010). The ionization dependence of the onset velocity of the BAL can be interpreted in the context of the disc+wind model.

4.5 Torus

Due to limited access to the mid-infrared range from the ground and to the low resolution of mid-infrared instrumentation in the last decades, observations of lensed quasars in the mid-infrared have been rather limited. Agol et al. (2000) reports the first measurements of the flux ratios in that range for the Einstein Cross. The flux ratios measured at 8.9 and 11.7 μm were found to be in agreement with those measured at radio wavelengths, but relative uncertainties reach ∼20%. Nevertheless, this observation has been used to rule out the presence of (compact) synchrotron emission in the mid-infrared, and support the idea that the emission in that range is dominated by dust arising from a region too large to be microlensed (Agol et al. 2000; Wyithe et al. 2002). Several additional observations of the Einstein Cross have been performed over the years (see Vives-Arias et al. 2016, for a census), in broad agreement with the early measurements of Agol et al. (2000). The absence of microlensing yields $R_{1/2} \gtrsim 200\sqrt{< M > /0.3 M_{\odot}}$ lt-days at about 11 μm (Vives-Arias et al. 2016). Due to this large size, the few other observations of lensed quasars in the mid-infrared have instead been used to search for flux ratio anomalies, interpreted as microlensing by substructures of mass $M \sim 10^6 - 10^8 M_{\odot}$ (e.g. Chiba et al. 2005; Minezaki et al. 2009).

Although microlensing in the mid-infrared range is small, it may not always be negligible. Observations of 6 lensed quasars at 2.2 μm and 3.8 μm by Fadely and Keeton (2011) show that microlensing can occur. Since this typically corresponds to ∼ 1 μm rest-frame, this may be explained by the yet substantial contribution of accretion disc emission. Observations of unlensed quasars as well as simulations however suggest that, for lower luminosity systems, the hot component of the torus (peaking around 2.2 μm rest-frame) may be microlensed in some systems (Stalevski et al. 2012; Sluse et al. 2013). Upcoming observations of lensed quasars with the James-Webb Space Telescope is expected to shed new lights on the presence of microlensing in wavelengths ranges dominated by dust emission.

4.6 Scattered and Polarized Emission

Polarization of UV-optical light in AGNs is observed at levels ranging between zero and 4-5%, rare systems sometimes reaching polarization levels of 10-20%. Broad absorption line AGNs, radio-loud systems and blazar generally display the larger linear polarization degrees. In the UV-optical, the polarization arises predominantly from scattering of the light by electrons, and maybe dust. The exact location of that region (equatorial/polar, inside/outside the BLR, etc), and even its ubiquity in all AGN are yet totally opened questions. Because microlensing re-weight the spatial distribution of the emission in the inner regions of AGNs, it is expected to produce (de-)polarization of AGNs, hence modifying both the polarization degree and angle of microlensed images (Belle and Lewis 2000; Hales and Lewis 2007; Kedziora et al. 2011).

The first observational evidence for polarization differences induced by microlensing required the high resolution capabilities of the HST targeting the BAL quasar H1413+117 (Chae et al. 2001). It took more than a decade before obtaining resolved polarimetric and spectropolarimetric observations of individual images of lensed quasars from the ground, again for H1413+117 (Hutsemékers et al. 2010, 2015; Sluse et al. 2015). Those studies have shown a twist of the polarisation angle of image D, which displays slowly varying microlensing since about 25 years. This observation can be explained with the presence in that BAL of two spatially separated regions producing orthogonal polarization: an equatorial region which is microlensed, and a more extended polar region. Resolved polarised observations of the two large separation lensed quasars J1004+4112 and Q0957+561 have been presented by Popović et al. (2020, 2021). Similarly, changes in the polarisation degree and angles are observed in those objects due to microlensing. This supports the existence of a compact a equatorial scattering region in systems which are not BALs.

In addition to an imprint of microlensing in the polarisation signal, spectroscopic data of microlensed BAL have revealed the presence of extended diffuse emission separated from the accretion disc (Sluse et al. 2015; Hutsemékers et al. 2020). This discovery has been facilitated by the presence of almost black absorption covering the emission from the accretion disc but not the most extended continuum. Differential microlensing has revealed non-microlensed continuum emission contributing to up to 40% of the observed continuum. Despite of contributing to a large fraction of the observed continuum, this emission is insufficient to explain the larger size of the accretion disc unveiled by microlensing studies (Sect. 4.3; Sluse et al. 2015). The presence of non-microlensed (pseudo-)continuum emission in non-BAL systems is yet to be demonstrated, as well as its exact origin. If this emission is generically present in AGNs, a small flickering of microlensed lightcurves could be expected at wavelengths free of broad emission lines. While such flickering is observed in numerous systems (see Sect. 3.4.2), it is not yet demonstrated that it is caused by extended continuum emission.

5 Lensing Galaxy Results

What can be constrained uniquely by microlensing about the lens galaxy is the partition of its mass into baryons (compact) and dark (smooth) matter, and how the baryonic component is further partitioned between the microlenses, i.e. its mass function. As we have already seen in Sect. 2.8 (see also Fig. 12), the values of κ, γ, κ_*[17] at the positions of the multiple images, which are directly linked to the local values of the second derivatives of the lens potential (see Saha et al. 2024), define microlensing variability. The total mass in baryons, defined by κ_*, can be further distributed in compact objects in different ways. Microlensing magnification is generally more sensitive to the mean microlens mass, but information on a mixture of masses and the IMF can be extracted under specific conditions.

[17] Separating the shear into a smooth and compact component does not make any difference in the lens equation (see Eq. (4)), as opposed to the mass density. Note that if there is an external shear component then it would have to be subtracted in order to link the local values of the shear at the positions of the multiple images to the second derivative of the lens mass distribution.

5.1 Measuring the Baryonic/Dark Matter Content Under Different Constraints from Macromodels

In principle, the κ, γ, κ_* could all be constrained as free parameters of a microlensing model fitted to observables, but in practice, only a single free parameter is used. This can be either κ_* itself, with κ, γ derived separately from a macromodel, or a parameter partitioning the total mass between a stellar and dark mass profile (usually the mass-to-light ratio), leading to a sequence of covariant κ, γ, κ_*. A lens macromodel is a mass distribution/lens potential fitted to the imaging data of a lensed quasar. Lens modelling requires some form of the mass profile to be chosen first, e.g. SIE, power-law, etc, and proceeds by making specific assumptions for the (unknown) extended source brightness profile, including the quasar point source, and solving the lens equation to fit the lensed light components to the data (see Shajib et al. 2024). It is straightforward to calculate the values of κ, γ at the multiple image positions from the resulting lens potential. One could additionally make use of the fact that light traces mass, fit a two-dimensional brightness profile to the observed lens light (e.g. a Sersic profile, spline polynomials, etc), assume some mass-to-light ratio, and thus compute a stellar-mass component to incorporate in the mass model. In combination with a dark matter component, e.g. a Navarro-Frenk-White profile, κ_* can be provided straightforwardly, albeit not in a completely unbiased way (see below).

Both flux ratio and light curve microlensing data can be used to constrain the local graininess of matter, i.e. the value of κ_*, and both require simulated magnification maps. These maps are generated for given κ, γ, κ_* combinations derived from some macromodel and used to extract observable quantities. In this case, all the other parameters of the model – accretion disc profile, effective velocity, microlens mass – are treated as nuisance parameters and marginalised over. By fitting the model to the data, the most likely κ_*, or more precisely κ, γ, κ_* combination, can be found. We present the results on κ_* from flux ratios and light curves below.

5.1.1 Flux Ratios

The idea of constraining κ_* from microlensing flux ratios was originally proposed by Schechter and Wambsganss (2002), who found that a smooth mass component in addition to microlenses can increase variability especially for saddle-points.[18] Bate et al. (2007) and Congdon et al. (2007) found that a finite source can in fact cause the same broadening of microlensing magnification distributions, requiring the two parameters – source size and κ_* – to be studied simultaneously. Mediavilla et al. (2009) studied flux ratios in 29 image pairs obtaining their κ, γ from a SIS plus external shear macromodel and found an overall 5^{+9}_{-3} per cent of matter in stars (at 90 per cent confidence) for a given source size (their result is positively correlated with the source size and probably contains various other systematic effects and biases, see the discussion in Mediavilla et al. 2009). In a later study, Jiménez-Vicente et al. (2015a) and Jiménez-Vicente et al. (2015b) took into account the source size simultaneously with κ_* and found consistently 20 per cent of matter in stars at $1-2$ effective radii for a sample of 18 lenses from flux ratios in the optical and X-rays respectively. They used the more realistic SIE plus external shear macromodels from Schechter et al. (2014), which are known to produce smaller magnifications. Bate et al. (2011) found $20-50$ per cent of matter in stars within $5-10$ kpc of the lens center for two systems after marginalizing over the source properties (size and temperature profile) and κ_* for Q 2237+0305. This

[18] Schechter and Wambsganss (2002) showed this for a single κ, γ combination. This result was later extended to the entire parameter space by Vernardos et al. (2014).

was an expected outcome for this system because the multiple images are located so close to the lens center that the stellar mass component is comparable to, or even dominates over, smoothly distributed dark matter (see references in Bate et al. 2011, for the macromodels they used). Pooley et al. (2012) found 7 per cent of matter in stars at a mean distance of 6.6 kpc from the lens center by analyzing an ensemble of 14 quadruply lensed quasars and a decreasing trend of κ_* with distance. They derived their κ, γ from a SIS plus external shear macromodel (see also Pooley et al. 2007; Blackburne et al. 2011).

The results summarized above are prone to systematic errors in measurements and degeneracies in modelling that are specifically associated with flux ratio snapshots. As discussed previously (see Sect. 4.3), analyzing flux ratios with microlensing is more robust for close image pairs that have almost the same macro-magnification and small time delays. But properly deblending the image fluxes can be tricky precisely because of the proximity of the images, while unaccounted for intrinsic source variability can lead to κ_* overestimates (Mediavilla et al. 2009). Any overlap of the observed wavelengths (e.g. for a broad band) with emission lines, especially broad lines that are known to undergo microlensing as well, could contaminate the measured flux ratios. The extent of the chromatic variation and the level of the no-microlensing baseline, which are known to affect source measurements (Bate et al. 2018), could also play a role for measuring κ_*, but this has not been investigated yet. Finally, there are cases where the flux ratio data are simply not sufficient to lead to any conclusive κ_* results[19] (e.g. Bate et al. 2008; Floyd et al. 2009).

5.1.2 Light Curves

Constraining κ_* from light curve data has proceeded largely by applying the fitting method of Kochanek (2004). Studies using this approach have adopted a different strategy to calculating κ, γ from a macromodel and having κ_* as a free parameter. Their mass model consists of a combination of a NFW halo and a concentric de Vaucouleurs component of the mass in stars, obtained by fitting the light of the lensing galaxy and a constant mass-to-light ratio (see also Lehar et al. 2000). The total mass being conserved, a single free parameter, f_{ML}, describes the combination of the two components, from purely stars ($f_{\mathrm{ML}} = 1$) to just a NFW halo ($f_{\mathrm{ML}} = 0$). Varying this parameter leads to different κ, γ, κ_* combinations, not just κ_* (and also a varying mass-to-light ratio across the lensing galaxy). Examples of such families of models are shown in Fig. 13 – the points from Morgan et al. (2008), Dai et al. (2010), and MacLeod et al. (2015).

Light curve data analysis favours low values of κ_*, albeit not decisively. Chartas et al. (2009) and Dai et al. (2010) find $f_{\mathrm{ML}} = 0.2$ and 0.3 for lensed quasars HE1104−1805 and RXJ 1131−1231 respectively. Modest trends supporting dark matter dominated models are found for PG1115+080 (Morgan et al. 2008), QJ0158−4325 (Morgan et al. 2012), and WFI2026−4536 (Cornachione et al. 2020a), while measurements for Q0957+561 (Hainline et al. 2012), SBS0909+532 (Hainline et al. 2013), and J0924+0219 (Morgan et al. 2006; MacLeod et al. 2015) have been inconclusive.

One of the biggest caveats of analyzing light curves is the very demanding computations (Kochanek 2004; Poindexter and Kochanek 2010) that have resulted in the analysis of only ≈ 10 systems (Cornachione and Morgan 2020). However, the generally weak trends in κ_* are probably due to insufficient uncorrelated variability in the light curves to constrain the

[19] For the accretion disc, Guerras et al. (2020) find that at least 4 measurements (or 4 pairs of images) should be used to obtain reliable estimations. Although not explicitly studied, something similar probably holds for measuring κ_* too.

dark matter fraction (Hainline et al. 2012). Including the time dimension, as opposed to snapshots, comes at the cost of additional parameters, mainly in relation to the effective velocity, but this alone does not seem to affect the measurements of κ_*.

5.1.3 Impact of Macromodel Uncertainty on Measuring κ_*

Apart from the possible systematic biases and statistical uncertainties inherent to the analysis methods described above, any uncertainty of the macromodel itself can potentially have an impact on κ_* measurements. Vernardos and Fluke (2014b) have shown that $\Delta\kappa$, $\Delta\gamma$ as small as 0.02 can lead to significant differences in the properties of microlensing magnification. Macromodel biases and uncertainties of this magnitude are not uncommon and can be due to three reasons that we briefly list here. First, they can be due to missing ingredients of the mass model, like perturbers, disc- or bar-like structures, or higher order moments in the potential (e.g. see Van de Vyvere et al. 2022, for an example of how the latter affect time delays). For example, not accounting for the nearby galaxy G2 and a few close-by perturbers in the model of WFI2033−4723 yields a relative change of κ by up to 60 per cent, and of the absolute macro-magnification by up to a factor 2 for some of the lensed images (Rusu et al. 2020). The presence of any other massive substructure ($> 10^6 M_\odot$), whose impact on flux ratios can be dramatic (Mao and Schneider 1998; Dalal and Kochanek 2002; Metcalf and Zhao 2002), can bias the values of κ, γ locally, albeit with different characteristics compared to microlensing (Inoue 2016). Second, there is a fundamental degeneracy between the initial mass function (IMF) and the dark matter fraction (e.g. see Oguri et al. 2014; Foxley-Marrable et al. 2018), which, in short, states that the same lens potential and light distribution can be produced by a low dark matter fraction and many low-mass (and less luminous) stars, or a high dark matter fraction and fewer but brighter stars. Finally, the mass-sheet degeneracy (Falco et al. 1985; Gorenstein et al. 1988), i.e. scaling the lens galaxy mass distribution and adding a constant surface mass density (mass-sheet), leaves observables such as the image positions, shapes, and flux ratios, unchanged but affects the κ, γ. Although there are ways to mitigate these effects, e.g. by breaking the mass-sheet degeneracy through modelling of kinematics data (see Shajib et al. 2024) or explicitly modelling substructures (see Vegetti et al. 2024), the complexity of the true lens potential and the corresponding quality of a macromodel will eventually affect microlensing measurements.

5.2 Microlens Mass and the IMF

Microlensing variability depends on the relative size of micro-caustics on the source plane with respect to the emitting source. A typical value for this size is $R_{\rm E}$ that is $\propto \mathrm{M}/\mathrm{M}_\odot{}^{1/2}$ (Eq. (7)), where M is the microlens mass. The masses for a population of microlenses are, in principle, free to vary in the broad range roughly from small star clusters (or even intermediate mass black holes) to sub-stellar objects like planets and moons. In the latter case, one can speak of 'nano-lensing',[20] which is still relevant for quasars as long as there is a wavelength to observe in which their emitting region is accordingly small. It turns out that microlensing variability (at a fixed source size) depends primarily on the microlens mean mass and less on the shape of the mass spectrum. However, the IMF of the lensing galaxy can still be probed by measuring the microlens mean mass across different wavelengths (or

[20] The other extreme is usually referred to as 'milli-lensing'. This is out of the scope of this review because the multiple images produced can be either separated enough to be detected or because any variability takes place on timescales of centuries or more. The reader is referred to Vegetti et al. (2024) for more details.

by indirectly measuring an overall mass-to-light ratio for the galaxy lens, see below). This is suggested by observations as well, for example, by the lack of correlation between the microlensing variability in optical and X-ray observations found by Mosquera et al. (2013) (see also Pooley et al. 2007), where the former could be due to stellar microlenses and the later due to planetary mass nanolenses (e.g. Guerras et al. 2020).

It has been known from early on (Wambsganss 1992; Lewis and Irwin 1995; Wyithe and Turner 2001; Congdon et al. 2007) that the expected microlensing magnification shows very little dependence on the shape of the mass spectrum of the microlenses. Consequently, a widely adopted approach has been using a fixed-mass microlens population to produce magnification maps. Schechter et al. (2004) found that magnification histograms can be substantially different in the case of a mix of two such populations whose masses differ by more than a factor of ten but have a comparable total mass. Esteban-Gutiérrez et al. (2020) further examine such bi-modal mass distributions and find that they can be replaced by a single population with mass equal to their geometric mean. Both studies admit that they intentionally focus on extreme cases and that for extended sources the low mass component can effectively behave like a mass-sheet – a case equivalent to a single fixed-mass population with a different κ_* and the low-mass objects absorbed into a smooth component.

The source appears the smallest in the X-rays and allows one to probe the sub-stellar mass range for the microlenses. The innermost accretion disc is known to emit the relativistically blurred iron Fe Kα line in the X-rays (Reynolds and Nowak 2003). Lensed quasars can display shifts in the peak energy of this line over time, attributed to microlensing (Chartas et al. 2017). Dai and Guerras (2018) and Bhatiani et al. (2019) modelled such shifts in three systems and interpreted them as due to the presence of a population of sub-stellar mass objects ($10^{-8} - 10^{-3}$ $M_\odot$). They base their argument on the fact that the cross section of an X-ray source is too small for microlensing by stellar-mass objects. This can be understood by considering the caustic density resulting from the same total mass density divided into a few stellar-mass microlenses compared to many more smaller mass ones. As the caustic density increases with smaller masses so does the corss section for microlensing of an X-ray (very small) source (see Fig. 2 in Bhatiani et al. 2019). The study of Guerras et al. (2020) similarly supports planetary masses for the microlenses using the standard deviation of distributions of uncorrelated X-ray flux ratios over time (i.e. a collapsed light curve).

Observations in the optical are mostly affected by stellar-mass microlenses. Jiménez-Vicente and Mediavilla (2019) calculate the microlens mean mass of a sample of 24 lensed quasars to be $0.08 < M/M_\odot < 0.21$ (at 68 per cent confidence) using optical and X-ray flux ratios. They used fixed-mass simulations and marginalized over all the other free parameters of their model. Nevertheless, there was still a degeneracy with the source size that was broken by using a size prior from reverberation mapping (Mediavilla et al. 2017). Using light curve data for Q 2237+0305, Kochanek (2004) and Poindexter and Kochanek (2010) measured $0.006 < M/M_\odot < 0.2$ and $0.12 < M/M_\odot < 1.94$ (at 68 per cent confidence) respectively. Both studies highlight the importance of the velocity priors used to get these marginally consistent measurements. The latter uses longer light curves and a velocity model that includes the random motions of the microlenses, which could explain the higher mass values found. Under similar dependence on velocity priors, MacLeod et al. (2015) and Morgan et al. (2018) found $0.016 < M/M_\odot < 0.8$ and $0.03 < M/M_\odot < 0.44$ (at 68 per cent confidence) for J0924+0219 and WFI2033−4723 respectively, while Cornachione et al. (2020b) found $0.1 < M/M_\odot < 1$ for Q0957+561 and a somewhat unexpected order of magnitude smaller mass range for SBS0909+532.

The interest in probing higher masses (up to 100 $M_\odot$) has been kept high due to observations towards the Galactic bulge by the MACHO experiment (Alcock et al. 2000) and the recent detection of gravitational waves by LIGO/Virgo, attributed to high-mass (≈ 40 $M_\odot$) black hole mergers. Interpreting the former observations is arguably model-dependent and can allow for more mass in the form of dark, compact objects Hawkins (2015). Hawkins (2020) reached this conclusion in order to explain microlensing variability in lensed quasar light curves as well. However, only theoretical arguments were set forward to attribute this to stellar-mass black holes (and not e.g. stars) and microlensing observables were not modelled explicitly using such mass distributions. A re-analyis of the same data by Awad et al. (2023) suggested that a standard explanation, where galaxies are composed of a population of stars with a standard IMF and smoothly distributed dark matter, cannot be ruled out. Similarly, Esteban-Gutiérrez et al. (2022) found a negligible fraction of mass in ≈ 30 $M_\odot$ objects by explicitly using a mass spectrum in their magnification map simulations.

In addition to directly embedding a mass spectrum in microlensing models, there is also an indirect way the IMF can be probed through the total mass in microlenses. This is based on breaking the IMF-dark matter fraction degeneracy (e.g. see Oguri et al. 2014), i.e. the lens total mass (κ, coming from the macromodel) and the lens light distribution can be attributed to either a low dark matter fraction and many low-mass (and less luminous) stars, or a high dark matter fraction and fewer but brighter stars. An example is the stellar mass fundamental plane (Hyde and Bernardi 2009), which is obtained by converting surface brightness to stellar mass through an assumption on the IMF (e.g. see Kauffmann et al. 2003). The spectral data used to achieve this light-to-mass conversion suffer from negligible contributions from stellar remnants, brown dwarfs, and red dwarfs that are too faint, hence placing loose constraints on the lower mass part of the IMF. Schechter et al. (2014) recalibrate the mean of the IMF (a Salpeter one in this case) to a higher mass based on microlensing measurements of X-ray flux ratios in 10 lensed quasars. We note that this was an indirect measurement; although the magnification maps used had a microlens mass spectrum (assumed to be a Kroupa IMF, e.g. Blackburne et al. 2011), its form was fixed and the free parameter of the model was in fact κ_*. Oguri et al. (2014) followed a similar approach to constrain the shape of the IMF, finding a preference for Salpeter over Chabrier (see Schechter et al. 2014, for a detailed comparison of the methodology of the two studies). Finally, Vernardos (2019) presented a theoretical study of how optical flux ratios could be used to constrain the IMF. They found that the IMF could be measured better in a sample of doubly-lensed quasars (due to their higher numbers) that additionally have higher ratios of Einstein to effective radius (i.e. their multiple images appear further away from the lens-galaxy light).

5.3 Convergence and Shear, κ, γ

As mentioned in the beginning of 5.1, the values of κ, γ can in principle be constrained alongside κ_* and the mean microlens mass from microlensing data. However, in practice this is almost never the case due to the very high computational cost of creating and sampling from many magnification maps. An exception is the study of MG0414+0534 by Vernardos (2018), where κ, γ were treated as free parameters and constrained (over an area) from flux ratio data independently from any macromodel. Vernardos and Tsagkatakis (2019) tested a proof-of-concept that the values of κ, γ could be measured from microlensing light curves as well. Although promising, these studies have their own limitations and their applicability to a wide range of systems/data remains to be explored. Achieving such measurements will constitute an entirely new probe of matter in galaxies and provide constraints to macromodels – κ, γ are directly linked to the second derivatives of the lens potential – with potential application to galaxy evolution, time delay cosmography, etc.

6 Future Prospects and Open Questions

From a curiosity applicable to only a handful of systems, microlensing has steadily matured over the years, both in terms of observations and methodology. It can now complement well-established fields like reverberation mapping (for AGN structure e.g. see Cackett et al. 2021) and galaxy-galaxy lensing (for the mass partition between baryons/dark matter and the IMF, e.g. see Shajib et al. 2024 and Vegetti et al. 2024). Microlensing constitutes an additional observational probe based on an "orthogonal" set of assumptions, e.g. it is not sensitive to the brightness profile of AGN emission in the same way as reverberation techniques, nor does it critically depend on the shape of the mass profile of the lens and the macroscopically observed brightness distribution of the lensed source (shape of the arcs, rings, etc). As such, it can be combined with more traditional methods, for example, to study the kinematics and morphology of the BLR (Garsden et al. 2011) and to break the degeneracy between the stellar IMF and the dark matter fraction (Oguri et al. 2014). Moreover, microlensing provides the additional advantage of probing objects that may be hard to study with standard methods. For instance, microlensing measurements are independent of the level of intrinsic variability of the source (a needed prior for the reverberation method to work), naturally probes objects at cosmic-noon (i.e. $1.5 < z < 2.5$), and thanks to macro-magnification, enables to study even faint systems at high redshift.

Currently, several hundreds of lensed quasars are known (see Lemon et al. 2024) but suitable microlensing data have been obtained for only a few tens of them. In the next decade, the field will be revolutionised by the advent of all-sky surveys, like LSST and Euclid, which are not only expected to discover thousands of new systems but will also provide high-quality data to perform microlensing analyses. Microlensing holds a great potential in addressing major science questions by facing the challenge of the upcoming avalanche of data and subsequent adaptation and refinement of its analysis techniques. The following subsections outline these three points in detail.

6.1 Science Potential

There is a treasure trove of information to be mined for the central quasar engine from monitoring data and in particular from HMEs. Once thousands of lensed quasars are discovered and regularly observed, we will be poised to start observing tens or even hundreds of such events. During their peaks, the inner accretion disc and the flow of material through the ISCO are magnified and encoded in the microlensing signal. With high cadence (<nightly) and signal-to-noise ratio data, especially in the X-rays and optical-UV wavelengths (could also include radio, see Sect. 4.6), we will be able to perform a variety of tests on accretion disc theory and the theory of relativity.

Two main microlensing results for the accretion disc are its size and temperature profile, which are respectively found larger and shallower than expected. More insights into these somewhat puzzling results would be gained by simply increasing the number of studied systems from the current 15 or so. Even without new and better-adapted to "big-data" methods, simply turning the crank with existing ones would suffice for this goal.

Microlensing has enabled the measurement of the BLR half-light radius, in general agreement with expectations from the $R - L$ relations from reverberation mapping. It has additionally confirmed its increase in size with ionization degree for tens of systems. For a small subsample of objects, constraints on the BLR kinematics have been set by modeling the broad line deformations. There is a great potential in expanding such works to more systems. The best observational setup for such work consists in spatially resolved spectroscopy, ideally from space or from the ground in combination with adaptive optics. The modeling of the kinematics is computationally more demanding, and may benefit from development of innovative modeling strategies to be massively applied to order(s) of magnitude more systems.

Directly measuring the graininess of matter and by extension the mass density in stars can shed light into the interplay of dark matter with baryons throughout galaxy evolution. This critical information is very hard to obtain with other methods and prone to degeneracies and assumptions. The goal to achieve is measuring κ_* as a function of galactic radius within the lens or per galaxy type (e.g. Pooley et al. 2012). Such microlensing results could be combined with other modelling methods as priors, e.g. galaxy-galaxy lens modelling (see Shajib et al. 2024), used to break degeneracies (e.g. Oguri et al. 2014), or recalibrate scaling relations like the fundamental plane (e.g. Schechter et al. 2014). Again, just the sheer number of new measurements will be enough for gaining important new knowledge.

The expected large number of new observations will enable slicing into subcategories sharing similar physical properties (e.g. quasar black hole mass and Eddington ratio, lens galaxy type and/or age, etc). This will be crucial for drawing a re-fined picture of AGN structure and to understand how their properties depend on key physical characteristics. This will be particularly powerful in combination with improved methods (see below), which would make the most out of every single measurement, i.e. not just counting on large number statistics.

Finally, we would like to make a brief comment on the potential of performing cosmological studies with microlensing. Firstly, understanding their central engine can clarify whether quasars can be used as standard candles, e.g. with relations such as the one between the luminosity in the X-rays and UV light (Risaliti and Lusso 2019). These are exactly the wavelengths where the quasar is the most prone and sensitive to microlensing. And secondly, microlensing is the only probe for peculiar velocities of galaxies at high redshifts ($z > 0.1$, Mediavilla et al. 2016). Such velocities can in turn be used to constrain the growth of structure (e.g. Koda et al. 2014) and shed more light into the σ_8 tension between early and late Universe (Di Valentino et al. 2021).

6.2 Future Observations

The two main and invaluable sources for microlensing data in the next decade will be the all-sky surveys by the Vera Rubin observatory and the Euclid satellite, both starting to provide data within 2023. The former, will provide ~ 800 epochs across 6 bands over a decade for each patch of sky (on average), while the latter will carry out a survey at a much higher spatial resolution (space-based). Both will provide data suitable for microlensing studies for thousands of systems without explicit need for follow-up: LSST will provide 10-year long light curves and Euclid flux ratios in the optical and infrared. The LSST light curves in particular will not only be suitable for long-term microlensing variability studies (see Sect. 3.4) but they could also be used to predict the onset of HMEs.

Imaging and high cadence photometry will play a key role in probing the accretion disc, but a lot may be gained by obtaining spatially resolved spectroscopy. This will be possible for much fainter than current systems, owing to the upcoming 30-m class telescopes. Such data may be instrumental in deblending the disc emission from any other superimposed source of extended (pseudo)-continuum emission. At the same time, BLR deformations may be studied in detail.

In X-rays, almost all of the progress made in constraining the size of the corona, the inner disc radius, and the dark matter content of the lensing galaxies has been a result of the ability to determine the X-ray fluxes of each image of a lensed quasar. This task requires sub-arcsecond spatial resolution in X-rays, and *Chandra* has been the only observatory with this capability. Launched in 1999, the mission has made incredible contributions to this field, but the satellite is aging and the main detector has lost most of its effective area below 1 keV. Currently, there is no mission with similar or better spatial resolution in X-rays that has been approved by any of the major space agencies. The Astro2020 Decadal recommended that NASA plan for the launch of the next flagship X-ray mission no earlier than the late 2040s. Such a mission would likely have *Chandra*-like or better spatial resolution.

In the interim, NASA Astrophysics has introduced a new line of satellites (Probes) the first of which is to be launched in the early 2030s and will be either a far-infrared mission or an X-ray mission. One of the X-ray mission concepts (AXIS) aims to have spatial resolution similar to *Chandra* but with an order of magnitude more collecting area. If such a mission was to meet these specifications, it would be able to make substantial progress in this field both by observing lensed quasars at higher redshifts / lower fluxes and by extending the X-ray light curves of *Chandra*-observed lensed quasars to over 30- and likely over 40-year baselines.

The presence of microlensing at radio wavelengths is still debated with only two claimed signatures in existing data. There is a possible confirmation bias in not finding more cases. A dedicated observing program is required to investigate the question in a systematic way. In the near future, the Square Kilometer Array may become an ideal facility to confirm or rule out the existence of microlensing in the radio domain, and probe radio emission at even higher scales than possible with interferometry.

The impact of microlensing on the polarisation of the quasar light has not yet been explored much, but important results on the presence of scattered light have been obtained. A polarimetric imaging survey with a 2-m class telescope may enable to systematically study the presence of scattered light in a larger number of microlensed quasars, possibly contributing to understanding oversized accretion discs. Spectropolarimatry may naturally complement photometric data, in particular for the study of the structure of BAL quasars. Tools for simulating polarized signal are basically in place, but data are required to develop the field.

6.3 Developing New Methods

For all microlensing applications, the increase in sample size will require novel analysis approaches. New technologies like GPUs have already helped in understanding the magnification map parameter space that would have otherwise been inaccessible due to the high computational cost. This remains true for the study of systematic biases in the analysis methods; a few works have explored this, albeit in a limited fashion, for the single-epoch method, while a similar study for the light curve method is hindered by its very resource-demanding computations. This is especially true when long light curves most likely containing more complex signals are used; one such example could be the role of the dynamic nature of the microlensing map (individual $\boldsymbol{v}_\star$ terms per microlens in Eq. (57)). Identifying and mitigating such biases will be essential for delivering groundbreaking science with microlensing. One way to achieve this is by understanding how the length of observations, the amount of microlensing amplitude variations within it, and errors in the macromodel κ, γ and magnification can affect resulting quasar structure measurements and correct for it, and similarly for determining the stellar matter fraction, κ_*. The alternative possibility of simply allowing for large number statistics to improve the precision of the results (without a guaranteed improvement on the accuracy due to possible systematic biases), although more straightforward to achieve, would prevent reaching the full potential of microlensing by, for example, studying subcategories of objects with key properties of interest (e.g. as a function of redshift).

Because the BLR covers a larger fraction of the microlensing map than the disc, its modeling might be affected by different systematic uncertainties than the latter. The modeling of the BLR together with the accretion disc limits the range of events reproducing the data. It may be worth investigating if this is associated with a gain in precision and accuracy on the retrieved sizes. A drawback of the BLR modeling strategy is the need for microlensing maps that are sufficiently large for modeling the BLR, and sufficiently high resolution to constrain the disc. Another challenge that arises with the increase in number of observational constraints (e.g. the microlensing in many velocity bins in the case of the BLR) is the difficulty for the models to reproduce the full signal down to the noise. It often happens that only a few tens of realisations out of millions of trials successfully reproduce the data. Simulations may help finding out if this is evidence for the need of more complex models, or if the observed events are simply rare. Whatever the answer, the computational cost is large, and alternatives to the existing "Monte-Carlo" sampling are desired. We note that this problem is not specific to BLR modeling, for instance, the models of high photometric precision light curves are generally unable to reproduce the data down to the noise. In the latter case, however, there are good reasons to think that it is not the statistics to blame, but the model that ignores non-disc contributions to the broad band flux (e.g. BLR emission blended with disc emission).

Machine Learning is another novel technology that could in fact help with achieving control over the systematics. However, apart from a handful of exploratory works, machine learning methods applied to microlensing have remained largely untapped. Key applications where such methods could make a difference would be fast generation of magnification maps and/or light curves for any κ, γ, κ_*, and source size, or direct inference of the model parameters (source size and geometry, κ_*, etc) from the data, avoiding expensive Bayesian likelihood estimations.

Last but not least, the unprecedented amount of data could be mined for the target information in novel ways that go beyond standard approaches, e.g. population studies, inclusion of more complex signals like reverberated flux and/or binary black holes, etc. Finally, HMEs were mentioned before in the context of machine learning, however, regardless of implementation, a robust predictor and follow-up trigger for such events should be developed and put to the task – there is a unique chance to capitalize LSST data for this.

6.4 Closing Remarks

We believe that with this review we have drawn a coherent and detailed picture of the current state of the field of quasar microlensing. We would argue that both data and methods have evolved from circumstantial and basic to targeted and advanced, but not yet fully mature – the field is in its "adolescence". This means that although first results are coherent and exciting, there is more work needed to control systematic biases and to be able to address the avalanche of data from all-sky surveys in the next decade. We are confident that this will happen and microlensing will achieve its full potential to deliver groundbreaking scientific results.

Acknowledgements We thank the International Space Science Institute in Bern (ISSI) for their hospitality and the conveners for organizing the stimulating workshop on "Strong Gravitational Lensing". GV has received funding from the European Union's Horizon 2020 research and innovation programme under the Marie Sklodovska-Curie grant agreement No 897124. GV's and MO's research was made possible by the generosity of Eric and Wendy Schmidt by recommendation of the Schmidt Futures program. DS would like to thank

Damien Hutsemékers and Lorraine Braibant for discussions and material that helped shaping Fig. 23. MM acknowledges the support of the Swiss National Science Foundation under grant P500PT_203114. VM acknowledges partial support from *Centro de Astrofísica de Valparaíso* and ANID Fondecyt Regular 1231418. TA acknowledges support from ANID-FONDECYT Regular Project 1190335, the Millennium Science Initiative ICN12_009 and the ANID-BASAL project FB210003. Thanks to F. Dux for producing Fig. 20.

Funding Open access funding provided by EPFL Lausanne.

Declarations

Competing Interests The authors have no conflicts of interest to report.

Open Access This article is licensed under a Creative Commons Attribution 4.0 International License, which permits use, sharing, adaptation, distribution and reproduction in any medium or format, as long as you give appropriate credit to the original author(s) and the source, provide a link to the Creative Commons licence, and indicate if changes were made. The images or other third party material in this article are included in the article's Creative Commons licence, unless indicated otherwise in a credit line to the material. If material is not included in the article's Creative Commons licence and your intended use is not permitted by statutory regulation or exceeds the permitted use, you will need to obtain permission directly from the copyright holder. To view a copy of this licence, visit http://creativecommons.org/licenses/by/4.0/.

References

Abajas C, Mediavilla E, Muñoz JA, Popović LČ, Oscoz A (2002) The influence of gravitational microlensing on the broad emission lines of quasars. Astrophys J 576(2):640–652. https://doi.org/10.1086/341793

Abajas C, Mediavilla E, Muñoz JA, Gómez-Álvarez P, Gil-Merino R (2007) Microlensing of a biconical broad-line region. Astrophys J 658(2):748–762. https://doi.org/10.1086/511023

Abolmasov P, Shakura NI (2012a) Microlensing evidence for super-Eddington disc accretion in quasars. Mon Not R Astron Soc 427:1867–1876. https://doi.org/10.1111/j.1365-2966.2012.21881.x

Abolmasov P, Shakura NI (2012b) Resolving the inner structure of QSO discs through fold-caustic-crossing events. Mon Not R Astron Soc 423:676–693. https://doi.org/10.1111/j.1365-2966.2012.20904.x

Abramowicz MA, Fragile CP (2013) Foundations of black hole accretion disk theory. Living Rev Relativ 16:1. https://doi.org/10.12942/lrr-2013-1

Abramowicz MA, Czerny B, Lasota JP, Szuszkiewicz E (1988) Slim accretion discs. Astron J 332:646

Agol E, Krolik J (1999) Imaging a quasar accretion disk with microlensing. Astrophys J 524:49–64. https://doi.org/10.1086/307800

Agol E, Jones B, Blaes O (2000) Keck mid-infrared imaging of QSO 22370305. Astrophys J 545:657–663

Alcock C, Allsman RA, Alves DR, Axelrod TS, Becker AC, Bennett DP, Cook KH, Dalal N, Drake AJ, Freeman KC, Geha M, Griest K, Lehner MJ, Marshall SL, Minniti D, Nelson CA, Peterson BA, Popowski P, Pratt MR, Quinn PJ, Stubbs CW, Sutherland W, Tomaney AB, Vandehei T, Welch D (2000) The MACHO project: microlensing results from 5.7 years of Large Magellanic Cloud observations. Astrophys J 542:281–307. https://doi.org/10.1086/309512

Allen JT, Hewett PC, Maddox N, Richards GT, Belokurov V (2011) A strong redshift dependence of the broad absorption line quasar fraction. Mon Not R Astron Soc 410:860–884. https://doi.org/10.1111/j.1365-2966.2010.17489.x

Amorim A, Bauböck M, Bentz MC, Brandner W, Bolzer M, Clénet Y, Davies R, de Zeeuw PT, Dexter J, Drescher A, Eckart A, Eisenhauer F, Förster Schreiber NM, Garcia PJV, Genzel R, Gillessen S, Gratadour D, Hönig S, Kaltenbrunner D, Kishimoto M, Lacour S, Lutz D, Millour F, Netzer H, Onken CA, Ott T, Paumard T, Perraut K, Perrin G, Petrucci PO, Pfuhl O, Prieto MA, Rouan D, Shangguan J, Shimizu T, Stadler J, Sternberg A, Straub O, Straubmeier C, Street R, Sturm E, Tacconi LJ, Tristram KRW, Vermot P, von Fellenberg S, Widmann F, Woillez J (GRAVITY Collaboration) (2021) The central parsec of NGC 3783: a rotating broad emission line region, asymmetric hot dust structure, and compact coronal line region. Astron Astrophys 648:A117. https://doi.org/10.1051/0004-6361/202040061

Angonin M-C, Remy M, Surdej J, Vanderriest C (1990) First spectroscopic evidence of microlensing on a BAL quasar? The case of H1413+117. Astron Astrophys 233:5

Anguita T, Schmidt RW, Turner EL, Wambsganss J, Webster RL, Loomis KA, Long D, Mcmillan R (2008) The multiple quasar Q2237+0305 under a microlensing caustic. Astron Astrophys 480:327–334. https://doi.org/10.1051/0004-6361:20078221

Antonucci R (1993) Unified models for active galactic nuclei and quasars. Annu Rev Astron Astrophys 31:473–521. https://doi.org/10.1146/annurev.aa.31.090193.002353

Awad P, Chan JHH, Millon M, Courbin F, Paic E (2023) Probing compact dark matter objects with microlensing in gravitationally lensed quasars. Astron Astrophys 673:A88. arXiv:2304.01320. https://doi.org/10.1051/0004-6361/202245615

Barsdell BR, Barnes DG, Fluke CJ (2010) Analysing astronomy algorithms for graphics processing units and beyond. Mon Not R Astron Soc 408:1936–1944. https://doi.org/10.1111/j.1365-2966.2010.17257.x

Barth AJ, Pancoast A, Bennert VN, Brewer BJ, Canalizo G, Filippenko AV, Gates EL, Greene JE, Li W, Malkan MA, Sand DJ, Stern D, Treu T, Woo JH, Assef RJ, Bae HJ, Buehler T, Cenko SB, Clubb KI, Cooper MC, Diamond-Stanic AM, Hönig SF, Joner MD, Laney CD, Lazarova MS, Nierenberg AM, Silverman JM, Tollerud EJ, Walsh JL (2013) The Lick AGN monitoring project 2011: Fe II reverberation from the outer broad-line region. Astrophys J 769:128. https://doi.org/10.1088/0004-637X/769/2/128

Barvainis R (1987) Hot dust and the near-infrared bump in the continuum spectra of quasars and active galactic nuclei. Astrophys J 320:537. https://doi.org/10.1086/165571

Bate NF, Fluke CJ (2012) A graphics processing unit-enabled, high-resolution cosmological microlensing parameter survey. Astrophys J 744:90. https://doi.org/10.1088/0004-637X/744/2/90

Bate NF, Webster RL, Wyithe JSB (2007) Smooth matter and source size in microlensing simulations of gravitationally lensed quasars. Mon Not R Astron Soc 381:1591–1596. https://doi.org/10.1111/j.1365-2966.2007.12330.x

Bate NF, Floyd DJE, Webster RL, Wyithe JSB (2008) A microlensing study of the accretion disc in the quasar MG 0414+0534. Mon Not R Astron Soc 391:1955–1960. https://doi.org/10.1111/j.1365-2966.2008.14020.x

Bate NF, Floyd DJE, Webster RL, Wyithe JSB (2011) A microlensing measurement of dark matter fractions in three lensing galaxies. Astrophys J 731:71. https://doi.org/10.1088/0004-637X/731/1/71

Bate NF, Vernardos G, O'Dowd MJ, Neri-Larios MD (2018) HST imaging of four lensed quasars. Mon Not R Astron Soc 479:4796–4814. https://doi.org/10.1093/mnras/sty1793

Belle KE, Lewis GF (2000) Microlensing of broad absorption line quasars: polarization variability. Publ Astron Soc Pac 112:320. https://doi.org/10.1086/316523

Bennert N, Falcke H, Schulz H, Wilson AS, Wills BJ (2002) Size and structure of the narrow-line region of quasars. Astrophys J 574:105–109

Benning M, Burger M (2018) Modern regularization methods for inverse problems. Acta Numerica 27:1–111. https://doi.org/10.1017/S0962492918000016. arXiv:1801.09922

Bentz MC, Katz S (2015) The agn black hole mass database. Publ Astron Soc Pac 127:67

Bentz MC, Denney KD, Grier CJ, Barth AJ, Peterson BM, Vestergaard M, Bennert VN, Canalizo G, Rosa GD, Filippenko AV, Gates EL, Greene JE, Li W, Malkan MA, Pogge RW, Stern D, Treu T, Woo JH (2013) The low-luminosity end of the radius-luminosity relationship for active galactic nuclei. Astrophys J 767:149. https://doi.org/10.1088/0004-637X/767/2/149

Bentz MC, Williams PR, Street R, Onken CA, Valluri M, Treu T (2021) A detailed view of the broad-line region in NGC 3783 from velocity-resolved reverberation mapping. Astrophys J 920:112. https://doi.org/10.3847/1538-4357/ac19af

Best H, Fagin J, Vernardos G, O'Dowd M (2022) Resolving the vicinity of supermassive black holes with gravitational microlensing. http://arxiv.org/abs/2210.10500

Bhatiani S, Dai X, Guerras E (2019) Confirmation of planet-mass objects in extragalactic systems. Astrophys J 885:77. https://doi.org/10.3847/1538-4357/ab46ac

Biggs AD (2023) A vla monitoring study of JVAS B1422+231: investigation of time delays and detection of extrinsic variability. Mon Not R Astron Soc. https://doi.org/10.1093/mnras/stad870

Biggs AD, Browne IWA (2018) A revised lens time delay for JVAS B0218+357 from a reanalysis of VLA monitoring data. Mon Not R Astron Soc 476:5393–5407. https://doi.org/10.1093/MNRAS/STY565

Birrer S et al (2024) Time-delay cosmography: measuring the Hubble constant and other cosmological parameters with strong gravitational lensing. Space Sci Rev 220

Blackburne JA, Kochanek CS (2010) The effect of a time-varying accretion disk size on quasar microlensing light curves. Astrophys J 718:1079–1084. https://doi.org/10.1088/0004-637X/718/2/1079

Blackburne JA, Pooley D, Rappaport S, Schechter PL (2011) Sizes and temperature profiles of quasar accretion disks from chromatic microlensing. Astrophys J 729:34. https://doi.org/10.1088/0004-637X/729/1/34

Blackburne JA, Kochanek CS, Chen B, Dai X, Chartas G (2014) The optical, ultraviolet, and X-ray structure of the quasar HE 0435–1223. Astrophys J 789:125. https://doi.org/10.1088/0004-637X/789/2/125

Blandford RD, Mckee CF (1982) Reverberation mapping of the emission line regions of Seyfert galaxies and quasars. Astrophys J 255:419–439

Bogdanov MB, Cherepashchuk AM (2002) Reconstruction of the strip brightness distribution in a quasar accretion disk from gravitational microlensing data. Astron Rep 46:626–633

Borguet B, Hutsemékers D (2010) A polar+equatorial wind model for broad absorption line quasars. Astron Astrophys 515:A22. https://doi.org/10.1051/0004-6361/200913255

Bourassa RR, Kantowski R (1975) The theory of transparent gravitational lenses. Astrophys J 195:13–21

Bourassa RR, Kantowski R, Norton TD (1973) The spheroidal gravitational lens. Astrophys J 185:747–756

Bozza V (2010) Microlensing with an advanced contour integration algorithm: Green's theorem to third order, error control, optimal sampling and limb darkening. Mon Not R Astron Soc 408:2188–2200. https://doi.org/10.1111/j.1365-2966.2010.17265.x

Braibant L, Hutsemékers D, Sluse D, Anguita T, García-Vergara CJ (2014) Microlensing of the broad-line region in the quadruply imaged quasar HE0435-1223. Astron Astrophys 565:L11. https://doi.org/10.1051/0004-6361/201423633

Braibant L, Hutsemékers D, Sluse D, Anguita T (2016) The different origins of high- and low-ionization broad emission lines revealed by gravitational microlensing in the Einstein cross. Astron Astrophys 592:A23. https://doi.org/10.1051/0004-6361/201628594

Braibant L, Hutsemékers D, Sluse D, Goosmann R (2017) Constraining the geometry and kinematics of the quasar broad emission line region using gravitational microlensing: I. Models and simulations. Astron Astrophys 607:A32. https://doi.org/10.1051/0004-6361/201731086

Brenneman LW, Reynolds CS (2006) Constraining black hole spin via X-ray spectroscopy. Astrophys J 652(2):1028–1043. https://doi.org/10.1086/508146

Bruce A, Lawrence A, MacLeod C, Elvis M, Ward MJ, Collinson JS, Gezari S, Marshall PJ, Lam MC, Kotak R, Inserra C, Polshaw J, Kaiser N, Kudritzki RP, Magnier EA, Waters C (2017) Spectral analysis of four 'hypervariable' AGN: a microneedle in the haystack? Mon Not R Astron Soc 467:1259–1280. https://doi.org/10.1093/mnras/stx168

Cackett EM, Gelbord J, Li Y-R, Horne K, Wang J-M, Barth AJ, Bai J-M, Bian W-H, Carroll RW, Du P, Edelson R, Goad MR, Ho LC, Hu C, Khatu VC, Luo B, Miller J, Yuan Y-F (2020) Supermassive black holes with high accretion rates in active galactic nuclei. XI. Accretion disk reverberation mapping of MRK 142. Astrophys J 896:1. https://doi.org/10.3847/1538-4357/ab91b5

Cackett EM, Bentz MC, Kara E (2021) Reverberation mapping of active galactic nuclei: from X-ray corona to dusty torus. iScience 24(6):102557. https://doi.org/10.1016/j.isci.2021.102557

Canizares CR (1982) Manifestations of a cosmological density of compact objects in quasar light. Astrophys J 263:508

Cappellari M (2016) Structure and kinematics of early-type galaxies from integral field spectroscopy. Annu Rev Astron Astrophys 54:597–665. https://doi.org/10.1146/annurev-astro-082214-122432

Cardelli JA, Savage BD (1988) Two lines of sight with exceedingly anomalous ultraviolet interstellar extinction. Astrophys J 325:864–879

Chae K-H, Turnshek DA, Schulte-Ladbeck RE, Rao SM, Lupie OL (2001) Hubble Space Telescope observations of the gravitationally lensed cloverleaf broad absorption line QSO H14131143: imaging polarimetry and evidence for microlensing of a scattering region. Astrophys J 561:653–659

Chan JHH, Rojas K, Millon M, Courbin F, Bonvin V, Jauffret G (2021) Measuring accretion disk sizes of lensed quasars with microlensing time delay in multi-band light curves. Astron Astrophys 647:A115. https://doi.org/10.1051/0004-6361/202038971

Chang K, Refsdal S (1979) Flux variations of QSO 0957+561 A, B and image splitting by stars near the light path. Nature 282:561

Chang K, Refsdal S (1984) Star disturbances in gravitational lens galaxies. Astron Astrophys 132:168

Chartas G, Agol E, Eracleous M, Garmire G, Bautz MW, Morgan ND (2002) Caught in the act: Chandra observations of microlensing of the radio-loud quasar MG J0414+0534. Astrophys J 568(2):509–521. https://doi.org/10.1086/339162

Chartas G, Eracleous M, Agol E, Gallagher Chandra SC (2004) Observations of the cloverleaf quasar H1413+117: a unique laboratory for microlensing studies of a LoBAL quasar. Astrophys J 606(1):78–84. https://doi.org/10.1086/382743

Chartas G, Eracleous M, Dai X, Agol E, Gallagher S (2007) Discovery of probable relativistic Fe emission and absorption in the cloverleaf quasar H 1413+117. Astrophys J 661(2):678–692. https://doi.org/10.1086/516816

Chartas G, Kochanek CS, Dai X, Poindexter S, Garmire G (2009) X-ray microlensing in RXJ1131-1231 and HE1104-1805. Astrophys J 693:174–185. https://doi.org/10.1088/0004-637X/693/1/174

Chartas G, Kochanek CS, Dai X, Moore D, Mosquera AM, Blackburne JA (2012) Revealing the structure of an accretion disk through energy-dependent X-ray microlensing. Astrophys J 757(2):137. https://doi.org/10.1088/0004-637X/757/2/137

Chartas G, Krawczynski H, Zalesky L, Kochanek CS, Dai X, Morgan CW, Mosquera A (2017) Measuring the innermost stable circular orbits of supermassive black holes. Astrophys J 837:26. https://doi.org/10.3847/1538-4357/aa5d50

Chartas G, Cappi M, Vignali C, Dadina M, James V, Lanzuisi G, Giustini M, Gaspari M, Strickland S, Bertola E (2021) Multiphase powerful outflows detected in high-z quasars. Astrophys J 920:24. https://doi.org/10.3847/1538-4357/ac0ef2

Chelouche D (2005) Gravitational microlensing and the structure of quasar outflows. Astrophys J 629:667

Chen B, Dai X, Kochanek CS, Chartas G, Blackburne JA, Morgan CW (2012) X-ray monitoring of gravitational lenses with Chandra. Astrophys J 755(1):24. https://doi.org/10.1088/0004-637X/755/1/24

Chen GCF, Suyu SH, Wong KC, Fassnacht CD, Chiueh T, Halkola A, Hu IS, Auger MW, Koopmans LVE, Lagattuta DJ, McKean JP, Vegetti S (2016) Sharp – III. First use of adaptive-optics imaging to constrain cosmology with gravitational lens time delays. Mon Not R Astron Soc 462:3457–3475. https://doi.org/10.1093/mnras/stw991

Chen B, Kantowski R, Dai X, Baron E, der Mark PV (2017) Accelerating gravitational microlensing simulations using the Xeon Phi coprocessor. Astron Comput 19:60–65. https://doi.org/10.1016/j.ascom.2017.03.005

Chiba M, Minezaki T, Kashikawa N, Kataza H, Inoue KT (2005) Subaru mid-infrared imaging of the quadruple lenses PG 1115+080 and B1422+231: limits on substructure lensing 1. Astrophys J 627:53

Collin S, Boisson C, Mouchet M, Dumont AM, Coupé S, Porquet D, Rokaki E (2002) Are quasars accreting at super-Eddington rates? Astron Astrophys 388:771–786. https://doi.org/10.1051/0004-6361:20020550

Congdon AB, Keeton CR, Osmer SJ (2007) Microlensing of an extended source by a power-law mass distribution. Mon Not R Astron Soc 376:263–272. https://doi.org/10.1111/j.1365-2966.2007.11426.x

Cornachione MA, Morgan CW (2020) Quasar microlensing variability studies favor shallow accretion disk temperature profiles. Astrophys J 895:93. https://doi.org/10.3847/1538-4357/ab8aed

Cornachione MA, Morgan CW, Burger HR, Shalyapin VN, Goicoechea LJ, Vrba FJ, Dahm SE, Tilleman TM (2020b) Near-infrared and optical continuum emission region size measurements in the gravitationally lensed quasars Q0957+561 and SBS0909+532. Astrophys J 905:7. https://doi.org/10.3847/1538-4357/abc25d

Cornachione MA, Morgan CW, Millon M, Bentz MC, Courbin F, Bonvin V, Falco EE (2020a) A microlensing accretion disk size measurement in the lensed quasar WFI 2026–4536. Astrophys J 895:125. https://doi.org/10.3847/1538-4357/ab557a

Czerny B, Li Y-R, Hryniewicz K, Panda S, Wildy C, Sniegowska M, Wang J-M, Sredzinska J, Karas V (2017) Failed radiatively accelerated dusty outflow model of the broad line region in active galactic nuclei. I. Analytical solution. Astrophys J 846:154. https://doi.org/10.3847/1538-4357/aa8810

Dai X, Guerras E (2018) Probing extragalactic planets using quasar microlensing. Astrophys J 853:27. https://doi.org/10.3847/2041-8213/aaa5fb

Dai L, Pascale M (2021) New approximation of magnification statistics for random microlensing of magnified sources. http://arxiv.org/abs/2104.12009

Dai X, Chartas G, Agol E, Bautz MW, Garmire Chandra GP (2003) Observations of QSO 2237+0305. Astrophys J 589(1):100–110. https://doi.org/10.1086/374548

Dai X, Kochanek CS, Chartas G, Kozłowski S, Morgan CW, Garmire G, Agol E (2010) The sizes of the X-ray and optical emission regions of RXJ 1131-1231. Astrophys J 709:278–285. https://doi.org/10.1088/0004-637X/709/1/278

Dalal N, Kochanek CS (2002) Direct detection of cold dark matter substructure. Astrophys J 572:25–33

Davis SW, Laor A (2011) The radiative efficiency of accretion flows in individual active galactic nuclei. Astrophys J 728:98. https://doi.org/10.1088/0004-637X/728/2/98

Dempsey R, Zakamska NL (2018) The size-luminosity relationship of quasar narrow-line regions. Mon Not R Astron Soc 477:4615–4626. https://doi.org/10.1093/mnras/sty941

Dexter J, Agol E (2011) Quasar accretion disks are strongly inhomogeneous. Astrophys J Lett 727:24. https://doi.org/10.1088/2041-8205/727/1/L24

Di Matteo T (1998) Magnetic reconnection: flares and coronal heating in active galactic nuclei. Mon Not R Astron Soc 299(1):15–20. https://doi.org/10.1046/j.1365-8711.1998.01950.x

Di Valentino E, Anchordoqui LA, Akarsu Ö, Ali-Haimoud Y, Amendola L, Arendse N, Asgari M, Ballardini M, Basilakos S, Battistelli E, Benetti M, Birrer S, Bouchet FR, Bruni M, Calabrese E, Camarena D, Capozziello S, Chen A, Chluba J, Chudaykin A, Colgáin EÓ, Cyr-Racine F-Y, de Bernardis P, de Cruz Pérez J, Delabrouille J, Dunkley J, Escamilla-Rivera C, Ferté A, Finelli F, Freedman W, Frusciante N, Giusarma E, Gómez-Valent A, Handley W, Harrison I, Hart L, Heavens A, Hildebrandt H, Holz D, Huterer D, Ivanov MM, Joudaki S, Kamionkowski M, Karwal T, Knox L, Kumar S, Lamagna L, Lesgourgues J, Lucca M, Marra V, Masi S, Matarrese S, Mazumdar A, Melchiorri A, Mena O, Mersini-Houghton L, Miranda V, Moreno-Pulido C, Mota DF, Muir J, Mukherjee A, Niedermann F, Notari A, Nunes RC, Pace F, Paliathanasis A, Palmese A, Pan S, Paoletti D, Pettorino V, Piacentini F, Poulin V, Raveri M, Riess AG, Salzano V, Saridakis EN, Sen AA, Shafieloo A, Shajib AJ, Silk J, Silvestri A, Sloth MS, Smith TL, Solà Peracaula J, van de Bruck C, Verde L, Visinelli L, Wandelt BD, Wang D, Wang J-M, Yadav AK, Yang W (2021) Cosmology intertwined III: $f\sigma_8$ and S_8. Astropart Phys 131:102604. https://doi.org/10.1016/j.astropartphys.2021.102604

Diego JM, Hannuksela OA, Kelly PL, Pagano G, Broadhurst T, Kim K, Li TGF, Smoot GF (2019) Observational signatures of microlensing in gravitational waves at LIGO/Virgo frequencies. Astron Astrophys 627:A130. https://doi.org/10.1051/0004-6361/201935490

Dobler G, Keeton CR, Wambsganss J (2007) Microlensing of central images in strong gravitational lens systems. Mon Not R Astron Soc 377:977–986. https://doi.org/10.1111/j.1365-2966.2007.11695.x

Edelson R, Gelbord JM, Horne K, McHardy IM, Peterson BM, Arévalo P, Breeveld AA, Rosa GD, Evans PA, Goad MR, Kriss GA, Brandt WN, Gehrels N, Grupe D, Kennea JA, Kochanek CS, Nousek JA, Papadakis I, Siegel M, Starkey D, Uttley P, Vaughan S, Young S, Barth AJ, Bentz MC, Brewer BJ, Crenshaw DM, Cáceres EDBADL, Denney KD, Dietrich M, Ely J, Fausnaugh MM, Grier CJ, Hall PB, Kaastra J, Kelly BC, Korista KT, Lira P, Mathur S, Netzer H, Pancoast A, Pei L, Pogge RW, Schimoia JS, Treu T, Vestergaard M, Villforth C, Yan H, Zu Y (2015) Space telescope and optical reverberation mapping project. II. Swift and HST reverberation mapping of the accretion disk of NGC 5548. Astrophys J 806:129. https://doi.org/10.1088/0004-637X/806/1/129

Eigenbrod A, Courbin F, Meylan G, Agol E, Anguita T, Schmidt RW, Wambsganss J (2008a) Microlensing variability in the gravitationally lensed quasar QSO 2237+0305 ≡ the Einstein cross. II. Energy profile of the accretion disk. Astron Astrophys 490:933–943. https://doi.org/10.1051/0004-6361

Eigenbrod A, Courbin F, Sluse D, Meylan G, Agol E (2008b) Microlensing variability in the gravitationally lensed quasar QSO 2237+0305 ≡ the Einstein cross. Astron Astrophys 480:647–661. https://doi.org/10.1051/0004-6361:20078703

Elitzur M, Ho LC, Trump JR (2014) Evolution of broad-line emission from active galactic nuclei. Mon Not R Astron Soc 438:3340–3351. https://doi.org/10.1093/mnras/stt2445

Esteban-Gutiérrez A, Agües-Paszkowsky N, Mediavilla E, Jiménez-Vicente J, Muñoz JA, Heydenreich S (2020) The impact of the mass spectrum of lenses in quasar microlensing studies. Constraints on a mixed population of primordial black holes and stars. Astrophys J 904:176. https://doi.org/10.3847/1538-4357/abbdf7

Esteban-Gutiérrez A, Agües-Paszkowsky N, Mediavilla E, Jiménez-Vicente J, Muñoz JA, Heydenreich S (2022) Abundance of LIGO/Virgo black holes from microlensing observations of quasars with reverberation mapping size estimates. Astrophys J 929:123. https://doi.org/10.3847/1538-4357/ac57c5

Event Horizon Telescope Collaboration (2019) First M87 Event Horizon Telescope results. IV. Imaging the central supermassive black hole. Astrophys J Lett 875:L4. https://doi.org/10.3847/2041-8213/ab0e85

Event Horizon Telescope Collaboration (2022) First Sagittarius A* Event Horizon Telescope results. III. Imaging of the galactic center supermassive black hole. Astrophys J Lett 930:L14. https://doi.org/10.3847/2041-8213/ac6429

Fabian AC, Rees MJ, Stella L, White NE (1989) X-ray fluorescence from the inner disc in Cygnus X-1. Mon Not R Astron Soc 238:729–736. https://doi.org/10.1093/mnras/238.3.729

Fabian AC, Zoghbi A, Ross RR, Uttley P, Gallo LC, Brandt WN, Blustin AJ, Boller T, Caballero-Garcia MD, Larsson J, Miller JM, Miniutti G, Ponti G, Reis RC, Reynolds CS, Tanaka Y, Young AJ (2009) Broad line emission from iron K- and L-shell transitions in the active galaxy 1H0707-495. Nature 459(7246):540–542. https://doi.org/10.1038/nature08007

Fabian AC, Kara E, Walton DJ, Wilkins DR, Ross RR, Lozanov K, Uttley P, Gallo LC, Zoghbi A, Miniutti G, Boller T, Brandt WN, Cackett EM, Chiang C-Y, Dwelly T, Malzac J, Miller JM, Nardini E, Ponti G, Reis RC, Reynolds CS, Steiner JF, Tanaka Y, Young AJ (2013) XMM observation of the narrow-line Seyfert 1 galaxy IRAS 13224-3809: rapid variability, high spin and a soft lag. Mon Not R Astron Soc 429(4):2917–2923. https://doi.org/10.1093/mnras/sts504

Fadely R, Keeton CR (2011) Near-infrared K and L' flux ratios in six lensed quasars. Astron J 141:101. https://doi.org/10.1088/0004-6256/141/3/101

Falco EE, Gorenstein VM, Shapiro II (1985) On model-dependent bounds on h0 from gravitational images: application to Q0957+561A, B. Astrophys J 289:1

Falco EE, Impey CD, Kochanek CS, Leha J, Mcleod BA, Rix H-W, Keeton CR, Mun JA, Peng CY (1999) Dust and extinction curves in galaxies with $z > 0$: the interstellar medium of gravitational lens galaxies. Astrophys J 523:617–632

Ferland GJ, Hu C, Wang JM, Baldwin JA, Porter RL, Hoof PAMV, Williams RJR (2009) Implications of infalling fe ii-emitting clouds in active galactic nuclei: anisotropic properties. Astrophys J 707:L82. https://doi.org/10.1088/0004-637X/707/1/L82

Ferland GJ, Porter RL, van Hoof PAM, Williams RJR, Abel NP, Lykins ML, Shaw G, Henney WJ, Stancil PC (2013) The 2013 release of cloudy. Rev Mex Astron Astrofís 49:137–163. https://doi.org/10.48550/arXiv.1302.4485

Fian C, Mediavilla E, Hanslmeier A, Oscoz A, Serra-Ricart M, Muñoz JA, Jiménez-Vicente J (2016) Size of the accretion disk in the graviationally lensed quasar SDSS J1004+4112 from the statistics of microlensing magnifications. Astrophys J 830(2):149. https://doi.org/10.3847/0004-637X/830/2/149

Fian C, Mediavilla E, Jiménez-Vicente J, Muñoz JA, Hanslmeier A (2018) Estimate of the accretion disk size in the gravitationally lensed quasar HE 0435-1223 using microlensing magnification statistics. Astrophys J 869:132. https://doi.org/10.3847/1538-4357/aaeed5

Fian C, Mediavilla E, Jiménez-Vicente J, Motta V, Muñoz JA, Chelouche D, Goméz-Alvarez P, Rojas K, Hanslmeier A (2021b) Revealing the structure of the lensed quasar Q 0957+561: I. Accretion disk size. Astron Astrophys 654:A70. https://doi.org/10.1051/0004-6361/202039854

Fian C, Mediavilla E, Motta V, Jiménez-Vicente J, Muñoz JA, Chelouche D, Hanslmeier A (2021a) Microlensing of the broad emission lines in 27 gravitationally lensed quasars: broad line region structure and kinematics. Astron Astrophys 653:A109. https://doi.org/10.1051/0004-6361/202039829

Floyd DJE, Bate NF, Webster RL (2009) The accretion disc in the quasar SDSS J0924+0219. Mon Not R Astron Soc 398:233–239. https://doi.org/10.1111/j.1365-2966.2009.15045.x

Fluke CJ, Webster RL (1999) Investigating the geometry of quasars with microlensing. Mon Not R Astron Soc 302:68–74. https://doi.org/10.1046/j.1365-8711.1999.02109.x

Fluke CJ, Barnes DG, Barsdell BR, Hassan AH (2011) Astrophysical supercomputing with GPUs: critical decisions for early adopters. Publ Astron Soc Aust 28:15–27

Foxley-Marrable M, Collett TE, Vernardos G, Goldstein DA, Bacon D (2018) The impact of microlensing on the standardisation of strongly lensed type IA supernovae. Mon Not R Astron Soc 478:5081–5090. http://arxiv.org/abs/1802.07738

Galeev AA, Rosner R, Vaiana GS (1979) Structured coronae of accretion disks. Astrophys J 229:318–326. https://doi.org/10.1086/156957

Gallagher SC, Brandt WN, Chartas G, Priddey R, Garmire GP, Sambruna RM (2006) An exploratory Chandra survey of a well-defined sample of 35 large bright quasar survey broad absorption line quasars. Astrophys J 644(2):709–724. https://doi.org/10.1086/503762

Gallo LC, Wilkins DR, Bonson K, Chiang C-Y, Grupe D, Parker ML, Zoghbi A, Fabian AC, Komossa S, Longinotti AL (2015) Suzaku observations of MRK 335: confronting partial covering and relativistic reflection. Mon Not R Astron Soc 446(1):633–650. https://doi.org/10.1093/mnras/stu2108

Gardner E, Done C (2017) The origin of the uv/optical lags in NGC 5548. Mon Not R Astron Soc 470:3591–3605. https://doi.org/10.1093/mnras/stx946

Garsden H, Lewis GF (2010) Gravitational microlensing: a parallel, large-data implementation. New Astron 15:181

Garsden H, Bate NF, Lewis GF (2011) Gravitational microlensing of a reverberating quasar broad-line region – I. Method and qualitative results. Mon Not R Astron Soc 418(2):1012–1027. https://doi.org/10.1111/j.1365-2966.2011.19552.x

Gaudi BS, Petters AO (2002a) Gravitational microlensing near caustics. I. Folds. Astrophys J 574:970–984. https://doi.org/10.1086/341063

Gaudi BS, Petters AO (2002b) Gravitational microlensing near caustics. II. Cusps. Astrophys J 580:468

Gil-Merino R, Goicoechea LJ, Shalyapin VN, Oscoz A (2018) New database for a sample of optically bright lensed quasars in the northern hemisphere. Astron Astrophys 616:A118. https://doi.org/10.1051/0004-6361/201832737

Goicoechea LJ, Artamonov BP, Shalyapin VN, Sergeyev AV, Burkhonov OA, Akhunov TA, Asfandiyarov IM, Bruevich VV, Ehgamberdiev SA, Shimanovskaya EV, Zheleznyak AP (2020) Liverpool-maidanak monitoring of the Einstein cross in 2006-2019. Astron Astrophys 637:A89. https://doi.org/10.1051/0004-6361/202037902

Gorenstein VM, Falco EE, Shapiro II (1988) Degeneracies in parameter estimates for models of gravitational lens systems. Astrophys J 327:693

Gott JR (1981) Are heavy halos made of low mass stars? A gravitational lens test. Astrophys J 243:140–146

Gould A, Miralda-Escude J (1997) Signatures of accretion disks in quasar microlensing. Astrophys J 483:13

Graham AW (2007) The black hole mass – spheroid luminosity relation. Mon Not R Astron Soc 379:711–722. https://doi.org/10.1111/j.1365-2966.2007.11950.x

Graham AW (2016) Galaxy bulges and their massive black holes: a review. In: Laurikainen E, Peletier R, Gadotti D (eds) Galactic bulges. Astrophysics and Space Science Library, vol 418. Springer, Cham, pp 263–313. https://doi.org/10.1007/978-3-319-19378-6_11

Granot J, Schechter PL, Wambsganss J (2003) The mean number of extra microimage pairs for macrolensed quasars. Astrophys J 583:575–583

Green PJ (2006) Lens-aided multi-angle spectroscopy (LAMAS) reveals small-scale outflow structure in quasars. Astrophys J 644:733. https://doi.org/10.1086/503760

Greencard L, Rokhlin V (1987) A fast algorithm for particle simulations. J Comput Phys 73:315–348

Grieger B, Kayser R, Refsdal S (1988) Gravitational micro-lensing as a clue to quasar structure. Astron Astrophys 194:54

Grieger B, Kayser R, Schramm T (1991) The deconvolution of the quasar structure from microlensing light curves. Astron Astrophys 252:508

Grier CJ, Pancoast A, Barth AJ, Fausnaugh MM, Brewer BJ, Treu T, Peterson BM (2017) The structure of the broad-line region in active galactic nuclei. II. Dynamical modeling of data from the AGN10 reverberation mapping campaign. Astrophys J 849:146. https://doi.org/10.3847/1538-4357/aa901b

Grier CJ, Shen Y, Horne K, Brandt WN, Trump JR, Hall PB, Kinemuchi K, Starkey D, Schneider DP, Ho LC, Homayouni Y, Li JI-H, McGreer ID, Peterson BM, Bizyaev D, Chen Y, Dawson KS, Eftekharzadeh S, Guo Y, Jia S, Jiang L, Kneib J-P, Li F, Li Z, Nie J, Oravetz A, Oravetz D, Pan K, Petitjean P, Ponder KA, Rogerson J, Vivek M, Zhang T, Zou H (2019) The Sloan Digital Sky Survey reverberation mapping project: initial C IV lag results from four years of data. Astrophys J 887:38. https://doi.org/10.3847/1538-4357/ab4ea5

Grzędzielski M, Janiuk A, Czerny B, Wu Q (2017) Modified viscosity in accretion disks: application to galactic black hole binaries, intermediate mass black holes, and active galactic nuclei. Astron Astrophys 603:A110. https://doi.org/10.1051/0004-6361/201629672

Guerras E, Mediavilla E, Jiménez-Vicente J, Kochanek CS, Muñoz Ja, Falco E, Motta V (2013a) Microlensing of quasar broad emission lines: constraints on broad line region size. Astrophys J 764:160. https://doi.org/10.1088/0004-637X/764/2/160

Guerras E, Mediavilla E, Jimenez-Vicente J, Kochanek CS, Muñoz JA, Falco E, Motta V, Rojas K (2013b) Microlensing of quasar ultraviolet iron emission. Astrophys J 778:123. https://doi.org/10.1088/0004-637X/778/2/123

Guerras E, Dai X, Mediavilla E (2020) A second-order moment of microlensing variability as a novel tool to constrain source emission size or discrete lens demographics in extragalactic research. Astrophys J 896:111. https://doi.org/10.3847/1538-4357/ab76b9

Guilbert PW, Rees MJ (1988) 'Cold' material in non-thermal sources. Mon Not R Astron Soc 233:475–484. https://doi.org/10.1093/mnras/233.2.475

Guo W-J, Li Y-R, Zhang Z-X, Ho LC, Wang J-M (2022) Accretion disk size measurements of active galactic nuclei monitored by the Zwicky Transient Facility. Astrophys J 929:19. https://doi.org/10.3847/1538-4357/ac4e84

Haardt F, Maraschi L (1991) A two-phase model for the X-ray emission from Seyfert galaxies. Astrophys J Lett 380:51. https://doi.org/10.1086/186171

Haardt F, Maraschi L (1993) X-ray spectra from two-phase accretion disks. Astrophys J 413:507. https://doi.org/10.1086/173020

Haardt F, Maraschi L, Ghisellini G (1994) A model for the X-ray and ultraviolet emission from Seyfert galaxies and galactic black holes. Astrophys J Lett 432:95. https://doi.org/10.1086/187520

Hainline LJ, Morgan CW, Beach JN, Kochanek CS, Harris HC, Tilleman T, Fadely R, Falco EE, Le TX (2012) A new microlensing event in the doubly imaged quasar Q 0957+561. Astrophys J 744:104. https://doi.org/10.1088/0004-637X/744/2/104.

Hainline LJ, Morgan CW, Macleod CL, Landaal ZD, Kochanek CS, Harris HC, Tilleman T, Goicoechea LJ, Shalyapin VN, Falco EE (2013) Time delay and accretion disk size measurements in the lensed quasar SBS 0909+532 from multiwavelength microlensing analysis. Astrophys J 774:69. https://doi.org/10.1088/0004-637X/774/1/69

Hales CA, Lewis GF (2007) Resolving the structure at the heart of bal quasars through microlensing induced polarisation variability. Publ Astron Soc Aust 24:30–40. https://doi.org/10.1071/AS07002

Hall PB, Noordeh ES, Chajet LS, Weiss E, Nixon CJ (2014) Modelling spikes in quasar accretion disc temperature. Mon Not R Astron Soc 442:1090–1109. https://doi.org/10.1093/mnras/stu890

Hall PB, Sarrouh GT, Horne K (2018) Non-blackbody disks can help explain inferred AGN accretion disk sizes. Astrophys J 854:93. https://doi.org/10.3847/1538-4357/aaa768

Hamann F, Chartas G, McGraw S, Hidalgo PR, Shields J, Capellupo D, Charlton J, Eracleous M (2013) Extreme-velocity quasar outflows and the role of X-ray shielding. Mon Not R Astron Soc 435:133–148. https://doi.org/10.1093/mnras/stt1231

Hartley P, Jackson N, Sluse D, Stacey HR, Vives-Arias H (2019) Strong lensing reveals jets in a sub-microJy radio-quiet quasar. Mon Not R Astron Soc 485(3):3009–3023. https://doi.org/10.1093/mnras/stz510

Hartley P, Jackson N, Badole S, McKean JP, Sluse D, Vives-Arias H (2021) Using strong lensing to understand the microjy radio emission in two radio quiet quasars at redshift 1.7. Mon Not R Astron Soc 508:4625–4638. https://doi.org/10.1093/mnras/stab2758

Hawkins MRS (2015) A new look at microlensing limits on dark matter in the galactic halo. Astron Astrophys 575:A107. https://doi.org/10.1051/0004-6361/201425400

Hawkins MRS (2020) The signature of primordial black holes in the dark matter halos of galaxies. Astron Astrophys 633:107. https://doi.org/10.1051/0004-6361/201936462

Hawkins MRS (2022) New evidence for a cosmological distribution of stellar mass primordial black holes. Mon Not R Astron Soc 512:5706–5714. https://doi.org/10.1093/mnras/stac863

Heyrovský D, Loeb A (1997) Microlensing of an elliptical source by a point mass. Astrophys J 490(1):38–50. https://doi.org/10.1086/304855

Higginbottom N, Proga D, Knigge C, Long KS, Matthews JH, Sim SA (2014) Line-driven disk winds in active galactic nuclei: the critical importance of ionization and radiative transfer. Astrophys J 789:19. https://doi.org/10.1088/0004-637X/789/1/19

Homayouni Y, Trump JR, Grier CJ, Horne K, Shen Y, Brandt WN, Dawson KS, Alvarez GF, Green PJ, Hall PB, Santisteban JVH, Ho LC, Kinemuchi K, Kochanek CS, Li JI-H, Peterson BM, Schneider DP, Starkey DA, Bizyaev D, Pan K, Oravetz D, Simmons A (2020) The Sloan Digital Sky Survey reverberation mapping project: Mg II lag results from four years of monitoring. Astrophys J 901:55. https://doi.org/10.3847/1538-4357/ababa9

Horne K, Welsh WF, Peterson BM (1991) Echo mapping of broad Hβ emission in NGC 5548. Astrophys J 367:5

Hu C, Du P, Lu KX, Li YR, Wang F, Qiu J, Bai JM, Kaspi S, Ho LC, Netzer H, Wang JM (2015) Supermassive black holes with high accretion rates in active galactic nuclei. III. Detection of Fe II reverberation in nine narrow-line Seyfert 1 galaxies. Astrophys J 804:138. https://doi.org/10.1088/0004-637X/804/2/138

Hubeny I, Agol E, Blaes O, Krolik JH (2000) Non-LTE models and theoretical spectra of accretion disks in active galactic nuclei. III. Integrated spectra for hydrogen-helium disks. Astrophys J 533:710–728

Huchra J, Gorenstein M, Kent S, Shapiro I, Smith G, Horine E, Whipple FL, Perley R (1985) 2237+0305: a new and unusual gravitational lens. Astron J 90:691

Hutsemékers D (1993) Selective gravitational microlensing and line profile variations in the BAL quasar H1413+117. Astron Astrophys 280:435

Hutsemékers D (1994) The use of gravitational microlensing to scan the structure of BAL QSOS. Astrophys Space Sci 216:361

Hutsemékers D, Sluse D (2021) Geometry and kinematics of the broad emission line region in the lensed quasar Q2237+0305. Astron Astrophys 654:A155. https://doi.org/10.1051/0004-6361/202141820

Hutsemékers D, Borguet B, Sluse D, Riaud P, Anguita T (2010) Microlensing in H1413+117: disentangling line profile emission and absorption in a broad absorption line quasar. Astron Astrophys 519:A103. https://doi.org/10.1051/0004-6361/200913247

Hutsemékers D, Sluse D, Braibant L, Anguita T (2015) Polarization microlensing in the quadruply imaged broad absorption line quasar H1413+117. Astron Astrophys 584:A61. https://doi.org/10.1051/0004-6361/201527243

Hutsemékers D, Braibant L, Sluse D, Anguita T, Goosmann R (2017) New constraints on quasar broad absorption and emission line regions from gravitational microlensing. Front Astron Space Sci 4:18. https://doi.org/10.3389/fspas.2017.00018

Hutsemekers D, Braibant L, Sluse D, Goosmann R (2019) Constraining the geometry and kinematics of the quasar broad emission line region using gravitational microlensing: II. Comparing models with observations in the lensed quasar HE0435-1223. Astron Astrophys 629:A43. https://doi.org/10.1051/0004-6361/201731087

Hutsemékers D, Sluse D, Kumar P (2020) Spatially separated continuum sources revealed by microlensing in the gravitationally lensed broad absorption line quasar SDSS J081830.46+060138.0. Astron Astrophys 633:A101. https://doi.org/10.1051/0004-6361/201936973

Hutsemékers D, Sluse D, Savic D, Richards GT (2023) Microlensing of the broad emission line region in the lensed quasar J1004+4112. Astron Astrophys. https://doi.org/10.1051/0004-6361/202245490

Hyde JB, Bernardi M (2009) The luminosity and stellar mass fundamental plane of early-type galaxies. Mon Not R Astron Soc 396:1171–1185. https://doi.org/10.1111/j.1365-2966.2009.14783.x

Ichimaru S (1977) Bimodal behavior of accretion disks: theory and application to Cygnus X-1 transitions. Astrophys J 214:840–855. https://doi.org/10.1086/155314

Inoue KT (2016) On the origin of the flux ratio anomaly in quadruple lens systems. Mon Not R Astron Soc 461:164–175. https://doi.org/10.1093/mnras/stw1270

Irwin MJ, Webster RL, Hewett PC, Corrigan RT, Jedrzejewski RI (1989) Photometric variations in the Q2237+0305 system: first detection of a microlensing event. Astron J 98:1989

Jiang Y-F, Davis SW, Stone JM (2016) Iron opacity bump changes the stability and structure of accretion disks in active galactic nuclei. Astrophys J 827:10. https://doi.org/10.3847/0004-637x/827/1/10

Jiménez-Vicente J, Mediavilla E (2019) The initial mass function of lens galaxies from quasar microlensing. Astrophys J 885:75. https://doi.org/10.3847/1538-4357/ab46b8

Jiménez-Vicente J, Mediavilla E (2022) Fast multipole method for gravitational lensing: application to high-magnification quasar microlensing. Astrophys J 941:80. https://doi.org/10.3847/1538-4357/ac9e59

Jiménez-Vicente J, Mediavilla E, Muñoz JA, Kochanek CS (2012) A robust determination of the size of quasar accretion disks using gravitational microlensing. Astrophys J 751:106. https://doi.org/10.1088/0004-637X/751/2/106

Jiménez-Vicente J, Mediavilla E, Kochanek CS, Muñoz JA, Motta V, Falco E, Mosquera AM (2014) The average size and temperature profile of quasar accretion disks. Astrophys J 783:47

Jiménez-Vicente J, Mediavilla E, Kochanek CS, Muñoz JA (2015a) Dark matter mass fraction in lens galaxies: new estimates from microlensing. Astrophys J 799:149. https://doi.org/10.1088/0004-637X/783/1/47

Jiménez-Vicente J, Mediavilla E, Kochanek CS, Muñoz JA (2015b) Probing the dark matter radial profile in lens galaxies and the size of X-ray emitting region in quasars with microlensing. Astrophys J 806:251. https://doi.org/10.1088/0004-637X/799/2/149

Jorstad S et al (2023) The Event Horizon Telescope image of the quasar NRAO 530. Astrophys J 943:170. https://doi.org/10.3847/1538-4357/acaea8

Jovanović P, Popović LČ, Simić S (2009) Influence of gravitational microlensing on broad absorption lines of QSOs: the case of the Fe K α line. New Astron Rev 53(7–10):156–161. https://doi.org/10.1016/j.newar.2009.07.008

Kaspi S, Smith PS, Netzer H, Maoz D, Jannuzi BT, Giveon U (2000) Reverberation measurements for 17 quasars and the size-mass-luminosity relations in active galactic nuclei. Astrophys J 533:631–649

Kaspi S, Brandt WN, Maoz D, Netzer H, Schneider DP, Shemmer O (2007) Reverberation mapping of high-luminosity quasars: first results. Astrophys J 659:997

Katz JI (1976) Nonrelativistic Compton scattering and models of quasars. Astrophys J 206:910–916. https://doi.org/10.1086/154455

Katz N, Balbus S, Paczynski B (1986) Random scattering approach to gravitational microlensing. Astrophys J 306:2

Kauffmann G, Heckman TM, White SDM, Charlot S, Tremonti C, Brinchmann J, Bruzual G, Peng EW, Seibert M, Bernardi M, Blanton M, Brinkmann J, Castander F, Csábai I, Fukugita M, Ivezic Z, Munn JA, Nichol RC, Padmanabhan N, Thakar AR, Weinberg DH, York D (2003) Stellar masses and star formation histories for 10^5 galaxies from the sloan digital sky survey. Mon Not R Astron Soc 341:33–53

Kayser R (1992) Gravitational microlensing. In: Kayser R, Schramm T, Nieser L (eds) Gravitational lenses. Lecture notes in physics, vol 406. Springer, Heidelberg, pp 143–155. https://doi.org/10.1007/3-540-55797-0_94

Kayser R, Refsdal S, Stabell R (1986) Astrophysical aplications of gravitational micro-lensing. Astron Astrophys 166:36

Kedziora DJ, Garsden H, Lewis GF (2011) Gravitational microlensing as a probe of the electron-scattering region in Q2237+0305. Mon Not R Astron Soc 415:1409–1418. https://doi.org/10.1111/j.1365-2966.2011.18787.x

Keeton CR, Burles S, Schechter PL, Wambsganss J (2006) Differential microlensing of the continuum and broad emission lines in SDSS J0924 + 0219, the most anomalous lensed quasar. Astrophys J 639:1–6

Khavinson D, Neumann G (2004) On the number of zeros of certain rational harmonic functions. https://doi.org/10.48550/arXiv.math/0401188

Kishimoto M, Hönig SF, Beckert T, Weigelt G (2007) The innermost region of AGN tori: implications from the HST/NICMOS type 1 point sources and near-ir reverberation. Astron Astrophys 476:713–721. https://doi.org/10.1051/0004-6361:20077911

Kishimoto M, Hönig SF, Antonucci R, Barvainis R, Kotani T, Tristram KRW, Weigelt G, Levin K (2011a) The innermost dusty structure in active galactic nuclei as probed by the Keck interferometer. Astron Astrophys 527:A121. https://doi.org/10.1051/0004-6361/201016054

Kishimoto M, Hönig SF, Antonucci R, Millour F, Tristram KRW, Weigelt G (2011b) Mapping the radial structure of AGN tori. Astron Astrophys 536:A78. https://doi.org/10.1051/0004-6361/201117367

Kochanek CS (2004) Quantitative interpretation of quasar microlensing light curves. Astrophys J 605:58–77. https://doi.org/10.1086/382180

Koda J, Blake C, Davis T, Magoulas C, Springob CM, Scrimgeour M, Johnson A, Poole GB, Staveley-Smith L (2014) Are peculiar velocity surveys competitive as a cosmological probe? Mon Not R Astron Soc 445(4):4267–4286. https://doi.org/10.1093/mnras/stu1610

Kollmeier JA, Onken CA, Kochanek CS, Gould A, Weinberg DH, Dietrich M, Cool R, Dey A, Eisenstein DJ, Jannuzi BT, Floc' EL, Stern D (2006) Black hole masses and Eddington ratios at $0.3 < z < 4$. Astrophys J 648:128

Koopmans LVE, Bruyn AGD (2000) Microlensing of multiply-imaged compact radio sources evidence for compact halo objects in the disk galaxy of B1600+434. Astron Astrophys 358:793–811

Koptelova E, Shimanovskaya E, Artamonov B, Yagola A (2007) Analysis of the Q2237+0305 light-curve variability with regularization technique. Mon Not R Astron Soc 381:1655–1662. https://doi.org/10.1111/j.1365-2966.2007.12335.x

Korista KT, Goad MR (2001) The variable diffuse continuum emission of broad-line clouds. Astrophys J 553:695–708

Korista KT, Alloin D, Barr P, Clavel J, Cohen RD, Crenshaw DM, Evans IN, Horne K, Koratkar AP, Kriss GA, Krolik JH, Malkan MA, Morris SL, Netzer H, O'Brien PT, Peterson BM, Reichert GA, Rodriguez-Pascual PM, Wamsteker W, Anderson KSJ, Axon DJ, Benitez E, Berlind P, Bertram R, Blackwell JH

Jr, Bochkarev NG, Boisson C, Carini M, Carrillo R, Carone TE, Cheng F-Z, Christensen JA, Chuvaev KK, Dietrich M, Dokter JJ, Doroshenko V, Dultzin-Hacyan D, England MN, Espey BR, Filippenko AV, Gaskell CM, Goad MR, Ho LC, Huchra JP, Jiang XJ, Kaspi S, Kollatschny W, Laor A, Luminet J-P, MacAlpine GM, MacKenty JW, Malkov YF, Maoz D, Martin PG, Matheson T, McCollum B, Merkulova N, Metik L, Mignoli M, Miller HR, Pastoriza MG, Pelat D, Penfold J, Perez M, Perola GC, Persaud JL, Peters J, Pitts R, Pogge RW, Pronik I, Pronik VI, Ptak RL, Rawley L, Recondo-Gonzalez MC, Rodriguez-Espinosa JM, Romanishin W, Sadun AC, Salamanca I, Santos-Lleo M, Sekiguchi K, Sergeev SG, Shapovalova AI, Shields JC, Shrader C, Shull JM, Silbermann NA, Sitko ML, Skillman DR, Smith HA, Smith SM, Snijders MAJ, Sparke LS, Stirpe GM, Stoner RE, Sun W-H, Thiele U, Tokarz S, Tsvetanov ZI, Turnshek DA, Veilleux S, Wagner RM, Wagner SJ, Wanders I, Wang T, Welsh WF, Weymann RJ, White RJ, Wilkes BJ, Wills BJ, Winge C, Wu H, Zou ZL (1995) Steps toward determination of the size and structure of the broad-line region in active galactic nuclei. VIII. An intensive HST, IUE, and ground-based study of NGC 5548. Astrophys J Suppl Ser 97:13

Koshida S, Minezaki T, Yoshii Y, Kobayashi Y, Sakata Y, Sugawara S, Enya K, Suganuma M, Tomita H, Aoki T, Peterson BA (2014) Reverberation measurements of the inner radius of the dust torus in 17 Seyfert galaxies. Astrophys J 788:159. https://doi.org/10.1088/0004-637X/788/2/159

Krawczynski H, Chartas G (2017) Simulations of the Fe Kα energy spectra from gravitationally microlensed quasars. Astrophys J 843(2):118. https://doi.org/10.3847/1538-4357/aa7896

Kundic T, Wambsganss J (1993) Gravitational microlensing: the effect of random motion of individual stars in the lensing galaxy. Astrophys J 404:455. https://doi.org/10.1192/bjp.111.479.1009-a

Lamer G, Schwope A, Wisotzki L, Christensen L (2006) Strange magnification pattern in the large separation lens SDSS J1004+4112 from optical to X-rays. Astron Astrophys 454:493–501. https://doi.org/10.1051/0004-6361:20064934

Lančová D, Abarca D, Kluźniak W, Wielgus M, Sądowski A, Narayan R, Schee J, Török G, Abramowicz M (2019) Puffy accretion disks: sub-Eddington, optically thick, and stable. Astrophys J 884:37. https://doi.org/10.3847/2041-8213/ab48f5

Laor A (1991) Line profiles from a disk around a rotating black hole. Astrophys J 376:90. https://doi.org/10.1086/170257

Laor A, Behar E (2008) On the origin of radio emission in radio-quiet quasars. Mon Not R Astron Soc 390:847–862. https://doi.org/10.1111/j.1365-2966.2008.13806.x

Lasota JP (2023) Active galactic nuclei. Wiley, New York, pp 1–320. https://doi.org/10.1002/9781394163724

Lasota JP, King AR, Dubus G (2015) X-ray transients: hyper- or hypo-luminous? Astrophys J Lett 801:L4. https://doi.org/10.1088/2041-8205/801/1/L4

Lawrence A (2012) The uv peak in active galactic nuclei: a false continuum from blurred reflection? Mon Not R Astron Soc 423:451–463. https://doi.org/10.1111/j.1365-2966.2012.20889.x

Lawther D, Vestergaard M, Raimundo S, Koay JY, Peterson BM, Fan X, Grupe D, Mathur S (2023) Flares in the changing look AGN MRK 590 – I. The UV response to X-ray outbursts suggests a more complex reprocessing geometry than a standard disc. Mon Not R Astron Soc 519:3903–3922. https://doi.org/10.1093/mnras/stac3515

Ledvina L, Heyrovský D, Dovčiak M (2018) X-ray line profile variations during quasar microlensing. Astrophys J 863(1):66. https://doi.org/10.3847/1538-4357/aad0f3

Lehar J, Falco EE, Kochanek CS, McLeod BA, Muñoz JA, Impey CD, Rix H-W, Keeton CR, Peng CY (2000) Hubble space telescope observations of 10 two-image gravitational lenses. Astrophys J 536:584

Lemon C et al (2024) Searching for strong gravitational lenses. Space Sci Rev 220. https://doi.org/10.1007/s11214-024-01042-9

Lewis GF (2020) Gravitational microlensing time delays at high optical depth: image parities and the temporal properties of fast radio bursts. Mon Not R Astron Soc 497:1583–1589. https://doi.org/10.1093/mnras/staa2044

Lewis GF, Belle KE (1998) Microlensing of broad absorption line quasars. Mon Not R Astron Soc 297:69–76

Lewis GF, Ibata RA (1998) Quasar image shifts resulting from gravitational microlensing. Astrophys J 501(2):478–485. https://doi.org/10.1086/305860

Lewis GF, Ibata RA (2004) Gravitational microlensing of quasar broad-line regions at large optical depths. Mon Not R Astron Soc 348(1):24–33. https://doi.org/10.1111/j.1365-2966.2004.07349.x

Lewis GF, Irwin MJ (1995) The statistics of micolensing light curves – I. Amplification probability distributions. Mon Not R Astron Soc 276:103

Lewis GF, Miralda-Escudé J, Richardson DC, Wambsganss J (1993) Microlensing light curves: a new and efficient numerical method. Mon Not R Astron Soc 261:647

Lewis GF, Irwin MJ, Hewett PC, Foltz CB (1998) Microlensing-induced spectral variability in Q2237+0305. Mon Not R Astron Soc 295:573–586

Li YR, Wang JM, Ho LC, Du P, Bai JM (2013) A Bayesian approach to estimate the size and structure of the broad-line region in active galactic nuclei using reverberation mapping data. Astrophys J 779:110. https://doi.org/10.1088/0004-637X/779/2/110

Li YP, Yuan F, Dai X (2019) Reconciling the quasar microlensing disc size problem with a wind model of active galactic nucleus. Mon Not R Astron Soc 483:2275–2281. https://doi.org/10.1093/mnras/sty3245

Lightman AP, White TR (1988) Effects of cold matter in active galactic nuclei: a broad hump in the X-ray spectra. Astrophys J 335:57. https://doi.org/10.1086/166905

Luhtaru R, Schechter PL, de Soto KM (2021) What makes quadruply lensed quasars quadruple? Astrophys J 915:4. https://doi.org/10.3847/1538-4357/abfda1

Lynden-Bell D (1969) Galactic nuclei as collapsed old quasars. Nature 223:690–694. https://doi.org/10.1038/223690a0

MacLeod CL, Morgan CW, Mosquera AM, Kochanek CS, Tewes M, Courbin F, Meylan G, Chen B, Dai X, Chartas G (2015) A consistent picture emerges: a compact X-ray continuum emission region in the gravitationally lensed quasar SDSS J0924+0219. Astrophys J 806:258. https://doi.org/10.1088/0004-637X/806/2/258

Makarov VV, Secrest NJ (2022) Quasars with proper motions and the link to double and multiple AGNs. Astrophys J 933(1):28. https://doi.org/10.3847/1538-4357/ac7047

Mao S (1993) Gravitational microlensing of gamma-ray bursts. Astrophys J 402:382–386

Mao S, Schneider P (1998) Evidence for substructure in lens galaxies? Mon Not R Astron Soc 295:587–594

Marconi A, Hunt LK (2003) The relation between black hole mass, bulge mass, and near-infrared luminosity. Astrophys J 589:21–24

Marziani P, Sulentic JW, Zwitter T, Dultzin-Hacyan D, Calvani M (2001) Searching for the physical drivers of the eigenvector 1 correlation space. Astrophys J 558:553–560

McGill P, Anderson J, Casertano S, Sahu KC, Bergeron P, Blouin S, Dufour P, Smith LC, Evans NW, Belokurov V, Smart RL, Bellini A, Calamida A, Dominik M, Kains N, Klüter J, Nielsen MB, Wambsganss J (2023) First semi-empirical test of the white dwarf mass-radius relationship using a single white dwarf via astrometric microlensing. Mon Not R Astron Soc 520(1):259–280. https://doi.org/10.1093/mnras/stac3532

Mediavilla E, Muñoz JA, Kochanek CS, Falco EE, Arribas S, Motta V (2005) The first precise determination of an optical-far-ultraviolet extinction curve beyond the local group ($z = 0.83$). Astrophys J 619:749

Mediavilla E, Munoz JA, Lopez P, Mediavilla T, Abajas C, Gonzalez-Morcillo C, Gil-Merino R (2006) A fast and very accurate approach to the computation of microlensing magnification patterns based on inverse polygon mapping. Astrophys J 653:942–953. https://doi.org/10.1086/508796

Mediavilla E, Muñoz JA, Falco E, Motta V, Guerras E, Canovas H, Jean C, Oscoz A, Mosquera AM (2009) Microlensing-based estimate of the mass fraction in compact objects in lens galaxies. Astrophys J 706:1451–1462. https://doi.org/10.1088/0004-637X/706/2/1451

Mediavilla E, Mediavilla T, Muñoz JA, Ariza O, Lopez P, Gonzalez-Morcillo C, Jiménez-Vicente J (2011) New developments on inverse polygon mapping to calculate gravitational lensing magnification maps: optimized computations. Astrophys J 741:42. https://doi.org/10.1088/0004-637X/741/1/42

Mediavilla E, Jiménez-Vicente J, Muñoz JA, Mediavilla T (2015) Resolving the innermost region of the accretion disk of the lensed quasar Q2237+0305 through gravitational microlensing. Astrophys J Lett 814:L26. https://doi.org/10.1088/2041-8205/814/2/L26

Mediavilla E, Jiménez-Vicente J, Muñoz JA, Battaner E (2016) Peculiar transverse velocities of galaxies from quasar microlensing. Tentative estimate of the peculiar velocity dispersion at $Z \sim 0.5$. Astrophys J 832(1):46. https://doi.org/10.3847/0004-637X/832/1/46

Mediavilla E, Jiménez-Vicente J, Muñoz JA, Vives-Arias H, Calderón-Infante J (2017) Limits on the mass and abundance of primordial black holes from quasar gravitational microlensing. Astrophys J 836:18. https://doi.org/10.3847/2041-8213/aa5dab

Mediavilla E, Jiménez-vicente J, Mejía-restrepo J, Motta V, Falco E, Muñoz JA, Fian C, Guerras E (2020) Individual estimates of the virial factor in 10 quasars: implications on the kinematics of the broad-line region. Astrophys J 895:111. https://doi.org/10.3847/1538-4357/ab8ae0

Melo A, Motta V, Godoy N, Mejia-Restrepo J, Assef RJ, Mediavilla E, Falco E, ávila-Vera F, Jerez R (2021) First black hole mass estimation for the quadruple lensed system WGD2038-4008. Astron Astrophys 656:A108. https://doi.org/10.1051/0004-6361/202141869

Merloni A, Fabian AC (2001) Thunderclouds and accretion discs: a model for the spectral and temporal variability of Seyfert 1 galaxies. Mon Not R Astron Soc 328(3):958–968. https://doi.org/10.1046/j.1365-8711.2001.04925.x

Metcalf RB, Zhao H (2002) Flux ratios as a probe of dark substructures in quadruple-image gravitational lenses. Astrophys J Lett 567:5–8

Middleton MJ, Heil L, Pintore F, Walton DJ, Roberts TP (2015) A spectral-timing model for ULXs in the supercritical regime. Mon Not R Astron Soc 447:3243–3263. https://doi.org/10.1093/mnras/stu2644

Millon M, Courbin F, Bonvin V, Paic E, Meylan G, Tewes M, Sluse D, Magain P, Chan JHH, Galan A, Joseph R, Lemon C, Tihhonova O, Anderson RI, Marmier M, Chazelas B, Lendl M, Triaud AHMJ, Wyttenbach A (2020a) Cosmograil. XIX. Time delays in 18 strongly lensed quasars from 15 years of optical monitoring. Astron Astrophys 640:105. https://doi.org/10.1051/0004-6361/202037740

Millon M, Galan A, Courbin F, Treu T, Suyu SH, Ding X, Birrer S, Chen GCF, Shajib AJ, Sluse D, Wong KC, Agnello A, Auger MW, Buckley-Geer EJ, Chan JHH, Collett T (2020b) TDCOSMO: I. An exploration of systematic uncertainties in the inference of H_0 from time-delay cosmography. Astron Astrophys 639:1–19. https://doi.org/10.1051/0004-6361/201937351

Millon M, Dalang C, Lemon C, Sluse D, Paic E, Chan JHH, Courbin F (2022) Evidence for a milliparsec-separation supermassive binary black hole with quasar microlensing. Astron Astrophys 668:A77. https://doi.org/10.1051/0004-6361/202244440

Mineshige S, Yonehara A (1999) Gravitational microlens mapping of a quasar accretion disk. Publ Astron Soc Jpn 51:497

Minezaki T, Chiba M, Kashikawa N, Inoue KT, Kataza H (2009) Subaru mid-infrared imaging of the quadruple lenses. II. Unveiling lens structure of MG0414+0534 and Q2237+030. Astrophys J 697:610–618. https://doi.org/10.1088/0004-637X/697/1/610

Moore GE (1965) Cramming more components onto integrated circuits. Electron Mag 38:4

Morgan CW, Kochanek CS, Morgan ND, Falco EE (2006) Microlensing of the lensed quasar SDSS 0924+0219. Astrophys J 647:874–885. https://doi.org/10.1086/505569

Morgan CW, Kochanek CS, Dai X, Morgan ND, Falco EE, Al MET (2008) X-ray and optical microlensing in the lensed quasar PG 1115 + 080 1. Astrophys J 689:755

Morgan CW, Kochanek CS, Morgan ND, Falco EE (2010) The quasar accretion disk size-black hole mass relation. Astrophys J 712:1129–1136. https://doi.org/10.1088/0004-637X/712/2/1129

Morgan CW, Hainline LJ, Chen B, Tewes M, Kochanek CS, Dai X, Kozlowski S, Blackburne JA, Mosquera AM, Chartas G, Courbin F, Meylan G (2012) Further evidence that quasar X-ray emitting regions are compact: X-ray and optical microlensing in the lensed quasar Q J0158-4325. Astrophys J 756:52. https://doi.org/10.1088/0004-637X/756/1/52

Morgan CW, Hyer GE, Bonvin V, Mosquera AM, Cornachione M, Courbin F, Kochanek CS, Falco EE (2018) Accretion disk size measurement and time delays in the lensed quasar WFI 2033–4723. Astrophys J 869:106. https://doi.org/10.3847/1538-4357/aaed3e

Mortonson MJ, Schechter PL, Wambsganss J (2005) Size is everything: universal features of quasar microlensing with extended sources. Astrophys J 628:594–603. https://doi.org/10.1086/431195

Mosquera AM, Kochanek CS (2011) The microlensing properties of a sample of 87 lensed quasars. Astrophys J 738:96. https://doi.org/10.1088/0004-637X/738/1/96

Mosquera AM, Muñoz JA, Mediavilla E (2009) Detection of chromatic microlensing in Q 2237 + 0305 A. Astrophys J 691:1292–1299. https://doi.org/10.1088/0004-637X/691/2/1292

Mosquera AM, Muñoz JA, Mediavilla E, Kochanek CS (2011) A study of gravitational lens chromaticity using ground-based narrowband photometry. Astrophys J 728:145. https://doi.org/10.1088/0004-637X/728/2/145

Mosquera AM, Kochanek CS, Chen B, Dai X, Blackburne JA, Chartas G (2013) The structure of the X-ray and optical emitting regions of the lensed quasar Q 2237+0305. Astrophys J 769:53

Motta V, Mediavilla E, Oz JAM, Falco E, Kochanek CS, Arribas S, García-Lorenzo B, Oscoz A, Serra-Ricart M (2002) Detection of the 2175 a ° extinction feature at $z = 0.83$. Astrophys J 574:719

Motta V, Mediavilla E, Falco E, Munoz JA (2012) Measuring microlensing using spectra of multiply lensed quasars. Astrophys J 755:82. https://doi.org/10.1088/0004-637X/755/1/82

Moustakas L, O'Dowd M, Anguita T, Webster R, Chartas G, Cornachione M, Dai X, Fian C, Hutsemekers D, Jimenez-Vicente J, Labrie K, Lewis G, Macleod C, Mediavilla E, Morgan CW, Motta V, Nierenberg A, Pooley D, Rojas K, Sluse D, Vernardos G, Wambsganss J, Yong SY (2019) Astro2020 science white paper – quasar microlensing: revolutionizing our understanding of quasar structure and dynamics. arXiv e-prints. arXiv:1904.12967. https://doi.org/10.48550/arXiv.1904.12967

Muñoz JA, Mediavilla E, Kochanek CS, Falco EE, Mosquera AM (2011) A study of gravitational lens chromaticity with the Hubble space telescope. Astrophys J 742:67. https://doi.org/10.1088/0004-637X/742/2/67

Muñoz JA, Vives-Arias H, Mosquera AM, Jiménez-Vicente J, Kochanek CS, Mediavilla E (2016) Structure of the accretion disk in the lensed quasar Q2237+0305 from multi-epoch and multi-wavelength narrowband photometry. Astrophys J 817:155. https://doi.org/10.3847/0004-637x/817/2/155

Murray N, Chiang J (1997) Disk winds and disk emission lines. Astrophys J 474:91–103

Murray N, Chiang J, Grossman SA, Voit GM (1995) Accretion disk winds from active galactic nuclei. Astrophys J 451:498

Mushotzky RF (1984) X-ray spectra and time variability of active galactic nuclei. Adv Space Res 3(10–12):157–165. https://doi.org/10.1016/0273-1177(84)90081-4

Nagao T, Marconi A, Maiolino R (2006) The evolution of the broad-line region among sdss quasars. Astron Astrophys 447:157–172. https://doi.org/10.1051/0004-6361:20054024

Narayan R, Yi I (1994) Advection-dominated accretion: a self-similar solution. Astrophys J Lett 428:13–16

Neira F, Anguita T, Vernardos G (2020) A quasar microlensing light-curve generator for LSST. Mon Not R Astron Soc 495:544–553. https://doi.org/10.1093/mnras/staa1208

Neronov A, Vovk I (2016) Test of relativistic gravity using microlensing of relativistically broadened lines in gravitationally lensed quasars. Phys Rev D 93(2):023006. https://doi.org/10.1103/PhysRevD.93.023006

Netzer H (2015) Revisiting the unified model of active galactic nuclei. Annu Rev Astron Astrophys 53:365–408. https://doi.org/10.1146/annurev-astro-082214-122302

Nixon C, King A, Price D, Frank J (2012) Tearing up the disk: how black holes accrete. Astrophys J Lett 757:L24. https://doi.org/10.1088/2041-8205/757/2/L24

Novikov ID, Thorne KS (1973) Astrophysics of black holes, DeWitt C, DeWitt BS (eds). Gordon & Breach, New York, pp 343–450

O'Dowd M, Bate NF, Webster RL, Wayth R, Labrie K (2011) Differential microlensing measurements of quasar broad-line kinematics in Q2237+0305. Mon Not R Astron Soc 415:1985–1998. https://doi.org/10.1111/j.1365-2966.2010.18119.x

O'Dowd MJ, Bate NF, Webster RL, Labrie K, Rogers J (2015) Microlensing constraints on broad absorption and emission line flows in the quasar H1413+117. Astrophys J 813:62. https://doi.org/10.1088/0004-637X/813/1/62

Oguri M, Marshall PJ (2010) Gravitationally lensed quasars and supernovae in future wide-field optical imaging surveys. Mon Not R Astron Soc 405:2579. https://doi.org/10.1111/j.1365-2966.2010.16639.x

Oguri M, Rusu CE, Falco EE (2014) The stellar and dark matter distributions in elliptical galaxies from the ensemble of strong gravitational lenses. Mon Not R Astron Soc 439:2494–2504. https://doi.org/10.1093/mnras/stu106

Oshima T, Mitsuda K, Fujimoto R, Iyomoto N, Futamoto K, Hattori M, Ota N, Mori K, Ikebe Y, Miralles JM, Kneib J-P (2001) Detection of an iron emission feature from the lensed broad absorption line QSO H1413+117 at $z = 2.56$. Astrophys J Lett 563(2):103–106. https://doi.org/10.1086/338653

Ostensen R, Refsdal S, Stabell R, Teuber J, Emanuelsen PI, Festin L, Florentin-Nielsen R, Gahm G (1996) Astron Astrophys 309:59.

Ota N, Inada N, Oguri M, Mitsuda K, Richards GT, Suto Y, Brandt WN, Castander FJ, Fujimoto R, Hall PB, Keeton CR, Nichol RC, Schneider DP, Eisenstein DE, Frieman JA, Turner EL, Minezaki T, Yoshii Chandra Y (2006) Observations of SDSS J1004+4112: constraints on the lensing cluster and anomalous X-ray flux ratios of the quadruply imaged quasar. Astrophys J 647(1):215–221. https://doi.org/10.1086/505385

Paczynski B (1986) Gravitational microlensing at large optical depth. Astrophys J 301:503

Padovani P, Alexander DM, Assef RJ, Marco BD, Giommi P, Hickox RC, Richards GT, Smolčić V, Hatziminaoglou E, Mainieri V, Salvato M (2017) Active galactic nuclei: what's in a name? Astron Astrophys Rev 25:2. https://doi.org/10.1007/s00159-017-0102-9

Paic E, Vernardos G, Sluse D, Millon M, Courbin F, Chan JH, Bonvin V (2022) Constraining quasar structure using high-frequency microlensing variations and continuum reverberation. Astron Astrophys 659:A21. https://doi.org/10.1051/0004-6361/202141808

Pancoast A, Brewer BJ, Treu T (2011) Geometric and dynamical models of reverberation mapping data. Astrophys J 730:139. https://doi.org/10.1088/0004-637X/730/2/139

Pancoast A, Brewer BJ, Treu T (2014) Modelling reverberation mapping data – I. Improved geometric and dynamical models and comparison with cross-correlation results. Mon Not R Astron Soc 445:3055–3072. https://doi.org/10.1093/mnras/stu1809

Papadakis IE, DovĆiak M, Kammoun ES (2022) X-ray illuminated accretion discs and quasar microlensing disc sizes. Astron Astrophys 666:A11. https://doi.org/10.1051/0004-6361/202142962

Parker ML, Tomsick JA, Miller JM, Yamaoka K, Lohfink A, Nowak M, Fabian AC, Alston WN, Boggs SE, Christensen FE, Craig WW, Fürst F, Gandhi P, Grefenstette BW, Grinberg V, Hailey CJ, Harrison FA, Kara E, King AL, Stern D, Walton DJ, Wilms J, Zhang WW (2015) NuSTAR and Suzaku observations of the hard state in Cygnus X-1: locating the inner accretion disk. Astrophys J 808(1):9. https://doi.org/10.1088/0004-637X/808/1/9

Pashchenko IN, Plavin AV, Kutkin AM, Kovalev YY (2020) A bias in VLBI measurements of the core shift effect in AGN jets. Mon Not R Astron Soc 499(3):4515–4525. https://doi.org/10.1093/mnras/staa3140

Peterson BM, Wandel A (1999) Keplerian motion of broad-line region gas as evidence for supermassive black holes in active galactic nuclei. Astrophys J 521:95–98

Peterson BM, Crenshaw DM, Meyers KA (1985) Variability of the emission-line spectra and optical continua of seyfert galaxies. III. Results for a homogeneous sample. Astrophys J 298:283–291

Peterson BM, Bentz MC, Desroches L-B, Filippenko AV, Ho LC, Kaspi S, Laor A, Maoz D, Moran EC, Pogge RW, Quillen AC (2005) Multiwavelength monitoring of the dwarf Seyfert 1 galaxy NGC 4395. I. A reverberation-based measurement of the black hole mass. Astrophys J 632(2):799–808. https://doi.org/10.1086/444494

Petters AO (1992) Morse theory and gravitational microlensing. J Math Phys 33:1915–1931. https://doi.org/10.1063/1.529667

Petters AO, Levine H, Wambsganss J (2001) Singularity theory and gravitational lensing. Progress in Mathematical Physics, vol 21. Birkhäuser, Boston, MA. https://doi.org/10.1007/978-1-4612-0145-8

Petters AO, Rider B, Teguia AM (2009a) A mathematical theory of stochastic microlensing. I. Random time-delay functions and lensing maps. J Math Phys 50:072503. https://doi.org/10.1063/1.3158854

Petters AO, Rider B, Teguia AM (2009b) A mathematical theory of stochastic microlensing. II. Random images, shear, and the Kac-Rice formula. J Math Phys 50:122501. https://doi.org/10.1063/1.3267859

Pfuhl O, Davies R, Dexter J, Netzer H, Hönig S, Lutz D, Schartmann M, Sturm E, Amorim A, Brandner W, Clénet Y, Zeeuw PTD, Eckart A, Eisenhauer F, Schreiber NMF, Gao F, Garcia PJV, Genzel R, Gillessen S, Gratadour D, Kishimoto M, Lacour S, Millour F, Ott T, Paumard T, Perraut K, Perrin G, Peterson BM, Petrucci PO, Prieto MA, Rouan D, Shangguan J, Shimizu T, Sternberg A, Straub O, Straubmeier C, Tacconi LJ, Tristram KRW, Vermot P, Waisberg I, Widmann F, Woillez J (GRAVITY Collaboration) (2020) An image of the dust sublimation region in the nucleus of NGC 1068. Astron Astrophys 634:A1. https://doi.org/10.1051/0004-6361/201936255

Poindexter S, Kochanek CS (2010) The transverse peculiar velocity of the Q2237+0305 lens galaxy and the mean mass of its stars. Astrophys J 712:658–667. https://doi.org/10.1088/0004-637X/712/1/658

Poindexter S, Morgan ND, Kochanek CS (2008) The spatial structure of an accretion disk. Astrophys J 673:34–38. https://doi.org/10.1086/524190

Pooley D, Blackburne JA, Rappaport S, Schechter PL (2007) X-ray and optical flux ratio anomalies in quadruply lensed quasars. I. Zooming in on quasar emission regions. Astrophys J 661:19–29

Pooley D, Rappaport S, Blackburne JA, Schechter PL, Wambsganss J (2012) X-ray and optical flux ratio anomalies in quadruply lensed quasars. II. Mapping the dark matter content in elliptical galaxies. Astrophys J 744:111. https://doi.org/10.1088/0004-637X/744/2/111

Popović LĆ, Chartas G (2005) The influence of gravitational lensing on the spectra of lensed quasi-stellar objects. Mon Not R Astron Soc 357:135–144. https://doi.org/10.1111/j.1365-2966.2004.08619.x

Popović LČ, Jovanović P, Mediavilla E, Zakharov AF, Abajas C, Muñoz JA, Chartas G (2006) A study of the correlation between the amplification of the Fe Kα line and the X-ray continuum of quasars due to microlensing. Astrophys J 637(2):620–630. https://doi.org/10.1086/498558

Popović LA, Afanasiev VL, Moiseev A, Smirnova A, Simić S, Savić D, Mediavilla EG, Fian C (2020) Spectroscopy and polarimetry of the gravitationally lensed quasar SDSS J1004+4112 with the 6m SAO RAS telescope. Astron Astrophys 634:A27. https://doi.org/10.1051/0004-6361/201936088

Popović LC, Afanasiev VL, Shablovinskaya ES, Ardilanov VI, Savić DJ (2021) Spectroscopy and polarimetry of the gravitationally lensed quasar Q0957+561. Astron Astrophys 647:A98. https://doi.org/10.1051/0004-6361/202039914

Pozdnyakov LA, Sobol IM, Syunyaev RA (1976) Multiple Compton scattering by relativistic electrons – Monte Carlo calculations of the emission spectrum. Sov Astron Lett 2:55–57

Press WH, Gunn JE (1973) Method for detecting a cosmological density of condensed objects. Astrophys J 185:397–412

Rauch KP, Blandford RD (1991) Microlensing and the structure of active galactic nucleus accretion discs. Astrophys J 381:39

Rauch KP, Mao S, Wambsganss J, Paczynski B (1992) Caustic-induced features in microlensing magnification probability distributions. Astrophys J 386:30

Refsdal S, Stabell R (1997) Gravitational microlensing of large sources including shear term effects. Astron Astrophys 325:877–880

Reynolds CS, Nowak MA (2003) Fluorescent iron lines as a probe of astrophysical black hole systems. Phys Rep 377:389

Richards GT, Keeton CR, Pindor B, Hennawi JF, Hall PB, Turner EL, Inada N, Oguri M, Ichikawa S-I, Becker RH, Gregg MD, White RL, Stuart J, Wyithe B, Schneider DP, Johnston DE, Frieman JA, Brinkmann J (2004) Microlensing of the broad emission line region in the quadruple lens SDSS J1004+4112. Astrophys J 610:679

Risaliti G, Lusso E (2019) Cosmological constraints from the Hubble diagram of quasars at high redshifts. Nat Astron 3:272–277. https://doi.org/10.1038/s41550-018-0657-z

Rojas K, Motta V, Mediavilla E, Falco E, Jiménez-Vicente J, Muñoz JA (2014) Strong chromatic microlensing in HE0047-1756 and SDSS1155+6346. Astrophys J 797(1):61

Rothschild RE, Mushotzky FR, Baity WA, Gruber DE, Matteson JL, Peterson LE (1983) 2-165 keV observations of active galaxies and the diffuse background. Astrophys J 269:423–437. https://doi.org/10.1086/161053

Rusu CE, Wong KC, Bonvin V, Sluse D, Suyu SH, Fassnacht CD, Chan JHH, Hilbert S, Auger MW, Sonnenfeld A, Birrer S, Courbin F, Treu T, Chen GC-F, Halkola A, Koopmans LVE, Marshall PJ, Shajib AJ (2020) H0LiCOW. XII. Lens mass model of WFI2033-4723 and blind measurement of its time-delay distance and H_0. Mon Not R Astron Soc 498:1440–1468. https://doi.org/10.1093/mnras/stz3451

Sądowski A, Abramowicz M, Bursa M, Kluźniak W, Lasota J-P, Różańska A (2011) Relativistic slim disks with vertical structure. Astron Astrophys 527:17. https://doi.org/10.1051/0004-6361/201015256

Saha P (2000) Lensing degeneracies revisited. Astron J 120:1654–1659

Saha P, Williams LLR (2011) Understanding micro-image configurations in quasar microlensing. Mon Not R Astron Soc 411:1671–1677. https://doi.org/10.1111/j.1365-2966.2010.17797.x

Saha P, Sluse D, Wagner J, Williams LLR (2024) Essentials of strong gravitational lensing. Space Sci Rev 220:12. https://doi.org/10.1007/s11214-024-01041-w

Sarkar A, Ferland GJ, Chatzikos M, Guzmán F, van Hoof PAM, Smyth RT, Ramsbottom CA, Keenan FP, Ballance CP (2021) Improved Fe II emission-line models for AGNs using new atomic data sets. Astrophys J 907:12. https://doi.org/10.3847/1538-4357/abcaa6

Schechter PL, Wambsganss J (2002) Quasar microlensing at high magnification and the role of dark matter: enhanced fluctuations and suppressed saddle points. Astrophys J 580:685–695. https://doi.org/10.1086/343856

Schechter PL, Udalski A, Szymański M, Kubiak M, Pietrzyński G, Soszyński I, Woźniak P, Żebruń K, Szewczyk O, Wyrzykowski Ł (2003) Microlensing of relativistic knots in the quasar HE 1104-1805 AB. Astrophys J 584:657

Schechter PL, Wambsganss J, Lewis GF (2004) Qualitative aspects of quasar microlensing with two mass components: magnification patterns and probability distributions. Astrophys J 613:77–85

Schechter PL, Pooley D, Blackburne JA, Wambsganss J (2014) A calibration of the stellar mass fundamental plane at $z \sim 0.5$ using the micro-lensing-induced flux ratio anomalies of macro-lensed quasars. Astrophys J 793:96. https://doi.org/10.1088/0004-637X/793/2/96

Schild RE (1996) Microlensing variability of the gravitationally lensed quasar Q0957+561 A, B. Astrophys J 464:125

Schmidt RW, Wambsganss J (2010) Quasar microlensing. Gen Relativ Gravit 42:2127–2150. https://doi.org/10.1007/s10714-010-0956-x

Schneider P (1993) Upper bounds on the cosmological density of compact objects with sub-solar masses from the variability of QSOS. Astron Astrophys 279:1–20

Schneider P, Wambsganss J (1990) Are the broad emission lines of quasars affected by gravitational microlensing? Astron Astrophys 237:42–53

Schneider P, Weiss A (1987) A gravitational lens origin for agn-variability? Consequences of micro-lensing. Astron Astrophys 171:49

Schneider P, Ehlers J, Falco EE (1992) Gravitational lenses. Springer, Berlin. https://doi.org/10.1007/978-3-662-03758-4

Schneider P, Kochanek CS, Wambsganss J (2006) Gravitational lensing: strong, weak, micro, Meylan G, Jetzer P, North P (eds). Springer, Berlin. https://doi.org/10.1007/978-3-540-30310-7

Shajib AJ et al (2024) Strong lensing by galaxies. Space Sci Rev 220

Shakura NI, Sunyaev RA (1973) Black holes in binary systems. Observational appearance. Astron Astrophys 24:337

Shalyapin VN, Goicoechea LJ (2014) Deep optical imaging and spectroscopy of the lens system SDSS J1339+1310. Astron Astrophys 568:A116. https://doi.org/10.1051/0004-6361/201323360

Shalyapin VN, Gil-Merino R, Goicoechea LJ (2021) Fast simulations of extragalactic microlensing. Astron Astrophys 653:A121. https://doi.org/10.1051/0004-6361/202140527

Shapiro SL, Lightman AP, Eardley DM (1976) A two-temperature accretion disk model for Cygnus X-1: structure and spectrum. Astrophys J 204:187–199. https://doi.org/10.1086/154162

Shen Y, Ho LC (2014) The diversity of quasars unified by accretion and orientation. Nature 513:210–213. https://doi.org/10.1038/nature13712

Shen Y, Greene JE, Strauss MA, Richards GT, Schneider DP (2008) Biases in virial black hole masses: an SDSS perspective. Astrophys J 680:169

Silpa S, Kharb P, Ho LC, Ishwara-Chandra CH, Jarvis ME, Harrison C (2020) Probing the origin of low-frequency radio emission in PG quasars with the uGMRT – I. Mon Not R Astron Soc 499:5826–5839. https://doi.org/10.1093/mnras/staa2970

Sluse D, Tewes M (2014) Imprints of the quasar structure in time-delay light curves: microlensing-aided reverberation mapping. Astron Astrophys 571:1–10. https://doi.org/10.1051/0004-6361/201424776

Sluse D, Claeskens JF, Hutsemékers D, Surdej J (2007) Multi-wavelength study of the gravitational lens system RXS J1131-1231. III. Long slit spectroscopy: micro-lensing probes the qso structure. Astron Astrophys 468:885–901. https://doi.org/10.1051/0004-6361:20066821

Sluse D, Schmidt R, Courbin F, Hutsemékers D, Meylan G, Eigenbrod A, Anguita T, Agol E, Wambsganss J (2011) Zooming into the broad line region of the gravitationally lensed quasar QSO 2237 + 0305 ≡ the Einstein cross. Astron Astrophys 528:100. https://doi.org/10.1051/0004-6361/201016110

Sluse D, Hutsemékers D, Courbin F, Meylan G, Wambsganss J (2012) Microlensing of the broad line region in 17 lensed quasars. Astron Astrophys 544:62. https://doi.org/10.1051/0004-6361/201219125

Sluse D, Kishimoto M, Anguita T, Wucknitz O, Wambsganss J (2013) Mid-infrared microlensing of accretion disc and dusty torus in quasars: effects on flux ratio anomalies. Astron Astrophys 553:A53. https://doi.org/10.1051/0004-6361/201220843

Sluse D, Hutsemékers D, Anguita T, Braibant L, Riaud P (2015) Evidence for two spatially separated UV continuum emitting regions in the cloverleaf broad absorption line quasar. Astron Astrophys 582:A109. https://doi.org/10.1051/0004-6361/201526832

Smith JE, Robinson A, Young S, Axon DJ, Corbett EA (2005) Equatorial scattering and the structure of the broad-line region in Seyfert nuclei: evidence for a rotating disc. Mon Not R Astron Soc 359:846–864. https://doi.org/10.1111/j.1365-2966.2005.08895.x

Speranza G, Balmaverde B, Capetti A, Massaro F, Tremblay G, Marconi A, Venturi G, Chiaberge M, Baldi RD, Baum S, Grandi P, Meyer ET, O'Dea C, Sparks W, Terrazas BA, Torresi E (2021) The MURALES survey. IV. Searching for nuclear outflows in 3C radio galaxies at $z < 0.3$ with MUSE observations. Astron Astrophys 653:A150. https://doi.org/10.1051/0004-6361/202140686

Stalevski M, Jovanović P, Popović LĆ, Baes M (2012) Gravitational microlensing of active galactic nuclei dusty tori. Mon Not R Astron Soc 425:1576–1584. https://doi.org/10.1111/j.1365-2966.2012.21611.x

Suganuma M, Yoshii Y, Kobayashi Y, Minezaki T, Enya K, Tomita H, Aoki T, Koshida S, Peterson BA (2006) Reverberation measurements of the inner radius of the dust torus in nearby Seyfert 1 galaxies. Astrophys J 639:46

Sunyaev RA, Titarchuk LG (1980) Comptonization of X-rays in plasma clouds – typical radiation spectra. Astron Astrophys 86:121

Sunyaev RA, Truemper J (1979) Hard X-ray spectrum of Cyg X-1. Nature 279:506–508. https://doi.org/10.1038/279506a0

Suyu SH et al (2024) Strong gravitational lensing and microlensing of supernovae. Space Sci Rev 220. https://doi.org/10.1007/s11214-024-01044-7

Tewes M, Courbin F, Meylan G (2013) Cosmograil: the cosmological monitoring of gravitational lenses. XI. Techniques for time delay measurement in presence of microlensing. Astron Astrophys 553:120. https://doi.org/10.1051/0004-6361/201220123

Thompson AC, Fluke CJ, Barnes DG, Barsdell BR (2010) Teraflop per second gravitational lensing ray-shooting using graphics processing units. New Astron 15:16

Thompson AC, Vernardos G, Fluke CJ, Barsdell BR (2014) GPU-D: generating cosmological microlensing magnification maps. Astrophysics Source Code Library ascl:1403.001

Thorne KS (1974) Disk-accretion onto a black hole. II. Evolution of the hole. Astrophys J 191:507–519

Tie SS, Kochanek CS (2018) Microlensing makes lensed quasar time delays significantly time variable. Mon Not R Astron Soc 473:80–90. https://doi.org/10.1093/mnras/stx2348

Treyer M, Wambsganss J (2004) Astrometric microlensing of quasars. Dependence on surface mass density and external shear. Astron Astrophys 416:19–34. https://doi.org/10.1051/0004-6361:20034284

Turner TJ, Pounds KA (1989) The EXOSAT spectral survey of AGN. Mon Not R Astron Soc 240:833–880. https://doi.org/10.1093/mnras/240.4.833

Udalski A, Szymanski M, Kubiak M et al (2006) The Optical Gravitational Lensing Experiment. OGLE-III long term monitoring of the gravitational lens QSO 2237+0305. Acta Astronomica 56:293–305

Urry CM, Padovani P (1995) Unified schemes for radio-loud active galactic nuclei. Publ Astron Soc Pac 107:803–845

Van de Vyvere L, Gomer MR, Sluse D, Xu D, Birrer S, Galan A, Vernardos G (2022) TDCOSMO: VII. Boxyness/discyness in lensing galaxies: detectability and impact on H_0. Astron Astrophys 659:A127. https://doi.org/10.1051/0004-6361/202141551

Vegetti S et al (2024) Strong gravitational lensing as a probe of dark matter. Space Sci Rev 220

Venumadhav T, Dai L, Miralda-Escudé J (2017) Microlensing of extremely magnified stars near caustics of galaxy clusters. Astrophys J 850:49. https://doi.org/10.3847/1538-4357/aa9575

Vernardos G (2018) A joint microlensing analysis of lensing mass and accretion disc models. Mon Not R Astron Soc 480:4675

Vernardos G (2019) Microlensing flux ratio predictions for Euclid. Mon Not R Astron Soc 483:5583

Vernardos G (2022) Simualting time-vayring strong lenses. Mon Not R Astron Soc 511:4417

Vernardos G, Fluke CJ (2013) A new parameter space study of cosmological microlensing. Mon Not R Astron Soc 434:832

Vernardos G, Fluke CJ (2014a) Adventures in the microlensing cloud: large datasets, eresearch tools, and GPUs. Astron Comput 6:1

Vernardos G, Fluke CJ (2014b) The effect of macromodel uncertainties on microlensing modelling of lensed quasars. Mon Not R Astron Soc 445:1223

Vernardos G, Tsagkatakis G (2019) Quasar microlensing light-curve analysis using deep machine learning. Mon Not R Astron Soc 486:1944–1952. https://doi.org/10.1093/mnras/stz868

Vernardos G, Fluke CJ, Bate NF, Croton DJ (2014) Gerlumph data release 1: high-resolution cosmological microlensing magnification maps and eresearch tools. Astrophys J Suppl Ser 211:16
Vernardos G, Fluke CJ, Bate NF, Croton DJ, Vohl D (2015) Gerlumph data release 2: 2.5 bilion microlensing light curves. Astrophys J Suppl Ser 217:23
Villafaña L, Williams PR, Treu T, Brewer BJ, Barth AJ, U V, Bennert VN, Vogler HA, Guo H, Bentz MC, Canalizo G, Filippenko AV, Gates E, Hamann F, Joner MD, Malkan MA, Woo J-H, Abolfathi B, Abramson LE, Armen SF, Bae H-J, Bohn T, Boizelle BD, Bostroem A, Brandel A, Brink TG, Channa S, Cooper MC, Cosens M, Donohue E, Fillingham SP, Gonzalez-Buitrago D, Halevi G, Halle A, Hood CE, Horne K, Horst JC, de Kouchkovsky M, Kuhn B, Kumar S, Leonard DC, Loveland D, Manzano-King C, McHardy I, Michel R, Olaes MKB, Park D, Park S, Pei L, Ross TW, Runco JN, Samuel J, Sanchez J, Scott B, Sexton RO, Shin J, Shivvers I, Spencer CL, Stahl BE, Stegman S, Stomberg I, Valenti S, Walsh JL, Yuk H, Zheng W (2022) The Lick AGN monitoring project 2016: dynamical modeling of velocity-resolved Hβ lags in luminous Seyfert galaxies. Astrophys J 930:52. https://doi.org/10.3847/1538-4357/ac6171
Vives-Arias H, Muñoz JA, Kochanek CS, Mediavilla E, Jiménez-Vicente J (2016) Observations of the lensed quasar Q2237+0305 with CanariCam at GTC. Astrophys J 831:43. https://doi.org/10.3847/0004-637x/831/1/43
Walsh D, Carswell RF, Weymann RJ (1979) 0957+ 561 A, B – twin quasistellar objects or gravitational lens. Nature 279:381
Wambsganss J (1990) Gravitational microlensing. PhD thesis. Ludwig Maximilians University, Munich
Wambsganss J (1992) Probability distributions for the magnification of quasars due to microlensing. Astrophys J 386:19
Wambsganss J (1999) Gravitational lensing: numerical simulations with a hierarchical tree code. J Comput Appl Math 109:353
Wambsganss J, Paczynski B, Schneider P (1990) Interpretation of the microlensing event in QSO2237+0305. Astrophys J 358:33
White SV, Jarvis MJ, Häußler B, Maddox N (2015) Radio-quiet quasars in the video survey: evidence for agn-powered radio emission at $s_{1.4\,\mathrm{GHz}} < 1$ mJy. Mon Not R Astron Soc 448:2665–2686. https://doi.org/10.1093/mnras/stv134
Wielgus M, Lančová D, Straub O, Kluźniak W, Narayan R, Abarca D, Różańska A, Vincent F, Török G, Abramowicz M (2022) Observational properties of puffy discs: radiative GRMHD spectra of mildly sub-Eddington accretion. Mon Not R Astron Soc 514:780–789. https://doi.org/10.1093/mnras/stac1317
Wilkins DR, Gallo LC (2015) Driving extreme variability: the evolving corona and evidence for jet launching in Markarian 335. Mon Not R Astron Soc 449(1):129–146. https://doi.org/10.1093/mnras/stv162
Witt HJ (1990) Investigation of high amplification events in light curves of gravitationally lensed quasars. Astron Astrophys 236:311
Witt HJ (1991) Der Mikrogravitationslinseneffekt – Theorie und Anwendungen. Dissertation. Universität Hamburg
Witt HJ (1993) An efficient method to compute microlensed light curves for point sources. Astrophys J 403:530
Witt HJ, Mao S (1994) Interpretation of microlensing events in Q2237+0305. Astrophys J 429:66
Witt HJ, Mao S, Schechter PL (1995) On the universality of microlensing in quadruple gravitational lenses. Astrophys J 443:18
Woźniak PR, Udalski A, Szymański M, Kubiak M, Pietrzyński G, Soszyński I, Żebruń K (2000) The optical gravitational lensing experiment: a hunt for caustic crossings in QSO 2237+0305. Astrophys J 540:65–67
Wyithe JSB, Loeb A (2002) Measuring the size of quasar broad-line clouds through time-delay light-curve anomalies of gravitational lenses. Astrophys J 577:615–625
Wyithe JSB, Turner EL (2001) Determining the microlens mass function from quasar microlensing statistics. Mon Not R Astron Soc 320:21–30. https://doi.org/10.1046/j.1365-8711.2001.03917.x
Wyithe JSB, Webster RL, Turner EL (2000a) The distribution of microlensed light-curve derivatives: the relationship between stellar proper motions and transverse velocity. Mon Not R Astron Soc 312:843–852
Wyithe JSB, Webster RL, Turner EL (2000b) Limits on the microlens mass function of Q2237+0305. Mon Not R Astron Soc 315:51–61
Wyithe JSB, Agol E, Turner EL, Schmidt RW (2002) Constraints on the mass profile of the lens galaxy G2237+0305. Mon Not R Astron Soc 330:575
Yang Q, Shen Y, Liu X, Aguena M, Annis J, Avila S, Banerji M, Bertin E, Brooks D, Burke D, Rosell AC, Kind MC, da Costa L, Vicente JD, Desai S, Diehl HT, Doel P, Flaugher B, Fosalba P, Frieman J, Garcia-Bellido J, Gerdes D, Gruen D, Gruendl R, Gschwend J, Gutierrez G, Hinton S, Hollowood DL, Honscheid K, Kuropatkin N, Maia M, March M, Marshall J, Martini P, Melchior P, Menanteau F, Miquel R, Paz-Chinchon F, Malagón AP, Romer K, Sanchez E, Scarpine V, Schubnell M, Serrano S,

Sevilla I, Smith M, Suchyta E, Tarle G, Varga TN, Wilkinson R (2020) Dust reverberation mapping in distant quasars from optical and mid-infrared imaging surveys. Astrophys J 900:58. https://doi.org/10.3847/1538-4357/aba59b

Yonehara A, Hirashita H, Richter P (2008) Origin of chromatic features in multiple quasars. Variability, dust, or microlensing. Astron Astrophys 478(1):95–109. https://doi.org/10.1051/0004-6361:20067014

Young P (1981) Q0957+561 – effects of random stars on the gravitational lens. Astrophys J 244:756–767

Yu Z, Martini P, Penton A, Davis TM, Kochanek CS, Lewis GF, Lidman C, Malik U, Sharp R, Tucker BE, Aguena M, Annis J, Bertin E, Bocquet S, Brooks D, Rosell AC, Carollo D, Kind MC, Carretero J, Costanzi M, da Costa LN, Pereira MES, Vicente JD, Diehl HT, Doel P, Everett S, Ferrero I, García-Bellido J, Gatti M, Gerdes DW, Gruen D, Gruendl RA, Gschwend J, Gutierrez G, Hinton SR, Hollowood DL, Honscheid K, James DJ, Kuehn K, Mena-Fernández J, Menanteau F, Miquel R, Nichol B, Paz-Chinchón F, Pieres A, Malagón AAP, Raveri M, Romer AK, Sanchez E, Scarpine V, Sevilla-Noarbe I, Smith M, Suchyta E, Swanson MEC, Tarle G, Vincenzi M, Walker AR, Weaverdyck N (2023) OzDES reverberation mapping program: Mg II lags and R-L relation. Mon Not R Astron Soc 522(3):4132–4147. https://doi.org/10.1093/mnras/stad1224

Zackrisson E, Bergvall N, Marquart T, Helbig P (2003) Can microlensing explain the long-term optical variability of quasars? Astron Astrophys 408:17–25. https://doi.org/10.1051/0004-6361:20030895

Zamaninasab M, Clausen-Brown E, Savolainen T, Tchekhovskoy A (2014) Dynamically important magnetic fields near accreting supermassive black holes. Nature 510:126–128. https://doi.org/10.1038/nature13399

Zdziarski AA, You B, Szanecki M (2022) Corrections to estimated accretion disk size due to color correction, disk truncation, and disk wind. Astrophys J Lett 939:2. https://doi.org/10.3847/2041-8213/ac9474

Zheng W, Chen X, Li G, Chen H-Z (2022) An improved GPU-based ray-shooting code for gravitational microlensing. Astrophys J 931:114. https://doi.org/10.3847/1538-4357/ac68ea

Publisher's Note Springer Nature remains neutral with regard to jurisdictional claims in published maps and institutional affiliations.

Authors and Affiliations

G. Vernardos[1,2,3] · D. Sluse[4] · D. Pooley[5,6] · R.W. Schmidt[7] · M. Millon[1,8] · L. Weisenbach[9] · V. Motta[10] · T. Anguita[11,12] · P. Saha[13] · M. O'Dowd[2,3,14] · A. Peel[1] · P.L. Schechter[15,16]

✉ G. Vernardos
georgios.vernardos@epfl.ch

1 Institute of Physics, Laboratory of Astrophysics, Ecole Polytechnique Fédérale de Lausanne (EPFL), Observatoire de Sauverny, 1290 Versoix, Switzerland

2 Department of Astrophysics, American Museum of Natural History, Central Park West and 79th Street, NY 10024, USA

3 Department of Physics and Astronomy, Lehman College of the City University of New York, Bronx, NY 10468, USA

4 STAR Institute, Quartier Agora – Allée du six Aout 19c, B-4000 Liége, Belgium

5 Department of Physics and Astronomy, Trinity University, San Antonio, TX 78212, USA

6 Eureka Scientific, Oakland, CA 94602, USA

7 Astronomisches Rechen-Institut, Zentrum für Astronomie der Universität Heidelberg, Mönchhofstr. 12-14, D-69120 Heidelberg, Germany

8 Kavli Institute for Particle Astrophysics and Cosmology and Department of Physics, Stanford University, CA 94305, Stanford, USA

9 Institute of Cosmology and Gravitation, University of Portsmouth, Dennis Sciama Building, Burnaby Road, Portsmouth, PO1 3FX, UK

[10] Instituto de Física y Astronomía, Facultad de Ciencias, Universidad de Valparaíso, Avda. Gran Bretaña 1111, Valparaíso, Chile

[11] Instituto de Astrofísica, Facultad de Ciencias Exactas, Universidad Andres Bello, Av. Fernandez Concha 700, Las Condes, Santiago, Chile

[12] Millennium Institute of Astrophysics, Nuncio Monseñor Sótero Sanz 100, Providencia, Santiago, Chile

[13] Physik-Institut, University of Zurich, Winterthurerstrasse 190, 8057 Zurich, Switzerland

[14] The Graduate Center of the City University of New York, 365 Fifth Avenue, New York, NY 10016, USA

[15] Massachusetts Institute of Technology, Department of Physics, Cambridge, MA 02139, USA

[16] Massachusetts Institute of Technology, Kavli Institute for Astrophysics and Space Research, Cambridge, MA 02139, USA

Space Science Reviews (2024) 220:13
https://doi.org/10.1007/s11214-024-01044-7

Strong Gravitational Lensing and Microlensing of Supernovae

Sherry H. Suyu[1,2,3] · **Ariel Goobar**[4] · **Thomas Collett**[5] · **Anupreeta More**[6,7] · **Giorgos Vernardos**[8]

Received: 2 June 2023 / Accepted: 30 December 2023 / Published online: 5 February 2024
© The Author(s) 2024

Abstract
Strong gravitational lensing and microlensing of supernovae (SNe) are emerging as a new probe of cosmology and astrophysics in recent years. We provide an overview of this nascent research field, starting with a summary of the first discoveries of strongly lensed SNe. We describe the use of the time delays between multiple SN images as a way to measure cosmological distances and thus constrain cosmological parameters, particularly the Hubble constant, whose value is currently under heated debates. New methods for measuring the time delays in lensed SNe have been developed, and the sample of lensed SNe from the upcoming Rubin Observatory Legacy Survey of Space and Time (LSST) is expected to provide competitive cosmological constraints. Lensed SNe are also powerful astrophysical probes. We review the usage of lensed SNe to constrain SN progenitors, acquire high-z SN spectra through lensing magnifications, infer SN sizes via microlensing, and measure properties of dust in galaxies. The current challenge in the field is the rarity and difficulty in finding lensed SNe. We describe various methods and ongoing efforts to find these spectacular explosions, forecast the properties of the expected sample of lensed SNe from upcoming surveys particularly the LSST, and summarize the observational follow-up requirements to enable the various scientific studies. We anticipate the upcoming years to be exciting with a boom in lensed SN discoveries.

Keywords Gravitational lensing: strong · Gravitational lensing: micro · Supernovae: general · (Cosmology:) distance scale · (Cosmology:) cosmological parameters · (ISM:) dust, extinction

1 Brief History

In an insightful and pioneering publication, Refsdal (1964) pointed out that supernovae (SNe) would be particularly interesting sources for studies involving strong gravitational lensing. This was arguably among the biggest early milestones in the field of gravitational lensing, following the realisation by Zwicky (1937) that the scenario, first proposed by Einstein in 1936, had realistic applications in extra-galactic astronomy. Refsdal pointed out that the measurement of the time delay between the arrival of SN images could be used to infer the Hubble constant (H_0).

Extended author information available on the last page of the article

1.1 History of Searches for Lensed SNe Behind Clusters

The idea to use "gravitational telescopes", i.e., known lensing galaxy clusters, to boost the faint signals from distant supernovae started to get traction about thirty years ago (Kovner and Paczynski 1988; Sullivan et al. 2000; Gunnarsson and Goobar 2003). Since the lensing magnification boosts the signal from the faint distant source behind the lens by a factor of μ, but leaves the dominant foreground sky noise unaffected, the signal-to-noise ratio scales as $SNR \propto \mu\sqrt{t}$ (where t is the exposure time), hence leading to a gain factor μ^2 in exposure length. However, the solid angle at the source planes shrinks by a factor μ behind the lens. Thus, the net gain/loss of searching for supernovae behind massive clusters is therefore a non-trivial combination of field of view, limiting depth, and supernova luminosity functions. Although not spectroscopically confirmed, several SN candidates were eventually found in cadenced observations in the Near-IR at the Very Large Telescopes (VLT) of the European Southern Observatory (ESO) (Stanishev et al. 2009; Goobar et al. 2009), including a $z = 1.7$ core-collapse SN behind Abell 1689, with a lens model inferred magnification $\mu = 4.3 \pm 0.3$ (Amanullah et al. 2011). In spite of the limitations of the survey, Petrushevska et al. (2016) demonstrated the power of gravitational lensing to place meaningful limits on the rate of core-collapse supernovae at very high redshifts.

1.2 Discoveries!

PS1-10afx was first reported by Chornock et al. (2013) as an unusual superluminous SN. However, shortly thereafter, Quimby et al. (2013) showed that it was a perfect match to a lensed SN Ia at redshift $z = 1.388$, with a large amplification, $\mu \sim 30$. Quimby et al. (2014) eventually also identified a foreground lens, at $z = 1.117$. Even if the putative lensing object was too close to the SN, it was consistent with lensing models and existing data. Since the lensed SN Ia classification only gained acceptance four years after the explosion, further investigation of the lensing nature of the SN or its type was not possible. Moreover, a subsequent single-band HST imaging (PI: Chornock) proved insufficient to verify the presence of any lensed images of the SN host galaxy.

The first detection of multiple images from a supernova came with SN Refsdal at redshift $z = 1.49$ (Kelly et al. 2015), found in Hubble Space Telescope (HST) surveys of the massive lensing galaxy cluster MACS J1149.6+2223 at $z = 0.54$. Based on both light curve shapes and spectroscopy, SN Refsdal was classified as a Type II SN, resembling the iconic SN 1987A in the LMC (Kelly et al. 2016b). This is a peculiar class of faint SNe, quite rare in the local universe. At the time of discovery, 4 multiple images of the SN were visible with HST (images S1-S4, see Fig. 1 taken from Grillo et al. 2018), in a cross configuration around one of the foreground cluster galaxies. There are two other images of the background spiral galaxy hosting the SN, labelled as SX (inset (a) in Fig. 1) and SY (the northern-most image, just outside the top-right corner of inset (a)).

Using mass models of the cluster, the discovery team (Kelly et al. 2015) predicted that SY happened before S1-S4, whereas SX would appear after S1-S4. However, the time of SX reappearance was uncertain from their model, ranging from 1 to 10 years. To refine the delay prediction, Grillo et al. (2016) used the Multi Unit Spectroscopic Explorer (MUSE; Bacon et al. 2010) on ESO's VLT to obtain spectroscopic observations of the field. The spectroscopic redshift measurements helped to separate foreground lenses from background sources. This spectroscopic data set was shared with multiple modelling teams, who attempted to predict the reappearance of image SX using the new observational data. Teams predicted their delays before the reappearance (Treu et al. 2016; Jauzac et al. 2016; Grillo

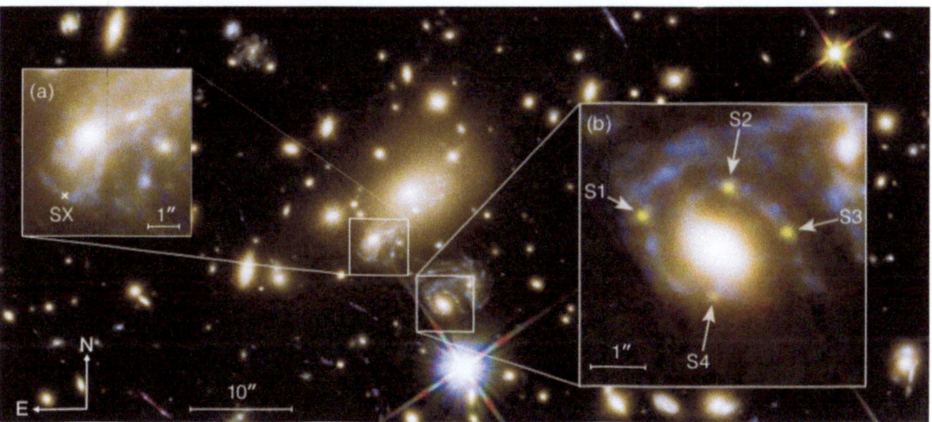

Fig. 1 Hubble Space Telescope image of SN Refsdal, the first strongly lensed SN system with spatially resolved images. Inset (a) shows the image SX that was detected in December 2015 (Kelly et al. 2016a), and inset (b) shows the multiple images S1, S2, S3 and S4 that were first discovered in November 2014 (Kelly et al. 2015). There is another image SY located in the northern most image of the spiral host galaxy, next to the top-right corner of the inset (a). Image taken from Grillo et al. (2018). Original image credit: NASA, ESA/Hubble

et al. 2016; Kawamata et al. 2016), providing a true blind test of the predictions and the modelling capabilities. Most teams predicted a short time delay of SX with respect to S1, within a year's time. Kelly et al. (2016a) detected the reappearance of SX in December 2015, and the resulting constraint on the time delay and magnification of SX agreed well with the predictions from two teams (Grillo et al. 2016; Kawamata et al. 2016).

Through the monitoring of the multiple images of SN Refsdal, Rodney et al. (2016) have measured the time delays and magnification ratios among images S1, S2, S3 and S4. The time-delay measurement for image SX from subsequent monitoring is $376.0^{+5.6}_{-5.5}$ days relative to image S1 (Kelly et al. 2023b). This time-delay measurement with a relative uncertainty of 1.5% provides the first opportunity for a precise H_0 measurement from lensed SNe, which we describe more in Sect. 2.3.

The first spatially resolved multiply-imaged Type Ia supernova, iPTF16geu, was detected by the intermediate Palomar Transient Factory (Goobar et al. 2017). As shown in Fig. 2, the ground-based imager at iPTF could not spatially resolve the very compact system with Einstein radius $\theta_{\rm E} = 0.3''$. However, because of the "standard candle" nature of SNe Ia, it became clear that strong lensing was the most likely explanation, since the SN was more than 30 standard deviations brighter than the expectations for a SN Ia at its measured redshift, $z = 0.409$. The spectra used to classify the supernova showed spectral lines from both the host galaxy at the same redshift, but also the deflecting galaxy at $z = 0.216$. Thanks to multi-band follow-up with HST, an accurate (model independent) measurement of the magnification was made, $\mu = 67.8^{+2.6}_{-2.9}$ (Dhawan et al. 2020), after correction for non-negligible extinction by dust in both the host and lens galaxies. The time delays between the SN images for this system were very small, about a day or less (More et al. 2017; Dhawan et al. 2020). The flux ratios between the supernova images (see Fig. 2) were not consistent with expectations from lensing, hinting contributions from stellar microlensing (More et al. 2017). This was further confirmed after differential dust extinction from within the lensing galaxy was also accounted for (Mörtsell et al. 2020).

In 2021, another strongly lensed SN, SN Requiem, was discovered in archival HST imaging of the galaxy cluster MACSJ0138.0−2155 by Rodney et al. (2021). A bright quiescent

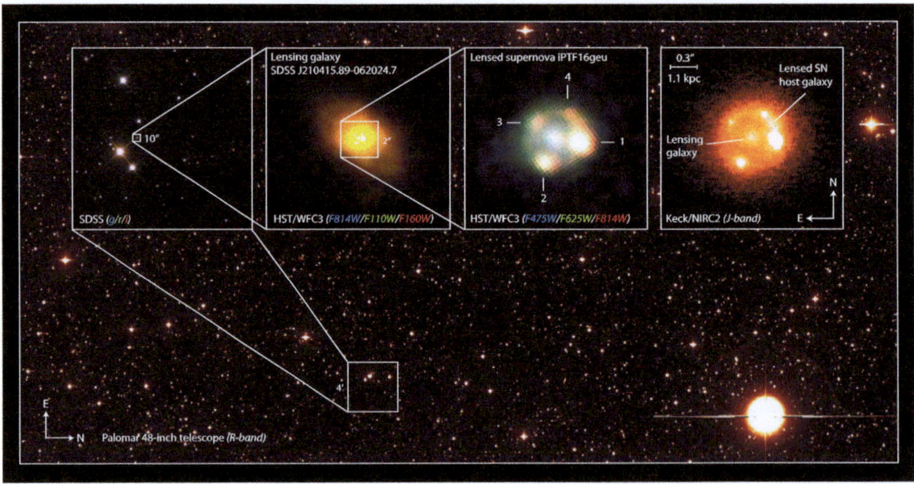

Fig. 2 Wide-field image from iPTF showing the portion of the CCD camera where iPTF16geu was found in modest seeing ($2''$). The insets show how the strong lensing nature, a quadruple lens with an Einstein radius of only $0.3''$, could be verified using HST imaging in the optical, and laser-guide-star adaptive-optics imaging from Keck in the Near-IR. Image credit: J. Johansson

and evolved elliptical galaxy at redshift $z = 1.95$ is lensed into giant arcs by the galaxy cluster as shown in Fig. 3, making this system one of the brightest NIR objects (e.g., Newman et al. 2018). Near the giant arcs are three point sources that were present in the single-epoch image in 2016 (Newman et al. 2018), but absent in the 2019 HST image (HST Program - REQUIEM; HST-GO-15663; PI:Akhshik). These 3 point sources are identified as multiple images of a SN hosted in the $z = 1.95$ galaxy, and the SN is likely to be of Type Ia. Based on the differences in the colour of the 3 SN images and SN Ia templates, the time delays could be estimated from the single-epoch observation in 2016, although with large uncertainties. The predicted time delays suggest that the fourth image of SN Requiem will appear in the future (circa 2037).

Recently, Goobar et al. (2023) discovered another strongly lensed SNe Ia, named Supernova Zwicky (a.k.a. SN 2022qmx), in the Zwicky Transient Facility (ZTF; Bellm et al. 2019). The discovery of this system is similar to the case of iPTF16geu – the multiple lensed SN Ia images are closely separated (with an Einstein radius of $\theta_{\rm E} \sim 0.17''$) and not spatially resolved by ZTF, thus resulting in a brightness through lensing magnification that is substantially higher than expected for a SN Ia. Pierel et al. (2023) obtained HST imaging of this system, showing clearly the four multiple images of the SN Ia in a symmetric configuration. Based on the single-epoch photometry of the SN Ia images, Pierel et al. (2023) measured short time delays of < 1 day between the multiple images, which are consistent with the predictions from their multiple lens mass models obtained using different lensing software. Both Goobar et al. (2023) and Pierel et al. (2023) find anomalous flux ratios of the SN images compared to the smooth mass model predictions, indicating the presence of dust, millilensing and/or microlensing.

The fifth system of spatially-resolved lensed SN is reported very recently by Chen et al. (2022). A SN was lensed by one of the galaxies in the galaxy cluster Abell 370 in 2010 into three multiple images, and Chen et al. (2022) discovered it in archival HST image, similar to the way that SN Resfdal was found. Based on their cluster lens mass modeling,

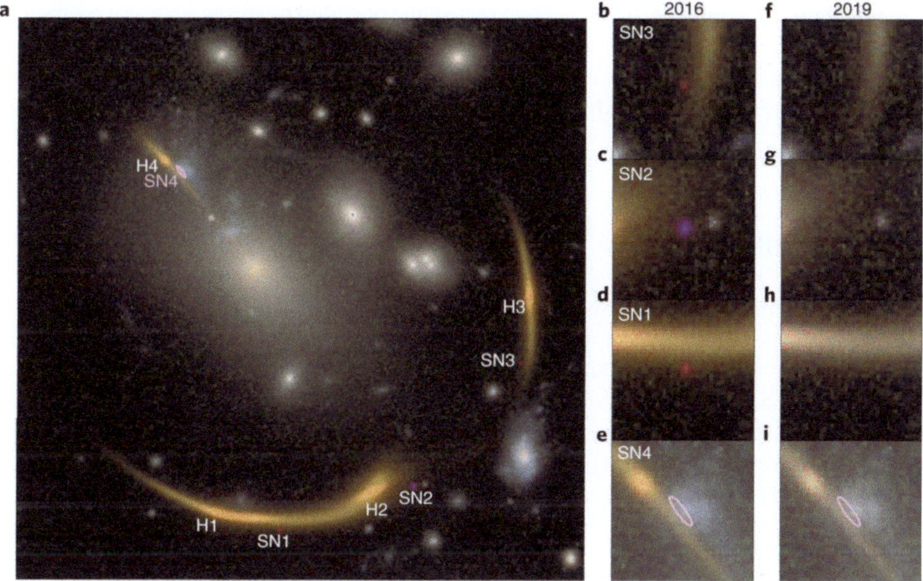

Fig. 3 Panel a): Color image of SN Requiem showing arc-like images of the distant host galaxy (H1-H4), the three SN images (SN1-SN3) and an ellipse pointing out the expected location of SN4. Panels b-i): zoomed-in regions around the SN locations from the data taken in 2016 (b-e) and 2019 (f-i). Figure taken from Rodney et al. (2021)

Chen et al. (2022) estimated the rest-frame age of the SN for each of the three SN images, with the youngest one being a mere ~ 6 hours after explosion. Using the early-phase light curve obtained in the single-epoch HST observations with multiple filters, Chen et al. (2022) classified the SN as a core-collapse SN and measured its pre-explosion radius of $\sim 500R_{\odot}$, consistent with a red supergiant. The estimated photometric redshift of the SN host galaxy is ~ 3.

The sixth system is the first strongly lensed SN detected with the James Webb Space Telescope (JWST) in the imaging of the galaxy cluster PLCK-G165.7+67.0, named SN H0pe (Frye et al. 2023; Polletta et al. 2023). This strong lensing cluster was discovered through the detection and follow-up of the brightest, strongly lensed dusty star-forming galaxies on the sky from Planck and Herschel (Cañameras et al. 2015; Planck Collaboration et al. 2015). This cluster and a lensed dusty galaxy have been observed and studied by Cañameras et al. (2015), Harrington et al. (2016), Cañameras et al. (2018), Frye et al. (2019), Pascale et al. (2022) prior to the JWST observations. Upon the discovery of SN H0pe (occurring in a background elliptical galaxy that is strongly lensed into a prominent arc), Frye et al. (2023) obtained additional JWST imaging and spectroscopy of the system in order to constrain the time delays of the SN images through photometry (Pierel et al., submitted) and spectroscopy (Chen et al., in prep.). Several papers are forthcoming to use this system to study SN Ia, SN host galaxy, dusty star-forming galaxy and cosmology (e.g., Johansson et al., Caminha et al., Kamieneski et al., Pascale et al., in prep.).

These first few strongly lensed SNe have opened a new window to probe cosmology and supernovae through these spectacular phenomena. Two reviews (Oguri 2019; Liao et al. 2022) on gravitationally lensed transients have illustrated the exciting utilities of lensed transients. Compared to these two reviews, we focus and dive deeper into the specific case of strongly lensed supernovae. In Sect. 2, we review the potential of lensed SNe as a cosmological probe. In Sect. 3, we describe the use of lensed SNe as an astrophysical probe, such as constraining the SN progenitors and dust in galaxies. In Sect. 4, we present various searches of lensed SNe and the expected rates in current and upcoming surveys. We summarise in Sect. 5.

2 Cosmological Probe

While the idea of using lensed SNe for measuring the Hubble constant dates back to Refsdal (1964), this time-delay method was first realised with lensed quasars that were discovered decades before lensed SNe, starting with the first lensed quasar system found by Walsh et al. (1979). The use of lensed quasars as a cosmological probe is reviewed by Birrer et al. (2024) in this collection (see also, e.g., Treu and Marshall 2016; Suyu et al. 2018; Treu et al. 2022). We briefly summarise the approach of time-delay cosmography in Sect. 2.1, and focus on SNe as lensed background sources, which are "standardisable candles" and a relative distance indicator. In Sect. 2.2, we describe approaches to measure time delays of lensed SNe, particularly novel ones through spectroscopic observations of SNe, which complement conventional techniques using photometric light curves. In Sect. 2.3, we show the present cosmological constraints and future forecasts from lensed SNe.

2.1 Jackpot: Two Cosmological Probes in One

We describe the various distance measurements that we can obtain from strongly lensed SNe, through the lensing time delays and the SN light curves. We briefly summarise the determination of the lensing distances that are covered in detail in the parallel reviews by, e.g., Birrer et al. (2024), and focus particularly on the new aspects that SNe bring.

The expression for the time delay between images i and j of a lensed SN is

$$\Delta t_{ij} = \frac{D_{\Delta t}}{c} \Delta \tau_{ij}, \tag{1}$$

where $D_{\Delta t}$ is the time-delay distance, c is the speed of light, and τ is the Fermat potential. The time-delay distance (e.g., Refsdal 1964; Suyu et al. 2010) is defined by

$$D_{\Delta t} \equiv (1 + z_{\mathrm{d}}) \frac{D_{\mathrm{d}} D_{\mathrm{s}}}{D_{\mathrm{ds}}}, \tag{2}$$

where z_{d} is the deflector (lens) redshift, and D_{d}, D_{s}, and D_{ds} are the angular diameter distance to the deflector, to the source, and between the deflector and the source, respectively. The Fermat potential difference between two SN image positions $\boldsymbol{\theta}_i$ and $\boldsymbol{\theta}_j$ (with corresponding source position $\boldsymbol{\beta}$) is

$$\Delta \tau_{ij} = \tau(\boldsymbol{\theta}_i; \boldsymbol{\beta}) - \tau(\boldsymbol{\theta}_j; \boldsymbol{\beta}), \tag{3}$$

where the Fermat potential is

$$\tau(\boldsymbol{\theta}; \boldsymbol{\beta}) = \frac{1}{2}(\boldsymbol{\theta} - \boldsymbol{\beta})^2 - \psi(\boldsymbol{\theta}), \tag{4}$$

and $\psi(\boldsymbol{\theta})$ is the lens potential. The dimensionless surface mass density $\kappa(\boldsymbol{\theta})$, a.k.a. lensing convergence, is related to the lens potential via the Poisson equation,

$$2\kappa(\boldsymbol{\theta}) = \nabla^2\psi(\boldsymbol{\theta}) \tag{5}$$

From Eq. (1), we see that by measuring the time delays Δt_{ij} and modelling the deflection and line-of-sight mass distributions to get $\Delta\tau_{ij}$, we can infer the time-delay distance $D_{\Delta t}$. Since $D_{\Delta t}$ is a combination of three angular diameter distances (Eq. (2)) and is thus proportional to H_0^{-1}, time-delay lenses allow us to measure directly H_0 with weak dependence on other cosmological parameters.

When the lens galaxy stellar velocity dispersion is measured, then the combination of time delays, velocity dispersion and high-resolution imaging of the lens system allows $D_{\rm d}$ to be measured (e.g., Paraficz and Hjorth 2009; Jee et al. 2015, 2019), in addition to $D_{\Delta t}$. The joint constraint of $D_{\Delta t}$ and $D_{\rm d}$ provides even more leverage on determining cosmological parameters (e.g., Jee et al. 2016; Yıldırım et al. 2020).

It is clear from Eq. (4) that in order to constrain H_0 with strong lens time delays, one needs to know both the 2D lens potential and the unlensed source position, neither of which are directly observable. The use of lens modelling is required to infer these quantities. However, strong lens models are subject to degeneracies, which are a major source of uncertainty for time-delay cosmography (Schneider and Sluse 2014). The largest problem is the mass-sheet degeneracy: re-scaling any model of the lensing convergence κ and adding a constant sheet of surface mass density leaves the predicted images unchanged, but alters the predicted time delay between the images (Falco et al. 1985). Breaking the mass-sheet degeneracy is therefore necessary to constrain H_0 from any lens with observations of image positions and time delays. In order to break the mass-sheet degeneracy, additional information is required, either non-lensing information (e.g., the dynamical mass of the system), the presence of sources at multiple redshifts (though see Schneider 2014) or knowledge of the intrinsic luminosity or size of the unlensed source (Kolatt and Bartelmann 1998). It is on this final point – a known luminosity of the source – that strongly lensed SNe shine as a powerful potential cosmological probe.

In this context, Type Ia SNe play a special role. Thanks to the homogeneity of their luminosity, these thermonuclear explosions could be used as sharp distance estimators to measure the accelerated expansion of the universe (Riess et al. 1998; Perlmutter et al. 1999), leading to the discovery of dark energy. State-of-the-art SN Ia cosmological surveys have demonstrated that the intrinsic luminosity of SNe Ia only varies by about 0.1 mag (Betoule et al. 2014; Scolnic et al. 2018). Birrer et al. (2022) have shown the great benefit of these sharp distance indicators to break the mass-sheet degeneracy to improve the accuracy of the measurements of the Hubble constant from time delays.

Despite the great opportunity presented by having a lensed standard candle, the exploitation of lensed SNe poses an additional challenge: microlensing by stars (Dobler and Keeton 2006). The intrinsic size of a SN is comparable to the Einstein radius of an individual star in the lens galaxy, a case analogous to microlensing of lensed quasars (Vernardos et al. 2023, in prep.). This means that observed magnification is sensitive to the – essentially unobservable – positions of stars in the lensing galaxy. Worse, as the supernova expands, the SN atmosphere crosses microlensing caustics. This causes a change in total magnification of the supernova and differential magnification across the atmosphere (Bagherpour et al. 2006; Goldstein et al. 2018; Huber et al. 2019; Mörtsell et al. 2020); this means that the light curves of each image can look quite different even though they are formed from the same background source, as illustrated in Fig. 4. This makes it a challenge for measuring accurate

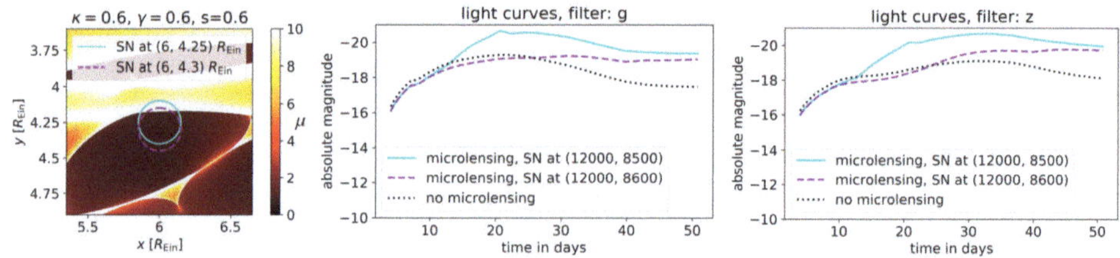

Fig. 4 Examples of Type Ia SNe light curves with significant microlensing. Left panel: a map of microlensing magnifications (μ, indicated by the color bar) on the source plane as a result from stars in the foreground lens galaxy. The map is for a lensed SN image with convergence $\kappa = 0.6$, shear $\gamma = 0.6$, and smooth dark matter fraction $s = 0.6$. In this example, the Einstein radius for the mean microlens is $R_{\rm Ein} = 7.2 \times 10^{-3}$ pc. The two circles in solid cyan and dashed magenta indicate the size of a SN Ia at 21 rest-frame days after explosion. Middle and right panels: microlensed light curves in the g-band (middle panel) and z-band (right panel) corresponding to the SN positions shown in the left panel (solid cyan and dashed magenta). The intrinsic light curve without microlensing is shown in dotted black. When the expanding photosphere of the SN crosses a microlensing caustic with high μ, its light curve changes substantially relative to the no-microlensing case. Figure taken from Huber et al. (2019)

time delays between images, though early colour curves can overcome this due to an achromatic phase in the microlensing at early times (Goldstein et al. 2018; Huber et al. 2021b) as can spectroscopic observations of absorption lines in the SN atmosphere (Johansson et al. 2021; Bayer et al. 2021).

The SNe never reach a large enough size for the microlensing effect of multiple stars to average out, so microlensing makes it hard to standardise lensed SNe. The scale of this problem depends in detail on the magnification of the image, the type of the image (time-delay minimum, maximum or saddle) and fraction of density in stars at the location of the image (e.g., Schechter and Wambsganss 2002; Huber et al. 2021b). For high macro-magnification images, the scatter can be as large as 1 magnitude in flux (Schechter and Wambsganss 2002; Yahalomi et al. 2017; Diego et al. 2022), entirely washing out the possibility of using lensed SNe as a standard candle. For images forming further from the Einstein Radius, and particularly for large mass systems where the dark matter fraction is higher at the Einstein radius, the scatter is much smaller and can be less than the intrinsic scatter of a Type Ia SN.

Foxley-Marrable et al. (2018) found that 20 percent of lensed SNe will contain an image with scatter comparable to the intrinsic scatter of a Type Ia supernova luminosity, assuming the dark matter fractions of elliptical galaxies derived in Auger et al. (2010) and a Saltpeter IMF. However, these standardisable images only form far beyond the Einstein radius of the lens. This means that the counter images are very close to the centre of the lens and demagnified. Measuring time delays will therefore be challenging, although the faint images arrive second so follow-up with a larger telescope may ameliorate this problem.

The 20 percent standardisable fraction is subject to two key caveats: the stellar initial mass function is assumed to be Salpeter-like and the ellipticity of the lensing mass follows that of the light. The assumption of a particular IMF sets the stellar-to-dark matter fraction. Figure 5 shows the importance of this assumption: at very high stellar fraction there are no strongly lensed images with minimal microlensing scatter (Weisenbach et al. 2021). With a population of lensed SNe, Foxley-Marrable et al. (2018) found that observing the amount of microlensing induced scatter can be used to constrain the IMF of the lens galaxy, with strong sensitivity for lenses with Einstein radius between $0.2''$ and $0.5''$. The assumption that mass follows light produces more elliptical lens mass distributions than is seen in cold-dark-matter-only simulations. Increasing the ellipticity of the lens moves the dashed lines in Fig. 5 to the left. Strongly lensed images can form at lower magnification, where microlensing

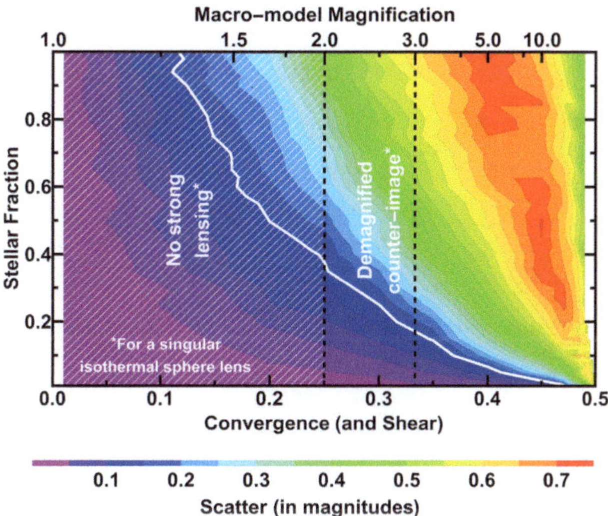

Fig. 5 Microlensing induced scatter (indicated by the color bar) for a point source lensed by a Singular Isothermal Sphere (SIS) macromodel as a function of stellar fraction and macro-magnification. For an SIS, the convergence and shear magnitude (shown in the x-axis) are the same at any given location on the lens plane. Scatter is lower for the lensed SN image that is far from the Einstein radius (with lower convergence/shear), but there is a limit since a convergence below 0.25 produces only single images (i.e., no strong lensing) and convergence below $\sim$0.35 produces a faint counterimage (close to lens center) that makes time-delay measurements more challenging. Image credit: Luke Weisenbach, modified based on Weisenbach et al. (2021)

is less significant. If real lenses are more spherical, then the standardisable fraction will fall. Therefore, the standardisation of lensed supernovae is subject to somewhat uncertain astrophysics. Exactly what we observe from a large sample of lensed supernovae will inform not just cosmology but also the astrophysics of matter in the strong lensing galaxy.

Despite microlensing distortions on the light curves of SNe and reductions in the number of standardisable SN, lensed SNe still have various advantages over the more conventional lensed quasars for time-delay cosmography. The drastically varying brightness of lensed SNe enable shorter monitoring campaigns (months) to obtain the light curves and thus more efficient measurements of time delays. Simulations of microlensed SN light curves with realistic photometric uncertainties showed that time delays can still be recovered accurately and precisely (Huber et al. 2019; Pierel and Rodney 2019), and the dominant source of uncertainty in the time delays is typically photometric uncertainties rather than microlensing distortions (Huber et al. 2021a). Another advantage of lensed SNe is that SNe fade after several months, revealing both the lens galaxy and the lensed SN host galaxy more clearly that enable more accurate lens mass modelling (e.g., Ding et al. 2021). In particular, spatially-resolved stellar kinematics of the foreground lens galaxy can be more readily acquired after the SN images fade, and the combination of lensing and kinematic data allows one to break the mass-sheet degeneracy (e.g., Barnabè et al. 2012; Yıldırım et al. 2020, 2023; Birrer et al. 2020; Shajib et al. 2023). Stellar kinematic maps of the lensed SN host galaxy would further constrain the lens mass distribution (Chirivì et al. 2020). We therefore anticipate lensed SNe to be an efficient cosmological probe in the upcoming era of time-domain astronomy when hundreds of such lensed SN events are expected to happen (e.g., Oguri and Marshall 2010; Quimby et al. 2014; Goldstein et al. 2019; Wojtak et al. 2019).

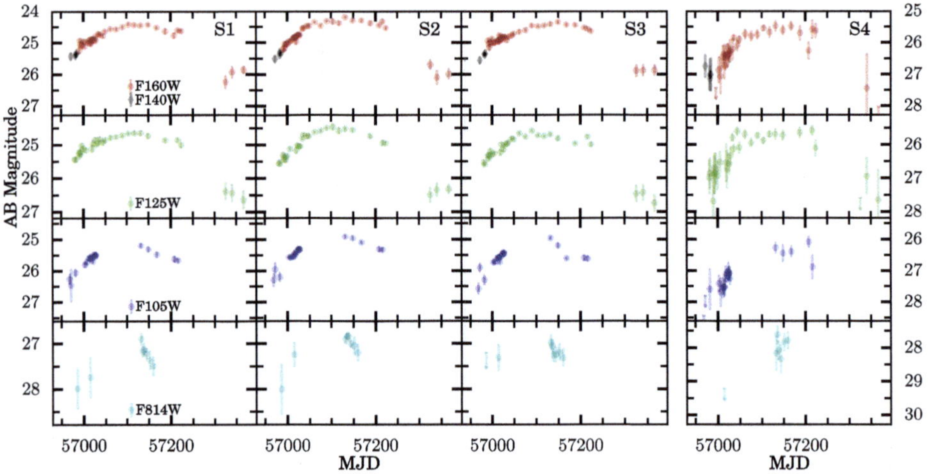

Fig. 6 Light curves of SN Refsdal from Hubble Space Telescope imaging in multiple wavelength filters. Each row consists of the light curves obtained in the specific filter that is indicated in the leftmost panel. Each column shows one of the four SN images, S1-S4 (left to right), that are indicated on the top panels. Each panel shows the observed AB magnitude as a function of the observer-frame days. Figure taken from Rodney et al. (2016)

2.2 Time-Delay Measurements

The time delays between multiple lensed images are a key ingredient for time-delay cosmography. The fractional uncertainties in the time delays translate directly to the fractional uncertainties in the $D_{\Delta t}$ and $D_{\rm d}$ measurements (Eq. (1)).

Traditionally, the measurements of time delays involve monitoring the lens system to acquire the light curves of either lensed quasars (e.g., Fassnacht et al. 2002; Courbin et al. 2018; Millon et al. 2020a,b) or lensed supernovae (e.g., Rodney et al. 2016; Dhawan et al. 2020). Figure 6 shows an example of the light curves of the four images (S1, S2, S3 and S4) of SN Refsdal obtained by Rodney et al. (2016). Curve-shifting techniques that account for microlensing distortions are then applied to these light curves to extract the time delays (e.g., Tewes et al. 2013; Pierel and Rodney 2019; Millon et al. 2020c). By making use of the characteristic SN Ia SEDs, Dhawan et al. (2020) used SN Ia template light curves to fit to the HST images of iPTF16geu and measured the 3 independent time delays between the four SN images. New machine learning approaches such as random forests are also being developed to infer the delays of microlensed SN Ia systems from the light curves (e.g., Huber et al. 2021a).

SNe are not only drastically changing their brightness, but their spectra also evolve substantially. The spectroscopic evolution of supernovae offers a new avenue to measure the time delays that are complementary and competitive to the light-curve techniques, as demonstrated by Johansson et al. (2021) for the case of iPTF16geu. For the most common type of core-collapse supernovae, SNe IIP, where the light curves rise fast to peak and then reach a plateau phase that could last for hundreds of days, spectral time delays may be the best way forward as there is a tight correlation between the velocity and time (e.g., $v(t) \propto t^{-0.464\pm0.017}$; Nugent et al. 2006). Figure 7 shows an example of the spectra of Type IIP SN1999em from the TARDIS simulation (Kerzendorf and Sim 2014; Vogl et al. 2019, 2020) at multiple epochs after explosion (Bayer et al. 2021). These epochs (in rest-frame days) are in the

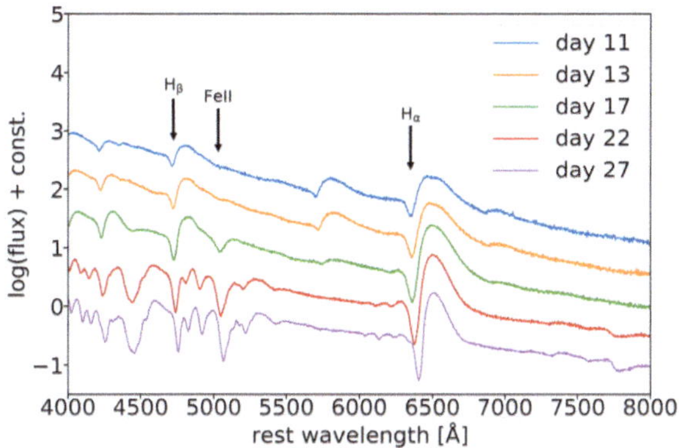

Fig. 7 Spectral evolution of core-collapse SN1999em based on the TARDIS simulation from Vogl et al. (2019, 2020). Three prominent spectral features, Hβ, FeII and Hα, are labelled. The SN phases are indicated by their rest-frame days after explosion. As the SN phase increases, the spectral features become stronger and the absorption wavelengths increase. Such a sequence of spectra of the first-appearing SN image provides the wavelength-phase relation, and a measurements of the absorption wavelength of a trailing SN image therefore provides information of its SN phase and thus its time delay relative to the first-appearing SN image. With each spectrum of signal-to-noise of 20, Bayer et al. (2021) showed that the time delays can be measured with uncertainties of ~ 2 days per spectral feature, even after accounting for the effects of microlensing. Figure taken from Bayer et al. (2021)

plateau phase of this Type IIP SN. The spectra show prominent absorption lines, especially Hβ, FeII and Hα that are indicated on the figure, and the absorption wavelength increases as a function of time. It is precisely such evolutions in the spectral features that allow us to infer the time delays between two SN images.

In more detail, a sequence of spectra of the first appearing SN image, say SN image A (such as those shown in Fig. 7) yields the absorption wavelength of each spectral feature as a function of the SN phase (days after explosion). When we obtain a single spectrum of a trailing SN image, say SN image B, and measure the absorption wavelength of the same spectral feature, we can determine the phase of SN B by using the relation between the absorption wavelength and phase from SN A. The determined phase of SN B relative to SN A, together with the known observational times of the epochs, then allows us to compute the time delays between SN images A and B. Microlensing of SN introduces scatter in the relation between the absorption wavelength and phase, and this effect can be quantified using microlensing maps such as those from the GERLUMPH software (Vernardos and Fluke 2014; Vernardos et al. 2014, 2015). Accounting for the effect of microlensing that distorts spectra, Bayer et al. (2021) have shown that this technique can yield time delays with uncertainties of $\sim$2 days per spectral feature with signal-to-noise in the spectra of $\sim$20 per wavelength bin of 3Å width. With multiple spectral features and epochs, the inferred time delays from such spectral approach will be even more precise and accurate.

2.3 Cosmography with Lensed SNe: Present Cosmological Constraints and Future Forecasts

Of the six known lensed SN systems with spatially resolved SN images, SN Refsdal and SN H0pe with time delays of tens to hundreds of days are suitable in delivering a precise and accurate H_0 measurement (with $<10\%$ uncertainty). As mentioned in Sect. 1, the time delay of image SX relative to the first image S1 of SN Refsdal has recently been measured

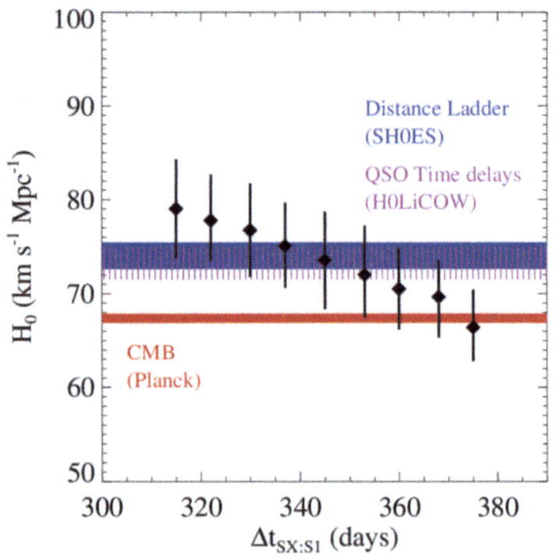

Fig. 8 Forecast of H_0 from SN Refsdal as a function of the time delay between SN images SX and S1, using the reference mass model of Grillo et al. (2020). The median values of H_0 (diamonds) with the 1σ uncertainties in flat ΛCDM are shown for 9 different hypothetical SX-S1 delays, each with an assumed uncertainty of 10 days. For comparison, the blue, magenta and red bands show, respectively, the 1σ credible intervals from SH0ES (Riess et al. 2019), H0LiCOW (Wong et al. 2020) and Planck (Planck Collaboration et al. 2020). Figure taken from Grillo et al. (2020)

with 1.5% uncertainty by Kelly et al. (2023b). Combining the time delays with the two best-fitting cluster mass models that are weighted by a selection of observables, Kelly et al. (2023a) obtained $H_0 = 66.6^{+4.1}_{-3.3}\,\mathrm{km\,s^{-1}\,Mpc^{-1}}$. This is in good agreement with the forecast (both in value and uncertainties) shown in Fig. 8 that is taken from Grillo et al. (2020), who predicted the cosmological constraints on H_0 for a range of SX time delays before the delay was measured.

Various studies indicate that current and future surveys would discover dozens to hundreds of lensed SNe (Goobar et al. 2002; Oguri and Marshall 2010; Goldstein et al. 2019; Wojtak et al. 2019; Pierel et al. 2021). In particular, Oguri and Marshall (2010) anticipated thousands of lensed quasars and $\sim$100 lensed SNe to be detected in the Rubin Observatory Legacy Survey of Space and Time (LSST; LSST Science Collaboration et al. 2009). Using a mock sample of $\sim$1500 well-observed lenses (consisting of 1476 lensed quasars and 66 lensed SNe) and assuming Planck priors in a flat Universe, Oguri and Marshall (2010) expected the following 1σ uncertainties on cosmological parameters: $\sigma(w_0) = 0.15$, $\sigma(w_a) = 0.41$ and $\sigma(h) = 0.017$ where (w_0, w_a) are the time-independent and time-dependent components of the dark energy equation-of-state parameter, and h is H_0 in units of $100\,\mathrm{km\,s^{-1}\,Mpc^{-1}}$.

Huber et al. (2019) analyzed in detail the number of lensed SNe Ia that could yield time-delay measurements with precisions better than 5% and accuracies better than 1%, with realistic microlensed SN Ia light curves and LSST observing strategies. While LSST is efficient at detecting lensed SNe given its wide survey area and depth, the observing cadence per filter is not rapid enough to map out light curves for precise delay measurements. Follow-up observations at the cadence of at least one epoch every 2 days would drastically increase the number of lensed SNe with "good" time delays (i.e., those delays with uncertainties $<$5% in terms of precision and $<$1% in terms of accuracy). Based on the results of Huber et al. (2019), Suyu et al. (2020) simulated a sample of 20 lensed SNe Ia from LSST that are expected to have good time delays. Assuming that these systems would have high-resolution imaging and spatially resolved kinematic measurements to break the mass-sheet degeneracy and yield 6.6% uncertainty on $D_{\Delta t}$ and 5% uncertainty on D_{d}, this modest sample of 20 lensed SNe Ia could yield constraints on H_0 and Ω_{m} of 1.3% and 19%, respectively, in flat ΛCDM, as illustrated in Fig. 9. In an open ΛCDM cosmology (i.e., allowing for a spatially curved Universe), a similar constraint on H_0 compared to flat ΛCDM is achievable, while in

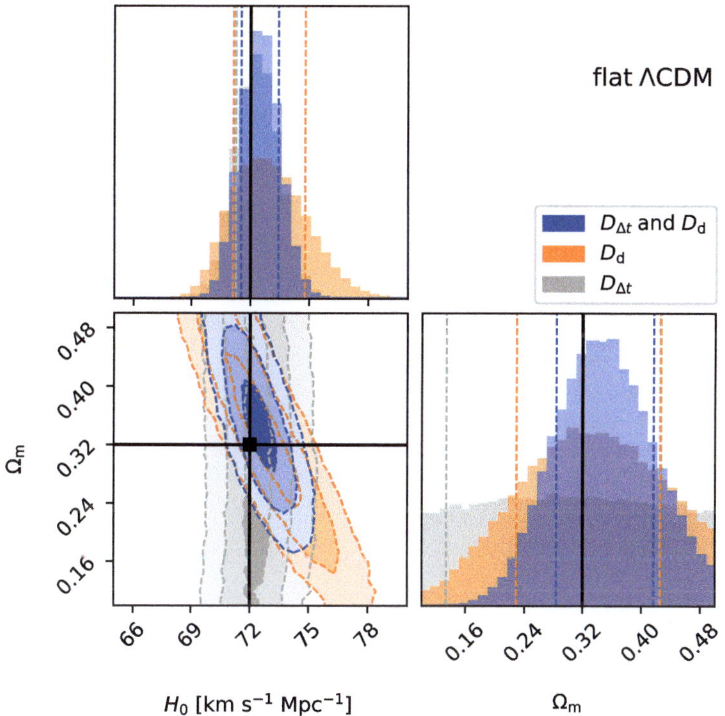

Fig. 9 Forecast of constraints on H_0 and Ω_m in flat ΛCDM from a sample of 20 lensed SNe Ia from LSST with precise and accurate time-delay measurements. Assuming distance measurements of $D_{\Delta t}$ and D_d with 6.6% and 5% uncertainties, respectively, for each lensed SN Ia, this modest sample is expected to yield H_0 and Ω_m with precisions of 1.3% and 19%, respectively. Figure taken from Suyu et al. (2020)

the flat wCDM cosmology (where the dark energy equation-of-state parameter w is allowed to vary), the H_0 constraint degrades to 3% (Suyu et al. 2020).

The forecast made by Suyu et al. (2020) did not make use of the standardisable nature of SNe Ia, due to the effects of microlensing and millilensing as mentioned in Sect. 2.1. Nonetheless, Foxley-Marrable et al. (2018) showed that lensed SNe Ia with asymmetric configurations, i.e., with a SN image located far outside the Einstein radius of the foreground lens, have low microlensing scatter of $\lesssim$0.15 mag for the outer SN image, which is comparable to the intrinsic dispersion of a typical SN Ia. These systems are thus standardisable in terms of overcoming microlensing perturbations, and Foxley-Marrable et al. (2018) estimated that about $\sim$22% of the $\sim$ 930 LSST systems predicted by Goldstein et al. (2018) would be standardisable. However, Goldstein et al. (2018) estimated that only $\sim$650 systems would be detected early enough by LSST to deliver reliable time delays. This implies a sample of $\sim$140 lensed SNe Ia that are standardisable and with potential time-delay measurement.

Following the estimates of Foxley-Marrable et al. (2018) and by making use of 144 standardisable lensed SNe Ia, Birrer et al. (2022) combined non-lensed SNe Ia with lensed SNe Ia via a Bayesian hierarchical framework to infer the constraints on H_0. The aim is to use the standardisable nature of SNe Ia to break the mass-sheet degeneracy, without using spatially resolved kinematics of the lens galaxy. Assuming optimistically that the time delays between SN image pairs can be measured with a precision of 2 days for all these systems and that the typical Einstein radius is $\sim 1''$, Birrer et al. (2022) found that H_0 can be measured with an uncertainty of 1.5%.

The various cosmological forecasts depend sensitively on assumptions of follow-up observations in order to classify the SN type, measure the time delays, and acquire high-resolution imaging and spectroscopy. As described in Sect. 2.1, these are necessary for measuring absolute distances to lensed SNe and using them as a cosmological probe. In Sect. 4.4, we discuss in more detail the follow-up requirements for lensed SNe.

3 Astrophysical Probe

Lensed SNe are powerful astrophysical tools to study supernovae and galaxies, in addition to probing cosmology through their time delays.

In Sect. 3.1, we explain how lensed SNe provide an excellent opportunity to constrain SN progenitors. In Sect. 3.2, we describe how lensing magnifications enable the acquisition of high-redshift SN spectra that are crucial for SN cosmology. In Sect. 3.3, we outline the use of microlensing to probe SN structure. In Sect. 3.4, we show lensed SNe as a probe of dust in the foreground lens galaxy.

3.1 SN Physics and Progenitors

Whilst it is accepted that a Type Ia SN is a thermonuclear detonation of a white dwarf (WD), the cause of the detonations remains uncertain even after decades of study (Maoz et al. 2014; Livio and Mazzali 2018).

The classic progenitor model (often referred to as the single degenerate channel) for a Type Ia SN is a close binary of a white dwarf and a post main sequence star (Whelan and Iben 1973). As the star expands, it overflows its Roche lobe and its outer layers are accreted onto the WD until the WD reaches the Chandrasekhar mass. At this point, electron degeneracy pressure can no longer support the WD, and gravitational collapse causes thermonuclear detonation of the WD. This model is attractive, since it naturally explains the standardisability of SNe Ia, with the progenitor always having the same mass and near identical composition. However the lack of evidence for surviving companions suggest that this channel cannot account for all SNe Ia (González Hernández et al. 2012; Lundqvist et al. 2015).

An alternative progenitor model is the double degenerate channel. Here two white dwarfs merge, exceeding the Chandrasekhar mass (Iben and Tutukov 1984; Webbink 1984). This mechanism leaves behind no companion star, but since each pair of white dwarfs will sum to a different mass, there is no obvious reason why this channel would produce a good standard candle. In addition to these two classic channels of single- and double-degenerate systems, there are other explosion mechanisms that have been explored, including sub-Chandrasekhar explosions (e.g., Sim et al. 2010), delayed detonations (e.g., Röpke et al. 2012), and double detonations of sub-Chandrasekhar WDs (e.g., Fink et al. 2007).

Early observations of SN light curves are critical in constraining the properties of SN progenitor systems (e.g. Kasen 2010, Piro et al. 2010, Rabinak and Waxman 2011, Nugent et al. 2011, Bloom et al. 2012, Goobar et al. 2015, Piro and Morozova 2016, Noebauer et al. 2017, Kochanek 2019, Fausnaugh et al. 2021, Yao et al. 2019, Miller et al. 2020, Bulla et al. 2020). If non-degenerate matter is close to the WD, then it should be shock heated by the explosion, producing excess high energy flux in the first few hours to days of the light curve. In addition to SNe Ia, early observations of core-collapse SNe are also helpful in constraining the properties, such as the sizes, of the progenitor stars (e.g., González-Gaitán et al. 2015; Chen et al. 2022). Even with the development of wide-field optical surveys, observing these earliest moments of SNe is heavily reliant on chance. Strong lensing gives us the opportunity to predict the precise reappearance time of a SN, such that very early data can be gathered (Suwa 2018; Suyu et al. 2020).

There are multiple challenges to this approach: early detection of the first SN image, possible image demagnification, distortions of SN light curves and spectra by microlensing, and time delay predictions of insufficient precision. We discuss each of these in turn.

Detection of a lensed SN based on only the appearance of the first image is more difficult, due to possible confusions between a lensed and an unlensed SN without the multiple SN images present. This confusion can be overcome when the lensed SN host galaxy is visible as lensed arc features in addition to the foreground lens galaxy.

Whilst lensed SNe will be discovered from highly magnified images, images occurring with significant time delay after the bright "discovery image" are typically demagnified in the case of two-image systems (Foxley-Marrable et al. 2020). The situation is better for four-image systems where some of the trailing SN images can be brighter than the discovery SN image. We expect $\sim 2/3$ of the lensed SNe to be two-image systems, and $\sim 1/3$ to be four-image systems (Oguri and Marshall 2010).

As seen in Fig. 4 and shown in Huber et al. (2019), microlensing distorts the SN light curves and spectra, potentially garbling the information on progenitors. Encouragingly, Suyu et al. (2020) have investigated the microlensing effects on four different explosion models of SNe Ia and demonstrated that spectral distortions due to microlensing are $\lesssim 1\%$ at the 1σ level within 10 rest-frame days after explosions. Therefore, microlensing is expected to have negligible impact on early-phase spectra (within $\sim$10 rest-frame days) for deciphering SN progenitors.

Finally, time delays cannot be precisely estimated from lens modelling alone. A 5-10 percent precision is typical from the best lens models (e.g., Shajib et al. 2019), and the uncertain value of the Hubble constant given the discrepant H_0 measurements (e.g., Verde et al. 2019; Di Valentino et al. 2021) adds several more percent of uncertainty in the predicted time delays. For systems with delays $\gtrsim 20$ days that are sufficiently long to catch the trailing SN images from their beginnings (given the time it takes to detect the first SN image and to arrange follow-up observations), a 10% uncertainty would translate to $\gtrsim 2$ days. Given this uncertainty, the predicted delay can easily miss the observations within ~ 2 days after explosion, the crucially early moments for constraining SN progenitors.

To overcome these challenges, deep and high-cadence (ideally daily) monitorings with imagers after the detection of the first appearing SN image would greatly help. Such a monitoring would acquire the early-phase light curves of trailing SN images, and spectroscopic observations of the early phases can be triggered as soon as a trailing SN image appears in the monitoring. Early-phase spectroscopic observations especially in the rest-frame UV are unprecedented and important for constraining SN progenitors (Suyu et al. 2020). The time delays of lensed SNe provide a unique and exciting avenue to acquire such spectroscopic observations for studying SN progenitors.

3.2 High-Z SN Spectra Through Gravitational Telescopes

The precision of SNe Ia as distance indicators, and thus their use to accurately constrain the nature of dark energy, is ultimately limited by our understanding of progenitor systems, e.g., the potential evolution of the SN properties, especially their absolute magnitude, over cosmic time. Thus, confirming the standard-candle nature through spectroscopic comparisons

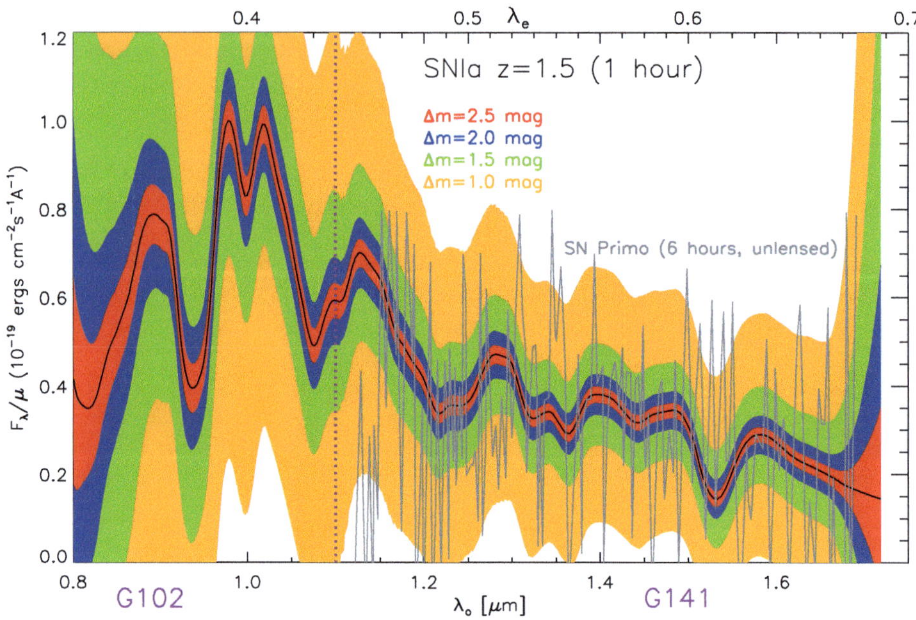

Fig. 10 Expected signal for HST observations through grisms G102 and G141 of a SN Ia at $z = 1.5$ for different magnifications provided by lensing. The shaded regions in [orange, green, blue, red] indicate the 1σ uncertainty per pixel in one-hour-long exposures for $\Delta m = [1, 1.5, 2.0, 2.5]$ mag of magnification, respectively. The observed spectrum of "SN Primo" (Rodney et al. 2012, six hours exposure time) is also shown for comparison

between SNe near and far is essential. With current instruments, detailed comparisons beyond $z > 1$ are not possible, due to the low SNR. This will remain challenging in the JWST era. However, multiple spectra of SNe at high redshift are required to check if the progenitor population evolves with redshift. Hence, the magnification of the signal from gravitational lensing could become essential to constrain one of the biggest systematic uncertainties in SN Ia cosmology (Petrushevska et al. 2017; Johansson et al. 2021). Figure 10 shows the noisy HST spectrum, based on six hours of observations of SN "Primo" at $z = 1.55$ observed by Rodney et al. (2012) compared with the expected signal if observed for shorter time through gravitational telescopes of different strengths, $\Delta m = 1, 1.5, 2.0$ and 2.5 mag, where Δm is the difference in magnitude coming from lensing magnification. These magnifications are expected to be typical for lensed SNe (Goldstein et al. 2019).

3.3 SN Structure with Microlensing

When a background source crosses a caustic network produced by microlenses, this allows one to estimate the physical dimensions of the light emitting surface, most robustly its half-light radius (Mortonson et al. 2005; Vernardos and Tsagkatakis 2019). For the case of SNe, their expanding ejecta lead to an increasing size that is directly linked to the expansion velocity. Because the (unobservable) distance from a caustic/high magnification region on the source plane is inversely proportional to changes in brightness due to microlensing, deformed SN light curves, like the ones shown in Fig. 4, can be used to constrain the size evolution, and subsequently the expansion velocity. In combination with the measurements of the photospheric velocity of the SN ejecta from spectroscopic observations, the SN size evolution can provide information on SN explosion models. Furthermore, microlensing is a potential way to probe asymmetries in the SN photosphere, since each of the multiple SN images will have different parts of its photosphere being microlensed.

We illustrate the size measurement via microlensing by the following toy model. Let us assume that the brightness of a lensed SN image at any given time t and wavelength λ (ignoring time delays and macromagnification for simplicity) is:

$$I(t,\lambda) = \int \mathrm{d}x \int \mathrm{d}y\, S(x,y;t,\lambda)\, \mu(x,y), \tag{6}$$

where S is the two dimensional intrinsic SN brightness profile, μ is the microlensing magnification on the source plane, and the integrals are performed over the extent of the source brightness profile. Assuming, for illustrative purposes, that the intrinsic surface brightness profile (S) is a circle of radius $R(t)$ with a known, constant surface brightness $i(t,\lambda)$, we get:

$$I(t,\lambda) = i(t,\lambda) \int_0^{2\pi} \int_0^R \mu(r,\theta)\, r\, \mathrm{d}r\, \mathrm{d}\theta = i(t,\lambda) F(R) \tag{7}$$

in polar coordinates (r,θ), where the value of the integral can be written as a radius and time-dependent factor, $F(R)$, and whose probability distribution can be calculated numerically from magnification maps. In fact, we can drop the assumption on a uniform profile and perform the convolution in Eq. (6) for different profile shapes. Therefore, if we know $S(x,y;t,\lambda)$ from a standardised SN type and calculate F, then we can measure $R(t)$. If the SN brightness profile shape varies over time, ratios of $I(t,\lambda)$ can be considered for separations in time Δt small enough so that the profile does not change by much. Finally, we note that Dobler and Keeton (2006) have used a similar approach, the time-weighted light curve derivative, albeit to measure the stellar mass fraction of the lensing galaxy.

In an analogous way, the relative size, or more specifically the half-light radius, of the SN ejecta at different wavelengths can be estimated from Eq. (6). Actually, the shape of the SN profile and/or the expansion velocity may vary in different wavelengths (see Figs. 11 and 12; Huber et al. 2019; Bayer et al. 2021), leading to different half-light radii (i.e., $R = R(t,\lambda)$ instead of $R(t)$). In the context of our toy model, the brightness ratio between two wavelengths at a given time t can be used to constrain the relative half-light radii:

$$\frac{I(t,\lambda_1)}{I(t,\lambda_2)} = \frac{i(t,\lambda_1)}{i(t,\lambda_2)} \frac{F(R_1)}{F(R_2)}. \tag{8}$$

Such measurements of radii across wavelengths and time not only provide a check on whether the expansion is homologous (when comparing the radii to the measured velocities from spectra), but also information on the explosion models that alter the dependence of radius on wavelength. It remains to be seen whether microlensing constraints on explosion models are competitive to the existing constraints based on SN spectral evolution modelling and analysis.

3.4 Dust Properties in Distant Galaxies

The different sightlines of multipe images offer unique opportunities to explore the properties of the interstellar medium of the deflecting galaxy, as was first shown for strongly

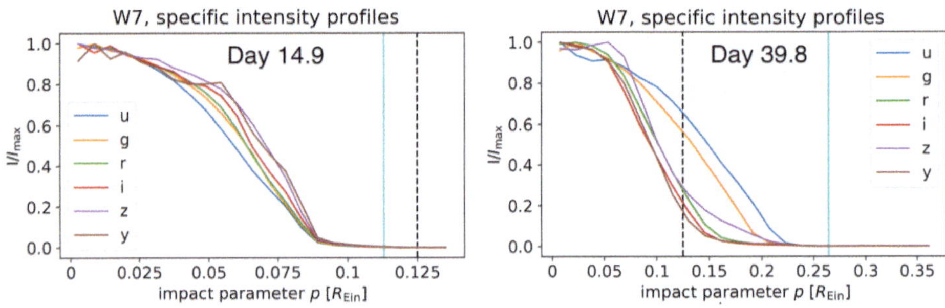

Fig. 11 Specific intensity profile of the W7 explosion model of SNe Ia. Left: radial intensity profiles in different filters 14.9 rest-frame days after explosion, in units of the Einstein radius of the microlenses ($R_{\mathrm{Ein}} = 2.2 \times 10^{16}$ cm for this specific case). The vertical solid cyan lines indicate the radius that encloses 99.9% of the total projected specific intensity. The vertical black dashed lines are random locations of caustics – effects of microlensing are strong when the specific intensity of the SN crosses a caustic. Right panel: same as the left panel but for a SN Ia 39.8 rest-frame days after explosion. Figure extracted and modified from Huber et al. (2019)

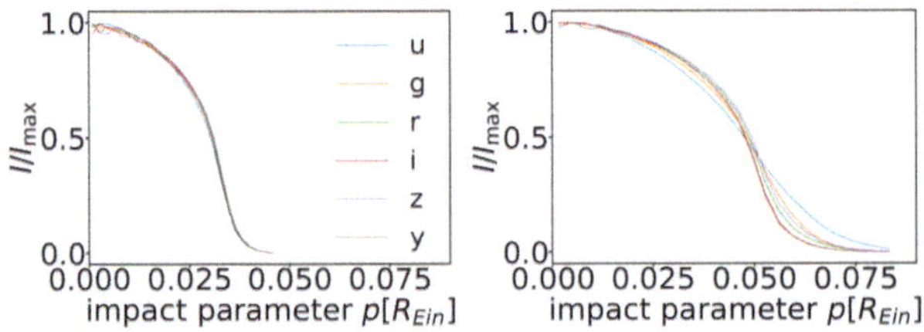

Fig. 12 Specific intensity profiles from TARDIS (Kerzendorf and Sim 2014; Vogl et al. 2019) modelling of SN 1999em, a Type IIP SN (Vogl et al. 2020) projected on the plane of the sky (Bayer et al. 2021). Left: radial intensity profiles in different filters 11 rest-frame days after explosion, in units of the Einstein radius of the microlenses ($R_{\mathrm{Ein}} = 2.9 \times 10^{16}$ cm for this specific case). Right panel: same as the left panel but for 27 rest-frame days after explosion. Figure taken from Bayer et al. (2021)

lensed QSOs by Falco et al. (1999) and in subsequent studies (e.g., Elíasdóttir et al. 2006; Østman et al. 2008; Hjorth et al. 2013). This is important as accurate measurements of e.g., the total-to-selective extinction parameter $R_V \equiv A_{\mathrm{V}}/E(B-V)$, a key number for many areas of astronomy, are very hard to establish outside the Milky Way and the Magellanic Clouds. The *extinction* in the V-band, A_{V}, is defined as

$$A_{\mathrm{V}} := m_{\mathrm{V}} - m_{\mathrm{V},0}, \tag{9}$$

where m_{V} is the apparent magnitude with dust extinction and $m_{\mathrm{V},0}$ is the intrinsic magnitude without dust extinction. The extinction for any other wavelength band is defined in a similar way. The *color excess* (or reddening) between the B and V band is

$$E(B-V) := A_{\mathrm{B}} - A_{\mathrm{V}} \tag{10}$$

$$= (m_{\mathrm{B}} - m_{\mathrm{V}}) - (m_{\mathrm{B},0} - m_{\mathrm{V},0}). \tag{11}$$

In the case of no dust extinction, then $A_{\mathrm{V}} = 0$, $A_{\mathrm{B}} = 0$ and $E(B-V) = 0$. For a fixed R_{V} value, higher A_{V} values (or higher $E(B-V)$ values) correspond to more dust along the sightline. The observed colors in multiple bands of the images are used to infer the

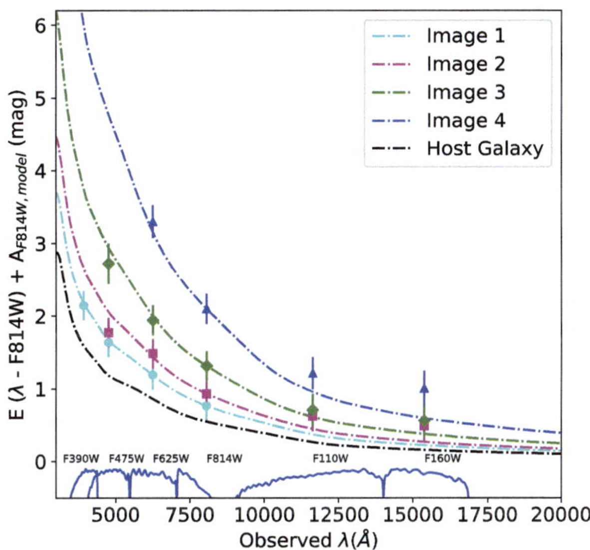

Fig. 13 The observed color excess for the resolved images from HST of iPTF16geu as a function of wavelength. The absorption from the host galaxy dust grains is plotted in black. For Image 1 we can see that the host galaxy is the dominant source of extinction, and for images 2, 3, 4 there is a progressively larger contribution from the dust in the lens galaxy, with correspondingly higher values of the color excess. Figure taken from Dhawan et al. (2020)

differential reddening as the various images travel through different regions of the lensing galaxy (i.e., infer the differences in A_V between the multiple SN images). Multi-wavelength observations of the lens system will also allow one to probe variations in R_V for different sightlines traversed by the multiple SN images (our own Milky Way shows variations in R_V; Fitzpatrick and Massa 2007; Schlafly et al. 2016). Corrections for differential extinctions in systems of strongly lensed SNe Ia is crucial to also make use of their standard candle nature to infer the lensing magnification.

Exploiting the well-known color evolution of SNe Ia, Dhawan et al. (2020) were able to infer both the extinction in the lensing galaxy for iPTF16geu, along with the common color excess from attenuation in the host galaxy using multi-band images from HST, as shown in Fig. 13. For iPTF16geu, with an impact parameter of about 1 kpc, two of the images (3 and 4) suffered significant extinction and the total magnification inferred from spatially unresolved images was significantly underestimated. Intriguingly, the best fit values of R_V for the lensing galaxy at $z = 0.216$ were much lower than what is observed for studies of stars in the Milky-Way (Schlafly et al. 2017), but similar to what is found in studies of well-measured, highly-reddened SNe Ia (Amanullah et al. 2015).

4 Searches and Rates

4.1 Methods

There are multiple ways to find lensed supernovae by making use of their image morphology, magnification, image multiplicity and/or time evolution. We briefly describe several approaches in this subsection.

4.1.1 Search Through Known Lensed Galaxies

A straightforward approach to find lensed SNe is to monitor known lensed galaxies and wait for a SN to explode in one of the lensed galaxies (Shu et al. 2018a). Using the sample of 128 galaxy-scale strong-lens systems from the Sloan Lens ACS Survey (SLACS; Bolton et al. 2006, 2008), the SLACS for the Masses Survey (S4TM; Shu et al. 2017), and the Baryon Oscillation Spectroscopic Survey Emission-Line Lens Survey (BELLS; Brownstein et al. 2012; Bolton et al. 2012), Shu et al. (2021, 2018a) estimated that the rates of strongly lensed SNe Ia and core-collapse SNe are 0.34 ± 0.03 and 2.8 ± 0.2 events per year, respectively.

One can either monitor known lens systems (especially ones with the highest star-formation and SN rates) through dedicated observing programs (Craig et al. 2021), or through wide-field imaging surveys such as the ongoing ZTF (Bellm et al. 2019). In the latter case, lens systems and lens candidates can be cross matched to transient alerts from ZTF through various brokers such as AMPEL (Nordin et al. 2019), ANTARES (Saha et al. 2014; Narayan et al. 2018; Lee et al. 2020; Matheson et al. 2021) and Lasair (Smith et al. 2019). Multi-epoch images of lens candidates can also be used to find lensed SNe based on temporal and spatial information of transients occurring near the lens candidates (Sheu et al. 2023). There are now thousands of confirmed and candidate lens systems from a wide range of lens searches, notably through machine learning approaches in recent years (e.g., Bolton et al. 2006; Limousin et al. 2009; Gavazzi et al. 2012; Brownstein et al. 2012; Vieira et al. 2013; Cañameras et al. 2015; Jacobs et al. 2017, 2019; Marshall et al. 2016; More et al. 2016; Sonnenfeld et al. 2018, 2020; Petrillo et al. 2019; Cañameras et al. 2020, 2021; Huang et al. 2020, 2021; Savary et al. 2022; Rojas et al. 2022; Shu et al. 2022; Tran et al. 2022). However, given the relatively shallow limiting depth of $r = 20.6\,\mathrm{mag}$ (5σ) of the ZTF survey (Bellm et al. 2019), individual lensed SN images associated with the lensed galaxies are likely below the detection threshold (Oguri and Marshall 2010; Wojtak et al. 2019). Nonetheless, the combination of the flux from all lensed SN images can be above the detection threshold, which motivates the next search approach through magnifications.

4.1.2 Search Through Lensing Magnification

While targeted searches for lensed SNe in known strongly lensed systems is straightforward, it is also constraining given the limited number of such systems. Since supernovae, especially SNe Ia, have a known narrow range of intrinsic brightness, it is feasible to identify particularly bright SNe as potential lensing candidates.

Quimby et al. (2014) proposed a technique to find unresolved lensed images of SNe Ia based on their high magnification and colors. Lensed SNe are expected to come from high redshifts where they will be in larger numbers owing to a larger volume and the lensing probability being higher for distant sources. The lensing magnification will make them appear brighter but their colors will largely be unaffected. Hence, in a color-magnitude diagram, the lensed SNe Ia will appear redder, at given magnitude, since they will be originating from higher redshifts compared to the unlensed population (see the blue circles above the black solid line in Fig. 14). Subluminous SNe, such as SN 1991bg, tend to be red as well (e.g., Miller et al. 2017) and would contaminate lensed SN candidates selected via this color-magnitude approach.

Another technique is based on having some knowledge of the redshifts of the galaxies involved, and it is interesting since it does not require multiple images to be spatially resolved. Thus it can in principle be used to detected arbitrarily compact systems, complementing other techniques, and allowing to probe the entire angular separation distribution of strongly lensed systems.

Figure 15 from Goobar et al. (2017) shows a practical realisation of detecting unresolved strongly lensed supernova: iPTF16geu was found to be a 30σ outlier when compared with other SNe Ia at the same redshift. Thus, even with the very modest $\sim 2''$ spatial resolution

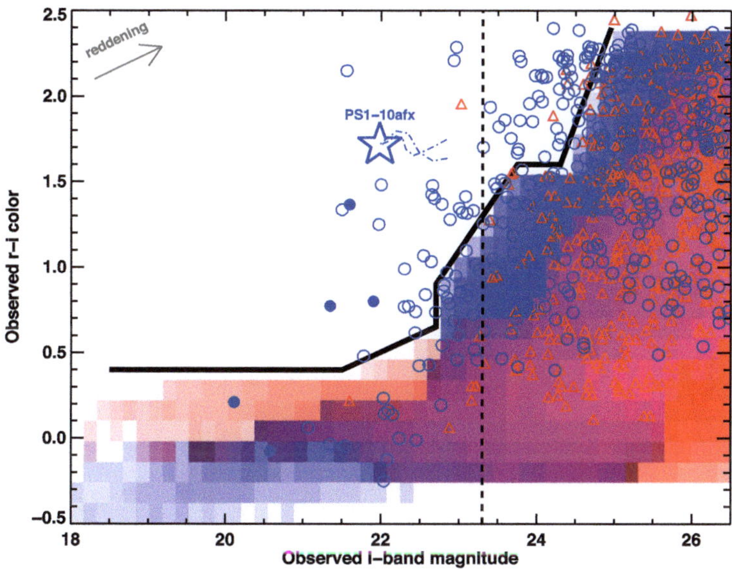

Fig. 14 The expected distribution of unlensed SNe population (SNe Ia in blue shaded, and core-collapse SNe in red shaded) is enclosed by thick black line showing the red limit. The lensed population predicted by Monte Carlo simulations is shown for SNe Ia (blue circles) and core-collapse SNe (red triangles). Vertical line marks the single epoch limit for LSST. Figure taken from Quimby et al. (2014)

at the 48-inch telescope at the Palomar Observatory, a strongly lensed system with $\theta_E = 0.3''$ could be identified. The difficulty with this approach is that spectroscopic observations are needed for classification of the transients, along with obtaining reliable redshifts of the galaxies, potentially a daunting task for large surveys with tens or even hundreds of new astrophysical transients found every night.

Goldstein and Nugent (2017) proposed a methodology to lower the rate of false positives: only considering transients spatially associated with elliptical galaxies, which make up about $\sim 80\%$ of the galaxy lenses and typically only host SNe Ia, including the sub-luminous ones. Furthermore, thanks to prominent 4000Å breaks, robust photometric redshifts can be computed for elliptical galaxies. Hence, if the transient appears to be too bright for being a SN Ia at the (photometric) redshift of the nearest galaxy, there is a good chance that it is a deflecting galaxy, as opposed to the host. Upon a secure identification, targeted high spatial-resolution imaging can be used to try to resolve the system.

4.1.3 Search Through Multiplicity of Images

For imaging surveys with faint limiting depth and good angular resolution such as the upcoming LSST, the multiple SN images of a lensed SN can be resolved and detected individually. Oguri and Marshall (2010) and Wojtak et al. (2019) investigated the number of lensed SNe detectable from such an approach. Specifically, Wojtak et al. (2019) imposed the following conditions for detecting a lensed SN via its image multiplicity: (1) the maximum image separation between the multiple SN images is between $0.5''$ and $4''$, where the lower limit is set by the expected seeing of LSST and the upper limit puts focus on galaxy-scale lens systems, (2) the flux ratio between the two images of a two-image (double) system is larger than 0.1, in order to have good contrast and clear identification of the images, and (3)

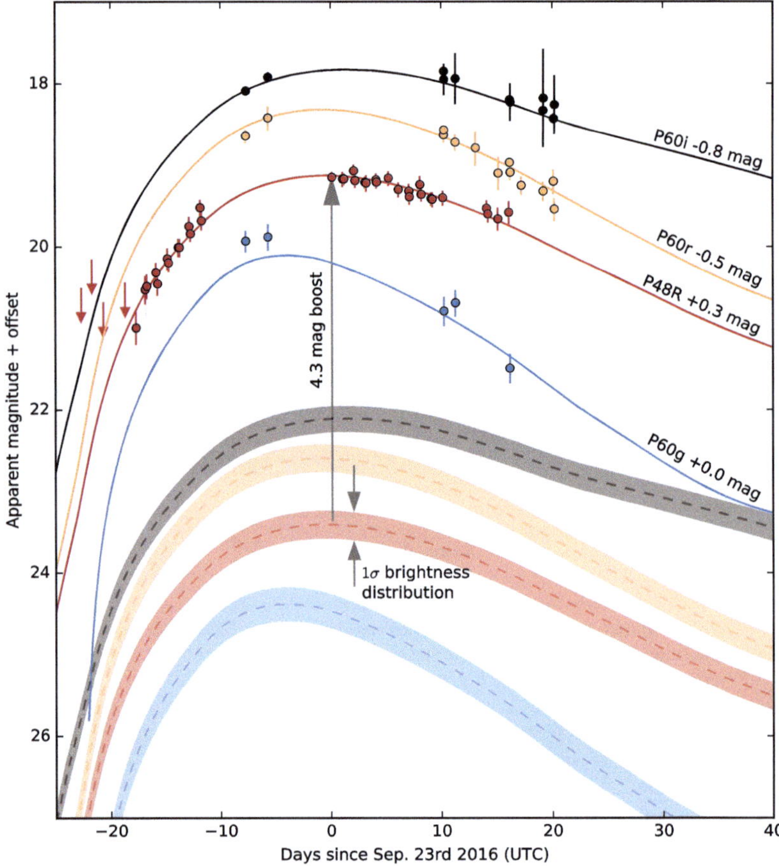

Fig. 15 Multicolor light curve of iPTF16geu showing that the unresolved supernova was 4.3 magnitudes (30 standard deviations) brighter than expected for its redshift. The magnitudes are measured with respect to time of maximum light in the R-band at P48 and in the g-, r-, and i-bands with the P60 telescope. The solid lines show the best-fitted SN Ia model to the data, while the dashed lines indicate the expected light curves at $z = 0.409$ (without lensing). The bands represent the standard deviation of the brightness distribution for SNe Ia. To fit the observed light curves, a brightness boost from gravitational lensing of 4.3 magnitudes is required. Figure from Goobar et al. (2017)

at least three of the four images in a quad system are detected, and both images of a double system are detected.

For surveys with limiting magnitude that are fainter than ∼23 mag in g, r or i-band, the detection via image multiplicity is expected to detect more lensed SN systems than the magnification approach. We illustrate this in Fig. 16 for lensed SNe Ia, and refer to Wojtak et al. (2019) for other types of lensed SNe that show similar trends. A combination of the two complementary approaches (magnification and multiplicity), as labelled by "hybrid" in Fig. 16, delivers more lens systems than the individual approaches on their own. In terms of cosmological applications, the multiplicity approach tends to detect lensed SN systems with longer time delays and larger image separations, which are more suitable for time-delay cosmography.

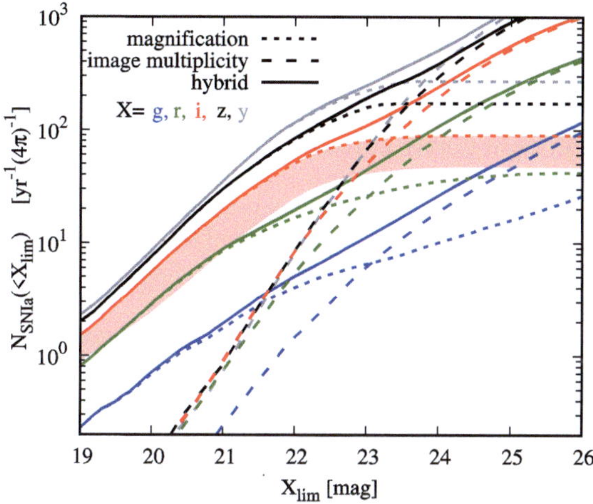

Fig. 16 The number of lensed SNe Ia expected per year over the whole sky as a function of the imaging survey limiting magnitude depth. The different colors correspond to detection in different filters (g, r, i, z or y). The dotted (dashed) lines show the expected number of lensed SNe Ia detected through the magnification (image multiplicity) approach. The hybrid approach, a combination of magnification and image multiplicity, is indicated by the solid curve. For shallow image surveys with limiting depth brighter than ~22 such as the ZTF survey, the magnification approach dominates in providing most of the expected lensed SNe. For deep image surveys with limiting depth fainter than ~23 such as the LSST survey, the image multiplicity approach start to dominate. Figure taken from Wojtak et al. (2019)

4.1.4 Search Through Spatio-Temporal Images

Most of the lensed SN searches in ZTF are based on the approaches of cross matches to lens systems/candidates (Sect. 4.1.1) or magnification (Sect. 4.1.2) since the angular resolution of ZTF (with $1''$ pixel sizes) does not resolve the multiple SN images of typical lens systems. However, in the advent of LSST, there will be high angular-resolution and deep images with at least hundreds of epochs per filter at each sky location. Building upon the image multiplicity method described in Sect. 4.1.3, Kodi Ramanah et al. (2022) proposed a machine-learning approach to classify an object based on images from cadenced surveys like the Young Supernova Experiment (YSE; Jones et al. 2021) and LSST. Machine learning techniques such as convolutional neural networks (CNN) perform well and are efficient at processing large amounts of imaging data for finding gravitational lenses (e.g., Jacobs et al. 2017, 2019; Lanusse et al. 2018; Metcalf et al. 2019; Petrillo et al. 2019; Cañameras et al. 2020, 2021). The spatio-temporal method of Kodi Ramanah et al. (2022) builds upon the success of the CNN and incorporate also time-domain information.

Rather than providing a neural network with the multi-band stacked (static) images of objects for classification, Kodi Ramanah et al. (2022) developed a network to take in a temporal series of images. The network architecture is a CNN that encodes long short-term memory (LSTM, a type of recurrent network; Sherstinsky 2020). The concept is that the temporal series of images will show the multiple SN images appearing at different epochs. Even if the multiple SN images are not well resolved, the centroids of the distribution of light would change as the multiple SN images brighten and dim at different times. Such change in the features of the object helps the neural network to distinguish lensed SN (with multiple SN images) from non-lensed SN (with single SN image). Using simulated images of YSE, Kodi Ramanah et al. (2022) demonstrated that the spatio-temporal network improves the classification accuracy by ~20% compared to networks that use static (e.g., single-epoch) images. The new development spatio-temporal network is very promising for application to the LSST.

4.1.5 Search Through Light Curves of Unresolved Multiple Images

For lens systems with small image separations between the multiple SN images or poor seeing conditions, the multiple images of the SN will be blended together. The light curve from such an "unresolved" lensed SN system will, therefore, be a superposition of the light curves of the individual SN images.

Bag et al. (2021) developed a new method that uses an orthogonal basis expansion around a trial light curve to fit an observed light curve in order to 1) determine whether the light curve is stemming from a lensed SN, and if so, 2) measure the time delays from the blended light curve. Therefore, the method is independent of any SN model assumptions. When tested on unresolved lensed SN Ia systems with two images, the method could accurately classify and measure the time delays from the blended light curves that have a time delay greater than 10 days. For systems with shorter delays, the method was unable to discern or infer the time delays robustly.

Denissenya and Linder (2022) have further expanded on this approach. They use a CNN (convolutional neural network), instead of the forward modeling approach, to determine the number of multiple (delayed) light curves that constitute a given observed light curve. A resulting value of one means that the light curve is not from a lensed SN, whereas a value of two or more indicates that the light curve stems from a lensed SN with two or more blended images. Using the CNN, Denissenya and Linder (2022) can accurately classify blended light curves coming from lensed SN Ia systems with time delays systematically longer than ~6 days and measure their delays with an uncertainty of ~1 day.

Both the studies of Bag et al. (2021) and Denissenya and Linder (2022) have not included microlensing effects, which are deferred to future work. Nonetheless, these first studies based on light curves show promise in finding lensed SNe with blended SN images.

4.2 Expected Number of Lensed SN Events

Strongly lensed transients are very rare. Hence, deep, wide-field time-domain optical and Near-IR surveys have the best chances to find samples of lensed SNe. Among these, the (optical) LSST survey at the Vera C. Rubin Observatory to see first light in ~2025, and the (NIR) Roman satellite, planned for launch some years later offer the best opportunities. In both cases, between several tens to hundreds of strongly lensed supernovae may be expected throughout the multi-year surveys (Goobar et al. 2002; Oguri and Marshall 2010; Quimby et al. 2014; Goldstein et al. 2019; Wojtak et al. 2019; Pierel et al. 2021). The cumulative number of SNe, Type Ia and core-collapse, expected to be found every year by LSST as a function of the detection threshold is shown in Fig. 17. A potential concern is the low cadence of the observations, which may require follow-up observations to measure time delays accurately (Huber et al. 2019).

4.3 Expected Lens Properties

Figure 18 shows simulations by Goldstein et al. (2019) of the distribution of system properties for strongly lensed SNe Ia in the LSST survey. The typical source redshift is $z_s \sim 0.9$ with time delays between 2 and 3 weeks. With a median image separation close to $1''$ and the expected good seeing conditions at the Rubin Observatory coupled with the plate-scale of the camera, most events would be spatially resolved.

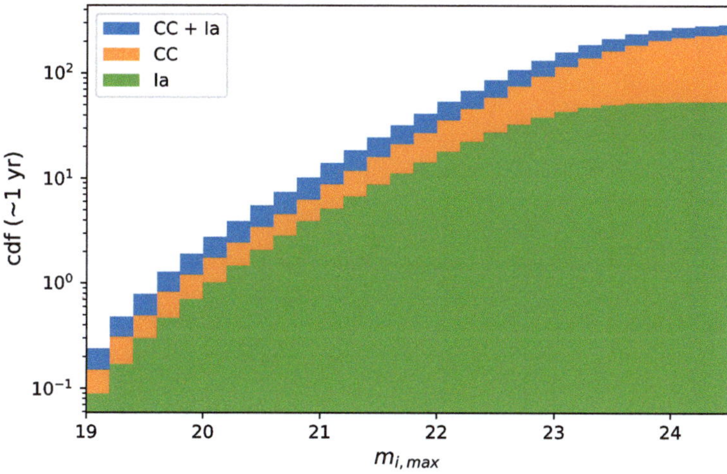

Fig. 17 Cumulative number of lensed Type Ia and core-collapse supernovae detections by the LSST survey per year as a function of their peak i-band magnitude, as computed in Goldstein et al. (2019). Figure credit: Ana Sagués Carracedo

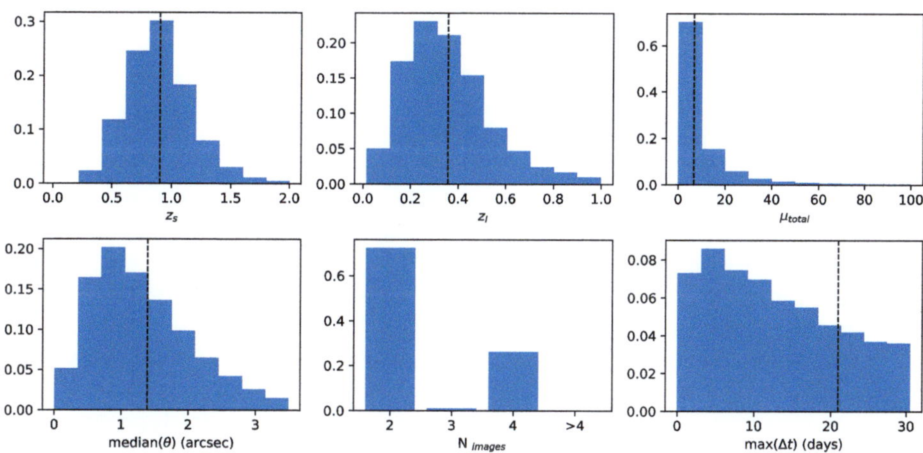

Fig. 18 Predicted system configurations for strongly lensed Type Ia supernovae in LSST as computed in Goldstein et al. (2019). Figure credit: Ana Sagués Carracedo

4.4 Requirements on Follow-up Observations

Full scientific exploitation of strongly lensed SNe depends on 1) early transient detection, ideally well before light curve peak; 2) rapid spectroscopic classification of the transient; 3) well-sampled, spatially resolved light curves; and 4) good wavelength coverage, necessary for accurate extinction corrections. From Fig. 17 one can see that the vast majority of the lensed SNe in forthcoming surveys will have (summed) peak magnitudes fainter than 21 mag. Hence, spectroscopic classification requires 8-m class telescopes, in most cases, potentially a major challenge. As argued in Sect. 3.4, an accurate measurement of the image magnifications requires a wide lever arm in wavelength coverage of the spatially resolved images, for at least a handful of epochs along the light curve. The new space telescope, JWST, would be ideally suited for follow-up. Its Near-IR sensitivity is an excellent match to the source redshifts shown in Fig. 18.

5 Summary

The first discoveries of strongly lensed SNe in recent years are opening a new window of exploration for cosmological and astrophysical studies. In this review, we have provided an overview of the analysis and results from these first lensed SN systems, and the future prospects for this exciting new field. The main takeaway points are as follows.

- There are currently six published strongly lensed SN systems with spatially resolved SN images. Four systems are lensed by galaxy clusters and two lensed by individual galaxies.
- Time delays between the multiple SN images of a lensed SN allow direct measurements of (1) the time-delay distance, (2) the angular-diameter distance to the lens in the case where stellar kinematic measurements of the lens galaxy are available, and (3) the luminosity distance in the case of a Type Ia SN that is a standardisable candle, as long as the SN magnifications due to microlensing and millilensing can be accurately accounted for. These distance measurements provide competitive constraints on cosmological parameters, particularly H_0.
- Several new methods to infer the time delays of lensed SNe have been developed in recent years, employing different kinds of data such as light curves, color curves or spectral evolution of SNe. Time-delay measurements with uncertainties of $\sim$1 day are achievable with real and mock data. The fractional uncertainty in the delays contribute directly to the fractional uncertainty on the time-delay distance and the lens angular-diameter distance.
- SNe Refsdal and H0pe are promising for delivering competitive H_0 measurements, and the first measurement from SN Refsdal with $\sim$6% uncertainty has been obtained. We expect that a modest sample of 20 lensed SNe Ia from the upcoming LSST to yield H_0 with nearly 1% precision in flat ΛCDM cosmology.
- The time delays of lensed SNe provide a unique opportunity to acquire very early-phase observations of SNe, especially in the rest-frame UV, and probe SN progenitors.
- Lensing magnifications allow the acquisition of spectra of SN above $z > 1$, that are crucial to reduce systematic uncertainties in SN Ia cosmology.
- Microlensing of SNe provides an avenue to constrain the sizes of the SN at different wavelengths.
- The dust properties in the (foreground lens) galaxies can be measured through the multiple SN sight lines in lensed SN systems.
- Various methods have been developed to search for lensed SNe, through e.g., the monitoring of known lens systems, magnification, image multiplicity, spatio-temporal evolution in the imaging, and light curves of unresolved SN images. Depending on the selection criteria, we expect at least dozens of lensed SNe to explode in the upcoming LSST, with their properties summarized in Sect. 4.3.

We are entering a new era of lensed SNe with the upcoming wide-field cadenced imaging surveys. Rapid follow-up observations including spectroscopic typing and light curve monitoring, will be crucial and necessary to make the best use of such events for cosmological and astrophysical studies.

Acknowledgements We thank Luke Weisenbach for producing Fig. 5, and Satadru Bag, Raoul Cañameras and Paul Schechter for discussions and feedback on the manuscript. We also thank the two anonymous reviewers for their constructive comments that improved our article. We are grateful to the International Space Science Institute (ISSI) in Bern for the hospitality and the stimulating workshop on "Strong Gravitational Lensing". SHS thanks the Max Planck Society for support through the Max Planck Research Group and the Max Planck Fellowship.

Funding Open Access funding enabled and organized by Projekt DEAL. This project has received funding from the European Research Council (ERC) under the European Union's Horizon 2020 research and innovation programme (LENSNOVA: grant agreement No 771776). This research is supported in part by the Excellence Cluster ORIGINS which is funded by the Deutsche Forschungsgemeinschaft (DFG, German Research Foundation) under Germany's Excellence Strategy – EXC-2094 – 390783311. AG acknowledges support from the Swedish National Space Board and *Vetenskapsrådet*, the Swedish Research Council. TEC is funded by a Royal Society University Research Fellowship. GV has received funding from the European Union's Horizon 2020 research and innovation programme under the Marie Sklodovska-Curie grant agreement No 897124. GV's research was made possible by the generosity of Eric and Wendy Schmidt by recommendation of the Schmidt Futures program.

Declarations

Competing Interests The authors declare that they have no competing interests. SHS is a guest editor of the Space Science Reviews topical collection "Strong Gravitational Lensing", which includes this review article.

Open Access This article is licensed under a Creative Commons Attribution 4.0 International License, which permits use, sharing, adaptation, distribution and reproduction in any medium or format, as long as you give appropriate credit to the original author(s) and the source, provide a link to the Creative Commons licence, and indicate if changes were made. The images or other third party material in this article are included in the article's Creative Commons licence, unless indicated otherwise in a credit line to the material. If material is not included in the article's Creative Commons licence and your intended use is not permitted by statutory regulation or exceeds the permitted use, you will need to obtain permission directly from the copyright holder. To view a copy of this licence, visit http://creativecommons.org/licenses/by/4.0/.

References

Amanullah R, Goobar A, Clément B, Cuby J-G, Dahle H, Dahlén T, Hjorth J, Fabbro S, Jönsson J, Kneib J-P, Lidman C, Limousin M, Milvang-Jensen B, Mörtsell E, Nordin J, Paech K, Richard J, Riehm T, Stanishev V, Watson D (2011) A highly magnified supernova at z = 1.703 behind the massive galaxy cluster A1689. Astrophys J Lett 742(1):7. https://doi.org/10.1088/2041-8205/742/1/L7

Amanullah R, Johansson J, Goobar A, Ferretti R, Papadogiannakis S, Petrushevska T, Brown PJ, Cao Y, Contreras C, Dahle H, Elias-Rosa N, Fynbo JPU, Gorosabel J, Guaita L, Hangard L, Howell DA, Hsiao EY, Kankare E, Kasliwal M, Leloudas G, Lundqvist P, Mattila S, Nugent P, Phillips MM, Sandberg A, Stanishev V, Sullivan M, Taddia F, Östlin G, Asadi S, Herrero-Illana R, Jensen JJ, Karhunen K, Lazarevic S, Varenius E, Santos P, Sridhar SS, Wallström SHJ, Wiegert J (2015) Diversity in extinction laws of type Ia supernovae measured between 0.2 and 2 μm. Mon Not R Astron Soc 453(3):3300–3328. https://doi.org/10.1093/mnras/stv1505

Auger MW, Treu T, Bolton AS, Gavazzi R, Koopmans LVE, Marshall PJ, Moustakas LA, Burles S (2010) The sloan lens ACS survey. X. Stellar, dynamical, and total mass correlations of massive early-type galaxies. Astrophys J 724(1):511–525. https://doi.org/10.1088/0004-637X/724/1/511

Bacon R, Accardo M, Adjali L, Anwand H, Bauer S, Biswas I, Blaizot J, Boudon D, Brau-Nogue S, Brinchmann J, Caillier P, Capoani L, Carollo CM, Contini T, Couderc P, Daguisé E, Deiries S, Delabre B, Dreizler S, Dubois J, Dupieux M, Dupuy C, Emsellem E, Fechner T, Fleischmann A, François M, Gallou G, Gharsa T, Glindemann A, Gojak D, Guiderdoni B, Hansali G, Hahn T, Jarno A, Kelz A, Koehler C, Kosmalski J, Laurent F, Le Floch M, Lilly SJ, Lizon J-L, Loupias M, Manescau A, Monstein C, Nicklas H, Olaya J-C, Pares L, Pasquini L, Pécontal-Rousset A, Pelló R, Petit C, Popow E, Reiss R, Remillieux A, Renault E, Roth M, Rupprecht G, Serre D, Schaye J, Soucail G, Steinmetz M, Streicher O, Stuik R, Valentin H, Vernet J, Weilbacher P, Wisotzki L, Yerle N (2010) The MUSE second-generation VLT instrument. In: McLean IS, Ramsay SK, Takami H (eds) Ground-based and airborne instrumentation for astronomy III. Society of photo-optical instrumentation engineers (SPIE) conference series, vol 7735, p 773508. https://doi.org/10.1117/12.856027

Bag S, Kim AG, Linder EV, Shafieloo A (2021) Be it unresolved: measuring time delays from lensed supernovae. Astrophys J 910(1):65. https://doi.org/10.3847/1538-4357/abe238

Bagherpour H, Branch D, Kantowski R (2006) Effects of gravitational microlensing on P Cygni profiles of type Ia supernovae. Astrophys J 638(2):946–950. https://doi.org/10.1086/498889

Barnabè M, Dutton AA, Marshall PJ, Auger MW, Brewer BJ, Treu T, Bolton AS, Koo DC, Koopmans LVE (2012) The SWELLS survey - IV. Precision measurements of the stellar and dark matter distributions in a spiral lens galaxy. Mon Not R Astron Soc 423(2):1073–1088. https://doi.org/10.1111/j.1365-2966.2012.20934.x

Bayer J, Huber S, Vogl C, Suyu SH, Taubenberger S, Sluse D, Chan JHH, Kerzendorf WE (2021) HOLISMOKES. V. Microlensing of type II supernovae and time-delay inference through spectroscopic phase retrieval. Astron Astrophys 653:A29. https://doi.org/10.1051/0004-6361/202040169. arXiv:2101.06229

Bellm EC, Kulkarni SR, Graham MJ, Dekany R, Smith RM, Riddle R, Masci FJ, Helou G, Prince TA, Adams SM, Barbarino C, Barlow T, Bauer J, Beck R, Belicki J, Biswas R, Blagorodnova N, Bodewits D, Bolin B, Brinnel V, Brooke T, Bue B, Bulla M, Burruss R, Cenko SB, Chang C-K, Connolly A, Coughlin M, Cromer J, Cunningham V, De K, Delacroix A, Desai V, Duev DA, Eadie G, Farnham TL, Feeney M, Feindt U, Flynn D, Franckowiak A, Frederick S, Fremling C, Gal-Yam A, Gezari S, Giomi M, Goldstein DA, Golkhou VZ, Goobar A, Groom S, Hacopians E, Hale D, Henning J, Ho AYQ, Hover D, Howell J, Hung T, Huppenkothen D, Imel D, Ip W-H, Ivezić Ž, Jackson E, Jones L, Juric M, Kasliwal MM, Kaspi S, Kaye S, Kelley MSP, Kowalski M, Kramer E, Kupfer T, Landry W, Laher RR, Lee C-D, Lin HW, Lin Z-Y, Lunnan R, Giomi M, Mahabal A, Mao P, Miller AA, Monkewitz S, Murphy P, Ngeow C-C, Nordin J, Nugent P, Ofek E, Patterson MT, Penprase B, Porter M, Rauch L, Rebbapragada U, Reiley D, Rigault M, Rodriguez H, van Roestel J, Rusholme B, van Santen J, Schulze S, Shupe DL, Singer LP, Soumagnac MT, Stein R, Surace J, Sollerman J, Szkody P, Taddia F, Terek S, Van Sistine A, van Velzen S, Vestrand WT, Walters R, Ward C, Ye Q-Z, Yu P-C, Yan L, Zolkower J (2019) The Zwicky Transient Facility: system overview, performance, and first results. Publ Astron Soc Pac 131(995):018002. https://doi.org/10.1088/1538-3873/aaecbe

Betoule M, Kessler R, Guy J, Mosher J, Hardin D, Biswas R, Astier P, El-Hage P, Konig M, Kuhlmann S, Marriner J, Pain R, Regnault N, Balland C, Bassett BA, Brown PJ, Campbell H, Carlberg RG, Cellier-Holzem F, Cinabro D, Conley A, D'Andrea CB, DePoy DL, doi M, Ellis RS, Fabbro S, Filippenko AV, Foley RJ, Frieman JA, Fouchez D, Galbany L, Goobar A, Gupta RR, Hill GJ, Hlozek R, Hogan CJ, Hook IM, Howell DA, Jha SW, Le Guillou L, Leloudas G, Lidman C, Marshall JL, Möller A, Mourão AM, Neveu J, Nichol R, Olmstead MD, Palanque-Delabrouille N, Perlmutter S, Prieto JL, Pritchet CJ, Richmond M, Riess AG, Ruhlmann-Kleider V, Sako M, Schahmaneche K, Schneider DP, Smith M, Sollerman J, Sullivan M, Walton NA, Wheeler CJ (2014) Improved cosmological constraints from a joint analysis of the SDSS-II and SNLS supernova samples. Astron Astrophys 568:22. https://doi.org/10.1051/0004-6361/201423413

Birrer S, Shajib AJ, Galan A, Millon M, Treu T, Agnello A, Auger M, Chen GC-F, Christensen L, Collett T, Courbin F, Fassnacht CD, Koopmans LVE, Marshall PJ, Park J-W, Rusu CE, Sluse D, Spiniello C, Suyu SH, Wagner-Carena S, Wong KC, Barnabè M, Bolton AS, Czoske O, Ding X, Frieman L, Van de Vyvere JA (2020) TDCOSMO. IV. Hierarchical time-delay cosmography – joint inference of the Hubble constant and galaxy density profiles. Astron Astrophys 643:165. https://doi.org/10.1051/0004-6361/202038861

Birrer S, Dhawan S, Shajib AJ (2022) The Hubble constant from strongly lensed supernovae with standardizable magnifications. Astrophys J 924:2. https://doi.org/10.3847/1538-4357/ac323a. arXiv:2107.12385

Birrer S, Millon M, Sluse D, Shajib AJ, Courbin F, Koopmans LVE, Suyu SH, Treu T (2024) Time-Delay Cosmography: Measuring the Hubble Constant and other cosmological parameters with strong gravitational lensing. Space Sci Rev 220. arXiv:2210.10833

Bloom JS, Kasen D, Shen KJ, Nugent PE, Butler NR, Graham ML, Howell DA, Kolb U, Holmes S, Haswell CA, Burwitz V, Rodriguez J, Sullivan M (2012) A compact degenerate primary-star progenitor of SN 2011fe. Astrophys J Lett 744(2):17. https://doi.org/10.1088/2041-8205/744/2/L17

Bolton AS, Burles S, Koopmans LVE, Treu T, Moustakas LA (2006) The Sloan Lens ACS survey. I. A large spectroscopically selected sample of massive early-type lens galaxies. Astrophys J 638(2):703–724. https://doi.org/10.1086/498884

Bolton AS, Burles S, Koopmans LVE, Treu T, Gavazzi R, Moustakas LA, Wayth R, Schlegel DJ (2008) The sloan lens ACS survey. V. The full ACS strong-lens sample. Astrophys J 682(2):964–984. https://doi.org/10.1086/589327

Bolton AS, Brownstein JR, Kochanek CS, Shu Y, Schlegel DJ, Eisenstein DJ, Wake DA, Connolly N, Maraston C, Arneson RA, Weaver BA (2012) The BOSS emission-line lens survey. II. Investigating mass-density profile evolution in the SLACS+BELLS strong gravitational lens sample. Astrophys J 757(1):82. https://doi.org/10.1088/0004-637X/757/1/82

Brownstein JR, Bolton AS, Schlegel DJ, Eisenstein DJ, Kochanek CS, Connolly N, Maraston C, Pandey P, Seitz S, Wake DA, Wood-Vasey WM, Brinkmann J, Schneider DP, Weaver BA (2012) The BOSS emission-line lens survey (BELLS). I. A large spectroscopically selected sample of lens galaxies at redshift ~0.5. Astrophys J 744(1):41. https://doi.org/10.1088/0004-637X/744/1/41

Bulla M, Miller AA, Yao Y, Dessart L, Dhawan S, Papadogiannakis S, Biswas R, Goobar A, Kulkarni SR, Nordin J, Nugent P, Polin A, Sollerman J, Bellm EC, Coughlin MW, Dekany R, Golkhou VZ, Graham MJ, Kasliwal MM, Kupfer T, Laher RR, Masci FJ, Porter M, Rusholme B, Shupe DL (2020) ZTF Early Observations of Type Ia Supernovae III: Early-Time Colors as a Test for Explosion Models and Multiple Populations. Astrophys J 902:48. https://doi.org/10.3847/1538-4357/abb13c. arXiv:2001.00587

Cañameras R, Nesvadba NPH, Guery D, McKenzie T, König S, Petitpas G, Dole H, Frye B, Flores-Cacho I, Montier L, Negrello M, Beelen A, Boone F, Dicken D, Lagache G, Le Floc'h E, Altieri B, Béthermin M, Chary R, de Zotti G, Giard M, Kneissl R, Krips M, Malhotra S, Martinache C, Omont A, Pointecouteau E, Puget J-L, Scott D, Soucail G, Valtchanov I, Welikala N, Yan L (2015) Planck's dusty GEMS: the brightest gravitationally lensed galaxies discovered with the Planck all-sky survey. Astron Astrophys 581:105. https://doi.org/10.1051/0004-6361/201425128

Cañameras R, Nesvadba NPH, Limousin M, Dole H, Kneissl R, Koenig S, Le Floc'h E, Petitpas G, Scott D (2018) Planck's dusty GEMS. V. Molecular wind and clump stability in a strongly lensed star-forming galaxy at z = 2.2. Astron Astrophys 620:60. https://doi.org/10.1051/0004-6361/201833679

Cañameras R, Schuldt S, Suyu SH, Taubenberger S, Meinhardt T, Leal-Taixé L, Lemon C, Rojas K, Savary E (2020) HOLISMOKES. II. Identifying galaxy-scale strong gravitational lenses in Pan-STARRS using convolutional neural networks. Astron Astrophys 644:163. https://doi.org/10.1051/0004-6361/202038219

Cañameras R, Schuldt S, Shu Y, Suyu SH, Taubenberger S, Meinhardt T, Leal-Taixé L, Chao DC-Y, Inoue KT, Jaelani AT, More A (2021) HOLISMOKES. VI. New galaxy-scale strong lens candidates from the HSC-SSP imaging survey. Astron Astrophys 653:L6. https://doi.org/10.1051/0004-6361/202141758. arXiv:2107.07829

Chen W, Kelly PL, Oguri M, Broadhurst TJ, Diego JM, Emami N, Filippenko AV, Treu TL, Zitrin A (2022) Shock cooling of a red-supergiant supernova at redshift 3 in lensed images. Nature 611(7935):256–259. https://doi.org/10.1038/s41586-022-05252-5

Chirivì G, Yıldırım A, Suyu SH, Halkola A (2020) Gravitational lensing and dynamics (GLaD): combined analysis to unveil properties of high-redshift galaxies. Astron Astrophys 643:135. https://doi.org/10.1051/0004-6361/202037929

Chornock R, Berger E, Rest A, Milisavljevic D, Lunnan R, Foley RJ, Soderberg AM, Smartt SJ, Burgasser AJ, Challis P, Chomiuk L, Czekala I, Drout M, Fong W, Huber ME, Kirshner RP, Leibler C, McLeod B, Marion GH, Narayan G, Riess AG, Roth KC, Sanders NE, Scolnic D, Smith K, Stubbs CW, Tonry JL, Valenti S, Burgett WS, Chambers KC, Hodapp KW, Kaiser N, Kudritzki R-P, Magnier EA, Price PA (2013) PS1-10afx at z = 1.388: Pan-STARRS1 discovery of a new type of superluminous supernova. Astrophys J 767(2):162. https://doi.org/10.1088/0004-637X/767/2/162

Courbin F, Bonvin V, Buckley-Geer E, Fassnacht CD, Frieman J, Lin H, Marshall PJ, Suyu SH, Treu T, Anguita T, Motta V, Meylan G, Paic E, Tewes M, Agnello A, Chao DC-Y, Chijani M, Gilman D, Rojas K, Williams P, Hempel A, Kim S, Lachaume R, Rabus M, Abbott TMC, Allam S, Annis J, Banerji M, Bechtol K, Benoit-Lévy A, Brooks D, Burke DL, Carnero Rosell A, Carrasco Kind M, Carretero J, D'Andrea CB, da Costa LN, Davis C, DePoy DL, Desai S, Flaugher B, Fosalba P, García-Bellido J, Gaztanaga E, Goldstein DA, Gruen D, Gruendl RA, Gschwend J, Gutierrez G, Honscheid K, James DJ, Kuehn K, Kuhlmann S, Kuropatkin N, Lahav O, Lima M, Maia MAG, March M, Marshall JL, McMahon RG, Menanteau F, Miquel R, Nord B, Plazas AA, Sanchez E, Scarpine V, Schindler R, Schubnell M, Sevilla-Noarbe I, Smith M, Soares-Santos M, Sobreira F, Suchyta E, Tarle G, Tucker DL, Walker AR, Wester W (2018) COSMOGRAIL: the COSmological MOnitoring of GRAvItational lenses. XVI. Time delays for the quadruply imaged quasar DES J0408-5354 with high-cadence photometric monitoring. Astron Astrophys 609:71. https://doi.org/10.1051/0004-6361/201731461

Craig P, O'Connor K, Chakrabarti S, Rodney SA, Pierel JR, McCully C, Perez-Fournon I (2021) A targeted search for strongly lensed supernovae and expectations for targeted searches in the Rubin era. ArXiv e-prints, arXiv:2111.01680

Denissenya M, Linder EV (2022) Deep learning unresolved lensed light curves. Mon Not R Astron Soc 515(1):977–983. https://doi.org/10.1093/mnras/stac1726

Dhawan S, Johansson J, Goobar A, Amanullah R, Mörtsell E, Cenko SB, Cooray A, Fox O, Goldstein D, Kalender R, Kasliwal M, Kulkarni SR, Lee WH, Nayyeri H, Nugent P, Ofek E, Quimby R (2020) Magnification, dust, and time-delay constraints from the first resolved strongly lensed type Ia supernova iPTF16geu. Mon Not R Astron Soc 491(2):2639–2654. https://doi.org/10.1093/mnras/stz2965

Di Valentino E, Mena O, Pan S, Visinelli L, Yang W, Melchiorri A, Mota DF, Riess AG, Silk J (2021) In the realm of the Hubble tension-a review of solutions. Class Quantum Gravity 38(15):153001. https://doi.org/10.1088/1361-6382/ac086d

Diego JM, Bernstein G, Chen W, Goobar A, Johansson JP, Kelly PL, Mörtsell E, Nightingale JW (2022) Microlensing and the type Ia supernova iPTF16geu. Astron Astrophys 662:34. https://doi.org/10.1051/0004-6361/202143009

Ding X, Liao K, Birrer S, Shajib AJ, Treu T, Yang L (2021) Improved time-delay lens modelling and H_0 inference with transient sources. Mon Not R Astron Soc 504(4):5621–5628. https://doi.org/10.1093/mnras/stab1240

Dobler G, Keeton CR (2006) Microlensing of lensed supernovae. Astrophys J 653(2):1391–1399. https://doi.org/10.1086/508769

Einstein A (1936) Lens-like action of a star by the deviation of light in the gravitational field. Science 84(2188):506–507. https://doi.org/10.1126/science.84.2188.506

Elíasdóttir Á, Hjorth J, Toft S, Burud I, Paraficz D (2006) Extinction curves of lensing galaxies out to z = 1. Astrophys J Suppl Ser 166(2):443–469. https://doi.org/10.1086/507131

Falco EE, Gorenstein MV, Shapiro II (1985) On model-dependent bounds on H 0 from gravitational images: application to Q 0957+561 A, B. Astrophys J Lett 289:1–4. https://doi.org/10.1086/184422

Falco EE, Impey CD, Kochanek CS, Lehár J, McLeod BA, Rix H-W, Keeton CR, Muñoz JA, Peng CY (1999) Dust and extinction curves in galaxies with z>0: the interstellar medium of gravitational lens galaxies. Astrophys J 523(2):617–632. https://doi.org/10.1086/307758

Fassnacht CD, Xanthopoulos E, Koopmans LVE, Rusin D (2002) A determination of H_0 with the CLASS gravitational lens B1608+656. III. A significant improvement in the precision of the time delay measurements. Astrophys J 581(2):823–835. https://doi.org/10.1086/344368

Fausnaugh MM, Vallely PJ, Kochanek CS, Shappee BJ, Stanek KZ, Tucker MA, Ricker GR, Vanderspek R, Latham DW, Seager S, Winn JN, Jenkins JM, Daylan T, Doty JP, Furesz G, Levine AM, Morris R, Pal A, Sha L, Ting EB, Wohler B (2021) Early-time light curves of type Ia supernovae observed with TESS. Astrophys J 908:51. https://doi.org/10.3847/1538-4357/abcd42. arXiv:1904.02171

Fink M, Hillebrandt W, Röpke FK (2007) Double-detonation supernovae of sub-Chandrasekhar mass white dwarfs. Astron Astrophys 476(3):1133–1143. https://doi.org/10.1051/0004-6361:20078438

Fitzpatrick EL, Massa D (2007) An analysis of the shapes of interstellar extinction curves. V. The IR-through-UV curve morphology. Astrophys J 663(1):320–341. https://doi.org/10.1086/518158

Foxley-Marrable M, Collett TE, Vernardos G, Goldstein DA, Bacon D (2018) The impact of microlensing on the standardization of strongly lensed type Ia supernovae. Mon Not R Astron Soc 478(4):5081–5090. https://doi.org/10.1093/mnras/sty1346

Foxley-Marrable M, Collett TE, Frohmaier C, Goldstein DA, Kasen D, Swann E, Bacon D (2020) Observing the earliest moments of supernovae using strong gravitational lenses. Mon Not R Astron Soc 495(4):4622–4637. https://doi.org/10.1093/mnras/staa1289

Frye BL, Pascale M, Qin Y, Zitrin A, Diego J, Walth G, Yan H, Conselice CJ, Alpaslan M, Bauer A, Busoni L, Coe D, Cohen SH, Dole H, Donahue M, Georgiev I, Jansen RA, Limousin M, Livermore R, Norman D, Rabien S, Windhorst RA (2019) analysis of a massive lensing cluster in a Hubble space telescope census of submillimeter giant arcs selected using Planck/Herschel. Astrophys J 871(1):51. https://doi.org/10.3847/1538-4357/aaeff7

Frye BL, Pascale M, Pierel J, Chen W, Foo N, Leimbach R, Garuda N, Cohen S, Kamieneski P, Windhorst R, Koekemoer AM, Kelly P, Summers J, Engesser M, Liu D, Furtak L, Polletta M, Harrington K, Willner S, Diego JM, Jansen R, Coe D, Conselice CJ, Dai L, Dole H, D'Silva JCJ, Driver S, Grogin N, Marshall MA, Meena A, Nonino M, Ortiz Rafael I, Pirzkal N, Robotham A, Ryan RE, Strolger L, Tompkins S, Trussler J, Willmer C, Yan H, Yun MS, Zitrin A (2023) The JWST Discovery of the Triply-imaged Type Ia "Supernova H0pe" and Observations of the Galaxy Cluster PLCK G165.7+67.0. ArXiv e-prints, arXiv:2309.07326

Gavazzi R, Treu T, Marshall PJ, Brault F, Ruff A (2012) The SL2S galaxy-scale gravitational lens sample. I. The alignment of mass and light in massive early-type galaxies at z = 0.2-0.9. Astrophys J 761(2):170. https://doi.org/10.1088/0004-637X/761/2/170

Goldstein DA, Nugent PE (2017) How to find gravitationally lensed type Ia supernovae. Astrophys J Lett 834(1):5. https://doi.org/10.3847/2041-8213/834/1/L5

Goldstein DA, Nugent PE, Kasen DN, Collett TE (2018) Precise time delays from strongly gravitationally lensed type Ia supernovae with chromatically microlensed images. Astrophys J 855(1):22. https://doi.org/10.3847/1538-4357/aaa975

Goldstein DA, Nugent PE, Goobar A (2019) Rates and properties of supernovae strongly gravitationally lensed by elliptical galaxies in time-domain imaging surveys. Astrophys J Suppl Ser 243(1):6 https://doi.org/10.3847/1538-4365/ab1fe0

González Hernández JI, Ruiz-Lapuente P, Tabernero HM, Montes D, Canal R, Méndez J, Bedin LR (2012) No surviving evolved companions of the progenitor of SN 1006. Nature 489(7417):533–536. https://doi.org/10.1038/nature11447

González-Gaitán S, Tominaga N, Molina J, Galbany L, Bufano F, Anderson JP, Gutierrez C, Förster F, Pignata G, Bersten M, Howell DA, Sullivan M, Carlberg R, de Jaeger T, Hamuy M, Baklanov PV, Blinnikov SI (2015) The rise-time of type II supernovae. Mon Not R Astron Soc 451(2):2212–2229. https://doi.org/10.1093/mnras/stv1097

Goobar A, Mörtsell E, Amanullah R, Nugent P (2002) Cosmological parameters from lensed supernovae. Astron Astrophys 393:25–32. https://doi.org/10.1051/0004-6361:20020987

Goobar A, Paech K, Stanishev V, Amanullah R, Dahlén T, Jönsson J, Kneib JP, Lidman C, Limousin M, Mörtsell E, Nobili S, Richard J, Riehm T, von Strauss M (2009) Near-IR search for lensed supernovae behind galaxy clusters. II. First detection and future prospects. Astron Astrophys 507(1):71–83. https://doi.org/10.1051/0004-6361/200811254

Goobar A, Kromer M, Siverd R, Stassun KG, Pepper J, Amanullah R, Kasliwal M, Sollerman J, Taddia F (2015) Constraints on the origin of the first light from SN 2014J. Astrophys J 799(1):106. https://doi.org/10.1088/0004-637X/799/1/106

Goobar A, Amanullah R, Kulkarni SR, Nugent PE, Johansson J, Steidel C, Law D, Mörtsell E, Quimby R, Blagorodnova N, Brandeker A, Cao Y, Cooray A, Ferretti R, Fremling C, Hangard L, Kasliwal M, Kupfer T, Lunnan R, Masci F, Miller AA, Nayyeri H, Neill JD, Ofek EO, Papadogiannakis S, Petrushevska T, Ravi V, Sollerman J, Sullivan M, Taddia F, Walters R, Wilson D, Yan L, Yaron O (2017) IPTF16geu: a multiply imaged, gravitationally lensed type Ia supernova. Science 356(6335):291–295. https://doi.org/10.1126/science.aal2729

Goobar A, Johansson J, Schulze S, Arendse N, Sagués Carracedo A, Dhawan S, Mörtsell E, Fremling C, Yan L, Perley D, Sollerman J, Joseph R, Hinds K-R, Meynardie W, Andreoni I, Bellm E, Bloom J, Collett TE, Drake A, Graham M, Kasliwal M, Kulkarni S, Miller A, Neill JD, Nordin J, Pierel J, Richard J, Riddle R, Rigault M, Rusholme B, Sharma Y, Stein R, Stewart G, Townsend A, Vinko J, Wheeler JC, Wold A (2023) SN Zwicky: uncovering a population of gravitational lens galaxies with magnified "standard candles". Nat Astron 7:1098–1107. https://doi.org/10.1038/s41550-023-01981-3. arXiv:2211.00656

Grillo C, Karman W, Suyu SH, Rosati P, Balestra I, Mercurio A, Lombardi M, Treu T, Caminha GB, Halkola A, Rodney SA, Gavazzi R, Caputi KI (2016) The story of supernova "refsdal" told by muse. Astrophys J 822(2):78. https://doi.org/10.3847/0004-637X/822/2/78

Grillo C, Rosati P, Suyu SH, Balestra I, Caminha GB, Halkola A, Kelly PL, Lombardi M, Mercurio A, Rodney SA, Treu T (2018) Measuring the value of the Hubble constant "à la Refsdal". Astrophys J 860(2):94. https://doi.org/10.3847/1538-4357/aac2c9

Grillo C, Rosati P, Suyu SH, Caminha GB, Mercurio A, Halkola A (2020) On the accuracy of time-delay cosmography in the frontier fields cluster MACS J1149.5+2223 with supernova Refsdal. Astrophys J 898(1):87. https://doi.org/10.3847/1538-4357/ab9a4c

Gunnarsson C, Goobar A (2003) Massive galaxy clusters as gravitational telescopes for distant supernovae. Astron Astrophys 405:859–866. https://doi.org/10.1051/0004-6361:20030648

Harrington KC, Yun MS, Cybulski R, Wilson GW, Aretxaga I, Chavez M, De la Luz V, Erickson N, Ferrusca D, Gallup AD, Hughes DH, Montaña A, Narayanan G, Sánchez-Argüelles D, Schloerb FP, Souccar K, Terlevich E, Terlevich R, Zeballos M, Zavala JA (2016) Early science with the large millimeter telescope: observations of extremely luminous high-z sources identified by Planck. Mon Not R Astron Soc 458(4):4383–4399. https://doi.org/10.1093/mnras/stw614

Hjorth J, Vreeswijk PM, Gall C, Watson D (2013) On inferring extinction laws in $z \sim 6$ quasars as signatures of supernova dust. Astrophys J 768(2):173. https://doi.org/10.1088/0004-637X/768/2/173

Huang X, Storfer C, Ravi V, Pilon A, Domingo M, Schlegel DJ, Bailey S, Dey A, Gupta RR, Herrera D, Juneau S, Landriau M, Lang D, Meisner A, Moustakas J, Myers AD, Schlafly EF, Valdes F, Weaver BA, Yang J, Yèche C (2020) Finding strong gravitational lenses in the DESI DECam legacy survey. Astrophys J 894(1):78. https://doi.org/10.3847/1538-4357/ab7ffb

Huang X, Storfer C, Gu A, Ravi V, Pilon A, Sheu W, Venguswamy R, Banka S, Dey A, Landriau M, Lang D, Meisner A, Moustakas J, Myers AD, Sajith R, Schlafly EF, Schlegel DJ (2021) Discovering new strong gravitational lenses in the DESI legacy imaging surveys. Astrophys J 909(1):27. https://doi.org/10.3847/1538-4357/abd62b

Huber S, Suyu SH, Noebauer UM, Bonvin V, Rothchild D, Chan JHH, Awan H, Courbin F, Kromer M, Marshall P, Oguri M, Ribeiro T, LSST dark energy science collaboration (2019) Strongly lensed SNe Ia in the era of LSST: observing cadence for lens discoveries and time-delay measurements. Astron Astrophys 631:161. https://doi.org/10.1051/0004-6361/201935370

Huber S, Suyu SH, Ghoshdastidar D, Taubenberger S, Bonvin V, Chan JHH, Kromer M, Noebauer UM, Sim SA, Leal-Taixé L (2021a) HOLISMOKES. VII. Time-delay measurement of strongly lensed SNe Ia using machine learning. Astron Astrophys 658:A157. https://doi.org/10.1051/0004-6361/202141956. arXiv:2108.02789

Huber S, Suyu SH, Noebauer UM, Chan JHH, Kromer M, Sim SA, Sluse D, Taubenberger S (2021b) HOLISMOKES. III. Achromatic phase of strongly lensed type Ia supernovae. Astron Astrophys 646:110. https://doi.org/10.1051/0004-6361/202039218

Iben I Jr, Tutukov AV (1984) Supernovae of type I as end products of the evolution of binaries with components of moderate initial mass (M $\lesssim$ 9 $M_\odot$). Astrophys J Suppl Ser 54:335–372. https://doi.org/10.1086/190932

Jacobs C, Glazebrook K, Collett T, More A, McCarthy C (2017) Finding strong lenses in CFHTLS using convolutional neural networks. Mon Not R Astron Soc 471(1):167–181. https://doi.org/10.1093/mnras/stx1492

Jacobs C, Collett T, Glazebrook K, Buckley-Geer E, Diehl HT, Lin H, McCarthy C, Qin AK, Odden C, Caso Escudero M, Dial P, Yung VJ, Gaitsch S, Pellico A, Lindgren KA, Abbott TMC, Annis J, Avila S, Brooks D, Burke DL, Carnero Rosell A, Carrasco Kind M, Carretero J, da Costa LN, De Vicente J, Fosalba P, Frieman J, García-Bellido J, Gaztanaga E, Goldstein DA, Gruen D, Gruendl RA, Gschwend J, Hollowood DL, Honscheid K, Hoyle B, James DJ, Krause E, Kuropatkin N, Lahav O, Lima M, Maia MAG, Marshall JL, Miquel R, Plazas AA, Roodman A, Sanchez E, Scarpine V, Serrano S, Sevilla-Noarbe I, Smith M, Sobreira F, Suchyta E, Swanson MEC, Tarle G, Vikram V, Walker AR, Zhang Y, DES Collaboration (2019) An extended catalog of galaxy-galaxy strong gravitational lenses discovered in DES using convolutional neural networks. Astrophys J Suppl Ser 243(1):17. https://doi.org/10.3847/1538-4365/ab26b6

Jauzac M, Richard J, Limousin M, Knowles K, Mahler G, Smith GP, Kneib J-P, Jullo E, Natarajan P, Ebeling H, Atek H, Clément B, Eckert D, Egami E, Massey R, Rexroth M (2016) Hubble frontier fields: predictions for the return of SN Refsdal with the MUSE and GMOS spectrographs. Mon Not R Astron Soc 457(2):2029–2042. https://doi.org/10.1093/mnras/stw069

Jee I, Komatsu E, Suyu SH (2015) Measuring angular diameter distances of strong gravitational lenses. J Cosmol Astropart Phys 2015(11):033. https://doi.org/10.1088/1475-7516/2015/11/033

Jee I, Komatsu E, Suyu SH, Huterer D (2016) Time-delay cosmography: increased leverage with angular diameter distances. J Cosmol Astropart Phys 2016(4):031. https://doi.org/10.1088/1475-7516/2016/04/031

Jee I, Suyu SH, Komatsu E, Fassnacht CD, Hilbert S, Koopmans LVE (2019) A measurement of the Hubble constant from angular diameter distances to two gravitational lenses. Science 365(6458):1134–1138. https://doi.org/10.1126/science.aat7371

Johansson J, Goobar A, Price SH, Sagués Carracedo A, Della Bruna L, Nugent PE, Dhawan S, Mörtsell E, Papadogiannakis S, Amanullah R, Goldstein D, Cenko SB, De K, Dugas A, Kasliwal MM, Kulkarni SR, Lunnan R (2021) Spectroscopy of the first resolved strongly lensed type Ia supernova iPTF16geu. Mon Not R Astron Soc 502(1):510–520. https://doi.org/10.1093/mnras/staa3829

Jones DO, Foley RJ, Narayan G, Hjorth J, Huber ME, Aleo PD, Alexander KD, Angus CR, Auchettl K, Baldassare VF, Bruun SH, Chambers KC, Chatterjee D, Coppejans DL, Coulter DA, DeMarchi L, Dimitriadis G, Drout MR, Engel A, French KD, Gagliano A, Gall C, Hung T, Izzo L, Jacobson-Galán WV, Kilpatrick CD, Korhonen H, Margutti R, Raimundo SI, Ramirez-Ruiz E, Rest A, Rojas-Bravo C, Siebert MR, Smartt SJ, Smith KW, Terreran G, Wang Q, Wojtak R, Agnello A, Ansari Z, Arendse N, Baldeschi A, Blanchard PK, Brethauer D, Bright JS, Brown JS, de Boer TJL, Dodd SA, Fairlamb JR, Grillo C, Hajela A, Hede C, Kolborg AN, Law-Smith JAP, Lin C-C, Magnier EA, Malanchev K, Matthews D, Mockler B, Muthukrishna D, Pan Y-C, Pfister H, Ramanah DK, Rest S, Sarangi A, Schrøder SL, Stauffer C, Stroh MC, Taggart KL, Tinyanont S, Wainscoat RJ (2021) young supernova experiment, the young supernova experiment: survey goals, overview, and operations. Astrophys J 908(2):143. https://doi.org/10.3847/1538-4357/abd7f5

Kasen D (2010) Seeing the collision of a supernova with its companion star. Astrophys J 708(2):1025–1031. https://doi.org/10.1088/0004-637X/708/2/1025

Kawamata R, Oguri M, Ishigaki M, Shimasaku K, Ouchi M (2016) Precise strong lensing mass modeling of four Hubble frontier field clusters and a sample of magnified high-redshift galaxies. Astrophys J 819(2):114. https://doi.org/10.3847/0004-637X/819/2/114

Kelly PL, Rodney SA, Treu T, Foley RJ, Brammer G, Schmidt KB, Zitrin A, Sonnenfeld A, Strolger L-G, Graur O, Filippenko AV, Jha SW, Riess AG, Bradac M, Weiner BJ, Scolnic D, Malkan MA, von der Linden A, Trenti M, Hjorth J, Gavazzi R, Fontana A, Merten JC, McCully C, Jones T, Postman M, Dressler A, Patel B, Cenko SB, Graham ML, Tucker BE (2015) Multiple images of a highly magnified supernova formed by an early-type cluster galaxy lens. Science 347(6226):1123–1126. https://doi.org/10.1126/science.aaa3350

Kelly PL, Brammer G, Selsing J, Foley RJ, Hjorth J, Rodney SA, Christensen L, Strolger L-G, Filippenko AV, Treu T, Steidel CC, Strom A, Riess AG, Zitrin A, Schmidt KB, Bradač M, Jha SW, Graham ML, McCully C, Graur O, Weiner BJ, Silverman JM, Taddia F (2016b) SN Refsdal: classification as a luminous and blue SN 1987A-like type II supernova. Astrophys J 831(2):205. https://doi.org/10.3847/0004-637X/831/2/205

Kelly PL, Rodney SA, Treu T, Strolger L-G, Foley RJ, Jha SW, Selsing J, Brammer G, Bradač M, Cenko SB, Graur O, Filippenko AV, Hjorth J, McCully C, Molino A, Nonino M, Riess AG, Schmidt KB, Tucker B, von der Linden A, Weiner BJ, Zitrin A (2016a) Deja vu all over again: the reappearance of supernova Refsdal. Astrophys J Lett 819(1):8. https://doi.org/10.3847/2041-8205/819/1/L8

Kelly PL, Rodney S, Treu T, Birrer S, Bonvin V, Dessart L, Foley RJ, Filippenko AV, Gilman D, Jha S, Hjorth J, Mandel K, Millon M, Pierel J, Thorp S, Zitrin A, Broadhurst T, Chen W, Diego JM, Dressler A, Graur O, Jauzac M, Malkan MA, McCully C, Oguri M, Postman M, Schmidt KB, Sharon K, Tucker BE, von der Linden A, Wambsganss J (2023b) The magnificent five images of supernova Refsdal: time delay and magnification measurements. Astrophys J 948(2):93. https://doi.org/10.3847/1538-4357/ac4ccb

Kelly PL, Rodney S, Treu T, Oguri M, Chen W, Zitrin A, Birrer S, Bonvin V, Dessart L, Diego JM, Filippenko AV, Foley RJ, Gilman D, Hjorth J, Jauzac M, Mandel K, Millon M, Pierel J, Sharon K, Thorp S, Williams L, Broadhurst T, Dressler A, Graur O, Jha S, McCully C, Postman M, Schmidt KB, Tucker BE, von der Linden A (2023a) Constraints on the Hubble constant from supernova Refsdal's reappearance. Science 380(6649):1322. https://doi.org/10.1126/science.abh1322

Kerzendorf WE, Sim SA (2014) A spectral synthesis code for rapid modelling of supernovae. Mon Not R Astron Soc 440(1):387–404. https://doi.org/10.1093/mnras/stu055

Kochanek CS (2019) The physics of flash (supernova) spectroscopy. Mon Not R Astron Soc 483(3):3762–3772. https://doi.org/10.1093/mnras/sty3363

Kodi Ramanah D, Arendse N, Wojtak R (2022) AI-driven spatio-temporal engine for finding gravitationally lensed supernovae. Mon Not R Astron Soc 512(4):5404–5417. https://doi.org/10.1093/mnras/stac838. arXiv:2107.12399

Kolatt TS, Bartelmann M (1998) Gravitational lensing of type IA supernovae by galaxy clusters. Mon Not R Astron Soc 296(3):763–772. https://doi.org/10.1046/j.1365-8711.1998.01466.x

Kovner I, Paczynski B (1988) Supernovae in luminous arcs. Astrophys J Lett 335:9. https://doi.org/10.1086/185328

Lanusse F, Ma Q, Li N, Collett TE, Li C-L, Ravanbakhsh S, Mandelbaum R, Póczos B (2018) CMU DeepLens: deep learning for automatic image-based galaxy-galaxy strong lens finding. Mon Not R Astron Soc 473(3):3895–3906. https://doi.org/10.1093/mnras/stx1665

Lee C-H, Matheson T, Saha A, Narayan G, Soraisam M, Stubens C, Wolf N, Snodgrass R, Kececioglu J, Scheidegger C (2020) ANTARES: a gateway to ZTF and LSST alerts. In: Society of photo-optical instrumentation engineers (SPIE) conference series, vol 11449. https://doi.org/10.1117/12.2560698

Liao K, Biesiada M, Zhu Z-H (2022) Strongly lensed transient sources: a review. Chin Phys Lett 39(11):119801. https://doi.org/10.1088/0256-307X/39/11/119801

Limousin M, Cabanac R, Gavazzi R, Kneib J-P, Motta V, Richard J, Thanjavur K, Foex G, Pello R, Crampton D, Faure C, Fort B, Jullo E, Marshall P, Mellier Y, More A, Soucail G, Suyu S, Swinbank M, Sygnet J-F, Tu H, Valls-Gabaud D, Verdugo T, Willis J (2009) A new window of exploration in the mass spectrum: strong lensing by galaxy groups in the SL2S. Astron Astrophys 502(2):445–456. https://doi.org/10.1051/0004-6361/200811473

Livio M, Mazzali P (2018) On the progenitors of type Ia supernovae. Phys Rep 736:1–23. https://doi.org/10.1016/j.physrep.2018.02.002

LSST Science Collaboration, Abell PA, Allison J, Anderson SF, Andrew JR, Angel JRP, Armus L, Arnett D, Asztalos SJ, Axelrod TS, Bailey S, Ballantyne DR, Bankert JR, Barkhouse WA, Barr JD, Barrientos LF, Barth AJ, Bartlett JG, Becker AC, Becla J, Beers TC, Bernstein JP, Biswas R, Blanton MR, Bloom JS, Bochanski JJ, Boeshaar P, Borne KD, Bradac M, Brandt WN, Bridge CR, Brown ME, Brunner RJ, Bullock JS, Burgasser AJ, Burge JH, Burke DL, Cargile PA, Chandrasekharan S, Chartas G, Chesley SR, Chu Y-H, Cinabro D, Claire MW, Claver CF, Clowe D, Connolly AJ, Cook KH, Cooke J, Cooray A, Covey KR, Culliton CS, de Jong R, de Vries WH, Debattista VP, Delgado F, Dell'Antonio IP, Dhital S, Di Stefano R, Dickinson M, Dilday B, Djorgovski SG, Dobler G, Donalek C, Dubois-Felsmann G, Durech J, Eliasdottir A, Eracleous M, Eyer L, Falco EE, Fan X, Fassnacht CD, Ferguson HC, Fernandez YR, Fields BD, Finkbeiner D, Figueroa EE, Fox DB, Francke H, Frank JS, Frieman J, Fromenteau S, Furqan M, Galaz G, Gal-Yam A, Garnavich P, Gawiser E, Geary J, Gee P, Gibson RR, Gilmore K, Grace EA, Green RF, Gressler WJ, Grillmair CJ, Habib S, Haggerty JS, Hamuy M, Harris AW, Hawley SL, Heavens AF, Hebb L, Henry TJ, Hileman E, Hilton EJ, Hoadley K, Holberg JB, Holman MJ, Howell SB, Infante L, Ivezic Z, Jacoby SH, Jain B, Jedicke R, Jee MJ, Garrett Jernigan J, Jha SW, Johnston KV, Jones RL, Juric M, Kaasalainen S, Kafka M (2009) LSST Science Book, Version 2.0. ArXiv e-prints, arXiv:0912.0201

Lundqvist P, Nyholm A, Taddia F, Sollerman J, Johansson J, Kozma C, Lundqvist N, Fransson C, Garnavich PM, Kromer M, Shappee BJ, Goobar A (2015) No trace of a single-degenerate companion in late spectra of supernovae 2011fe and 2014J. Astron Astrophys 577:39. https://doi.org/10.1051/0004-6361/201525719

Maoz D, Mannucci F, Nelemans G (2014) Observational clues to the progenitors of type Ia supernovae. Annu Rev Astron Astrophys 52:107–170. https://doi.org/10.1146/annurev-astro-082812-141031

Marshall PJ, Verma A, More A, Davis CP, More S, Kapadia A, Parrish M, Snyder C, Wilcox J, Baeten E, Macmillan C, Cornen C, Baumer M, Simpson E, Lintott CJ, Miller D, Paget E, Simpson R, Smith AM, Küng R, Saha P, Collett TE (2016) SPACE WARPS - I. Crowdsourcing the discovery of gravitational lenses. Mon Not R Astron Soc 455(2):1171–1190. https://doi.org/10.1093/mnras/stv2009

Matheson T, Stubens C, Wolf N, Lee C-H, Narayan G, Saha A, Scott A, Soraisam M, Bolton AS, Hauger B, Silva DR, Kececioglu J, Scheidegger C, Snodgrass R, Aleo PD, Evans-Jacquez E, Singh N, Wang Z, Yang S, Zhao Z (2021) The ANTARES astronomical time-domain event broker. Astron J 161(3):107. https://doi.org/10.3847/1538-3881/abd703

Metcalf RB, Meneghetti M, Avestruz C, Bellagamba F, Bom CR, Bertin E, Cabanac R, Courbin F, Davies A, Decencière E, Flamary R, Gavazzi R, Geiger M, Hartley P, Huertas-Company M, Jackson N, Jacobs C, Jullo E, Kneib J-P, Koopmans LVE, Lanusse F, Li C-L, Ma Q, Makler M, Li N, Lightman M, Petrillo CE, Serjeant S, Schäfer C, Sonnenfeld A, Tagore A, Tortora C, Tuccillo D, Valentín MB, Velasco-Forero S, Verdoes Kleijn GA, Vernardos G (2019) The strong gravitational lens finding challenge. Astron Astrophys 625:119. https://doi.org/10.1051/0004-6361/201832797

Miller AA, Kasliwal MM, Cao Y, Adams SM, Goobar A, Knežević S, Laher RR, Lunnan R, Masci FJ, Nugent PE, Perley DA, Petrushevska T, Quimby RM, Rebbapragada UD, Sollerman J, Taddia F, Kulkarni SR (2017) Color me intrigued: the discovery of iPTF 16fnm, an SN 2002cx-like object. Astrophys J 848(1):59. https://doi.org/10.3847/1538-4357/aa8c7e

Miller AA, Yao Y, Bulla M, Pankow C, Bellm EC, Cenko SB, Dekany R, Fremling C, Graham MJ, Kupfer T, Laher RR, Mahabal AA, Masci FJ, Nugent PE, Riddle R, Rusholme B, Smith RM, Shupe DL, van Roestel J, Kulkarni SR (2020) ZTF Early Observations of Type Ia Supernovae II: First Light, the Initial Rise, and Time to Reach Maximum Brightness. Astrophys J 902:47. https://doi.org/10.3847/1538-4357/abb13b. arXiv:2001.00598

Millon M, Courbin F, Bonvin V, Buckley-Geer E, Fassnacht CD, Frieman J, Marshall PJ, Suyu SH, Treu T, Anguita T, Motta V, Agnello A, Chan JHH, Chao DC-Y, Chijani M, Gilman D, Gilmore K, Lemon C, Lucey JR, Melo A, Paic E, Rojas K, Sluse D, Williams PR, Hempel A, Kim S, Lachaume R, Rabus M (2020b) TDCOSMO. II. Six new time delays in lensed quasars from high-cadence monitoring at the MPIA 2.2 m telescope. Astron Astrophys 642:193. https://doi.org/10.1051/0004-6361/202038698

Millon M, Courbin F, Bonvin V, Paic E, Meylan G, Tewes M, Sluse D, Magain P, Chan JHH, Galan A, Joseph R, Lemon C, Tihhonova O, Anderson RI, Marmier M, Chazelas B, Lendl M, Triaud AHMJ, Wyttenbach A (2020a) COSMOGRAIL. XIX. Time delays in 18 strongly lensed quasars from 15 years of optical monitoring. Astron Astrophys 640:105. https://doi.org/10.1051/0004-6361/202037740

Millon M, Tewes M, Bonvin V, Lengen B, Courbin F (2020c) PyCS3: a python toolbox for time-delay measurements in lensed quasars. J Open Sour Softw 5(53):2654. https://doi.org/10.21105/joss.02654

More A, Verma A, Marshall PJ, More S, Baeten E, Wilcox J, Macmillan C, Cornen C, Kapadia A, Parrish M, Snyder C, Davis CP, Gavazzi R, Lintott CJ, Simpson R, Miller D, Smith AM, Paget E, Saha P, Küng R, Collett TE (2016) SPACE WARPS- II. New gravitational lens candidates from the CFHTLS discovered through citizen science. Mon Not R Astron Soc 455(2):1191–1210. https://doi.org/10.1093/mnras/stv1965

More A, Suyu SH, Oguri M, More S, Lee C-H (2017) Interpreting the strongly lensed supernova iPTF16geu: time delay predictions, microlensing, and lensing rates. Astrophys J Lett 835(2):25. https://doi.org/10.3847/2041-8213/835/2/L25

Mortonson MJ, Schechter PL, Wambsganss J (2005) Size is everything: universal features of quasar microlensing with extended sources. Astrophys J 628(2):594–603. https://doi.org/10.1086/431195

Mörtsell E, Johansson J, Dhawan S, Goobar A, Amanullah R, Goldstein DA (2020) Lens modelling of the strongly lensed type Ia supernova iPTF16geu. Mon Not R Astron Soc 496(3):3270–3280. https://doi.org/10.1093/mnras/staa1600

Narayan G, Zaidi T, Soraisam MD, Wang Z, Lochner M, Matheson T, Saha A, Yang S, Zhao Z, Kececioglu J, Scheidegger C, Snodgrass RT, Axelrod T, Jenness T, Maier RS, Ridgway ST, Seaman RL, Evans EM, Singh N, Taylor C, Toeniskoetter J, Welch E, Zhu S, ANTARES Collaboration (2018) Machine-learning-based brokers for real-time classification of the LSST alert stream. Astrophys J Suppl Ser 236(1):9. https://doi.org/10.3847/1538-4365/aab781

Newman AB, Belli S, Ellis RS, Patel SG (2018) Resolving quiescent galaxies at $z \gtrsim 2$. I. Search for gravitationally lensed sources and characterization of their structure, stellar populations, and line emission. Astrophys J 862(2):125. https://doi.org/10.3847/1538-4357/aacd4d

Noebauer UM, Kromer M, Taubenberger S, Baklanov P, Blinnikov S, Sorokina E, Hillebrandt W (2017) Early light curves for type Ia supernova explosion models. Mon Not R Astron Soc 472(3):2787–2799. https://doi.org/10.1093/mnras/stx2093

Nordin J, Brinnel V, van Santen J, Bulla M, Feindt U, Franckowiak A, Fremling C, Gal-Yam A, Giomi M, Kowalski M, Mahabal A, Miranda N, Rauch L, Reusch S, Rigault M, Schulze S, Sollerman J, Stein R,

Yaron O, van Velzen S, Ward C (2019) Transient processing and analysis using AMPEL: alert management, photometry, and evaluation of light curves. Astron Astrophys 631:147. https://doi.org/10.1051/0004-6361/201935634

Nugent P, Sullivan M, Ellis R, Gal-Yam A, Leonard DC, Howell DA, Astier P, Carlberg RG, Conley A, Fabbro S, Fouchez D, Neill JD, Pain R, Perrett K, Pritchet CJ, Regnault N (2006) Toward a cosmological Hubble diagram for type II-P supernovae. Astrophys J 645(2):841–850. https://doi.org/10.1086/504413

Nugent PE, Sullivan M, Cenko SB, Thomas RC, Kasen D, Howell DA, Bersier D, Bloom JS, Kulkarni SR, Kandrashoff MT, Filippenko AV, Silverman JM, Marcy GW, Howard AW, Isaacson HT, Maguire K, Suzuki N, Tarlton JE, Pan Y-C, Bildsten L, Fulton BJ, Parrent JT, Sand D, Podsiadlowski P, Bianco FB, Dilday B, Graham ML, Lyman J, James P, Kasliwal MM, Law NM, Quimby RM, Hook IM, Walker ES, Mazzali P, Pian E, Ofek EO, Gal-Yam A, Poznanski D (2011) Supernova SN 2011fe from an exploding carbon-oxygen white dwarf star. Nature 480(7377):344–347. https://doi.org/10.1038/nature10644

Oguri M (2019) Strong gravitational lensing of explosive transients. Rep Prog Phys 82(12):126901. https://doi.org/10.1088/1361-6633/ab4fc5

Oguri M, Marshall PJ (2010) Gravitationally lensed quasars and supernovae in future wide-field optical imaging surveys. Mon Not R Astron Soc 405(4):2579–2593. https://doi.org/10.1111/j.1365-2966.2010.16639.x

Østman L, Goobar A, Mörtsell E (2008) Extinction properties of lensing galaxies. Astron Astrophys 485(2):403–415. https://doi.org/10.1051/0004-6361:20079187

Paraficz D, Hjorth J (2009) Gravitational lenses as cosmic rulers: Ω_m, Ω_Λ from time delays and velocity dispersions. Astron Astrophys 507(3):49–52. https://doi.org/10.1051/0004-6361/200913307

Pascale M, Frye BL, Dai L, Foo N, Qin Y, Leimbach R, Bauer AM, Merlin E, Coe D, Diego J, Yan H, Zitrin A, Cohen SH, Conselice CJ, Dole H, Harrington K, Jansen RA, Kamieneski P, Windhorst RA, Yun MS (2022) Possible ongoing merger discovered by photometry and spectroscopy in the field of the galaxy cluster PLCK G1657+67 0. Astrophys J 932(2):85. https://doi.org/10.3847/1538-4357/ac6ce9

Perlmutter S, Aldering G, Goldhaber G, Knop RA, Nugent P, Castro PG, Deustua S, Fabbro S, Goobar A, Groom DE, Hook IM, Kim AG, Kim MY, Lee JC, Nunes NJ, Pain R, Pennypacker CR, Quimby R, Lidman C, Ellis RS, Irwin M, McMahon RG, Ruiz-Lapuente P, Walton N, Schaefer B, Boyle BJ, Filippenko AV, Matheson T, Fruchter AS, Panagia N, Newberg HJM, Couch WJ, Project TSC (1999) Measurements of Ω and Λ from 42 high-redshift supernovae. Astrophys J 517(2):565–586. https://doi.org/10.1086/307221

Petrillo CE, Tortora C, Vernardos G, Koopmans LVE, Verdoes Kleijn G, Bilicki M, Napolitano NR, Chatterjee S, Covone G, Dvornik A, Erben T, Getman F, Giblin B, Heymans C, de Jong JTA, Kuijken K, Schneider P, Shan H, Spiniello C, Wright AH (2019) LinKS: discovering galaxy-scale strong lenses in the kilo-degree survey using convolutional neural networks. Mon Not R Astron Soc 484(3):3879–3896. https://doi.org/10.1093/mnras/stz189

Petrushevska T, Amanullah R, Goobar A, Fabbro S, Johansson J, Kjellsson T, Lidman C, Paech K, Richard J, Dahle H, Ferretti R, Kneib JP, Limousin M, Nordin J, Stanishev V (2016) High-redshift supernova rates measured with the gravitational telescope A 1689. Astron Astrophys 594:54. https://doi.org/10.1051/0004-6361/201628925

Petrushevska T, Amanullah R, Bulla M, Kromer M, Ferretti R, Goobar A, Papadogiannakis S (2017) Testing for redshift evolution of type Ia supernovae using the strongly lensed PS1-10afx at z = 1.4. Astron Astrophys 603:136. https://doi.org/10.1051/0004-6361/201730989

Pierel JDR, Rodney S (2019) Turning gravitationally lensed supernovae into cosmological probes. Astrophys J 876(2):107. https://doi.org/10.3847/1538-4357/ab164a

Pierel JDR, Rodney S, Vernardos G, Oguri M, Kessler R, Anguita T (2021) Projected cosmological constraints from strongly lensed supernovae with the roman space telescope. Astrophys J 908(2):190. https://doi.org/10.3847/1538-4357/abd8d3

Pierel JDR, Arendse N, Ertl S, Huang X, Moustakas LA, Schuldt S, Shajib AJ, Shu Y, Birrer S, Bronikowski M, Hjorth J, Suyu SH, Agarwal S, Agnello A, Bolton AS, Chakrabarti S, Cold C, Courbin F, Della Costa JM, Dhawan S, Engesser M, Fox OD, Gall C, Gomez S, Goobar A, Jimenez C, Johansson J, Li G, Marques-Chaves R, Mao S, Mazzali PA, Perez-Fournon I, Petrushevska T, Poidevin F, Rest A, Sheu W, Shirley R, Silver E, Storfer C, Treu T, Wojtak R, Zenati Y (2023) LensWatch: I. Resolved HST Observations and Constraints on the Strongly-Lensed Type Ia Supernova 2022qmx ("SN Zwicky"). Astrophys J 948:115. https://doi.org/10.3847/1538-4357/acc7a6. arXiv:2211.03772

Piro AL, Morozova VS (2016) Exploring the potential diversity of early type ia supernova light curves. Astrophys J 826(1):96. https://doi.org/10.3847/0004-637X/826/1/96

Piro AL, Nakar E (2014) Constraints on shallow ^{56}Ni from the early light curves of type Ia supernovae. Astrophys J 784(1):85. https://doi.org/10.1088/0004-637X/784/1/85

Piro AL, Chang P, Weinberg NN (2010) Shock breakout from type Ia supernova. Astrophys J 708(1):598–604. https://doi.org/10.1088/0004-637X/708/1/598

Planck Collaboration, Aghanim N, Altieri B, Arnaud M, Ashdown M, Aumont J, Baccigalupi C, Banday AJ, Barreiro RB, Bartolo N, Battaner E, Beelen A, Benabed K, Benoit-Lévy A, Bernard J-P, Bersanelli M, Bethermin M, Bielewicz P, Bonavera L, Bond JR, Borrill J, Bouchet FR, Boulanger F, Burigana C, Calabrese E, Canameras R, Cardoso J-F, Catalano A, Chamballu A, Chary R-R, Chiang HC, Christensen PR, Clements DL, Colombi S, Couchot F, Crill BP, Curto A, Danese L, Dassas K, Davies RD, Davis RJ, de Bernardis P, de Rosa A, de Zotti G, Delabrouille J, Diego JM, Dole H, Donzelli S, Doré O, Douspis M, Ducout A, Dupac X, Efstathiou G, Elsner F, Enßlin TA, Falgarone E, Flores-Cacho I, Forni O, Frailis M, Fraisse AA, Franceschi E, Frejsel A, Frye B, Galeotta S, Galli S, Ganga K, Giard M, Gjerløw E, González-Nuevo J, Górski KM, Gregorio A, Gruppuso A, Guéry D, Hansen FK, Hanson D, Harrison DL, Helou G, Hernández-Monteagudo C, Hildebrandt SR, Hivon E, Hobson M, Holmes WA, Hovest W, Huffenberger KM, Hurier G, Jaffe AH, Jaffe TR, Keihänen E, Keskitalo R, Kisner TS, Kneissl R, Knoche J, Kunz M, Kurki-Suonio H, Lagache G, Lamarre J-M, Lasenby A, Lattanzi M, Lawrence CR, Le Floc'h E, Leonardi R, Levrier F, Liguori M, Lilje PB, Linden-Vørnle M, López-Caniego M, Lubin PM, Macías-Pérez JF, MacKenzie T, Maffei B, Mandolesi N, Maris M, Martin PG, Martinache C, Martí nez-González E, Masi S, Matarrese S, Mazzotta P, Melchiorri A, Mennella A, Migliaccio M, Moneti A, Montier L, Morgante G, Mortlock D, Munshi D, Murphy JA, Natoli P, Negrello M, Nesvadba NPH, Novikov D, Novikov I, Omont A, Pagano L, Pajot F, Pasian F, Perdereau O, Perotto L, Perrotta F, Pettorino V, Piacentini F, Piat M, Plaszczynski S, Pointecouteau E, Polenta G, Popa L, Pratt GW, Prunet S, Puget J-L, Rachen JP, Reach WT, Reinecke M, Remazeilles M, Renault C, Ristorcelli I, Rocha G, Roudier G, Rusholme B, Sandri M, Santos D, Savini G, Scott D, Spencer LD, Stolyarov V, Sunyaev R, Sutton D, Sygnet J-F, Tauber JA, Terenzi L, Toffolatti L, Tomasi M, Tristram M, Tucci M, Umana G, Valenziano L, Valiviita J, Valtchanov I, Van Tent B, Vieira JD, Vielva P, Wade LA, Wandelt BD, Wehus IK, Welikala N, Zacchei A, Zonca A (2015) Planck intermediate results. XXVII. High-redshift infrared galaxy overdensity candidates and lensed sources discovered by Planck and confirmed by Herschel-SPIRE. Astron Astrophys 582:30. https://doi.org/10.1051/0004-6361/201424790

Planck Collaboration, Aghanim N, Akrami Y, Ashdown M, Aumont J, Baccigalupi C, Ballardini M, Banday AJ, Barreiro RB, Bartolo N, Basak S, Battye R, Benabed K, Bernard J-P, Bersanelli M, Bielewicz P, Bock JJ, Bond JR, Borrill J, Bouchet FR, Boulanger F, Bucher M, Burigana C, Butler RC, Calabrese E, Cardoso J-F, Carron J, Challinor A, Chiang HC, Chluba J, Colombo LPL, Combet C, Contreras D, Crill BP, Cuttaia F, de Bernardis P, de Zotti G, Delabrouille J, Delouis J-M, Di Valentino E, Diego JM, Doré O, Douspis M, Ducout A, Dupac X, Dusini S, Efstathiou G, Elsner F, Enßlin TA, Eriksen HK, Fantaye Y, Farhang M, Fergusson J, Fernandez-Cobos R, Finelli F, Forastieri F, Frailis M, Fraisse AA, Franceschi E, Frolov A, Galeotta S, Galli S, Ganga K, Génova-Santos RT, Gerbino M, Ghosh T, González-Nuevo J, Górski KM, Gratton S, Gruppuso A, Gudmundsson JE, Hamann J, Handley W, Hansen FK, Herranz D, Hildebrandt SR, Hivon E, Huang Z, Jaffe AH, Jones WC, Karakci A, Keihänen E, Keskitalo R, Kiiveri K, Kim J, Kisner TS, Knox L, Krachmalnicoff N, Kunz M, Kurki-Suonio H, Lagache G, Lamarre J-M, Lasenby A, Lattanzi M, Lawrence CR, Le Jeune M, Lemos P, Lesgourgues J, Levrier F, Lewis A, Liguori M, Lilje PB, Lilley M, Lindholm V, López-Caniego M, Lubin PM, Ma Y-Z, Macías-Pérez JF, Maggio G, Maino D, Mandolesi N, Mangilli A, Marcos-Caballero A, Maris M, Martin PG, Martinelli M, Martí nez-González E, Matarrese S, Mauri N, McEwen JD, Meinhold PR, Melchiorri A, Mennella A, Migliaccio M, Millea M, Mitra S, Miville-Deschênes M-A, Molinari D, Montier L, Morgante G, Moss A, Natoli P, Nørgaard-Nielsen HU, Pagano L, Paoletti D, Partridge B, Patanchon G, Peiris HV, Perrotta F, Pettorino V, Piacentini F, Polastri L, Polenta G, Puget J-L, Rachen JP, Reinecke M, Remazeilles M, Renzi A, Rocha G, Rosset C, Roudier G, Rubiño-Martí n JA, Ruiz-Granados B, Salvati L, Sandri M, Savelainen M, Scott D, Shellard EPS, Sirignano C, Sirri G, Spencer LD, Sunyaev R, Suur-Uski A-S, Tauber JA, Tavagnacco D, Tenti M, Toffolatti L, Tomasi M, Trombetti T, Valenziano L, Valiviita J, Van Tent B, Vibert L, Vielva P, Villa F, Vittorio N, Wandelt BD, Wehus IK, White M, White SDM, Zacchei A, Zonca A (2020) Planck 2018 results. VI. Cosmological parameters. Astron Astrophys 641:6. https://doi.org/10.1051/0004-6361/201833910

Polletta M, Nonino M, Frye B, Gargiulo A, Bisogni S, Garuda N, Thompson D, Lehnert M, Pascale M, Willner SP, Kamieneski P, Leimbach R, Cheng C, Coe D, Cohen SH, Conselice CJ, Dai L, Diego J, Dole H, Driver SP, D'Silva JCJ, Fontana A, Foo N, Furtak LJ, Grogin NA, Harrington K, Hathi NP, Jansen RA, Kelly P, Koekemoer AM, Mancini C, Marshall MA, Pierel JDR, Pirzkal N, Robotham A, Rutkowski MJ, Ryan RE, Snigula JM, Summers J, Tompkins S, Willmer CNA, Windhorst RA, Yan H, Yun MS, Zitrin A (2023) Spectroscopy of the supernova H0pe host galaxy at redshift 1.78. Astron Astrophys 675:4. https://doi.org/10.1051/0004-6361/202346964

Quimby RM, Werner MC, Oguri M, More S, More A, Tanaka M, Nomoto K, Moriya TJ, Folatelli G, Maeda K, Bersten M (2013) Extraordinary magnification of the ordinary type Ia supernova PS1-10afx. Astrophys J Lett 768(1):20. https://doi.org/10.1088/2041-8205/768/1/L20

Quimby RM, Oguri M, More A, More S, Moriya TJ, Werner MC, Tanaka M, Folatelli G, Bersten MC, Maeda K, Nomoto K (2014) Detection of the gravitational lens magnifying a type Ia supernova. Science 344(6182):396–399. https://doi.org/10.1126/science.1250903

Rabinak I, Waxman E (2011) The early UV/optical emission from core-collapse supernovae. Astrophys J 728(1):63. https://doi.org/10.1088/0004-637X/728/1/63

Refsdal S (1964) On the possibility of determining Hubble's parameter and the masses of galaxies from the gravitational lens effect. Mon Not R Astron Soc 128:307. https://doi.org/10.1093/mnras/128.4.307

Riess AG, Filippenko AV, Challis P, Clocchiatti A, Diercks A, Garnavich PM, Gilliland RL, Hogan CJ, Jha S, Kirshner RP, Leibundgut B, Phillips MM, Reiss D, Schmidt BP, Schommer RA, Smith RC, Spyromilio J, Stubbs C, Suntzeff NB, Tonry J (1998) Observational evidence from supernovae for an accelerating universe and a cosmological constant. Astron J 116(3):1009–1038. https://doi.org/10.1086/300499

Riess AG, Casertano S, Yuan W, Macri LM, Scolnic D (2019) Large Magellanic Cloud Cepheid standards provide a 1% foundation for the determination of the Hubble constant and stronger evidence for physics beyond ΛCDM. Astrophys J 876(1):85. https://doi.org/10.3847/1538-4357/ab1422

Rodney SA, Riess AG, Dahlen T, Strolger L-G, Ferguson HC, Hjorth J, Frederiksen TF, Weiner BJ, Mobasher B, Casertano S, Jones DO, Challis P, Faber SM, Filippenko AV, Garnavich P, Graur O, Grogin NA, Hayden B, Jha SW, Kirshner RP, Kocevski D, Koekemoer A, McCully C, Patel B, Rajan A, Scarlata C (2012) A type Ia supernova at redshift 1.55 in Hubble Space Telescope infrared observations from CANDELS. Astrophys J 746(1):5. https://doi.org/10.1088/0004-637X/746/1/5

Rodney SA, Strolger L-G, Kelly PL, Bradač M, Brammer G, Filippenko AV, Foley RJ, Graur O, Hjorth J, Jha SW, McCully C, Molino A, Riess AG, Schmidt KB, Selsing J, Sharon K, Treu T, Weiner BJ, Zitrin A (2016) SN refsdal: photometry and time delay measurements of the first Einstein cross supernova. Astrophys J 820(1):50. https://doi.org/10.3847/0004-637X/820/1/50

Rodney SA, Brammer GB, Pierel JDR, Richard J, Toft S, O'Connor KF, Akhshik M, Whitaker K (2021) A gravitationally lensed supernova with an observable two-decade time delay. Nat Astron 5:1118. https://doi.org/10.1038/s41550-021-01450-9. arXiv:2106.08935

Rojas K, Savary E, Clément B, Maus M, Courbin F, Lemon C, Chan JHH, Vernardos G, Joseph R, Cañameras R, Galan A (2022) Search of strong lens systems in the dark energy survey using convolutional neural networks. Astron Astrophys 668:73. https://doi.org/10.1051/0004-6361/202142119

Röpke FK, Kromer M, Seitenzahl IR, Pakmor R, Sim SA, Taubenberger S, Ciaraldi-Schoolmann F, Hillebrandt W, Aldering G, Antilogus P, Baltay C, Benitez-Herrera S, Bongard S, Buton C, Canto A, Cellier-Holzem F, Childress M, Chotard N, Copin Y, Fakhouri HK, Fink M, Fouchez D, Gangler E, Guy J, Hachinger S, Hsiao EY, Chen J, Kerschhaggl M, Kowalski M, Nugent P, Paech K, Pain R, Pecontal E, Pereira R, Perlmutter S, Rabinowitz D, Rigault M, Runge K, Saunders C, Smadja G, Suzuki N, Tao C, Thomas RC, Tilquin A, Wu C (2012) Constraining type Ia supernova models: SN 2011fe as a test case. Astrophys J Lett 750(1):19. https://doi.org/10.1088/2041-8205/750/1/L19

Saha A, Matheson T, Snodgrass R, Kececioglu J, Narayan G, Seaman R, Jenness T, Axelrod T (2014) ANTARES: a prototype transient broker system. In: Peck AB, Benn CR, Seaman RL (eds) Observatory operations: strategies, processes, and systems V. Society of photo-optical instrumentation engineers (SPIE) conference series, vol 9149, p 914908. https://doi.org/10.1117/12.2056988

Savary E, Rojas K, Maus M, Clément B, Courbin F, Gavazzi R, Chan JHH, Lemon C, Vernardos G, Cañameras R, Schuldt S, Suyu SH, Cuillandre J-C, Fabbro S, Gwyn S, Hudson MJ, Kilbinger M, Scott D, Stone C (2022) Strong lensing in UNIONS: toward a pipeline from discovery to modeling. Astron Astrophys 666:1. https://doi.org/10.1051/0004-6361/202142505

Schechter PL, Wambsganss J (2002) Quasar microlensing at high magnification and the role of dark matter: enhanced fluctuations and suppressed saddle points. Astrophys J 580(2):685–695. https://doi.org/10.1086/343856

Schlafly EF, Meisner AM, Stutz AM, Kainulainen J, Peek JEG, Tchernyshyov K, Rix H-W, Finkbeiner DP, Covey KR, Green GM, Bell EF, Burgett WS, Chambers KC, Draper PW, Flewelling H, Hodapp KW, Kaiser N, Magnier EA, Martin NF, Metcalfe N, Wainscoat RJ, Waters C (2016) The optical-infrared extinction curve and its variation in the Milky Way. Astrophys J 821(2):78. https://doi.org/10.3847/0004-637X/821/2/78

Schlafly EF, Peek JEG, Finkbeiner DP, Green GM (2017) Mapping the extinction curve in 3d: structure on kiloparsec scales. Astrophys J 838(1):36. https://doi.org/10.3847/1538-4357/aa619d

Schneider P (2014) Can one determine cosmological parameters from multi-plane strong lens systems? Astron Astrophys 568:2. https://doi.org/10.1051/0004-6361/201424450

Schneider P, Sluse D (2014) Source-position transformation: an approximate invariance in strong gravitational lensing. Astron Astrophys 564:103. https://doi.org/10.1051/0004-6361/201322106

Scolnic DM, Jones DO, Rest A, Pan YC, Chornock R, Foley RJ, Huber ME, Kessler R, Narayan G, Riess AG, Rodney S, Berger E, Brout DJ, Challis PJ, Drout M, Finkbeiner D, Lunnan R, Kirshner RP, Sanders NE, Schlafly E, Smartt S, Stubbs CW, Tonry J, Wood-Vasey WM, Foley M, Hand J, Johnson E, Burgett WS,

Chambers KC, Draper PW, Hodapp KW, Kaiser N, Kudritzki RP, Magnier EA, Metcalfe N, Bresolin F, Gall E, Kotak R, McCrum M, Smith KW (2018) The complete light-curve sample of spectroscopically confirmed SNe ia from pan-STARRS1 and cosmological constraints from the combined pantheon sample. Astrophys J 859(2):101. https://doi.org/10.3847/1538-4357/aab9bb

Shajib AJ, Birrer S, Treu T, Auger MW, Agnello A, Anguita T, Buckley-Geer EJ, Chan JHH, Collett TE, Courbin F, Fassnacht CD, Frieman J, Kayo I, Lemon C, Lin H, Marshall PJ, McMahon R, More A, Morgan ND, Motta V, Oguri M, Ostrovski F, Rusu CE, Schechter PL, Shanks T, Suyu SH, Meylan G, Abbott TMC, Allam S, Annis J, Avila S, Bertin E, Brooks D, Carnero Rosell A, Carrasco Kind M, Carretero J, Cunha CE, da Costa LN, De Vicente J, Desai S, Doel P, Flaugher B, Fosalba P, García-Bellido J, Gerdes DW, Gruen D, Gruendl RA, Gutierrez G, Hartley WG, Hollowood DL, Hoyle B, James DJ, Kuehn K, Kuropatkin N, Lahav O, Lima M, Maia MAG, March M, Marshall JL, Melchior P, Menanteau F, Miquel R, Plazas AA, Sanchez E, Scarpine V, Sevilla-Noarbe I, Smith M, Soares-Santos M, Sobreira F, Suchyta E, Swanson MEC, Tarle G, Walker AR (2019) Is every strong lens model unhappy in its own way? Uniform modelling of a sample of 13 quadruply+ imaged quasars. Mon Not R Astron Soc 483(4):5649–5671. https://doi.org/10.1093/mnras/sty3397

Shajib AJ, Mozumdar P, Chen GC-F, Treu T, Cappellari M, Knabel S, Suyu SH, Bennert VN, Frieman JA, Sluse D, Birrer S, Courbin F, Fassnacht CD, Villafaña L, Williams PR (2023) TDCOSMO. XIII. Improved Hubble constant measurement from lensing time delays using spatially resolved stellar kinematics of the lens galaxy. Astron Astrophys 673:A9. https://doi.org/10.1051/0004-6361/202345878. arXiv:2301.02656

Sherstinsky A (2020) Fundamentals of recurrent neural network (RNN) and long short-term memory (LSTM) network. Phys D: Nonlinear Phenom 404:132306. https://doi.org/10.1016/j.physd.2019.132306

Sheu W, Huang X, Cikota A, Suzuki N, Schlegel D, Storfer C (2023) Retrospective search for strongly lensed supernovae in the DESI Legacy Imaging Surveys. Astrophys J 952(1):10. https://doi.org/10.3847/1538-4357/acd1e4. arXiv:2301.03578

Shu Y, Brownstein JR, Bolton AS, Koopmans LVE, Treu T, Montero-Dorta AD, Auger MW, Czoske O, Gavazzi R, Marshall PJ, Moustakas LA (2017) The sloan lens ACS survey. XIII. Discovery of 40 new galaxy-scale strong lenses. Astrophys J 851(1):48. https://doi.org/10.3847/1538-4357/aa9794

Shu Y, Bolton AS, Mao S, Kang X, Li G, Soraisam M (2018a) Prediction of supernova rates in known galaxy-galaxy strong-lens systems. Astrophys J 864(1):91. https://doi.org/10.3847/1538-4357/aad5ea

Shu Y, Bolton AS, Mao S, Kang X, Li G, Soraisam M (2021) Erratum: "Prediction of supernova rates in known galaxy-galaxy strong-lens systems" (2018, ApJ, 864, 91). Astrophys J 919(1):67. https://doi.org/10.3847/1538-4357/ac24a4

Shu Y, Cañameras R, Schuldt S, Suyu SH, Taubenberger S, Inoue KT, Jaelani AT, High-redshift HOLISMOKESVIII (2022) Strong-lens search in the hyper suprime-cam subaru strategic program. Astron Astrophys 662:4. https://doi.org/10.1051/0004-6361/202243203

Sim SA, Röpke FK, Hillebrandt W, Kromer M, Pakmor R, Fink M, Ruiter AJ, Seitenzahl IR (2010) Detonations in sub-Chandrasekhar-mass C+O white dwarfs. Astrophys J Lett 714(1):52–57. https://doi.org/10.1088/2041-8205/714/1/L52

Smith KW, Williams RD, Young DR, Ibsen A, Smartt SJ, Lawrence A, Morris D, Voutsinas S, Nicholl M (2019) Lasair: the transient alert broker for LSST:UK. Res. Notes Am. Astron. Soc. 3(1):26. https://doi.org/10.3847/2515-5172/ab020f

Sonnenfeld A, Chan JHH, Shu Y, More A, Oguri M, Suyu SH, Wong KC, Lee C-H, Coupon J, Yonehara A, Bolton AS, Jaelani AT, Tanaka M, Miyazaki S, Komiyama Y (2018) Survey of gravitationally-lensed objects in HSC imaging (SuGOHI). I. Automatic search for galaxy-scale strong lenses. Publ Astron Soc Jpn 70:29. https://doi.org/10.1093/pasj/psx062

Sonnenfeld A, Verma A, More A, Baeten E, Macmillan C, Wong KC, Chan JHH, Jaelani AT, Lee C-H, Oguri M, Rusu CE, Veldthuis M, Trouille L, Marshall PJ, Hutchings R, Allen C, O'Donnell J, Cornen C, Davis CP, McMaster A, Lintott C, Miller G (2020) Survey of gravitationally-lensed objects in HSC imaging (SuGOHI). VI. Crowdsourced lens finding with space warps. Astron Astrophys 642:148. https://doi.org/10.1051/0004-6361/202038067

Stanishev V, Goobar A, Paech K, Amanullah R, Dahlén T, Jönsson J, Kneib JP, Lidman C, Limousin M, Mörtsell E, Nobili S, Richard J, Riehm T, von Strauss M (2009) Near-IR search for lensed supernovae behind galaxy clusters. I. Observations and transient detection efficiency. Astron Astrophys 507(1):61–69. https://doi.org/10.1051/0004-6361/200911982

Sullivan M, Ellis R, Nugent P, Smail I, Madau P (2000) A strategy for finding gravitationally lensed distant supernovae. Mon Not R Astron Soc 319(2):549–556. https://doi.org/10.1046/j.1365-8711.2000.03875.x

Suwa Y (2018) Supernova forecast with strong lensing. Mon Not R Astron Soc 474(2):2612–2616. https://doi.org/10.1093/mnras/stx2953

Suyu SH, Marshall PJ, Auger MW, Hilbert S, Blandford RD, Koopmans LVE, Fassnacht CD, Treu T (2010) Dissecting the gravitational lens B1608+656. II. Precision measurements of the Hubble constant, spatial curvature, and the dark energy equation of state. Astrophys J 711(1):201–221. https://doi.org/10.1088/0004-637X/711/1/201

Suyu SH, Chang T-C, Courbin F, Okumura T (2018) Cosmological distance indicators. Space Sci Rev 214(5):91. https://doi.org/10.1007/s11214-018-0524-3

Suyu SH, Huber S, Cañameras R, Kromer M, Schuldt S, Taubenberger S, Yıldırım A, Bonvin V, Chan JHH, Courbin F, Nöbauer U, Sim SA, Sluse D (2020) HOLISMOKES. I. Highly optimised lensing investigations of supernovae, microlensing objects, and kinematics of ellipticals and spirals. Astron Astrophys 644:162. https://doi.org/10.1051/0004-6361/202037757

Tewes M, Courbin F, Meylan G (2013) COSMOGRAIL: the COSmological MOnitoring of GRAvItational lenses. XI. Techniques for time delay measurement in presence of microlensing. Astron Astrophys 553:120. https://doi.org/10.1051/0004-6361/201220123

Tran K-VH, Harshan A, Glazebrook K, Keerthi Vasan GC, Jones T, Jacobs C, Kacprzak GG, Barone TM, Collett TE, Gupta A, Henderson A, Kewley LJ, Lopez S, Nanayakkara T, Sanders RL, Sweet SM (2022) The AGEL survey: spectroscopic confirmation of strong gravitational lenses in the DES and DECaLS fields selected using convolutional neural networks. Astron J 164(4):148. https://doi.org/10.3847/1538-3881/ac7da2

Treu T, Marshall PJ (2016) Time delay cosmography. Astron Astrophys Rev 24(1):11. https://doi.org/10.1007/s00159-016-0096-8

Treu T, Brammer G, Diego JM, Grillo C, Kelly PL, Oguri M, Rodney SA, Rosati P, Sharon K, Zitrin A, Balestra I, Bradač M, Broadhurst T, Caminha GB, Halkola A, Hoag A, Ishigaki M, Johnson TL, Karman W, Kawamata R, Mercurio A, Schmidt KB, Strolger L-G, Suyu SH, Filippenko AV, Foley RJ, Jha SW, Patel B (2016) "Refsdal" meets popper: comparing predictions of the re-appearance of the multiply imaged supernova behind MACSJ1149.5+2223. Astrophys J 817(1):60. https://doi.org/10.3847/0004-637X/817/1/60

Treu T, Suyu SH, Marshall PJ (2022) Strong lensing time-delay cosmography in the 2020s. Astron Astrophys Rev 30(1):8. https://doi.org/10.1007/s00159-022-00145-y

Verde L, Treu T, Riess AG (2019) Tensions between the early and late universe. Nat Astron 3:891–895. https://doi.org/10.1038/s41550-019-0902-0

Vernardos G, Fluke CJ (2014) Adventures in the microlensing cloud: large datasets, eResearch tools, and GPUs. Astron Comput 6:1–18. https://doi.org/10.1016/j.ascom.2014.05.002

Vernardos G, Tsagkatakis G (2019) Quasar microlensing light-curve analysis using deep machine learning. Mon Not R Astron Soc 486(2):1944–1952. https://doi.org/10.1093/mnras/stz868

Vernardos G, Fluke CJ, Bate NF, Croton D (2014) GERLUMPH data release 1: high-resolution cosmological microlensing magnification maps and eResearch tools. Astrophys J Suppl Ser 211(1):16. https://doi.org/10.1088/0067-0049/211/1/16

Vernardos G, Fluke CJ, Bate NF, Croton D, Vohl D (2015) GERLUMPH data release 2: 2.5 billion simulated microlensing light curves. Astrophys J Suppl Ser 217(2):23. https://doi.org/10.1088/0067-0049/217/2/23

Vernardos G et al (2023) Microlensing of strongly lensed quasars. arXiv:2312.00931

Vieira JD, Marrone DP, Chapman SC, De Breuck C, Hezaveh YD, Weiß A, Aguirre JE, Aird KA, Aravena M, Ashby MLN, Bayliss M, Benson BA, Biggs AD, Bleem LE, Bock JJ, Bothwell M, Bradford CM, Brodwin M, Carlstrom JE, Chang CL, Crawford TM, Crites AT, de Haan T, Dobbs MA, Fomalont EB, Fassnacht CD, George EM, Gladders MD, Gonzalez AH, Greve TR, Gullberg B, Halverson NW, High FW, Holder GP, Holzapfel WL, Hoover S, Hrubes JD, Hunter TR, Keisler R, Lee AT, Leitch EM, Lueker M, Luong-van D, Malkan M, McIntyre V, McMahon JJ, Mehl J, Menten KM, Meyer SS, Mocanu LM, Murphy EJ, Natoli T, Padin S, Plagge T, Reichardt CL, Rest A, Ruel J, Ruhl JE, Sharon K, Schaffer KK, Shaw L, Shirokoff E, Spilker JS, Stalder B, Staniszewski Z, Stark AA, Story K, Vanderlinde K, Welikala N, Williamson R (2013) Dusty starburst galaxies in the early universe as revealed by gravitational lensing. Nature 495(7441):344–347. https://doi.org/10.1038/nature12001

Vogl C, Sim SA, Noebauer UM, Kerzendorf WE, Hillebrandt W (2019) Spectral modeling of type II supernovae. I. Dilution factors. Astron Astrophys 621:29. https://doi.org/10.1051/0004-6361/201833701

Vogl C, Kerzendorf WE, Sim SA, Noebauer UM, Lietzau S, Hillebrandt W (2020) Spectral modeling of type II supernovae. II. A machine-learning approach to quantitative spectroscopic analysis. Astron Astrophys 633:88. https://doi.org/10.1051/0004-6361/201936137

Walsh D, Carswell RF, Weymann RJ (1979) 0957+561 A, B: twin quasistellar objects or gravitational lens? Nature 279:381–384. https://doi.org/10.1038/279381a0

Webbink RF (1984) Double white dwarfs as progenitors of R coronae borealis stars and type I supernovae. Astrophys J 277:355–360. https://doi.org/10.1086/161701

Weisenbach L, Schechter PL, Pontula S (2021) "Worst-case" microlensing in the identification and modeling of lensed quasars. Astrophys J 922(1):70. https://doi.org/10.3847/1538-4357/ac2228

Whelan J, Iben I Jr (1973) Binaries and supernovae of type I. Astrophys J 186:1007–1014. https://doi.org/10.1086/152565

Wojtak R, Hjorth J, Gall C (2019) Magnified or multiply imaged? - search strategies for gravitationally lensed supernovae in wide-field surveys. Mon Not R Astron Soc 487(3):3342–3355. https://doi.org/10.1093/mnras/stz1516

Wong KC, Suyu SH, Chen GC-F, Rusu CE, Millon M, Sluse D, Bonvin V, Fassnacht CD, Taubenberger S, Auger MW, Birrer S, Chan JHH, Courbin F, Hilbert S, Tihhonova O, Treu T, Agnello A, Ding X, Jee I, Komatsu E, Shajib AJ, Sonnenfeld A, Blandford RD, Koopmans LVE, Marshall PJ, Meylan G (2020) H0LiCOW – XIII. A 2.4 per cent measurement of H_0 from lensed quasars: 5.3σ tension between early- and late-Universe probes. Mon Not R Astron Soc 498(1):1420–1439. https://doi.org/10.1093/mnras/stz3094

Yahalomi DA, Schechter PL, Wambsganss J (2017) A Quadruply Lensed SN Ia: Gaining a Time-Delay...Losing a Standard Candle. ArXiv e-prints, arXiv:1711.07919

Yao Y, Miller AA, Kulkarni SR, Bulla M, Masci FJ, Goldstein DA, Goobar A, Nugent P, Dugas A, Blagorodnova N, Neill JD, Rigault M, Sollerman J, Nordin J, Bellm EC, Cenko SB, De K, Dhawan S, Feindt U, Fremling C, Gatkine P, Graham MJ, Graham ML, Ho AYQ, Hung T, Kasliwal MM, Kupfer T, Laher RR, Perley DA, Rusholme B, Shupe DL, Soumagnac MT, Taggart K, Walters R, Yan L (2019) ZTF early observations of type Ia supernovae. I. Properties of the 2018 sample. Astrophys J 886(2):152. https://doi.org/10.3847/1538-4357/ab4cf5

Yıldırım A, Suyu SH, Halkola A (2020) Time-delay cosmographic forecasts with strong lensing and JWST stellar kinematics. Mon Not R Astron Soc 493(4):4783–4807. https://doi.org/10.1093/mnras/staa498

Yıldırım A, Suyu SH, Chen GC-F, Komatsu E (2023) TDCOSMO VIII: Cosmological distance measurements in light of the mass-sheet degeneracydegeneracy: Forecasts from strong lensing and integral field unit stellar kinematics. Astron Astrophys 675:A21. https://doi.org/10.1051/0004-6361/202142318. arXiv:2109.14615

Zwicky F (1937) Nebulae as gravitational lenses. Phys Rev 51:290–290. https://doi.org/10.1103/PhysRev.51.290

Publisher's Note Springer Nature remains neutral with regard to jurisdictional claims in published maps and institutional affiliations.

Authors and Affiliations

Sherry H. Suyu[1,2,3] · Ariel Goobar[4] · Thomas Collett[5] · Anupreeta More[6,7] · Giorgos Vernardos[8]

✉ S.H. Suyu
suyu@mpa-garching.mpg.de

A. Goobar
ariel@fysik.su.se

T. Collett
thomas.collett@port.ac.uk

A. More
anupreeta@iucaa.in

G. Vernardos
georgios.vernardos@epfl.ch

1 TUM School of Natural Sciences, Department of Physics, Technical University of Munich, James-Franck-Straße 1, 85748 Garching, Germany

2 Max-Planck-Institut für Astrophysik, Karl-Schwarzschild-Str. 1, 85748 Garching, Germany

3 Institute of Astronomy and Astrophysics, Academia Sinica, 11F of ASMAB, No.1, Sect. 4, Roosevelt Road, Taipei 10617, Taiwan

4 The Oskar Klein Centre, Department of Physics, Stockholm University, Albanova University Center, SE-106 91 Stockholm, Sweden

[5] Institute of Cosmology and Gravitation, University of Portsmouth, Dennis Sciama Building, Burnaby Road, Portsmouth, PO1 3FX, UK

[6] The Inter-University Centre for Astronomy and Astrophysics (IUCAA), Post Bag 4, Ganeshkhind, Pune 411007, India

[7] Kavli Institute for the Physics and Mathematics of the Universe (IPMU), 5-1-5 Kashiwanoha, Kashiwa-shi, Chiba 277-8583, Japan

[8] Institute of Physics, Laboratory of Astrophysics, Ecole Polytechnique Fédérale de Lausanne (EPFL), Observatoire de Sauverny, 1290 Versoix, Switzerland

Space Science Reviews (2024) 220:57
https://doi.org/10.1007/s11214-024-01088-9

Microlensing Near Macro-Caustics

Luke Weisenbach[1] · Timo Anguita[2,3] · Jordi Miralda-Escudé[4,5,6] · Masamune Oguri[7,8] · Prasenjit Saha[9] · Paul L. Schechter[10]

Received: 12 October 2023 / Accepted: 1 July 2024 / Published online: 30 July 2024
© The Author(s) 2024

Abstract

Microlensing near macro-caustics is a complex phenomenon in which swarms of micro-images produced by micro-caustics form on both sides of a macro-critical curve. Recent discoveries of highly magnified images of individual stars in massive galaxy cluster lenses, predicted to be formed by these micro-image swarms, have stimulated studies on this topic. In this article, we explore microlensing near macro-caustics using both simulations and analytic calculations. We show that the mean total magnification of the micro-image swarms follows that of an extended source in the absence of microlensing. Micro-caustics join into a connected network in a region around the macro-critical line of a width proportional to the surface density of microlenses; within this region, the increase of the mean magnification toward the macro-caustic is driven by the increase of the number of micro-images rather than individual magnifications of micro-images. The maximum achievable magnification in micro-caustic crossings decreases with the mass fraction in microlenses. We conclude with a review of applications of this microlensing phenomenon, including limits to the fraction of dark matter in compact objects, and searches of Population III stars and dark matter subhalos. We argue that the discovered highly magnified stars at cosmological distances already imply that less than $\sim 10\%$ of the dark matter may be in the form of compact objects with mass above $\sim 10^{-6}\ M_\odot$.

✉ J. Miralda-Escudé
miralda@icc.ub.edu

1 Institute of Cosmology and Gravitation, University of Portsmouth, Portsmouth PO1 3FX, UK

2 Instituto de Astrofisica, Departamento de Ciencias Fisicas, Universidad Andrés Bello, Santiago, Chile

3 Millennium Institute of Astrophysics, Santiago, Chile

4 Institute of Cosmos Sciences, University of Barcelona, Barcelona 08028, Spain

5 Institució Catalana de Recerca i Estudis Avançats, Barcelona 08010, Spain

6 Institut d'Estudis Espacials de Catalunya, Castelldefels 08860, Spain

7 Center for Frontier Science, Chiba University, 1-33 Yayoi-cho, Inage-ku, Chiba 263-8522, Japan

8 Department of Physics, Graduate School of Science, Chiba University, 1-33 Yayoi-Cho, Inage-Ku, Chiba 263-8522, Japan

9 Physik-Institut, University of Zurich, Winterthurerstr. 190, 8057 Zurich, Switzerland

10 MIT Kavli Institute 37-635, Cambridge MA 02139, USA

Keywords Gravitational lensing · Strong gravitational lensing · Gravitational microlensing · Caustic crossing

1 Introduction

Paczynski (1986) showed how a screen of point mass micro-lenses (stars) of sufficiently high surface density produces an extended swarm of micro-images, the sum of whose fluxes is, on average, equal to the flux expected for a point source lensed by a smooth mass distribution with the same surface density (see also Katz et al. 1986; Granot et al. 2003; Saha and Williams 2011).

A caustic is a closed one-dimensional locus in the source plane of a lens system that maps onto a one-dimensional locus in the image plane – a critical curve – along which images of point sources undergo infinite magnification in the geometric optics approximation. A physical source has a finite angular size that implies a maximum magnification, which is reached when the source straddles a caustic and gives rise to two merging mirror images straddling the corresponding critical curve. When a source crosses a caustic, a pair of mirror images either annihilates at or emerges from the critical curve (see e.g., Schneider et al. 1992).

These two distinct phenomena combine when one has micro-lensing near a caustic, with the emergence or annihilation of a pair of elongated swarms of micro-images at a critical curve. The combination of these two phenomena was considered by Wambsganss (1990) as a possible explanation for BL Lac systems. The discovery of a bright transient, thought to be a highly magnified star at redshift 1.5, near the critical curve of the lensing cluster MACS J1149.5+2223 (Kelly et al. 2018) led to renewed interest, triggering theoretical studies of the "trains" of micro-images, or elongated swarms, formed on the two sides of a macro-critical curve (Venumadhav et al. 2017; Diego et al. 2018; Oguri et al. 2018; Dai and Pascale 2021) as well as observational searches of more such events (Rodney et al. 2018; Chen et al. 2019; Kaurov et al. 2019; Dai et al. 2020).

In what follows, after reviewing the macro-lens model near macro-caustics (Sect. 2), we briefly review the behavior of extended sources in the vicinity of a caustic (Sect. 3). This is then combined with the behavior of micro-image swarms in regions of high optical depth discussed in the article by Vernardos et al. (2024, this collection) to show how these two phenomena modify the predictions for each phenomenon considered separately and how these can be used to interpret transients near macro-caustics (Sect. 4). Finally, we review recent observations of micro-caustic crossing near macro-caustics and discuss some applications of these microlensing phenomena, including the possibility of searching for more exotic substructure (Sect. 5).

2 The Macro Model

To study microlensing near macro-caustics, we define a macro mass model (i.e., the lens model without microlenses) that contains a caustic and a critical curve, and set the coordinate system origin of the image and source planes at a point on the critical curve and the caustic, respectively. The Fermat time-delay surface that satisfies this condition is written as

$$\tau = \frac{1}{2}(\boldsymbol{\beta} - \boldsymbol{\theta})^2 - \psi(\boldsymbol{\theta}) \, , \tag{1}$$

where $\boldsymbol{\beta}$ and $\boldsymbol{\theta}$ are two-dimensional vectors of the angular positions in the source plane and the image plane, respectively, with coordinates (β_1, β_2) and (θ_1, θ_2). The Taylor-expanded deflection potential, up to third order at the origin, is

$$\begin{aligned}\psi(\boldsymbol{\theta}) =& \frac{1}{2}(\psi_{11}\theta_1^2 + 2\psi_{12}\theta_1\theta_2 + \psi_{22}\theta_2^2) \\ &+ \frac{1}{6}(\psi_{111}\theta_1^3 + 3\psi_{112}\theta_1^2\theta_2 + 3\psi_{122}\theta_1\theta_2^2 + \psi_{222}\theta_2^3) \ .\end{aligned} \tag{2}$$

Note that the first order terms of the expansion disappear because of the requirement that the origin of the image plane maps to the origin of the source plane. Denoting convergence at the origin as κ_0, i.e.,

$$\psi_{11} + \psi_{22} = 2\kappa_0 \ , \tag{3}$$

and using the condition that the critical curve and the caustic pass through the origin, the second derivatives of the deflection potential can be described as

$$\begin{aligned}\psi_{11} &= \kappa_0 + (1 - \kappa_0)\cos\omega \ , \\ \psi_{22} &= \kappa_0 - (1 - \kappa_0)\cos\omega \ , \\ \psi_{12} &= -(1 - \kappa_0)\sin\omega \ ,\end{aligned} \tag{4}$$

where ω is an arbitrary constant parameter.

Equations (2) and (4) form a general expression of the lens model near the macro-caustic. We have the additional freedom to rotate the coordinate system to simplify it further. For instance, choosing the coordinate system with $\omega = 0$ realizes the locally orthogonal coordinate system around a fold caustic, which is our main interest here. We could also choose to instead rotate the coordinate system to eliminate one of the third derivatives of the deflection potential, although such a transformation makes discussion regarding the direction and extent of the image trains less intuitive.

It is also helpful to consider more restrictive cases for the microlensing analysis. One possible choice is to make convergence constant across the field, for which we can easily attain microlensing simulations with a constant stellar mass fraction to the total matter density. Since convergence κ is computed from Eq. (2) as

$$\kappa = \kappa_0 + \frac{1}{2}(\psi_{111} + \psi_{122})\theta_1 + \frac{1}{2}(\psi_{112} + \psi_{222})\theta_2 \ , \tag{5}$$

we need $\psi_{111} = -\psi_{122}$ and $\psi_{112} = -\psi_{222}$ to realize constant convergence. Setting $\omega = 0$ and denoting $\psi_{111} = -1/d_{\parallel}$ and $\psi_{112} = -1/d_{\perp}$, the lens equation becomes

$$\begin{aligned}\beta_1 =& \frac{\theta_1^2}{2d_{\parallel}} + \frac{\theta_1\theta_2}{d_{\perp}} - \frac{\theta_2^2}{2d_{\parallel}} \ , \\ \beta_2 =& 2(1 - \kappa_0)\theta_2 + \frac{\theta_1^2}{2d_{\perp}} - \frac{\theta_1\theta_2}{d_{\parallel}} - \frac{\theta_2^2}{2d_{\perp}} \ ,\end{aligned} \tag{6}$$

which, apart from notation, is the same as the macro lens model adopted in e.g., Dai and Pascale (2021).

The parameters $d_{\parallel}$ and $d_{\perp}$ have dimensions of the angular variable θ, and are on the order of the Einstein radius of the macro lens model, which is $\mathcal{O}(1'')$ for galaxy-scale strong lensing and $\mathcal{O}(10'')$ for cluster-scale strong lensing. In this paper we will use a dimensionless variable defined as the angular coordinates divided by the Einstein radius of each microlens, $\theta_\star$, which is $\theta_\star = \mathcal{O}(10^{-6''})$ for typical stars acting as microlenses at cosmological distances. Thus $d_\parallel$ and $d_\perp$ are on the order of $\sim 10^6 - 10^7$ in this dimensionless variable.

Alternatively, we consider a restricted case where the critical curve and a train of micro-images (which follows the direction of the principal axis that has zero eigenvalue at the critical curve) are perpendicular to each other. This is realized by setting $\psi_{112} = \psi_{122} = \psi_{222} = 0$ along with $\omega = 0$. Denoting $\psi_{111} = -1/E$, the lens equation becomes

$$\begin{aligned} \beta_1 =& \frac{\theta_1^2}{2E} , \\ \beta_2 =& 2(1-\kappa_0)\theta_2 . \end{aligned} \tag{7}$$

The variable E again has dimensions of the angle θ, and is on the order of $\sim 10^6 - 10^7$ in the unit defined by $\theta_\star$.

While this macro lens model is a restricted case, it makes the analysis much easier and hence is instructive; we will use it in Sect. 3. From our experience in modeling such systems, introducing other degrees of freedom related to further derivatives of the potential introduces modest quantitative changes but not major qualitative differences. In this model, convergence changes near the critical line as

$$\kappa = \kappa_0 - \frac{\theta_1}{2E} . \tag{8}$$

3 Extended Sources in the Vicinity of a Caustic

In this Section, we review the behavior of magnification of extended sources in the vicinity of a macro caustic in the absence of microlensing. We use the simple macro model described by Eq. (7). At an image plane position $\boldsymbol{\theta} = (\theta_1, \theta_2)$, magnification factors along the θ_1 and θ_2 directions, which we denote μ_1 and μ_2 respectively (equal to the inverse eigenvalues of the magnification matrix), are easily computed as

$$\begin{aligned} \mu_1 =& \frac{E}{\theta_1} , \\ \mu_2 =& \frac{1}{2(1-\kappa_0)} . \end{aligned} \tag{9}$$

The magnification of a point image at that point is $\mu = \mu_1\mu_2$. A point source at $\boldsymbol{\beta} = (\beta_1, \beta_2)$ has two images on both sides of the critical curve at $\theta_1 = \pm\sqrt{2E\beta_1}$ if and only if $\beta_1 > 0$. Thus the magnification of these two images of a point source at $\boldsymbol{\beta}$ is

$$\mu(\boldsymbol{\beta}) = \frac{1}{(1-\kappa_0)}\sqrt{\frac{E}{2\beta_1}} , \tag{10}$$

for $\beta_1 > 0$.

For an extended source with a Gaussian surface brightness profile with standard deviation σ, the magnification factor of the source at $\boldsymbol{\beta}$ is given by

$$\mu(\boldsymbol{\beta}) = \frac{\sqrt{E}}{2\sqrt{\pi}\sigma(1-\kappa_0)} \int_0^\infty \frac{\mathrm{d}\beta_1'}{\sqrt{\beta_1'}} \exp\left[-\frac{(\beta_1'-\beta_1)^2}{2\sigma^2}\right] . \tag{11}$$

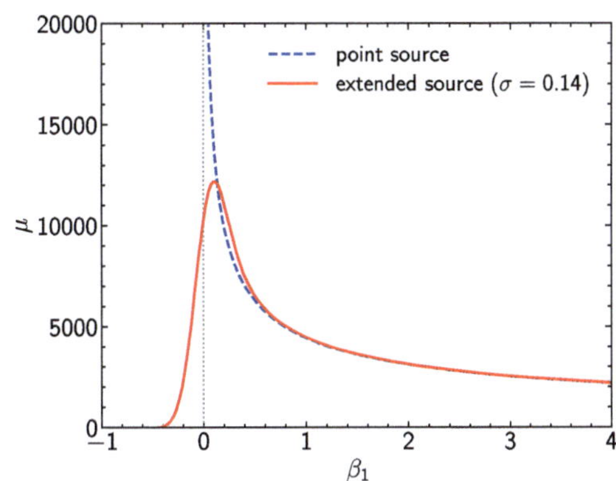

Fig. 1 Magnification factors as a function of the source position β_1, both for the point source (*dashed*) and the extended source (*solid*), in absence of microlensing. We adopt the macro model of Eq. (7) with $\kappa_0 = 0.5$ and $E = 10^7$. The extended source follows the Gaussian surface brightness distribution with $\sigma = 0.14$

As an example, Fig. 1 shows magnification factors as a function of the source position β_1, both for the point source and the extended source, for the case $E = 10^7$, $\kappa_0 = 0.5$ and $\sigma = 0.14$. The magnification of the point source diverges as the source approaches the caustic at $\beta_1 = 0$, while the magnification of the extended source saturates at a maximum value that can be estimated as the magnification of the point source (Eq. (10)) at $\beta_1 = \sigma$ (e.g., Miralda-Escudé 1991), i.e.,

$$\mu_{\max} \approx \frac{1}{(1-\kappa_0)}\sqrt{\frac{E}{2\sigma}} \,. \tag{12}$$

For the example in Fig. 1, this estimate yields $\mu_{\max} \approx 12{,}000$, close to the maximum magnification directly derived from Eq. (11), $\mu_{\max} \approx 12{,}200$.

4 Microlensing Near Macro-Caustics

We now describe the impact of microlensing on the magnification of sources.

4.1 Average Magnification Profile Interpreted as an Effectively Extended Source

As shown in Neindorf (2003) and Venumadhav et al. (2017), and discussed in the article by Vernardos et al. (2024, this collection) for quasars, the total magnification of a source in the presence of microlensing, averaged over all random point mass realizations of the microlenses, is the magnification under the macro-lens model of the source profile convolved with the probability distribution function (PDF) of the microlens deflection angle:

$$\langle \mu(\boldsymbol{\beta}) \rangle = \int I_\star(\boldsymbol{\beta}')\mu_{\rm B}(\boldsymbol{\beta}+\boldsymbol{\beta}')\,{\rm d}^2\boldsymbol{\beta}' \,, \tag{13}$$

where $\boldsymbol{\beta}$ is the position of the source, $\boldsymbol{\beta}'$ is a dummy variable of integration over the source plane, $I_\star(\boldsymbol{\beta}')$ is the source profile $I(\boldsymbol{\beta}')$ convolved with the PDF of the microlens deflection angle $p(\boldsymbol{\alpha}_\star)$, and $\mu_{\rm B}(\boldsymbol{\beta})$ denotes the magnification factor of the background macro-model (i.e., the mass model after the point mass lenses are smoothed) for a source at $\boldsymbol{\beta}$. The mean effect of microlenses is encapsulated in the convolved source profile $I_\star(\boldsymbol{\beta}')$. The PDF of the deflection angle has a bivariate Gaussian core with a width of

$$\sigma_\star = \theta_\star \kappa_\star^{1/2}\left[\ln(2e^{1-\gamma_E} N_\star^{1/2})\right]^{1/2} \,, \tag{14}$$

Fig. 2 Magnification of a point source as a function of source position β_1 near the macro-caustic in the presence of microlenses (*blue solid line*), using the macro model of Eq. (7) with $\kappa_0 = 0.5$ and $E = 10^7$, and convergence of point masses of $\kappa_\star = 0.0025$. For comparison, magnification of an extended source with a Gaussian surface brightness distribution with $\sigma = \sigma_{\alpha} = 0.14$ (see also Eq. (14)) in the absence of microlenses is plotted by a dashed line

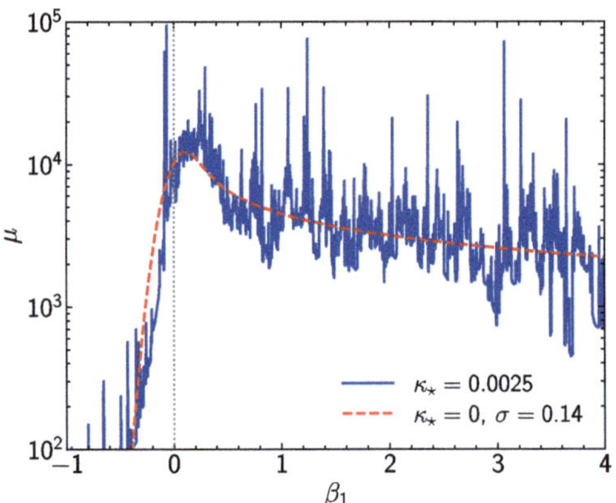

where $\kappa_\star$ is the convergence of point mass lenses (stars), $\theta_\star$ is the Einstein radius of each point mass, $\gamma_E \approx 0.577$ is the Euler-Mascheroni constant, and $N_\star$ is the number of point masses.

Although the deflection angle distribution $p(\boldsymbol{\alpha}_\star)$ is not precisely a Gaussian, it can be approximated this way whenever the deflection is contributed by many microlenses, according to the central limit theorem (this fails, of course, for the tails of this distribution caused by large deflections near a single microlens). The behavior of the mean magnification of point sources near a macro-caustic should then follow that of an extended Gaussian source in the absence of microlensing as studied in Sect. 3. To explicitly check this point, we make numerical simulations of the magnification of a point source in the simple macro model of Eq. (7), with $\kappa_0 = 0.5$ and $E = 10^7\theta_\star$, adding randomly distributed microlenses of a single mass corresponding to an Einstein radius $\theta_\star$, within a circle of radius $\theta_{\max} = 15{,}000\,\theta_\star$. A surface density contributing a convergence $\kappa_\star = 0.0025$ in this macro model is replaced with microlenses. This stellar mass fraction is similar to the expected value for the first discovery of a micro-caustic crossing event due to stars making up the intracluster light (see Sect. 5.1). With the choice of these parameters, from Eq. (14) we obtain the effective width of the Gaussian as $\sigma_\star \approx 0.14\,\theta_\star$. We solve the lens equation using an adaptive mesh method implemented in the `glafic` software (Oguri 2010) together with a hierarchical tree algorithm to speed up calculations of deflection angles from an ensemble of point mass lenses (see e.g., Wambsganss 1999).

Figure 2 shows the result of moving the source along the β_1 axis, fixing $\beta_2 = 0$ and computing the total magnification as a function of the source position, or lightcurve. While many sharp peaks corresponding to micro-caustic crossings are present, the mean magnification closely follows the prediction of the total magnification for an extended source without microlenses as studied in Sect. 3, shown by a dashed line in Fig. 2. Note that the size of the extended source is not fitted to the lightcurve, but determined by Eq. (14). This confirms the average effect of microlensing in macro-caustic crossings can indeed be interpreted as an effectively extended source, as discussed in the literature (e.g., Dai and Pascale 2021).

The lightcurve in Fig. 2 exhibits large fluctuations originating from a random realization of positions of point masses. A better comparison with the model of an effectively extended source is obtained by averaging the calculation of the lightcurve over independent realizations of the point mass positions. The mean of 10 realizations is shown in Fig. 3, indicating a converging match of the lightcurve of an effectively extended source with the numerical result.

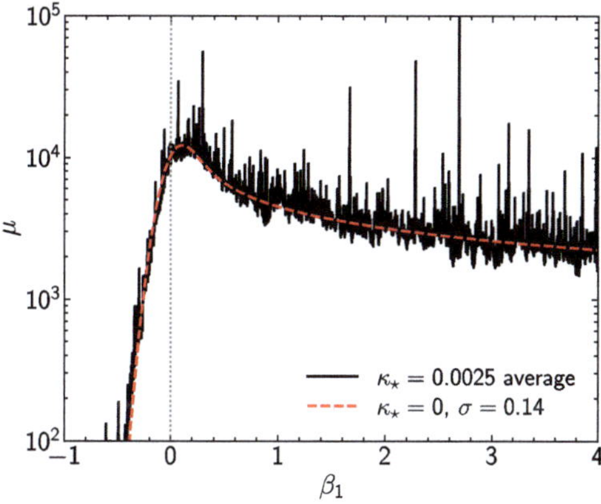

Fig. 3 Similar to Fig. 2, but averaging over 10 realizations of point mass positions

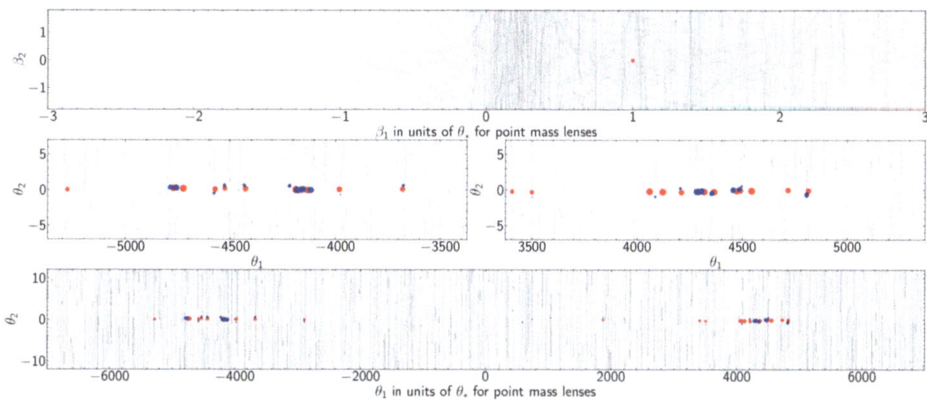

Fig. 4 An example of the micro-image swarms (or "trains"). Upper and lower panels show the source and image planes, respectively, and the middle panel shows enlarged view of micro-images in the plane around the swarms. The macro model caustic and critical curve (i.e., those in absence of microlenses) are vertical lines at $\beta_1 = \theta_1 = 0$. Micro-caustics and critical curves are shown by gray thin lines. The source position is $(\beta_1, \beta_2) = (1, 0)$, and corresponding multiple images in the image plane are shown by filled circles in the middle and lower panels. The size of the circles scales with the logarithm of the magnification factor of each image, and red and blue circles indicate micro-minimum and micro-saddle point images. In total there are 12 micro-minimum images on the negative parity side ($\theta_1 < 0$), and 15 micro-minimum images on the positive parity side ($\theta_1 > 0$). The set-up of the calculation is same as the one in Fig. 2

4.2 Effective Sizes

Accounting for the high-angle tail of the PDF of the microlens deflection angle, about 99% of the flux of a point source should be found within a region of size $10 \cdot \theta_\star \kappa_\star^{1/2}$ in the source plane, when considering the smooth mapping to the source place caused only by the smoothed lens and including the average effect of microlenses as an effective source, as explained above. Transforming to the image plane, one gets an ellipse of semi-major axes $r_1 \approx 10 \cdot \theta_\star \kappa_\star^{1/2} \mu_1$ and $r_2 \approx 10 \cdot \theta_\star \kappa_\star^{1/2} \mu_2$. This effective size of the micro image swarm is discussed as well in Vernardos et al. (2024) for the case of quasars.

Figure 4 shows an example of the micro image swarm (or "train") from calculations given in Sect. 4.1. The Figure indicates that there are two trains of the micro images on both sides of the macro critical curve at $\theta_1 = 0$ such that they are highly spread along horizontal di-

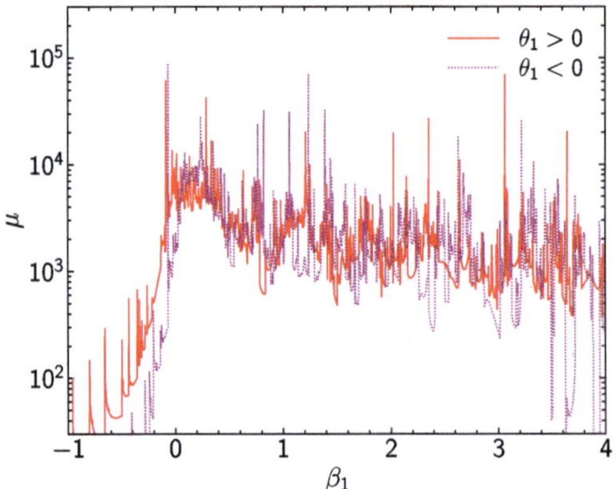

Fig. 5 Similar to Fig. 2, but magnification factors on the positive parity side ($\theta_1 > 0$) and on the negative parity side ($\theta_1 < 0$) are plotted separately by solid and dotted lines, respectively

rections. Since in our macro mass model $\mu_2 = 1$ and $\mu_1(\theta_1 = 4000) = 2500$, the discussion above suggests that the size of the swarm are $r_1 \approx 1250$ and $r_2 \approx 0.5$ in the dimensionless unit. The distribution of micro images shown in Fig. 4 appears to be consistent with this estimate.

4.3 Parity Dependence

As seen in Fig. 4, unless β_1 is small enough that the two swarms of images are merged, there are two image swarms appearing on each side of the macro-critical line. If the two swarms are well separated, they can be observationally resolved and their total magnification can be separately measured. To check the different behavior of each swarm, Fig. 5 shows the total magnifications of the two swarms located on the positive and negative parity sides as a function of the source position. A noticeable difference is that the total magnification of the negative parity side reaches significantly smaller values. As noted in Diego et al. (2018) and Oguri et al. (2018), this is because in the negative parity region, a source can be more strongly de-magnified by the microlenses de-magnified compared with the original macro model magnification (see also, e.g., Schechter and Wambsganss 2002). This effect becomes less pronounced as β_1 decreases because of averaging over many micro-images (see also Sect. 4.5).

4.4 Thickness of the Corrugated Macro-Caustic

The micro-critical lines are corrugated, that is to say, typically joined together in a large-scale network, within a width θ_w around the original macro-critical curve, where the characteristic size of the micro-critical curve of a single microlens is equal to the mean separation between microlenses (Venumadhav et al. 2017). For microlenses with Einstein radius $\theta_\star$ and convergence contribution $\kappa_\star$, their mean separation is given by $\theta_\star \kappa_\star^{-1/2}$. At a separation θ from the macro-critical curve, the magnification eigenvalue of the smooth lens is θ/E, and the micro-critical curve is boosted to a size $\theta_\star(\theta/E)^{-1/2}$. Equating these two quantities we derive the thickness of the corrugated micro-critical curves, θ_w, as

$$\theta_w = E\kappa_\star \, . \tag{15}$$

For our specific example shown in this paper, we find $\theta_w = 25{,}000$, which indicates that results shown in previous subsections are those for micro-images well within the corrugated

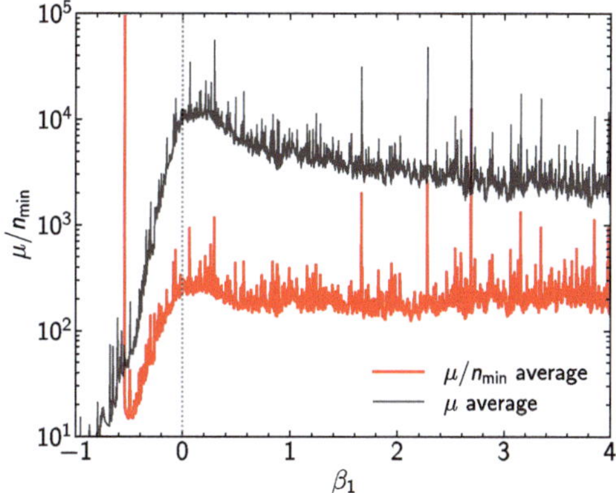

Fig. 6 Magnification factors divided by the number of micro-minimum images ($n_{\rm min}$), averaged over 10 realizations, are plotted as a function of the source position β_1 near the macro-caustic. For comparison, the average total magnification factors plotted in Fig. 3 are also shown by a thin line. The set-up of the calculation is same as the one in Fig. 2

band. Within this thickness $\theta_{\rm w}$, the magnification of the individual micro-images is roughly constant, and given by roughly the magnification at the separation $\theta_{\rm w}$ i.e.,

$$\mu_i \sim \frac{1}{2(1-\kappa_0)\kappa_\star}, \tag{16}$$

which reduces to $\mu_i \sim 400$ in our example. Note that this typical magnification of an individual micro-image is not the same as the total magnification, which increases as we approach the macro-critical line owing to the increasing number of micro-images in the two swarms. As we will see below, the maximum magnification achieved in microcaustic crossings is also constant within the region of width r_w, in the same way as the typical magnification of a single micro-image.

4.5 Mean Number of Extra Micro-Image Pairs

Petters et al. (2009) provide a generic formula for the expected number of positive parity images in lensing systems (see their Eq. 11). In general, the formula relies on knowing the expected value of the magnification as a function of image plane position, conditional upon the fact that image plane positions are mapped to a particular source-plane position. In the case of constant surface mass density and shear, the formula simplifies greatly due to the fact that the expected value of the magnification is independent of image plane position. This is not the case near a macro-caustic, and the matter of the distribution of the magnification is currently an open question. We cannot therefore provide analytic estimates of the mean number of micro-minima, nor the mean magnification per micro-minima, as we could with constant surface mass density and shear. However, we can provide some commentary based on arguments from the simpler case.

For constant convergence and shear that produces an image of formally infinite magnification, the equations in Vernardos et al. (2024) are found to produce an infinite number of expected micro-minima. However, the random shear due to the micro-lenses combines in such a way that the mean magnification per minimum is finite. We know that near a macro-caustic, the magnification will not in fact diverge – as the discussion of the previous sections show, the point mass behaves as a an extended source. The number of micro-minima will not diverge then either.

Figure 6 shows the magnification divided by the number of micro-minima from calculations given in Sect. 4.1. The result indicates that the increase of the mean magnification as approaching the macro-caustic is achieved mainly by the increased number of micro-images, rather than the increase of magnifications of individual micro-images. It is found that the average magnification of micro-minima is $\sim 200-300$, which we find is reasonably close to a rough estimation given in equation (16).

The maximum number of images is reached when the two swarms merge, over a region in the source plane with the width of the effective source size induced by microlensing, $\theta_\star \kappa_\star^{1/2} \sim 10^{-7}$ arcsec for typical values. The time it would take for a source to cross this region is of the order of a year, so if a supermagnified source is close to the position of the smooth model caustic, it should be possible to measure the variation in the frequency of caustic crossings and average magnification that occurs at this position over this timescale.

4.6 Maximum Achievable Magnification

The maximum magnification in a microimage is reached when the source crosses a micro-caustic, which occurs over a timescale determined by the source angular radius and the transverse velocities of source and lens relative to us (Miralda-Escudé 1991). The maximum magnification is determined by the eigenvalue gradient E at the critical line, and its characteristic value is limited by the presence of compact microlenses within the large-scale gravitational lens. The practical case this has been observed in is when the large-scale lens is a cluster of galaxies or a galaxy, and the compact microlenses are individual stars in cluster galaxies or the intracluster population.

As shown in Sect. 4.4, within the corrugated band, the average magnification of any random point (i.e., microimage) in the image plane is $\bar{\mu} \sim [2\kappa_\star(1-\kappa_0)]^{-1}$. This can also be interpreted as the eigenvalue with smallest absolute value fluctuating with a typical value of $\kappa_\star$ with the other eigenvalue being fixed to $2(1-\kappa_0)$. The scale of variation of the small eigenvalue in the image plane is the mean distance between microlenses, $\theta_\star \kappa_\star^{-1/2}$, which in the source plane results in a typical separation between micro-caustics of $\theta_\star \kappa_\star^{1/2}$, implying that the maximum magnification that is typically reached for individual images of a source of angular radius θ_s in micro-critical lines is

$$\mu_{\text{peak}} \sim \frac{(\theta_\star \kappa_\star^{1/2}/\theta_s)^{1/2}}{2\kappa_\star(1-\kappa_0)} = \frac{\theta_\star^{1/2}\kappa_\star^{-3/4}}{2\theta_s^{1/2}(1-\kappa_0)} . \tag{17}$$

An interesting observation from this result is that the maximum achievable magnification decreases with increasing stellar mass density $\kappa_\star$.

5 Applications of Observations

5.1 Summary of Observations

The first discovery of a micro-caustic crossing near a macro-caustic was reported by Kelly et al. (2018), as the rapid transient MACS J1149 Lensed Star 1 (also known as 'Icarus') near the macro critical curve of the massive cluster MACS J1149.5+2223 at $z=0.54$, interpreted as a highly magnified individual star at $z=1.49$. The timescale of the lightcurve near the peak ($\lesssim 10$ days) constrains the size of the background source to be $\lesssim 200\ R_\odot$ for typical transverse velocities expected from large-scale structure (Oguri et al. 2018). Together with the spectral energy distribution and peak magnitude, this led Kelly et al. (2018) to conclude that the transient is an image of a blue supergiant at $z=1.49$ magnified by more than a factor of 2000. Stars making up the intracluster light can fully account for the observed microlensing event rate. Kelly et al. (2018) also reported a separate short transient event detected 0.26″ from Lensed Star 1, which can be a counterimage of Lensed Star 1.

Rodney et al. (2018) reported two peculiar fast transients (collectively nicknamed 'Spock') in a giant arc at $z = 1.0054$ lensed by the massive cluster MACS J0416.1−2403 at $z = 0.397$. While they interpret the transients as eruptions of a luminous blue variable star or a recurrent nova, they also noted these events might be micro-caustic crossings near the macro-critical curve. Improved mass modeling of this cluster to constrain the shape of the critical curves would be helpful to discriminate these possibilities.

The discovery of Icarus has triggered searches of micro-caustic crossings in archival *Hubble Space Telescope* images. Chen et al. (2019) and Kaurov et al. (2019) reported the discovery of a highly magnified star (blue supergiant) at redshift $z = 0.94$ in a giant arc behind the massive cluster MACS J0416.1−2403. The possible counterimage of the background star at an offset of $\sim 0.1''$ is also reported. The discovery suggests that micro-caustic crossing events may be ubiquitously found in deep imaging of massive clusters of galaxies.

Dai et al. (2020) reported several asymmetric surface brightness features in a giant arc at $z = 2.93$ produced by the galaxy cluster SDSS J1226+2152 at $z = 0.43$. While such asymmetric features can be produced by micro-caustic crossings, Dai et al. (2020) concluded that they are more likely to be produced by subhalos with masses of $\sim 10^6 - 10^8\ M_\odot$, based on the absence of notable time variation over a six-year baseline.

Several other observations of supermagnified stars in lensing clusters have been reported recently, some of which are being given special names, and are continuing to be discovered with the ongoing observations with JWST: an additional star called *Mothra* in MACSJ0416 (Diego et al. 2023b), the star *Earendel* in WHL0137-08 (Welch et al. 2022b,a), Godzilla in PSZ1 G311.65 (Diego et al. 2022), *Quyllur* in the El Gordo cluster (Diego et al. 2023a), and several other stars in Abell 2744 (Chen et al. 2022), MACSJ0647 (Meena et al. 2023b; Furtak et al. 2024), and Abell 370 (Kelly et al. 2022; Meena et al. 2023a). In this short paper we cannot do justice to all the wonderful science that is made possible by this blossoming of discoveries.

5.2 Some Applications

We now discuss some applications that the phenomenon of super-magnification in the micro-caustics of lensing clusers of galaxies has already demonstrated or may be developed in the future. This list is by no means complete and is limited by the length of this paper; in general any source, such as supernovae or quasars, may be studied in special ways when highly magnified; in the case of gravitational waves, diffraction effects mean that the interest of lensing observations shifts to masses larger than stellar ones, corresponding to a Schwarzshild radius larger than the observed wavelengths.

5.2.1 Compact Objects in the Dark Matter

As discussed in Sect. 4, one of the most important model parameters that control the property of microlensing near macro-caustics is the surface density of microlenses, $\kappa_\star$. Normally only stars in lensing clusters or galaxies are considered as microlenses, whose abundance near macro-caustics can be inferred from observations of intracluster light. However, if a fraction of dark matter is composed of dark compact objects, such as primordial black holes (Carr and Hawking 1974), the surface density of microlenses may be much higher. This, in turn, means we can constrain the abundance of compact objects from observations of micro-caustic crossings in a manner complementary to other constraints (see e.g., Sasaki et al.

2018, for a review of constraints on primordial black holes), including quasar microlensing (Mediavilla et al. 2017).

Constraints on compact objects from the Icarus highly magnified star have extensively been explored in Kelly et al. (2018) as well as in follow-up papers by Venumadhav et al. (2017), Diego et al. (2018), and Oguri et al. (2018). For instance, as explained in Sect. 4.6 the maximum achievable magnification is reduced as $\kappa_\star^{-3/4}$. Following this idea, Oguri et al. (2018) adopted a simple analytic model to derive constraints on the compact object abundance for a wide range of mass ranging from $10^{-5}\ M_\odot$ to $10^2\ M_\odot$. In addition, a large $\kappa_\star$ reduces the time variability of lensed sources due to averaging effects, which also yields constraints (Kelly et al. 2018; Diego et al. 2018).

Furthermore, as discussed in Sect. 4.4, the compact object abundance determines the width of the corrugated band around the macro-critical curve within which most highly magnified images appear, so an independent constraint is obtained from the spread of micro-caustic crossings around macro critical curves (Venumadhav et al. 2017). A limit can then be derived on $\kappa_\star$ in compact objects from the set of highly magnified stars at cosmological distances reported so far. The width of the corrugated band is $\theta_w = d_{\|}\kappa_\star$, where $d_{\|}$ is the gradient of the small magnification eigenvalue. In the first event named Icarus, the macro model predicts $d_{\|} \simeq 4\,\mathrm{arcmin}^{-1}$, and the separation of the image from the predicted critical curve is ~ 0.13 arcsec (Kelly et al. 2018; Venumadhav et al. 2017). In the MACS J0416.1-2403 event, the macro model predicts $d_{\|} = 7\,\mathrm{arcmin}^{-1}$ and the separation of the image is 0.1 arcsec (Diego et al. 2018; Oguri et al. 2018). The ratios of the observed image separations compared to the maximum separation where images with the highest magnifications are observed is therefore $0.012\kappa_\star$ and $0.009\kappa_\star$, respectively. The random probability these two values are observed at least as small as they are is $\sim 10^{-4}\kappa_\star^2$. Requiring this probability to be at least larger than 1% we obtain the limit $\kappa_\star < 0.1$. This is valid for a broad range of microlens mass, roughly $10^{-6}\ M_\odot < M$, over which other effects like diffraction or source size are not important (see Fig. 9 in Venumadhav et al. 2017). Naturally, as the number of discovered events increases in cluster lenses that can be adequately modeled, this limit on any objects that are sufficiently compact to act as microlenses near macro-caustics will improve essentially to the value of the microlenses accounted by intracluster stars.

Recently, this argument has been developed in detail using most of the discovered supermagnified stars by Vall Müller and Miralda-Escudé (2024), and in the future this limit should rapidly improve as more examples of these stars with extreme magnifications are revealed by the potential of JWST and other telescopes.

5.2.2 Observing Population III Stars

Microlensing near the macro-caustics provides a means of observing distant individual stars that cannot be observed without the high magnification achieved in micro-caustic crossings. Windhorst et al. (2018) explored the possibility of observing individual Population III (Pop III) stars, which are metal-free stars formed from pristine gas, at $z \simeq 7-17$, and concluded that direct observations of highly magnified Pop III stars would be possible by monitoring a few to a few tens of mass clusters over a decade with the James Webb Space Telescope.

5.2.3 Dark Matter Subhalos

Dai et al. (2018) proposed a novel method to constrain the abundance of dark matter subhalos with masses of $10^6 - 10^8\ M_\odot$, whose existence is naturally predicted by the standard cold dark matter paradigm (e.g., Diemand et al. 2008), using observations of image pairs of sources near caustics. The method uses astrometry of these highly magnified image pairs of stars or other luminous objects (star clusters or star-forming regions) in caustic-straddling giant arcs and searches for perturbations of the shape of macro critical curves, which are expected to be smooth on small scales in the absence of subhalos. Subhalos with masses of $10^6 - 10^8\ M_\odot$ produce irregularities on the critical curve shape (see also Abe et al. 2024), and the astrometric perturbations of image pairs, at the level of $20 - 80$ mas, which can be detected by accurate centroid measurements of several magnified pairs by the James Webb Space Telescope. Known cluster galaxies produce irregularities on larger scales only, corresponding to their higher masses.

5.2.4 Axion Minihalos

Finally, axion minihalos may also be detectable using highly magnified stars. Axion minihalos are expected in many variations of the theory of axions for dark matter that solve the strong QCD problem, with masses of $\sim 10^{-12}\ M_\odot$. These are extremely difficult to detect, not only because of their tiny size but also because their surface density is highly subcritical to lensing. However, when a star crosses one of the micro-caustics affected by micro-lensing from intracluster stars, axion minihalos would create a secondary level of perturbations on the magnification at smaller scales that would cause perturbations of the lightcurve of micro-caustic crossing events on small time scales. This was predicted in Dai and Miralda-Escudé (2020), proposing a possible route to the detection of axion dark matter using only astrophysical methods.

6 Summary

In this artcile we have explored the expected microlensing magnification of sources close to macro-caustics. For this purpose, we have performed simulations and compared the results to the analytical expectations of the behavior of microimages, also described in Venumadhav et al. (2017), Dai and Pascale (2021), Vernardos et al. (2024, this collection).

We have shown that the mean magnification due to lens plane microlenses of point sources in the vicinity of the macro-caustic follows that of an extended source in the vicinity of a macro-caustic in the absence of microlensing. This is precisely correct when the source profile matches the probability distribution function of the microlens deflection angle.

Similarly, we have shown that the two swarms of microimages at each side of the macro-caustic are spread within an ellipse that results from mapping the microlens deflection angle PDF, or the corresponding "extended source", to the lens plane. We note that even though the sizes of the two image swarms are symmetric, the swarm of images at the negative parity side of the macro-caustic can show significantly lower magnification values. This is consistent with the findings of Schechter and Wambsganss (2002) who show that microlensed sources in macro-saddle points are more prone to be de-magnified than in macro-minima.

We have reviewed the characteristic number and average magnification of the micro-images within the microlens-distorted (or corrugated) macro-caustic. We observe that the magnification per micro-image remains roughly constant within this corrugated caustic and with a value given by the separation of micro-image pairs from micro-caustics. The maximum individual micro-image magnification is limited by the angular size of the source, and decreases with increasing stellar mass density. The increase of the mean magnification when approaching the position of the macro-caustic is driven by the increase in the number of micro-images rather than their individual magnifications. Analogous to the fact that micro-images behave as a lensed extended source and thus the total magnification is not infinite when crossing the macro-caustic, the number of micro-images reaches a maximum when the two image swarms merge.

The microlensing phenomena occurring near macro-caustics that we have reviewed in this Chapter have not been observed until recently. Since the dawn of the era of time-domain astronomy, in the past few years several transients have been shown to be consistent with such microlensing in the vicinity of macro-caustics. These discoveries imply that microcaustic crossing events should be ubiquitous when monitoring observations of massive galaxy clusters, and will surely be increased by observations with JWST reaching for fainter magnitudes. With this in mind, we have non-exhaustively reviewed some future (and current) applications of the different observational signatures, ranging from identifying population III stars from individual caustic crossing events to constraining the abundance and nature of primordial black holes and dark matter subhalos from the statistical properties of observed caustic crossing events and probing the distorted macro-caustic from astrometric perturbations in magnified background stars.

Funding Open Access funding provided thanks to the CRUE-CSIC agreement with Springer Nature. This work was supported in part by JSPS KAKENHI Grant Number JP22H01260, JP20H05856, and by Spanish grant PID2019-108122GB-C32. TA acknowledges support the Millennium Science Initiative ICN12_009 and the ANID BASAL project FB210003.

Declarations

Competing Interests The authors declare no competing interests.

Open Access This article is licensed under a Creative Commons Attribution 4.0 International License, which permits use, sharing, adaptation, distribution and reproduction in any medium or format, as long as you give appropriate credit to the original author(s) and the source, provide a link to the Creative Commons licence, and indicate if changes were made. The images or other third party material in this article are included in the article's Creative Commons licence, unless indicated otherwise in a credit line to the material. If material is not included in the article's Creative Commons licence and your intended use is not permitted by statutory regulation or exceeds the permitted use, you will need to obtain permission directly from the copyright holder. To view a copy of this licence, visit http://creativecommons.org/licenses/by/4.0/.

References

Abe KT, Kawai H, Oguri M (2024) Analytic approach to astrometric perturbations of critical curves by substructures. Phys Rev D 109:083517. https://doi.org/10.1103/PhysRevD.109.083517. arXiv:2311.18211 [astro-ph.CO]

Carr BJ, Hawking SW (1974) Black holes in the early universe. Mon Not R Astron Soc 168:399–416. https://doi.org/10.1093/mnras/168.2.399

Chen W, Kelly PL, Diego JM, Oguri M, Williams LLR, Zitrin A, Treu TL, Smith N, Broadhurst TJ, Kaiser N, Foley RJ, Filippenko AV, Salo L, Hjorth J, Selsing J (2019) Searching for highly magnified stars at cosmological distances: discovery of a redshift 0.94 blue supergiant in archival images of the galaxy cluster MACS J0416.1-2403. Astrophys J 881(1):8. https://doi.org/10.3847/1538-4357/ab297d

Chen W, Kelly PL, Treu T, Wang X, Roberts-Borsani G, Keen A, Windhorst RA, Zhou R, Bradac M, Brammer G, Strait V, Broadhurst TJ, Diego JM, Frye BL, Meena AK, Zitrin A, Pascale M, Castellano M, Marchesini D, Morishita T, Yang L (2022) Early results from GLASS-JWST. VIII. An extremely magnified blue supergiant star at redshift 2.65 in the A2744 cluster field. Astrophys J Lett 940(2):54. https://doi.org/10.3847/2041-8213/ac9585

Dai L, Miralda-Escudé J (2020) Gravitational lensing signatures of axion dark matter minihalos in highly magnified stars. Astron J 159(2):49. https://doi.org/10.3847/1538-3881/ab5e83

Dai L, Pascale M (2021) New Approximation of Magnification Statistics for Random Microlensing of Magnified Sources. arXiv e-prints arXiv:2104.12009 [astro-ph.GA]

Dai L, Venumadhav T, Kaurov AA, Miralda-Escudé J (2018) Probing dark matter subhalos in galaxy clusters using highly magnified stars. Astrophys J 867(1):24. https://doi.org/10.3847/1538-4357/aae478

Dai L, Kaurov AA, Sharon K, Florian M, Miralda-Escudé J, Venumadhav T, Frye B, Rigby JR, Bayliss M (2020) Asymmetric surface brightness structure of caustic crossing arc in SDSS J1226+2152: a case for dark matter substructure. Mon Not R Astron Soc 495(3):3192–3208. https://doi.org/10.1093/mnras/staa1355

Diego JM, Kaiser N, Broadhurst T, Kelly PL, Rodney S, Morishita T, Oguri M, Ross TW, Zitrin A, Jauzac M, Richard J, Williams L, Vega-Ferrero J, Frye B, Filippenko AV (2018) Dark matter under the microscope: constraining compact dark matter with caustic crossing events. Astrophys J 857(1):25. https://doi.org/10.3847/1538-4357/aab617

Diego JM, Pascale M, Kavanagh BJ, Kelly P, Dai L, Frye B, Broadhurst T (2022) Godzilla, a monster lurks in the sunburst galaxy. Astron Astrophys 665:134. https://doi.org/10.1051/0004-6361/202243605

Diego JM, Meena AK, Adams NJ, Broadhurst T, Dai L, Coe D, Frye B, Kelly P, Koekemoer AM, Pascale M, Willner SP, Zackrisson E, Zitrin A, Windhorst RA, Cohen SH, Jansen RA, Summers J, Tompkins S, Conselice CJ, Driver SP, Yan H, Grogin N, Marshall MA, Pirzkal N, Robotham A, Ryan RE, Willmer CNA, Bradley LD, Caminha G, Caputi K, Carleton T, Kamieneski P (2023a) JWST's PEARLS: a new lens model for ACT-CL J0102−4915, "El Gordo," and the first red supergiant star at cosmological distances discovered by JWST. Astron Astrophys 672:3. https://doi.org/10.1051/0004-6361/202245238

Diego JM, Sun B, Yan H, Furtak LJ, Zackrisson E, Dai L, Kelly P, Nonino M, Adams N, Meena AK, Willner SP, Zitrin A, Cohen SH, Silva JCJD, Jansen RA, Summers J, Windhorst RA, Coe D, Conselice CJ, Driver SP, Frye B, Grogin NA, Koekemoer AM, Marshall MA, Pirzkal N, Robotham A, Rutkowski MJ, Ryan Russell E J, Tompkins S, Willmer CNA, Bhatawdekar R (2023b) JWST's PEARLS: Mothra, a new kaiju star at z = 2.091 extremely magnified by MACS0416, and implications for dark matter models. Astron Astrophys 679:A31. https://doi.org/10.1051/0004-6361/202347556. arXiv:2307.10363 [astro-ph.CO]

Diemand J, Kuhlen M, Madau P, Zemp M, Moore B, Potter D, Stadel J (2008) Clumps and streams in the local dark matter distribution. Nature 454(7205):735–738. https://doi.org/10.1038/nature07153

Furtak LJ, Meena AK, Zackrisson E, Zitrin A, Brammer GB, Coe D, Diego JM, Eldridge JJ, Jiménez-Teja Y, Kokorev V, Ricotti M, Welch B, Windhorst Abdurro'uf RA, Andrade-Santos F, Bhatawdekar R, Bradley LD, Broadhurst T, Chen W, Conselice CJ, Dayal P, Frye BL, Fujimoto S, Hsiao TY-Y, Kelly PL, Mahler G, Mandelker N, Norman C, Oguri M, Pirzkal N, Postman M, Ravindranath S, Vanzella E, Wilkins SM (2024) Reaching for the stars - JWST/NIRSpec spectroscopy of a lensed star candidate at z = 4.76. Mon Not R Astron Soc 527(1):7–13. https://doi.org/10.1093/mnrasl/slad135

Granot J, Schechter PL, Wambsganss J (2003) The mean number of extra microimage pairs for macrolensed quasars. Astrophys J 583(2):575–583. https://doi.org/10.1086/345447

Katz N, Balbus S, Paczynski B (1986) Random scattering approach to gravitational microlensing. Astrophys J 306:2. https://doi.org/10.1086/164313

Kaurov AA, Dai L, Venumadhav T, Miralda-Escudé J, Frye B (2019) Highly magnified stars in lensing clusters: new evidence in a galaxy lensed by MACS J0416.1-2403. Astrophys J 880(1):58. https://doi.org/10.3847/1538-4357/ab2888

Kelly PL, Diego JM, Rodney S, Kaiser N, Broadhurst T, Zitrin A, Treu T, Pérez-González PG, Morishita T, Jauzac M, Selsing J, Oguri M, Pueyo L, Ross TW, Filippenko AV, Smith N, Hjorth J, Cenko SB, Wang X, Howell DA, Richard J, Frye BL, Jha SW, Foley RJ, Norman C, Bradac M, Zheng W, Brammer G, Benito AM, Cava A, Christensen L, de Mink SE, Graur O, Grillo C, Kawamata R, Kneib J-P, Matheson T, McCully C, Nonino M, Pérez-Fournon I, Riess AG, Rosati P, Schmidt KB, Sharon K, Weiner BJ (2018) Extreme magnification of an individual star at redshift 1.5 by a galaxy-cluster lens. Nat Astron 2:334–342. https://doi.org/10.1038/s41550-018-0430-3

Kelly PL, Chen W, Alfred A, Broadhurst TJ, Diego JM, Emami N, Filippenko AV, Keen A, Li Kei S, Lim J, Meena AK, Oguri M, Scarlata C, Treu T, Williams H, Williams LLR, Zhou R, Zitrin A, Foley RJ, Jha SW, Kaiser N, Mehta V, Rieck S, Salo L, Smith N, Weisz DR (2022) Flashlights: More than a Dozen High-Significance Microlensing Events of Extremely Magnified Stars in Galaxies at Redshifts z = 0.7-1.5. arXiv e-prints arXiv:2211.02670 [astro-ph.CO]. https://doi.org/10.48550/arXiv.2211.02670

Mediavilla E, Jiménez-Vicente J, Muñoz JA, Vives-Arias H, Calderón-Infante J (2017) Limits on the mass and abundance of primordial black holes from quasar gravitational microlensing. Astrophys J Lett 836(2):18. https://doi.org/10.3847/2041-8213/aa5dab

Meena AK, Chen W, Zitrin A, Kelly PL, Golubchik M, Zhou R, Alfred A, Broadhurst T, Diego JM, Filippenko AV, Li SK, Oguri M, Smith N, Williams LLR (2023a) Flashlights: an off-caustic lensed star at redshift z = 1.26 in Abell 370. Mon Not R Astron Soc 521(4):5224–5231. https://doi.org/10.1093/mnras/stad869

Meena AK, Zitrin A, Jiménez-Teja Y, Zackrisson E, Chen W, Coe D, Diego JM, Dimauro P, Furtak LJ, Kelly PL, Oguri M, Welch Abdurro'uf B, Andrade-Santos F, Adamo A, Bhatawdekar R, Bradač M, Bradley LD, Broadhurst T, Conselice CJ, Dayal P, Donahue M, Frye BL, Fujimoto S, Hsiao TY-Y, Kokorev V, Mahler G, Vanzella E, Windhorst RA (2023b) Two lensed star candidates at z $\simeq$ 4.8 behind the galaxy cluster MACS J0647.7+7015. Astrophys J Lett 944(1):6. https://doi.org/10.3847/2041-8213/acb645

Miralda-Escudé J (1991) The magnification of stars crossing a caustic. I. Lenses with smooth potentials. Astrophys J 379:94. https://doi.org/10.1086/170486

Neindorf B (2003) A probability theoretical access to extragalactic microlensing. Astron Astrophys 404:83–92. https://doi.org/10.1051/0004-6361:20030098

Oguri M (2010) The mass distribution of SDSS J1004+4112 revisited. Publ Astron Soc Jpn 62:1017. https://doi.org/10.1093/pasj/62.4.1017

Oguri M, Diego JM, Kaiser N, Kelly PL, Broadhurst T (2018) Understanding caustic crossings in giant arcs: characteristic scales, event rates, and constraints on compact dark matter. Phys Rev D 97(2):023518. https://doi.org/10.1103/PhysRevD.97.023518

Paczynski B (1986) Gravitational microlensing at large optical depth. Astrophys J 301:503. https://doi.org/10.1086/163919

Petters AO, Rider B, Teguia AM (2009) A mathematical theory of stochastic microlensing. II. Random images, shear, and the Kac-Rice formula. J Math Phys 50(12):122501. https://doi.org/10.1063/1.3267859

Rodney SA, Balestra I, Bradac M, Brammer G, Broadhurst T, Caminha GB, Chirivı G, Diego JM, Filippenko AV, Foley RJ, Graur O, Grillo C, Hemmati S, Hjorth J, Hoag A, Jauzac M, Jha SW, Kawamata R, Kelly PL, McCully C, Mobasher B, Molino A, Oguri M, Richard J, Riess AG, Rosati P, Schmidt KB, Selsing J, Sharon K, Strolger L-G, Suyu SH, Treu T, Weiner BJ, Williams LLR, Zitrin A (2018) Two peculiar fast transients in a strongly lensed host galaxy. Nat Astron 2:324–333. https://doi.org/10.1038/s41550-018-0405-4

Saha P, Williams LLR (2011) Understanding micro-image configurations in quasar microlensing. Mon Not R Astron Soc 411(3):1671–1677. https://doi.org/10.1111/j.1365-2966.2010.17797.x

Sasaki M, Suyama T, Tanaka T, Yokoyama S (2018) Primordial black holes—perspectives in gravitational wave astronomy. Class Quantum Gravity 35(6):063001. https://doi.org/10.1088/1361-6382/aaa7b4

Schechter PL, Wambsganss J (2002) Quasar microlensing at high magnification and the role of dark matter: enhanced fluctuations and suppressed saddle points. Astrophys J 580(2):685–695. https://doi.org/10.1086/343856

Schneider P, Ehlers J, Falco EE (1992) Gravitational Lenses. Astronomy and Astrophysics Library. Springer, Berlin, Heidelberg. https://doi.org/10.1007/978-3-662-03758-4

Vall Müller C, Miralda-Escudé J (2024) Limits on Dark Matter Compact Objects implied by Supermagnified Stars in Lensing Clusters. arXiv e-prints arXiv:2403.16989 [astro-ph.CO]. https://doi.org/10.48550/arXiv.2403.16989

Venumadhav T, Dai L, Miralda-Escudé J (2017) Microlensing of extremely magnified stars near caustics of galaxy clusters. Astrophys J 850(1):49. https://doi.org/10.3847/1538-4357/aa9575

Vernardos G, Sluse D, Pooley D, Schmidt RW, Millon M, Weisenbach L, Motta V, Anguita T, Saha P, O'Dowd M, Peel A, Schechter PL (2024) Microlensing of strongly lensed quasars. Space Sci Rev 220(1):14. https://doi.org/10.1007/s11214-024-01043-8

Wambsganss J (1990) Gravitational microlensing. PhD thesis. Ludwig-Maximilians-Universität, Munich

Wambsganss J (1999) Gravitational lensing: numerical simulations with a hierarchical tree code. J Comput Appl Math 109(1):353–372

Welch B, Coe D, Diego JM, Zitrin A, Zackrisson E, Dimauro P, Jiménez-Teja Y, Kelly P, Mahler G, Oguri M, Timmes FX, Windhorst R, Florian M, de Mink SE, Avila RJ, Anderson J, Bradley L, Sharon K, Vikaeus A, McCandliss S, Bradač M, Rigby J, Frye B, Toft S, Strait V, Trenti M, Sharma S, Andrade-Santos F, Broadhurst T (2022a) A highly magnified star at redshift 6.2. Nature 603(7903):815–818. https://doi.org/10.1038/s41586-022-04449-y

Welch B, Coe D, Zackrisson E, de Mink SE, Ravindranath S, Anderson J, Brammer G, Bradley L, Yoon J, Kelly P, Diego JM, Windhorst R, Zitrin A, Dimauro P, Jiménez-Teja Abdurro'uf Y, Nonino M, Acebron A, Andrade-Santos F, Avila RJ, Bayliss MB, Benítez A, Broadhurst T, Bhatawdekar R, Bradač M, Caminha GB, Chen W, Eldridge J, Farag E, Florian M, Frye B, Fujimoto S, Gomez S, Henry A, Hsiao TY-Y, Hutchison TA, James BL, Joyce M, Jung I, Khullar G, Larson RL, Mahler G, Mandelker N, McCandliss S, Morishita T, Newshore R, Norman C, O'Connor K, Oesch PA, Oguri M, Ouchi M, Postman M, Rigby JR, Ryan REJ, Sharma S, Sharon K, Strait V, Strolger L-G, Timmes FX, Toft S, Trenti M, Vanzella E, Vikaeus A (2022b) JWST imaging of earendel, the extremely magnified star at redshift z = 6.2. Astrophys J Lett 940(1):1. https://doi.org/10.3847/2041-8213/ac9d39

Windhorst RA, Timmes FX, Wyithe JSB, Alpaslan M, Andrews SK, Coe D, Diego JM, Dijkstra M, Driver SP, Kelly PL, Kim D (2018) On the observability of individual population III stars and their stellar-mass black hole accretion disks through cluster caustic transits. Astrophys J Suppl Ser 234(2):41. https://doi.org/10.3847/1538-4365/aaa760

Publisher's Note Springer Nature remains neutral with regard to jurisdictional claims in published maps and institutional affiliations.

Space Science Reviews (2025) 221:29
https://doi.org/10.1007/s11214-025-01157-7

Wave Optics, Interference, and Decoherence in Strong Gravitational Lensing

Calvin Leung[1,2] · **Dylan Jow**[3,4] · **Prasenjit Saha**[5] · **Liang Dai**[6] · **Masamune Oguri**[7] · **Léon V.E. Koopmans**[8,9]

Received: 23 April 2023 / Accepted: 28 February 2025 / Published online: 13 March 2025
© The Author(s) 2025

Abstract
Wave-optical effects in gravitational lensing have long been predicted, and with the discovery of populations of compact transients such as gravitational wave events and fast radio bursts, may soon be observed. We present an observer's review of the relevant theory underlying wave-optical effects in gravitational lensing. Starting from the curved-spacetime scalar wave equation, we derive the Fresnel-Kirchoff diffraction integral, and analyze it in the eikonal and wave optics regimes. We answer the question of what makes interference effects observable in some systems but not in others, and how interference effects allow for complementary information to be extracted from lensing systems as compared to traditional measurements. We end by discussing how diffraction effects affect optical depth forecasts and lensing near caustics, and how compact, low-frequency transients like gravitational waves and fast radio bursts provide promising paths to open up the frontier of coherent gravitational lensing.

Keywords Gravitational lensing · Wave optics · Gravitational waves · Transients

1 Introduction

Gravitational lensing in the geometric optics approximation has enjoyed immense success as a unique and precise probe of cosmology and astrophysics. However, electromagnetic and gravitational radiation both obey wave equations, have a wave-like nature, and interfere and diffract as they propagate to Earth from the cosmos. While interference phenomena may seem of little relevance outside a controlled laboratory, it is worth remembering that the first theoretical predictions of gravitational lensing (Chwolson 1924) were made decades before the first observation of an Einstein ring (Hewitt et al. 1988). Similarly, wave effects in gravitational lensing were predicted as far back as 1973 (Ohanian 1973): decades before detections were on the horizon.

Now, there is hope on multiple fronts: Lensed supernovae (Kelly et al. 2015) have already been detected. In principle, wave-optical effects should be observable in the lensing of gravitational wave (GW) signals from binary black holes (Abbott et al. 2016), binary neutron stars (Abbott et al. 2017a), and even black hole-neutron star mergers (Abbott et al. 2021c), as well as a large population of fast radio bursts (FRBs) at low frequencies (CHIME/FRB

Extended author information available on the last page of the article

Collaboration et al. 2021). Growing observational efforts in these areas puts astronomers well on their way to detecting interference effects in gravitational lensing. Measuring these interference effects enables unique tests of gravity: for example, testing the equivalence principle using precise measurements of time lags between electromagnetic and gravitational-wave images (Takahashi 2017). It is also worth mentioning that Wheeler's cosmic delayed-choice experiment, concerning the interference of single photons over long astronomical path lengths, remains a classic test of the foundations of quantum mechanics (Miller and Wheeler 1997; Jacques et al. 2007; Doyle and Carico 2009; Leung et al. 2018).

Wave-optical effects in gravitational lensing also offer a significant enhancement of gravitational lensing as an astrophysical tool. In traditional lensing, only a handful of numbers (the flux magnification, time delay, and image separations/distortions) can be measured, leading to difficult degeneracies in lensing-based inference: the mass-sheet degeneracy is perhaps the best-known example (Falco et al. 1985; Saha 2000). In the wave-optics limit, much more information is present, since the amplitude and phase of the magnification have nontrivial wavelength dependences which may be observable in the near future. For instance, lensing time delays may be extracted coherently with dramatically-increased precision. At radio wavelengths these time delays may be measured coherently with a precision of $\sim 10^{-9} - 10^{-6}$ seconds: several orders of magnitude more precise than any lensing delay ($\sim$ 1 millisecond) inferred incoherently from FRB light curves (Wucknitz et al. 2021; Jow et al. 2020), and orders of magnitude better than other lensed transients.

Wave-optical effects may also shed light on the emission physics of compact objects (such as GWs and FRBs), by using gravitational lenses as astrophysical-scale interferometers. This concept has been demonstrated in the study of pulsar scattering in the ionized interstellar medium to resolve the nanoarcsecond-scale emission region of the Crab pulsar (Lin et al. 2023). The observation of interference effects in gravitational lensing is a new frontier in using lensing to learn about our universe.

The theory of interference effects in gravitational lensing was developed since as far back as 1973 (Ohanian 1973, 1974; Peters 1974). Further developments have mainly focused on GW lensing (Deguchi and Watson 1986a,b; Nakamura 1998; Nakamura and Deguchi 1999; Baraldo et al. 1999; Takahashi and Nakamura 2003; Macquart 2004) and because of the relative simplicity of the gravitational waveforms produced by compact binary coalescences, and the low (kilohertz) emission frequencies at which interference effects are most pronounced. However, wave-optical effects in lensing are of growing importance at frequencies far higher than gravitational waves. An early example was in using "femtolensing" to probe extremely low mass scales by analyzing the spectra of gamma-ray bursts (GRBs) (Gould 1992; Stanek et al. 1993; Nemiroff et al. 1993) More recently, various other applications using lensing to search for exotic phenomena have had to contend with wave-optical effects, which limit the reach of the search. Some examples include optical microlensing searches for primordial black holes (PBHs; see e.g. Sugiyama et al. (2020), searches for PBHs using lensed FRBs (Katz et al. 2020; Kader et al. 2022; Leung et al. 2022), searches for small-scale density perturbations in minihalos or e.g. axion dark matter (Venumadhav et al. 2017; Dai and Miralda-Escudé 2020), and even searching for lensing by cosmic strings (Xiao et al. 2022).

In this review, we aim to present the fundamental concepts underlying wave interference in gravitational lensing. We begin with the Fresnel-Kirchoff diffraction integral (though a first-principles derivation is provided in the Appendix), and analyze it in two opposite regimes. We first analyze the stationary-phase (a.k.a. eikonal) regime in Sect. 4. Then, we present some analytical solutions from the diffractive (a.k.a. wave optics) regime (Sect. 5) and comment on numerical solution methods for this regime, in which the stationary phase

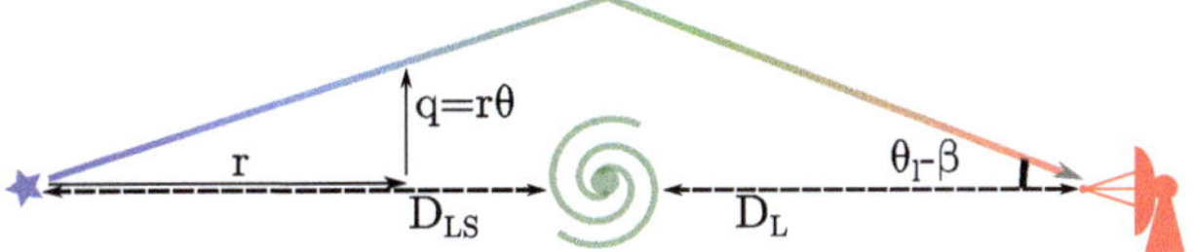

Fig. 1 A schematic illustrating the source-frame coordinates (r) used to derive the Fresnel-Kirchoff diffraction integral (see the Appendix), as well as the observer-frame distances and angles commonly used in the literature and throughout the main text

approximation fails. We analyze how the finite angular sizes of sources pose a challenge to detecting interference effects (Sect. 6) for sources which are physically coherent (e.g. gravitational waves) and sources which are point-like (spatially coherent), and how wave optics in the vicinity of a caustic limits the maximum magnification attainable (Sect. 7) and qualitatively alters the coherence requirements for interference effects to be observed.

Finally, in anticipation of detections of wave-optical effects, we discuss practical ramifications. One novel result is a calculation of the optical depth to lensing in wave optics, which increases by $\sim 50\%$ relative to that of geometric optics (Sect. 8). several proposed application of wave optics in lensing (Sect. 9). We finally comment on how various compact transients offer opportunities for detections in the future (Sect. 10).

2 The Fresnel-Kirchoff Diffraction Integral

Many key results in wave-optical gravitational lensing can be derived starting with the Fresnel-Kirchoff diffraction integral.

$$F(\omega, \boldsymbol{\beta}) \propto \int d^2\boldsymbol{\theta}_l \exp\left\{i\omega\left[\frac{D_{\rm eff}}{2c}|\boldsymbol{\beta} - \boldsymbol{\theta}_l|^2 - \hat{\psi}(D_L\boldsymbol{\theta}_l)\right]\right\} \tag{1}$$

This can be derived from the propagation equation in general relativity; we refer the motivated reader to the Appendix. The derivation relies on several assumptions which hold well in most astrophysical scenarios. First, that the lens is thin with respect to the propagation path (Kling and Frittelli 2008); second, that the light takes piecewise-straight lines; and third, that gravity along the line of sight is sufficiently weak and static that the Newtonian potential suffices to describe its effects; and fourth, that the deflections of light rays are small compared to the path lengths involved.

In Eq. (1), ω is the observing frequency and $\boldsymbol{\beta}$ is the source's angular position with respect to the lens $\boldsymbol{\theta}_l$ (see Fig. 1). We also have defined $D_{\rm eff} = D_L D_S / D_{LS}$ where the distances D_S, D_L, and D_{LS} are the angular diameter distances between the observer and the source, between the observer and the lens, and between the source and lens respectively. In a cosmological context, one must also account for the redshifted frequency of the light as it encounters the lens: $\omega \to \omega(1 + z_l)$ (Schneider et al. 1992, Chap. 4.6).

It is convenient to define a characteristic scale to measure deflection angles. While any scale may be chosen, each lens model comes with a different "natural" choice of the angular scale. For instance, the angular scale is often chosen such that the dimensionless angular radius of the Einstein ring is 1. For example, for a singular isothermal sphere characterized by its velocity dispersion σ, the deflection angle is a constant and defines the angular scale θ_E by which to normalize Eq. (1):

$$\theta_E = 4\pi \frac{\sigma^2}{c^2} \frac{D_{LS}}{D_S} \tag{2}$$

While the singular isothermal sphere is common in geometric optics, its solution in wave optics is quite complicated. Instead, in wave optics, the point mass model is much more commonly analyzed. We write Eq. (1) in terms of dimensionless variables using a notional point-mass lens model. However, this non-dimensionalization scheme is not restricted to a point-mass lens; The constant M can be understood to be some characteristic mass scale of the lens. Similarly θ_E need not be the Einstein radius, but is understood to be some characteristic scale of deflections.

$$\theta_E = \sqrt{\frac{4GMD_{\text{eff}}}{c^2}} \tag{3}$$

$$\boldsymbol{\theta}_l \rightarrow \mathbf{x}\theta_E \tag{4}$$

$$\boldsymbol{\beta} \rightarrow \mathbf{y}\theta_E \tag{5}$$

$$\hat{\psi}(\boldsymbol{\theta}_l) \rightarrow \psi(\boldsymbol{x})\theta_E^2 D_{\text{eff}} \tag{6}$$

$$\omega \rightarrow \frac{\Omega c}{D_{\text{eff}}\theta_E^2} \tag{7}$$

Note that

$$\Omega = 4GM\omega/c^3 = 4\pi R_s/\lambda \tag{8}$$

has a physical interpretation as the ratio of one wavelength to the Schwarzschild radius $R_s = 2GM/c^2$. The dimensionless frequency, Ω, therefore plays a crucial role in determining whether geometric optics is a valid approximation. Equation (1) becomes

$$F(\Omega, \boldsymbol{y}) = \frac{\Omega}{2\pi i}\int d^2\boldsymbol{x}\exp(i\Omega\tau(\boldsymbol{x}, \boldsymbol{y})) \tag{9}$$

where we have also introduced a normalization constant into $F(\Omega, \boldsymbol{y})$ such that when $\psi \rightarrow 0$, $F \rightarrow 1$. The dimensionless time-delay potential is

$$\tau(\boldsymbol{x}, \boldsymbol{y}) = \frac{|\boldsymbol{x} - \boldsymbol{y}|^2}{2} - \psi(\boldsymbol{x}). \tag{10}$$

In interstellar optics, there is an independent length scale in addition to the Einstein radius. The Fresnel scale is the transverse length scale over which interference phenomena occur; we define it as $R_F = \sqrt{\lambda D}$. To see why $R_F = \sqrt{\lambda D}$, consider a spherical wavefront of wavelength λ emanating from a point source towards a plane at some distance $D \gg \lambda$. The shortest straight-line path from the source to the screen is of length D, while a slightly longer straight-line path of length $D + \lambda/2$ hits the screen a transverse distance $\sqrt{\lambda D}$ away from the origin and destructively interferes. We see that wave optics effects manifest on small angular deflections where $\sqrt{\lambda/D} \ll 1$, justifying the small-angle approximation above.

3 Different Regimes in Wave Optical Gravitational Lensing

Equation (9) looks simple, but describes extremely diverse phenomena, including all of standard (geometric optics-based) gravitational lensing phenomena. This is because the complex-valued amplification F (not to be confused for the positive-valued flux magnification from geometric optics) carries both amplitude and phase information, which can result in constructive and destructive interference. For an amplification factor of F, the flux of the source gets magnified by $|F|^2$, in analogy with the Born rule from quantum mechanics. To extend the analogy, the overall phase of F is arbitrary, as the choice of the origin of the source plane $\boldsymbol{y}$ was arbitrary.

Just like in traditional geometric optics, we care about the stationary points of the lens potential $\tau(\boldsymbol{x}, \boldsymbol{y})$:

$$\frac{\partial \tau}{\partial x_a} = y_a - x_a - \partial_a \psi = 0. \tag{11}$$

However, in wave optics, we sum over all possible paths through the lens plane: the resulting amplification is proportional to the size of the coherent "patches," or Fresnel zones, surrounding the stationary points in the lens plane. The first Fresnel zone is defined here as the patch surrounding each stationary point in the lens plane for which the phase has not changed by $2\,\pi$; its size is related to the magnification of the image. Higher-order Fresnel zones (corresponding to a phase difference of $2\,\pi$-$4\,\pi$, 4π-6π, etc.) also contribute to the signal and are responsible for the phenomenology of wave optics.

In the top row of Fig. 2 we illustrate the phase delay potential $\Omega\tau(\boldsymbol{x}, \boldsymbol{y})$ modulo 2π over the lens plane for several values of Ω. We hold a source fixed at $\boldsymbol{y}_0 = (1, 0)$ (one Einstein radius away from the origin of the source plane) for various choices of Ω, with a point-mass lens ($\psi(\boldsymbol{x}) = \log(|\boldsymbol{x}|)$. To compare the coherent patches with traditional geometric optics, we place dotted circles centered at the two stationary points of τ, at $\boldsymbol{x} = (x_\pm, 0)$ for

$$x_\pm = \frac{y \pm \sqrt{y^2 + 4}}{2}. \tag{12}$$

The radii of the circles are chosen according to their geometric-optics magnifications, and are scaled as $1/\Omega$ to ensure that at high Ω, due to the prefactor in Eq. (9), the flux magnification $|F|^2$ stays constant.

The phase over the lens plane varies significantly as a function of Ω, suggesting that the amplification, and therefore the lens magnification, can become frequency-dependent, unlike in traditional gravitational lensing, where magnification ratios are completely achromatic. This frequency dependence arises because Eq. (8) introduces a characteristic length scale (the wavelength), and therefore chromaticity, into gravitational lensing. We qualitatively depict the frequency dependence of $|F|^2$ as a function of Ω in Fig. 3.

There are three qualitative regimes of gravitational lensing: wave optics, eikonal optics, and geometric optics. Mathematically, they are classified based on how the integrand in Eq. (9) behaves.

The behavior of the integrand is controlled by ΩA, which sets the size of a coherent patch in the lens plane around the individual stationary points of $\tau(\boldsymbol{x}, \boldsymbol{y})$ (Jow et al. 2023; Berry 2021). Here, A is the determinant of the magnification matrix, which is proportional to the area of the Fresnel zone surrounding the stationary point in the time delay potential.

$$A_{ab} = \frac{\partial^2 \tau}{\partial \boldsymbol{x}_a \partial \boldsymbol{x}_b} = 1 - \frac{\partial^2 \psi}{\partial \boldsymbol{x}_a \partial \boldsymbol{x}_b}. \tag{13}$$

If ΩA is small, then the light does not "feel" the lens and not much lensing occurs. The first Fresnel zones of each stationary points are overlapping, and in this regime the images are inseparably blurred together on the sky, and are separated by less than a wavelength in the time domain. There are still multiple stationary points of the time-delay potential, so

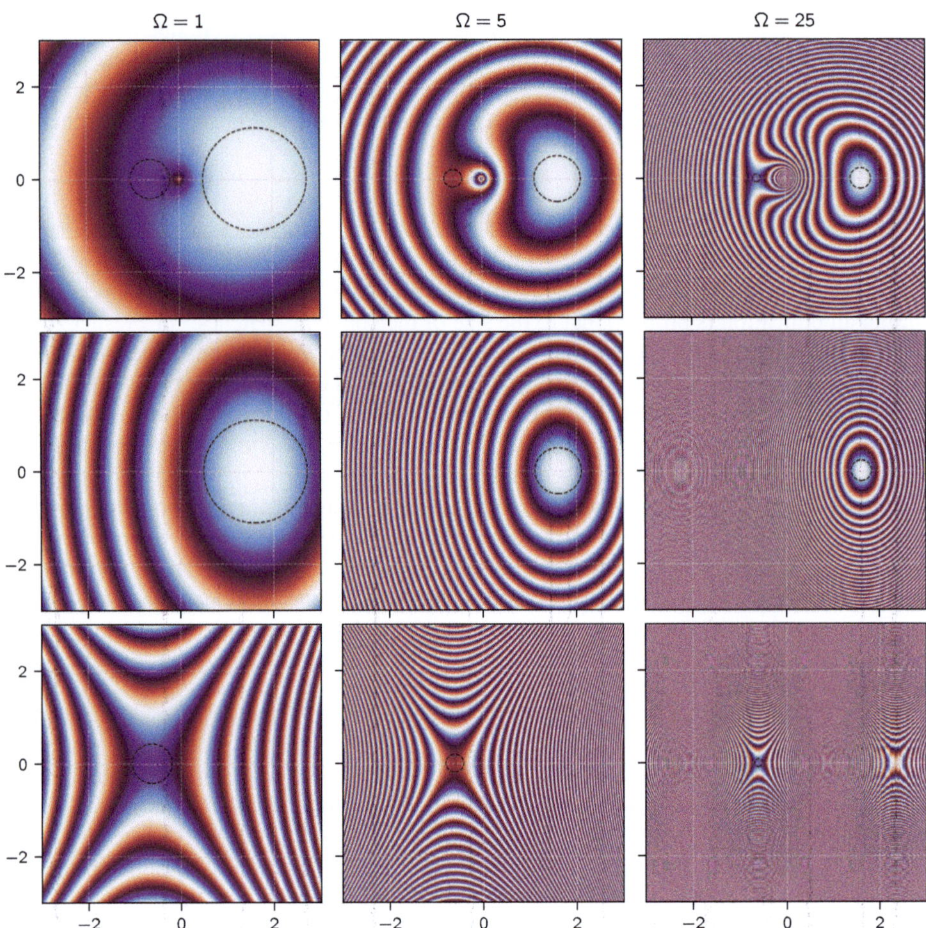

Fig. 2 Top row: The phase $\Omega\tau(\boldsymbol{x})$ corresponding to the lensing potential $\tau(\boldsymbol{x}, \boldsymbol{y}) = \frac{1}{2}|\boldsymbol{x} - \boldsymbol{y}|^2 - \ln(|x|)$, for three values of $\Omega = 1$ (heavily diffracted), $\Omega = 5$, $\Omega = 25$. We fix the source at $\boldsymbol{y} = (1, 0)$ and evaluate the phase as a function of position in the lens plane, visualized with a 2π-periodic colormap. To guide the eye, we have drawn two circles centered on the loci of the image positions $\boldsymbol{x}_\pm$ evaluated in the geometric optics limit. The radii of the pairs of circles are proportional to the flux magnification ratios $\mu_\pm$, and are scaled as $1/\Omega$. Top left: In the heavily-diffracted regime ($\Omega = 1$), the time delay between images is less than 2π radians. The images are superimposed in the time domain within one wavelength, and are rendered indistinguishable due to diffraction. Top center: As the frequency increases ($\Omega = 5$), the images are separated by more than one wavelength, and there are several oscillations between the two stationary points of the lens. The images are therefore no longer blended by diffraction; they become distinguishable in both the time domain and the angular domain. The interference leads to constructive and destructive interference of the phase of the integrand when integrated over the whole lens plane, and shows a simple sinusoidal functional form when plotted against Ω. Top right: In the high-frequency limit (here shown as $\Omega = 25$ for visualization purposes), the two images in the time domain are very well-resolved. The total magnification may be treated as a sum of discrete images – the discrete stationary points of the lens potential – using geometric optics. Middle/bottom rows: The stationary phase approximation to the "plus"/"minus" image individually, for the same three values of Ω. It can be seen that the stationary phase approximation works better and better for higher values of Ω, as the relevant parts of the integrand become increasingly confined around the stationary points. Note that a second saddle point appears on the right side of the bottom right panel; this is a image rendering artifact

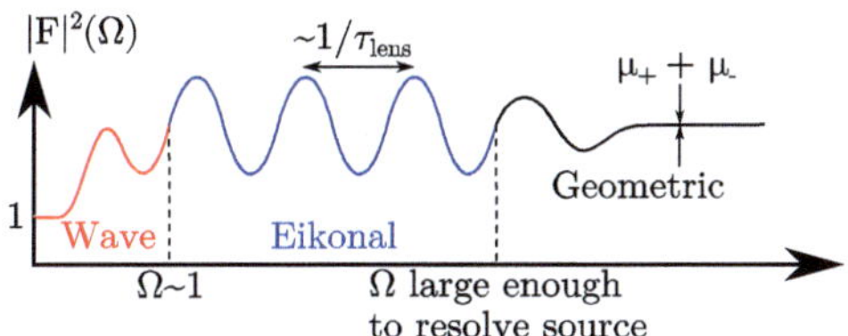

Fig. 3 A schematic depicting different regimes in gravitational lensing as a function of Ω. Left red: When $\Omega \sim 1$, we are in the diffractive or wave optics regime where the lensing is present but suppressed ($|F|^2 \to 1$) for $\Omega \ll 1$. Center blue: In the eikonal regime, the interference fringe between the two images, here represented as two amplitudes F_+ and F_-, is clearly visible. The total magnification has a perfect sinusoidal dependence (see Eq. (19)) with a characteristic spectral oscillation scale of $1/\tau_{lens}$. Right black: Finally, at high frequencies, we enter the geometric optics (traditional lensing) regime when the lens's resolving power is sufficiently great to resolve the finite extent of the source. In this limit, the interference effects become washed out, and the lensing magnification becomes achromatic. In this limit the total flux magnification is the sum of the image flux magnifications: $|F_+|^2 + |F_-|^2$

the lensing is still "strong lensing", but it is impossible to associate a particular photon to a particular image. The wave optics regime can be thought of as the transition between no lensing at all and the presence of an interference pattern caused by gravitational lensing, as seen in Fig. 3.

When the phase of the integrand varies rapidly (large values of ΩA), the individual stationary points are separable into discrete images on the sky (and into discrete events in the time domain, for transient events). Each image is observationally separable from the others, and they interfere with each other to form a characteristic interference pattern. This interference regime is the *eikonal* limit. When the observing frequency is so high (extremely large values of ΩA), the oscillations are quenched by the finite extent of the source in the *source* plane, and ordinary geometric optics applies.

The geometric optics regime is the one in which most gravitational lensing observations have been conducted. In this work, we make the distinction between the eikonal limit (i.e. the stationary-phase approximation of Eq. (9)) and the geometric optics regime (which we take to be the regime in which the eikonal images are added together incoherently at the observer). Some authors do not make this distinction and simply refer to the eikonal limit as geometric optics as well. Grillo and Cordes (2018) distinguish the incoherent and coherent regimes as "zeroth" and "first-order" geometric optics.

These three different regimes can be organized as a function of Ω and A. With modern observational capabilities ranging from ground-based (and soon, space-based) gravitational wave detectors to gamma-ray telescopes, Ω can vary by 30 orders of magnitude for different lenses and wavelengths, and A can vary from order unity to $\sim 10^3$ in rare extreme-magnification scenarios. Since these are rare we will focus on Ω. Figure 4 shows Ω as a function of the mass scale of the lens and the wavelength of the source, to illustrate how the different lensing regimes apply in different physical scenarios.

Within Fig. 4, we have also delineated a second boundary, which denotes whether the radiation is detected incoherently (without phase information, e.g. with a flux detector) or coherently (with phase information, e.g. with an antenna). At long wavelengths (radio and gravitational waves), it is possible with current technology to detect the phase of the radiation field using electromagnetic/gravitational antennas. These situations where we have access to the phase of the electromagnetic/gravitational field are extremely powerful and enable qualitively different observations to be carried out. For example, recording the field amplitude instead of the intensity (voltages instead of light-curves) allows antennas to reach

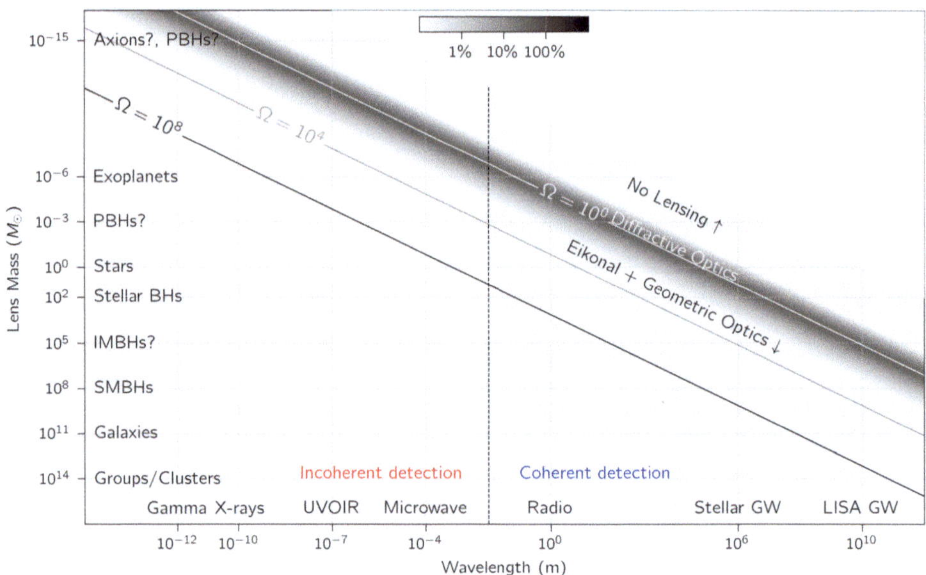

Fig. 4 The transition from geometric to diffractive optics, as quantified by $\Omega = 4\pi R_S/\lambda$, as function of lens mass and observing wavelength. The gray band quantifies, using a perturbative expansion which is accurate for eikonal optics and no-lensing limits but not the transition between the two (Jow et al. 2023), the breakdown of eikonal optics at $\Omega \sim 1$: the regime where wave optics effects matter the most. The gray color scheme depicts the fractional residual (Eq. 16 in Jow et al. 2023) in this "diffractive" regime between a full wave-optics calculation and perturbative expansion. Right half: At longer wavelengths, advances in instrumentation (e.g. wideband voltage recording in radio telescopes and gravitational wave detectors) have enabled measurements of the amplitude (coherent detection) instead of the flux (incoherent detection). Left half: At high Ω, the stationary-phase approximation holds, and geometric/eikonal optics applies. In the $\Omega \ll 1$ regime, the lenses can be thought of as being "too small" to impart any phase on the passing wave

Nyquist-limited time resolution for resolving shorter lensing delays, regardless of the flux variability timescale of the source. This is similar to radio interferometry, where coherent voltage recording allows precise time-delay measurements between different receivers even when the sources do not exhibit time variability in their light curves. Improving time resolution through coherent detectors and recording instrumentation therefore opens up new phase space towards smaller time delays (corresponding to smaller mass scales) in gravitational lensing.

4 Eikonal Optics

Lensing in eikonal optics is similar to treating the source as a plane wave hitting an opaque screen with a set of well-separaged pinholes of negligible size. Each pinhole's size and phase relative to the other pinholes is set by the traditionally-calculated flux magifications and time delays. To see this mathematically, we simplify Eq. (9) using the stationary phase approximation. Conceptually, this collapses the integral over the lens plane to a coherent superposition of a small number of paths defined by the stationary points of the lens potential. Taking the limit as $\Omega \to \infty$ results in rapid oscillations of the phase of the integrand as a function of $\boldsymbol{x}$. In this limit, only the stationary points of $\Omega\tau(x, y)$ as a function of x, which we index by j, contribute constructively to the integrand. These positions $\boldsymbol{x}_j$ correspond to the image positions as calculated with geometric optics. Near the jth stationary point, the time delay potential can be approximated by a Taylor expansion:

$$\tau_j(\boldsymbol{x},\boldsymbol{y}) \approx \tau(\boldsymbol{x}_j) + \frac{1}{2}A^j_{ab}(\boldsymbol{x}-\boldsymbol{x}_j)_a(\boldsymbol{x}-\boldsymbol{x}_j)_b + \cdots \tag{14}$$

where $A^j_{ab} = \partial_a\partial_b\tau(\boldsymbol{x}_j)$ is the inverse magnification matrix evaluated at the jth image. In Fig. 2, we plot in the top row the phase $\Omega\tau(\boldsymbol{x},\boldsymbol{y})$ modulo 2π, and in the bottom two rows the stationary phase approximation of the phase potential about each of the two stationary points. We approximate the Fresnel-Kirchhoff integral as the coherent superposition of both of these expansions. This is called the stationary phase (or saddle-point) approximation, which relies on the fact that contributions far away from the stationary points cancel each other out.

$$\int d^2\boldsymbol{x}\exp(i\Omega\tau(\boldsymbol{x})) \approx \sum_j \int d^2\boldsymbol{x}\exp(i\Omega\tau_j(\boldsymbol{x})) \tag{15}$$

$$= \sum_j \exp(i\Omega\tau(\boldsymbol{x}_j)) \int d^2\boldsymbol{x}\exp(i\Omega A^j_{ab}(\boldsymbol{x}-\boldsymbol{x}_j)_a(\boldsymbol{x}-\boldsymbol{x}_j)_b/2) \tag{16}$$

We evaluate the above expression analytically, being a sum of complex Gaussian integrals. Since A^j_{ab} is a symmetric matrix, it is diagonalizable via a rotation of $\boldsymbol{x}$; we call its eigenvalues λ^j_1 and λ^j_2. The Gaussian integral associated with each image contributes a factor of $\frac{2\pi i}{\Omega\sqrt{\lambda^j_1}\sqrt{\lambda^j_2}}$. The eigenvalues λ^j_1, λ^j_2 may be both positive if the image is a minimum of the time-delay potential, both negative if the image is a maximum of the time-delay potential, or they may have different signs if the image is at a saddle point of the potential. If we define F_j by

$$F_j^{-1} = \left|\sqrt{\lambda^j_1}\sqrt{\lambda^j_2}\right| \tag{17}$$

the amplification becomes

$$F(\Omega) = \sum_j F_j \exp(i\Omega\tau(\boldsymbol{x}_j) - i\pi n_j). \tag{18}$$

where $n_j = 0, 1/2, 1$ keeps track of the extra factor of $-i$ arising from the sign of the eigenvalues. This phase shift arises from the image parity and is called the Morse phase; in the eikonal limit it is discretely-valued.[1] It has been known for decades (Blandford and Narayan 1986; Nakamura and Deguchi 1999) but only recently have the observational ramifications become observationally relevant, in particular in the parameter space of gravitational-wave lensing (Dai and Venumadhav 2017; Dai et al. 2020). The overall Morse phase is not predicted by waveform models. In the specific setting of gravitational waves generated by the nearly-circular inspirals of compact binaries, an overall Morse phase shift of 180° is degenerate with a 180° orbital phase shift of the binary as long as we restrict our analysis to the dominant quadrupole mode of the radiation. Orbital eccentricity or the inclusion of

[1] However, this is not necessarily true in the diffractive regime when the stationary phase approximation breaks down – see Appendix A of (Jow et al. 2023).

higher-order modes in model waveforms can also break this degeneracy (Janquart et al. 2021; Ezquiaga et al. 2021).

The relative Morse phase between two images of the same waveform can be used to reduce backgrounds in searches for lensed GWs (Dai et al. 2020). In the time domain, a relative Morse phase of 180° induces a sign flip of the electromagnetic/gravitational waveform (Type II images; $n_j = 1$) relative to a Type I image. In the case of saddle points (Type III images; $n_j = 1/2$), the Morse phase applies a Hilbert transform to the unlensed waveform. This latter distortion is non-trivial and leads to biases in parameter estimation unless this is properly taken into account in search templates (Vijaykumar et al. 2023).

Interference effects in lensing are imprinted not only in the phase, but also the amplitude of the observed lensing event. As shown in Fig. 3, interference induces modulations of the intrinsic spectrum of a source whose period is $\Delta\Omega = 2\pi/(\tau(\boldsymbol{x}_j) - \tau(\boldsymbol{x}_k))$ in the eikonal regime. This is observationally interesting because the cross-correlation between a bright and faint image is easier to detect than the faint image in autocorrelation at fixed detection signal-to-noise (Jow et al. 2020; Takahashi and Nakamura 2003). To see this, we start by expanding Eq. (18):

$$|F|^2 = \sum_j |F_j|^2 + 2\sum_{j<k} F_j F_k \cos(\Omega(\tau(\boldsymbol{x}_j) - \tau(\boldsymbol{x}_k)) - \pi(n_j - n_k)) \tag{19}$$

where we recognize $|F_j|^2 = 1/\det A^j_{ab}$ as the flux magnification of each image present in geometric optics. This is analogous to an experiment allowing diffraction and interference of light through pinholes of varying sizes and positions. Now suppose that we detect the brightest image, which has magnification $|F_j|^2$, and that the noise floor is $\varepsilon^2|F_j|^2$ with $\varepsilon < 1$. To check for the presence of a fainter image k in auto-correlation, its flux must be $|F_k|^2 \gtrsim |F_j|^2\varepsilon^2$. However, detecting the cross-correlation term requires that $|F_j F_k| \sim |F_j|^2\varepsilon^2$. In other words, we are sensitive to dimmer secondary images at fixed signal-to-noise ratio ε^{-1}.

The eikonal limit, or stationary-phase approximation, is a particularly useful approximation because the behaviour of the stationary-phase points, a.k.a. the images, is described by the rich mathematics of catastrophe theory. Catastrophe theory characterizes the topology of caustics, which are the boundaries in parameter space at which the number of stationary-phase points changes. The essence of what makes catastrophe theory so powerful is the result that the topology of caustics can be classified by a small number of "elementary catastrophes" (Thom 1967). The result is that lenses will behave qualitatively similarly based on which elementary catastrophes characterize the lenses, even though they may differ significantly in functional form (Nye 1999). While, technically, the eikonal limit breaks down at caustics (at caustics, two or more of the inverse magnification matrices are degenerate, leading to formally infinite magnifications), the area of the region around the caustic for which the eikonal approximation fails goes to zero as $\Omega \to \infty$. Moreover, the full wave optics diffraction patterns of some of the elementary catastrophes are known (Berry and Upstill 1980). For example, as we will see in Sect. 7.2, the diffraction pattern near a fold catastrophe is given exactly by the Airy function (see also Nakamura and Deguchi 1999). Therefore, the known diffraction patterns near caustics coupled with the eikonal limit away from the caustics can provide a full description of a generic lens. This approach is taken in Grillo and Cordes (2018).

The effects mentioned so far – Morse phase shifts, spectral modulation of the amplification factor, and favorable scaling of lensing cross-sections – arise as generic interference phenomena present in eikonal optics. Going into the fully diffractive regime adds additional features which we will discuss next.

5 Diffractive Optics ($\Omega \gtrsim 1$)

In the case where Ω is not much greater than 1, Eq. (9) cannot be simplified using the stationary phase approximation. For an axisymmetric lens, the potential $\psi(\boldsymbol{x}) \to \psi(x)$, and the amplification can be written as

$$F(\Omega, y) = -i\Omega \exp(i\Omega y^2/2) \times \tag{20}$$

$$\int_0^\infty dx x J_0(\Omega x y) \exp\left[i\Omega(x^2/2 - \psi(x))\right] \tag{21}$$

where J_0 is the zeroth-order Bessel function. In certain cases this may be computed analytically. For a point mass where $\psi(x) = \log(x)$ this simplifies to

$$F(\Omega, y) = \exp\left[\frac{\pi\Omega}{4} + i\frac{\Omega}{2}\left(\ln\left(\frac{\Omega}{2}\right)\right)\right] \times \tag{22}$$

$$\Gamma\left(1 - \frac{i\Omega}{2}\right) {}_1F_1\left(\frac{i\Omega}{2}, 1; \frac{i\Omega y^2}{2}\right). \tag{23}$$

In this equation (see also Eq. 21 of Deguchi and Watson 1986a, Eq. 2.16 of Katz et al. 2018 for slightly different forms), Γ is the gamma function and ${}_1F_1$ is the hypergeometric function.[2]

The point mass is the simplest analytical solution for the amplification factor in wave optics in gravitational lensing. However, other analytical solutions exist which we briefly mention: the solution for a Kerr (rotating) black hole is very similar to that of the point mass; it includes a shift in the point-mass lensing potential $\psi(\boldsymbol{x}) = \log(x) - \frac{\boldsymbol{x} \cdot \boldsymbol{\alpha}}{x^2}$, where the shift $\vec{\alpha} = \hat{n} \times \vec{J}/(M R_E)$ depends on the physical Einstein radius R_E and the lens's angular momentum vector per unit mass $\vec{J}/M$ relative to the line-of-sight unit vector $\hat{n}$. The amplification is the same as that of Eq. (23), being circularly-symmetric, albeit shifted by an angle α (Baraldo et al. 1999). This is similar to the Sagnac effect for a rotating optical interferometer. Analytic solutions for the singular isothermal sphere (Matsunaga and Yamamoto 2006), an infinite cosmic string (Suyama et al. 2006), a one-dimensional sinusoidal phase screen (Beach and Lovelace 1997), and the solar gravitational lens (Turyshev 2017; Engeli and Saha 2022) are also known.

Solutions to the diffraction integral beyond the analytical ones presented here are often computed numerically. However, this can be difficult because the phase in integrals such as Eq. (9) oscillates faster for larger values of the argument $\boldsymbol{x}$. Some of the first numerical efforts based on contour integration were developed for the calculation of femtolensing spectra (Ulmer and Goodman 1995); this method has also been applied successfully to gravitational waves (Mishra et al. 2021) and in the limit of extended sources (Tambalo et al. 2023a). In addition, approximate methods have been developed (Grillo and Cordes 2018) which go beyond the eikonal approximation and which remain effective near caustics (see also discussion of Sect. 1 of Jow et al. 2021)). In addition, the Fresnel integral can be written as a convolution and evaluated via an FFT. This performs well near caustics, but requires the lens plane to be densely sampled with a finite and regular two-dimensional grid fine enough to capture the details of the oscillatory integrand.

[2]Wolfram MathWorld: Hypergeometric ${}_1F_1$.

More recently, Picard-Lefschetz theory has been used to assist in the analysis and evaluation of Fresnel integrals (Feldbrugge et al. 2023; Feldbrugge and Turok 2020; Jow et al. 2021; Shi and Xu 2021; Jow et al. 2023; Tambalo et al. 2023a). These methods essentially involve deforming the two lens plane integrals into a set of carefully-chosen contours in the complex plane along which contour integrals are non-oscillatory and absolutely convergent. One feature of Picard-Lefschetz integration is that while it converges slowly for $\Omega \sim 1$, performance improves as the integrands become *more* oscillatory ($\Omega \gg 1$), so it is complementary to many other techniques. A limitation is that the lens phase function $\phi(\boldsymbol{x})$ (e.g. $\phi(\boldsymbol{x}) = \Omega\tau(\boldsymbol{x}, \boldsymbol{y})$) must be able to be evaluated when applying analytic continuation to the integrand, "complexifying" it from $\mathbb{R}^2 \to \mathbb{C}^2$. This is straightforward for the broad class of rational functions (quotients of polynomials), which have no branch cuts (also a desirable property) and a small number of isolated poles. The point mass lens has $\psi(x) = \log(|x|)$ which *does* have a branch cut; in a follow-up work to (Feldbrugge et al. 2023), a technique to evaluate the isolated and the binary point lens potentials by reparametrizing the radial integral was reported (Feldbrugge and Turok 2020). However, well-known lens models like the singular isothermal sphere ($\psi(x) = |x|$) have not yet been evaluated using this method. Despite this limitation, the utility of Picard-Lefschetz theory is not limited to its numerical application, but also includes its conceptual power. By extending the analysis of optics into the complex plane, Picard-Lefschetz theory gives a well-defined prescription for separating contributions to the total observed flux from individual geometric images, even at low frequencies. Picard-Lefschetz theory can be seen as a way to extend the "discrete images" description from geometric optics lensing deep into the wave regime. In this way, Picard-Lefschetz theory can conceptually bridge the gap between these two regimes (Jow et al. 2023).

In Fig. 5 we show several magnification maps, evaluated using the Picard-Lefschetz method for several values of Ω. The map coordinates $\boldsymbol{\mu} = (\mu_x, \mu_y)$ are related to the source plane and observer plane coordinates ($\boldsymbol{y}$ and $\boldsymbol{y}_{\text{obs}}$ respectively) by $\boldsymbol{\mu} = \boldsymbol{y}\frac{D_L}{D_S} + \boldsymbol{y}_{\text{obs}}\frac{D_{LS}}{D_S}$. This parametrization reflects the symmetry of the source and the observer in optics: in gravitational lensing we may think of fixing $\boldsymbol{y}_{\text{obs}} = 0$, and treat $\boldsymbol{\mu}$ as a source plane coordinate, while in a laboratory diffraction experiment we may instead fix $\boldsymbol{y} = 0$ and observe the diffraction pattern on the observer's screen, which spans many values of $\boldsymbol{y}_{\text{obs}}$. The Fresnel integral evaluated is that of the Chang-Refsdal lens (Chang and Refsdal 1979), for which

$$\tau(\boldsymbol{x}, \boldsymbol{y}) = \frac{1}{2}|\boldsymbol{x} - \boldsymbol{\mu}|^2 - \log(|\boldsymbol{x}|) - \frac{\gamma}{2}(x_1^2 - x_2^2) \tag{24}$$

whose single free parameter $\gamma > 0$ describes the external shear.

5.1 A Useful Analytical Approximation

The point mass model is extremely useful, but in practice, hypergeometric functions with complex arguments can be cumbersome to numerically evaluate (Pearson et al. 2014). There is a useful analytic approximation to Eq. (23) which addresses three shortcomings of the eikonal approximation (Eq. (18)): namely, that in full wave optics, the first few crests in Eq. (23) are not regularly spaced, and are not of equal height, and that it is still possible to get infinite magnification ratios. This analytical approximation is fast, accurate, and reproduces a finite magnification near the point caustic at $y = 0$ (see Fig. 8), making it useful in studying the high-magnification limit. Its shortcoming is that it leaves out a factor of $1 - \exp(-\pi\Omega)$ in

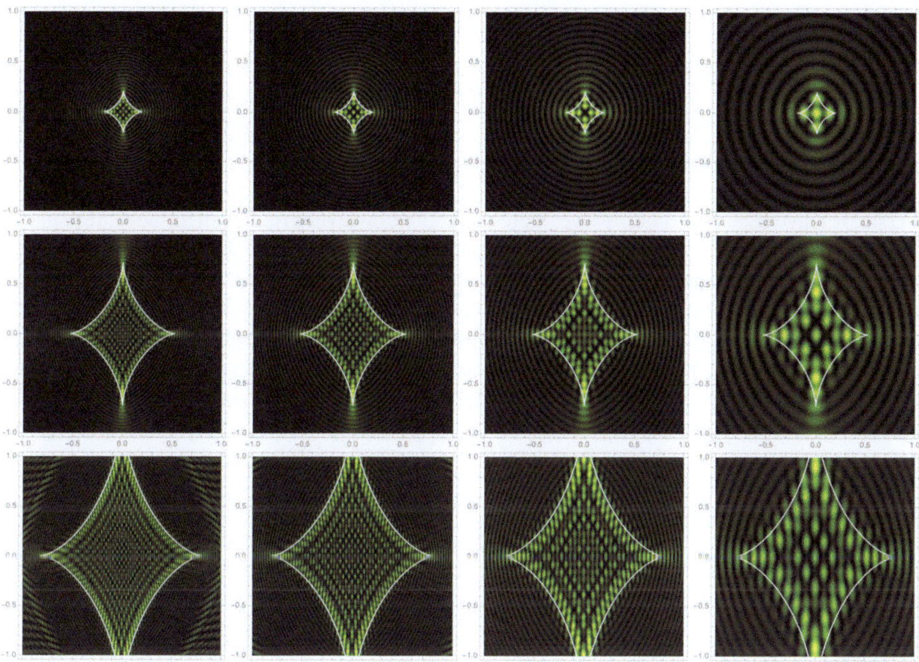

Fig. 5 A map of $|F|^2(\Omega, \boldsymbol{\mu})$ for a point mass+external shear model (Eq. (24)), where the map coordinates $\boldsymbol{\mu}$ are the scaled source plane coordinates $\boldsymbol{y}$. Columns: From left to right, each column corresponds to $\Omega = 100, 75, 50, 25$. Rows: From top to bottom, each row corresponds to $\gamma = 0.1, 0.3, 0.5$. The diamond caustic, at which the geometric optics magnification diverges, is overlaid as a white diamond on each map. Reproduced with permission from Feldbrugge and Turok (2020)

the denominator which accounts for the suppression of lensing as $\Omega \to 0$. It can be derived by solving for the stationary points of

$$\tau(x) = \frac{x^2}{2} - \psi(x) \tag{25}$$

which gives the geometric-optics image positions x_E such that $x_E = \psi'(x_E)$. To second order in $\Delta x = x - x_E$,

$$\tau(x) = \tau(x_E) + \frac{1}{2}(1 - \psi''(x_E))\Delta x^2 = \tau(x_E) + (1 - \kappa)\Delta x^2. \tag{26}$$

The region of the lens plane where the rays interfere constructively is defined by

$$2\pi \gtrsim (1 - \psi''(x_E))\Delta x^2 \Omega, \tag{27}$$

and this domain has a width of $\Delta x \sim (\Omega|1 - \psi''(x_E)|)^{-1/2}$. Within this domain, the Bessel function argument of Eq. (21) changes by $\Omega \Delta x y$, and outside this domain, it oscillates quickly. If $y \lesssim \Omega^{-1/2}$ and ψ'' is of order unity and slowly varying over Δx, then $1 \gtrsim \Omega \Delta x y$, and the Bessel function does not change over the important range of the integration. Pulling $x J_0(\Omega x y)$ out of the integral sign leaves a Gaussian integral from 0 to ∞:

$$|F_{\text{wave}}|^2 \approx \pi \Omega x_E^2 |1 - \psi''(x_E)|^{-1} J_0^2(\Omega x y) \tag{28}$$

$$\leq \pi \Omega x_E^2 |1 - \psi''(x_E)|^{-1} = 2\pi \Omega x_E^2 |1 - \kappa(x_E)|^{-1}.$$

where we have used $2(1 - \kappa(x_E)) = 1 - \psi''(x_E)$.

6 Transition to Geometric Optics: Finite Source Size

So far, we have assumed a point-like source of monochromatic radiation in the source plane. In reality, we observe radiation from a finite patch of $\boldsymbol{y}$-values in the source plane. In general, increasing the observing frequency increases the angular resolution and washes out the chromatic interference patterns seen in both the eikonal and wave optics regimes, e.g. Eqs. (19) and (23). This has been the main factor suppressing interference effects in gravitationally-lensed GRB systems, though recently-discovered transients like gravitational waves, pulsars, and FRBs may overcome this limitation. Before we dive into quantitative treatment of decoherence, it is worth defining two notions of coherence relevant for interstellar optics: physical coherence and spatial coherence.

6.1 Physical Coherence

We call emission *physically-coherent* if the phase of the emitted waves is predictable. The most relevant example of this is in gravitational wave astronomy, where waveform models predict the phase and amplitude of gravitational wave signals.

Another example of a situation where temporal coherence is necessary is in astrophysical masers, where the maser emission is in phase. In addition, proposals exist to use the sun as a gravitational lens (Maccone 2010) to search for optical laser pulses from hypothetical extrasolar civilizations. The idea is that if the emission region is phase-coherent, the field may be constructively magnified by the collecting area of the sun's gravitational field, enabling sensitive technosignature searches with modest telescope resources. In this case, the maximum magnification is limited by how point-like the laser is, and the size of a coherent patch in the lens plane, which governs the "collecting area" of the solar gravitational lens.

Lensing must preserve the physical coherence of the source: in the gravitational-wave case, certain lensing events may go undetected by e.g. matched-filtering search pipelines (Dai et al. 2018). In the other case, the maximum magnification is not successfully attained.

6.2 Spatial Coherence

In most scenarios, we do not work directly with the wave field, since we cannot predict its phase. In these cases, which constitute the majority of astronomy, we consider predictions for second-order correlations in the wavefield as a function of time and frequency. For instance, since the overall phase of the incoming radiation is unknown, we make predictions for the impact of interference effects on the spectrum of a star, or the output of the lag correlation of time series data in order to search for lensing.

When studying interference and wave optics phenomena using second-order correlations, the source is required to be *spatially* coherent. In the terminology of telescopes, this is equivalent to the source's finite angular size being unresolved by an interferometer. In the language of interstellar optics one might analogously say that the source is unresolved by the inhomogeneous media through which the waves travel. This in turn places stringent limits on the apparent angular size of the source. In some cases this corresponds to the true size of the emission region; in others, relativistic beaming, interstellar scattering, or atmospheric fluctuations can modify the apparent size of the source.

6.3 Source-Size Requirements for Coherence

Having qualitatively described different kinds of coherence involved in studying wave optics, we derive the corresponding requirements on the source size for both types of coherence in a consistent framework. We use eikonal/geometric optics, which covers most of the aforementioned scenarios, but caution that a wave-optics treatment is necessary in the vicinity of caustics. We consider Eq. (10) in the vicinity of one of the images, which we label j. The time delay close to $\boldsymbol{x}_j$ can be expanded as

$$\delta\tau_j = \tau(\boldsymbol{x}, \boldsymbol{y}) - \tau(\boldsymbol{x}_j, \boldsymbol{y}) \tag{29}$$

As we vary the position in the source plane about the central position, the geometric-optics image positions will shift. To take this into account we need to take the total derivative of the time delay in $\overline{\boldsymbol{y}} = \overline{y}_a = \boldsymbol{y} - \boldsymbol{y}_0$ subject to the lens equation constraint.

$$\delta\tau_j = \frac{\partial\tau}{\partial y_a}\overline{y}_a + \frac{1}{2}\frac{\partial^2\tau}{\partial x_c \partial x_d}\frac{\partial x_c}{\partial y_a}\frac{\partial x_d}{\partial y_b}\overline{y}_a\overline{y}_b \tag{30}$$

$$+\frac{1}{2}\frac{\partial^2\tau_j}{\partial x_c \partial y_b}\frac{\partial x_c}{\partial y_a}\overline{y}_a\overline{y}_b + \frac{1}{2}\frac{\partial^2\tau_j}{\partial y_a \partial y_b}\overline{y}_a\overline{y}_b \tag{31}$$

For lensing to preserve the temporal coherence of the signal while only modulating the amplitude, we need $\Omega\delta\tau_j \leq 1$. Using $A_{ab} = \dfrac{\partial x_a}{\partial y_b}$ and $y_a = x_a - \partial\psi/\partial x_a$ we have

$$\delta\tau_j = -\partial_a\psi_j\overline{y}_a + \delta_{ab}\overline{y}_a\overline{y}_b/2 \tag{32}$$

where the value of the lens potential at x_j is ψ_j. This simplification arises because the images are points of stationary phase, so delays are stationary under small shifts of the image positions. The maximum allowed source size for temporal coherence depends on the details of the potential ψ and is generally not isotropic in the lens plane, but a convenient estimate which is independent of ϕ can be obtained by choosing the y_1 direction to be perpendicular to $\partial_a\psi_j$. Then, the size of the source along the y_1 direction must satisfy

$$|\overline{y}_1| \leq (2\Omega)^{-1/2} \tag{33}$$

or in physical units,

$$R_{src} \leq \sqrt{\frac{\lambda D_S D_{LS}}{4\pi D_L}} \tag{34}$$

Equation (34), or the Fresnel scale, is plotted in Fig. 7 using the order-of-magnitude approximation $D_S D_{LS}/D_L \sim D_{\rm eff}$, and ignoring cosmological factors of $1+z$.

For spatial coherence, the time delay between two images $\delta\tau_j - \delta\tau_k = \delta\tau_{jk}$ must remain constant over the finite extent of the source. We require the phase shift to be $\Omega\delta\tau_{jk} < 1$. Using Eq. (31) with images j and k gives to first order

$$\delta\tau_{jk} = -(x_j - x_k)_a\overline{y}_a \tag{35}$$

We see that in the eikonal optics limit, the source size requirements on observing interference effects in lensing are independent of the image magnifications and depend only on the

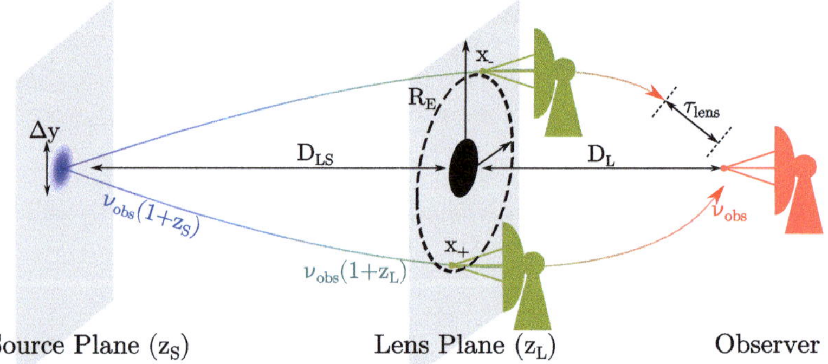

Fig. 6 In the eikonal limit, a lensed source remains coherent if the spectral oscillations do not significantly change over the source's finite angular size, here shown as a blue patch. When the images are well-separated, "rays" from the source traverse the lens plane through the image positions $\boldsymbol{x}_\pm$ calculated via geometric optics. For the point-mass lens, one image passes outside the Einstein ring at a transverse physical scale R_E, and the other passes inside the Einstein ring on the opposite side of the lens. The interference fringes from lensing are present if the source looks point-like as observed by an interferometer with a baseline of Eq. (38) (i.e. a pair of antennas placed in the lens plane at the apparent locations of the gravitational lens images in the lens plane)

image positions, in analogy with the standard double-slit experiment. Then the size of the resolution element in the source plane can be calculated by evaluating Eq. (35) by conveniently choosing the y_1 direction to be oriented parallel to $\boldsymbol{x}_j - \boldsymbol{x}_k$. Then $\Omega|\tau_{jk}| \leq 1$ implies

$$|y_1| \leq \Omega^{-1}|\boldsymbol{x}_j - \boldsymbol{x}_k|^{-1}. \tag{36}$$

This can be translated back to physical units using Eq. (7):

$$R_{src} \leq \frac{D_S \lambda}{2\pi D_{\text{slit}}}. \tag{37}$$

with

$$D_{\text{slit}} = D_{\text{eff}} \theta_E |\boldsymbol{x}_j - \boldsymbol{x}_k|. \tag{38}$$

D_{slit} is the effective baseline that goes into the Rayleigh resolution limit for light of wavelength λ passing through a double-slit (Schneider 1983) located at the lens plane whose separation is D_{slit}, as shown in Fig. 6. As an order of magnitude estimate we can assume $|\boldsymbol{x}_j - \boldsymbol{x}_k| \sim 1$, getting

$$R_{src} \leq \frac{D_S \lambda}{2\pi D_{\text{eff}} \theta_E} \tag{39}$$

Equation (39) is plotted in Fig. 7 using the approximation $D_S \sim D_{\text{eff}}$ and ignoring cosmological factors of $1 + z$.

In some scenarios, eikonal optics breaks down, such as near caustics and wave optics is necessary. For physically-incoherent finite sources, which constitute the majority of astrophysical scenarios, analytical solutions are available from Matsunaga and Yamamoto (2006) (point-mass lens), and Zabel and Peterson (2003) (fold caustic crossing).

These works explore how fringes are lost as the source size is increased, or equivalently, how Ω is increased. As Ω is increased, the fringe spacing in frequency will decrease. When

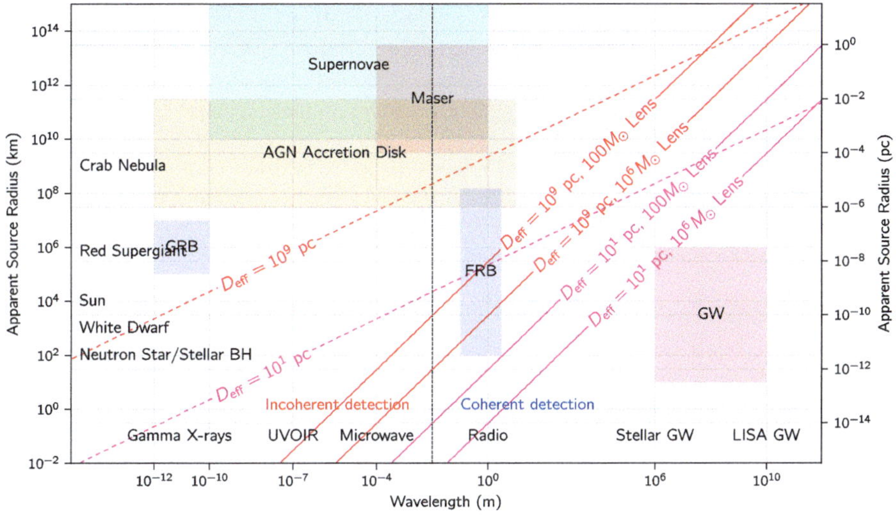

Fig. 7 Regimes of coherence in gravitational lensing. We plot the apparent radius of the source against the observing wavelength (ignoring redshift effects). Solid lines: To preserve spatial coherence, the source must lie below the solid lines corresponding to Eq. (39). Dashed lines: To preserve temporal coherence, in the diffractive/wave optics regime, the source size must be below the dashed lines corresponding to Eq. (34). For reference, we have roughly labeled the apparent sizes of various sources and wavelengths at which they have been observed in the various boxes. Sizes of supernovae, GRBs, FRBs, and GWs are taken from Oguri (2019); estimates for scatter-broadenened FRB sizes are from Leung et al. (2022), Katz et al. (2018)

this happens, there must be sufficiently high spectral resolution, or else the interference feature may be washed out by the finite frequency resolution of the detector. More fundamentally, if the frequency of light is increased further, the angular resolution of the interferometer is enhanced until the interference feature decreases in contrast. When either the finite spectral resolution or finite-source size degrade the interference pattern, ordinary geometric optics takes over.

7 Finite Magnification in Wave Optics

Having generically analyzed eikonal and wave optics and the transition to incoherent (traditional) gravitational lensing, we can now turn to studying more specific phenomena which arise in wave optics. One failure of geometric optics is that for certain regions of the source plane, magnification ratios become infinite. These regions – called caustics – are typically isolated points or one-dimensional curves in the source plane where geometric-optics rays "cross" and as such are regions where interference effects are most crucial (Ohanian 1983).

7.1 The Einstein Ring

the most famous example of a caustic is the Einstein ring for a point-mass lens. Rays emanating from the point caustic at $\boldsymbol{y}_0 = 0$ pierce the lens plane at a ring of points called the "critical curve" (here, the critical curve is defined by $|\boldsymbol{x}_0| = 1$) on their way to the observer.

The infinite magnification of geometric optics lensing is removed by two distinct effects: finite source size (only an infinitely small patch is magnified infinitely), and diffraction (Herlt and Stephani 1976; Benson and Cooke 1979). In other words, to truly attain infinite magnification, one would need an infinitely-small point source which radiates at an infinitely-short wavelength. In Fig. 8 we calculate magnification maps for both a point source close to the caustic, and a finite-sized source centered on the caustic, in the case of infinite and finite frequency (geometric optics and wave optics respectively).

The magnification profiles in geometric optics and diffractive optics can be computed as follows. In geometric optics the total magnification of a point source at some impact parameter y is

$$|F_{\text{geo}}|^2 = \frac{y^2+2}{y\sqrt{y^2+4}} \tag{40}$$

To obtain the finite-size result, we average Eq. (40) over a finite-sized top-hat of radius $\Delta y > 0$ in the source plane centered at $y = 0$. This gives a finite total brightness. The amplification factor for a circular, on-axis source of radius Δy is

$$\langle |F_{\text{geo}}|^2 \rangle_{\Delta y} = \sqrt{1 + \frac{4}{\Delta y^2}} \tag{41}$$

In wave optics, we repeat the top-hat averaging over the point-mass magnification map derived from wave optics (Eq. (28)). This yields

$$\langle |F_{\text{wave}}(\Omega)|^2 \rangle_{\Delta y} = \frac{\pi \Omega x_E^2 [J_0(\Omega x_E \Delta y)^2 + J_1(\Omega x_E \Delta y)^2]}{|1 - \psi''(x_E)|} \tag{42}$$

In Fig. 8, we plot Eqs. (40) and (28) in the top panel, and Eqs. (42) and (41) in the bottom panel. While both diffraction and finite source size effects modify the behavior of $|F|^2$ near $y = 0$, they have very different qualitatitive features – even a tiny departure from a point source washes out the large oscillations characteristic of diffractive optics. It is also evident that when the source size is on the caustic, the finite source size solution deviates considerably from the point source solution (Witt and Mao 1994; Sugiyama et al. 2020). The discrepancy between using a point-source formula and the extended source matter most when the source dimension $\Delta y \lesssim \Omega^{-1}$; this is most relevant for gravitational waves, which have low values of Ω and have source sizes $\lesssim 100$ km (Oguri 2019).

7.2 Fold Caustics

The well-known Einstein ring is an example of a point caustic: an isolated point in the source plane where the geometric-optics magnification diverges. However, caustics need not be isolated; they can be one-dimensional curves in the source plane. An example is the diamond caustic for the Chang-Refsdal lens, as shown in Fig. 5 in its full wave-optics glory. We will see how the wave-optical treatment regularizes the divergences in geometric optics. The infinite magnifications in geometric optics arise at stationary points of the lensing potential where one of the eigenvalues of the Hessian goes to zero. In wave optics this means that the size of the coherent patch in the lens plane is limited by the third derivative of the potential.

Following Nakamura and Deguchi (1999), we outline the calculation of the magnification map of the source plane in scalar wave optics. We start with the Fresnel integral (Eq. (9)) where a source at $\boldsymbol{y}_0$ lies directly on the caustic with a corresponding point on the critical curve $\boldsymbol{x}_0$. We can choose coordinate systems in both the source and lens planes with unit vectors $(\hat{\boldsymbol{y}}_1, \hat{\boldsymbol{y}}_2)$, $(\hat{\boldsymbol{x}}_1, \hat{\boldsymbol{x}}_2)$.

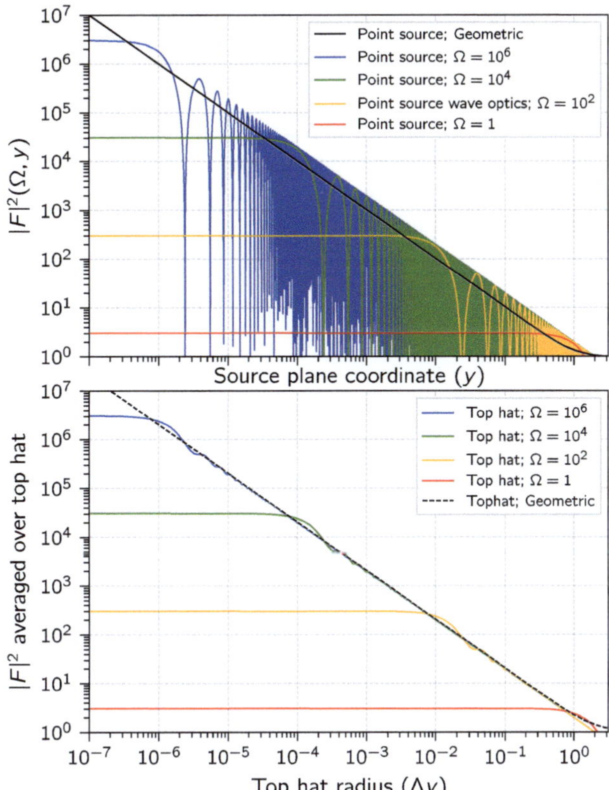

Fig. 8 The behavior of the magnification $|F|^2$ taking into account finite source size and wave optics effects near the Einstein ring ($y \to 0$). Top: The magnification map as a function of $\boldsymbol{y}$ near the origin saturates at $\boldsymbol{y} \sim 1/\Omega$ at a magnification of $\pi\Omega$. Bottom: the magnification map for an on-axis source, now as a function of source radius Δy

We begin with a third-order Taylor expansion about $\boldsymbol{x}_0$ and reason through each term. Here, $\overline{\boldsymbol{x}} = \boldsymbol{x} - \boldsymbol{x}_0$, and $\overline{\boldsymbol{y}} = \boldsymbol{y} - \boldsymbol{y}_0$, and the indices a, b, c run over the two dimensions of the lens plane.

$$\tau(\boldsymbol{x}, \boldsymbol{y}) - \tau(\boldsymbol{x_0}, \boldsymbol{y}) = \frac{\partial \tau}{\partial x_a}\overline{\boldsymbol{x}}_{\boldsymbol{a}} + \frac{1}{2}\frac{\partial^2 \tau}{\partial x_a \partial x_b}\overline{\boldsymbol{x}}_{\boldsymbol{a}}\overline{\boldsymbol{x}}_{\boldsymbol{b}} \tag{43}$$

$$+ \frac{1}{6}\frac{\partial^3 \tau}{\partial x_a \partial x_b \partial x_c}\overline{\boldsymbol{x}}_{\boldsymbol{a}}\overline{\boldsymbol{x}}_{\boldsymbol{b}}\overline{\boldsymbol{x}}_{\boldsymbol{c}} \tag{44}$$

As we approach the caustic in the source plane, the magnification diverges and one of the eigenvalues of the Hessian matrix is zero. The potential is therefore free of curvature along the direction corresponding to that eigenvector, which we will call the $\hat{\boldsymbol{x}}_1$ direction. The time delay potential is flat to second order along $\hat{\boldsymbol{x}}_1$, and we include the dominant third-order term along this direction. Along the non-divergent ($\hat{\boldsymbol{x}}_2$) direction orthogonal to $\hat{\boldsymbol{x}}_1$, the curvature is $1 - \psi_{22}$, where $\psi_{22} = \partial^2_{x_2}\psi$.

$$\tau(\boldsymbol{x}, \boldsymbol{y}) - \tau(\boldsymbol{x_0}, \boldsymbol{y}) \approx -\overline{\boldsymbol{x}} \cdot \overline{\boldsymbol{y}} + \frac{1 - \psi_{22}(\boldsymbol{x_0})}{2}x_2^2 - \frac{\psi_{111}}{6}x_1^3 \tag{45}$$

As $\boldsymbol{y}$ approaches $\boldsymbol{y}_0$, the first derivative term vanishes, creating an image at $\boldsymbol{x}_0$ whose magnification is roughly the size of the patch in the lens plane over which the phase changes by 2π. Traversing the $\hat{\boldsymbol{x}}_2$ direction (fixing $x_1 = 0$), the second derivative of the potential sets the width of the coherent patch. Along the $\hat{\boldsymbol{x}}_1$ axis (fixing $x_2 = 0$), since the potential has no second derivative, the *third* derivative sets the extent of the coherent patch along the $\hat{\boldsymbol{x}}_1$ direction:

$$\partial^3\tau/\partial x_1^3 = -\psi_{111}(\boldsymbol{x}_0). \tag{46}$$

If we define $p = (1 - \psi_{22}(\boldsymbol{x}_0))^{-1}$ and $q = -2/\psi_{111}(\boldsymbol{x}_0) > 0$, the integrand is a quadratic along the x_2 direction and cubic along the x_1 direction:

$$F = \frac{\Omega}{2\pi i}\int d\overline{x}_1 \exp\left[i\Omega\left(-\overline{x}_1\overline{y}_1 + \frac{\overline{x}_1^3}{3q}\right)\right] \times \tag{47}$$

$$\int d\overline{x}_2 \exp\left[i\Omega\left(-\overline{x}_2\overline{y}_2 + \frac{\overline{x}_2^2}{2p}\right)\right] \tag{48}$$

Taking $\mu_* = |p|q^{2/3}\Omega^{1/3}$, $Y = q^{-1/3}\Omega^{-2/3}$, and letting Ai denote the Airy function,[3] the amplification factor is

$$|F| = 2\pi\mu_*\mathrm{Ai}(-\overline{y}_1/Y) \tag{49}$$

The characteristic fringe spacing near the fold caustic is $\sim Y \propto \Omega^{-2/3}$; Jaroszynski and Paczynski (1995) points out that it is possible to observe this fringing during a caustic crossing to measure the finite linear size of a source.

The resolution element is also different near the fold caustic. The intuition behind this is that near a fold caustic, the slit separation vanishes and the magnification increases, making the higher-order terms of Eq. (35) important. We start with the fold-caustic potential (Eq. (45)) and see that the image separation is $|x_j - x_k| = (\psi_{111}y_1/2)^{1/2}$. Using Eq. (35) and assuming that all third- and fourth-order derivatives of ψ vanish except ψ_{111} leaves only one relevant term in the sum ($a = b = c = d = e = f = 1$), where we have $A_{ad} = A_{be} = A_{cf} = (\psi_{111}x_1)^{-1}$. The time delay for different patches of the source is

$$\delta\tau_{jk} = \frac{2^{3/2}}{|\psi_{111}|^{1/2}}|y_1|^{1/2}\overline{y}_1 - \frac{2^{-3/2}}{6|\psi_{111}|^{1/2}}\frac{\overline{y}_1^3}{|y_1|^{3/2}} \tag{50}$$

and we can see that the second term dominates when $|y_1|$ is sufficiently small (i.e. $|y_1| \lesssim \overline{y}_1$).

8 Modified Optical Depths in Wave Optics

Modifying the amplification factor's dependence on the impact parameter near the Einstein ring modifies any optical depth calculations for which the observations are flux-limited. For forecasts and constraints, it is important to characterize the conditions under which diffraction modifies the optical depth relative to the standard case of geometric optics. One effect, discussed previously in Sect. 4, is that interference terms, if they are not washed out, are easier to detect than autocorrelation terms. With a search method that utilizes spectral information, this leads to an increase in the optical depth (Katz et al. 2020; Jow et al. 2020; Kader et al. 2022) which becomes significant when the source is much brighter than the limiting flux threshold.

[3] see Wolfram MathWorld

In Sect. 7, we discussed how the infinite magnifications in the point-mass model which appear in geometric optics get regularized by wave optics. One might initially suspect that this decreases the optical depth in a wave optics calculation relative to the geometric-optics calculation, but the reality is that it depends on the magnification scale (Sugiyama et al. 2020), and the angular motion of the source over the observation period. Diffraction eliminates the possibility of chance alignments creating extremely high magnification events ($|F|^2 \gtrsim \pi\Omega$) (Bontz and Haugan 1981; Nakamura and Deguchi 1999; De Paolis et al. 2002), as can be seen by the flattening of the wave optics magnification curves at small y in the bottom panel of Fig. 8 (for compact sources, magnifications of up to $|F|^2 \approx \pi\Omega$ can be achieved). However, for more modest amplification factors, the sidelobes of $F(\Omega, y)$ contribute non-negligibly to the cross-section. Let us take a graphical approach to understand this. Note the rapidly-varying oscillations in the top panel of Fig. 8. In a typical optical depth calculation with a certain sensitivity threshold $|F|^2_{\rm min}$, we integrate over all y in the source plane for which $|F(y)|^2 > |F|^2_{\rm min}$ to obtain the optical depth $\sigma(> |F_{\rm min}|^2)$. In the geometric optics case, $|F_{\rm geo}|^2(y)$ is a decreasing function. Therefore, the region of the source plane that counts towards the optical depth is a circle of radius y_* such that $|F_{\rm geo}(y_*)|^2 = |F|^2_{\rm min}$. Its area, the cross-section to lensing, is $\sigma(> |F_{\rm min}|^2) = \sigma_{\rm geo} = \pi y_*^2$. In the wave optics case, it is not as straightforward to calculate a threshold value of y, because the amplification factor is oscillatory, as seen in the top panel of Fig. 8. We numerically evaluate Eq. (28) on a dense grid of points $\{y_i\}$, checking for each y_i whether $|F(y_i)|^2 > |F_{\rm min}|^2$ and accumulating $2\pi y_i \Delta y_i$ to $\sigma(> |F_{\rm min}|^2)$ if the magnification is above the threshold. Repeating this as a function of $|F_{\rm min}|^2$, for $|F_{\rm geo}|^2$ (Eq. (40)), as well as $|F_{\rm wave}|^2$ (we use the approximation in Eq. (28)) allows us to calculate the optical depth in wave optics and in geometric optics as a function of the sensitivity cutoff $|F_{\rm min}|^2$; the ratio is plotted in Fig. 9 (see also Appendix B of Sammons et al. 2022, which studies the effect of high magnification lensing events on fast transient luminosity functions).

Finally, for $\Omega \lesssim 1$, the lensing is somewhat suppressed. This was considered in the context of kilohertz gravitational waves, which do not "feel" mass distributions smaller than $\sim 10^2 M_\odot$ (Nakamura 1998). The bad news is that this limits the mass reach of searches for gravitational wave lensing. The good news is that existing search pipelines are reasonably sensitive to diffractively-lensed gravitational waves, and that the subtle distortions due to diffractive lensing will be detectable for louder events (Dai et al. 2018). Moreover, the chromaticity induces a mass scale dependence of the lensing effect. This enables lensing magnifications to probe intervening density fluctuations as a function of mass scale (i.e. the matter power spectrum) (Takahashi 2006; Oguri and Takahashi 2020) and dark matter on dwarf galaxy scales (Choi et al. 2021; Guo and Lu 2022). In addition, diffraction suppresses the effect of microlensing on the measured magnification ratios of gravitational waves macro-lensed by galaxies or clusters, unless the macro-magnifications $\gtrsim 15$. This can offer cleaner time-delay cosmography measurements and allow gravitational waves to be localized (Mishra et al. 2021; Cheung et al. 2021).

Since $F(\Omega, \mathbf{y})$ is oscillatory as a function of Ω, the cross-section of microlensing is higher at low frequencies. However, as the frequency approaches $\Omega \leq 1$, no microlensing occurs (see next section). According to Fig. 4, the reach of optical microlensing is also cut off at low masses due to diffractive effects. These two effects lead to overestimating cross-sections calculated in the geometric optics limit. Due to finite source size effects (Witt and Mao 1994) and wave effects, the maximum magnification is less than infinity but drops off more slowly. See e.g. Fig. 5 of Sugiyama et al. (2020) for a detailed study of the effect of finite source size effect on the light curve of a transiting optical source.

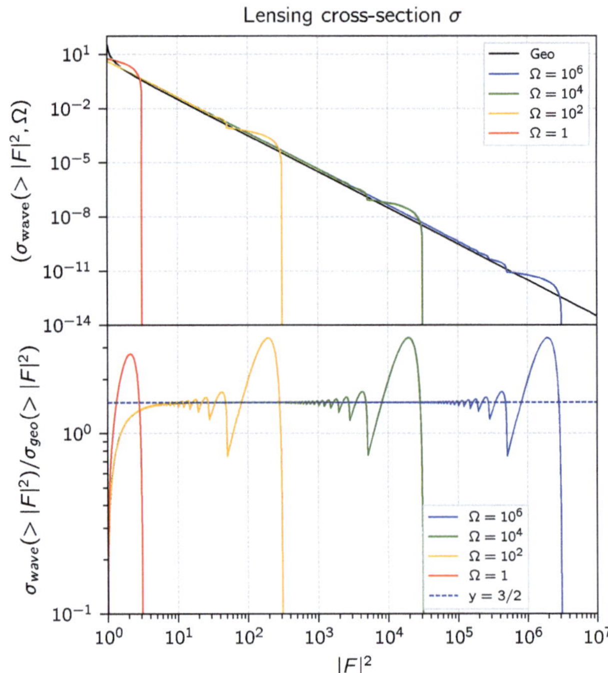

Fig. 9 The lensing cross-section σ for a minimum detectable amplification factor $|F|^2$. Top: we plot several values of Ω as well as the cross-section calculated in geometric optics. Bottom: the ratio of $\sigma_{\rm wave}/\sigma_{\rm geo}$ is well-approximated by a constant between $|F|^2 = [1, 2\pi\Omega]$ with a height of $y = 3/2$. The conclusion is that interference effects enhances lensing cross-sections for high magnifications by a factor of $\approx 3/2$ while cutting off huge amplifications greater than $\approx \pi\Omega$. This result has the caveat that Eq. (28) is used to approximate Eq. (23), which is an excellent approximation for $\Omega \gtrsim 1$

9 Possible Applications of Wave-Optics

As illustrated in Fig. 3, the takeaway of wave optics in lensing is that the chromatic amplification yields interesting opportunities to extract information. We now highlight several proposed applications of wave optics and interference effects.

9.1 The Lens as an Astronomical Interferometer

Most fundamentally, the interference between multiple images allows each image to be used as a path in a virtual interferometer. This adds transverse spatial resolution which can easily exceed that of ground-based instrumentation, and precise measurements of the time delay between each path. The increased angular resolution has been used to resolve pulsar and FRB emission regions with exquisite transverse spatial resolution in the context of plasma lensing ($\sim 10^4$ km Main et al. 2018; Kumar et al. 2024; Nimmo et al. 2025). This can also be done with gravitationally-lensed wavefronts, but these systems have not yet been discovered.

The time delay itself can also be used as a powerful probe. As discussed previously, the quasi-sinusoidal ripple generated by this interference enables a precise, phase-coherent measurement of lensing time delays independent of intrinsic source variability. Measuring time delays this way is a ubiquitous idea that has been proposed for various physical systems, from sub-Hertz gravitational waves generated by chirping binaries lensed by $\sim 10^8 M_\odot$ objects (Takahashi and Nakamura 2003). to radio pulsar/fast radio bursts lensed by sub-solar mass compact objects and planets (Jow et al. 2020). We emphasize that the interference fringes rely on correlated fluctuations in the wave field, not the total power. This means that for sufficiently compact sources, precise lensing delays may be measured towards non-transient sources. However, the stringent size requirements on preserving the interference fringes excludes most, if not all, currently known continuum sources from this type of measurement. Transient signals from compact objects are precisely those which are known to be sufficiently point-like to realize the interferometric measurement of time delays. In transient lensing as compared with continuum-source lensing, we lose our ability to monitor the Einstein crossing timescale, but we have gained a single-shot measurement of the lens delay timescale free of uncertainties from estimating the relative velocities of the source and lens. Using a simple lens model, the lensing delay and the flux ratio can completely determine the mass and instantaneous impact parameter of an unresolved lens (Itoh et al. 2009, see also). In the case of compact yet persistent emitters or recurring transient emission such as pulsations from pulsars, measurement of the Einstein crossing timescale can be used to extract additional information such as the relative motion of the lens and the source.

9.2 Breaking the Mass-Sheet Degeneracy

The chromaticity in lensing is not strictly a time-delay measurement. It can also be understood as a rough mass measurement: if the mass is smaller than one wavelength ($\Omega \sim 1$) we should expect to see suppression of the lensing due to wave optics (Cremonese et al. 2021). When in the eikonal or wave-optics regimes, this insight can be used to break the mass-sheet degeneracy, which allows mass to be freely exchanged between the compact lens and an extended, uniform sheet even when the image positions, magnifications, and time delays are fixed.

9.3 Multi-Messenger Tests of the Equivalence-Principle

One particularly elegant test of fundamental physics is to use a multi-messenger observation of a lensed gravitational-wave + electromagnetic source to test gravity. This allows for exquisite measurements of wave optical effects, since due to their low Ω, gravitational waves are potent probes of diffractive effects, while electromagnetic waves are typically in the eikonal or the geometric optics regimes. Observing simultaneously-emitted low-frequency (gravitational) images and high-frequency (electromagnetic) images can expose wave optics effects such as modifications to the phase velocity of diffractively-lensed images (Takahashi 2017), though analyses of the causal "wavefront" (Suyama 2020; Ezquiaga et al. 2020) reveals that no violation of causality occurs, though an apparent violation of causality may arise from incorrect separation of the observed image superposition into individual images (Tanaka and Suyama 2023).

9.4 Gravitational Scintillation and the Small-Scale Matter Power Spectrum

It would be remiss to omit in this review the similarities between wave optics/interference effects and scintillation (see e.g. Narayan 1992; Goodman et al. 1987). The interference pattern in Fig. 3 is similar to diffractive scintillation in pulsar astronomy, where inhomogeneous plasma in front of a compact background source causes an intrinsically featureless spectrum to exhibit stochastic ripples (or scintles) over some characteristic bandwidth. By watching the scintles evolve over time, it is possible to constrain the motion of the intervening plasma screens (see e.g. Walker et al. 2004). A drifting gravitational lens can similarly produce a slowly-evolving amplification factor as it transits (see upper left panel of Fig. 10). A collection of moving lenses causing gravitational scintillation has also been considered. Macquart (2004) derives analytical expressions for the phase perturbations sourced by clumpy matter distributions; Congedo et al. (2006) considers the particular case of gravitational scintillation by stellar clusters in the Galactic center. Takahashi et al. (2005) considers the validity of the Born approximation and the impact of chromatic (diffractive) effects in gravitational scintillation. Diffractive effects as a probe of the small-scale matter power spectrum are considered in Takahashi (2006), Oguri and Takahashi (2020), Inamori and Suyama (2021), Oguri and Takahashi (2022).

9.5 Beyond Scalar Wave Optics

Scalar wave optics already captures much of the relevant phenomenology of interference effects in gravitational lensing. From an observational standpoint, the eikonal and wave optics regimes are still uncharted territory. Virtually all observations of gravitational lensing thus far fall squarely in the geometric optics regime. However, in an era of gravitational-wave observations from the ground and from space, there are compelling observational reasons to go beyond scalar wave optics.

Notably, we have worked with scalar waves in the weak gravity limit, ignoring cases where $U \sim 1$, for which Eq. (1) is not valid. Numerically simulating wave propagation in strongly-curved spacetimes becomes necessary and is an active area of research in general relativity (Zenginoğlu and Galley 2012; Yang et al. 2014). These numerical solutions of scalar wave propagation reveal rich phenomenology, particularly near caustics, where numerical simulations predict multipath propagation for integer numbers of cycles around the photon ring. The simulated "caustic echo" has a period which is in good agreement with the light-crossing time of the photon ring (Stewart 1989).

Other aspects of lensing are not captured by the scalar theory, such as polarization-dependent effects. Two examples from general relativity are the gravitational Faraday rotation induced by rotating gravitational potentials (Ishihara et al. 1988; Nouri-Zonoz 1999; Asada and Kasai 2000; Li et al. 2022) and the gravitational spin-Hall effect, in which the trajectories of waves through a gravitational potential becomes frequency- and polarization-dependent (Gosselin et al. 2007; Oancea et al. 2020), leading to dispersive wave propagation and different arrival times than predicted by geometric optics (Oancea et al. 2024).

Note also that while the scalar wave formalism for both gravitational and plasma lensing are effectively identical (the only difference being that gravitational potentials are frequency independent, but plasma lensing potentials have a $\sim \lambda^2$ dependence), magnetic fields in ionized plasma cause different polarizations to experience different refractive indices in a phenomenon known as birefringence (Li et al. 2019). Thus, while the scalar wave theory of gravitational lensing and plasma lensing share significant overlap, notable differences arise in the vector propagation.

A final limitation is that we have so far limited our discussion to the effects of scalar perturbations to the metric on the propagation of waves through space-time. In general, metric perturbations may themselves have vector and tensor components (e.g. gravitational waves). Gravitational waves will induce additional time delays in the propagation of light from background sources which may be observable (indeed, this is the basis for pulsar timing array experiments). Gravitational waves themselves may therefore lens coherent sources of radiation (Boyle and Pen 2012; Rahvar 2018) inducing characteristic interference patterns; this is an under-explored and active area of research. Polarization effects, such as the gravitational Faraday rotation, may also be more pronounced in lensing by tensor and vector perturbations, and may have observable impacts on CMB lensing (Dai 2014). However, in light of the complexity and magnitude of these effects, for the remainder of this review paper, we will focus on scalar wave optics in the aforementioned regimes: eikonal scalar

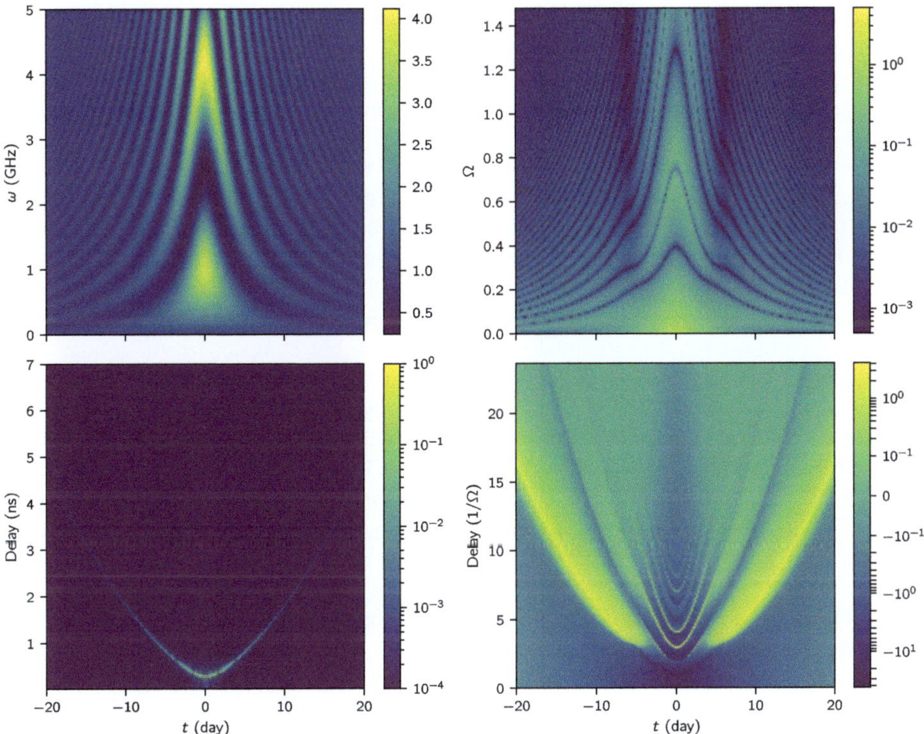

Fig. 10 Top left: Frequency-dependent amplification $|F_{\text{wave}}(\Omega, y)|^2$ spectrum of a compact, broadband radio sources as it transits behind a point-mass lens of $5M_{\otimes}$, $D = 1$ kpc with a minimum impact parameter of $y_0 = 0.5$. The amplification oscillates like $\cos^2(\Omega(\tau_+ - \tau_-))$ as a function of frequency for large impact parameters where the eikonal optics (see Eq. (19)) approximation holds. Top right: Fractional discrepancy between $|F_{\text{wave}}(\Omega, y)|^2$ and $|F_{\text{geo}}(\Omega, y)|^2$ for the point-mass lens. Diffractive effects distort the regular fringe pattern predicted from geometric optics. Bottom left: The Fourier transform of the top left panel along the frequency axis, i.e. the delay spectrum as a function of time through the transit. The arc tracks the instantaneous lensing delay $(\tau_+(t) - \tau_-(t))$ as it changes throughout the transit. Bottom right: The Fourier transform of the top right panel along the frequency axis. Adapted with permission from (Jow et al. 2020)

optics, diffractive scalar optics, and the transition from coherent to incoherent (traditional) gravitational lensing.

10 Observational Prospects and Discussion

We have presented a summary of the various aspects of interference in gravitational lensing. In some ways, interference in gravitational lensing makes lensing easier to detect. For example, in the eikonal limit, the presence of interference terms between images makes searching for the interference terms much more sensitive than searches for individual images: one simply searches for a regular sinusoidal pattern in the spectrum whose spacing allows for a direct measurement of the lensing time delay. For diffractive optics with a point-mass lens, the sinusoidal pattern becomes distorted at low frequencies as Ω tends to 1; the unique form of this spectral distortion can break degeneraces in microlensing, allowing for a measurement of the impact parameter and (redshifted) mass of the lens simultaneously.

In addition, diffraction limits the maximum attainable magnification near caustics, where interference effects become the most pronounced. We find in diffractive lensing that extremely large magnifications near cusps and folds are no longer possible, but that "sidelobes" in the magnification pattern can boost the cross-section to lensing. We calculate this boost, and find (in agreement with Sammons et al. (2022), which appeared while this manuscript was being prepared) that to a very good approximation, the lensing cross-section is higher by a factor of $3/2$ as compared to geometric optics for large ($1 \ll |F|^2 < \pi\Omega$) magnifications.

Finally, we characterized the primary challenge to observing interference in gravitational lensing: the paucity of point-like, or spatially coherent, sources of radiation. A true point source, even one that does not exhibit any significant light curve variations, could be used to measure lensing delays to Nyquist-limited time resolution (nanoseconds at radio frequencies). Such compact steady sources are not yet known, but compact *transient* sources at cosmological distances such as supernovae and fast radio bursts could be used. Gravitational waves from compact binary mergers observed by current and upcoming ground- and space-based detectors are particularly promising candidates for observing diffractive lensing. We comment briefly on the different observational prospects here, but refer the reader to Oguri (2019) for a comprehensive discussion of transient lensing.

In the field of lensed transients, supernovae have a head start since a handful of lensed supernovae have already been discovered (e.g. Kelly et al. 2015). Since Ω is large for optical and near-infrared supernovae, they can potentially be extremely highly magnified Diego (2019). However, according to Fig. 8, supernovae photospheres are not sufficiently compact, except at cosmological distances.

GRBs are among the most luminous and distant transients known, leading to favorable lensing rates and therefore some searches over the last few decades (Gould 1992), but because of the extremely short wavelengths involved, the emission is in most cases too extended, i.e. spatially incoherent (Stanek et al. 1993; Witt and Mao 1994; Matsunaga and Yamamoto 2006). While the apparent size of the emission is dependent on the beaming factor (Katz et al. 2018; Khangulyan et al. 2022), and can vary from burst to burst, most GRBs are not point-like enough to observe the spectral features of femtolensing; furthermore the range of lens sizes required to observe interference effects in GRBs may not be astrophysical (see Fig. 4).

FRBs are extremely compact, but their true angular sizes are highly uncertain. They may be angularly broadened (Lazio et al. 2008) into a "scattering disk" at the low (sub-GHz) radio frequencies at which they are typically observed (Katz et al. 2020; Kumar and Beniamini 2023). Nevertheless, spatial coherence in the sense of Eq. (36) is preserved for sub-solar mass lenses (Kader et al. 2022; Leung et al. 2022). The bulk of FRBs detected to date have been surveyed by the Canadian Hydrogen Intensity Mapping Experiment at low (sub-gigahertz) frequencies (CHIME/FRB Collaboration et al. 2021), but because of the strong frequency dependence of the size of the scattering disk ($R_{src} \propto \lambda^2$), these FRBs are often angularly-broadened by inhomogeneous plasma, which leads to a larger effective source size and potentially washes out interference fringes. This problem can be mitigated with FRBs surveyed at even slightly higher frequencies by upcoming experiments like CHORD and the DSA-2000 (Vanderlinde et al. 2019; Hallinan et al. 2019); these may enable observation of coherent interference patterns in FRBs lensed by stars (Connor and Ravi 2023).

Perhaps the most promising way to observe wave optics is to go to low frequencies is via gravitational wave detectors. The Laser Interferometer Space Antenna (LISA) mission is the way to do this, and many forecasts have been done with LISA-scale sources and lenses in mind, e.g. Çalışkan et al. (2023), Tambalo et al. (2023b). However, it is also possible to use existing ground-based gravitational wave detectors like LIGO (LIGO Scientific Collaboration et al. 2015), Virgo (Acernese et al. 2015), and KAGRA (KAGRA Collaboration et al. 2019), whose pristine sources are not contaminated by propagation effects, except perhaps by microlenses near the images of a macro-lens (Mishra et al. 2021). Gravitational-wave sources are extremely compact ($\sim$ 100 km Oguri 2019), and searches have already been designed and conducted in both the eikonal and diffractive regimes (Haris et al. 2018; Hannuksela et al. 2019; Dai et al. 2020; Abbott et al. 2021b), though no candidates have been conclusively confirmed. Prospects for detecting wave-optics effects in continuous GW sources have also been studied (Basak et al. 2023; Savastano et al. 2024); while dim, these sources could potentially be found in extremely long integrations (Liao et al. 2019; Suvorov 2022). A key observational challenge of GW lensing to overcome is that with the current generation of gravitational wave detectors (LIGO, Virgo, KAGRA, and GEO600), gravitational waves are difficult to localize to a host galaxy without an electromagnetic counterpart. However, rapid progress, particularly in FRBs and gravitational-wave instruments, will soon lead to detections of interference and diffractive effects in gravitational lensing, and eventually the creative application of lensing to unique measurements which can advance astronomy and fundamental physics.

Appendix: The Curved-Spacetime Scalar Wave Equation

In this section we derive the starting point of this paper from the scalar wave equation in a curved spacetime:

$$0 = g^{ab}\nabla_a \nabla_b \phi. \tag{51}$$

Scalar wave theory captures interference and diffraction phenomena in gravitational lensing,[4] and is sufficient for analyzing most interference phenomena in gravitational lensing. The starting point of scalar wave theory is the Fresnel-Kirchoff integral, from which most lensing phenomena can be derived. It requires a little bit of general relativity, but we will assume the reader has little background in that area.

The formalism of gravitational lensing is similar to the Newtonian dynamics of point masses orbiting in a potential. In the same way that the external potential is fixed for point masses, we consider the dynamics of a scalar field representing our wave, ϕ, on a static background spacetime, whose curvature is specified by the metric tensor g_{ab}. In a weak gravitational field [5] with only scalar degrees of freedom, the metric tensor can be written in terms of the gravitational potential as $g_{00} = -(1+2U)$ and $g_{ii} = 1-2U$ for $i = 1, 2, 3$, and where U is the Newtonian gravitational potential, and where we use units in which $c = 1$ for now.

In Eq. (51), ∇_a refers to the covariant derivative. The covariant derivative needs to be expanded into partial differential equations using several identities from general relativity. We introduce the notation $g = \det(g_{ab}) = g_{00}g_{11}g_{22}g_{33}$ for our diagonal metric. Since scalars are Lorentz-invariant, the covariant derivative of a scalar is the same as its partial derivative:

$$\nabla_b \phi = \partial_b \phi. \tag{52}$$

[4]but ignores e.g. polarization-dependent effects.

[5]Confusingly, “strong lensing”, which refers to the regime in lensing in which multiple images arise, occurs mostly in the weak-field limit; we discuss the strong-field limit briefly in Sect. 9.5.

The analogous rule for (the covariant divergence of) a vector is[6]

$$\nabla_a V^a = \frac{1}{\sqrt{-g}}\partial_a(\sqrt{-g}V^a). \tag{53}$$

Equation (53) implies

$$g^{ab}\nabla_a\nabla_b\phi = \nabla_a(g^{ab}\nabla_b\phi) = \nabla_a(g^{ab}\partial_b\phi) \tag{54}$$

$$= \frac{1}{\sqrt{-g}}\partial_a(\sqrt{-g}g^{ab}\partial_b\phi) = 0. \tag{55}$$

Equation (55) is the wave equation often used as a starting point in studying gravitational lensing in wave optics, e.g. Nakamura and Deguchi (1999), Macquart (2004). We now expand the four non-zero terms in U. Since we are working in the weak-field limit, $|U| \ll 1$; this justifies neglecting terms of $\mathcal{O}(U^2)$. Since the potential is static, the 00 term evaluates to

$$\frac{1}{\sqrt{-g}}\partial_0(\sqrt{-g}g^{00}\partial_0\phi) = -\frac{1}{1+2U}\partial_t^2\phi \approx -(1-2U)\partial_t^2\phi. \tag{56}$$

The $a = b = 1$ term to first order in U is

$$\frac{1}{\sqrt{-g}}\partial_1(\sqrt{-g}g^{11}\partial_1\phi) = (1+2U)\partial_1^2\phi. \tag{57}$$

The full equation is

$$0 = -(1-2U)\partial_0^2\phi + (1+2U)\sum_{i=1,2,3}\partial_i^2\phi + \mathcal{O}(U^2) \tag{58}$$

which can compactly be written as

$$0 = \nabla^2\phi - (1-4U)\partial_t^2\phi \tag{59}$$

where the ∇^2 (without the indices) refers to the flat-space Laplacian. There are various ways to solve Eq. (59), which is analogous to the Schrödinger equation (Davydov 1965; Peters 1974; Deguchi and Watson 1986b). In circumstances when the potential is not time-dependent, the most straightforward solution approach is to take the Fourier transform over time, letting $\phi(\vec{r}, t) \to \tilde{\phi}(\vec{r}, \omega)$. We obtain

$$4\omega^2 U\tilde{\phi} = (\nabla^2 + \omega^2)\tilde{\phi}. \tag{60}$$

Equation (60) encodes the physics of wave propagation in a Newtonian potential. To see this, we can plug in the Fourier mode solution $\tilde{\phi} = e^{i\vec{k}(\omega)\cdot\vec{r} - i\omega t}$ to obtain the dispersion relation for waves in a gravitational field, and the expression for the group velocity of light in a gravitational field, which encodes the Shapiro delay:

$$\omega(k) = \pm|k|(1-4U)^{-1/2} \approx \pm|k|(1+2U) \tag{61}$$

[6]This in turn arises from an identity for a contracted Christoffel symbol: $\Gamma^c_{ca} = \frac{\partial_a(\sqrt{-g})}{\sqrt{-g}}$; see e.g. Lightman et al. (1975) (Exercise 7.7j) for more details.

$$d\omega/dk = 1 + 2U = v_g \tag{62}$$

We can continue to solve Eq. (60) by considering its solution when $U = 0$. In a coordinate system centered on the source (see Fig. 1), the solution is $\tilde{\phi}_0(\vec{r}) = \exp(i\omega r)/r$. We make that ansatz $\tilde{\phi}(\vec{r}, \omega) = F(\vec{r})\phi_0$, where $F(\vec{r}, \omega)$ is the amplification factor of the wave field (as contrasted with the flux, i.e. the field value squared). Expanding the spherical Laplacian as

$$\nabla^2 = \frac{1}{r^2}\frac{\partial}{\partial r}\left(r^2\frac{\partial}{\partial r}\right) + \frac{1}{r^2}\nabla_\theta^2 \tag{63}$$

gives

$$4\omega^2 U F = 2i\omega\partial_r F + \partial_r^2 F + \frac{1}{r^2}\nabla_\theta^2 F. \tag{64}$$

We now study some properties of F to build some intuition with the goal to further simplify Eq. (64). Since have defined F to not capture the r^{-1} decay of the wave amplitude or the oscillation of the wavefront, we expect F to be slowly varying along the radial direction. If the total propagation path has length $\sim D$, dimensional analysis suggests that the radial variation of F is of order $|\partial_r^2 F|/|\partial_r F| \sim 1/D$. If we further make the approximation that the path length $D \gg \lambda$, then the $\partial_r^2 F$ term in Eq. (64) is much smaller than the $2i\omega\partial_r F$ term and can be neglected. The transverse variation of F encodes the interference pattern and is much more interesting. In fact, we visualize a transverse slice of $|F|^2$ every time we hold up a piece of paper to see the interference pattern behind a double slit laboratory experiment. As mentioned in the main text the relevant angular scale for the interference pattern is the Fresnel angular scale $\sqrt{\lambda/D} \ll 1$; working with these small angles enables us to replace the angular part of the Laplacian with its flat-sky equivalent. We parameterize the two angular dimensions using a flat-sky coordinate system $\boldsymbol{q} = r\boldsymbol{\theta}$ whose origin is the straight-line path to the observer, and define $\nabla_q^2 = \frac{1}{r^2}\nabla_\theta^2$. With all these simplifications, Eq. (64) becomes

$$i\partial_r F = -\frac{1}{2\omega}\nabla_q^2 F + 2\omega U F. \tag{65}$$

Equation (65) resembles the Schrödinger equation (Takahashi and Nakamura 2003), and we may solve it using the Feynman path integral. The Lagrangian which goes into the Feynman path integral can be derived from the classical Hamiltonian corresponding to Eq. (65), which governs transverse deflections of rays evolving in two transverse dimensions as a function of time, or equivalently, the radius r from the waves' origin. The evolution of the transverse trajectories of rays is specified by $\boldsymbol{q}$ and $\dot{\boldsymbol{q}} = r\dot{\boldsymbol{\theta}}$. The conjugate momentum is $\boldsymbol{p} = mr\dot{\boldsymbol{\theta}}$, where the analogous "mass" (not to be confused with the lens mass!) $m = \omega$ and the classical potential is $V = 2\omega U$. The Lagrangian corresponding to this Hamiltonian can be obtained by a Legendre transform:

$$L(\boldsymbol{q}, \dot{\boldsymbol{q}}, r) = \dot{\boldsymbol{q}} \cdot (m\dot{\boldsymbol{q}}) - H(\boldsymbol{q}, \boldsymbol{p}) = \omega\left[\frac{\dot{\boldsymbol{q}}^2}{2} - 2U\right] \tag{66}$$

where $H = \frac{\boldsymbol{p}^2}{2m} + V(r, \boldsymbol{q})$. Having written down the Lagrangian, solutions to Eq. (65) can be written as an integral over all possible ray trajectories $\boldsymbol{q}(r)$:

$$F(\omega, \vec{r}) = \int \mathcal{D}[\boldsymbol{q}(r)] \exp\left[i\int L(\boldsymbol{q}, \dot{\boldsymbol{q}}, r)\, dr\right] \tag{67}$$

The kinetic and potential terms in the Lagrangian correspond to the geometric time delay ($\tau_{\rm geo}$) and the Shapiro delay ($\tau_{\rm grav}$) in optics (Blandford and Narayan 1986). Putting the factors of c back in, the geometric delay along the line of sight is

$$\hat{\tau}_{\rm geo} = \int_0^{D_S} \frac{r^2 \dot{\boldsymbol{\theta}}(r)^2}{2}\, dr \tag{68}$$

To evaluate this integral we can make the simplifying assumption that the light paths are straight rays bent "instantly" once a ray hits a single lens plane as shown in Fig. 1. This relies on the lens to be thin enough along the propagation direction that the ray is not deflected appreciably relative to its "piecewise-straight" approximation (Schneider et al. 1992, Chap. 4.3). In this approximation the above integral reduces to

$$= \frac{D_L D_S}{2 D_{LS} c} |\boldsymbol{\theta}_l - \boldsymbol{\beta}|^2. \tag{69}$$

where $\boldsymbol{\theta}_l$ and $\boldsymbol{\beta}$ are the angular positions of the lens and the apparent source position (see Fig. 1). This is the same as assuming that far away from the lens plane, the index of refraction of the medium is homogeneous enough such that rays travel in approximately straight lines.

The gravitational part of the time delay is captured entirely by the Newtonian potential in the limit of weak gravity ($|U| \ll 1$):

$$\hat{\tau}_{\rm grav} = -\frac{2}{c^3} \int_0^{D_S} U(\boldsymbol{q}, r)\, dr. \tag{70}$$

In the lens frame, we approximate this integral by collapsing the gravitational potential over the z axis (the thin-lens approximation). Defining the physical impact parameter $\boldsymbol{b}$ ($= r_l \boldsymbol{\theta}$ from the source's perspective, and $\boldsymbol{b} = D_L \boldsymbol{\theta}_l$ from the observer's perspective) we define

$$\hat{\psi}(\boldsymbol{b}) = \frac{2}{c^3} \int U(\boldsymbol{b} + z\hat{z})\, dz \approx -\hat{\tau}_{\rm grav}. \tag{71}$$

The approximations in Eq. (68) and (71) reduce the infinite-dimensional Feynman path integral to the Fresnel-Kirchhoff diffraction integral, where $D_{\rm eff} = D_L D_S / D_{LS}$:

$$F(\omega, \boldsymbol{\beta}) \propto \int d^2\boldsymbol{\theta}_l \exp\left\{ i\omega \left[\frac{D_{\rm eff}}{2c} |\boldsymbol{\beta} - \boldsymbol{\theta}_l|^2 - \hat{\psi}(D_L \boldsymbol{\theta}_l) \right] \right\} \tag{72}$$

Acknowledgements We thank the International Space Science Institute in Bern (ISSI) for their hospitality and the conveners for organizing the workshop on Strong Gravitational Lensing where this paper was initially written, and Paul Schechter for helpful discussions throughout the preparation of this manuscript. C. L. is supported by a Miller Postdoctoral Fellowship at UC Berkeley. L.D. acknowledges research grant support from the Alfred P. Sloan Foundation (Award Number FG-2021-16495). M.O. was supported by JSPS KAKENHI Grant Number JP22K18720.

Funding Open access funding provided by University of Zurich.

Declarations

Competing Interests The authors declare no competing interests.

Open Access This article is licensed under a Creative Commons Attribution 4.0 International License, which permits use, sharing, adaptation, distribution and reproduction in any medium or format, as long as you give appropriate credit to the original author(s) and the source, provide a link to the Creative Commons licence, and indicate if changes were made. The images or other third party material in this article are included in the article's Creative Commons licence, unless indicated otherwise in a credit line to the material. If material is not included in the article's Creative Commons licence and your intended use is not permitted by statutory regulation or exceeds the permitted use, you will need to obtain permission directly from the copyright holder. To view a copy of this licence, visit http://creativecommons.org/licenses/by/4.0/.

References

Abbott BP, Abbott R, Abbott TD, et al LIGO Scientific Collaboration, Virgo Collaboration (2016) Observation of gravitational waves from a binary black hole merger. Phys Rev Lett 116(6):061102. https://doi.org/10.1103/PhysRevLett.116.061102

Abbott BP, Abbott R, Abbott TD, et al LIGO Scientific Collaboration, Virgo Collaboration (2017a) GW170817: observation of gravitational waves from a binary neutron star inspiral. Phys Rev Lett 119(16):161101. https://doi.org/10.1103/PhysRevLett.119.161101

Abbott R, Abbott TD, Abraham S, et al LIGO Scientific Collaboration, Virgo Collaboration (2021b) Search for lensing signatures in the gravitational-wave observations from the first half of LIGO-Virgo's third observing run. Astrophys J 923(1):14. https://doi.org/10.3847/1538-4357/ac23db

Abbott R, Abbott TD, Abraham S, et al Ligo Scientific Collaboration, VIRGO Collaboration, KAGRA Collaboration (2021c) Observation of gravitational waves from two neutron star-black hole coalescences. Astrophys J Lett 915(1):5. https://doi.org/10.3847/2041-8213/ac082e

Acernese F, Agathos M, Agatsuma K, et al (2015) Advanced Virgo: a second-generation interferometric gravitational wave detector. Class Quantum Gravity 32(2):024001. https://doi.org/10.1088/0264-9381/32/2/024001

Asada H, Kasai M (2000) Can we see a rotating gravitational lens? Prog Theor Phys 104(1):95–102. https://doi.org/10.1143/PTP.104.95

Baraldo C, Hosoya A, Nakamura TT (1999) Gravitationally induced interference of gravitational waves by a rotating massive object. Phys Rev D 59(8):083001. https://doi.org/10.1103/PhysRevD.59.083001

Basak S, Sharma AK, Kapadia SJ, Ajith P (2023) Astrophys J Lett 942:L31. https://doi.org/10.3847/2041-8213/acab50

Beach TL, Lovelace EV (1997) Diffraction by a sinusoidal phase screen. Radio Sci 32(3):913–921. https://doi.org/10.1029/97RS00063

Benson JR, Cooke JH (1979) High-intensification regions of gravitational lenses. Astrophys J 227:360–363. https://doi.org/10.1086/156739

Berry MV Scalings for diffraction-decorated caustics in gravitational lensing. J Opt 23(6):065604 (2021) https://doi.org/10.1088/2040-8986/abfee1

Berry MV, Upstill C (1980) Iv catastrophe optics: morphologies of caustics and their diffraction patterns. Prog Opt 18:257–346

Blandford R, Narayan R (1986) Fermat's principle, caustics, and the classification of gravitational lens images. Astrophys J 310:568. https://doi.org/10.1086/164709

Bontz RJ, Haugan MP (1981) A diffraction limit on the gravitational lens effect. Astrophys Space Sci 78(1):199–210. https://doi.org/10.1007/BF00654034

Boyle L, Pen U-L (2012) Pulsar timing arrays as imaging gravitational wave telescopes: angular resolution and source (de)confusion. Phys Rev D 86(12):124028. https://doi.org/10.1103/PhysRevD.86.124028

Çalışkan M, Ji L, Cotesta R, et al (2023) Phys Rev D 107:043029. https://doi.org/10.1103/PhysRevD.107.043029

Chang K, Refsdal S (1979) Flux variations of QSO 0957 + 561 A, B and image splitting by stars near the light path. Nature 282(5739):561–564. https://doi.org/10.1038/282561a0

Cheung YH, Gais J, Hannuksela OA, Li FG (2021) Stellar-mass microlensing of gravitational waves. Mon Not R Astron Soc 503(3):3326–3336. https://doi.org/10.1093/mnras/stab579

CHIME/FRB Collaboration, Amiri M, Andersen BC, Bandura K, et al (2021) The first CHIME/FRB fast radio burst catalog. Astrophys J Suppl Ser 257(2):59. https://doi.org/10.3847/1538-4365/ac33ab

Choi HG, Park C, Jung S (2021) Small-scale shear: peeling off diffuse subhalos with gravitational waves. Phys Rev D 104(6):063001. https://doi.org/10.1103/PhysRevD.104.063001

Chwolson O (1924) Über eine mögliche Form fiktiver Doppelsterne. Astron Nachr 221:329. https://doi.org/10.1002/asna.19242212003

Congedo G, de Paolis F, Longo P, Nucita AA, Vetrugno D, Qadir A (2006) Gravitational wave scintillation by a stellar cluster. Int J Mod Phys D 15(11):1937–1945. https://doi.org/10.1142/S0218271806009248
Connor L, Ravi V (2023) Stellar prospects for FRB gravitational lensing. Mon Not R Astron Soc 521(3):4024–4038. https://doi.org/10.1093/mnras/stad667
Cremonese P, Ezquiaga JM, Salzano V (2021) Breaking the mass-sheet degeneracy with gravitational wave interference in lensed events. Phys Rev D 104(2):023503. https://doi.org/10.1103/PhysRevD.104.023503
Dai L (2014) Rotation of the cosmic microwave background polarization from weak gravitational lensing. Phys Rev Lett 112(4):041303. https://doi.org/10.1103/PhysRevLett.112.041303
Dai L, Miralda-Escudé J (2020) Gravitational lensing signatures of axion dark matter minihalos in highly magnified stars. Astron J 159(2):49. https://doi.org/10.3847/1538-3881/ab5e83
Dai L, Venumadhav T (2017) On the waveforms of gravitationally lensed gravitational waves. arXiv:1702.04724
Dai L, Li S-S, Zackay B, Mao S, Lu Y (2018) Detecting lensing-induced diffraction in astrophysical gravitational waves. Phys Rev D 98(10):104029. https://doi.org/10.1103/PhysRevD.98.104029
Dai L, Zackay B, Venumadhav T, Roulet J, Zaldarriaga M (2020) Search for Lensed Gravitational Waves Including Morse Phase Information: an Intriguing Candidate in O2. arXiv:2007.12709
Davydov AS (1965) Quantum mechanics. International series in natural philosophy, vol 1. Elsevier, Amsterdam
De Paolis F, Ingrosso G, Nucita AA, Qadir A (2002) A note on gravitational wave lensing. Astron Astrophys 394:749–752. https://doi.org/10.1051/0004-6361:20021258
Deguchi S, Watson WD (1986a) Wave effects in gravitational lensing of electromagnetic radiation. Phys Rev D 34(6):1708–1718. https://doi.org/10.1103/PhysRevD.34.1708
Deguchi S, Watson WD (1986b) Diffraction in gravitational lensing for compact objects of low mass. Astrophys J 307:30. https://doi.org/10.1086/164389
Diego JM (2019) The universe at extreme magnification. Astron Astrophys 625:84. https://doi.org/10.1051/0004-6361/201833670
Doyle LR, Carico DP (2009) Quantum uncertainty considerations for gravitational lens interferometry. Open Astron J 2(1):63–71. https://doi.org/10.2174/1874381100902010063
Engeli S, Saha P (2022) Optical properties of the solar gravity lens. Mon Not R Astron Soc 516(4):4679–4683. https://doi.org/10.1093/mnras/stac2522
Ezquiaga JM, Hu W, Lagos M (2020) Apparent superluminality of lensed gravitational waves. Phys Rev D 102(2):023531. https://doi.org/10.1103/PhysRevD.102.023531
Ezquiaga JM, Holz DE, Hu W, Lagos M, Wald RM (2021) Phase effects from strong gravitational lensing of gravitational waves. Phys Rev D 103(6):064047. https://doi.org/10.1103/PhysRevD.103.064047
Falco EE, Gorenstein MV, Shapiro II (1985) On model-dependent bounds on H 0 from gravitational images : application to Q 0957+561 A, B. Astrophys J Lett 289:1–4. https://doi.org/10.1086/184422
Feldbrugge J, Turok N (2020) Gravitational lensing of binary systems in wave optics. arXiv:2008.01154
Feldbrugge J, Pen U-L, Turok N (2023) Oscillatory path integrals for radio astronomy. Ann Phys 451:169255. https://doi.org/10.1016/j.aop.2023.169255
Goodman JJ, Romani RW, Blandford RD, Narayan R (1987) The effects of caustics on scintillating radio sources. Mon Not R Astron Soc 229:73–102. https://doi.org/10.1093/mnras/229.1.73
Gosselin P, Bérard A, Mohrbach H (2007) Spin Hall effect of photons in a static gravitational field. Phys Rev D 75(8):084035. https://doi.org/10.1103/PhysRevD.75.084035
Gould A (1992) Femtolensing of Gamma-Ray bursters. Astrophys J Lett 386:5. https://doi.org/10.1086/186279
Grillo G, Cordes J (2018) Wave asymptotics and their application to astrophysical plasma lensing. arXiv:1810.09058
Guo X, Lu Y (2022) Probing the nature of dark matter via gravitational waves lensed by small dark matter halos. Phys Rev D 106(2):023018. https://doi.org/10.1103/PhysRevD.106.023018
Hallinan G, Ravi V, Weinreb S, et al (2019) The DSA-2000 — a radio survey camera. In: Bulletin of the American Astronomical Society, vol 51, p 255. https://doi.org/10.48550/arXiv.1907.07648
Hannuksela OA, Haris K, Ng YK, Kumar S, Mehta AK, Keitel D, Li FG, Ajith P (2019) Search for gravitational lensing signatures in LIGO-Virgo binary black hole events. Astrophys J Lett 874(1):2. https://doi.org/10.3847/2041-8213/ab0c0f
Haris K, Mehta AK, Kumar S, Venumadhav T, Ajith P (2018) Identifying strongly lensed gravitational wave signals from binary black hole mergers. arXiv:1807.07062
Herlt E, Stephani H (1976) Wave optics of the spherical gravitational lens. Part I: diffraction of a plane electromagnetic wave by a large star. Int J Theor Phys 15(1):45–65. https://doi.org/10.1007/BF01807086
Hewitt JN, Turner EL, Schneider DP, Burke BF, Langston GI (1988) Unusual radio source MG1131+0456: a possible Einstein ring. Nature 333(6173):537–540. https://doi.org/10.1038/333537a0

Inamori M, Suyama T (2021) Universal relation between the variances of distortions of gravitational waves owing to gravitational lensing. Astrophys J Lett 918(2):30. https://doi.org/10.3847/2041-8213/ac2142

Ishihara H, Takahashi M, Tomimatsu A (1988) Gravitational Faraday rotation induced by a Kerr black hole. Phys Rev D 38(2):472–477. https://doi.org/10.1103/PhysRevD.38.472

Itoh Y, Futamase T, Hattori M (2009) Method to measure a relative transverse velocity of a source-lens-observer system using gravitational lensing of gravitational waves. Phys Rev D 80(4):044009. https://doi.org/10.1103/PhysRevD.80.044009

Jacques V, Wu E, Grosshans F, Treussart F, Grangier P, Aspect A, Roch J-F (2007) Experimental realization of Wheeler's delayed-choice Gedanken experiment. Science 315(5814):966. https://doi.org/10.1126/science.1136303

Janquart J, Seo E, Hannuksela OA, Li FG, Van Den Broeck C (2021) On the identification of individual gravitational-wave image types of a lensed system using higher-order modes. Astrophys J Lett 923(1):1. https://doi.org/10.3847/2041-8213/ac3bcf

Jaroszynski M, Paczynski B (1995) Diffraction effects in microlensing of Q2237+0305. Astrophys J 455:443. https://doi.org/10.1086/176593

Jow DL, Foreman S, Pen U-L, Zhu W (2020) Wave effects in the microlensing of pulsars and FRBs by point masses. Mon Not R Astron Soc 497(4):4956–4969. https://doi.org/10.1093/mnras/staa2230

Jow DL, Lin FX, Tyhurst E, Pen U-L (2021) Imaginary images and Stokes phenomena in the weak plasma lensing of coherent sources. Mon Not R Astron Soc 507(4):5390–5402. https://doi.org/10.1093/mnras/stab2337

Jow DL, Pen U-L, Feldbrugge J Regimes in astrophysical lensing: refractive optics, diffractive optics, and the Fresnel scale. Mon Not R Astron Soc 525(2):2107–2124 (2023) https://doi.org/10.1093/mnras/stad2332

Kader Z, Leung C, Dobbs M, et al CHIME/FRB Collaboration (2022) High-time resolution search for compact objects using fast radio burst gravitational lens interferometry with CHIME/FRB. Phys Rev D 106(4):043016. https://doi.org/10.1103/PhysRevD.106.043016

KAGRA Collaboration, Akutsu T, Ando M, Arai K, et al (2019) KAGRA: 2.5 generation interferometric gravitational wave detector. Nat Astron 3:35–40. https://doi.org/10.1038/s41550-018-0658-y

Katz A, Kopp J, Sibiryakov S, Xue W (2018) Femtolensing by dark matter revisited. J Cosmol Astropart Phys 2018(12):005. https://doi.org/10.1088/1475-7516/2018/12/005

Katz A, Kopp J, Sibiryakov S, Xue W (2020) Looking for MACHOs in the spectra of fast radio bursts. Mon Not R Astron Soc 496(1):564–580. https://doi.org/10.1093/mnras/staa1497

Kelly PL, Rodney SA, Treu T, et al (2015) Multiple images of a highly magnified supernova formed by an early-type cluster galaxy lens. Science 347(6226):1123–1126. https://doi.org/10.1126/science.aaa3350

Khangulyan D, Barkov MV, Popov SB (2022) Fast radio bursts by high-frequency synchrotron maser emission generated at the reverse shock of a powerful magnetar flare. Astrophys J 927(1):2. https://doi.org/10.3847/1538-4357/ac4bdf

Kling TP, Frittelli S (2008) Astrophys J 675:115. https://doi.org/10.1086/527368

Kumar P, Beniamini P (2023) Mon Not R Astron Soc 520:247. https://doi.org/10.1093/mnras/stad160

Kumar P, Beniamini P, Gupta O, Cordes JM (2024) Mon Not R Astron Soc 527:457. https://doi.org/10.1093/mnras/stad3010

Lazio WJ, Ojha R, Fey AL, Kedziora-Chudczer L, Cordes JM, Jauncey DL, Lovell JE (2008) Angular broadening of intraday variable AGNs. II. Interstellar and intergalactic scattering. Astrophys J 672(1):115–121. https://doi.org/10.1086/520572

Leung C, Brown A, Nguyen H, Friedman AS, Kaiser DI, Gallicchio J (2018) Astronomical random numbers for quantum foundations experiments. Phys Rev A 97(4):042120. https://doi.org/10.1103/PhysRevA.97.042120

Leung C, Kader Z, Masui KW, et al (2022) Constraining primordial black holes using fast radio burst gravitational-lens interferometry with CHIME/FRB. Phys Rev D 106(4):043017. https://doi.org/10.1103/PhysRevD.106.043017

Li D, Lin FX, Main R, Pen U-L, van Kerkwijk MH, Yang I-S (2019) Constraining magnetic fields through plasma lensing: application to the Black Widow pulsar. Mon Not R Astron Soc 484(4):5723–5733. https://doi.org/10.1093/mnras/stz374

Li Z, Qiao J, Zhao W, Er X (2022) Gravitational Faraday rotation of gravitational waves by a Kerr black hole. J Cosmol Astropart Phys 2022(10):095. https://doi.org/10.1088/1475-7516/2022/10/095

Liao K, Biesiada M, Fan X-L (2019) Astrophys J 875:139. https://doi.org/10.3847/1538-4357/ab1087

Lightman AP, Press WH, Price RH, Teukolsky SA (1975) Problem book in relativity and gravitation. Princeton University Press, Princeton

LIGO Scientific Collaboration, Aasi J, Abbott BP, Abbott R, et al (2015) Advanced LIGO. Class Quantum Gravity 32(7):074001. https://doi.org/10.1088/0264-9381/32/7/074001

Lin R, van Kerkwijk MH, Main R, Mahajan N, Pen U-L, Kirsten F (2023) Resolving the emission regions of the crab pulsar's giant pulses. II. Evidence for relativistic motion. Astrophys J 945(2):115. https://doi.org/10.3847/1538-4357/acba95

Maccone C (2010) FOCAL mission to 550 thru 1000 AU: status review 2009. Acta Astronaut 67(5):521–525. https://doi.org/10.1016/j.actaastro.2010.03.013

Macquart J-P (2004) Scattering of gravitational radiation. Second order moments of the wave amplitude. Astron Astrophys 422:761–775. https://doi.org/10.1051/0004-6361:20034512

Main R, Yang IS, Chan V, et al (2018) Nature 557:522. https://doi.org/10.1038/s41586-018-0133-z

Matsunaga N, Yamamoto K (2006) The finite source size effect and wave optics in gravitational lensing. J Cosmol Astropart Phys 2006(1):023. https://doi.org/10.1088/1475-7516/2006/01/023

Miller WA, Wheeler JA (1997) Delayed-choice experiments and Bohr's elementary quantum phenomenon. In: Advanced series in applied physics, vol 4. World Scientific, Singapore, pp 72–84. https://doi.org/10.1142/9789812819895_0008

Mishra A, Meena AK, More A, Bose S, Bagla JS (2021) Gravitational lensing of gravitational waves: effect of microlens population in lensing galaxies. Mon Not R Astron Soc 508(4):4869–4886. https://doi.org/10.1093/mnras/stab2875

Nakamura TT (1998) Gravitational lensing of gravitational waves from inspiraling binaries by a point mass lens. Phys Rev Lett 80(6):1138–1141. https://doi.org/10.1103/PhysRevLett.80.1138

Nakamura TT, Deguchi S (1999) Wave optics in gravitational lensing. Prog Theor Phys Suppl 133:137–153. https://doi.org/10.1143/PTPS.133.137

Narayan R (1992) The physics of pulsar scintillation. Philos Trans R Soc Lond Ser A 341(1660):151–165. https://doi.org/10.1098/rsta.1992.0090

Nemiroff RJ, Norris JP, Wickramasinghe WADT, Horack JM, Kouveliotou C, Fishman GJ, Meegan CA, Wilson RB, Paciesas WS (1993) Searching Gamma-Ray bursts for gravitational lensing echoes: implications for compact dark matter. Astrophys J 414:36. https://doi.org/10.1086/173054

Nimmo K, Pleunis Z, Beniamini P, et al (2025) Nature 637:48. https://doi.org/10.1038/s41586-024-08297-w

Nouri-Zonoz M (1999) Gravitoelectromagnetic approach to the gravitational Faraday rotation in stationary spacetimes. Phys Rev D 60(2):024013. https://doi.org/10.1103/PhysRevD.60.024013

Nye JF (1999) Natural focusing and fine structure of light: caustics and wave dislocations. IOP Publishing, Bristol

Oancea MA, Joudioux J, Dodin IY, Ruiz DE, Paganini CF, Andersson L (2020) Gravitational spin Hall effect of light. Phys Rev D 102(2):024075. https://doi.org/10.1103/PhysRevD.102.024075

Oancea MA, Stiskalek R, Zumalacárregui M (2024) Frequency- and polarization-dependent lensing of gravitational waves in strong gravitational fields. Phys Rev D 109(12):124045. https://doi.org/10.1103/PhysRevD.109.124045

Oguri M (2019) Strong gravitational lensing of explosive transients. Rep Prog Phys 82(12):126901. https://doi.org/10.1088/1361-6633/ab4fc5

Oguri M, Takahashi R (2020) Probing dark low-mass halos and primordial black holes with frequency-dependent gravitational lensing dispersions of gravitational waves. Astrophys J 901(1):58. https://doi.org/10.3847/1538-4357/abafab

Oguri M, Takahashi R (2022) Amplitude and phase fluctuations of gravitational waves magnified by strong gravitational lensing. Phys Rev D 106(4):043532. https://doi.org/10.1103/PhysRevD.106.043532

Ohanian HC (1973) Focusing of gravitational radiation. Phys Rev D 8(8):2734–2735. https://doi.org/10.1103/PhysRevD.8.2734

Ohanian HC (1974) On the focusing of gravitational radiation. Int J Theor Phys 9(6):425–437. https://doi.org/10.1007/BF01810927

Ohanian HC (1983) The caustics of gravitational 'lenses'. Astrophys J 271:551–555. https://doi.org/10.1086/161221

Pearson JW, Olver S, Porter MA (2014) Numerical Methods for the Computation of the Confluent and Gauss Hypergeometric Functions. arXiv:1407.7786. https://doi.org/10.48550/arXiv.1407.7786

Peters PC (1974) Index of refraction for scalar, electromagnetic, and gravitational waves in weak gravitational fields. Phys Rev D 9(8):2207–2218. https://doi.org/10.1103/PhysRevD.9.2207

Rahvar S (2018) Gravitational grating. Mon Not R Astron Soc 479(1):406–414. https://doi.org/10.1093/mnras/sty1369

Saha P (2000) Lensing degeneracies revisited. Astron J 120(4):1654–1659. https://doi.org/10.1086/301581

Sammons MW, James CW, Trott CM, Walker M (2022) The effect of gravitational lensing on fast transient event rates. Mon Not R Astron Soc. https://doi.org/10.1093/mnras/stac3013

Savastano S, Vernizzi F, Zumalacárregui M (2024) Phys Rev D 109:024064. https://doi.org/10.1103/PhysRevD.109.024064

Schneider P (1983) The mutual coherence of the images of gravitationally lensed objects. In: Swings J-P (ed) Quasars and gravitational lenses. Liege international astrophysical colloquia, vol 24, pp 131–133

Schneider P, Ehlers J, Falco EE (1992) Derivation of the lens equation, pp 119–155. Springer, Berlin. https://doi.org/10.1007/978-3-662-03758-4_4

Shi X, Xu Z (2021) Plasma microlensing dynamic spectrum probing fine structures in the ionized interstellar medium. Mon Not R Astron Soc 506(4):6039–6051. https://doi.org/10.1093/mnras/stab2108

Stanek KZ, Paczynski B, Goodman J (1993) Features in the spectra of Gamma-Ray bursts. Astrophys J Lett 413:7. https://doi.org/10.1086/186946

Stewart JM (1989) Solutions of the wave equation on a Schwarzschild space-time with localized energy. Proc R Soc Lond Ser A 424(1866):239–244. https://doi.org/10.1098/rspa.1989.0078

Sugiyama S, Kurita T, Takada M (2020) On the wave optics effect on primordial black hole constraints from optical microlensing search. Mon Not R Astron Soc 493(3):3632–3641. https://doi.org/10.1093/mnras/staa407

Suvorov AG (2022) Astrophys J 930:13. https://doi.org/10.3847/1538-4357/ac5f45

Suyama T (2020) On arrival time difference between lensed gravitational waves and light. Astrophys J 896(1):46. https://doi.org/10.3847/1538-4357/ab8d3f

Suyama T, Tanaka T, Takahashi R (2006) Exact wave propagation in a spacetime with a cosmic string. Phys Rev D 73(2):024026. https://doi.org/10.1103/PhysRevD.73.024026

Takahashi R (2006) Amplitude and phase fluctuations for gravitational waves propagating through inhomogeneous mass distribution in the universe. Astrophys J 644(1):80–85. https://doi.org/10.1086/503323

Takahashi R (2017) Arrival time differences between gravitational waves and electromagnetic signals due to gravitational lensing. Astrophys J 835(1):103. https://doi.org/10.3847/1538-4357/835/1/103

Takahashi R, Nakamura T (2003) Wave effects in the gravitational lensing of gravitational waves from chirping binaries. Astrophys J 595(2):1039–1051. https://doi.org/10.1086/377430

Takahashi R, Suyama T, Michikoshi S (2005) Scattering of gravitational waves by the weak gravitational fields of lens objects. Astron Astrophys 438(1):5–8. https://doi.org/10.1051/0004-6361:200500140

Tambalo G, Zumalacárregui M, Dai L, Cheung MH-Y (2023a) Phys Rev D 108:043527. https://doi.org/10.1103/PhysRevD.108.043527

Tambalo G, Zumalacárregui M, Dai L, Cheung MH-Y (2023b) Phys Rev D 108:103529. https://doi.org/10.1103/PhysRevD.108.103529

Tanaka S, Suyama T (2023) Kramers-Kronig relation in gravitational lensing. Phys Rev D 108(4):044015. https://doi.org/10.1103/PhysRevD.108.044015

Thom R (1967) Structural stability and morphogenesis. CRC Press, Boca Raton. https://books.google.ca/books?id=YVNPDwAAQBAJ

Turyshev SG (2017) Wave-theoretical description of the solar gravitational lens. Phys Rev D 95(8):084041. https://doi.org/10.1103/PhysRevD.95.084041

Ulmer A, Goodman J (1995) Femtolensing: beyond the semiclassical approximation. Astrophys J 442:67. https://doi.org/10.1086/175422

Vanderlinde K, Liu A, Gaensler B, et al (2019) The Canadian hydrogen observatory and radio-transient detector (CHORD). In: Canadian astronomy long range plan 2020–2030. White papers, vol 2020, p. 028. https://doi.org/10.5281/zenodo.3765414

Venumadhav T, Dai L, Miralda-Escudé J Microlensing of extremely magnified stars near caustics of galaxy clusters. Astrophys J 850(1):49 (2017) https://doi.org/10.3847/1538-4357/aa9575

Vijaykumar A, Mehta AK, Ganguly A (2023) Detection and parameter estimation challenges of type-II lensed binary black hole signals. Phys Rev D 108(4):043036. https://doi.org/10.1103/PhysRevD.108.043036

Walker MA, Melrose DB, Stinebring DR, Zhang CM (2004) Interpretation of parabolic arcs in pulsar secondary spectra. Mon Not R Astron Soc 354(1):43–54. https://doi.org/10.1111/j.1365-2966.2004.08159.x

Witt HJ, Mao S (1994) Can lensed stars be regarded as pointlike for microlensing by MACHOs? Astrophys J 430:505. https://doi.org/10.1086/174426

Wucknitz O, Spitler LG, Pen U-L (2021) Cosmology with gravitationally lensed repeating fast radio bursts. Astron Astrophys 645:44. https://doi.org/10.1051/0004-6361/202038248

Xiao H, Dai L, McQuinn M Detecting cosmic strings with lensed fast radio bursts. Phys Rev D 106(10):103033 (2022) https://doi.org/10.1103/PhysRevD.106.103033

Yang H, Zhang F, Zimmerman A, Chen Y (2014) Scalar Green function of the Kerr spacetime. Phys Rev D 89(6):064014. https://doi.org/10.1103/PhysRevD.89.064014

Zabel SA, Peterson JB (2003) Astrophys J 594:456. https://doi.org/10.1086/376896

Zenginoğlu A, Galley CR (2012) Caustic echoes from a Schwarzschild black hole. Phys Rev D 86(6):064030. https://doi.org/10.1103/PhysRevD.86.064030

Publisher's Note Springer Nature remains neutral with regard to jurisdictional claims in published maps and institutional affiliations.

Authors and Affiliations

Calvin Leung[1,2] · **Dylan Jow**[3,4] · **Prasenjit Saha**[5] · **Liang Dai**[6] · **Masamune Oguri**[7] · **Léon V.E. Koopmans**[8,9]

✉ P. Saha
psaha@physik.uzh.ch

C. Leung
calvin_leung@berkeley.edu

1 Department of Astronomy, University of California, 205 Campbell Hall, Berkeley, CA, 94720, USA

2 MIT Kavli Institute for Astrophysics and Space Research, Massachusetts Institute of Technology, 77 Massachusetts Ave., Cambridge, MA, 02139, USA

3 Canadian Institute for Theoretical Astrophysics, 60 St. George Street, Toronto, ON M5S 3H8, Canada

4 Department of Physics, University of Toronto, 60 St. George Street, Toronto, ON M5S 1A7, Canada

5 Physik-Institut, University of Zurich, 8057, Zurich, Switzerland

6 Department of Physics, University of California, 366 Physics North MC 7300, Berkeley, CA 94720, USA

7 Kapteyn Astronomical Institute, University of Groningen, P.O. Box 800, 9700AV, Groningen, the Netherlands

8 Center for Frontier Science, Chiba University, 1-33 Yayoi-cho, Inage-ku, Chiba, 263-8522, Japan

9 Department of Physics, Graduate School of Science, Chiba University, 1-33 Yayoi-Cho, Inage-Ku, Chiba, 263-8522, Japan

MIX
Papier aus verantwortungsvollen Quellen
Paper from responsible sources
FSC® C105338

If you have any concerns about our products,
you can contact us on
ProductSafety@springernature.com

In case Publisher is established outside the EU,
the EU authorized representative is:
Springer Nature Customer Service Center GmbH
Europaplatz 3, 69115 Heidelberg, Germany

Printed by Libri Plureos GmbH
in Hamburg, Germany